Pediatric and Adult Celiac Disease

A Clinically Oriented Perspective

Pediatric and Adult Celiac Disease

A Clinically Oriented Perspective

Edited by

Gino Roberto Corazza
Department of Internal Medicine, University of Pavia, Fondazione IRCCS Policlinico San Matteo, Pavia, Italy

Riccardo Troncone
European Laboratory for the Investigation of Food-Induced Diseases, Department of Translational Medical Science, University Federico II of Naples, Naples, Italy

Marco Vincenzo Lenti
Department of Internal Medicine, University of Pavia, Fondazione IRCCS Policlinico San Matteo, Pavia, Italy

Marco Silano
Department of Cardiovascular and Endocrine-metabolic diseases and Ageing, Istituto Superiore di Sanità, Rome, Italy

ELSEVIER

ACADEMIC PRESS

An imprint of Elsevier

Academic Press is an imprint of Elsevier
125 London Wall, London EC2Y 5AS, United Kingdom
525 B Street, Suite 1650, San Diego, CA 92101, United States
50 Hampshire Street, 5th Floor, Cambridge, MA 02139, United States

ISBN: 978-0-443-13359-6

For Information on all Academic Press publications
visit our website at https://www.elsevier.com/books-and-journals

Publisher: Mica H. Haley
Acquisitions Editor: Stacy Masucci
Editorial Project Manager: Tracy I. Tufaga
Production Project Manager: Omer Mukthar
Cover Designer: Christian Bilbow

Typeset by MPS Limited, Chennai, India

This book is dedicated to Giovanni Gasbarrini and Salvatore Auricchio, our Mentors in medicine, whose invaluable teachings have enriched its content.

We honor the memory of Anne Ferguson and Monty Losowsky who continue to inspire us.

Contents

List of contributors

Cristina Aimo Rare Diseases Unit, Section of Dermatology, Department of Health Sciences, USL Toscana Centre (European Reference Network-Skin Member), University of Florence, Florence, Italy

Nicola Aronico Department of Internal Medicine, University of Pavia, Fondazione IRCCS Policlinico San Matteo, Pavia, Italy

Renata Auricchio Department of Translational Medical Science and European Laboratory for the Investigation of Food-Induced Diseases, University of Naples Federico II, Naples, Italy

Michele Barone Gastroenterology Unit, Department of Emergency and Organ Transplantation, University of Bari "Aldo Moro", Bari, Italy

Federico Biagi Department of Internal Medicine and Medical Therapeutics, University of Pavia; Gastroenterology Unit, IRCCS Istituti Clinici Scientifici Maugeri, Pavia, Italy

Paola Ilaria Bianchi Department of Internal Medicine, University of Pavia, Fondazione IRCCS Policlinico San Matteo, Pavia, Italy

Stefano Bibbò Digestive Disease Center, Fondazione Policlinico Universitario "A. Gemelli" IRCCS, Rome, Italy; Department of Translational Medicine and Surgery, Catholic University of Rome, Rome, Italy

Giacomo Broglio Department of Internal Medicine, University of Pavia, Fondazione IRCCS Policlinico San Matteo, Pavia, Italy

Giacomo Caio Department of Translational Medicine, University of Ferrara, Ferrara, Italy

Emma Calabrese Department of Systems Medicine, University of Rome "Tor Vergata", Rome, Italy; Gastroenterology Unit, Fondazione Policlinico "Tor Vergata", Rome, Italy

Giovanni Cammarota Digestive Disease Center, Fondazione Policlinico Universitario "A. Gemelli" IRCCS, Rome, Italy; Department of Translational Medicine and Surgery, Catholic University of Rome, Rome, Italy

Marzia Caproni Rare Diseases Unit, Section of Dermatology, Department of Health Sciences, USL Toscana Centre (European Reference Network-Skin Member), University of Florence, Florence, Italy

Antonio Carroccio Department of Health Promotion Sciences, Maternal and Infant Care, University of Palermo, Internal Medicine and Medical Specialties (PROMISE), Palermo, Italy

Stefania Castellaneta Interdisciplinary Department of Medicine, Pediatric Section, University of Bari "Aldo Moro," Bari, Italy

Carlo Catassi Department of Pediatrics, Polytechnic University of Marche, Ancona, Italy

Giulia Catassi Pediatric Gastroenterology and Liver Unit, Sapienza University of Rome—Umberto I Hospital, Rome, Italy

Carolina Ciacci Department of Medicine Surgery and Dentistry, Scuola Medica Salernitana University of Salerno, Baronissi, Salerno, Italy

Rachele Ciccocioppo Gastroenterology Unit, Department of Medicine, AOUI Policlinico GB Rossi & University of Verona, Verona, Italy

Roberto Copparoni Office V—Nutrition and Information to Consumers, Directorate General for Food Safety and Nutrition, Italian Ministry of Health, Rome, Italy

Gino Roberto Corazza Department of Internal Medicine, University of Pavia, Fondazione IRCCS Policlinico San Matteo, Pavia, Italy

Alberto Corrà Rare Diseases Unit, Section of Dermatology, Department of Health Sciences, USL Toscana Centre (European Reference Network-Skin Member), University of Florence, Florence, Italy

Fernanda Cristofori Interdisciplinary Department of Medicine, Pediatric Section, University of Bari "Aldo Moro," Bari, Italy

Vanessa Nadia Dargenio Interdisciplinary Department of Medicine, Pediatric Section, University of Bari "Aldo Moro," Bari, Italy

Roberto De Giorgio Department of Translational Medicine, University of Ferrara, Ferrara, Italy

Giuseppe De Lisi Pathology Unit, Department of Molecular Medicine, University of Pavia, Fondazione IRCCS Policlinico San Matteo, Pavia, Italy

Ida De Micco Department of Medicine Surgery and Dentistry, Scuola Medica Salernitana University of Salerno, Baronissi, Salerno, Italy

Antonio Di Sabatino Department of Internal Medicine, University of Pavia, Fondazione IRCCS Policlinico San Matteo, Pavia, Italy

Luigia Di Stasio Institute of Food Science, Department of Biology, Agriculture and Food Sciences, National Research Council of Italy, Avellino, Italy

Michele Di Stefano Department of Internal Medicine, University of Pavia, Fondazione IRCCS Policlinico San Matteo, Pavia, Italy

Valentina Discepolo Department of Translational Medical Science and European Laboratory for the Investigation of Food-Induced Diseases, University of Naples Federico II, Naples, Italy

Alessio Fasano T.H. Chan School of Public Health, Department of Nutrition, Boston, MA, United States; Massachusetts General Hospital, Harvard Medical School, Department of Pediatrics, Boston, MA, United States

Ruggiero Francavilla Interdisciplinary Department of Medicine, Pediatric Section, University of Bari "Aldo Moro," Bari, Italy

Antonio Gasbarrini Digestive Disease Center, Fondazione Policlinico Universitario "A. Gemelli" IRCCS, Rome, Italy; Department of Translational Medicine and Surgery, Catholic University of Rome, Rome, Italy

Giovanni Gasbarrini Digestive Disease Center, Fondazione Policlinico Universitario "A. Gemelli" IRCCS, Rome, Italy; Department of Translational Medicine and Surgery, Catholic University of Rome, Rome, Italy

Simona Gatti Department of Pediatrics, Polytechnic University of Marche, Ancona, Italy

Carmen Gianfrani Institute of Biochemistry and Cell Biology, Department of Biomedical Sciences, National Research Council of Italy, Naples, Italy

Alessandra Giuliano Department of Health Promotion Sciences, Maternal and Infant Care, University of Palermo, Internal Medicine and Medical Specialties (PROMISE), Palermo, Italy

Luigi Greco Department of Translational Medical Science and European Laboratory for the Investigation of Food-Induced Diseases, University of Naples Federico II, Naples, Italy

Stefano Guandalini Department of Pediatrics, Section of Gastroenterology, Hepatology and Nutrition, University of Chicago Medicine, Chicago, IL, United States

Gianluca Ianiro Digestive Disease Center, Fondazione Policlinico Universitario "A. Gemelli" IRCCS, Rome, Italy; Department of Translational Medicine and Surgery, Catholic University of Rome, Rome, Italy

Marco Vincenzo Lenti Department of Internal Medicine, University of Pavia, Fondazione IRCCS Policlinico San Matteo, Pavia, Italy

Elena Lionetti Department of Pediatrics, Polytechnic University of Marche, Ancona, Italy

Marco Lucioni Department of Experimental Medicine, University of Pavia, Fondazione IRCCS Policlinico San Matteo, Pavia, Italy

Stiliano Maimaris Department of Internal Medicine and Medical Therapeutics, University of Pavia; Gastroenterology Unit, IRCCS Istituti Clinici Scientifici Maugeri, Pavia, Italy

Gianfranco Mamone Institute of Food Science, Department of Biology, Agriculture and Food Sciences, National Research Council of Italy, Avellino, Italy

Roberta Mandile Department of Translational Medical Science and European Laboratory for the Investigation of Food-Induced Diseases, University of Naples Federico II, Naples, Italy

Pasquale Mansueto Department of Health Promotion Sciences, Maternal and Infant Care, University of Palermo, Internal Medicine and Medical Specialties (PROMISE), Palermo, Italy

Irene Marafini Department of Systems Medicine, University of Rome "Tor Vergata", Rome, Italy; Gastroenterology Unit, Fondazione Policlinico "Tor Vergata", Rome, Italy

Elena Biancamaria Mariotti Rare Diseases Unit, Section of Dermatology, Department of Health Sciences, USL Toscana Centre (European Reference Network-Skin Member), University of Florence, Florence, Italy

Caterina Mengoli Department of Internal Medicine, University of Pavia, Fondazione IRCCS Policlinico San Matteo, Pavia, Italy

Giovanni Monteleone Department of Systems Medicine, University of Rome "Tor Vergata", Rome, Italy; Gastroenterology Unit, Fondazione Policlinico "Tor Vergata", Rome, Italy

Ivan Monteleone Department of Biomedicine and Prevention, University of Rome "Tor Vergata", Rome, Italy

Susanna Neuhold Associazione Italiana Celiachia AIC—APS, Genoa, Italy

Luisa Novellino Associazione Italiana Celiachia AIC—APS, Genoa, Italy

Caterina Pilo Associazione Italiana Celiachia AIC—APS, Genoa, Italy

Giuseppe Plutino Office V—Nutrition and Information to Consumers, Directorate General for Food Safety and Nutrition, Italian Ministry of Health, Rome, Italy

Naire Sansotta Pediatric Hepatology Gastroenterology and Transplantation Hospital Papa Giovanni XXIII, Bergamo, Italy

Giovanni Santacroce Department of Internal Medicine, University of Pavia, Fondazione IRCCS Policlinico San Matteo, Pavia, Italy

Annalisa Schiepatti Department of Internal Medicine and Medical Therapeutics, University of Pavia; Gastroenterology Unit, IRCCS Istituti Clinici Scientifici Maugeri, Pavia, Italy

Aurelio Seidita Department of Health Promotion Sciences, Maternal and Infant Care, University of Palermo, Internal Medicine and Medical Specialties (PROMISE), Palermo, Italy

Marco Silano Department of Cardiovascular and Endocrine-Metabolic Diseases and Ageing, Istituto Superiore di Sanità, Rome, Italy

Edoardo Troncone Department of Systems Medicine, University of Rome "Tor Vergata", Rome, Italy; Gastroenterology Unit, Fondazione Policlinico "Tor Vergata", Rome, Italy

Riccardo Troncone Department of Translational Medical Science and European Laboratory for the Investigation of Food-Induced Diseases, University of Naples Federico II, Naples, Italy

Alessandro Vanoli Pathology Unit, Department of Molecular Medicine, University of Pavia, Fondazione IRCCS Policlinico San Matteo, Pavia, Italy

Serena Vitale Institute of Biochemistry and Cell Biology, Department of Biomedical Sciences, National Research Council of Italy, Naples, Italy

Umberto Volta Department of Medical and Surgical Sciences (DIMEC), University of Bologna, Bologna, Italy

Annalisa Zanini Department of Surgery, Oncology and Gastroenterology, University of Padua, Padua, Italy

Fabiana Zingone Department of Surgery, Oncology and Gastroenterology, University of Padua, Padua, Italy

Preface

For many years, Italian medicine has held a prominent position in the study of celiac disease in both children and adults. Italy has hosted important international symposia, and many Italian scientists have been involved in international consensus groups and guideline development. The amount of scientific research produced in Italy competes with that of the most productive countries. Hence, it is not surprising that this book was conceived and written entirely by Italian authors, including a few who hold prestigious positions in North American institutions.

To note, this book does not aim to provide an Italian view on celiac disease, simply because such a view does not exist. The authors belong to independent groups and have gained their expertise at different foreign institutions. As a proof, throughout the book, some controversial opinions on specific topics emerge, as underlined by the editors themselves with appropriate callouts, highlighting controversial areas that may be the object of future research.

The main goals of this book are twofold. First, we wanted to provide a clear and up-to-date understanding of childhood and adult celiac disease, useful for all professionals and scientists (e.g., pediatricians, gastroenterologists, internists, pathologists, immunologists and basic scientists, dermatologists, nutritionists, and dieticians) involved in its management. Second, we aimed at having a clinical perspective as a leitmotif, as indicated in the book subtitle. However, this did not deter us from deepening those subtle mechanisms that represent the bases of the disease, since their understanding undoubtedly improves clinical practice. Furthermore, exploring the modalities of presentation of a food antigen known in its molecular structure to the immune system, the enzymatic mechanisms underlying the strengthening of its immunogenicity, the ways of clustering across autoimmune diseases, and the progression from chronic inflammation to lymphoma have made celiac disease a model of paramount importance for understanding broader mechanisms involved in the breakdown of immunological tolerance.

Going back to the clinical ground, although the spectrum of clinical presentation is wider and clearer than in the past, this disease never ceases to amaze us. The vast majority of patients are still undiagnosed, or are diagnosed with years of delay, despite the better disease awareness and the availability of accurate diagnostic tools. Recent data suggest that this happens not only in the primary care but also in the secondary care. Also, due to their relevant role in promoting educational and awareness campaigns, in fundraising for research, in supporting patients and their families, and in dealing with stakeholders and socio-political institutions, we have included a chapter dedicated to the activity of the patient association "Associazione Italiana Celiachia" (AIC), member of the AOECS.

We think that there is a real need for a new book because, compared to the latest on this topic, a number of novel issues and data have emerged. We will discuss the epidemiology of this constantly increasing condition, along with the many uncertainties around its differentiation from a series of emergent disorders (e.g., nonceliac gluten/wheat sensitivity, olmesartan enteropathy, immunodeficiencies, seronegative celiac disease, wheat allergy) that are yet unclear to many physicians. Furthermore, new guidelines on pediatric celiac disease were published in 2020 and the appropriateness of a biopsy-free approach in adults is still disputed. Finally, we report here for the first time a critical reappraisal of nondietary treatments, as well as the results of the recent phase II–IV trials focusing on refractory celiac disease and enteropathy-associated T-cell lymphoma.

To conclude, the editors would like to express their deepest gratitude to Tracy Tufaga, Stacy Masucci, and all the Elsevier editorial staff for their valuable assistance.

Gino Roberto Corazza
Riccardo Troncone
Marco Vincenzo Lenti
Marco Silano

Chapter 1

The dynamic epidemiology of celiac disease

Giulia Catassi[1], Elena Lionetti[2], Simona Gatti[2] and Carlo Catassi[2]
[1]*Pediatric Gastroenterology and Liver Unit, Sapienza University of Rome—Umberto I Hospital, Rome, Italy,* [2]*Department of Pediatrics, Polytechnic University of Marche, Ancona, Italy*

1.1 Introduction

The distribution of celiac disease (CD) over space and time has attracted the interest of researchers since the early knowledge of this condition. During the first half of the 20th century, CD was regarded as a rare disease in most Western countries, with an estimated prevalence of 1:4000–1:8000 and the notable exception of Western Ireland where a frequency of one every 400 young children, for some unknown reason/s, was reported (Mylotte et al., 1973). At that time, it was postulated that the higher prevalence of CD in the North-West of Europe was somewhat related to the higher frequency of HLA-related predisposing genotypes, along a South-East/North-West gradient from the Middle East area (Cronin & Shanahan, 2001).

The subsequent evolution of research, particularly the development of sensitive and simple diagnostic tools, particularly the IgA class antitissue transglutaminase (tTG) antibodies, unmasked a completely different story: (1) CD is one of the commonest life-long disorders, with an average prevalence of 1% in the general population and a higher frequency in well-defined at risk groups, such as relatives of affected individuals and patients with autoimmune comorbidities; (2) the disease has a worldwide distribution, even though the local prevalence varies according to genetic and environmental factors; (3) still today, due to huge clinical variability, most CD cases escape diagnosis unless actively searched by proactive policies, such as mass screening in the pediatric age group (Catassi et al., 2022).

In this chapter we will review the epidemiology of CD in the world, as described by the incidence, the prevalence, and the changes over time of these parameters in the general population. The analysis of these dynamic data will show how the knowledge of CD epidemiology may contribute to the understanding of the etiology of the disease and to the development of preventive strategies.

1.1.1 Definitions

For the sake of clarity, we define here the terms that will be repeatedly used throughout this chapter.

CD. Patients who have been diagnosed with CD according to currently available diagnostic guidelines. In adults, the diagnostic algorithm include both the positivity of serum celiac autoantibodies and the finding of villus atrophy (grade 3A to 3C lesion according to the Marsh-Oberhuber criteria) at the small intestinal biopsy. In children, the biopsy may or may not be required, according to the levels of the celiac autoantibodies.

Potential CD. Positivity of serological CD markers, particularly of IgA anti-tTG and antiendomysial (EMA) antibodies, in a patient with a normal histological picture (grade 0 or grade 1 lesion) at the small intestinal biopsy (Catassi et al., 2022).

Celiac autoimmunity. Positivity of serological CD markers, particularly of IgA anti-tTG and antiendomysial (EMA) antibodies (small intestinal biopsy not performed).

Prevalence of CD. The total number of individuals in a population who have CD at a specific period of time, usually expressed as a percentage of the population.

Pediatric and Adult Celiac Disease. DOI: https://doi.org/10.1016/B978-0-443-13359-6.00006-6

Incidence of CD. The number of individuals who are newly diagnosed with CD during a particular time-period (usually a year).

CD detection rate. The proportion of CD cases that are diagnosed on clinical basis, as compared to the overall prevalence of the disease in a given population.

Screening. Presumptive identification of unrecognized disease by the application of tests, examinations, or other procedures which can be applied rapidly. Screening tests sort out apparently well persons who probably have a disease from those who probably do not. A screening test is not intended to be diagnostic. Persons with positive or suspicious findings must be referred for diagnosis and necessary treatment.

CD case-finding. The process of looking for disease identification among people who actively seek care due to symptoms compatible with CD. It is in principle, but not necessarily, a "patient initiated" pathway to diagnosis (Lönnroth et al., 2013). Passive Case Finding (PCF, from the patient to the doctor) is defined as a CD diagnosis at health facilities among individuals who are aware of their symptoms and present for evaluation of the same on their own, while Active Case Finding (ACF, from the doctor to the patient) is the systematic CD testing of all individuals who are at risk of developing this disorder (Saini & Garg, 2020), due to the presence of one or more symptoms/conditions/comorbidities correlated to CD.

1.2 Incidence of celiac disease

Although apparently simple to measure, the evaluation of CD incidence requires a well-organized system of data collection about new cases of disease and the referral population (an example of this is the Swedish National Celiac Disease Register established in 1998), which is not available in many countries. For this reason, CD incidence cannot be calculated in many areas of the world.

In Western countries, a recent systematic review and meta-analysis reported the pooled incidence of CD in women and men to be 17.4 (95% CI: 13.7, 21.1) and 7.8 (95% CI: 6.3, 9.2) per 100,000 person-years, respectively. CD incidence has largely increased during the last decades in these countries (King et al., 2020). During the period 2010–14, twenty times more CD patients were diagnosed than during 1975–79 in the United Kingdom. The largest increase in diagnosis rates occurred in young women, the elderly, and the Asian immigrants (Holmes & Muirhead, 2017). In children, large increases in the incidence of CD across Europe have reached 50 per 100,000 person-years in Scandinavia and Spain with stabilization in some (notably Sweden and Finland). Larger increases in incidence have been found in older age groups than in infants and ages <5 years (Roberts et al., 2021). In the United States (Olmsted County, Minnesota) the overall age- and sex-adjusted incidence of CD increased from 11.1 per 100,000 persons/year in 2000–01 to 17.3 in 2008–10. Interestingly, the relative frequency of classical CD among incident cases decreased over time between 2000 and 2010 (Ludvigsson et al., 2013). The previously mentioned meta-analysis also found that the average annual incidence of CD in Europe, North America and Oceania has been rising over the past several decades by 7.5% (95% CI: 5.8, 9.3) per year (King et al., 2020). On the other hand, epidemiological data on CD incidence are still lacking for many countries (e.g., India, China, and Russia).

The increasing incidence of CD should not be interpreted as an evidence of higher frequency of the disease. This finding is primarily related to the improvement of the diagnostic rate, which in turn depends on both improved diagnostics, particularly diffuse availability of sensitive and simple serological test such as the anti-tTG antibody determination, and increased awareness of the high clinical variability of the disease among physicians and the general population.

1.3 Prevalence of celiac disease in the world

The frequency of CD is properly described by its prevalence, as determined by studies on samples of the general population, for example, school-age children performing the first-level test at school. Fig. 1.1 shows data on CD prevalence in different countries.

A recent systematic review and meta-analysis found the pooled world prevalence of CD autoimmunity to be 1.4% [95% confidence interval (CI): 1.1%–1.7%], based on positive results from tests for IgA anti-tTG ab and/or EMA. The prevalence of CD autoimmunity varied from continent to continent, highest being in Europe and Asia. The pooled global prevalence of biopsy-confirmed CD was 0.7% (95% CI: 0.5%–0.9%) with regional variations (0.4% in South America, 0.5% in Africa and North America, 0.6% in Asia, and 0.8% in Europe and Oceania). The prevalence of CD was significantly higher in females than in males (0.6% vs 0.4%), and in children than in adults (0.9% vs 0.5%) (Makharia et al., 2022; Singh et al., 2018).

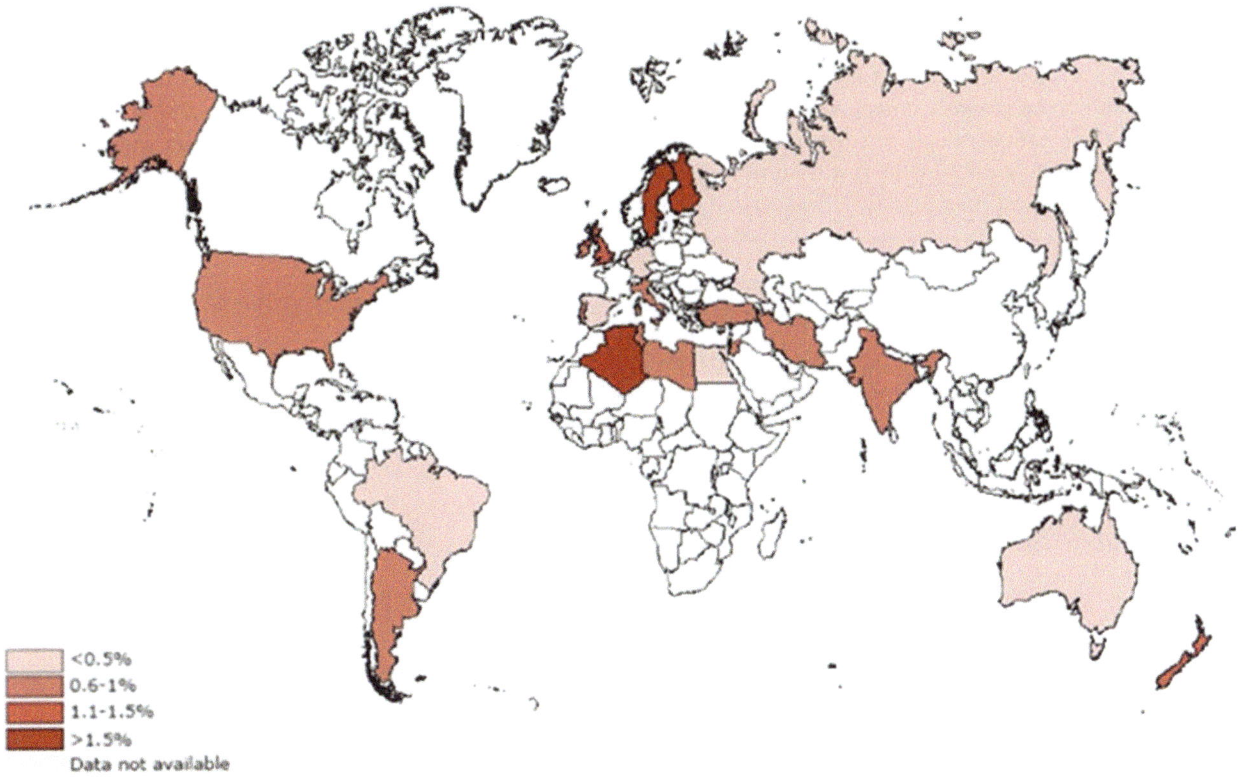

FIGURE 1.1 Prevalence of CD in different countries. Schematic representation of the prevalence of CD in different countries. *From Lionetti and Catassi (2014). Co-localization of gluten consumption and HLA-DQ2 and -DQ8 genotypes, a clue to the history of celiac disease. Digestive and Liver Disease, 46(12), 1057–1063. https://doi.org/10.1016/j.dld.2014.08.002. No changes were made to the Figure. Creative Common License found at https://www.sciencedirect.com/science/article/pii/S1590865814006008?via%3Dihub.*

Wide variations of CD prevalence in Europe were reported in the year 2010, with an overall estimate of 1.0% (95% CI 0.9–1.1). In subjects aged 30–64 years CD prevalence was 2.4% in Finland (2.0–2.8), 0.3% in Germany (0.1–0.4), and 0.7% in Italy (0.4–1.0) (Mustalahti et al., 2010). More recently, the prevalence of previously undiagnosed CD from screening surveys (histology based) in Europe ranged from 0.10% to 3.03% (median = 0.70%), with a significantly increasing annual trend. Prevalence since 2000 was significantly higher in northern Europe (1.60%) than in eastern (0.98%), southern (0.69%) and western (0.60%) Europe (Roberts et al., 2021). In two different birth cohorts of 12-years old Swedish students the overall prevalence of CD was 2.2 and 2.9%, respectively (Ivarsson et al., 2013). In Tromsø, Norway, 12,981 adults participated in a population-based CD serological screening. The prevalence of previously diagnosed CD was 0.37%. Additionally, the prevalence of previously undiagnosed CD was 1.10%. Thus 1.47% of this Norvegian population sample had CD, of whom 75% were previously undiagnosed (Kvamme et al., 2022). In a large pediatric CD screening performed in Italy on 9006 school age children the overall prevalence of CD was 1.62% (see details of this study in paragraph 5) (Lionetti et al., 2023).

In a large multicentre study performed in the United States in the early 2000s, Fasano and coworkers found a CD prevalence of 0.75% in a sample of 4126 subjects (Fasano et al., 2003). More recently, in a sample of 30,425 adults resident in Olmsted County (Minnesota, United States) the estimated prevalence of CD was 1.1% (95% CI: 1.0%–1.2%) (Choung et al., 2017). The analysis of data collected in the United States National Health and Nutrition Examination Survey, from 2009 through 2014, on 22,277 participants 6 years and older showed a 0.7% prevalence of CD. Interestingly, the disease was more common among individuals who lived at latitudes of 35–39 degrees North (odds ratio, 3.2; 95% CI: 1.4–7.1) or at latitudes of 40 degrees North or more (odds ratio, 5.4; 95% CI: 2.6–11.3) than individuals who lived at latitudes below 35 degrees North, independent of race or ethnicity, socioeconomic status, and body mass index (Unalp-Arida et al., 2017).

In Central and South America, Australia, North Africa and the Middle East areas, CD is as common as in the Western world (around 1% of the general population) (Ahadi et al., 2016; Alarida et al., 2011; Chin et al., 2009; Gandolfi et al., 2000; Israeli et al., 2010; Mora et al., 2012; Remes-Troche et al., 2006), while no data are available for

Sub-Saharan African countries. An impressive prevalence of CD autoimmunity (5.6%) has been documented in the Arab population of Saharawi, probably related to the abrupt and recent changes in the dietary pattern of this population (from a low- to a high-gluten diet) (Catassi et al., 1999).

Currently a continent under deep investigation is Asia (Singh et al., 2016). Based on available data, three different pattern of CD epidemiology can be recognized in Asia: (1) areas with Western-type prevalence (e.g., Turkey, North-Western India, North-Western China), (2) areas with very low CD prevalence (<0.2%) due to the rice-based dietary pattern and overall low prevalence of HLA-DQ2/DQ8, for example, Vietnam, Philippines and Japan, and (3) unknown CD prevalence, for example, Russia, Central Asia and South-East Asia. In China, CD was mostly unknown in the past, however two recent population-based studies showed that this disorder is not uncommon in this country, particularly in north China, where wheat is a staple food (Fukunaga et al., 2020; Poddighe et al., 2019; Ramakrishna et al., 2016; Yuan et al., 2017; Zhou et al., 2020).

In summary, epidemiological studies performed during the last decades have shown that wherever gluten-containing cereals (particularly wheat) are staple food, which is the case in most world countries, CD is a common disease affecting around 1% of the general population, with significant variations between countries and a definitely higher prevalence in some areas, such as Sweden and Italy.

1.4 Factors influencing the prevalence of celiac disease

Generally speaking, the between-countries variability of CD prevalence may depend on "spurious" (1) or "true" (2) differences:

1. Spurious differences. Factors related to the study-design may heavily influence the prevalence of the disease, such as: (a) age of investigated subjects (higher prevalence in children than in adults); (b) gender distribution (CD is more common in females); (c) selection bias in the sampled population. For instance, the prevalence of CD may be underestimated in screening studies performed in blood donors, representing the "healthier" segment of the population, or overestimated in selected groups of subjects such as patients seen at the general gastroenterology clinic; (d) study-protocol and screening test. Pioneer CD screening studies using the antigliadin antibody (AGA) as first-level screening test underestimated CD prevalence, as compared to an anti-tTG based screening, by nearly 40% (Gatti et al., 2020); (e) diagnostic criteria. The prevalence of CD is usually lower in studies requiring evidence of the celiac enteropathy at the small intestinal biopsy for diagnosis (cases of potential CD may be excluded from prevalence calculation).
2. True differences. On the other hand, the true prevalence of CD is directly related to the population prevalence of the CD-predisposing genotypes HLA-DQ2 and -DQ8 (30%–40% in most countries) and to the average level of wheat consumed at the population level. This is clearly exemplified by data from India: CD is much more common (0.9%) in the Northern part of the country, where wheat is the staple food, than in the Southern part (0.01%) showing both a lower prevalence of HLA-DQ2 genotype and a lower wheat consumption (Ramakrishna et al., 2016; Verma et al., 2022). Likewise, higher consumption of gluten-containing food could explain the higher prevalence of CD found in the South of Italy (1.93%) as compared to Central Italy (1.36%) (Lionetti et al., 2023). Interestingly, on a global scale HLA-DQ2 prevalence and amount of gluten consumption apparently show a positive correlation throughout the world (for possible explanations of this finding see paragraph 6.1) (Lionetti & Catassi, 2014).

1.5 Pediatric screening for celiac disease: lessons from the Italian studies

Italy is not only the country where the first mass screening project for CD in children was implemented in the early 90 s (Catassi et al., 1994), but also the homeland of several CD pediatric screening performed during these last 30 years. In all these studies a first-level screening test was performed at school and a complete CD diagnostic algorithm was activated in screening-positive subjects. The analysis of these studies allows a depiction of the evolving epidemiology of this common condition.

In a multicenter, countrywide study performed in the years 1992–94, 17,201 students aged 6–15 years underwent CD screening by the combined determination of serum IgG- and IgA-AGA test. CD was diagnosed in 82 subjects. The prevalence of undiagnosed CD was 0.48%, 1 in 210 subjects. The overall prevalence of CD, including previously diagnosed CD cases, was 0.54%, that is, one in 184 subjects. The ratio of known to undiagnosed CD cases was 1–7 (Catassi et al., 1996). As shown many years later by a validation study, the IgA AGA test, as compared to an IgA anti-tTG screening, underestimated CD prevalence by 39%. After adjustment for this underestimation, the 1992–94 CD prevalence was 0.88% (Gatti et al., 2020). During the period 1999–2001, 3541 subjects (1500 from "continental" Italy and 2041 from Sardinia) were screened for CD using IgA anti-tTG antibodies as first-level tests. A diagnosis of CD was made in 25 subjects, that is, an overall prevalence of 1:126 individuals (0.79%) (Fabiani et al., 2004). Another anti-tTG–based screening study performed on Italian children in 1999–2000 reported a CD prevalence of 1.1%, slightly higher than the 1993–95 estimate but definitely lower than the CD prevalence found in recent studies (Tommasini et al., 2004). During the years 2015–20 a large nationwide multicenter CD screening was again performed in 9008 school-age children (5–11 years of age) screened at school in eight different towns. This time the first level screening test was the determination of HLA-related predisposing genotypes HLA-DQ2 and -DQ8 (Fig. 1.2).

In this recent study the overall prevalence of CD in Italy was 1.62% (95% CI,1.39–1.86), 1.62% (95% CI 1.25–2.06) in the North of Italy, 1.36% (95% CI 1.04–1.75) in the Center, and 1.93% (95% CI 1.50–2.45) in the South, with a statistically significant difference between Central and South Italy ($p = 0.0482$) (Fig. 1.3).

The percentage of CD cases diagnosed prior to the school screening, that is, the so-called "visible part of the CD iceberg," was only 35% (Lionetti et al., 2023).

Based on the above data, there is little doubt that a significant increase of childhood CD prevalence in childhood took place in Italy during the last 30 years. The reason/s for the linear increase of CD prevalence in Italian children during the last decades remain/s unclear. Changes of the major determinants of CD, namely host genetic make-up and gluten consumption, are unlikely to explain this finding. Although increasing immigration from extra-European countries occurred during these last decades in Italy, only 11.5% of the most recent study groups were nonwhite children. Furthermore, most of these non-European children originated from Asian and South American countries, that is, areas generally showing a prevalence of CD lower than Europe (Lionetti et al., 2023). As for consumption of wheat derivatives, food supply data, if anything, indicate a decreasing trend in the consumption of gluten-containing products by the Italian population (Data Available at http://Www.Fao.Org/Faostat, n.d.). Likewise, we are unaware of any recent change of other environmental CD potential risk factors, for example, early dietary habits, early gluten composition, dough fermentation, intestinal infections, and drug use/abuse.

The recent Italian CD screening studies also shed a light on the sensitivity of the CD case-finding approach, as well as the different performances of an active vs a passive case-finding policy. In this population sample there were 157 children with CD: (1) 55 (35%) had received a clinical diagnosis of CD prior to the screening (by means of a PCF process); (2) 59 cases (38%) were found among screened students presenting at least one of the condition listed in the criteria for CD case-finding (an ACF process); (3) 43 cases (27%) were detected among screened children showing no indication to CD case-finding (diagnosis only by mass screening) (Fig. 1.4).

Positivity of at least one criterion for CD case finding was found in 53% of all screened subjects. There was only a minimal, statistically not significant difference in the prevalence of CD between HLA positive children with at least one (3.39%) and those with no criterion for CD case-finding (2.83%) ($p = 0.419$).

Put in the context of a country characterized by high awareness of CD clinical polymorphism among doctors and free availability of a primary care pediatrician for all children, these simple findings highlight that: (1) a PCF policy detects only about one third of CD cases; (2) a proactive CF policy (ACF) still does not intercept >25% of CD cases; (3) even more importantly, the ACF is not selective (requires testing more than 50% of the general pediatric population)

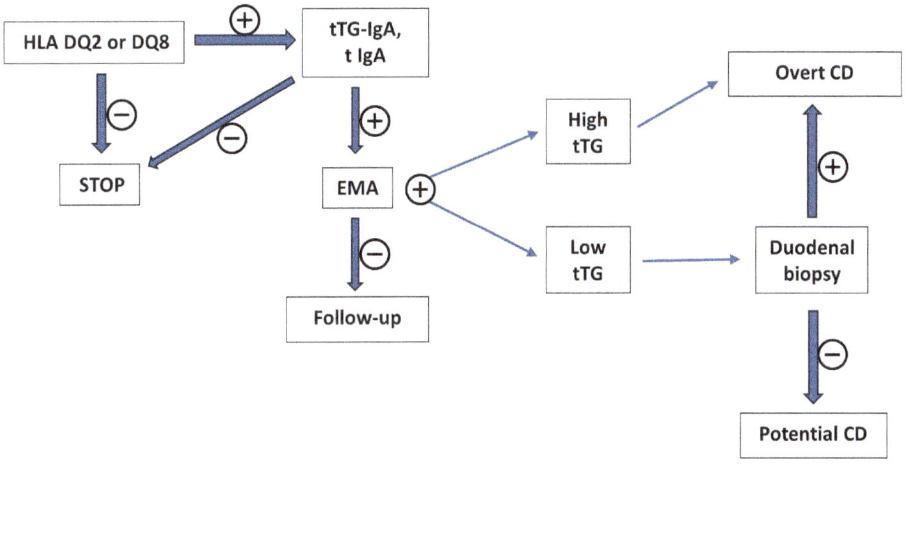

FIGURE 1.2 Diagnostic algorithm used in recent CD screening studies in Italy. Flow-chart showing the diagnostic algorithm used in recent CD screening studies in Italy. tTG-IgA = antitissue transglutaminase of IgA class; t IgA = total serum IgA; CD = celiac disease; EMA = antiendomysial antibodies.

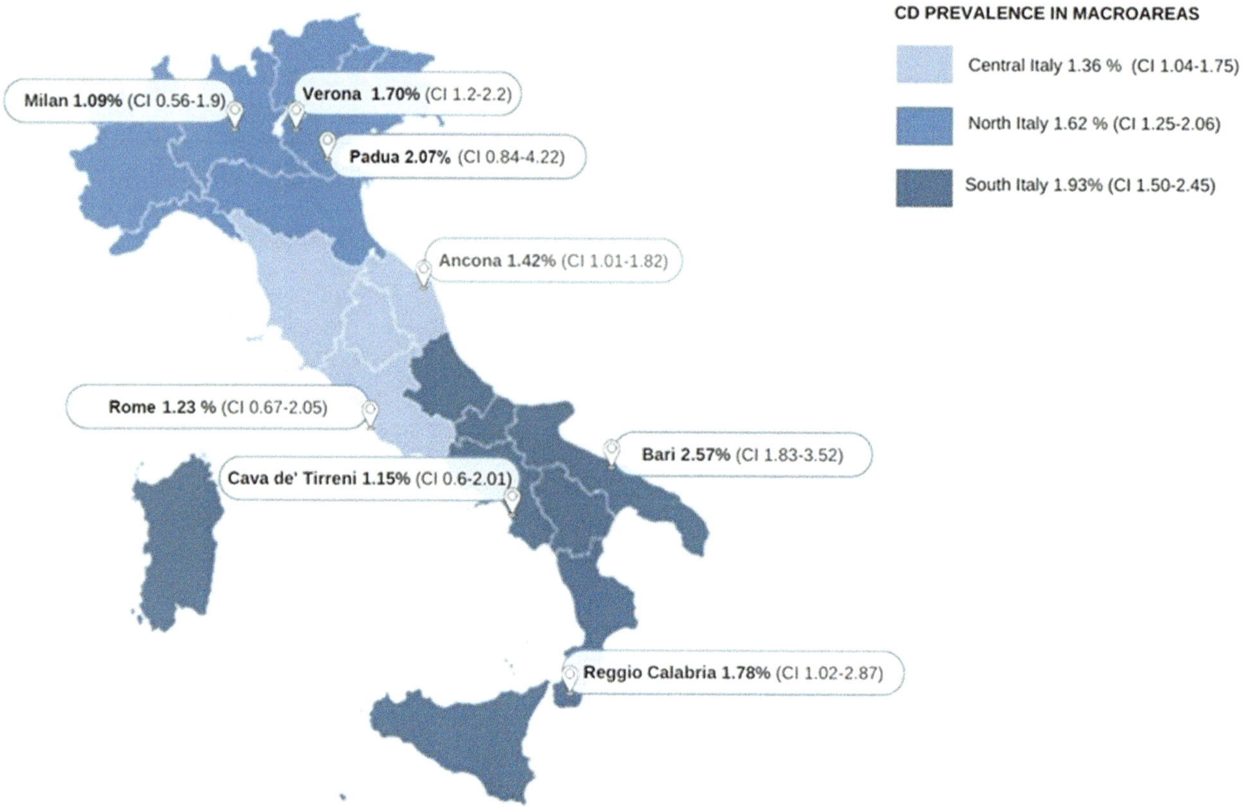

FIGURE 1.3 Prevalence of CD in Italy. Prevalence of CD in different areas of Italy, in a sample of 9006 school age children screened by the algorithm shown in the previous figure. *From Lionetti et al. (2023). Prevalence and detection rate of celiac disease in Italy: Results of a SIGENP multicenter screening in school-age children. Dig Liver Dis, 2023(22), 871–874. No changes were made to the Figure. Creative Common License found at: https://www.sciencedirect.com/science/article/pii/S1590865822008714?via%3Dihub.*

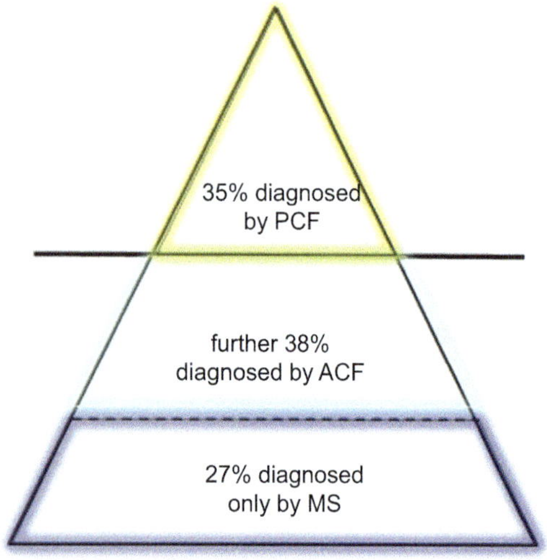

FIGURE 1.4 Representation of the "celiac" iceberg. The celiac iceberg in Italy, as seen by the 2015–20 studies. PCF = passive case-finding; ACF = active case-finding; MS = mass screening. *Data from Catassi et al. (2023). Celiac Disease detection strategies: poor performance of the case-finding policy. Am J Gastroenterol, 2023.*

and is not significantly more efficient than a by-chance testing. Due to the possible negative consequences of delayed/missed CD diagnoses (Al-Toma et al., 2019) we believe that these results suggest considering the implementation of simple mass CD screening programs, particularly focused on the pediatric age-group (Catassi et al., 2023).

1.6 Is the prevalence of celiac disease changing over time?

1.6.1 In the ancient time

The history of Homo sapiens dates back to about 200,000 years ago (200 KYA), however CD is a "modern" disease appearing on the stage only after gluten-containing cereals were introduced in the human diet around 12 KYA with the agricoltural revolution. Aretaeus of Cappadocia, living in the 2nd century AD, recorded a malabsorptive syndrome with chronic diarrhea, causing a debilitation of the whole body that he named the "Celiac affection" (Catassi et al., 2022). However there are no reports on CD in the ancient time, apart from the anecdotal description of a possible disease-case in a young woman died in Italy during the first century AD. Her mummy presented evidence of CD manifestations, particularly short height (140 cm), dental enamel hypoplasia, skeletal signs of anemia (cribra orbitalia), and a decreased bone mass with osteoporosis and bone fragility (Gasbarrini et al., 2010). Interestingly, the analysis of DNA extracted from a bone sample and a tooth of this mummy displayed the haplotype HLA-DQ 2.5 associated to the highest risk of CD (Gasbarrini et al., 2012).

The extreme rarity of CD in the ancient time is indirectly suggested by the analysis of the current geographical distribution of HLA CD-predisposing genotypes and the world distribution of wheat consumption. As previously anticipated, domestication of gluten-containing cereals (particularly ancient varieties of wheat and barley) began approximately 12 KYA in Neolithic settlements in the north-eastern (Turkey, Iran, and Iraq) and in the south-western (Palestine, Syria, and Lebanon) regions of the so-called "Fertile Crescent" area. Cultivation of wheat and barley slowly spread northwest ward across Europe to reach Western countries, as Britain, "only" 4 KYA (Ammerman & Cavalli-Sforza, 1984). The agriculture revolution paved the way to new diseases, such as CD. Since treatment with the gluten-free diet was not available in the ancient time, the CD predisposing genotype could exert a negative impact on human reproductive fitness and survival over thousand years. As a consequence of a prolonged negative selective pressure, CD and its predisposing genotypes should be less common in Middle East than in European countries (as originally theorized by Simoons) (Lionetti & Catassi, 2014; Simoons, 1981). On the other hand, nowadays CD is equally common, for instance, in United Kingdom and Turkey, and the frequency of HLA-DQ2 (the major CD-predisposing haplotype) shows a positive correlation with the amount of gluten consumed in different world countries (Fig. 1.5)

and is even higher in Turkey and Iran, that is, countries consuming more wheat and for a longer duration of time, than in Finland and Ireland, that is, countries consuming gluten during a shorter time span. To explain these counterintuitive findings, we suggested that the diffusion of HLA CD-predisposing genotypes and that of CD occurred in different time periods. The currently high prevalence of CD is most likely a recent event that took place only in the last centuries/decades and is possibly explained by environmental conditions, such as higher susceptibility toward autoimmune diseases caused by reduced frequency and severity of infections (so-called hygiene hypothesis) and/or changes in the intestinal microbiome (Lionetti & Catassi, 2014).

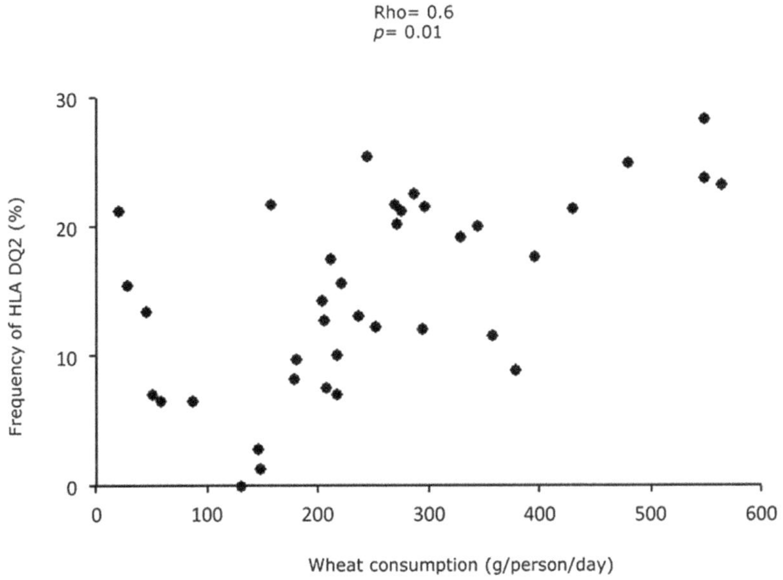

FIGURE 1.5 Relation between HLA-DQ2 and gluten consumption. There is a positive linear correlation between the prevalence of HLA-DQ2 genotype and the mean gluten consumption in different countries *From Lionetti and Catassi (2014). Co-localization of gluten consumption and HLA-DQ2 and -DQ8 genotypes, a clue to the history of celiac disease. Digestive and Liver Disease, 46(12), 1057–1063. https://doi.org/10.1016/j.dld.2014.08.002. No changes were made to the Figure. Creative Common License found at: https://www.sciencedirect.com/science/article/pii/S1590865814006008?via%3Dihub.*

1.6.2 In the modern time

The pediatrician Samuel Gee gave the first modern-day description of CD in children in a lecture at Hospital for Sick Children, Great Ormond Street, London, in 1887. However, the first data on CD prevalence in the general population became available only one hundred years later, following the introduction of simple and sensitive serological diagnostic tests for CD.

We already discussed in paragraph 5 the evidences showing that in Italy the prevalence of CD almost doubled over the last 30 years. What has happened in other parts of the world? Data collected at different time-points are available for some countries. In the United States, we analyzed sera from 3511 adults resident in Washington County, Maryland, with matched samples from 1974 (CLUE study I) and 1989 (CLUE study II). CD autoimmunity was detected in seven subjects in 1974 (prevalence 1:501) and in an additional nine subjects in 1989 (prevalence 1:219). During a 15-year period CD prevalence increased twofold in the CLUE cohort and fivefold in the United States from 1974 to the year 2000 (from 0.2% to 1%) (Catassi et al., 2010). Recently, 23,400 children from general population in Denver metropolitan area were enrolled in the ASK study from 2016 to 2019, and 23,319 participant children had the IgA antitransglutaminase tested. Of these, 2% showed positivity of celiac autoimmunity by both the chemiluminescence and the radioimmunoassay antibody determination (He et al., 2020). Although the prevalence of celiac autoimmunity (instead than biopsy-proven CD) tends to overestimate the prevalence of CD, it is easy to notice a linear trend of a marked increase in the prevalence of CD in the United States during the last 50 years.

An interesting clue to the understanding of the causes of the recent surge of CD prevalence observed in some countries is provided by a CD screening study performed in school-age children living in adjacent Russian Karelia ($n = 1988$) and Finland ($n = 3654$). These populations are equally exposed to grain products and partly share the same ancestry, but live in completely different socioeconomic environments. Anti-tTG antibodies were less frequent in Russian Karelia than in Finland (0.6% vs 1.4%, $p = 0.005$). CD was confirmed by duodenal biopsy in four of the eight anti-tTG IgA positive Karelian children, giving a prevalence of 1 in 496 compared to 1 in 107 children in Finland. Authors suggested that the lower prevalence of CD autoimmunity and of CD itself in Russian Karelia than in Finland may be the consequence of a protective environment characterized by inferior prosperity and standard of hygiene in Karelia (Kondrashova et al., 2008).

On the other hand, stabilization or even decrease in CD prevalence has also been observed in other places. As previously mentioned, between 1984 and 1986 Sweden experienced an "epidemic" of clinical CD in young children, attributed partly to changes in infant feeding. Later on, a two-phase cross sectional screening study was performed in which 13,279 12-years old children from two birth cohorts participated, born in 1993 and 1997, respectively. The total prevalence of CD was 2.9% and 2.2% for the 1993 and 1997 cohorts, respectively. The significant reduction of CD prevalence in the 1997 cohort was attributed to changes in infant feeding practices (Ivarsson et al., 2013). As a consequence of interventions aimed to primarily prevent CD development, the overall prevalence of CD could hopefully decrease in many countries.

1.7 How to improve the celiac disease detection rate (unraveling the celiac iceberg)

There is no doubt that the CD detection rate has increased in many countries over the last decades, particularly due to the adoption of case-finding policies and greater disease awareness. For instance, in Italy the CD detection rate was only 11% in the years 1992–94 and rose to 40% in 2015–20 (Gatti et al., 2020) (Lionetti et al., 2023). However, the percentage of cases that are currently diagnosed on clinical basis remains much smaller than the overall CD prevalence. This concept is often expressed by the picture of an iceberg that shows a visible part (the one diagnosed on clinical ground) and an even bigger submerged portion (the proportion of cases remaining undiagnosed) (Catassi, 2017). Even in countries showing a high level of CD awareness, on average 60%–70% of cases still remain undiagnosed and, being untreated, are exposed to the risk of long-term complications (Kvamme et al., 2022). In other countries, for example, India, the visible/diagnosed part of the CD iceberg is smaller than 5% (Ramakrishna et al., 2016).

How to approach the celiac iceberg is still a matter of debate, as discussed also in Chapter 8 of this book. Mass screening is highly sensitive and allows the detection of the large majority of CD cases together with an early treatment. If the screening is performed in the pediatric age group, it may prevent long-term complications (Catassi et al., 2022). However there are ethical and practical arguments against CD mass screening: (1) the advantages of an early diagnosis have not been entirely clarified; (2) life-long CD treatment by the gluten-free diet (GFD) may have a negative impact on the psycho-social aspects of daily life; (3) the GFD is associated with increased economic costs, both at the individual and societal level; (4) it remains arbitrary to establish the age of subjects at screening given that CD may develop at any age, as clearly shown by the previously mentioned American CLUE study (Catassi et al., 2010). Late CD

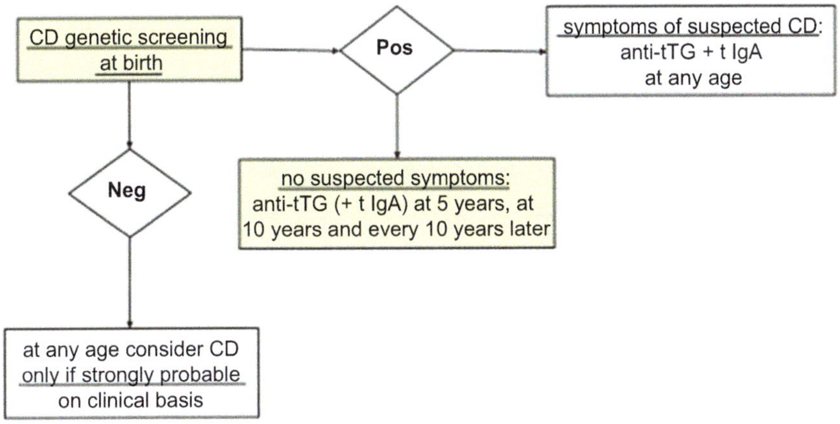

FIGURE 1.6 CD screening algorithm. Schematic representation of a proposed CD screening algorithm.

sero-conversion in previously negative subjects has been shown by long-term follow-up studies of screened subjects, a finding that is however more common in at-risk individuals (e.g., relatives of CD patients) (Paavola et al., 2022) than in the general population (Sandström et al., 2022).

Most experts and international societies currently agree that CD case-finding (testing only cases belonging to at-risk groups) is the best option to detect subclinical CD (Husby et al., 2020; Ludvigsson et al., 2014; Rubio-Tapia et al., 2023). For instance, the recently updated guidelines for the diagnosis and management of CD of the American College of Gastroenterology state that mass screening of asymptomatic individuals is not supported because CD does not fulfill some of the major World Health Organization criteria for mass screening, specifically the criteria that the natural history of the condition, including development of latent-to-declared disease, should be adequately understood and there should be an agreed policy on whom to treat as patients. Case finding is the currently preferred strategy to increase detection of cases, although the pros and cons of this approach are still a "matter of debate" (Rubio-Tapia et al., 2023).

Case-finding is indeed ethically sound, cheap and cost-effective. However case-finding also has important limitations: (1) low sensitivity; (2) it may have a socio-economic bias; (3) up to 85% of patients eventually found by screening have suffered from unrecognized symptoms for some time—even for many years—before diagnosis (Popp et al., 2019). Furthermore, as discussed in paragraph 5, even a policy of active case-finding is not only difficult to implement in the context of many national health systems but also poorly sensitive and specific.

In order to definitely improve the CD diagnostic rate, it is our opinion that new policies should be tested in the near future, for instance based on: (1) early identification (e.g., at birth) of HLA genetically at-risk individuals with stratification of HLA risk, and (2) targeted periodic serological CD testing of genetically predisposed individuals, for example, at 5 and 10 years of age and then every 10 years (or whenever symptoms of suspected CD do appear) (Fig. 1.6).

1.8 Conclusions

In most areas of the world, CD is one of the commonest life-long disorders showing a prevalence of 1%–2% in the general population. Even in areas currently characterized by a lower prevalence of disease, such as Eastern Asia, CD could become a frequent disorder due to the progressive "Westernization" of the diet on a worldwide basis. The reason/s of this high disease prevalence is/are still unclear, but could be related to the general increase of autoimmune diseases related to shifts in the immune response secondary to reduced frequency and severity of infectious diseases (so called "hygiene" hypothesis) and/or changes in the intestinal microbiome. Despite a significant increase in the incidence of clinically diagnosed CD, largely related to the diffusion of simple and sensitive diagnostic tools and increased awareness of CD clinical polymorphism, still a significant proportion of cases (> 50% on average with spikes of >90% in some Asian areas) remains undiagnosed in many countries.

A careful policy of systematic case-finding is currently considered the best buy for improving the CD detection rate, however this policy has significant limitations. Targeted screening of genetically predisposed individuals could represent the road ahead for an efficient intervention of secondary prevention (early diagnosis) of CD.

References

Ahadi, Z., Shafiee, G., Razmandeh, R., Keshtkar, A. A., Sani, M. N., Azemati, B., Sanaei, M., & Heshmat, R. (2016). Prevalence of celiac disease among the Iranian population: A systematic review and meta-analysis of observational studies. *Turkish Journal of Gastroenterology*, *27*(2), 122–128. Available from https://doi.org/10.5152/tjg.2015.150191.

Alarida, K., Harown, J., Ahmaida, A., Marinelli, L., Venturini, C., Kodermaz, G., Tozzoli, R., Mandolesi, A., Bearzi, I., & Catassi, C. (2011). Coeliac disease in Libyan children: A screening study based on the rapid determination of anti-transglutaminase antibodies. *Digestive and Liver Disease*, *43*(9), 688–691. Available from https://doi.org/10.1016/j.dld.2011.01.002.

Al-Toma, A., Volta, U., Auricchio, R., Castillejo, G., Sanders, D. S., Cellier, C., Mulder, C. J., & Lundin, K. E. A. (2019). European Society for the Study of Coeliac Disease (ESsCD) guideline for coeliac disease and other gluten-related disorders. *United European Gastroenterology Journal*, *7* (5), 583–613. Available from https://doi.org/10.1177/2050640619844125.

Ammerman, A. J., & Cavalli-Sforza, L. L. (1984). *The Neolithic transition and the genetics of populations in Europe*. Princeton University Press.

Catassi, C. (2017). New celiac icebergs are spotted, other are slowly emerging. *Journal of Pediatric Gastroenterology and Nutrition*, *65*(6), 601–602. Available from https://doi.org/10.1097/MPG.0000000000001698.

Catassi, C., Fabiani, E., Rätsch, I. M., Coppa, G. V., Giorgi, P. L., Pierdomenico, R., Alessandrini, S., Iwanejko, G., Domenici, R., Mei, E., Miano, A., Marani, M., Bottaro, G., Spina, M., Dotti, M., Montanelli, A., Barbato, M., Viola, F., Lazzari, R., & Ventura, A. (1996). The coeliac iceberg in Italy. A multicentre antigliadin antibodies screening for coeliac disease in school-age subjects. *Acta Paediatrica, International Journal of Paediatrics, Supplement*, *85*(Issue 412), 29–35.

Catassi, C., Kryszak, D., Bhatti, B., Sturgeon, C., Helzlsouer, K., Clipp, S. L., Gelfond, D., Puppa, E., Sferruzza, A., & Fasano, A. (2010). Natural history of celiac disease autoimmunity in a USA cohort followed since 1974. *Annals of Medicine*, *42*(7), 530–538. Available from https://doi.org/10.3109/07853890.2010.514285.

Catassi, C., Rätsch, I. M., Fabiani, E., Rossini, M., Coppa, G. V., Giorgi, P. L., Bordicchia, F., & Candela, F. (1994). Coeliac disease in the year 2000: Exploring the iceberg. *The Lancet*, *343*(8891), 200–203. Available from https://doi.org/10.1016/S0140-6736(94)90989-X.

Catassi, C., Rätsch, I. M., Gandolfi, L., Pratesi, R., Fabiani, E., El Asmar, R., Frijia, M., Bearzi, I., & Vizzoni, L. (1999). Why is coeliac disease endemic in the people of the Sahara? *Lancet*, *354*(9179), 647–648. Available from https://doi.org/10.1016/S0140-6736(99)02609-4.

Catassi, C., Verdu, E. F., Bai, J. C., & Lionetti, E. (2022). Coeliac disease. *The Lancet*, *399*(10344), 2413–2426. Available from https://doi.org/10.1016/S0140-6736(22)00794-2.

Catassi, G., Pjetraj, D., Gatti, L., & Catassi, C. (2023). Celiac Disease detection strategies: Poor performance of the case-finding policy. The American Journal of Gastroenterology, 118, 1702-1703.

Chin, M. W., Mallon, D. F., Cullen, D. J., Olynyk, J. K., Mollison, L. C., & Pearce, C. B. (2009). Screening for coeliac disease using anti-tissue transsglutaminase antibody assays, and prevalence of the disease in an Australian community. *Medical Journal of Australia*, *190*(8), 429–432. Available from https://doi.org/10.5694/j.1326-5377.2009.tb02491.x.

Choung, R. S., Larson, S. A., Khaleghi, S., Rubio-Tapia, A., Ovsyannikova, I. G., King, K. S., Larson, J. J., Lahr, B. D., Poland, G. A., Camilleri, M. J., & Murray, J. A. (2017). Prevalence and morbidity of undiagnosed celiac disease from a community-based study. *Gastroenterology*, *152*(4), 830–839. Available from https://doi.org/10.1053/j.gastro.2016.11.043, e5.

Cronin, C. C., & Shanahan, F. (2001). Why is celiac disease so common in Ireland? *Perspectives in Biology and Medicine*, *44*(3), 342–352. Available from https://doi.org/10.1353/pbm.2001.0045.

Fabiani, E., Peruzzi, E., Mandolesi, A., Garbuglia, G., Fanciulli, G., D'Appello, A. R., Gasparin, M., Bravi, E., Bearzi, I., Galeazzi, R., & Catassi, C. (2004). Anti-human versus anti-guinea pig tissue transglutaminase antibodies as the first-level serological screening test for coeliac disease in the general population. *Digestive and Liver Disease*, *36*(10), 671–676. Available from https://doi.org/10.1016/j.dld.2004.05.008.

Fasano, A., Berti, I., Gerarduzzi, T., Not, T., Colletti, R. B., Drago, S., Elitsur, Y., Green, P. H. R., Guandalini, S., Hill, I. D., Pietzak, M., Ventura, A., Thorpe, M., Kryszak, D., Fornaroli, F., Wasserman, S. S., Murray, J. A., & Horvath, K. (2003). Prevalence of Celiac disease in at-risk and not-at-risk groups in the United States: A large multicenter study. *Archives of Internal Medicine*, *163*(3), 286–292. Available from https://doi.org/10.1001/archinte.163.3.286.

Fukunaga, M., Ishimura, N., Abe, T., Takeda, M., Isomura, M., Kinoshita, Y., & Ishihara, S. (2020). Serological screening for celiac disease in adults in Japan: Shimane CoHRE study. *JGH Open*, *4*(4), 558–560. Available from https://doi.org/10.1002/jgh3.12334.

Gandolfi, L., Pratesi, R., Cordoba, J. C. M., Tauil, P. L., Gasparin, M., & Catassi, C. (2000). Prevalence of celiac disease among blood donors in Brazil. *American Journal of Gastroenterology*, *95*(3), 689–692. Available from https://doi.org/10.1016/S0002-9270(99)00906-5.

Gasbarrini, G., Miele, L., Corazza, G. R., & Gasbarrini, A. (2010). When was celiac disease born?: The italian case from the archeologic site of cosa. *Journal of Clinical Gastroenterology*, *44*(7), 502–503. Available from https://doi.org/10.1097/MCG.0b013e3181d345a5.

Gasbarrini, G., Rickards, O., Martínez-Labarga, C., Pacciani, E., Chilleri, F., Laterza, L., Marangi, G., Scaldaferri, F., & Gasbarrini, A. (2012). Origin of celiac disease: How old are predisposing haplotypes? *World Journal of Gastroenterology*, *18*(37), 5300–5304. Available from https://doi.org/10.3748/wjg.v18.i37.5300.

Gatti, S., Lionetti, E., Balanzoni, L., Verma, A. K., Galeazzi, T., Gesuita, R., Scattolo, N., Cinquetti, M., Fasano, A., Catassi, C., Annibali, R., Del Baldo, G., Franceschini, E., Palpacelli, A., Monachesi, C., Catassi, G. N., Trevisan, M. T., Anton, G., & Colombari, A. M. (2020). Increased Prevalence of Celiac Disease in School-age Children in Italy. *Clinical Gastroenterology and Hepatology*, *18*(3), 596–603. Available from https://doi.org/10.1016/j.cgh.2019.06.013.

He, L., Jia, X., Gu, Y., Miao, D., Waugh, K., Geno, C., Liu, E., Rewers, M., & Yu, L. (2020). Large-scale screening in general population children for celiac disease with a multiplex electrochemiluminescence (ECL) assay. *Journal of Immunology Research, 2020*. Available from https://doi.org/10.1155/2020/8897656.

Holmes, G. K. T., & Muirhead, A. (2017). Epidemiology of coeliac disease in a single centre in Southern Derbyshire 1958–2014. *BMJ Open Gastroenterology*, *4*(1), e000137. Available from https://doi.org/10.1136/bmjgast-2017-000137.

Husby, S., Koletzko, S., Korponay-Szabó, I., Kurppa, K., Mearin, M. L., Ribes-Koninckx, C., Shamir, R., Troncone, R., Auricchio, R., Castillejo, G., Christensen, R., Dolinsek, J., Gillett, P., Hróbjartsson, A., Koltai, T., Maki, M., Nielsen, S. M., Popp, A., Størdal, K., & Wessels, M. (2020). European society paediatric gastroenterology, hepatology and nutrition guidelines for diagnosing coeliac disease 2020. *Journal of Pediatric Gastroenterology and Nutrition*, *70*(1), 141–156. Available from https://doi.org/10.1097/MPG.0000000000002497.

Israeli, E., Hershcovici, T., Grotto, I., Rouach, Z., Branski, D., & Goldin, E. (2010). Prevalence of celiac disease in an adult Jewish population in Israel. *Israel Medical Association Journal*, *12*(5), 266–269. Available from http://www.ima.org.il/imaj/ar10may-03.pdf.

Ivarsson, A., Myléus, A., Norström, F., van der Pals, M., Rosén, A., Högberg, L., Danielsson, L., Halvarsson, B., Hammarroth, S., Hernell, O., Karlsson, E., Stenhammar, L., Webb, C., Sandström, O., & Carlsson, A. (2013). Prevalence of childhood celiac disease and changes in infant feeding. *Pediatrics*, *131*(3), e687–e694. Available from https://doi.org/10.1542/peds.2012-1015.

King, J. A., Jeong, J., Underwood, F. E., Quan, J., Panaccione, N., Windsor, J. W., Coward, S., Debruyn, J., Ronksley, P. E., Shaheen, A. A., Quan, H., Godley, J., Veldhuyzen Van Zanten, S., Lebwohl, B., Ng, S. C., Ludvigsson, J. F., & Kaplan, G. G. (2020). Incidence of celiac disease is increasing over time: A systematic review and meta-analysis. *American Journal of Gastroenterology*, *115*(4), 507–525. Available from https://doi.org/10.14309/ajg.0000000000000523.

Kondrashova, A., Mustalahti, K., Kaukinen, K., Viskari, H., Volodicheva, V., Haapala, A. M., Ilonen, J., Knip, M., Mäki, M., Hyöty, H., Hyöty, H., Knip, M., Viskari, H., Ilonen, J., Reunanen, A., Uibo, R., Salur, L., Ludvigsson, J., Marciulionyte, D., & Romanov, A. (2008). Lower economic status and inferior hygienic environment may protect against celiac disease. *Annals of Medicine*, *40*(3), 223–231. Available from https://doi.org/10.1080/07853890701678689.

Kvamme, J. M., Sørbye, S., Florholmen, J., & Halstensen, T. S. (2022). Population-based screening for celiac disease reveals that the majority of patients are undiagnosed and improve on a gluten-free diet. *Scientific Reports*, *12*(1). Available from https://doi.org/10.1038/s41598-022-16705-2.

Lionetti, E., Pjetraj, D., Gatti, Catassi, G., Bellantoni, A., Boffardi, M., Cananzi, M., Cinquetti, M., Francavilla, R., Malamisura, B., Montuori, M., Zuccotti, G., Cristofori, F., Gaio, P., Passaro, T., Penagini, F., Testa, A., Trovato, C., & Catassi, C. (2023) Prevalence and detection rate of celiac disease in Italy: Results of a SIGENP multicenter screening in school-age children. Digestive and Liver Disease: Official Journal of the Italian Society of Gastroenterology and the Italian Association for the Study of the Liver, 22, 871–874.

Lionetti, E., & Catassi, C. (2014). Co-localization of gluten consumption and HLA-DQ2 and -DQ8 genotypes, a clue to the history of celiac disease. *Digestive and Liver Disease*, *46*(12), 1057–1063. Available from https://doi.org/10.1016/j.dld.2014.08.002.

Lönnroth, K., Corbett, E., Golub, J., Godfrey-Faussett, P., Uplekar, M., Weil, D., & Raviglione, M. (2013). Systematic screening for active tuberculosis: Rationale, definitions and key considerations. *International Journal of Tuberculosis and Lung Disease*, *17*(3), 289–298. Available from https://doi.org/10.5588/ijtld.12.0797.

Ludvigsson, J. F., Bai, J. C., Biagi, F., Card, T. R., Ciacci, C., Ciclitira, P. J., Green, P. H. R., Hadjivassiliou, M., Holdoway, A., Van Heel, D. A., Kaukinen, K., Leffler, D. A., Leonard, J. N., Lundin, K. E. A., McGough, N., Davidson, M., Murray, J. A., Swift, G. L., Walker, M. M., & Sanders, D. S. (2014). Diagnosis and management of adult coeliac disease: Guidelines from the British society of gastroenterology. *Gut*, *63*(8), 1210–1228. Available from https://doi.org/10.1136/gutjnl-2013-306578.

Ludvigsson, J. F., Rubio-Tapia, A., Van Dyke, C. T., Melton, L. J., Zinsmeister, A. R., Lahr, B. D., & Murray, J. A. (2013). Increasing incidence of celiac disease in a north American Population. *American Journal of Gastroenterology*, *108*(5), 818–824. Available from https://doi.org/10.1038/ajg.2013.60.

Makharia, G. K., Singh, P., Catassi, C., Sanders, D. S., Leffler, D., Ali, R. A. R., & Bai, J. C. (2022). The global burden of coeliac disease: Opportunities and challenges. *Nature Reviews Gastroenterology and Hepatology*, *19*(5), 313–327. Available from https://doi.org/10.1038/s41575-021-00552-z.

Mora, M., Litwin, N., Toca, M. D. C., Azcona, M. I., Neffa, R. S., Battiston, F., Solaegui, M., Ortiz, G., Wagener, M., Olivera, J., Marchisone, S., Oropeza, G., Bastianelli, C., González, A., & Rezzónico, G. (2012). Prevalence of celiac disease: A multicenter trial in pediatric population from five urban districts of Argentina. *Archivos Argentinos de Pediatria*, *110*(6), 490–496. Available from https://doi.org/10.5546/aap.2012.490.

Mustalahti, K., Catassi, C., Reunanen, A., Fabiani, E., Heier, M., McMillan, S., Murray, L., Metzger, M. H., Gasparin, M., Bravi, E., & Mki, M. (2010). The prevalence of celiac disease in Europe: Results of a centralized, international mass screening project. *Annals of Medicine*, *42*(8), 587–595. Available from https://doi.org/10.3109/07853890.2010.505931.

Mylotte, M., Egan-Mitchel, B., McCarthy, C. F., & McNicholl, B. (1973). Incidence of coeliac disease in the west of ireland. *British Medical Journal*, *1*(5855), 703–705. Available from https://doi.org/10.1136/bmj.1.5855.703.

Paavola, S., Kurppa, K., Huhtala, H., Saavalainen, P., Lindfors, K., & Kaukinen, K. (2022). Coeliac disease re-screening among once seronegative at-risk relatives: A long-term follow-up study. *United European Gastroenterology Journal*, *10*(6), 585–593. Available from https://doi.org/10.1002/ueg2.12255.

Poddighe, D., Rakhimzhanova, M., Marchenko, Y., & Catassi, C. (2019). Pediatric celiac disease in Central and East Asia: Current knowledge and prevalence. *Medicina*, *55*(1), 11. Available from https://doi.org/10.3390/medicina55010011.

Popp, A., Kivelä, L., Fuchs, V., & Kurppa, K. (2019). Diagnosing celiac disease: Towards wide-scale screening and serology-based criteria. *Gastroenterology Research and Practice*, 2019. Available from https://doi.org/10.1155/2019/2916024.

Ramakrishna, B. S., Makharia, G. K., Chetri, K., Dutta, S., Mathur, P., Ahuja, V., Amarchand, R., Balamurugan, R., Chowdhury, S. D., Daniel, D., Das, A., George, G., Gupta, S. D., Krishnan, A., Prasad, J. H., Kaur, G., Pugazhendhi, S., Pulimood, A., Ramakrishna, K., & Verma, A. K. (2016). Prevalence of adult celiac disease in India: Regional variations and associations. *American Journal of Gastroenterology*, *111*(1), 115–123. Available from https://doi.org/10.1038/ajg.2015.398.

Remes-Troche, J. M., Ramírez-Iglesias, M. T., Rubio-Tapia, A., Alonso-Ramos, A., Velazquez, A., & Uscanga, L. F. (2006). Celiac disease could be a frequent disease in Mexico: Prevalence of tissue transglutaminase antibody in healthy blood donors. *Journal of Clinical Gastroenterology*, *40*(8), 697–700. Available from https://doi.org/10.1097/00004836-200609000-00007.

Roberts, S. E., Morrison-Rees, S., Thapar, N., Benninga, M. A., Borrelli, O., Broekaert, I., Dolinsek, J., Martin-de-Carpi, J., Mas, E., Miele, E., Pienar, C., Ribes-Koninckx, C., Thomassen, R. A., Thomson, M., Tzivinikos, C., Thorne, K., John, A., & Williams, J. G. (2021). Systematic review and meta-analysis: the incidence and prevalence of paediatric coeliac disease across Europe. *Alimentary Pharmacology and Therapeutics*, *54*(2), 109–128. Available from https://doi.org/10.1111/apt.16337.

Rubio-Tapia, A., Hill, I. D., Semrad, C., Kelly, C. P., Greer, K. B., Limketkai, B. N., & Lebwohl, B. (2023). American College of Gastroenterology Guidelines Update: Diagnosis and management of celiac disease. *American Journal of Gastroenterology*, *118*(1), 59–76. Available from https://doi.org/10.14309/ajg.0000000000002075.

Saini, V., & Garg, K. (2020). Case finding strategies under National Tuberculosis Elimination Programme (NTEP. *Indian Journal of Tuberculosis*, *67*(4), S101–S106. Available from https://doi.org/10.1016/j.ijtb.2020.09.029.

Sandström, O., Norström, F., Carlsson, A., Högberg, L., van der Palz, M., Stenhammar, L., Webb, C., Ivarsson, A., & Myléus, A. (2022). Five-year follow-up of new cases after a coeliac disease mass screening. *Archives of Disease in Childhood*, *107*(6), 596–600. Available from https://doi.org/10.1136/archdischild-2021-322755.

Simoons, F. J. (1981). *Coeliac disease as a geographic problem* (pp. 179–199). Masson.

Singh, P., Arora, A., Strand, T. A., Leffler, D. A., Catassi, C., Green, P. H., Kelly, C. P., Ahuja, V., & Makharia, G. K. (2018). Global prevalence of celiac disease: Systematic review and meta-analysis. *Clinical Gastroenterology and Hepatology*, *16*(6), 823–836. Available from https://doi.org/10.1016/j.cgh.2017.06.037, e2.

Singh, P., Arora, S., Singh, A., Strand, T. A., & Makharia, G. K. (2016). Prevalence of celiac disease in Asia: A systematic review and meta-analysis. *Journal of Gastroenterology and Hepatology (Australia)*, *31*(6), 1095–1101. Available from https://doi.org/10.1111/jgh.13270.

Tommasini, A., Not, T., Kiren, V., Baldas, V., Santon, D., Trevisiol, C., Berti, I., Neri, E., Gerarduzzi, T., Bruno, I., Lenhardt, A., Zamuner, E., Spanò, A., Crovella, S., Martellossi, S., Torre, G., Sblattero, D., Marzari, R., Bradbury, A., & Ventura, A. (2004). Mass screening for coeliac disease using antihuman transglutaminase antibody assay. *Archives of Disease in Childhood*, *89*(6), 512–515. Available from https://doi.org/10.1136/adc.2003.029603.

Unalp-Arida, A., Ruhl, C. E., Choung, R. S., Brantner, T. L., & Murray, J. A. (2017). Lower prevalence of celiac disease and gluten-related disorders in persons living in Southern vs Northern Latitudes of the United States. *Gastroenterology*, *152*(8), 1922–1932. Available from https://doi.org/10.1053/j.gastro.2017.02.012, e2.

Verma, A. K., Mechenro, J., Monachesi, C., Venugopal, G., Catassi, G. N., Lionetti, E., Ramakrishna, B. S., & Catassi, C. (2022). Distribution of celiac disease predisposing genes HLA-DQ2 and HLA-DQ8 in the native population of Southern India. *Indian Journal of Gastroenterology*, *41*(3), 240–246. Available from https://doi.org/10.1007/s12664-022-01251-6.

Yuan, J., Zhou, C., Gao, J., Li, J., Yu, F., Lu, J., Li, X., Wang, X., Tong, P., Wu, Z., Yang, A., Yao, Y., Nadif, S., Shu, H., Jiang, X., Wu, Y., Gilissen, L., & Chen, H. (2017). Prevalence of celiac disease autoimmunity among adolescents and young adults in China. *Clinical Gastroenterology and Hepatology*, *15*(10), 1572–1579. Available from https://doi.org/10.1016/j.cgh.2017.04.025, e1.

Zhou, C., Gao, F., Gao, J., Yuan, J., Lu, J., Sun, Z., Xu, M., Engel, J., Hui, W., Gilissen, L., & Chen, H. (2020). Prevalence of coeliac disease in Northwest China: Heterogeneity across Northern Silk road ethnic populations. *Alimentary Pharmacology and Therapeutics*, *51*(11), 1116–1129. Available from https://doi.org/10.1111/apt.15737.

Chapter 2

Wheat gluten proteins: from taxonomy to toxic epitopes

Gianfranco Mamone[1], Luigia Di Stasio[1], Serena Vitale[2] and Carmen Gianfrani[2]
[1]*Institute of Food Science, Department of Biology, Agriculture and Food Sciences, National Research Council of Italy, Avellino, Italy,*
[2]*Institute of Biochemistry and Cell Biology, Department of Biomedical Sciences, National Research Council of Italy, Naples, Italy*

2.1 Introduction

Wheat is one of the most widely consumed cereals worldwide. However, wheat-based foods have to be strictly avoided by a substantial number of people suffering from celiac disease (CD), which affects approximately 1% of the general population. Gluten contains hundreds of homologous proteins characterized by high levels of glutamine and proline residues. This peculiar aminoacidic composition strongly hampers physiologic digestion by gastrointestinal proteases, and large peptides with immunogenic potential are delivered to the gut, eliciting mucosal inflammation in CD patients.

Gluten proteins represent the main storage nutrient of germinating plants and, albeit with a poor nutritional value for humans, they are largely used in the food industry for their viscoelasticity properties in baking products. The first evidence of settled wheat agriculture is documented about 10,000 years ago, with the cultivation of diploid (genome AA) and tetraploid (genome AABB) species, such as einkorn and emmer crops. Over the millennia, extensive breeding of these ancient grains led to the rise of modern tetraploid durum and hexaploid soft (genome AABBDD) species, which are among the most consumed cereals worldwide. However, almost concomitantly with the spreading of wheat cultivation, gluten intolerance arose among the wheat-consuming populations due to the large immunogenicity potential of gluten proteins.

By using sophisticated analytical and immunological approaches, including biochemistry, proteomic, mass spectroscopy, enzyme-linked immunosorbent assay (ELISA), and celiac-gut T-cell lines, more than 50 different gluten peptides have been identified so far that elicit an inflammatory T-cell response in CD patients. The comprehensive definition of the gluten epitope repertoire is crucial for designing immunological or enzymatic therapies as alternatives to the gluten-exclusion diet, as well as for obtaining wheat species naturally devoid of immunogenic sequences, or resulting from gene silencing molecular engineering. Furthermore, the availability of wheat species with low immunotoxic gluten proteins, although not suitable for the diet of celiacs, could have important implications for disease prevention in high genetic-risk subjects.

2.2 Wheat species: ancient heritage and modern cultivars

Wheat is among the most important cereal crops and a major source of carbohydrates and proteins in the human diet. The term "wheat" refers to several species and subspecies belonging to the botanical genus "*Triticum*." Although the genetics of wheat is a complex matter, due to the natural hybridization and polyploidization processes that occurred over millennia, *Triticum* (*T.*) species are classified according to the ploidy levels into: (1) diploid wheat (genome AA, BB, or DD), (2) tetraploid wheat (genome AABB), and (3) hexaploid wheat (genome AABBDD). The hexaploid *T. aestivum* and the tetraploid *T. turgidum* are the most economically important species. *T. aestivum*, also known as soft or bread wheat, is more suitable for the preparation of cakes and biscuits (bakery food) and contributes to about 95% of the world's wheat production. *T. turgidum* is grown predominantly in the Mediterranean area and represents about 5% of total wheat production, and it is used mainly for the production of pasta (De Vita, 2019).

Pediatric and Adult Celiac Disease. DOI: https://doi.org/10.1016/B978-0-443-13359-6.00017-0

T. species are also subdivided according to their historical origin into ancient, heritage, and modern wheat. "Ancient grains" represent early forms of landrace, originated through spontaneous plant hybridizations, and remained unchanged over the centuries. Ancient grains include einkorn (diploid *T. monococcum*, variety *monococcum*), emmer (tetraploid *T. turgidum*, variety *dicoccum*), and spelt (hexaploid *T. aestivum*, variety *spelta*). Hybridization events were responsible for the formation of "heritage varieties," which were selected before the mid-20th century. "Heritage varieties" refers to traditional *T.* ones that have kept unchanged their original characteristics. Several heritage cultivars of *T. aestivum* and *T. durum* remained unchanged over the years, that is, Senatore Cappelli (durum) and autonomia (aestivum) (Spisni et al., 2019). The "modern wheat" grain originated, particularly after World War II, as a consequence of the intensive scientific plant breeding campaign aimed at increasing the production of wheat crops with high yields and improved technological properties (Prandi et al., 2017).

Ancient, heritage, and modern varieties own different yields and morphological and physiological characteristics, including the height of the plants, the length of the phenological cycle, susceptibility to biotic and abiotic stresses, and rheological properties quality (Khan, 2016).

In the last decade, ancient and heritage grains have been reintroduced due to the growing awareness of alternative cereals considered natural and healthy (Bordoni et al., 2017). However, the scientific panorama still debates the benefits of ancient or heritage wheats, which are often supposed to have superior health-promoting properties compared to modern cultivars.

2.2.1 Genomic: the concept of ploidy and natural crossing of ancestral wheat species

The complex genome of modern wheat has its roots in wild and ancient hybridizations among species, and an evolutionary process occurred from ancient to modern wheat (Fig. 2.1). The domestication of wheat is dated to $\approx$ 10,000 BCE with the extensive cultivation of einkorn diploid species (*T. monococcum*, AA genome). Diploid species also exist in wild forms named *T. urartu* (AA), Ae. speltoides (BB), and *Aegilops tauschii* (DD) (Molberg et al., 2005).

Tetraploid wheat was likely domesticated about the same time as diploid wheat and contains two genomes designated as AABB, originating through natural cross-breeding of the einkorn donors of AA (*T. urartu*) with a related wild grass donor of the BB genome (probably an unknown form of modern *Aegilops speltoides*). The domesticated tetraploid species, *T. turgidum*, includes two subspecies/varieties named emmer and durum (Peng et al., 2011). The wild form of tetraploid grain is *T. dicoccoides*, while minor forms are Khorasan, *T. turgidum* ssp. *turanicum*, which includes the form known as Kamut.

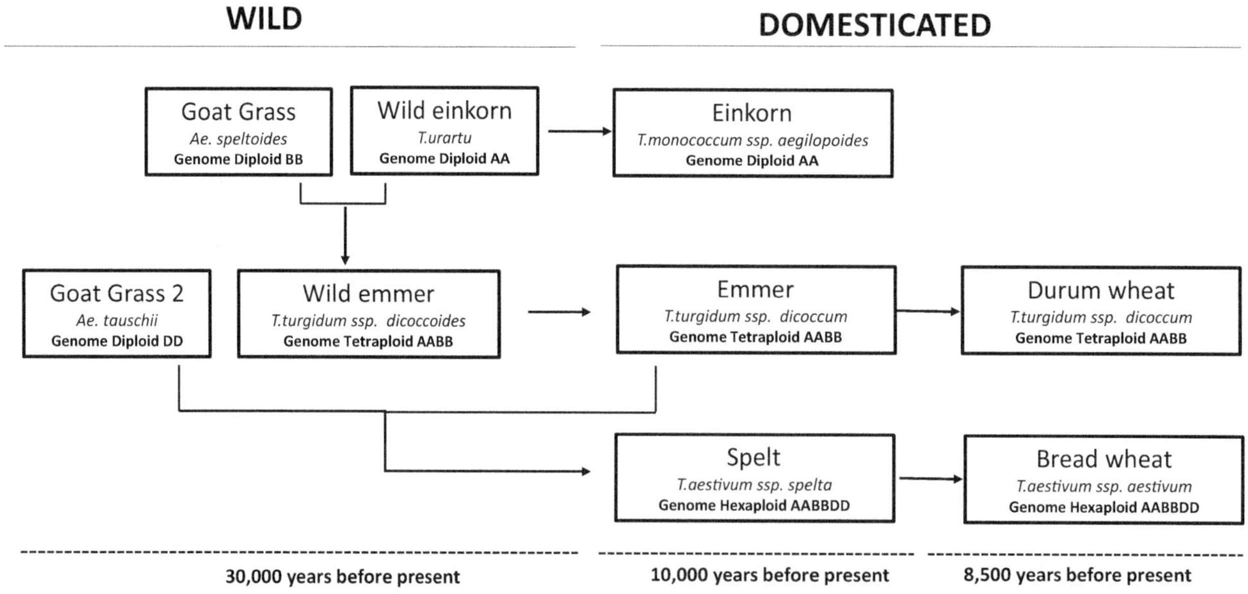

FIGURE 2.1 Evolution of the wheat species. Natural crossing of diploid wheat ancestors (with the AA and BB genomes) led to the development of tetraploid emmer wheat (AABB genomes), which subsequently became domesticated and diversified into durum wheat. Cultivated bread wheat (AABBDD genomes) arose from the crossing of the durum wheat (AABB) and diploid wild goat grass (DD).

The hexaploid wheat (genome AABBDD) has no wild-form equivalent and originated through crosslinking of durum wheat (AABB) and wild diploid species known as *Aegilops tauschii*, which donated the D genome (Peng et al., 2011).

All wheat species have chromosome numbers in multiples of seven (7n). More specifically, diploid wheat has 14 chromosomes (7 × 2) while tetraploid and hexaploid wheat have multiple of 14; more precisely, tetraploid wheat has 28 (7 × 4) chromosomes, and hexaploid wheat has 42 (7 × 6) chromosomes. Wheat chromosomes are named using a number, with the latter indicating to which genome they belong. Because of the close relationship between the donor species having the A, B, and D genomes, this polyploidy composition implies a high genetic homology among the different wheat species.

2.3 Protein wheat chemistry: albumin/globulin, gliadins, and glutenins

Wheat protein content accounts for about 10% to 15% of the dry weight of seed. Ancient species, such as *T. monococcum*, have a protein content sharply superior to that of modern ones, averaging 18% and often exceeding 20% (Borghi et al., 1996; Brandolini et al., 2008; Corbellini et al., 1999). In general, wheat proteins are classified according to Osborne's classification into soluble albumins/globulins proteins and insoluble gluten proteins. Gluten represents 80% of total grain protein content and is divided into alcohol-soluble monomeric gliadins and alcohol-insoluble polymeric glutenins. The former are monomeric proteins with molecular weights ranging from 28,000 to 55,000, subclassified into α/β-, γ-, and ω-gliadins types. The glutenin fractions consist of disulfide-linked proteins with molecular masses ranging from about 60,000 to more than 10 million. Following the reduction of inter- and intra-chain disulfide bonds, the glutenin subunits show solubility in aqueous alcohols similar to gliadins. Based on their electrophoretic mobility, glutenins are divided into high-molecular-weight (HMW-GS) and low-molecular-weight (LMW-GS) proteins (Wieser, 2007). LMW-GS are further subclassified into B-, C-, and D-types (Payne & Corfield, 1979).

2.4 Technological aspects: why gluten proteins are the dough treasure?

According to the Codex Alimentarius, "gluten is a protein fraction of wheat, insoluble in water and NaCl 0.5 mol/L", while according to the US Food and Drug Administration, gluten is defined as "proteins that occur naturally in these same grains and can cause adverse health effects in people with celiac disease" (Codex Alimentarius Commission, 2008). From a technological point of view, gluten is defined as the "cohesive, viscoelastic proteinaceous material prepared as a by-product of the isolation of starch from wheat flour" (Day et al., 2006; Lamacchia et al., 2014).

When wheat flour is added and mixed with water, gluten forms a dough with unique rheological properties. Notably, the chemistry behind these peculiar properties of gluten protein has been the subject of study for over 250 years (Bailey, 1941). Gluten improves the emulsifying properties during kneading, retains leavening gases, provides cohesiveness to dough during processing, fixes the crumb structure, and gives elasticity to the bread texture and more generally to bakery products (Krupa-Kozak & Drabińska, 2015). Furthermore, gliadins and glutenins are both responsible for rheological properties. In fact, gliadins take part in the development of the gluten network during mixing through the formation of intermolecular hydrogen bonds and hydrophobic bonds between nonpolar amino acid side chains, which also interact with the flour lipids. For these reasons, gliadins are important in determining the extensibility characteristics of dough (Khatkar & Schofield, 2002; Veraverbeke & Delcour, 2002), while glutenins confer strength and elasticity to the dough (Khatkar et al., 2002).

2.5 Large utilization of vital gluten in the food industry, as additive

Vital wheat gluten is a protein concentrate (approximately 75% protein, up to 8% moisture, and varying amounts of starch, lipids, and fibers), obtained from wheat flour starting with the separation of wheat flour into starch and gluten, its main constituents, and ending with a drying step (Ortolan & Steel, 2017; Van Der Borght et al., 2005). Given the extraordinary technological qualities of gluten, the vital wheat gluten is nowadays widely used in the baking industry as an ingredient (Marchetti et al., 2012; Melnyk et al., 2012). It is usually added to improve the viscoelastic properties of bakery products with poor bread quality or for the production of certain specialty breads, such as loaf bread with added fibers and/or germ and loaf bread obtained with long fermentation (Ortolan & Steel, 2017; Tedrus et al., 2001). The wide use of vital gluten has been hypothesized as a possible reason for the striking increase in the incidence of gluten-related disorders over the last decades. Interestingly, although it has been reported that the use of vital gluten as a food additive in different food products has tripled since 1977, how the amount of gluten consumption impacts the CD incidence is still controversial (Kasarda, 2013). Based on various assumptions, the author did not find clear evidence of an increase in the gluten content of wheat during the 20th century or the spread of CD. The author declared that wheat breeding to obtain higher gluten content or changes in the *pro-capite* intake of wheat might play a role, but there is a lack of suitable data on the incidence of CD to confirm those possibilities (Kasarda, 2013). Later studies done in the "Prevent Celiac Disease" study, a prospective randomized trial of infants with a family history of CD, tested the tolerance to gluten in this high-risk group through the introduction of small quantities of gluten beginning at 4 months of age. The authors showed negative results regarding the role of vital gluten in CD spread (Ludvigsson & Green, 2014; Vriezinga et al., 2014). In contrast, it was reported that the amount of gluten consumed within the first 2 years of age of Swedish genetically predisposed children increased CD risk by at least twofold (Andrén Aronsson et al., 2016).

Finally, to what extent a relatively small amount of vital wheat gluten is influential in CD is still a matter of debate.

2.6 Resistance of gluten proteins to gastrointestinal proteases: in vitro digestion model

Several in vitro digestion models have been proposed since the 1990s to study the changes of gluten proteins along with the gastrointestinal tract in relation to CD pathogenesis. It is important to emphasize that the gluten primary structure plays a fundamental role in determining resistance to gastrointestinal proteases. In fact, the high proline/glutamine residue content of gluten proteins implies a marked inaccessibility of proteases at the cleavage site. The resulting undigested oligopeptides can pass through the small intestine mucosal barrier until the lamina propria, where they are deamidated by tissue transglutaminase (tTG) and thus able to trigger the T-cell-mediated adaptive immune responses (Hausch et al., 2002; Shan et al., 2002, 2005). However, the early in vitro digestion studies were limited to pepsin and trypsin proteolysis to mimic the gastric and pancreatic phases, respectively. The use of a more accurate in vitro digestion strategy involving duodenal enzymes (trypsin, chymotrypsin, elastase, and carboxypeptidase) as well as the *eso-* and *endo-*peptidases of the brush-border membrane of villi enterocytes is to be attributed to Shan et al. (2002, 2005). By mimicking duodenal and intestinal digestion, the authors were able to demonstrate the digestive protease resistance of 33-mer from α-gliadins and 26-mer from γ-gliadins. A similar approach was employed to demonstrate the resistance to digestive proteases of a 25-mer peptide, which includes the toxic sequence P31–43 (Mamone et al., 2007).

Notably, by using this extensive digestive protocol, our group demonstrated that gluten proteins from *T. monococcum* are more susceptible to digestion compared to those from soft wheat *T. aestivum*, findings that imply a lower immunotoxicity on T cells (Gianfrani et al., 2005, 2012). In recent times, a more complex in vitro digestion model, named INFOGEST, has been applied by the scientific community, designed to reproduce the physiological process as close as possible to the in vivo one. The novelty of the INFOGEST model is to investigate the behavior of dietary components in the real food matrix without any previous protein extraction. This strategy has also been applied in recent years to the study of immunogenic peptides from wheat products. Ogilvie et al. (2020) assessed the immunogenicity of gluten residues in wheat-based food, demonstrating that gluten-toxic peptides were rapidly released from the food matrix during the intestinal phase. Boukid et al. (2019) successfully applied the INFOGEST method to detect and quantify gluten peptides relevant to CD pathogenesis.

2.7 Immunogenicity of gluten proteins: pathogenic implication for celiac disease

A combination of genetic and environmental causes—not yet fully understood—contributes to the onset of CD. The undisputed environmental factor that triggers the immune response is wheat gluten and related proteins in rye and barley. Upon dietary consumption, the gastrointestinal-resistant gliadin and glutenin fragments, ranging from 15 to 50 amino acid residues, are released in high concentration in the gut lumen, stimulating both innate and adaptive immune responses in CD patients (Gianfrani et al., 2005). Several studies have demonstrated that a key role in inflammatory responses in CD is played by intestinal CD4 + T cells (Camarca et al., 2009, 2017; Lundin et al., 1993; Tye-Din et al., 2010) that recognize gluten peptides when presented by disease-associated HLA-DQ2/DQ8 molecules expressed on the antigen-presenting cells (APCs). Following recognition of gluten-derived peptides, CD4 + T cells, mainly residing in the lamina propria, proliferate and produce inflammatory cytokines dominated by interferon-γ and interleukin-21. Gluten-specific CD4 + T cells infiltrate the intestinal mucosa of CD patients (Lundin et al., 1993), but not healthy subjects (Molberg et al., 1997), and, furthermore, they persist in the CD intestinal mucosa for decades as memory cells, prone to being activated by gluten peptides (Christophersen et al., 2016). Since these intestinal gluten-reactive T cells can be isolated from small bowel biopsies and in vitro expanded as T-cell lines, they represent an important resource for bioassays useful to detect the immunogenicity of many gluten peptides and to dissect CD pathogenesis.

TABLE 2.1 Immunodominant DQ2.5-restricted T-cell gluten peptides.[a]

Epitope	Previous peptide name	Peptide-binding motif 9-Mer core regions[b]								
α-Gliadins										
DQ2.5-glia-α1a	DQ2-α-I, α9	P	F	P	Q	P	**Q**	L	P	Y
DQ2.5-glia-α1b	DQ2-α-III	P	Y	P	Q	P	**Q**	L	P	Y
DQ2.5-glia-α2	DQ2-α-II, α2	P	Q	P	**Q**	L	P	Y	P	Q
DQ2.5-glia-α3	Glia-α20	F	R	P	**Q**	Q	P	Y	P	Q
γ-Gliadins										
DQ2.5-glia-γ1	DQ2-γ-I	P	Q	Q	S	F	P	**Q**	Q	**Q**
DQ2.5-glia-γ2	DQ2-γ-II	I	Q	P	**Q**	Q	P	A	Q	L
DQ2.5-glia-γ3	DQ2-γ-III	**Q**	Q	P	**Q**	Q	P	Y	P	**Q**
DQ2.5-glia-γ4a	DQ2-γ-IV	S	Q	P	**Q**	Q	**Q**	F	P	Q
DQ2.5-glia-γ4b	DQ2-γ-VIIc	P	Q	P	**Q**	Q	**Q**	F	P	Q
DQ2.5-glia-γ4c	DQ2-γ-VIIa	**Q**	Q	P	**Q**	Q	P	F	P	Q
DQ2.5-glia-γ4d	DQ2-γ-VIIb	P	Q	P	**Q**	Q	P	F	C	**Q**
DQ2.5-glia-γ4e		L	Q	P	**Q**	Q	P	F	P	**Q**
DQ2.5-glia-γ5	DQ2-γ-VI	**Q**	Q	P	F	P	**Q**	Q	P	Q
ω-Gliadins										
DQ2.5-glia-ω1	DQ2-ω-I	P	F	P	Q	P	**Q**	Q	P	F
DQ2.5-glia-ω2	DQ2-ω-II	P	Q	P	**Q**	Q	P	F	P	W
Glutenins										
DQ2.5-glut-L1	Glutenin-17	P	F	S	**Q**	Q	**Q**	Q	P	V
DQ2.5-glut-L2	Glutenin-156	F	S	**Q**	Q	Q	**Q**	S	P	F
Secalins										
DQ2.5-sec-1	Sec-α9	P	F	P	Q	P	**Q**	Q	P	F
DQ2.5-sec-2	Sec-α2	P	Q	P	**Q**	Q	P	F	P	Q
DQ2.5-sec-3	P	F	P	**Q**	Q	P	E	Q	I	
Hordeins										
DQ2.5-hor-1	Hor-α9	P	F	P	Q	P	**Q**	Q	P	F
DQ2.5-hor-2	Hor-α2	P	Q	P	**Q**	Q	P	F	P	Q
DQ2.5-hor-3a	Hor-I	P	I	P	**Q**	Q	P	Q	P	Y
DQ2.5-hor-3b		P	Y	P	**Q**	Q	P	Q	P	Y
Avenins										
DQ2.5-ave-1a	Av-α9A	P	Y	P	E	Q	**Q**	E	P	F
DQ2.5-ave-1b	Av-α9B	P	Y	P	E	Q	**Q**	Q	P	F
DQ2.5-ave-1c		P	Y	P	E	Q	**Q**	Q	P	I

Glutamine residues deamidated by tTG are shown in the table as **Q**.
[a]*References for current nomenclature: Sollid et al. (2012, 2020).*
[b]*The 9-mer core regions are reported for the most immunogenic peptides derived from wheat gluten (α-, γ-, and ω-gliadins), glutenins (low-molecular-weight glutenin: glut-L, and high-molecular-weight glutenin: glut-H), barley (hordeins), rye (secalins), and oat (avenins).*

TABLE 2.2 Immunodominant DQ8-restricted T-cell gluten peptides.[a]

Epitope	Previous peptide name	Peptide-binding motif								
		9-Mer core regions*								
α-Gliadins										
DQ8-glia-α1	DQ8-α-I	**Q**	G	S	F	Q	P	S	Q	**Q**
γ-Gliadins										
DQ8-glia-γ1a	DQ8-γ-Ia	**Q**	Q	P	**Q**	Q	P	F	P	Q
DQ8-glia-γ1b	DQ8-γ-Ib	**Q**	Q	P	**Q**	Q	P	Y	P	**Q**
DQ8-glia-γ2		P	Q	Q	S	F	P	**Q**	Q	**Q**
Glutenins										
DQ8-glut-H1	DQ8-GLT-I	**Q**	G	Y	Y	P	T	S	P	**Q**

*The 9-mer core regions are reported for the most immunogenic peptides derived from wheat gliadins and glutenins and restricted by HLA-DQ8 molecule. Glutamine residues deamidated by tTG are shown in the table as **Q**.
[a]*References: Sollid et al. (2012, 2020).*

So far, a large repertoire of more than 50 different gluten immunogenic peptides (epitopes) have been identified by using T-cell lines or peripheral blood mononuclear cells, these latter taken from CD volunteers after a brief (1–3 days) gluten consumption, as described more extensively below (Camarca, Radano, et al., 2012; Tye-Din et al., 2010).

This large repertoire of immune-toxic peptides can be explained by the high homology between the various gluten proteins, either in α-, γ-, and ω-gliadins or in LMW- and HMW glutenins (Wieser, 2007). This results in hundreds of sequences able to stimulate a proinflammatory response in CD patients (Tables 2.1 and 2.2). Furthermore, immunogenic peptides with high homology with the wheat T-cell epitopes were found in barley hordeins and rye secalins (Tye-Din et al., 2010) (Table 2.1). Although avenins from oats are considered mostly safe (Garsed & Scott, 2007), some CD patients are clinically sensitive to their consumption (Lundin et al., 2003).

So far, great effort has been made in the CD-scientific community to identify potentially immunotoxic gluten peptides. Different factors determine immunogenicity, first of all, the resistance to proteolytic degradation due to the peculiar aminoacidic sequence, but also the susceptibility to tTG-mediated deamidation and the capability to bind CD-associated HLA-DQ2 and DQ8 molecules (Camarca, del Mastro, et al., 2012; Sollid et al., 2012, 2020).

2.8 QXP aminoacidic consensus for deamidation by tissue transglutaminase

The binding of gluten peptides to CD-associated HLA-DQ2/DQ8 molecules expressed on APC is a fundamental step in the activation of T cells. However, it has remained unexplained for a long time how this specific binding could happen since these HLA molecules prefer to bind negatively charged peptides and gluten proteins constitutively lack them. In 1998 it was demonstrated that tTG specifically catalyzes a deamidation reaction that converts glutamine into glutamic acid (Molberg et al., 1998; Van De Wal et al., 1998). In particular, tTG preferably deamidates glutamines in the consensus sequence QXP, in which Q and P represent glutamine and proline, respectively, and X represents any other amino acid (Fleckenstein et al., 2002; Hardy et al., 2015; Ráki et al., 2017; Vader et al., 2002).

2.8.1 Aminoacidic sequence motifs for binding to HLA-DQ2 and HLA-DQ8 restriction molecules and prediction analysis

Crystallographic studies of peptide–HLA complexes indicated that HLA-DQ2-restricted gluten peptides have a glutamic acid (E) allocated in position (P) 4, P6, or P7, and proline residues preferentially accommodated in P1, P3, P6, and P8, but never in P2, P4, P7, and P9 (Fig. 2.2; Tables 2.1 and 2.3). With regard to HLA-DQ8-restricted gluten epitopes, the structural analysis pointed out that negatively charged amino acids are required at P1 and P9 for binding to HLA-DQ8 molecules, and E residues can also be located at the P4 site, while proline residues can be positioned in P3, P6, and P8 (Fig. 2.2 and Table 2.2). The selective deamidation of the partially digested gluten fragments increases their binding affinity to HLA molecules, resulting in the formation of stable peptide–HLA complexes.

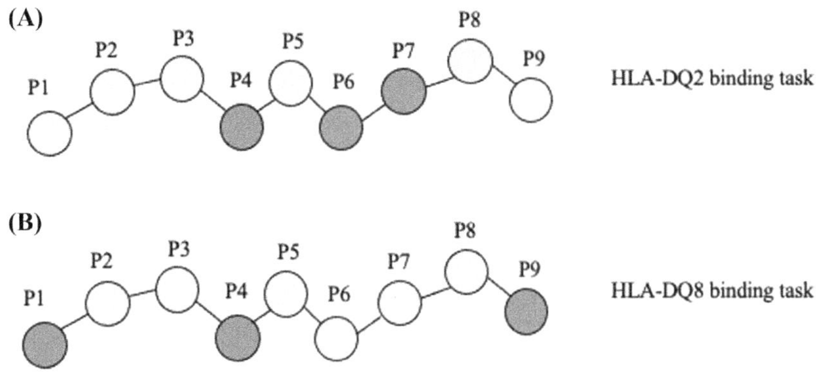

FIGURE 2.2 Peptide-binding task of gluten or related protein epitopes and HLA-DQ2/HLA-DQ8 molecules. Anchor amino acid residues of the HLA-binding task are shown in positions P1–P9. Glutamine residues deamidated by tTG2 and required for binding to HLA molecules are indicated with a dark gray circle. (A) HLA-DQ2-restricted epitopes have glutamine residues transformed into glutamic acid allocated in position (P) 4, P6, or P7. (B) HLA-DQ8-restricted epitopes have glutamine residues transformed into glutamic acid allocated in positions P1 and P9, and sometimes in P4.

TABLE 2.3 Immunodominant DQ2.2-restricted T-cell gluten peptides.[a]

	Previous peptide name	Peptide-binding motif								
		9-Mer core regions								
α-Gliadins										
DQ2.2-glut-α1		**Q**	G	S	V	Q	P	Q	**Q**	L
DQ2.2-glut-α2		Q	Y	S	Q	P	**Q**	Q	P	I
Glutenins										
DQ2.2-glut-L1	glutenin-17	P	F	S	**Q**	Q	**Q**	Q	P	V

The 9-mer core regions are reported for the most immunogenic peptides derived from wheat gliadins and glutenins and restricted by HLA-DQ2.2 molecule. Glutamine residues deamidated by tTG are shown in the table as **Q**.
[a]*References for current nomenclature: Sollid et al. (2012, 2020).*

Tables 2.1 and 2.2 contain the nine amino acid core sequences (positions 1–9, P1–P9) of the most known epitopes currently identified that bind to HLA-DQ2/DQ8, either in adult or pediatric CD patients (Camarca et al., 2017; Ráki et al., 2017; Sollid et al., 2012).

Of note, the peptides recognized by celiac CD4 + T cells are longer than nine amino acids, and *N*- and *C*-terminal flanking residues can influence the reactivity of specific T cells.

2.9 In vitro tools for the identification of T-cell immunogenic peptides in gluten proteins

As above mentioned, a great majority of known epitopes were identified by in vitro tools using gluten-specific polyclonal T-cell lines and T-cell clones generated from the jejunal mucosa of CD patients by cyclic stimulation with gluten and growth factors. The expanded T-cell cultures are assayed for their antigen-specificity by ELISA and T-cell proliferation. Other epitopes were identified by the short-term gluten oral challenge procedure. This is a minimally invasive in vivo strategy to detect gluten-specific T cells in the peripheral blood of treated celiac patients who consume gluten (approximately 6–16 gr gluten/die) for 1–3 days (Anderson et al., 2000, 2005; Camarca, Radano, et al., 2012; Picascia et al., 2015). Gluten-responsive T cells are transiently mobilized from gut mucosa to peripheral blood and can be detected by a sensitive immunospot assay (ELISPOT) (Picascia et al., 2015, 2020) or tetramer technology by multiparametric flow cytometry (Ráki et al., 2007).

2.10 Wheat breeding and gene editing procedures for wheat detoxification to obtain "safe" gluten

Recent studies have suggested that gluten proteins from the ancient *T. monococcum* landraces release a reduced number of immunogenic peptides capable of activating pathogenic CD4 + T cells due to their high in vitro digestibility compared to hexaploid soft wheat (Di Stasio et al., 2020; Gianfrani et al., 2005, 2012). These promising results are important for CD prevention in subjects at high risk of developing the disease, such as first-degree relatives of CD patients with genetic predisposition (Vriezinga et al., 2014). Indeed, although some wheat species are less immunogenic than others, to date, no existing wheat variety is completely safe for CD patients, with a gluten content of less than 20 mg/kg of gluten (20 PPM), and thus recommendable for their diet.

Alternative biotechnological approaches are currently under investigation aimed at creating wheat lines with reduced or null immunogenic potential and thus obtaining cereals with a "safe" gluten content for CD patients. These procedures utilize RNA interference (RNAi) and clustered regularly interspaced short palindromic repeats-associated nuclease 9 (CRISPR/Cas9) molecular technologies to silence immunogenic gluten sequences. Several studies have shown the transcript levels of different gliadin families can be downregulated by genetically modifying wheat using RNAi silencing, with a consequently reduced number of epitopes or even none detected in α-gliadins, γ-gliadins, and ω-gliadins (Altenbach et al., 2019; Barro et al., 2016; Becker et al., 2012; García-Molina et al., 2019; Gil-Humanes et al., 2008, 2010). Gil-Humanes et al. also demonstrated that the reduced-gliadin bread cooked with the flour obtained from transgenic wheat lines had a lower gliadin content, up to 97% compared to normal flour, but retained similar baking properties (Gil-Humanes et al., 2014). Similarly, Altenbach et al. genetically transformed with an RNAi plasmid the ω-1,2 gliadin genes from *T. aestivum* flour, reducing the levels of immunogenic sequences (Altenbach et al., 2019). Notwithstanding, in vivo clinical trials in CD volunteers would be needed to assess the real tolerance and safety of these RNAi-edited products. In addition, this approach introduces genetic modifications that remain in the wheat lines and currently make them unmarketable in many countries as they are considered genetically modified organisms (GMOs) (Jouanin et al., 2018).

By contrast, CRISPR/Cas9 gene editing is an alternative technology to edit and/or delete the epitopes in gliadin/glutenin genes, and to obtain wheat with hypoimmunogenic gluten, but with the same organoleptic properties required for viscoelasticity of dough and good end-use quality in wheat (Verma et al., 2021). Because the Cas9 construct used to generate the alterations is removed soon after, the mutations obtained are identical to those that could occur naturally, and the resulting CRISPR-edited wheat lines are not considered GMOs in most countries and therefore commercially more interesting (Jouanin et al., 2018). Sánchez-León et al. used CRISPR/Cas9 to modify the α-gliadin genes targeting two conserved sites adjacent to the epitope-containing region (Sánchez-León et al., 2018). The authors obtained up to 35 genes mutated from the 45 identified in the wild type in a single line, with a reduction in immunoreactivity of 85% assessed by R5 and G12 ELISA.

Based on these findings, although the lines modified using this promising CRISPR/Cas9 gene editing tool demonstrated a marked reduction of immunogenic activity of gluten in wheat varieties, there is still a long way to go, and further studies are needed to try to develop a completely safe wheat for CD patients (Jouanin et al., 2020).

2.11 Conclusion

Gluten contains a multitude of peptides able to trigger either innate or adaptive immune responses in CD. Overall, the definition of the entire repertoire of immunogenic and toxic gluten peptides at the molecular and structural level has added new and relevant information to better understand the mechanism that initiates the inflammatory response in CD patients and induces the loss of oral tolerance to gluten. In addition, the investigation of the metabolic fate of gluten, in particular the digestive process in the gastrointestinal tract, has improved the knowledge of CD pathogenesis. Recently, great attention has been paid to the nutritional properties of ancient diploid wheat species as well as the immune tolerability of gluten proteins for those suffering from gluten-induced disorders. However, although qualitative and quantitative differences in the presence of toxic epitopes between modern wheat and ancient wheat varieties have been demonstrated by plenty of studies, there is no direct evidence that the effects of wheat breeding have contributed to the increased prevalence of CD. More intervention studies involving multiple clinical centers are required to assess the efficacy of a diet based on more digestible and less immunotoxic wheat cultivars that may prevent CD in at-risk populations.

References

Altenbach, S. B., Chang, H. C., Yu, X. B., Seabourn, B. W., Green, P. H., & Alaedini, A. (2019). Elimination of omega-1,2 gliadins from bread wheat (Triticum aestivum) flour: Effects on immunogenic potential and end-use quality. *Frontiers in Plant Science*, *10*, 580. Available from https://doi.org/10.3389/fpls.2019.00580.

Anderson, R. P., Degano, P., Godkin, A. J., Jewell, D. P., & Hill, A. V. S. (2000). In vivo antigen challenge in celiac disease identifies a single transglutaminase-modified peptide as the dominant A-gliadin T-cell epitope. *Nature Medicine*, *6*(3), 337–342. Available from https://doi.org/10.1038/73200.

Anderson, R. P., Van Heel, D. A., Tye-Din, J. A., Barnardo, M., Salio, M., Jewell, D. P., & Hill, A. V. S. (2005). T cells in peripheral blood after gluten challenge in coeliac disease. *Gut*, *54*(9), 1217–1223. Available from https://doi.org/10.1136/gut.2004.059998.

Andrén Aronsson, C., Lee, H. S., Koletzko, S., Uusitalo, U., Yang, J., Virtanen, S. M., Liu, E., Lernmark, Å., Norris, J. M., & Agardh, D. (2016). Effects of gluten intake on risk of celiac disease: A case-control study on a Swedish Birth Cohort. *Clinical Gastroenterology and Hepatology*, *14* (3), 403–409. Available from https://doi.org/10.1016/j.cgh.2015.09.030, e3.

Bailey, C. H. (1941). A translation of Beccari's "Concerning grain" 1728. *Cereal Chemistry*, *18*, 555–561.

Barro, F., Iehisa, J. C. M., Giménez, M. J., García-Molina, M. D., Ozuna, C. V., Comino, I., Sousa, C., & Gil-Humanes, J. (2016). Targeting of prolamins by RNAi in bread wheat: Effectiveness of seven silencing-fragment combinations for obtaining lines devoid of coeliac disease epitopes from highly immunogenic gliadins. *Plant Biotechnology Journal*, *14*(3), 986–996. Available from https://doi.org/10.1111/pbi.12455.

Becker, D., Wieser, H., Koehler, P., Folck, A., Mühling, K. H., & Zörb, C. (2012). Protein composition and techno-functional properties of transgenic wheat with reduced a-gliadin content obtained by RNA interference. *Journal of Applied Botany and Food Quality*, *85*(1), 23–33.

Bordoni, A., Danesi, F., Di Nunzio, M., Taccari, A., & Valli, V. (2017). Ancient wheat and health: a legend or the reality? A review on KAMUT khorasan wheat. *International Journal of Food Sciences and Nutrition*, *68*(3), 278–286. Available from https://doi.org/10.1080/09637486.2016.1247434.

Borghi, B., Castagna, R., Corbellini, M., Heun, M., & Salamini, F. (1996). Breadmaking quality of Einkorn wheat (Triticum monococcum ssp. monococcum). *Cereal Chemistry*, *73*(2), 208–214.

Boukid, F., Prandi, B., Faccini, A., & Sforza, S. (2019). A complete mass spectrometry (MS)-based peptidomic description of gluten peptides generated during in vitro gastrointestinal digestion of durum wheat: Implication for celiac disease. *Journal of the American Society for Mass Spectrometry*, *30*(8), 1481–1490. Available from https://doi.org/10.1007/s13361-019-02212-8.

Brandolini, A., Hidalgo, A., & Moscaritolo, S. (2008). Chemical composition and pasting properties of einkorn (Triticum monococcum L. subsp. monococcum) whole meal flour. *Journal of Cereal Science*, *47*(3), 599–609. Available from https://doi.org/10.1016/j.jcs.2007.07.005.

Camarca, A., Anderson, R. P., Mamone, G., Fierro, O., Facchiano, A., Costantini, S., Zanzi, D., Sidney, J., Auricchio, S., Sette, A., Troncone, R., & Gianfrani, C. (2009). Intestinal T cell responses to gluten peptides are largely heterogeneous: Implications for a peptide-based therapy in celiac disease. *Journal of Immunology*, *182*(7), 4158–4166. Available from https://doi.org/10.4049/jimmunol.0803181.

Camarca, A., Auricchio, R., Picascia, S., Fierro, O., Maglio, M., Miele, E., Malamisura, B., Greco, L., Troncone, R., & Gianfrani, C. (2017). Gliadin-reactive T cells in Italian children from preventCD cohort at high risk of celiac disease. *Pediatric Allergy and Immunology*, *28*(4), 362–369. Available from https://doi.org/10.1111/pai.12720.

Camarca, A., del Mastro, A., & Gianfrani, C. (2012). Repertoire of gluten peptides active in celiac disease patients: Perspectives for translational therapeutic applications. *Endocrine, Metabolic and Immune Disorders - Drug Targets*, *12*(2), 207–219. Available from https://doi.org/10.2174/187153012800493549.

Camarca, A., Radano, G., Di Mase, R., Terrone, G., Maurano, F., Auricchio, S., Troncone, R., Greco, L., & Gianfrani, C. (2012). Short wheat challenge is a reproducible in-vivo assay to detect immune response to gluten. *Clinical and Experimental Immunology*, *169*(2), 129–136. Available from https://doi.org/10.1111/j.1365-2249.2012.04597.x.

Christophersen, A., Risnes, L. F., Bergseng, E., Lundin, K. E. A., Sollid, L. M., & Qiao, S. W. (2016). Healthy HLA-DQ2.5 + subjects lack regulatory and memory T cells specific for immunodominant gluten epitopes of celiac disease. *Journal of Immunology*, *196*(6), 2819–2826. Available from https://doi.org/10.4049/jimmunol.1501152.

Codex Alimentarius Commission (2008). https://www.fao.org/3/ca2329en/CA2329EN.pdf.

Corbellini, M., Empilli, S., Vaccino, P., Brandolini, A., Borghi, B., Heun, M., & Salamini, F. (1999). Einkorn characterization for bread and cookie production in relation to protein subunit composition. *Cereal Chemistry*, *76*(5), 727–733. Available from https://doi.org/10.1094/CCHEM.1999.76.5.727.

Day, L., Augustin, M. A., Batey, I. L., & Wrigley, C. W. (2006). Wheat-gluten uses and industry needs. *Trends in Food Science and Technology*, *17* (2), 82–90. Available from https://doi.org/10.1016/j.tifs.2005.10.003.

De Vita, P. (2019). *Durum wheat (Triticum turgidum ssp. durum) breeding to meet the challenge of climate change*, & (Eds.), *Advances in plant breeding strategies: Cereals* (Vol. 5, pp. 471–524). Springer International Publishing. Available from https://doi.org/10.1007/978-3-030-23108-8_13.

Di Stasio, L., Picascia, S., Auricchio, R., Vitale, S., Gazza, L., Picariello, G., Gianfrani, C., & Mamone, G. (2020). Comparative analysis of in vitro digestibility and immunogenicity of gliadin proteins from durum and einkorn wheat. *Frontiers in Nutrition*, *7*, 56. Available from https://doi.org/10.3389/fnut.2020.00056.

Fleckenstein, B., Molberg, Ø., Qiao, S. W., Schmid, D. G., Von Mülbe, F. D., Elgstøen, K., Jung, G., & Sollid, L. M. (2002). Gliadin T cell epitope selection by tissue transglutaminase in celiac disease. Role of enzyme specificity and pH influence on the transamidation versus deamidation reactions. *Journal of Biological Chemistry*, *277*(37), 34109–34116. Available from https://doi.org/10.1074/jbc.M204521200.

García-Molina, M. D., Giménez, M. J., Sánchez-León, S., & Barro, F. (2019). Gluten free wheat: Are we there? *Nutrients*, *11*(3), 487. Available from https://doi.org/10.3390/nu11030487.

Garsed, K., & Scott, B. B. (2007). Can oats be taken in a gluten-free diet? A systematic review. *Scandinavian Journal of Gastroenterology*, *42*(2), 171–178. Available from https://doi.org/10.1080/00365520600863944.

Gianfrani, C., Auricchio, S., & Troncone, R. (2005). Adaptive and innate immune responses in celiac disease. *Immunology Letters*, *99*(2), 141–145. Available from https://doi.org/10.1016/j.imlet.2005.02.017.

Gianfrani, C., Maglio, M., Rotondi Aufiero, V., Camarca, A., Vocca, I., Iaquinto, G., Giardullo, N., Pogna, N., Troncone, R., Auricchio, S., & Mazzarella, G. (2012). Immunogenicity of monococcum wheat in celiac patients. *The American Journal of Clinical Nutrition*, *96*(6), 1339–1345. Available from https://doi.org/10.3945/ajcn.112.040485.

Gil-Humanes, J., Pistón, F., Altamirano-Fortoul, R., Real, A., Comino, I., Sousa, C., Rosell, C. M., Barro, F., & Sestak, K. (2014). Reduced-gliadin wheat bread: An alternative to the gluten-free diet for consumers suffering gluten-related pathologies. *PLoS ONE*, *9*(3)e90898. Available from https://doi.org/10.1371/journal.pone.0090898.

Gil-Humanes, J., Pistón, F., Hernando, A., Alvarez, J. B., Shewry, P. R., & Barro, F. (2008). Silencing of γ-gliadins by RNA interference (RNAi) in bread wheat. *Journal of Cereal Science*, *48*(3), 565–568. Available from https://doi.org/10.1016/j.jcs.2008.03.005.

Gil-Humanes, J., Pistón, F., Tollefsen, S., Sollid, L. M., & Barro, F. (2010). Effective shutdown in the expression of celiac disease-related wheat gliadin T-cell epitopes by RNA interference. *Proceedings of the National Academy of Sciences of the United States of America*, *107*(39), 17023–17028. Available from https://doi.org/10.1073/pnas.1007773107.

Hardy, M. Y., Girardin, A., Pizzey, C., Cameron, D. J., Watson, K. A., Picascia, S., Auricchio, R., Greco, L., Gianfrani, C., La Gruta, N. L., Anderson, R. P., & Tye-Din, J. A. (2015). Consistency in polyclonal T-cell responses to gluten between children and adults with celiac disease. *Gastroenterology*, *149*(6), 1541–1552. Available from https://doi.org/10.1053/j.gastro.2015.07.013, e2.

Hausch, F., Shan, L., Santiago, N. A., Gray, G. M., & Khosla, C. (2002). Intestinal digestive resistance of immunodominant gliadin peptides. *American Journal of Physiology - Gastrointestinal and Liver Physiology*, *283*(4), G996–G1003. Available from https://doi.org/10.1152/ajpgi.00136.2002.

Jouanin, A., Boyd, L., Visser, R. G. F., & Smulders, M. J. M. (2018). Development of wheat with hypoimmunogenic gluten obstructed by the gene editing policy in Europe. *Frontiers in Plant Science*, *9*, 1523. Available from https://doi.org/10.3389/fpls.2018.01523.

Jouanin, A., Gilissen, L., Schaart, J., Leigh, F., Cockram, J., Wallington, E., Boyd, L. A., van den Broeck, H. C., van der Meer, I. M., America, A. H. P., Visser, R. G. F., & Smulders, M. J. M. (2020). CRISPR/Cas9 gene editing of gluten in wheat to reduce gluten content and exposure—Reviewing methods to screen for coeliac safety. *Frontiers in Nutrition*, *7*, 51.

Kasarda, D. D. (2013). Can an increase in celiac disease be attributed to an increase in the gluten content of wheat as a consequence of wheat breeding? *Journal of Agricultural and Food Chemistry*, *61*(6), 1155–1159. Available from https://doi.org/10.1021/jf305122s.

Khan, K. (2016). *Wheat: Chemistry and technology*. Elsevier.

Khatkar, B. S., Fido, R. J., Tatham, A. S., & Schofield, J. D. (2002). Functional properties of wheat gliadins. II. Effects on dynamic rheological properties of wheat gluten. *Journal of Cereal Science*, *35*(3), 307–313. Available from https://doi.org/10.1006/jcrs.2001.0430.

Khatkar, B. S., & Schofield, J. D. (2002). Dynamic rheology of wheat flour dough. II. Assessment of dough strength and bread-making quality. *Journal of the Science of Food and Agriculture*, *82*(8), 823–826. Available from https://doi.org/10.1002/jsfa.1111.

Krupa-Kozak, U., & Drabińska, N. (2015). *Technological and health-related aspects of gluten*.

Lamacchia, C., Camarca, A., Picascia, S., Di Luccia, A., & Gianfrani, C. (2014). Cereal-based gluten-free food: How to reconcile nutritional and technological properties of wheat proteins with safety for celiac disease patients. *Nutrients*, *6*(2), 575–590. Available from https://doi.org/10.3390/nu6020575.

Ludvigsson, J. F., & Green, P. H. R. (2014). The missing environmental factor in celiac disease. *New England Journal of Medicine*, *371*(14), 1341–1343. Available from https://doi.org/10.1056/NEJMe1408011.

Lundin, K. E. A., Nilsen, E. M., Scott, H. G., Løberg, E. M., Gjøen, A., Bratlie, J., Skar, V., Mendez, E., Loøvik, A., & Kett, K. (2003). Oats induced villous atrophy in coeliac disease. *Gut*, *52*(11), 1649–1652. Available from https://doi.org/10.1136/gut.52.11.1649.

Lundin, K. E. A., Scott, H., Hansen, T., Paulsen, G., Halstensen, T. S., Fausa, O., Thorsby, E., & Sollid, L. M. (1993). Gliadin-specific, HLA-DQ (α1*0501, β1*0201) restricted t cells isolated from the small intestinal mucosa of celiac disease patients. *Journal of Experimental Medicine*, *178*(1), 187–196. Available from https://doi.org/10.1084/jem.178.1.187.

Mamone, G., Ferranti, P., Rossi, M., Roepstorff, P., Fierro, O., Malorni, A., & Addeo, F. (2007). Identification of a peptide from α-gliadin resistant to digestive enzymes: Implications for celiac disease. *Journal of Chromatography B: Analytical Technologies in the Biomedical and Life Sciences*, *855*(2), 236–241. Available from https://doi.org/10.1016/j.jchromb.2007.05.009.

Marchetti, L., Cardós, M., Campaña, L., & Ferrero, C. (2012). Effect of glutens of different quality on dough characteristics and breadmaking performance. *LWT - Food Science and Technology*, *46*(1), 224–231. Available from https://doi.org/10.1016/j.lwt.2011.10.002.

Melnyk, J. P., Dreisoerner, J., Marcone, M. F., & Seetharaman, K. (2012). Using the Gluten Peak Tester as a tool to measure physical properties of gluten. *Journal of Cereal Science*, *56*(3), 561–567. Available from https://doi.org/10.1016/j.jcs.2012.07.015.

Molberg, Ø., Kett, K., Scott, H., Thorsby, E., Sollid, L. M., & Lundin, K. E. A. (1997). Gliadin specific, HLA DQ2-restricted T cells are commonly found in small intestinal biopsies from coeliac disease patients, but not from controls. *Scandinavian Journal of Immunology*, *46*(1), 103–108. Available from https://doi.org/10.1046/j.1365-3083.1997.d01-93.x-i2.

Molberg, Ø., Mcadam, S. N., Körner, R., Quarsten, H., Kristiansen, C., Madsen, L., Fugger, L., Scott, H., Norën, O., Roepstorff, P., Lundin, K. E. A., Sjöström, H., & Sollid, L. M. (1998). Tissue transglutaminase selectively modifies gliadin peptides that are recognized by gut-derived T cells in celiac disease. *Nature Medicine*, *4*(6), 713–717. Available from https://doi.org/10.1038/nm0698-713.

Molberg, Ø., Uhlen, A. K., Jensen, T., Flæte, N. S., Fleckenstein, B., Arentz-Hansen, H., Raki, M., Lundin, K. E. A., & Sollid, L. M. (2005). Mapping of gluten T-cell epitopes in the bread wheat ancestors: Implications for celiac disease. *Gastroenterology*, *128*(2), 393–401. Available from https://doi.org/10.1053/j.gastro.2004.11.003.

Ogilvie, O., Roberts, S., Sutton, K., Domigan, L., Larsen, N., Gerrard, J., & Demarais, N. (2020). Proteomic modelling of gluten digestion from a physiologically relevant food system: A focus on the digestion of immunogenic peptides from wheat implicated in celiac disease. *Food Chemistry*, *333*, 127466. Available from https://doi.org/10.1016/j.foodchem.2020.127466.

Ortolan, F., & Steel, C. J. (2017). Protein characteristics that affect the quality of vital wheat gluten to be used in baking: A review. *Comprehensive Reviews in Food Science and Food Safety*, *16*(3), 369–381. Available from https://doi.org/10.1111/1541-4337.12259.

Payne, P. I., & Corfield, K. G. (1979). Subunit composition of wheat glutenin proteins, isolated by gel filtration in a dissociating medium. *Planta, 145*(1), 83–88. Available from https://doi.org/10.1007/BF00379931.

Peng, J. H., Sun, D., & Nevo, E. (2011). Domestication evolution, genetics and genomics in wheat. *Molecular Breeding, 28*(3), 281–301. Available from https://doi.org/10.1007/s11032-011-9608-4.

Picascia, S., Mandile, R., Auricchio, R., Troncone, R., & Gianfrani, C. (2015). Gliadin-specific T-cells mobilized in the peripheral blood of coeliac patients by short oral gluten challenge: Clinical applications. *Nutrients, 7*(12), 10020–10031. Available from https://doi.org/10.3390/nu7125515.

Picascia, S., Camarca, A., Malamisura, M., Mandile, R., Galatola, M., Cielo, D., Gazza, L., Mamone, G., Auricchio, S., Troncone, R., Greco, L., Auricchio, R., & Gianfrani, C. (2020). In celiac disease patients the in vivo challenge with the diploid triticum monococcum elicits a reduced immune response compared to hexaploid wheat. *Molecular Nutrition & Food Research, 64*(11)1901032. Available from https://doi.org/10.1002/mnfr.201901032.

Prandi, B., Tedeschi, T., Folloni, S., Galaverna, G., & Sforza, S. (2017). Peptides from gluten digestion: A comparison between old and modern wheat varieties. *Food Research International, 91*, 92–102. Available from https://doi.org/10.1016/j.foodres.2016.11.034.

Ráki, M., Dahal-Koirala, S., Yu, H., Korponay-Szabó, I. R., Gyimesi, J., Castillejo, G., Jahnsen, J., Qiao, S. W., & Sollid, L. M. (2017). Similar responses of intestinal T cells from untreated children and adults with celiac disease to deamidated gluten epitopes. *Gastroenterology, 153*(3), 787–798. Available from https://doi.org/10.1053/j.gastro.2017.05.016, e4.

Ráki, M., Fallang, L. E., Brottveit, M., Bergseng, E., Quarsten, H., Lundin, K. E. A., & Sollid, L. M. (2007). Tetramer visualization of gut-homing gluten-specific T cells in the peripheral blood of celiac disease patients. *Proceedings of the National Academy of Sciences of the United States of America, 104*(8), 2831–2836. Available from https://doi.org/10.1073/pnas.0608610104.

Sánchez-León, S., Gil-Humanes, J., Ozuna, C. V., Giménez, M. J., Sousa, C., Voytas, D. F., & Barro, F. (2018). Low-gluten, nontransgenic wheat engineered with CRISPR/Cas9. *Plant Biotechnology Journal, 16*(4), 902–910. Available from https://doi.org/10.1111/pbi.12837.

Shan, L., Molberg, Ø., Parrot, I., Hausch, F., Filiz, F., Gray, G. M., Sollid, L. M., & Khosla, C. (2002). Structural basis for gluten intolerance in celiac sprue. *Science, 297*(5590), 2275–2279. Available from https://doi.org/10.1126/science.1074129.

Shan, L., Qiao, S. W., Arentz-Hansen, H., Molberg, Ø., Gray, G. M., Sollid, L. M., & Khosla, C. (2005). Identification and analysis of multivalent proteolytically resistant peptides from gluten: Implications for celiac sprue. *Journal of Proteome Research, 4*(5), 1732–1741. Available from https://doi.org/10.1021/pr050173t.

Sollid, L. M., Qiao, S. W., Anderson, R. P., Gianfrani, C., & Koning, F. (2012). Nomenclature and listing of celiac disease relevant gluten T-cell epitopes restricted by HLA-DQ molecules. *Immunogenetics, 64*(6), 455–460. Available from https://doi.org/10.1007/s00251-012-0599-z.

Sollid, L. M., Tye-Din, J. A., Qiao, S. W., Anderson, R. P., Gianfrani, C., & Koning, F. (2020). Update 2020: Nomenclature and listing of celiac disease–relevant gluten epitopes recognized by CD4 + T cells. *Immunogenetics, 72*(1–2), 85–88. Available from https://doi.org/10.1007/s00251-019-01141-w.

Spisni, E., Imbesi, V., Giovanardi, E., Petrocelli, G., Alvisi, P., & Valerii, M. C. (2019). Differential physiological responses elicited by ancient and heritage wheat cultivars compared to modern ones. *Nutrients, 11*(12), 2879. Available from https://doi.org/10.3390/nu11122879.

Tedrus, G., de, A. S., Ormenese, R., de, C. S. C., Speranza, S. M., Chang, Y. K., & Bustos, F. M. (2001). Estudo da adição de vital glúten à farinha de arroz, farinha de aveia e amido de trigo na qualidade de pães. *Ciência e Tecnologia de Alimentos, 21*(1), 20–25. Available from https://doi.org/10.1590/S0101-20612001000100006.

Tye-Din, J. A., Stewart, J. A., Dromey, J. A., Beissbarth, T., Van Heel, D. A., Tatham, A., Henderson, K., Mannering, S. I., Gianfrani, C., Jewell, D. P., Hill, A. V. S., McCluskey, J., Rossjohn, J., & Anderson, R. P. (2010). Comprehensive, quantitative mapping of T cell epitopes in gluten in celiac disease. *Science Translational Medicine, 2*(41). Available from https://doi.org/10.1126/scitranslmed.3001012.

Vader, W., Kooy, Y., Van Veelen, P., De Ru, A., Harris, D., Benckhuijsen, W., Pea, S., Mearin, L., Drijfhout, J. W., & Koning, F. (2002). The gluten response in children with celiac disease is directed toward multiple gliadin and glutenin peptides. *Gastroenterology, 122*(7), 1729–1737. Available from https://doi.org/10.1053/gast.2002.33606.

Van De Wal, Y., Kooy, Y., Van Veelen, P., Peña, S., Mearin, L., Papadopoulos, G., & Koning, F. (1998). Cutting edge: Selective deamidation by tissue transglutaminase strongly enhances gliadin-specific T cell reactivity. *Journal of Immunology, 161*(4), 1585–1588.

Van Der Borght, A., Goesaert, H., Veraverbeke, W. S., & Delcour, J. A. (2005). Fractionation of wheat and wheat flour into starch and gluten: Overview of the main processes and the factors involved. *Journal of Cereal Science, 41*(3), 221–237. Available from https://doi.org/10.1016/j.jcs.2004.09.008.

Veraverbeke, W. S., & Delcour, J. A. (2002). Wheat protein composition and properties of wheat glutenin in relation to breadmaking functionality. *Critical Reviews in Food Science and Nutrition, 42*(3), 179–208. Available from https://doi.org/10.1080/10408690290825510.

Verma, A. K., Mandal, S., Tiwari, A., Monachesi, C., Catassi, G. N., Srivastava, A., Gatti, S., Lionetti, E., & Catassi, C. (2021). Current status and perspectives on the application of CRISPR/Cas9 gene-editing system to develop a low-gluten, non-transgenic wheat variety. *Foods, 10*(10), 2351. Available from https://doi.org/10.3390/foods10102351.

Vriezinga, S. L., Auricchio, R., Bravi, E., Castillejo, G., Chmielewska, A., Crespo Escobar, P., Kolaček, S., Koletzko, S., Korponay-Szabo, I. R., Mummert, E., Polanco, I., Putter, H., Ribes-Koninckx, C., Shamir, R., Szajewska, H., Werkstetter, K., Greco, L., Gyimesi, J., Hartman, C., … Mearin, M. L. (2014). Randomized feeding intervention in infants at high risk for celiac disease. *New England Journal of Medicine, 371*(14), 1304–1315. Available from https://doi.org/10.1056/nejmoa1404172.

Wieser, H. (2007). Chemistry of gluten proteins. *Food Microbiology, 24*(2), 115–119. Available from https://doi.org/10.1016/j.fm.2006.07.004.

Chapter 3

Genetic predisposition

Luigi Greco and Renata Auricchio
Department of Translational Medical Science and European Laboratory for the Investigation of Food-Induced Diseases, University of Naples Federico II, Naples, Italy

3.1 Why genetics?

Celiac disease (CD) is an emerging epidemic of a multifactorial disorder with a complex genetic predisposition. CD has a strong heritable genetic component, which was definitely suggested by studies on twins obtained by matching celiac patients' registry with the Italian Registry of Twins (Greco et al., 2002). The concordance rates for the disease in monozygous twins (83%–86%) compared to that of dizygous twins (17%–20%) is the highest among autoimmune diseases and confirms the strong genetic component. The sibling relative risk is quite high, close to 20% (Nisticò et al., 2006). Obviously, the inheritance pattern does not follow a Mendelian inheritance rule, since multiple genes are involved in predisposing to this condition. In several autoimmune diseases, such as CD and type 1 diabetes the main genetic risk factor is represented by HLA class II genes. The encoded HLA molecules expressed on the surface of antigen presenting cells (APC) have a key role in presenting the antigens to inflammatory CD4 + T lymphocytes.

3.2 The human leukocyte antigen complex

The HLA complex occupies a 4 Mb region on chromosome 6p21. It is assembled by two subunits: the DQ2 dimer (named DQ2.5) is made by an "A" chain encoded by the DQA*0501 gene and a "B" chain encoded by the DQB*0201 gene. The HLA-DQ2 dimer can be encoded in cis on the same chromosome or in trans in two homologous chromosomes. The DQA*0301 and the DQB1*0302 chains aggregate in the DQ8 dimer. Fig. 3.1 shows the structure of the HLA dimer with two A chains and two B chains, forming the groove where the gliadin peptide is recognized and linked (Sollid & Lundin, 2014).

Although the cis and trans configurations of DQ2 differ at just one amino acid in each subunit, this does not influence antigen presentation as the variants are in noncritical regions and both confer similar disease risk. The alpha chain

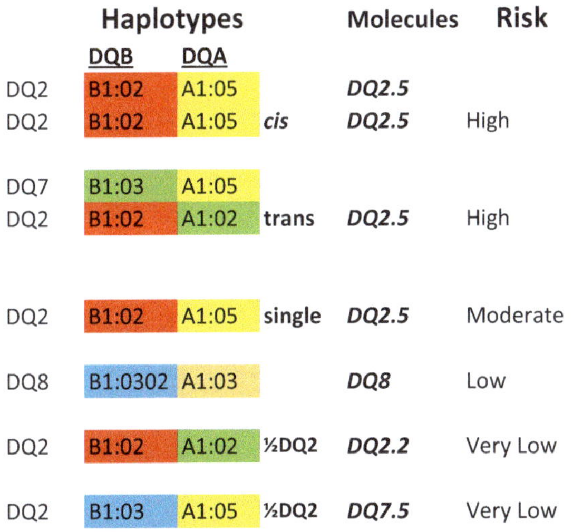

FIGURE 3.1 Structure of the HLA DQ Dimer. Each molecule is composed by a "B" unit and an "A" unit. The B1:02 (red) in combination with the A1:05 (yellow) assemble the DQ2.5 dimer, which is associated to the highest risk. It may be homozygous with two DQ2.5 in cis (first group), or it may be composed in trans on two different chromosomes (second group), with two B1:02 and B1:03 at risk molecules, conferring the highest risk also. DQ2.5 can be present alone = moderate risk (third group). DQ8 is composed by a B1:0302 (sky blue) and an A1:03 (orange) (fourth group). Different combination of B1 and A1 assemble dimers at low risk.

Pediatric and Adult Celiac Disease. DOI: https://doi.org/10.1016/B978-0-443-13359-6.00020-0

encoded by DQA1*05 allele and beta chain encoded by DQB1*02 allele (i.e., DQ2.5 in cis or trans) confer the risk of CD. Several studies have shown that the risk to develop CD is increased 4–6 times by homozygosity for the cis haplotype or for a second DQB1*02 allele on the other haplotype (i.e. DQ2.5/DQ2.5 or DQ2.5/DQ2.2) (Margaritte-Jeannin et al., 2004; Meijer et al., 2022).

Association of variants in the HLA complex with CD was first reported in 1972 using serological methods (Tosi et al., 1983). A strong linkage disequilibrium with Dr3 and Dr7 was observed since the early days, shared with other diseases: indeed, celiac patients are at substantially higher risk of having type I diabetes, autoimmune thyroid disease, and other immunological conditions that are also associated with Dr3-Dr4 and DQ2-DQ8 haplotypes. Later studies found that the strongest CD association was with HLA-DQ2, in particular with the combination of alleles encoding the alpha and beta chain variants DQA1*05 and DQB1*02 of the DQ2 heterodimer (Ek et al., 1978; Keuning et al., 1976). The genetic association data implicating DQ2 in CD is strengthened by functional studies identifying an immunodominant tissue transglutaminase modified gliadin epitope recognized by DQ2 restricted CD4+ T cells either in the intestine (Molberg et al., 1997) as well as in peripheral blood after antigen challenge (Camarca et al., 2021).

The main HLA haplotypes predisposing to CD are:

1. the DQ2.5 homozygote coding for two identical dimers DQ2.5 with higher capacity to recognize the antigen (where both DQA1*0501 and DQB1*0201 are located in cis on the same DQ2 haplotype)
2. the DQ2.5 dimers can be assembled in trans by one gene in a chromosome and the other on its homologous, where these two risk alleles are located on different haplotypes [in combination of DQ7(DQA1*0501-DQB1*0301) and DQ2(DQA1*0201-DQB1*0202) haplotypes]
3. the DQ2.5 can generate one single dimer in cis
4. the DQ8 dimer (DQA*0301-DQB1*0302) is produced alone
5. a combination of DQ2.5 and DQ8

In rare CD individuals the presence of only half of the above mentioned heterodimers (i.e., DQA*05 only) has been observed (Karell et al., 2003). Disease susceptibility depends on the dosage effect of the DQ2.5 heterodimer (Ploski et al., 1993; van Belzen et al., 2004). Homozygous individuals for the Dr3-DQ2 haplotype or heterozygous Dr3- DQ2/Dr7-DQ2 express the highest levels of DQ2.5 heterodimers (Margaritte-Jeannin et al., 2004).

Experimental evidence demonstrated that the expression of these molecules influences the achievement of the HLA-peptide complex threshold, needed for activation and proliferation of autoreactive CD4+ T cells. Pisapia demonstrated that in CD individuals DQA1*05 and DQB1*02 alleles, when carried by heterozygous APC, are differentially expressed (Pisapia et al., 2016). This marked expression of CD associated alleles in the APC of heterozygous patients resulted in the production of high concentration of DQ2.5 molecules, comparable to those produced in DQ2.5 homozygous cells. As a consequence, the strength of the gliadin-specific CD4+ T cells response is mainly dependent on the antigen dose.

3.2.1 Function

The primary function of the DQ molecules is to present exogenous peptide antigens (e.g., peptides from several gliadin molecules) to the T-cell receptor of helper T cells. The discovery that Human Tissue Transglutaminase 2 (tTG2) is able to convert, by deamidation, glutamine residues into negatively charged glutamate residues inside the gliadin molecules, gave the best explanation for the preferential recognition of these deamidated peptides by the HLA DQ2 and DQ8 molecules (Molberg et al., 1998). DQ2 and DQ8 molecules have positively charged pockets who attract negatively charged peptides. HLA-DQ2 has a lysine at position β71, with preference for binding negatively charged peptides residues at positions P4, P6, and P7. Hence DQ2-restricted T cells have a selective preference for deamidated peptides. DQ2 and DQ8 do not have an aspartic acid at position β57 which renders the P9 pocket basic, originating a preference for negatively charged residues at P9, which are present in several gluten epitopes. The α-chain of DQ2.5 has a tyrosine at position 22, which, in contrast to the phenylalanine of DQ2.2, forms hydrogen bonds with the peptide main chain so that DQ2.5 forms more stable complexes with gluten peptides than does DQ2.2, increasing the capacity of T cell to drive an autoimmune cascade. The reason of the strong association of specific HLA molecules to CD is explained by the unusual ability of DQ2 to bind the proline-rich gluten peptides remained after gastrointestinal digestion of gluten that have been deamidated by tissue transglutaminase (Fallang et al., 2009; Petersen et al., 2014).

The structure of DQ2 complexed with epitope from gluten shows that the glutamate at P6, which is formed by tissue transglutaminase deamidation, is an important anchor residue as it participates in an extensive hydrogen-bonding involving Lys-71 of DQ2. The core of the complex HLA groove-gliadin peptide is shown in Fig. 3.2 (Kim et al., 2004).

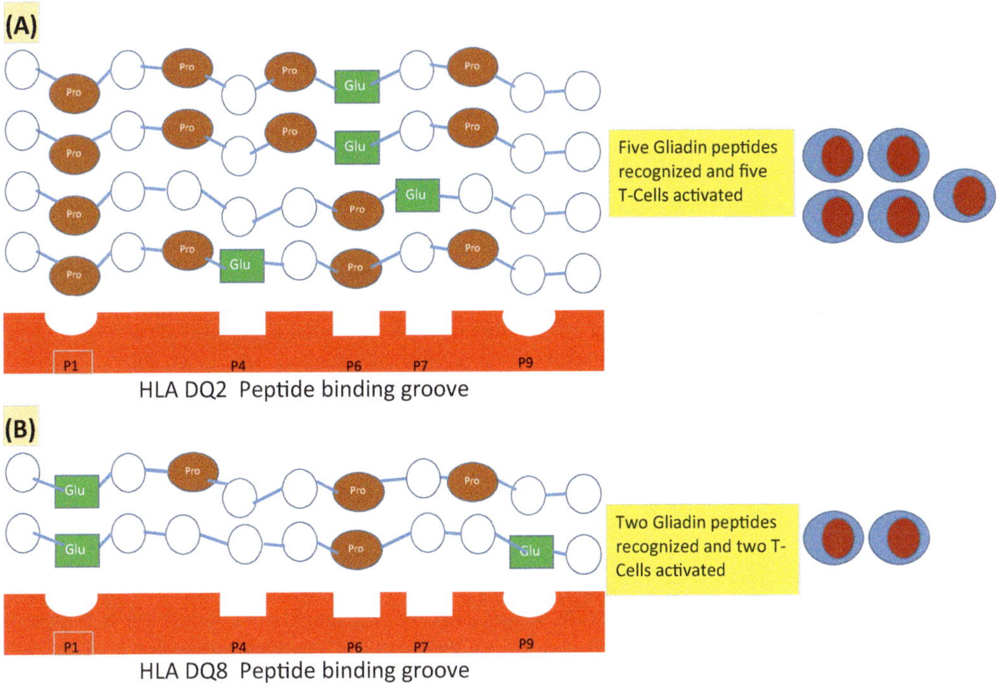

FIGURE 3.2 Structure of the HLA DQ2 and DQ8 binding grooves. (A) shows the HLA DQ2 binding sites for gliadin peptides. Five peptides are recognized in the P1, P4, P6 and P7 HLA pockets. The HLA DQ2 APC is so able to present five peptides to the T-cells, stimulating a strong activation. (B) shows the HLA DQ8 binding sites for just two gliadin peptides in the P1, P6 and P9 pockets, so that the presentation to the T-cells is significantly reduced, compared to DQ2.

HLA DQA1*05 and DQB1*02 alleles encoding the DQ2.5 molecules and HLA DQA1*0301 and DQB1*0302 alleles encoding DQ8 molecules are strongly associated with CD and type 1 diabetes (T1D). Farina previously demonstrated that DQA1*05 and DQB1*02 showed a higher expression with respect to non-CD associated alleles in heterozygous DQ2.5 positive (HLA Dr1/Dr3) APC of CD patients. This differential expression affected the level of the encoded DQ2.5 molecules on the APC surface and established the strength of gluten-specific CD4 + T cells response (Farina et al., 2019).

Apart from the genes encoding the DQ molecules, the HLA region also contains a large number of immune-related genes that are good candidates for susceptibility to CD. Dissecting the separate effects of other genes located in the HLA complex is a hard task because of the high linkage disequilibrium in the HLA region. Several studies have suggested the presence of other susceptibility genes for CD in the HLA region, including MICA, MICB and TNF genes (Fernandez et al., 2002; McManus et al., 1996), however, the majority of these studies were unable to perform a proper conditioning on the DQ2 haplotype.

3.2.2 HLA frequency in celiac disease

Around 90% of celiac patients possess the HLA-DQ2.5 molecules. Carriage of the HLA-DQ8 molecule (DQA1*0301-DQB1*0302) without DQ2 was found in 6% of celiac cases in a large study of 1008 European celiac patients (Bolognesi et al., 2003). Interestingly, in this European Study, a further 6% of celiac cases who did not carry classical DQ2.5 possessed either the alpha or beta chain of the heterodimer (Louka et al., 2003). Only 4 of 1008 celiac cases in the European cohort did not possess a subunit of either DQ2 or DQ8 (Karell et al., 2003).

3.2.3 Geographic distribution

The HLA-DQ2 allele is common in the healthy population, carried by approximately 30%–35% of Caucasians (Abadie et al., 2011; Bourgey et al., 2007). The presence of DQ2 is necessary, but not sufficient, to develop CD: the vast majority of DQ2-DQ8 carriers never develop the disease; this originated the search for other non-HLA genetic variants predisposing to disease. Since HLA DQ2–8 have the essential function to recognize and present the gluten antigen, it is unlikely that other genetic variants could contribute at similar level to genetic predisposition; thus, without these genes, CD is highly unlikely.

Regional variants have been observed in the frequency of these haplotypes, with the HLA-DQ8 ranging from about 2%–4% in Western countries to 25%–30% in the Iranian population and in the middle East (Liu et al., 2014; Makharia, 2015). Indeed, CD prevalence appears to be quite homogeneous worldwide in wheat consuming population, now ranging from 1% to 2,5%, where environmental factors, as exposure to gluten, contribute to the regional differences.

In the United States and Europe, where the prevalence of CD ranges between 1% and 2%, 25%–30% of individuals carry HLA-DQ2 and DQ8 susceptibility heterodimers (Liu et al., 2014), whereas the low prevalence of HLA predisposing DQ2/DQ8 molecules in far-East Asia could explain, in addition to the low wheat consumption, the anecdotal reports of CD in these countries (Makharia, 2015). On the other side in Sweden, where the prevalence of CD is estimated at 2.13%, the frequency of CD-predisposing HLA-DQ genotypes is 53% (Sandström et al., 2013).

This observation is corroborated by a world-wide gradient distribution of HLA-DQ molecules: 52.7% in Saudi Arabia (Al-Hussaini et al., 2018), 58% in Iran (Saberi-Firouzi et al., 2008), 31.9% in North India (Sood et al., 2006), and less than 10% in Japan and China (Yuan et al., 2013).

Looking to the world-wide perspective, no significant correlation was apparently present between the prevalence of CD and the levels of wheat consumption, the sum of the frequencies of Dr3-DQ2 and Dr4-DQ8, or the product of both factors. Nevertheless, excluding the outliers countries with extreme data, (Algeria, Finland, Mexico, north India, and Tunisia), a significant correlation was observed between the combination of both risk factors and the incidence of CD worldwide (correlation coefficient $R^2 = 0.4$; $p = 0.002$). It was observed a significant correlation between the prevalence of CD and wheat consumption ($R^2 = 0.14$, $p = 0.03$) and the prevalence of CD and the frequency of DQ2 + DQ8 haplotypes ($R^2 = 0.24$, $p = 0.03$) (Sollid & Lie, 2005).

3.2.4 Prediction of risk

A cohort of 188 Italian families with a symptomatic CD patient (the proband), at least one sib and both parents (798 individuals) were typed for HLA genotype (Bourgey et al., 2007). Of all 127 more probands with both parents were added to the sample to improve the robustness of the estimate. The parental haplotypes not transmitted to the proband are robust representatives of the general population (affected family based controls) and provide an unbiased estimate of the haplotype frequencies.

Risk of CD for individuals with genotype DQ listed in the Italian population is shown in Fig. 3.3. The corresponding haplotype combination and DR genotypes are shown for each group. The risk R is the ratio between the probability that an individual in the population has the defined genotype (control haplotype frequency) by the probability of CD in the population (disease prevalence) with that haplotype. It is clear that individuals carrying two copies of the DQB1*0201

DQ Genotype		Corresponding DR genotype	Genotypic Group	Risk
DQA1*0501-DQB1*0201	DQA1*0501-DQB1*0201	DR3/DR3	G1	0,21
DQA1*0501-DQB1*0201	DQA1*0201-DQB1*0202	DR3/DR7		
DQA1*0201-DQB1*0202	DQA1*0505-DQB1*0301	DR5/DR7	G2	0,17
DQA1*0501-DQB1*0201	DQA1*0505-DQB1*0301	DR3/DR5	G3	0,06
DQA1*0501-DQB1*0201	DQA1*0301-DQB1*0302	DR3/DR4		
DQA1*0501-DQB1*0201	Others	DR3/DRX		
DQA1*0201-DQB1*0202	DQA1*0201-DQB1*0202	DR7/DR7	G4	0,05
DQA1*0201-DQB1*0202	DQA1*0301-DQB1*0302	DR7/DR4		
DQA1*0301-DQB1*0302	DQA1*0301-DQB1*0302	DR4/DR4		
Others	Others	Others	G5	0,006

FIGURE 3.3 Risk of CD for individuals with genotype DQ listed in the Italian population.

(group G1) have a risk (21%) comparable to a heterozygous transmission, while those who have one single copy (group G3) have a much lower risk (6%).

The risk R that a new sib from a family with a proband case, would develop CD according to the known HLA genotype of the affected proband was estimated according to Mendelian segregation, assuming no recombination in the DR-DQ haplotypes. The typing of the parents does significantly increase the precision of the estimate of the familial recurrence risk for a newborn. Fig. 3.4 shows the distribution of the risk of sibs of a proband according to their own HLA genotype. About 40% of sibs have a very low risk (group G5, bleu bar), while about 30% are at high risk (group G1 and G2, red bars).

Fig. 3.5 shows the risk for a newborn across the parents' genotype. Crossing the genotype of the father (vertical) with that of the mother (horizontal) you may have an estimate of the familial recurrence risk for a new baby. Colors indicate the potential information available with parental genotyping. The risk for a new baby may be negligible (<1%, blue), moderate (1%–10%, green), intermediate (10%–15%, orange), high (15%–20%, yellow) or very high (>20%,

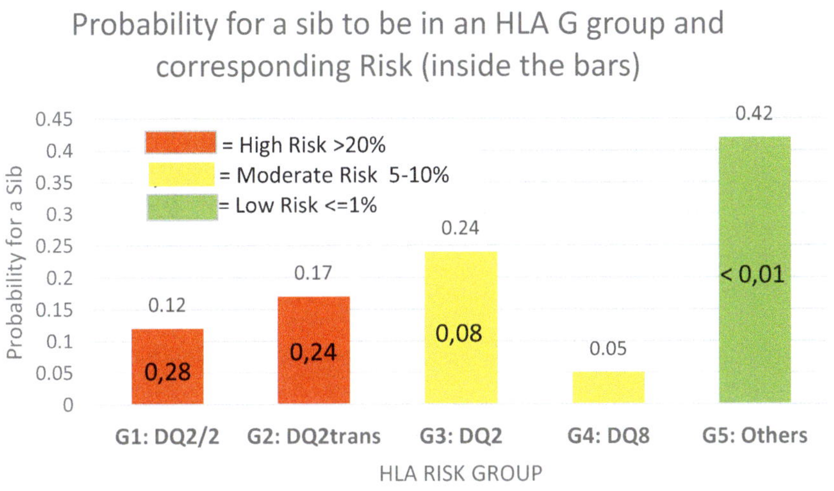

FIGURE 3.4 Risk for sibs of proband according to their own HLA genotype. The average estimated probability (on top of the bar) that sibs of proband will be in an HLA at risk "G" group and their corresponding risk (inside the bar) is shown. Red bars are at highest risk (Group G1 and G2) with Double DQ2.5. Yellow bars indicate a moderate risk (Group G3 and G4 with single DQ2 or DQ8). About 40% of sibs have a very low risk (green bar, group G5).

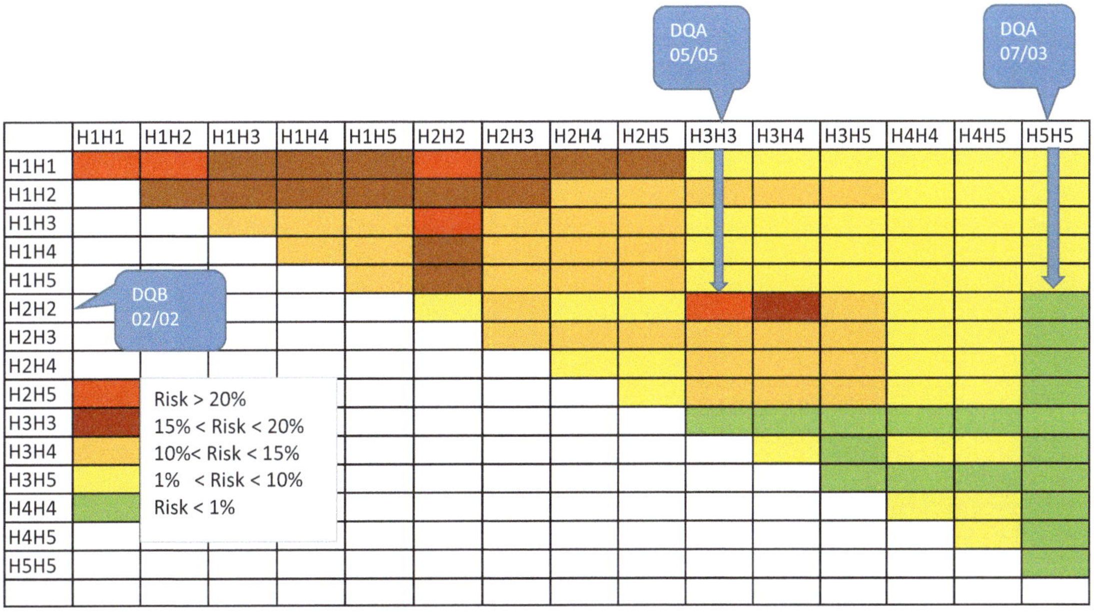

FIGURE 3.5 Risk for a sib of a proband according to the HLA DQ genotype of the parents. Crossing the father's genotype with mother's genotype you get the average estimate of the newborn. Red and Brown cells are associated to the highest risk. Green cells are at low risk. The orange color indicate a moderate risk, but the estimate has wide confidence intervals, suggesting the need to type directly the newborn. HAPLOTYPES: H1 = DQA1*05—DQB1*0201, H2 = DQA1*02—DQB1*0202, H3 = DQA1*05—DQB1*03, H4 = DQA1*0301—DQB1*0302, H5 = Others.

red). The orange and yellow boxes show that the risk estimate has a wide range by confidence intervals. In these families the genotyping of the newborn will considerably increase the precision of the estimate of the familial recurrence risk.

To show how the table works, let us consider:

1. a mother of a CD child, carrying the H2H2 (DQA1*02—DQB1*0202/DQA1*02—DQB1*0202) genotype, has a 29% risk for another CD baby if the father is H1H1 (DQA1*05—DQB1*0201/DQA1*05—DQB1*0201), but only a 7% risk if the father is H2/H2 (DQA1*02—DQB1*0202/DQA1*02—DQB1*0202).
2. a father of a CD child carrying an H2H5 (DQA1*02—DQB1*0202/Others) genotype has a small (2%) risk to have another CD baby if the mother is H2H5 (DQA1*02—DQB1*0202/Others) but a quite higher (12%) risk if she is H3H3 (DQA1*05—DQB1*03/DQA1*05—DQB1*03).

HLA was typed in 470 trio families with a CD proband from France (=117), Italy (=128) and Norway-Sweden (=225). For each country the relative risks associated with each genotype group were computed taking into account the control haplotype frequencies (Margaritte-Jeannin et al., 2004). Setting for each country the risk to develop CD at 1 for the individuals homozygous for DQ2, either made by DQ2.5 as well as by DQ2.2, the relative risk of the other groups is estimated. For the G1 group (DQ2/DQ2) the relative risk is higher for all countries and set to 1 (reference value). Then there is a higher risk (0,68) in Italy compared to France and Sweden (0,28–0,23) for the DQ2 in trans (group G2). Similarly, the risk is not homogeneous for the group G3 and G4. The gradient in the genotype group is not only related to different haplotype frequencies, but also to different genotype relative risk among populations, showing considerable differences in the relative risk north-south for each DQ genotype group (Sollid & Lundin, 2014).

The estimate of the risk to develop CD in a newborn from families with a confirmed CD case (familial recurrence risk) was fully confirmed by a large European cohort of the PREVENT-CD study (Meijer et al., 2022). Indeed, the frequency of CD was significantly higher in children homozygous for HLA-DQ2 (DQ2/DQ2 and DQ2/Dr7-DQ2), than children with HLA-DQ2 heterozygosity or DQ8 positivity, with a cumulative incidence at eight years of 35.4% ($n = 40$) versus 18.2% ($n = 14$). In girls, the risk to develop CD was significantly increased in both groups of HLA-DQ2 homozygosity, with hazard ratios of 14.8 and 12.5 for DQ2/DQ2 and for DQ2/DR7-DQ2, respectively. In boys, the risk to develop CD was also significantly increased in those with DQ2/ DQ2, but not in those with DQ2/DR7-DQ2 with hazard ratios of 5.0 and 1.0, respectively. Based on the variables' regression coefficients in a multivariate model, a risk stratification score was constructed for each child. Median (1.12) and first and third IQR (IQ1 = 0.90 and IQ3 = 1.44) were used as cut-off values for dividing the risk groups into low (0–0.90 points), low-medium (0.91–1.12 points), high-medium (1.13–1.44 points) and high (≥1.45 points) risk score. The total points score is mapped as a corresponding risk of CD probability shown in Fig. 3.6.

3.2.5 Age effect

In The Environmental Determinants of Diabetes in the Young study (TEDDY), 6,579 children with more than two tissue transglutaminase autoantibody (tTGA) measurements were followed from birth (Maziarz et al., 2022). After 12.5-year follow-up, 1276 (19.3%) were tTGA positive on more than two consecutive visits (defined as CD autoimmunity). HLA-DQ2/2 was associated with younger age of CD autoimmunity appearance: the hazard ratio to develop CD autoimmunity was 18 in children younger than 2 years, 10.1 in those between 2 and 4 years, 6.8 in those of 4–12,5 years. HLA-DQ2/8 was associated with HR of 9 while DQ8/8 gave a HR of 6.7.

Risk Group	Risk Score	HLA DQ/Dr Haplotype
group 1	high	DQ2/ DQ2 (*DQ2.5/DQ2.5*) and DQ2/DR7–DQ2 (*DQ2.5/DQ2.2*);
group 2	high medium	DQ2/ DQ7 (*DQ2.2/DQ7*);
group 3	high medium	DQ2/ DQ7 (*DQ2.5/DQ7*), DQ2/ DQ8 (*DQ2.5/DQ8*), and DQ2/other (*DQ2.5/other*);
group 4	low-medium	DQ2/ DQ2 (*DQ2.2/DQ2.2*), DQ2/DQ8 (*DQ2.2/DQ8*), and DQ8/ DQ8 (*DQ8/DQ8*);
group 5	low	DQ2/other (*DQ2.2/other*), DQ8/ DQ7 (*DQ8/DQ7*), and DQ8/other (*DQ8/other*);
"other"	very low	any HLA-DQ haplotype except DQ2, DR7–DQ2, DQ8, or DQ7.

FIGURE 3.6 Familial recurrence risk in the European cohort of the PREVENT-CD study. Risk of CD probability.

3.2.6 Refractory celiac disease

Interestingly, the homozygosity for the DQ2 haplotype is highly increased in a small subgroup of CD patients who did not respond to the gluten-free diet and who show aberrant intestinal T-cells (refractory CD type II patients) (see Chapter 14). 44%–62% of RCDII patients are homozygous for DQ2 compared to 20%–24% of uncomplicated CD patients (Wolters et al., unpublished data). This might reflect a correlation of the DQ2.5 dose with the severity of the disease.

3.2.7 When is HLA typing helpful?

Without HLA molecules that recognize a gluten peptide, the diagnosis of CD is highly unlikely, or impossible, since it does lack the starting mechanism that generate the abnormal immune response. Nevertheless, the high sensitivity and specificity of the anti-human tissue transglutaminase decrease significantly the diagnostic requirement of this relatively complex analysis: all celiac patients have one of the above listed HLA gene. HLA typing may add a precious diagnostic contribution in several cases:

1. when antibodies have borderline or uncertain values;
2. when the level of antibodies is low and the patient refuse a small bowel biopsy;
3. when small intestinal mucosa gives unspecific findings (Marsh 0–1 stages);
4. to estimate the genetic risk within first-degree relatives of CD subjects.

The overall risk range to develop CD in wheat consuming population is shown in Fig. 3.7.

3.3 Non-HLA gene

We understood that the strong HLA gene does explain no more than 40% of the genetic predisposition (heredity) of CD. Gusev and Sollid, by relating heredity to the amount of variance explained by the known associated genes (Gusev et al., 2013), estimated that most of the heredity has to be explained by a large amount of polymorphisms (up to 2550) in very common genes which bear a very small risk (1,001 to 1,1) with near identity with people not affected by CD, but may altogether modulate the "normal" immune response toward autoimmunity. To look for non-HLA gene, several Genome Wide Association Studies (GWAS) were performed in different populations in order to identify additional predisposing genes to the known HLA: 39 loci were associated with CD, but some loci included more than one associated gene, thus raising the number of the involved polymorphisms up to 57 (Caggiari et al., 2011). Nevertheless, as predicted by Sollid (Sollid & Lundin, 2014), the addition of these loci contributed to the estimation of the heredity no more than 10%–12%, so that the total heredity variance of HLA + associated SNPs raised from 40%–45% to about 54%.

Further GWAS raised the number of non-HLA CD associated gene to 115, many of them shared with several others immune-related disorders, contributing to the understanding of the common genetic background inducing an abnormal autoimmune response (Kumar et al., 2012). A group from Finland looked for specific set of CD associated loci to distinct clinical phenotypes: they tested 39 non-HLA CD SNPs in 625 celiac patients and found that the TLR7/TLR8 locus was associated with early (<7 years) disease onset and the SH2B3/ATXN2, ITGA4/UBE2E3 and IL2/IL21 with later onset (after 7 years of age). The latter three loci were associated with a more severe small bowel mucosal damage and SH2B3/ATXN2 with type 1 diabetes. Having more CD associated polymorphism was cumulatively associated to a more severe disease course (Cerqueira et al., 2021).

Expression quantitative trait loci (e-QTL) are genetic variants that affect gene expression levels: in CD e-QTLs are potential contributors to the disease pathology. 53% of the polymorphisms associated to CD are able to modify the

HLA Status	Risk range
HLA unknown and no familial case	1–2%
HLA DQ2 or DQ8 absent	0,001-0,002%
HLA DQ2 or DQ8 present and no familial case	2–4%
HLA unknown and familial case	10–12%
HLA DQ2 or DQ8 present and familial case	10–20%

FIGURE 3.7 Risk range associated to HLA genotype. The overall risk range to develop CD in wheat consuming population.

expression of the gene according to the genotype. This is a cis-expression of quantitative trait loci. Since the number of eQTL is larger than the number of gene expected by chance it is likely that these polymorphisms modulate gene expression either at the gene location as well as on distant gene. Dubois et al. identified a number of e-QTLs associated with CD susceptibility: genetic variants located in or near genes involved in immune regulation, such as the gene IL18RAP, were associated with altered gene expression in CD patients (Dubois et al., 2010). Raj et al. identified a number of e-QTLs associated with CD risk in individuals of European ancestry: these e-QTLs were enriched in genes related to immune regulation and intestinal function, such as the gene PTGER4 (Raj et al., 2013). Overall, these studies suggest that e-QTLs play an important role in the development and progression of CD, and may represent potential targets for therapeutic interventions in the future (Zhernakova et al., 2009). Only 5% of these new polymorphisms were present in coding regions, and about 5% and 9% in 5' and 3' untranslated regions, respectively. This means that most of the identified SNPs were either in intergenic or intronic regions, suggesting that their role is likely to be the regulation of gene expression by epigenetic pathways. Just three of the CD associated loci are in protein-coding regions. Hence, 90% of the polymorphisms associated with CD has to be related to epigenetic and regulating pathways.

Van der Graaf et al. attempted to integrate genetic data with a transcriptomic dataset from blood samples in order to identify eQTL gene (van der Graaf et al., 2020). They identified 118 genes across 50 CD-associated regions. Four pathways were associated to the identified gene polymorphisms:

1. cell adhesion and proliferation (Clorf106, FASLG);
2. T-cell activation (THEMIS, IL2, CD28, CTLA4, UBASH3A);
3. activation of monocytes and dendritic cells (CCR1, CC2, IL21);
4. transcriptomic factors (IRF4, FTS1).

By analyzing the coexpression and pathways of the 118 genes, three more domains were suggested:

1. INFγ signaling (STAT1, CD274, IL12A);
2. TCR activation (CD28, CTLA4, ICOS);
3. chemokine receptor binding (CCR1, CCR2, CCR3, IL12EA, IL21, IL18R1).

As expected, all genes associated to CD participate to the innate and induced immune response to antigen and to the following intra-cellular signaling. By integrating genetics and expression, 65 eQTL gene were found clustered together on a region of chromosome 12, where a TRAF-type zinc finger domain containing 1 (TRAFD1) was identified as a master regulator of genes involved in IFNγ signaling and MHC I antigen processing/presentation (van der Graaf et al., 2020). The TRAFD1 gene is the trans regulator of 41 CD associated genes. The TRAFD1 is also a regulator of the NFkB signaling pathway that is immediately activated by the exposure to gluten of the small intestinal mucosa of CD subjects (Maiuri et al., 2003). This master regulator, together with the other independently controlled genes, confirms the involvement either of the adaptive immunity, as well of the innate immunity, in the pathogenesis of CD. It appears that CD subjects have all normal and well-functioning genes, but the minor variations in their sequence do predispose to a specific and peculiar response to a common antigen, once some environmental cofactors ensue. This "abnormal" or "excessive" response is acted through the ordinary cell and molecular pathways of interaction between the host and the antigen.

3.3.1 Genetic risk score

An European Consortium attempted to improve the estimate of the genetic risk to develop CD by adding to the HLA associated risk the contribution brought by the addition of up to 57 CD-associated single nucleotide polymorphisms (SNPs), identifies in the GWAS, in 2675 cases and 2815 controls (Romanos et al., 2014). A Genetic Risk Score (GRS) was computed adding the number of CD-associated SNPs for each individual. Individuals in the top quintile of GRS had 2,5 times higher risk than those with a mean GRS and 7,2 times, than those in the bottom quintile. The combination of the HLA with 57 non-HLA SNP improved the discrimination between CD and controls producing an area under the curve of 0,854, much more better than the risk associated to HLA only. The GRS model with 57 SNPs allowed to reclassify more accurately 14,6% of the individuals, respect to HLA only. This model was especially efficacious for the individuals with an intermediate HLA based risk score: 15% of them were moved to the high risk category when non-HLA SNPs were considered.

In conclusion, the estimation of the non-HLA genes in addition to the known HLA risk score, is likely to predict since birth, within risk families, those babies who are more likely to develop CD. Such prediction allows to offer closer follow up and surveillance to the high-risk babies by the easy availability of the point-of-care antihuman tissue transglutaminase antibodies, much before clinical problems ensue.

3.4 Positive selection of CD-associated gene

The unexpected rise in the prevalence of CD in all wheat consuming populations rise the possibility of a positive selection of the CD associated gene over the more than 15,000 years of wheat consumption. If CD would have been present since the early days of development of agriculture, the vast majority of affected, especially children, would have had quite low chances to survive malnutrition, malabsorption, and infections. As other diseases CD would have disappeared over 12,000 years of wheat consumption if there had not been any protective factor to balance the burden of disease (Raj et al., 2013). Abadie et al. (2011) explored the chances of positive selection of CD associated genes and found that four genes, shared with other autoimmune diseases, showed a strong signature of positive selection. Homozygous carriers of the CD-associated SH2B3 allele produced, upon stimulation, larger quantities of proinflammatory cytokines (IL-1ß, IL-6 and IL-8). Hence, this entails a stronger and faster protection from common infections. Similarly, homozygous carriers of the IL18RAP risk allele have a lower expression of IL-18RAP protein, reducing the signaling in response to IL-18 and so producing less IFN-γ. The homozygous carriers of the CD associated CCR5-delta32 polymorphism significantly reduced the susceptibility to HIV-1. Homozygous carriers of the Delta32 mutation are resistant to HIV-1 infection because the mutation prevents functional expression of the CCR5 chemokine receptor normally used by HIV-1 to enter CD4 + T cells. Hence celiac individuals who carry two copies of this allele appears to be protected by the HIV infection, or they may have a very mild disease. Also heterozygous subjects have some level of protection against AIDS and rarely suffer from a severe disease (Ni et al., 2018). Heterozygosity for CCR5D32 was shown to be significantly associated with spontaneous hepatitis C viral clearance and with significantly lower hepatic inflammation (Goulding et al., 2005).

The expansion of agriculture (15–12,000 years ago) promoted a rapid expansion of the world population, with the related diffusion of infections and zoonoses: the presence of immunologically advantageous genetic profiles could have, at least partially, compensated the risk associated to the rise of CD in the wheat consuming populations. The very low frequency of the CD susceptibility alleles with positive selection in the African populations could be related to the relatively recent introduction of agriculture in this area (Abadie et al., 2011).

3.5 Shared loci and pathways with other autoimmune and inflammatory disease

There are several other autoimmune diseases that share common genetic risk factors with CD (Raj et al., 2013), including:

1. Type 1 diabetes: several genes are associated with an increased risk of both CD and type 1 diabetes, including the HLA-DQ2 and HLA-DQ8 genes;
2. Rheumatoid arthritis: the HLA-DRB1 gene has been identified as a shared genetic risk factor for both CD and rheumatoid arthritis;
3. Autoimmune thyroid diseases: CD and Hashimoto's thyroiditis and Graves' disease share a common genetic susceptibility region on chromosome 6q HLA Dr3, CTLA4, PTPN22;
4. Inflammatory bowel disease (IBD): CD and IBD share a common genetic susceptibility region on chromosome 3p21;
5. Multiple sclerosis: a higher prevalence of CD in patients with multiple sclerosis has been documented, as well as a shared susceptibility region on chromosome 6p21.

Overall, the HLA region on chromosome 6p21 has been identified as a common genetic risk factor for many autoimmune diseases, including CD. However, other genetic regions have also been implicated in the development of autoimmune diseases, and the precise mechanisms by which these genes contribute to disease development are not yet fully understood. Screening the list of GWAS gene associated to CD for links with other autoimmune and inflammatory disease Abadie (Abadie et al., 2011) found that 35% of the CD associated polymorphisms were shared with type I diabetes, sharing gene involved in the detection of viruses, cytokine and chemokine signaling and T-cell activation. The gene acting in the inflammasome (e.g., IL18RAP, IL18R1, IL1RL1 and IL1RL2) overlapped among CD, IBD and some autoimmune diseases, confirming the involvement of inflammation in the pathogenesis of CD. The comparison with other diseases enhanced the specific presence in CD of gene involved in the central role of T cell differentiation, activation, and function (Fernandez et al., 2002).

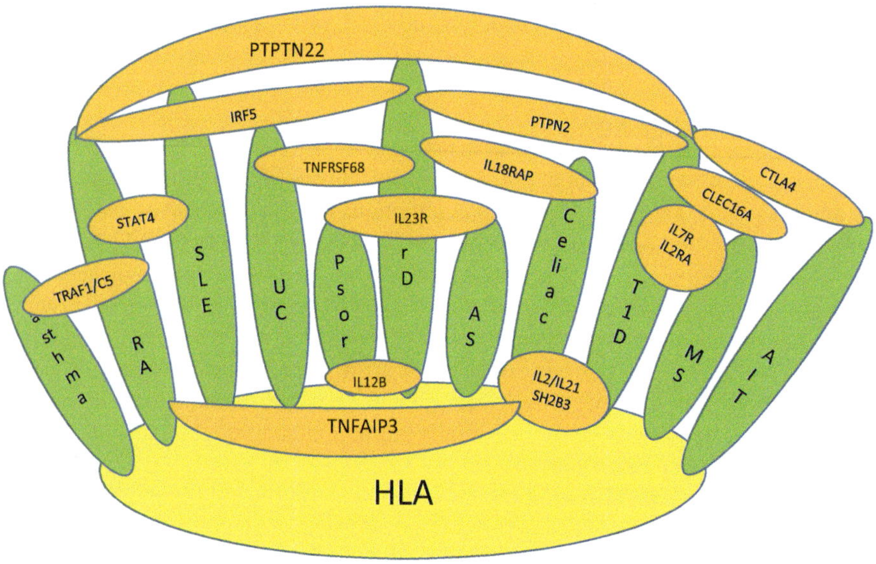

FIGURE 3.8 Gene shared among autoimmune and inflammatory diseases. HLA genes are shared among all the listed diseases. Several genes are shown in the orange ellipsis shared among two or more diseases. The majority of these are gene involved in inflammation and response to antigens. Asthma, Rheumatoid Arthritis (RA), Systemic lupus erythematosus (SLE), Ulcerative Colitis (UC), Psoriasis (Psor), Crohn's Disease (CrD), Ankylosing spondylitis (As)s, Celiac Disease (Celiac), Type I Diabetes (T1D), Multiple Sclerosis (MS), Autoimmune thyroid disease (AIT).

Zhernakova et al. (2010) using the data from GWAS studies and nonsynonymous SNPs, described the genetic scenario shared among auto immune and chronic inflammatory diseases. Fig. 3.8 shows that they share the HLA domains, on which the key molecular pathways involved in these disorders ensue. As expected, the functional pathways of inflammation and immune response are shared among these diseases. T-cell differentiation, immune-cell activation, signaling innate immunity and TNF signaling are frequently involved in several diseases, while specific cytokines or chemokines are more related to chronic inflammatory diseases. To note, IL21 and SH2B3 are shared between CD and type I diabetes (Dubois et al., 2010; Zhernakova et al., 2010).

3.6 Transcriptomics

The expression of gene associated to CD shall give an outlook of the phenotype of each single individual. Gene expression is tissue and cell specific, hence, data obtained on whole tissues may only give a global indication of gene expression. Total mRNA was screened in small bowel biopsies of 24 CD individuals and 21 controls finding that the IFN-gamma signaling pathway adequately distinguish cases from controls. Exploring the gene cluster, a transcriptomic profile was able to differentiate the degree of histologic damage of the small intestinal mucosa (Wolf et al., 2021): damage of the villi was related to the IFN-gamma signaling, which included INFG, GPB1, STAT1, Jak2 and HLA-DPA1 genes, the increased crypt proliferation was associated to CCNA2, CCNB1, CDK1, KIF20A and BIRCS genes, the activation of lamina propria was related to the activation of TLR7, CD81 and CDH2 genes. The structure and function of villi were connected to ABC1, ABCG2, EFNA1, CD36n and DPB4 genes, while crypts and inflammation showed differential activation of CCNA2, MYC, EPHB4, DUSP6, CASP1 and BIRC3. An increased expression of gene signature of T-cells, effector memory CD8 + T- and B-cells was also observed. Interestingly a genomic profile of the different pattern of the CD specific mucosal damage was indeed suggested, and this may accurately replace the traditional histology. Twenty-nine differentiated biomarkers were identified, confirmed by PCR and validated in a third group, in 20 biopsies from active CD (Bragde et al., 2018). The pathways of differential expression, compared to 20 controls, were the immune response, the response to infections, the intestinal barrier function and several metabolic pathways. Five biomarkers (CXCL10, GBP5, IN127 and UBD) were differentially expressed in healthy subjects that later manifested CD.

We analyzed the expression of nine CD associated genes in blood mononuclear cells of 22 HLA predisposed children from at-risk families for CD, followed from birth to 6 years of age (Galatola et al., 2017). Nine of them developed CD (patients) and 13 did not (controls). Gene expression was analyzed at 4–12 months of life (before diagnosis), at the time of CD diagnosis, and after at least 1 year of a gluten-free diet. Three genes (KIAA, TAGAP [T-cell Activation GTPase Activating Protein], and SH2B3 [SH2B Adapter Protein 3]) were overexpressed in patients, compared to controls, at least 9 months before CD diagnosis. At a stepwise discriminant analysis, four genes (RGS1 [Regulator of G-protein signaling 1], TAGAP, TNFSF14 [Tumor Necrosis Factor (Ligand) Superfamily member 14], and SH2B3) differentiate patients from controls before serum antibodies production and clinical symptoms ensued. Multivariate equation correctly distinguished 95.5% CD from non-CD children.

3.7 Gene regulation

3.7.1 Acetylation and methylation of histone

The modification of histones is one of the crucial mechanism to unroll DNA filaments to be expressed. A set of specialized enzyme are dedicated to this function, namely acetylation, phosphorylation, ubitiquination and methylation enzymes are the key operators. All these mechanisms are tissue and cell specific, in order to modulate the required specific functions.

Different levels of histone acetylation were found in T- cells from CD biopsies upon stimulation with IL-15, INFß and IL-21 (Zorro et al., 2020). It appeared that different stimulations enhanced differentiated profile of histone acetylation: IL-15 stimulated the expression of several noncoding RNA with potential regulatory effects on gene expression, while IFN-ß modulated the function of a main histone acetylation enzyme H3K27ac.

Methylation of histones is a further mechanism linked to repression of transcription: several CD associated SNPs, shared with rheumatoid arthritis, are able to modulate the activity of a main histone-methylation enzyme H3K4me3 (Gutierrez-Achury et al., 2016). The pathogenesis of crypts hyperplasia in CD was associated to the disruption of the Wnt/PRC2 (Polycomb repressive complex) axis since a histone methylating enzyme (H33K27me3) showed a different pattern according to the position along the villous-crypt axis, controlling gene related to proliferation and differentiation in the epithelium (Oittinen et al., 2017).

3.7.2 DNA methylation

DNA methyltransferase enzymes (DNMTs) overlay methyl groups to the C-5 residue of cytosine pyrimidine ring in the CpG-rich islands located in promoter and regulatory regions of DNA sequence. This methylation limits the transcription of gene, which are not expressed. Methylation of DNA sequences mostly happen in the early days of life and is largely heritable, with no modifications in the DNA sequence, but it may also happen during lifetime under specific stimuli. This fine regulation of gene expression is tissue and cell specific and is crucial to the development or protection of diseases.

Fernandez-Jimenez et al. (2014) compared the whole methylome of epithelial and immune cells from biopsies of CD and controls, and found a differential methylation pattern of epithelium compared to immune cells, confirming the cell specificity of the process. They found that CD patients on a gluten-free diet showed a levels of gene methylation of the NFκB pathway intermediate between that observed in active CD and healthy controls, suggesting that gene methylation might be modulated by the proline-rich peptide of the gliadin molecule (Fernandez-Jimenez et al., 2019). The DNA methylation pattern appears to be highly specific for a defined disease status, to the point that it can be used to diagnose CD and other diseases on a DNA from a saliva sample (Hearn et al., 2020). The fine regulation of these genes by methylation influences critical pathways like NFκB signaling pathway, IL17 signaling cascade, RIG-I like receptor signaling pathway, NOD-like receptor pathways, fully implicated in the abnormal immune response to gluten (Ghosh et al., 2022).

Our group (Cielo et al., 2019) separated the epithelium from the lamina propria of the small bowel mucosa of active CD biopsies and found a decreased methylation of SB2H3 and IL21 gene (strongly associated to CD) causing an over expression of these genes, so modifying the immune response to the gliadin antigen. In epithelial cells, SH2B3 was differentially methylated between CD patients and controls, whereas IL21, TNFAIP3 and cREL showed differences in methylation level between CD and controls in the lamina propria. Further investigation showed that IL21 expression in the lamina propria was greater in CD patients: this gene was 20% less methylated. Similarly, the lower methylation of the SH2B3 gene in CD epithelial cells was associated with higher expression of the gene. By using the Epigenome Roadmap tool, we identified the presence of several regulating elements for histone modifications: H3K27ac and H3K4me3 in the small intestine and H3K4me1, H3K4me3, and H3K9ac in duodenal mucosa and a DNAse hypersensitive tract. Bioinformatics analysis showed that the differential methylation is centered in a genomic area of a DNA regulatory element and involves four key histone modifications and a DNAse hypersensitive tract.

3.7.3 Micro-RNA

MicroRNAs (miRNAs) are short RNA sequences (20–25 nucleotides) accounting for about 30% of gene expression regulation, affecting various processes, including chromatin conformation and transcription, as well as mRNA stability and translation. Usually they present an inverse relationship with their targets, inducing their downregulation. miRNAs were screened in CD duodenal tissue to identify a specific signature representative of the events in the mucosa, and to determine whether the miRNAs "signature" was returning to normal levels in patients on a GFD (Tan et al., 2021).

Capuano et al. evaluated biopsies of children with CD at diagnosis and on a GFD, identifying about 20% miRNAs differentially regulated (Capuano et al., 2011). The upregulation of miR-449a (55 times over expressed in CD) targets Notch1 and KLF4 genes involved in intestinal cell proliferation and differentiation. Indeed, Notch1 expression was significantly downregulated in CD biopsies, confirming the regulatory role of miRNAs also on proliferation. Other miRNAs were also downregulated, that is, miR-124a, miR-189, miR-299−5p and miR-379, similarly to what was reported in Crohn's disease. However, a more comprehensive approach was needed, and a recent paper correlated, in the duodenal mucosa of celiac subjects, the miRNA and mRNA expression patterns obtained by sequencing (Magni et al., 2014). The analysis revealed the presence of a complex network involving various pathways known to be deregulated in CD, suggesting that CD-associated miRNAs play a central role in causing the intestinal damage.

Studies on adult CD patients identified different subset of miRNAs according to the clinical picture and the severity of intestinal damage. A significant downregulation of miR-31−5p was observed in all CD groups: indeed, miR-31−5p had as a target Foxp3, essential for T-Reg development; again, a significant inverse correlation was observed between the miRNA and the target mRNA. These changes were induced by gliadin, as demonstrated by in vitro stimulation of biopsies of patients on a GFD. It is interesting to note that the data on miR-192 are similar to those observed in IBD, in which miR-192 regulated NOD2 expression.

The identification of a specific miRNA plasma signature remains elusive at the present time. An analysis performed in pediatric CD revealed the presence of a trend similar to that observed in duodenal biopsies for miR-192−5p, miR-31−5p and miR-21−5p. However, none of these miRNAs could be defined as a marker able to evaluate the recovery of the mucosa while on a GFD, since there was no return to normal levels or a clear cut-off could not be established. The presence of miRNA in biological fluids stimulated the search for biomarkers able to identify CD patients on plasma or serum samples. Felli et al. found that miR-192−5p, miR-215−5p and miR-125b-5p were able to discriminate active CD, CD on a GFD and controls with high accuracy and specificity (Felli et al., n.d.). Bianchi et al. identified in sera 15 miRNAs able to discriminate adult patients according to two type of aggressive refractory disease (Bianchi et al., 2022). By bioinformatic they suggested the involvement of the pathway of TGF-β and cell-cell signaling by Wnt.

3.7.4 Long noncoding RNA

Long noncoding RNAs (lncRNAs) are RNA molecules longer than 200 nucleotides which do not code for proteins. They play regulatory roles in gene expression by interacting with DNA, RNA, and proteins. They may directly interact with DNA, or act by regulation of histone modification and chromatin rearrangement. Several lncRNAs are dysregulated in CD (Zorro et al., 2020).

Ricaño-Ponce et al. found that two CD associated SNPs (AP002954.4 and AC104820.2) regulate the expression of lncRNAs which have a crucial role in modulating the immune response, by a tissue and cell specific mechanism (Ricaño-Ponce et al., 2015). In another study, increased expression of Carl's lncRNA was observed in the cytoplasm followed by increased expression of NFkB pathway genes, suggesting that increased Carl's gene expression and its effect on cytoplasmic signaling NFkB is associated with the development of an inflammatory reaction in CD (Castellanos-Rubio et al., 2017).

One of the lncRNA involved in CD expression is lnc13, that binds the histone deacetylase 1 and the ribonucleoprotein HNRNPD to generate a complex that downregulates the expression of inflammatory gene. The activation of NFkB by inflammation reduces lnc13, causing the removal of the repression of gene expression (Castellanos-Rubio et al., 2017).

PTPN2 (protein tyrosine phosphatase nonreceptor type 2) is a CD associated gene involved in the activation and regulation of T-cell signaling. SNPs in the PTPN2 locus are associated with changes in levels of DNA methylation at four CpG sites 7.5 kb downstream of PTPN2 and changes in expression of the LINC01882, a long noncoding located on chr18 of PTPN2. Silencing this lncRNA modulates the transcription regulation pathway in T cells that could be involved in rheumatoid arthritis and other autoimmune diseases, where PTPN2 SNPs are known to be associated, thus providing a functional link between unexplained gene polymorphisms and disease (Castellanos-Rubio et al., 2016). Autoimmune diseases associated noncoding SNPs in the PTPN2 locus are linked to differential expression levels of the lncRNAs LINC01882 and RP11−973H7.1. These results show that autoimmune disease associated noncoding SNPs in the PTPN2 locus are also associated to differential expression levels of the lncRNAs LINC01882 and RP11−973H7.1.

The most up-regulated gene, upon LINC01882 knockdown, is BZRAP1 but no function for BZRAP1 has been reported for blood cells. The knockdown of LINC01882 also downregulated the expression of MAP2K4 and ZEB1, where MAP2K4 is a component of the kinase cascade that mediates cellular responses to cytokine signals and ZEB1 encodes a transcription factor that represses IL2 expression in T cells (Houtman et al., 2018).

In recent a study, the lncRNA expression was analyzed in 32 celiacs and 22 controls. The PTPN2-gene-associated lncRNA LINC01882 was more expressed in CD; as a consequence, PTPN2 and MAP2K4 genes show lower expression in CD. In conclusion LINC01882 plays a role in the development of CD, in association with rs12971201 PTPN2-SNP. Together, these results suggest that LINC01882 is involved in T-cell activation (Cielo et al., unpublished results). These studies demonstrate that lncRNAs are able to potentiate or suppress the inflammatory response in a specific gene. Increasing our understanding of exactly how lncRNAs function in regulating inflammation will provide new targets for the development of novel therapeutic strategies for many inflammatory diseases.

3.8 Conclusions

CD has the strongest heritability among autoimmune diseases, partially explained by the specific capacity of the HLA DQ2 and DQ8 dimers to recognize gluten peptides in order to present them to the T-cell receptor. Noteworthy, HLA genes are not different in CD compared to the general population and most of the HLA DQ2 DQ8 carriers do not develop the disease. Moreover, the expression of the DQ2 and DQ8 molecules is different in CD compared to controls, since they have the same structure but different regulation. Without this dimer that traps the gliadin peptide it is impossible to activate the abnormal immune response leading to autoimmunity and disease. The polymorphic nature of HLA is associated to a different capacity to recognize and present gluten peptides to T-cells, hence, there is a gradient of risk to develop CD according to the number and combinations of the DQ dimer.

HLA explains less than 50% of this heritability: more than 50 SNPs contribute to the development of the response to gliadin, activation of inflammation, and cellular damage. Altogether these SNPs add less than 10% to the heritability. So there is half of the genetic component that is not yet completely understood. Epigenetic mechanisms play a crucial role in the fine regulation of gene expression and to the working machinery leading to the disease. Gene expression is already defined in babies who will eventually develop the disease, before the exposure to gluten, and much before the appearance of specific antibodies and small intestinal damage. Fine regulation starts at the histone level, where acetylation and methylation enzymes control the unfolding of DNA strings; then methylation of the unfolded DNA sequences controls the possibility of expression of these CD associated genes. Once the mRNA is produced, there is a long way before it may reach the ribosome in order to start the protein production. A series of micro as well as long-noncoding RNA do regulate the chances to create the final product of the CD associated genes.

References

Abadie, V., Sollid, L. M., Barreiro, L. B., & Jabri, B. (2011). Integration of genetic and immunological insights into a model of celiac disease pathogenesis. *Annual Review of Immunology*, *29*(1), 493–525. Available from https://doi.org/10.1146/annurev-immunol-040210-092915.

Al-Hussaini, A., Alharthi, H., Osman, A., Eltayeb-Elsheikh, N., & Chentoufi, A. (2018). Genetic susceptibility for celiac disease is highly prevalent in the Saudi population. *Saudi Journal of Gastroenterology*, *24*(5), 268–273. Available from https://doi.org/10.4103/sjg.SJG_551_17.

Bianchi, N., Doneda, L., Elli, L., Taccioli, C., Vaira, V., Scricciolo, A., Lombardo, V., Terrazzan, A., Colapietro, P., Terranova, L., Bergamini, C., Vecchi, M., Scaramella, L., Nandi, N., & Roncoroni, L. (2022). Circulating microRNAs suggest networks associated with biological functions in aggressive refractory type 2 celiac disease. *Biomedicines*, *10*(6), 1408. Available from https://doi.org/10.3390/biomedicines10061408.

Bolognesi, E., Karell, K., Percopo, S., Coto, I., Greco, L., Mantovani, V., Suoraniemi, E., Partanen, J., Mustalahti, K., Mäki, M., & Momigliano-Richiardi, P. (2003). Additional factor in some HLA DR3/DQ2 haplotypes confers a fourfold increased genetic risk of celiac disease. *Tissue Antigens*, *61*(4), 308–316. Available from https://doi.org/10.1034/j.1399-0039.2003.00028.x.

Bourgey, M., Calcagno, G., Tinto, N., Gennarelli, D., Margaritte-Jeannin, P., Greco, L., Limongelli, M. G., Esposito, O., Marano, C., Troncone, R., Spampanato, A., Clerget-Darpoux, F., & Sacchetti, L. (2007). HLA related genetic risk for coeliac disease. *Gut*, *56*(8), 1054–1059. Available from https://doi.org/10.1136/gut.2006.108530.

Bragde, H., Jansson, U., Fredrikson, M., Grodzinsky, E., & Söderman, J. (2018). Celiac disease biomarkers identified by transcriptome analysis of small intestinal biopsies. *Cellular and Molecular Life Sciences*, *75*(23), 4385–4401. Available from https://doi.org/10.1007/s00018-018-2898-5.

Caggiari, L., Toffoli, G., De Re, V., Orzes, N., Spina, M., De Zorzi, M., Maiero, S., Cannizzaro, R., & Canzonieri, V. (2011). KIR/HLA combination associated with the risk of complications in celiac disease. *The International Journal of Biological Markers*, *26*(4), 221–228. Available from https://doi.org/10.5301/jbm.2011.8903.

Camarca, A., Radano, G., Mase, D., et al. (2021). Short whet challenge is a reproducible in-vivo assay to detect immune response to gluten. *Clinical and Experimental Immunology*, *169*, 129–360.

Capuano, M., Iaffaldano, L., Tinto, N., Montanaro, D., Capobianco, V., Izzo, V., Tucci, F., Troncone, G., Greco, L., Sacchetti, L., & Thi Thu Nguyen, H. (2011). MicroRNA-449a overexpression, reduced NOTCH1 signals and scarce goblet cells characterize the small intestine of celiac patients. *PLoS One*, *6*(12), e29094. Available from https://doi.org/10.1371/journal.pone.0029094.

Castellanos-Rubio, A., Fernandez-Jimenez, N., Kratchmarov, R., Luo, X., Bhagat, G., Green, P. H. R., Schneider, R., Kiledjian, M., Bilbao, J. R., & Ghosh, S. (2016). A long noncoding RNA associated with susceptibility to celiac disease. *Science (New York, N.Y.)*, *352*(6281), 91–95. Available from https://doi.org/10.1126/science.aad0467.

Castellanos-Rubio, A., Kratchmarov, R., Sebastian, M., Garcia-Etxebarria, K., Garcia, L., Irastorza, I., & Ghosh, S. (2017). Cytoplasmic form of carlr lncRNA facilitates inflammatory gene expression upon NF-kB activation. *Journal of Immunology*, *199*(2), 581–588. Available from https://doi.org/10.4049/jimmunol.1700023.

Cerqueira, J. X. M., Saavalainen, P., Kurppa, K., Laurikka, P., Huhtala, H., Nykter, M., Koskinen, L. E., Yohannes, L., Kilpeläinen, D. A., Shcherban, E., Palotie, A., Kaukinen, A., & Lindfors, K. (2021). Independent and cumulative coeliac disease-susceptibility loci are associated with distinct disease phenotypes. *Journal of Human Genetics*, *66*(6), 613–623. Available from https://doi.org/10.1038/s10038-020-00888-5.

Cielo, D., Galatola, M., Fernandez-Jimenez, N., De Leo, L., Garcia-Etxebarria, K., Loganes, C., Tommasini, A., Not, T., Auricchio, R., Greco, L., & Bilbao, J. R. (2019). Combined analysis of methylation and gene expression profiles in separate compartments of small bowel mucosa identified celiac disease patients' signatures. *Scientific Reports*, *9*(1). Available from https://doi.org/10.1038/s41598-019-46468-2.

Dubois, P. C. A., Trynka, G., Franke, L., Hunt, K. A., Romanos, J., Curtotti, A., Zhernakova, A., Heap, G. A. R., Ádány, R., Aromaa, A., Bardella, M. T., Van Den Berg, L. H., Bockett, N. A., De La Concha, E. G., Dema, B., Fehrmann, R. S. N., Fernández-Arquero, M., Fiatal, S., Grandone, E., & Van Heel, D. A. (2010). Multiple common variants for celiac disease influencing immune gene expression. *Nature Genetics*, *42*(4), 295–302. Available from https://doi.org/10.1038/ng.543.

Ek, J., Albrechtsen, D., Solheim, B. G., & Thorsby, E. (1978). Strong association between the HLA-dw3-related b cell alloantigen -DRw3* and coeliac disease. *Scandinavian Journal of Gastroenterology*, *13*(2), 229–233. Available from https://doi.org/10.3109/00365527809181753.

Fallang, L. E., Bergseng, E., Hotta, K., Berg-Larsen, A., Kim, C. Y., & Sollid, L. M. (2009). Differences in the risk of celiac disease associated with HLA-DQ2.5 or HLA-DQ2.2 are related to sustained gluten antigen presentation. *Nature Immunology*, *10*(10), 1096–1101. Available from https://doi.org/10.1038/ni.1780.

Farina, P., Pisapia, B., Vitale, F., Mozzillo, G., & Giovanna, D. P. (2019). HLA-DQA1 and HLA-DQB1 alleles, conferring susceptibility to celiac disease and type 1 diabetes, are more expressed than non-predisposing alleles and are coordinately regulated. *Cells*, *8*(7), 751. Available from https://doi.org/10.3390/cells8070751.

Felli, B., A., Uva, A. Cangelosi, Ancinelli, C., Paolini, M., A., Silano, M., Vincentini, O., Catassi, C., Lionetti, Gatti, S.F., & Masotti, A. (n.d.). Circulating microRNAs as novel non-invasive biomarkers of paediatric celiac disease and adherence to gluten-free diet. EBioMedicine, 76. https://doi.org/10.1016/j.ebiom.2022.103851.Epub.

Fernandez, L., Fernandez-Arquero, M., Gual, L., Lazaro, F., Maluenda, C., Polanco, I., Figueredo, M. A., & De La Concha, E. G. (2002). Triplet repeat polymorphism in the transmembrane region of the MICA gene in celiac disease. *Tissue Antigens*, *59*(3), 219–222. Available from https://doi.org/10.1034/j.1399-0039.2002.590307.x.

Fernandez-Jimenez, N., Castellanos-Rubio, A., Plaza-Izurieta, L., Irastorza, I., Elcoroaristizabal, X., Jauregi-Miguel, A., Lopez-Euba, T., Tutau, C., de Pancorbo, M. M., Vitoria, J. C., & Bilbao, J. R. (2014). Coregulation and modulation of NF B-related genes in celiac disease: Uncovered aspects of gut mucosal inflammation. *Human Molecular Genetics*, *23*(5), 1298–1310. Available from https://doi.org/10.1093/hmg/ddt520.

Fernandez-Jimenez, N., Garcia-Etxebarria, K., Plaza-Izurieta, L., Romero-Garmendia, I., Jauregi-Miguel, A., Legarda, M., Ecsedi, S., Castellanos-Rubio, A., Cahais, V., Cuenin, C., Degli Esposti, D., Irastorza, I., Hernandez-Vargas, H., Herceg, Z., & Bilbao, J. R. (2019). The methylome of the celiac intestinal epithelium harbours genotype-independent alterations in the HLA region. *Scientific Reports*, *9*(1). Available from https://doi.org/10.1038/s41598-018-37746-6.

Galatola, M., Cielo, D., Panico, C., Stellato, P., Malamisura, B., Carbone, L., Gianfrani, C., Troncone, R., Greco, L., & Auricchio, R. (2017). Presymptomatic diagnosis of celiac disease in predisposed children: The role of gene expression profile. *Journal of Pediatric Gastroenterology & Nutrition*, *65*(3), 314–320. Available from https://doi.org/10.1097/mpg.0000000000001519.

Ghosh, S., Khetarpal, P., & Senapati, S. (2022). Functional implications of the CpG island methylation in the pathogenesis of celiac disease. *Molecular Biology Reports*, *49*(10), 10051–10064. Available from https://doi.org/10.1007/s11033-022-07585-w.

Goulding, C. A., Murphy, A., MacDonald, G., Barrett, S., Crowe, J., Hegarty, J., McKiernan, S., Kelleher, D., & McManus, R. (2005). The CCR5-Δ32 mutation: Impact on disease outcome in individuals with hepatitis C infection from a single source. *Gut*, *54*(8), 1157–1161. Available from https://doi.org/10.1136/gut.2004.055699.

Greco, L., Romino, R., Coto, I., Cosmo, D. I., Percopo, N., Maglio, S., et al. (2002). Concordance, disease progression, and heritability of coeliac disease in Italian twins. *Gut*, *50*, 803–808.

Gusev, A., Bhatia, G., Zaitlen, N., Vilhjalmsson, B. J., Diogo, D., Stahl, E. A., Gregersen, P. K., Worthington, J., Klareskog, L., Raychaudhuri, S., Plenge, R. M., Pasaniuc, B., & Price, A. L. (2013). Quantifying missing heritability at known GWAS Loci. *PLoS Genetics*, *9*(12). Available from https://doi.org/10.1371/journal.pgen.1003993.

Gutierrez-Achury, J., Zorro, M. M., Ricaño-Ponce, I., Zhernakova, D. V., Diogo, D., Raychaudhuri, S., Franke, L., Trynka, G., Wijmenga, C., & Zhernakova, A. (2016). Functional implications of disease-specific variants in loci jointly associated with coeliac disease and rheumatoid arthritis. *Human Molecular Genetics*, *25*(1), 180–190. Available from https://doi.org/10.1093/hmg/ddv455.

Hearn, N. L., Chiu, C. L., & Lind, J. M. (2020). Comparison of DNA methylation profiles from saliva in Coeliac disease and non-coeliac disease individuals. *BMC Medical Genomics*, *13*(1). Available from https://doi.org/10.1186/s12920-020-0670-9.

Houtman, S. K., Chemin, A., Hensvold, R. K., Tandre, M.-L., Eloranta, L., Rönnblom, S., & Uebe. (2018). Anca Irinel Catrina, Vivianne Malmström, Leonid Padyukov. T Cells Are InfluenCD by a long non-coding RNA in the autoimmune associated PTPN2. *Locus J Autoimmun*, *90*, 28, –3.

Karell, K., Louka, A. S., Moodie, S. J., Ascher, H., Clot, F., Greco, L., Ciclitira, P. J., Sollid, L. M., & Partanen, J. (2003).). HLA types in celiac disease patients not carrying the DQA1 *05-DQB1 *02 (DQ2) heterodimer: Results from the European genetics cluster on celiac disease. *Human Immunology*, *64*(4), 469–477. Available from https://doi.org/10.1016/S0198-8859(03)00027-2.

Keuning, J. J., Peña, A. S., Van Hooff, J. P., Van Leeuwen, A., & Van Rood, J. J. (1976). HLA-DW3 associated with cœliac disease. *The Lancet*, *307* (7958), 506–508. Available from https://doi.org/10.1016/S0140-6736(76)90294-4.

Kim, C. Y., Quarsten, H., Bergseng, E., Khosla, C., & Sollid, L. M. (2004). Structural basis for HLA-DQ2-mediated presentation of gluten epitopes in celiac disease. *Proceedings of the National Academy of Sciences of the United States of America*, *101*(12), 4175–4179. Available from https://doi.org/10.1073/pnas.0306885101.

Kumar, V., Wijmenga, C., & Withoff, S. (2012). From genome-wide association studies to disease mechanisms: Celiac disease as a model for autoimmune diseases. *Seminars in Immunopathology*, *34*(4), 567–580. Available from https://doi.org/10.1007/s00281-012-0312-1.

Liu, E., Lee, H. S., Aronsson, C. A., Hagopian, W. A., Koletzko, S., Rewers, M. J., Eisenbarth, G. S., Bingley, P. J., Bonifacio, E., Simell, V., & Agardh, D. (2014). Risk of pediatric celiac disease according to HLA haplotype and country. *New England Journal of Medicine*, *371*(1), 42–49. Available from https://doi.org/10.1056/NEJMoa1313977.

Louka, A. S., Moodie, S. J., Karell, K., Bolognesi, E., Ascher, H., Greco, L., Momigliano-Richiardi, P., Partanen, J., Ciclitira, P. J., Sollid, L. M., Holopainen, P., Collin, P., Mustalahti, K., Mäki, M., Clerget-Darpoux, F., Babron, M. C., Clot, F., Hugot, J. P., D'Alfonso, S., & Vaughan, R. W. (2003). A collaborative European search for non-DQA1 *05-DQB1 *02 Celiac disease loci on HLA-Dr3 haplotypes: Analysis of transmission from homozygous parents. *Human Immunology*, *64*(3), 350–358. Available from https://doi.org/10.1016/S0198-8859(02)00822-4.

Magni, S., Comani, G. B., Elli, L., Vanessi, S., Ballarini, E., Nicolini, G., Rusconi, M., Castoldi, M., Meneveri, R., Muckenthaler, M. U., Bardella, M. T., & Barisani, D. (2014). MIRNAs affect the expression of innate and adaptive immunity proteins in celiac disease. *American Journal of Gastroenterology*, *109*(10), 1662. Available from https://doi.org/10.1038/ajg.2014.203.

Maiuri, M. C., De Stefano, D., Mele, G., Fecarotta, S., Greco, L., Troncone, R., & Carnuccio, R. (2003). Nuclear factor κB is activated in small intestinal mucosa of celiac patients. *Journal of Molecular Medicine*, *81*(6), 373–379. Available from https://doi.org/10.1007/s00109-003-0440-0.

Makharia, G. K. (2015). Celiac disease screening in Southern and East Asia. *Digestive Diseases*, *33*(2), 167–174. Available from https://doi.org/10.1159/000369537.

Margaritte-Jeannin, P., Babron, M. C., Bourgey, M., Louka, A. S., Clot, F., Percopo, S., Coto, I., Hugot, J. P., Ascher, H., Sollid, L. M., Greco, L., & Clerget-Darpoux, F. (2004). HLA-DQ relative risks for coeliac disease in European populations: A study of the European Genetics Cluster on Coeliac Disease. *Tissue Antigens*, *63*(6), 562–567. Available from https://doi.org/10.1111/j.0001-2815.2004.00237.x.

Maziarz, M., Lynch, K., Aronsson, L., K., Koletzko, Kurppa, K., Liu et al. (2022). Temporal associations of genetic and environmental factors on the risk of celiac disease autoimmunity in early childhood: the teddy study. In Proceedings of the ninteenth international celiac disease symposium.

McManus, R., Moloney, M., Borton, M., Finch, A., Chuan, Y. T., Lawlor, E., Weir, D. G., & Kelleher, D. (1996). Association of celiac disease with microsatellite polymorphisms close to the tumor necrosis factor genes. *Human Immunology*, *45*(1), 24–31. Available from https://doi.org/10.1016/0198-8859(95)00144-1.

Meijer, C. R., Auricchio, R., Putter, H., Castillejo, G., Crespo, P., Gyimesi, J., Hartman, C., Kolacek, S., Koletzko, S., Korponay-Szabo, I., Ojinaga, E. M., Polanco, I., Ribes-Koninckx, C., Shamir, R., Szajewska, H., Troncone, R., Villanacci, V., Werkstetter, K., & Mearin, M. L. (2022). Prediction models for celiac disease development in children from high-risk families: Data from the PreventCD cohort. *Gastroenterology*, *163*(2), 426–436. Available from https://doi.org/10.1053/j.gastro.2022.04.030.

Molberg, O., Kett, K., Scott, H., Thorsby, E., Sollid, L. M., & Lundin, K. E. (1997). Gliadin specific, HLA DQ2-restricted T cells are commonly found in small intestinal biopsies from coeliac disease patients, but not from controls. *Scandinavian Journal of Immunology*, *46*(1), 103–108.

Molberg, Ø., Mcadam, S. N., Körner, R., Quarsten, H., Kristiansen, C., Madsen, L., Fugger, L., Scott, H., Norën, O., Roepstorff, P., Lundin, K. E. A., Sjöström, H., & Sollid, L. M. (1998). Tissue transglutaminase selectively modifies gliadin peptides that are recognized by gut-derived T cells in celiac disease. *Nature Medicine*, *4*(6), 713–717. Available from https://doi.org/10.1038/nm0698-713.

Ni, J., Wang, D., & Wang, S. (2018). The CCR5-Delta32 genetic polymorphism and HIV-1 infection susceptibility: A meta-analysis. *Open Medicine (Poland)*, *13*(1), 467–474. Available from https://doi.org/10.1515/med-2018-0062.

Nisticò, L., Fagnani, C., Coto, I., Percopo, S., Cotichini, R., Limongelli, M. G., Paparo, F., D'Alfonso, S., Giordano, M., Sferlazzas, C., Magazzù, G., Momigliano-Richiardi, P., Greco, L., & Stazi, M. A. (2006). Concordance, disease progression, and heritability of coeliac disease in Italian twins. *Gut*, *55*(6), 803–804. Available from https://doi.org/10.1136/gut.2005.083964.

Oittinen, M., Popp, A., Kurppa, K., Lindfors, K., Mäki, M., Kaikkonen, M. U., & Viiri, K. (2017). Polycomb Repressive Complex 2 Enacts Wnt signaling in intestinal homeostasis and contributes to the instigation of stemness in diseases entailing epithelial hyperplasia or neoplasia. *Stem Cells*, *35*(2), 445–457. Available from https://doi.org/10.1002/stem.2479.

Petersen, J., Montserrat, V., Mujico, J. R., Loh, K. L., Beringer, D. X., Van Lummel, M., Thompson, A., Mearin, M. L., Schweizer, J., Kooy-Winkelaar, Y., Van Bergen, J., Drijfhout, J. W., Kan, W. T., La Gruta, N. L., Anderson, R. P., Reid, H. H., Koning, F., & Rossjohn, J. (2014). T-cell receptor recognition of HLA-DQ2-gliadin complexes associated with celiac disease. *Nature Structural and Molecular Biology*, *21*(5), 480–488. Available from https://doi.org/10.1038/nsmb.2817.

Pisapia, L., Camarca, A., Picascia, S., Bassi, V., Barba, P., Del Pozzo, G., & Gianfrani, C. (2016). HLA-DQ2.5 genes associated with celiac disease risk are preferentially expressed with respect to non-predisposing HLA genes: Implication for anti-gluten T cell response. *Journal of Autoimmunity*, *70*, 63–72. Available from https://doi.org/10.1016/j.jaut.2016.03.016.

Ploski, R., Ek, J., Thorsby, E., & Sollid, L. M. (1993). On the HLA-DQ(α1*0501, β1*0201)-associated susceptibility in celiac disease: A possible gene dosage effect of DQB1*0201. *Tissue Antigens*, *41*(4), 173–177. Available from https://doi.org/10.1111/j.1399-0039.1993.tb01998.x.

Raj, T., Kuchroo, M., Replogle, J. M., Raychaudhuri, S., Stranger, B. E., & De Jager, P. L. (2013). Common risk alleles for inflammatory diseases are targets of recent positive selection. *American Journal of Human Genetics*, *92*(4), 517–529. Available from https://doi.org/10.1016/j.ajhg.2013.03.001.

Ricaño-Ponce, I., Wijmenga, C., & Gutierrez-Achury, J. (2015). Genetics of celiac disease. *Best Practice & Research. Clinical Gastroenterology*, *29* (3), 399–412. Available from https://doi.org/10.1016/j.bpg.2015.04.004.

Romanos, J., Rosén, A., Kumar, V., Trynka, G., Franke, L., Szperl, A., Gutierrez-Achury, J., van Diemen, C. C., Kanninga, R., Jankipersadsing, S. A., Steck, A., Eisenbarth, G., van Heel, D. A., Cukrowska, B., Bruno, V., Mazzilli, M. C., Núñez, C., Bilbao, J. R., Mearin, M. L., & Wijmenga, C. (2014). Improving coeliac disease risk prediction by testing non-HLA variants additional to HLA variants. *Gut*, *63*(3), 415–422. Available from https://doi.org/10.1136/gutjnl-2012-304110.

Saberi-Firouzi, M., Omrani, G., Nejabat, M., Mehrabani, D., & Khademolhosseini, F. (2008). Prevalence of celiac disease in Shiraz, Southern Iran. *Saudi Journal of Gastroenterology*, *14*(3), 135–138. Available from https://doi.org/10.4103/1319-3767.41732.

Sandström, O., Rosén, A., Lagerqvist, C., Carlsson, A., Hernell, O., Högberg, L., & Ivarsson, A. (2013). Transglutaminase IgA antibodies in a celiac disease mass screening and the role of HLA-DQ genotyping and endomysial antibodies in sequential testing. *Journal of Pediatric Gastroenterology and Nutrition*, *57*(4), 472–476. Available from https://doi.org/10.1097/MPG.0b013e31829ef65d.

Sollid, L. M., & Lie, B. A. (2005). Celiac disease genetics: Current concepts and practical applications. *Clinical Gastroenterology and Hepatology*, *3* (9), 843–851. Available from https://doi.org/10.1016/S1542-3565(05)00532-X.

Sollid, L., & Lundin, K. (2014). *The autoimmune diseases*. Academic Press.

Sood, A., Midha, V., Sood, N., Avasthi, G., & Sehgal, A. (2006). Prevalence of celiac disease among school children in Punjab, North India. *Journal of Gastroenterology and Hepatology (Australia)*, *21*(10), 1622–1625. Available from https://doi.org/10.1111/j.1440-1746.2006.04281.x.

Tan, I. L., Barisani, D., Panceri, R., Modderman, R., Visschedijk, M., Weersma, R. K., Wijmenga, C., Jonkers, I., Coutinho de Almeida, R., & Withoff, S. (2021). A Combined mRNA- and miRNA-sequencing approach reveals miRNAs as potential regulators of the small intestinal transcriptome in celiac disease. *International Journal of Molecular Sciences*, *22*(21), 11382. Available from https://doi.org/10.3390/ijms222111382.

Tosi, R., Vismara, D., Tanigaki, N., Ferrara, G. B., Cicimarra, F., Buffolano, W., Follo, D., & Auricchio, S. (1983). Evidence that celiac disease is primarily associated with a DC locus allelic specificity. *Clinical Immunology and Immunopathology*, *28*(3), 395–404. Available from https://doi.org/10.1016/0090-1229(83)90106-X.

van Belzen, M. J., Koeleman, B. P. C., Crusius, J. B. A., Meijer, J. W. R., Bardoel, A. F. J., Pearson, P. L., Sandkuijl, L. A., Houwen, R. H. J., & Wijmenga, C. (2004). Defining the contribution of the HLA region to cis DQ2-positive coeliac disease patients. *Genes and Immunity*, *5*(3), 215–220. Available from https://doi.org/10.1038/sj.gene.6364061.

van der Graaf, A., Zorro, M., Claringbould, A., Vosa, U., Aguirre-Gamboa, R., Li, C., Mooiweer, J., Ricano-Ponce, I., Borek, Z., Koning, F., Kooy-Winkelaar, Y., Sollid, L., Qiao, S.W., Kumar, V., Li, Y., Franke, L., Withoff, S., Wijmenga, C., Sanna, S., & Jonkers, I. (2020). Systematic prioritization of candidate genes in disease loci identifies TRAFD1 as a master regulator of IFNγ signalling in celiac disease. bioRxiv. https://doi.org/10.1101/2020.03.04.973487

Wolf, J., Willscher, E., Loeffler-Wirth, H., Schmidt, M., Flemming, G., Zurek, M., Uhlig, H. H., Händel, N., & Binder, H. (2021). Deciphering the transcriptomic heterogeneity of duodenal coeliac disease biopsies. *International Journal of Molecular Sciences*, *22*(5), 2551. Available from https://doi.org/10.3390/ijms22052551.

Yuan, J., Gao, J., Li, X., Liu, F., Wijmenga, C., Chen, H., et al. (2013). The tip of the celiac iceberg in China: A systematic review and meta analysis. *PLoS One*, *8*.

Zhernakova, A., Elbers, C. C., Ferwerda, B., Romanos, J., Trynka, G., Dubois, P. C., de Kovel, C. G. F., Franke, L., Oosting, M., Barisani, D., Bardella, M. T., Joosten, L. A. B., Saavalainen, P., van Heel, D. A., Catassi, C., Netea, M. G., & Wijmenga, C. (2010). Evolutionary and functional analysis of celiac risk loci reveals SH2B3 as a protective factor against bacterial infection. *American Journal of Human Genetics*, *86*(6), 970–977. Available from https://doi.org/10.1016/j.ajhg.2010.05.004.

Zhernakova, A., Van Diemen, C. C., & Wijmenga, C. (2009). Detecting shared pathogenesis from the shared genetics of immune-related diseases. *Nature Reviews. Genetics*, *10*(1). Available from https://doi.org/10.1038/nrg2489.

Zorro, M. M., Aguirre-Gamboa, R., Mayassi, T., Ciszewski, C., Barisani, D., Hu, S., Weersma, R. K., Withoff, S., Li, Y., Wijmenga, C., Jabri, B., & Jonkers, I. H. (2020). Tissue alarmins and adaptive cytokine induce dynamic and distinct transcriptional responses in tissue-resident intraepithelial cytotoxic T lymphocytes. *Journal of Autoimmunity*, *108*, 102422. Available from https://doi.org/10.1016/j.jaut.2020.102422.

Chapter 4

Relevance of environmental cofactors

Ruggiero Francavilla[1], Vanessa Nadia Dargenio[1], Stefania Castellaneta[1], Michele Barone[2] and Fernanda Cristofori[1]

[1]*Interdisciplinary Department of Medicine, Pediatric Section, University of Bari "Aldo Moro," Bari, Italy,* [2]*Gastroenterology Unit, Department of Emergency and Organ Transplantation, University of Bari "Aldo Moro", Bari, Italy*

4.1 Introduction

Celiac disease (CD) is a common disease, with a prevalence of around 1.4%, which varies according to age, sex, and region (Singh et al., 2018). As described in other chapters (1 and 3), despite gluten being essential to develop CD, the disease only affects individuals carrying specific human leukocyte antigen (HLA)-risk haplotypes. HLA-DQ2 or DQ8 are necessary, but not sufficient, to develop CD, as approximately 40% of the general population have permissive genes, but only 3% will develop CD (Leonard et al., 2015). CD incidence is rising and is expected to triple by 2050 (West et al., 2014). Furthermore, the concordance between monozygotic twins is 80%. All these observations suggest that environmental cofactors other than gluten are contributing to CD pathogenesis. Factors such as mode of delivery, birth season, early life infections, infant feeding practices, amount and type of other wheat proteins consumed, and antibiotic exposure have all been implicated in triggering the immune response to dietary gluten, thus contributing to CD onset. Therefore CD results from a complex interplay among genetic, environmental, and immune factors leading to an inappropriate mucosal T-cell response to gluten. Potential environmental cofactors will be discussed in this chapter.

4.2 Pregnancy and birth

4.2.1 Prenatal and perinatal factors

Maternal smoking (Mårild et al., 2019), diet (Lund-Blix et al., 2019), infections, and drug use during pregnancy (e.g., antibiotics and paracetamol) have been speculated to influence the risk of CD (Mårild et al., 2017).

Ludvigsson et al. reported no association between CD and parental smoking habits during pregnancy or the first year of life (Ludvigsson & Ludvigsson, 2005). On the contrary, exploring the Norwegian Mother and Child Cohort (MoBa) data on 94,019 children, the authors showed that ongoing smoking during pregnancy was inversely associated with CD in childhood (Mårild et al., 2019). However, other studies did not reveal any association (Roberts et al., 2009). More recently published data from the same cohort demonstrated that children of mothers with high fiber intake during early pregnancy had a lower risk of CD. In contrast, maternal gluten intake was positively associated with the risk of offspring CD (Lund-Blix et al., 2020).

Some studies explored if maternal infection during pregnancy could influence the risk of developing CD during childhood (Kahrs et al., 2019; Kemppainen et al., 2017). In a Swedish study, the authors measured serum cytokines during early pregnancy from women who later gave birth to children with CD. The results suggest that increased levels of maternal Th1-mediated cytokines (TNF-alpha, INF-gamma; IL2; IL1-beta; IL12) during early pregnancy could be associated with the development of CD in the offspring. The authors speculated two possible explanations: either an infection as the trigger of the mother's immune response, which led to permanent changes in the fetus, or a noninfectious agent, such as psychological stress and/or a nutritional substance that can contribute to the immunological status (Lindehammer et al., 2011).

In a large Norwegian cohort study, including more than 80,000 children followed from pregnancy through childhood, the authors investigated whether maternal infections, antibiotics, and paracetamol in pregnancy were associated with CD risk in the offspring. Structured questionnaires were administered at weeks 18 and 30 of pregnancy and when

Pediatric and Adult Celiac Disease. DOI: https://doi.org/10.1016/B978-0-443-13359-6.00003-0

the child was six months old to acquire information on infective exposure and treatments. The authors found no clear association between maternal infectious episodes, paracetamol and antibiotic use and development of CD in children (Mårild et al., 2017). Their finding on maternal use of antibiotics is consistent with the results reported in a Swedish cohort of 8729 children that showed no statistically significant association between antibiotic exposure during pregnancy and CD in offspring. Risk estimates remained unchanged after adjustment for breastfeeding, age at gluten introduction and infections in the child's first year of life (Mårild et al., 2014).

In conclusion, a few studies with conflicting results reported that some mothers' lifestyles (smoking and diet) could negatively influence the risk of developing CD in the offspring. At the same time, no evidence supports the role of maternal infection or drug use during pregnancy as risk factors for CD.

Finally, some evidence suggests that being small for gestational age (SGA) might be related to CD (Mårild et al., 2012; Sandberg-Bennich et al., 2002; Steiner et al., 2019; Wingren et al., 2012).

4.2.2 Mode of delivery

In the past ten years, the global cesarean section (CS) rate increased to 21%, with an average annual increase of about 4%. Children delivered by CS are not exposed to maternal bacterial colonization compared to those born by vaginal birth, with a negative effect on the composition of the gut microbiota (Decker et al., 2011).

The influence of the mode of delivery as risk factor for CD has been evaluated by several authors and is summarized in Table 4.1.

In a Danish population-based cohort study on two million infants, Sevelsted et al. found no association between CS and CD (Sevelsted et al., 2015). Moreover, Koletzko et al. reported a lack of association between CS and the risk of CD or CD autoimmunity in 6087 genetically predisposed children (Koletzko et al., 2018). Identical results were reported by Sander et al. (Dydensborg Sander et al., 2018) in an observational register-based cohort study on 1.5 million children in Norway and Denmark, by Soullane et al. in Canada (Soullane et al., 2021) and by Roberts et al. in England (Roberts et al., 2009).

In our experience, examining data from a cohort of 553 children at CD risk, the prevalence of CD autoimmunity or overt CD was not different between children born by CS or vaginal delivery (Lionetti et al., 2017). On the other hand, a multicenter, case-control study showed that the mode of delivery might affect gut microbiota and increases the risk of CD (Decker et al., 2010). Other cohort studies support this findings (Andersen et al., 2020) (Adlercreutz et al., 2015). Interestingly, some studies report that elective, but not an emergency, CS was associated with a higher risk of CD (Mårild et al., 2012; Tanpowpong et al., 2023). In elective CS, the birth must occur before the onset of labor; thus children do not have contact with maternal vaginal microbiota in the birth canal. Moreover, mothers undergoing elective CS are more likely to receive antibiotic prophylaxis. Other studies have not confirmed the difference between elective and emergency CS (Emilsson et al., 2015) (Canova et al., 2014).

Finally, Yang et al. recently performed a systematic review with meta-analysis on 11 observational studies. They showed that CS was not associated with an increase in the risk of CD (OR = 1.03), regardless of whether it was in election or emergency (Yang et al., 2022).

4.2.3 Season of birth

It has been hypothesized that children who developed CD in early childhood are more likely to be born in summer or spring than in winter months. The seasonal effect might relate to different viral exposures during pregnancy or during first gluten exposure. We have studied a population of 596 CD children compared with a reference group that included all subjects born in the same period and in the same geographical area reporting a summer birth preponderance in CD patients compared to controls (28.2% vs 23.0%; OR 1.31) (Capriati et al., 2015). The season effect has been evaluated by several studies summarized in Table 4.2. While viral infections are the most substantial risk factor in favor of seasonal variation in CD risk, vitamin D level differences during pregnancy due to differences in sunlight and UV-B exposure might be a possible further explanation.

4.3 Geographic region (latitude)

The latitudinal gradient effect is described for several autoimmune diseases, including CD. In a recent meta-analysis and meta-regression including 128 studies, the authors confirmed that serology-based CD prevalence increases proceeding from the equator to the poles. Lack of sunlight exposure and subsequent vitamin D deficiency could explain this

TABLE 4.1 Available evidence on mode of delivery as risk factor for celiac disease.

Reference	Country	Study design	Population	Main results
Roberts et al. (2009)	England	Population-based case-control study	248.521 children	Reduced risk of CD for CS [OR = 0.29; (95% CI = 0.07–1.1)]
Decker et al. (2011)	Germany	Multicenter, case-control study	1.950 children	Higher CS rate was found in patients with CD [OR: 1.8 (95% CI: 1.13–2.88)]
Mårild et al. (2012)	Sweden	Population-based case-control study	11.749 children	Increased risk of CD after elective CS [aOR 1.15; (95% CI: 1.04–1.26)] No increased risk of CD after emergency CS [aOR 1.02; (95% CI: 0.92–1.13)] or any CS [aOR 1.02; (95% CI: 0.99–1.13)]
Canova et al. (2014)	Italy	Population-based Cohort study	203,000 children	CS have little influence on the risk of childhood CD Planned cesarean [IRR 1.11; (95% CI: 0.80–1.53)] Unplanned cesarean [IRR 1.1; (95% CI: 0.79, 1.55)]
Sevelsted et al. (2015)	Denmark	Observational register-based cohort study	Two million term children	CD is not associated with CS [IRR 0.99; (95%CI 0.87–1.14)]
Emilsson et al. (2015)	Norway	Population-based case-control study (MoBa)	114,500 children	CD was not associated with CS [OR, 0.84; (95% CI 0.65–1.09)], Elective CS [OR, 0.67; (95% CI 0.44–1.03)]
Adlercreutz et al. (2015)	Sweden	Observational register-based cohort study	811.599 children	CS is associated with a double diagnosis of type 1 diabetes and CD [OR 1.60; (95% CI: 1.07–2.39)]
Lionetti et al. (2017)	Italy	Cohort study (CELIPREV)	553 children (HLA DQ 2/8)	At 5 years of age, the prevalence of CD was not different between children born by CS or vaginally; 19% versus 14%, P = NS
Koletzko et al. (2018)	United States Finland Germany Sweden	Multicentre observational cohort study (TEDDY)	6.721 children	CS is not associated with increased risk for CD in the offspring [HR 0.85; (95% CI 0.65–1.11)]
Dydensborg Sander et al. (2018)	Denmark Norway	Observational register-based cohort study	1,051,028 children (Denmark) 537,457 children (Norway)	No association between the mode of delivery and the risk of diagnosed CD [aOR 1.11 (95% CI: 0.96–1.29)] (Denmark) [aOR 0.96 (95% CI: 0.84–1.09)] (Norway)
Andersen et al. (2020)	Denmark	Observational register-based cohort study	2.672.708 children	Being born by CS was associated with development of CD [HR 1.10 (95% CI: 1.03–1.18)]
Soullane et al. (2021)	Canada	Cohort study	934,873 children	CS is not associated with hospitalization for CD [HR 0.86: (95% CI 0.71–1.04)]
Tanpowpong et al. (2023)	United States	Cohort study	44.539 children	CS without labor may be associated with an increased risk for CD [aHR 1.56; (95%CI: 1.01–2.41)] CS with labor is not associated with an increased risk for CD [aHR 1.26; (95% CI: 0.83–1.93)]

TABLE 4.2 Birth season as risk factor for celiac disease.

Reference	Country	Study design	Population	Main results
Ivarsson et al. (2003)	Sweden	Register-based cohort study	2.151 CD children 4,5 milion controls	Higher risk for develop CD below 2 years of age if born during summer as compared with the winter [RR = 1.4; (95% CI 1.2–1.7)]
Lewy et al. (2009)	Israel	Retrospective medical record review	431 CD children	Girls with CD and patients of both sexes with a family history of CD have a different pattern of seasonality of birth from the general population
Lebwohl et al. (2013)	Sweden	Case control study	29.096 CD children 144.522 controls	Summer birth is associated with a small increased risk of later CD [OR 1.06; (95% CI 1.03–1.08)] highest risk for developing CD before age of 2 y [OR 1.17; (95% CI 1.10–1.26)]
Tanpowpong et al. (2013)	United States	Retrospective medical record review	1.964 CD patients	More patients were born in spring (27%) than summer (25%), fall (25%), and winter (23%)
Capriati et al. (2015)	Italy	Retrospective	596 CD children 439,990 controls	Summer birth preponderance in CD patients compared to controls (28.2% CD vs 23.0% controls population) [OR 1.31; (95% CI 1.10–1.57)]
Adlercreutz et al. (2015)	Sweden	Register-based cohort study	768.395 children	There is an association between being born in summer and having CD [OR 1.14; (95% CI 1.07–1.22)]
Namatovu et al. (2016)	Sweden	Prospective cohort longitudinal study	1.912.204 children	CD risk was higher for children born during spring [aHR 1.08; (95% CI 1.01–1.16)], summer [aHR 1.10; (95% CI 1.03–1.18)] and autumn [aHR 1.10; (95% CI 1.02 to 1.18)] as compared with children born during winter
Assa et al. (2018)	Israel	Population-based cohort study	10.566 CD patients	CD risk was significantly higher for subjects born in May [OR 1.07; (95% CI 1.00–1.15)] and June [OR 1.09; (95% CI 1.02–1.17)]
Daniel et al. (2019)	Israel	Population-based retrospective time series analysis study	308.903 children	Autumn births associated with a higher risk of CD [IRR 1.22; (95% CI 1.001–1.50)]
Moos et al. (2020)	Denmark	Register-based study; semiecological design	217.249 children	Season effect for children born in autumn [(OR 1,6); (95% CI 1.16–2.21)] and summer [(OR 1.5 95% CI 1.1; 2.1)]

association. On the other hand, latitude may influence also socioeconomic status or common viral infections (Celdir et al., 2022).

4.4 Diet

The Committee of Nutrition of the European Society for Pediatric Gastroenterology, Hepatology, and Nutrition (ESPGHAN) in 2017 released a position paper on complementary nutrition incorporating new evidence that had emerged from RCTs focusing primarily on the association between CD risk and breastfeeding and the quantity of gluten introduced to an infant's diet (Andrén Aronsson et al., 2019; Crespo-Escobar et al., 2017; Lund-Blix et al., 2019; Mårild, Dong, et al., 2019; Meijer et al., 2022). This position paper concluded that neither breastfeeding "per se" nor breastfeeding during gluten introduction reduces the risk of CD. Gluten should be introduced into the infant's diet anytime between 4 and 12 months of age. Large amounts of gluten should be avoided during the first weeks after gluten introduction. However, the optimal weaning gluten dose has not been determined. Since genetic risk alleles are rarely known in infants at the time of solid food introduction, the recommendations apply to all infants (Fewtrell et al., 2017).

4.4.1 Breastfeeding

According to the CELIPREV and PreventCD studies, no significant difference in the risk of CD or CD autoimmunity was found between children who were ever or exclusively breastfed and those who were never breastfed. Moreover, the duration of any breastfeeding did not influence the risk of CD (Lionetti et al., 2014; Vriezinga et al., 2014). A similar conclusion comes from the BABYDIET study (Sandra Hummel et al., 2011), cohort studies (Andrén Aronsson et al., 2019; Hummel et al., 2007, 2021; Jansen et al., 2014; Norris et al., 2005; Ziegler et al., 2003) and case-control studies (Bittker & Bell, 2019; Challacombe et al., 1997; Decker et al., 2011; Roberts et al., 2009) which found no correlation between the duration of exclusive lactation and CD risk. Only few studies have investigated the potential impact of breastfeeding on the risk of CD in individuals with an unknown genetic risk for CD (Jansen et al., 2014; Norris et al., 2005; Størdal et al., 2013; Welander et al., 2010) and found no influence on the incidence of CD risk.

In the PreventCD and CELIPREV studies, breastfeeding at the time of gluten introduction did not influence CD development (OR 1.34) (Lionetti et al., 2014; Vriezinga et al., 2014). The TEDDY study (Aronsson et al., 2015) did not find an increased risk of CD in children breastfed for fewer than one month during gluten introduction. However, a trend of increased CD autoimmunity risk was reported when breastfeeding lasted less than one month. In a meta-analysis of case-control studies (Szajewska et al., 2023), the group of children breastfed during gluten introduction had a decreased risk of CD compared to the control group (OR = 0.51).

4.4.2 Age at gluten introduction

None of the RCTs that studied the effect of age on gluten introduction found any association with CD risk (Beyerlein et al., 2014; Hummel et al., 2011; Lionetti et al., 2014; Sellitto et al., 2012; Vriezinga et al., 2014). In a meta-analysis of two trials (Szajewska et al., 2023), the authors reported a difference in CD autoimmunity risk at two years of age between children who introduced gluten at 12 or 6 months (RR 0.40).

The influence of gluten introduction timing on the development of CD autoimmunity in infants with an unknown genetic risk for CD was reported in the EAT study (Logan et al., 2020). The authors found that the risk was surprisingly lower in early (4 months) versus conventional (6 months) wheat introduction. The MoBA cohort (Lund-Blix et al., 2019) showed that introducing gluten between 4 and 6 months decrease the risk of CD as compared to 4 months. Moreover, a meta-analysis (Szajewska et al., 2023) found an increased risk of CD in infants who had received gluten at three months of age (OR 1.88).

4.4.3 Amount of gluten intake at weaning

Only two RCTs (CELIPREV and PreventCD) investigated the mean quantity of gluten consumed in populations at risk of CD (Lionetti et al., 2014; Vriezinga et al., 2014). The amount of gluten intake (between 18 and 36 months) was not associated with CD at five years of age in the PreventCD cohort. An updated analysis (with a longer follow-up and higher number of CD children) (Meijer et al., 2022) found that a higher amount of gluten intake during the first 3 years was associated with a higher risk of CD. Several studies confirmed these findings. The TEDDY study (Andrén Aronsson et al., 2019) found that a higher daily gluten intake was related to an elevated risk of CD and CD autoimmunity risk. The DAISY study (Mårild et al., 2019), found that a larger gluten intake throughout childhood was associated with an increased risk of CD development. The MoBA study indicated a greater risk of CD with increased cumulative gluten intake throughout childhood (Lund-Blix et al., 2019). Finally, a meta-analysis suggested that when gluten is introduced with a follow-up formula (but not solid foods), there is an increased risk of CD (Szajewska et al., 2023).

ESPGHAN recommends to limit the gluten intake at least for the first month after its introduction and throughout infancy.

4.5 Intestinal infections and CD

Several data support the hypothesis that various infectious agents, particularly viruses, can induce immune reactions associated with an increased risk of CD (Plot & Amital, 2009; Sarno et al., 2015). Large-scale, population-based cohort studies have reported that viral or bacterial infections early in life can promote disease, influencing the maturation of the mucosal immune system or predisposing to bacterial superinfections and derangement of commensal microbiota (Kemppainen et al., 2017; Mårild et al., 2015).

In a recent systematic review with a meta-analysis including 15 observational studies, it was observed that any infection might be associated with an increased risk of CD later in life (OR = 1.37) (Jiang et al., 2020).

Several mechanisms have been proposed to explain the immune responses that may promote CD. The first possible mechanism is "molecular mimicry," a condition that may favor autoimmunity when components of the pathogen's structure may be similar to the host's. Therefore immune cells that are activated in response to the pathogen can generate an autoimmune response through a mechanism of cross-reactivity versus self-components (Kim et al., 2006). A second possible hypothesis is that a nonspecific, hyperreactive antiviral immune response sustains a proinflammatory environment with a prolonged release of self-antigens from damaged tissue, triggering autoimmunity (Fujinami et al., 2006). Finally, it has been hypothesized that an "epitope spreading" secondary to the infectious damage can initiate the onset of autoreactive T-cells versus self-epitopes (Getts et al., 2013). These hypotheses still need to be confirmed.

On the contrary, some infections may play a protective role abrogating autoimmunity. Several hypotheses have been made: (1) posttranslational editing of proteins (switch from nonself to self), (2) modification of gluten epitopes (Lerner et al., 2017), (3) Th1 to Th2 immune response shift (Christen & von Herrath, 2005), and (4) apoptosis of immune cells, which dampens the immune system response and minimizes the progression of autoimmune disease (mainly secondary to Cytomegalovirus (CMV) and Epstein Barr Virus (EBV) infection) (Lerner et al., 2017).

The protective effect has also been observed in helminth infections. In particular, it has been reported that helminths modulate the immune system and generate a Th2 environment in the mouse model. This mechanism appears capable of protecting against the onset of autoimmune disorders or relieving the consequences of established autoimmune inflammation (Zandman-Goddard & Shoenfeld, 2009). The interest in helmints has led to their use as potential nondietetic therapy for CD. Identifying gut-resident bacteria species that may improve celiac symptoms and intestinal damage is of great interest.

4.5.1 Helicobacter pylori

Helicobacter pylori (HP) is the causative agent of the most common chronic bacterial infections affecting up to 50% of the general population. Two recent meta-analysis showed a negative relationship between HP and CD; however there was a significant statistical heterogeneity between the studies (Amlashi et al., 2021; Yue et al., 2022), and they did not establish a causal relationship.

HP evades human immune defense through several mechanisms that downregulate both innate and adaptive immune responses (Monack et al., 2004). Mechanisms by which HP may interfere with the development of CD are summarized in Fig. 4.1.

Overall, the quality of the evidence remains low. At the same time, most articles did not account for several confounding factors such as Marsh type, HP virulence, and IgA deficiency, which could have affected the results.

4.5.2 Campylobacter

Campylobacter is a major bacterial cause of enterocolitis worldwide, affecting individuals of all ages, with a higher incidence in children. Studies associating *Campylobacter* infection with CD are limited. A possible role of *Campylobacter* in CD development has been proposed through the mechanism of molecular mimicry in association with other genetic and environmental factors.

An extensive U.S. database was used to assess the risk of CD following gastrointestinal infection. During the study period the incidence of CD was 0.07 per 100,000 person-years among those who had experienced bacterial diarrhea and 0.04 in those who had not. The rate of CD was twice as high (0.15 per 100,000 person-years) among those who had been exposed to *Campylobacter* compared to those who had not. No association was found for any of the other pathogens investigated (Riddle et al., 2013).

4.5.3 Viral infection

It is well-established that viruses may trigger autoimmune conditions (Hussein & Rahal, 2019; Lernmark, 2016). Although a direct link between viruses and CD has not been demonstrated, numerous pieces of evidence suggest their role.

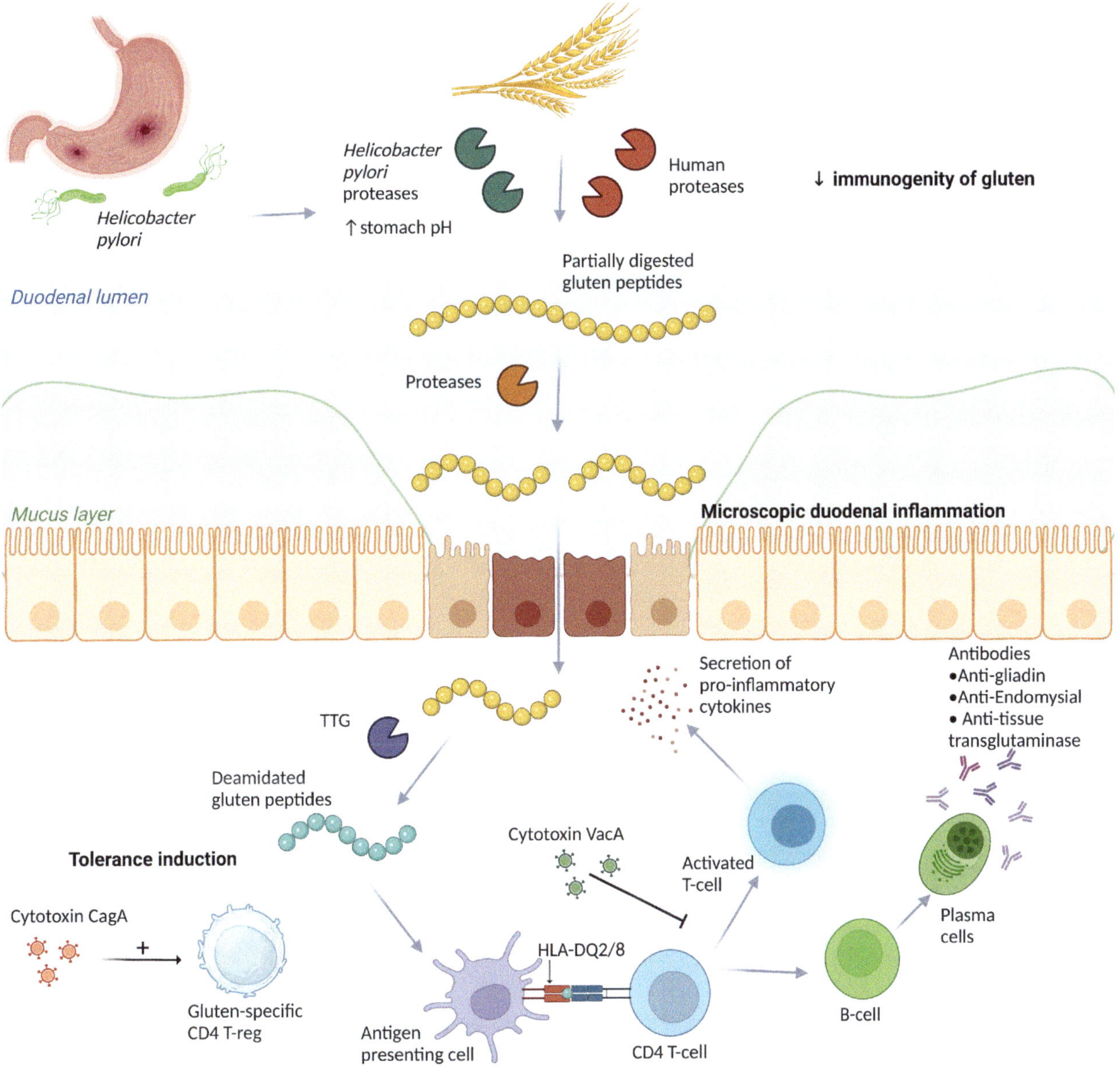

FIGURE 4.1 Mechanisms by which Helicobacter pylori (HP) may interfere with the development of celiac disease (CD). HP infection can affect immune responses in the small intestine to gluten and can lead to microscopic duodenal inflammation. Some studies have hypothesized that HP may also play a protective role by reducing the immunogenicity of gluten through secretion of proteases and increasing the pH of the gastric lumen. HP producing cytotoxin A-associated gene (CagA) appears to be involved in the immunoregulatory mechanism through its association with Treg cells. HP vacuolating cytotoxin (VacA) has an immunomodulatory role by inhibiting integrin-linked kinase (ILK), intervening in antigen presentation by MHC class II and directly suppressing T cells (Gebert et al., 2003; Yuan et al., 2009).

A viral infection can induce the up-regulation of IFN and IL-15 that may precipitate a proinflammatory Th1 response and, ultimately, promote the emergence of autoimmune diseases. Moreover, other concomitant exogenous factors can contribute to a proinflammatory milieu so that even before intestinal damage occurs, CD patients experience low-grade inflammation of the epithelium (Porpora et al., 2022).

It has been hypothesized that reovirus (the family to which rotavirus belongs) may be involved in the onset of CD (Bouziat et al., 2017). Higher antibodies to reovirus or human adenovirus serotype 2 have been found in individuals with CD, indicating a possible increased risk in genetically predisposed individuals (Kagnoff et al., 1987). Reoviruses disrupt the immunological response to dietary antigens, such as gluten, by abrogating antigen-specific regulatory T-cell answers and interfering with development of oral tolerance to food antigens. Finally, during reovirus infection, an elevated induction of type 1 IFN plays a crucial role in the CD susceptibility gene network (Bouziat et al., 2017).

4.5.3.1 Rotavirus

Rotavirus (RV) is the leading cause of acute gastroenteritis worldwide in infants and children under five. Most infections are caused by Group A strains and are transmitted by the fecal-oral route (Kirkwood, 2010). RV infects intestinal cells and causes various symptoms, from asymptomatic to severe diarrhea with dehydration; however, infection is not limited to the intestinal mucosa, and systemic spread of the virus has been demonstrated (Blutt & Conner, 2007). One possible mechanism of RV involvement in the pathogenesis of CD is molecular mimicry of human RV capsid VP7 antigen, which is recognized by a subset of IgA transglutaminase (TGAs) (Vaarala et al., 2017).

In a recent descriptive cross-sectional study conducted in Pakistan, 1,090 blood samples were screened for of TGAs, and fecal samples tested for RV infection. Of 112 TGAs IgA-positive patients, 66% were also infected with RV, showing a statistically significant association between RV infection and CD autoimmunity (Sadiq et al., 2022). Two prospective studies in genetically predisposed children showed that a RV infection could increase the risk of CD (Stene et al., 2006) and CD autoimmunity (Stewart et al., 2018).

Following the introduction of oral RV vaccines for the global pediatric population, a marked reduction in hospitalization rates has been observed (Hemming-Harlo et al., 2017). The impact of vaccination programs on the development of autoimmune diseases is still under study. In Finland, a population-based cohort study did not find a protective role of the vaccination program. However, longer follow-up might be needed to draw firm conclusions (Vaarala et al., 2017).

Finally, it should be also considered the potential molecular mimicry effect of the vaccines, which are both live attenuated, similar to the wild-type RV (Gómez-Rial et al., 2019); however, the lack of an exacerbated gut inflammatory component in RV vaccination could avoid the triggering effect provoked by natural infection.

Table 4.3 summarizes the evidence on the role of RV infection and vaccine as risk factor for CD.

4.5.3.2 Other viruses

Several studies have reported that herpesvirus infections are associated with an increased risk of CD (Kemppainen et al., 2017; Lerner et al., 2017) while others have reported a protective role (Christen & von Herrath, 2005; Plot & Amital, 2009). The defensive hypothesis is mainly based on a lower incidence of serum antibodies to CMV, EBV, and Rubella in CD patients compared with healthy controls (Plot, Amital, Barzilai, et al., 2009). Jansen et al. also found that levels of anti-CMV, EBV or Herpes simplex type 1 IgG in serum samples were inversely correlated with levels of IgA-type TGAs, suggesting a protective role of these viruses (Jansen et al., 2016).

A recent prospective birth cohort study found that enterovirus infections were more frequent in children who developed CD ($n = 41$) than in controls ($n = 53$) (OR 6.3) (Oikarinen et al., 2020). No difference was found in the frequency of adenovirus infections. Similar results were found in two case-control series (Lindfors et al., 2020) (Kahrs et al., 2019).

4.6 Drug exposure and CD risk

Among the environmental factors that may influence the development of CD, several drugs, particularly antibiotics and proton pump inhibitors (PPIs), and among micronutrients, vitamin D, have been implicated.

4.6.1 Antibiotic

Antibiotics are among the most commonly used drugs in infants and children. Today, about 10%–25% (Stephansson et al., 2011) of pregnant women use antibiotics, and more than two-thirds of children receive antibiotics before they reach age two. Antibiotics affect the gut microbiota of infants (Fouhy et al., 2012) (Fallani et al., 2010), and adults (Jakobsson et al., 2010), and these perturbations can persist for years.

Six studies have estimated the CD risk associated with antibiotic exposure and the results have been summarized in a meta-analysis showing that in childhood, antibiotic exposure was significantly associated with a higher risk of CD (OR, 1.15). Comparisons of the antibiotic class showed that penicillin V was associated, macrolide use was marginally associated while no relationship was found for cephalosporins or antifungal medications (Jiang et al., 2020). Four studies examined the association between exposure to antibiotics and the development of CD. Overall, 0.22% of the 531.197 antibiotic-exposed children developed CD, in contrast to 0.23% of the 1,160,442 nonexposed children. The research, thus did not support a link between antibiotic exposure and the development of CD (OR = 1.21) (Duong et al., 2022).

TABLE 4.3 Rotavirus (RV) infection and RV vaccine as risk factor for celiac disease (CD).

Reference	Country	Study design	Population	Study time/ Postvaccination Follow-up (y)	Main results
Stene et al. (2006)	United States	Prospective study	1931 children (HLA DQ2/ 8) 54/1,931 CD patients 18/54 CD patients with one RV infection 33/54 CD patients with ≥2 RV infections	From December 1993 to January 2003	RV predicted a higher risk of CD autoimmunity (for one infection) [RR 3.76; (95% CI 0.76–18.7)], for ≥2 infections, [RR for trend per increase in number of infections = 1.94; (95% CI 1.04–3.61)].
Rostami-Nejad et al. (2010)	Iran	Cross-sectional study	670 Adult patients 25/670 CD patients 8/25 CD patients with RV infection	From September 2006 to September 2007	The prevalence of RV was not statistically significant between individuals with positive and negative CD serology ($p = 0.2$).
Sadiq et al. (2022)	Pakistan	Cross-sectional study	1090 patients 112/1,090 CD patients 78/112 CD patients with RV infection	From January 2019 to December 2019	Association between RV infection and CD ($p < 0.05$).
Kemppainen et al. (2017)	United States Finland Germany Sweden	Multicentre observational cohort study (TEDDY)	6721 children 732/6,721 CD patients	From September 2004 to March 2015/0–4	Association between CD and GI infection was high in Sweden [HR, 1.45; (95% CI, 1.12–1.87)] and Germany [HR 1.56; (95% CI, 0.70–3.48)], where RV vaccination coverage was low, but not observed in Finland [HR, 0.99; (95% CI, 0.62–1.57)] or in United States [HR 1.22; (95% CI, 0.86–1.73)] where the RV vaccination coverage was high.
Vaarala et al. (2017)	Finland	Population-based cohort study	121,650 children 293 CD patients	4–6	Reduced risk of CD for RV vaccine [IRR 0.87; (95% CI 0.65–1.17)].
Hemming-Harlo et al. (2019)	Finland	Questionnaire study	5764 children 48 CD patients	11–14	Prevalence of CD was significantly lower in RV vaccine group [0.60%; (95% CI 0.38–0.93)] than in placebo group [1.11%; (95% CI 0.78–1.6)].
Inns et al. (2021)	United Kingdom	Prospective cohort study	880,629 children 1657 CD patients	0–5	[HR for the association between RV vaccination and CD was 1.05 (95% CI 0.86–1.28)].
Kim et al. (2022)	United States	Retrospective cohort study	2,109,225 children 1379/ 2,109,225 CD patients	0–17	Partial vaccination [HR 0.90; (95% CI 0.73–1.11)] and complete vaccination [HR 1.03; (95% CI 0.88–1.21)] were not associated with increased or decreased risk of CD.

No association was found in the Estonian birth cohort ($n = 5819$), in the Finnish cohort ($n = 3256$) (Simre et al., 2016) and in the TEDDY study. Overall, the available data are inconsistent, and the exact mechanism by which early antibiotic administration might predispose to CD remains unclear.

4.6.2 Proton pump inhibitors

Proton Pump Inhibitors (PPIs) are gastric acid suppression drugs often prescribed to pediatric patients. Their link to CD has been hypothesized because pH alterations affect food antigen degradation. It may affect the gut microbiome, intestinal permeability (Caio et al., 2019) and increases the risk of enteric infections (Leonard et al., 2007).

A population-based case-control study conducted in Sweden in 2014 which included 2934 subjects with CD and 14,584 controls, showed a strong positive association between PPI use and subsequent CD diagnosis (OR 4.79) (Lebwohl et al., 2014). The authors hypothesized that gut dysbiosis might modulate the inflammatory response, leading to increased immune reactivity toward ingested gluten.

Boechler et al. performed a retrospective cohort study, including 968,524 children, 1704 of whom had CD. PPI use during the first six months of life was associated with an increased risk of CD (H.R., 2.23) (Boechler et al., 2023).

Based on the results of these studies, future research is needed to clarify the mechanism(s) by which PPIs or acid suppression, in general, affect the pathogenesis of CD.

4.6.3 Vitamin D

The possible link between vitamin D and CD lies in its involvement in the regulation of both innate and adaptive immune systems (Pecora et al., 2020) and in the observation of an increased risk of developing CD in the presence of hypovitaminosis D (Akhshayaa et al., 2021; Lionetti et al., 2021; Shree et al., 2022), among people living in northern latitudes (Unalp-Arida et al., 2017).

Vitamin D might protect the intestinal mucosa from gluten-induced damage by down-regulating the Th1 response. In an vitro study, vitamin D increases the expression of tight junction-associated proteins, thus optimizing epithelial permeability (Lebwohl et al., 2018). A recent systematic review including 27 articles with 1137 patients with CD and 2613 controls (Lu et al., 2021), showed that CD patients, regardless of age, had significantly lower serum concentrations of vitamin D than controls. Whether vitamin D deficiency is involved in the pathogenesis of the disease or it just reflects a consequence of malabsorption still needs to be defined.

Interestingly, a significant increase in the incidence of CD in Sweden was observed precisely one year after the institution of a mandatory fortification program. Similarly, the TEDDY study has pointed out an association between vitamin D concentrations <30 nmol/L and >75 nmol/L in early childhood and an increased risk of CD in genetically at-risk children (Smith et al., 2014). Bittker et al. suggested that children who receive oral vitamin D for more than three months have a higher risk of developing CD (Bittker & Bell, 2019). The proposed explanation is that high levels of vitamin D up-regulate several inflammatory mediators and toll-like receptors (TLR2, TLR4, IL-17A, CCR4, CXCR4, IL12RB1, IL12RB2) involved in the immunological activation (Bittker, 2020).

Many medical authorities approve vitamin D supplementation during childhood. However, there is currently no clear consensus on the optimal vitamin D concentrations in the general pediatric population.

4.7 Pesticide's and persistent organic pollutant

The biological processes through which persistent organic pollutants (POPs) might impact CD remain to be defined. Animal models on endocrine-disrupting chemicals have shown that greater exposure to polychlorinated biphenyls (PCBs) is associated with disruption of tight junctions in the gut epithelium (Choi et al., 2010) and the ingestion of Bisphenol A (BPA) has also been shown to reduce tight junction proteins (Feng et al., 2019). These agents might determine a "leaky gut" condition, which allows the first passage of gliadin across the intestinal epithelium into the lamina propria, where a response from the immune system is initiated.

Since the late 1970s, soy, corn, and wheat farmers have employed glyphosate, a nonselective herbicide, to control weeds. It has also been hypothesized that elevated exposure to the herbicide might play a role in the rising incidence of CD, considering glyphosate's effects on intestinal microbiota modification, micronutrient intake, enzymatic detoxification and vitamin D3 activation in the liver (Samsel & Seneff, 2013). Nonetheless, its role in CD pathogenesis remains purely theoretical.

4.8 CD triggers in adulthood

Even though CD is recognized to arise at any age in genetically susceptible individuals on a gluten-containing diet (Al-Toma et al., 2019), little is known about the factors that trigger the disease in adulthood. An essential preliminary consideration is that the incidence of CD in adults is five times lower than in childhood, and whether this difference is due to environmental factors operating in infancy, remains to be demonstrated (Mariné et al., 2011). In this section, we will analyze all putative triggers leading the onset of the disease in adulthood, starting from the difference between the stronger predictors: the HLA pattern and the gluten-containing diet.

4.8.1 Different CD prevalence in children and adults

CD is undoubtedly distributed worldwide with a different geographical prevalence (Mariné et al., 2011). However, even if we consider epidemiologic studies based on identical analytical methods for CD screening, these differences could be attributed not only to environmental triggering risk factors but also to possible bias due to the age and gender of individuals included. One unsolved question concerns the considerable variability in age at disease onset (Al-Toma et al., 2019). Older studies report a female-to-male ratio ranging from 1:3 to 1.5:1; more than 70% of the new diagnoses are in adults (Fasano et al., 2003). A recent cross-sectional study evaluated the prevalence of CD in 4230 consecutive subjects, aged 1−80 years, reproducing the reference population by age and gender. In this study, 21 subjects developed CD. According to other data of the literature, there was a predominance of CD in female subjects (ratio 1:2.5) (Mariné et al., 2011), and the prevalence was higher in children (1:71) than in adults (1:357) ($p = 0.00005$). This latter finding was unexpected considering that CD is a long-lasting disease.

4.8.2 Genotypes and CD onset

A hierarchy of disease penetrance exists among the permissive HLA haplotypes, with over 90% of patients expressing HLA-DQ2.5 heterodimers and HLA-DQ8 heterodimers or HLA-DQ2.2 in the remaining patients (Levescot et al., 2022). Notably, the DQ2.5 homozygote pattern is the haplotype with the highest penetrance associated with earlier disease development. Only two studies have been performed in adults, one in combination with pediatric patients and one in adults alone (Espino & Núñez, 2021). While most studies show an association between HLA-DQ2.5 and an earlier CD onset in pediatric patients, this is not the case in others including adults.

A possible suggestion supporting the association between a specific haplotype and an adulthood onset of CD comes from a study (Biagi et al., 2012) showing that potential CD in adults is more common among subjects with DQ8. This finding suggests that DQ8-positive patients could experience a long latency period and evolve toward an overt disease in adulthood. This hypothesis appears reasonable since 90% of the CD patients have a DQ2.5 haplotype.

Another aspect to consider is the involvement of non-HLA loci with a mild association with CD (Romanos et al., 2014). However, the possibility that these autoimmune-associated genes could influence the appearance of CD at different ages is unknown.

4.8.3 Diet-related aspects and CD onset

Western-style diet (WD) is strongly associated with calorie-dense foods, rich in saturated fatty acids, simple carbohydrates, and animal proteins, and its relationship with noncommunicable diseases, is highly debated (García-Montero et al., 2021). There is increasing evidence that "Western" diets promote inflammation through direct or indirect modulation of the immune system (Tilg & Moschen, 2015). Theoretically, WD can have an impact on CD pathogenesis because it is responsible for dramatic changes in intestinal microbiota composition, increased mucosal inflammation, and intestinal permeability. This leads to systemic low-grade inflammation. In detail, high-fat diets enhance the permeation of luminal contents into the mucosal and submucosal layers promoting mucosal inflammation (Ji et al., 2011), enhancing lipopolysaccharide (LPS) uptake, and increasing concentrations of LPS in the circulatory system. Moreover, at the mucosal level, it can activate the NF-κB pathway increasing proinflammatory response and the release of proinflammatory cytokines and chemokines. Similarly, a high-sugar diet and a low intake of fresh fruits and vegetables are responsible for the decreased antiinflammatory and antioxidant nutrients (e.g., polyphenols, minerals, or vitamins) predisposing to low-grade chronic inflammation. Unfortunately, even if there are studies on the possible influence of dietary patterns on CD development in pediatric patients (Barroso et al., 2018), the precise mechanisms related to the diet are complex and challenging to quantify in adulthood. Therefore its contribution of this aspect remains unclear.

4.9 Conclusion

The finding that only 3%–4% of people who carry the significant risk genes and consume gluten develop CD indicates that unidentified environmental cofactors contribute to the pathogenesis of CD. Several factors (maternal, perinatal and neonatal) have been studied as possible risk factors for CD. Still, except for the season of birth, no definite role has been demonstrated for all the others. Long-term cohort studies reported no significant influence of timing or modality of gluten introduction to an infant's diet on CD development in children at risk. Clinical studies have focused on infections common in early childhood, indicating that viruses can impair oral tolerance to a dietary antigen. The ability of a virus to induce expression of inflammatory cytokines may be crucial to breaking oral tolerance and promoting CD. Finally, differences in microbiome composition have been identified between genetically susceptible individuals who develop CD and those who remain healthy.

Aside from potential viral/bacterial triggers, the microbiome has recently received much attention. Recent changes in the microbiota could have contributed to the increased prevalence of CD in the last 50 years (Cristofori et al., 2018). Several observational studies have reported differences in the microbiota of patients with CD versus controls, but with little consistency, except for an increase in proteobacteria and a decrease in species with anti-inflammatory properties, a finding, however, reported in several chronic inflammatory intestinal conditions.

References

Adlercreutz, E. H., Wingren, C. J., Vincente, R. P., Merlo, J., & Agardh, D. (2015). Perinatal risk factors increase the risk of being affected by both type 1 diabetes and coeliac disease. *Acta Paediatrica (Oslo, Norway : 1992)*, *104*(2), 178–184. Available from https://doi.org/10.1111/apa.12836.

Akhshayaa, G., Seth, A., Kumar, P., & Jain, A. (2021). Prevalence and management of vitamin D deficiency in children with newly diagnosed coeliac disease: Cohort study. *Paediatrics and International Child Health*, *41*(4), 247–252. Available from https://doi.org/10.1080/20469047.2021.1996089.

Al-Toma, A., Volta, U., Auricchio, R., Castillejo, G., Sanders, D. S., Cellier, C., Mulder, C. J., & Lundin, K. E. A. (2019). European Society for the Study of Coeliac Disease (ESsCD) guideline for coeliac disease and other gluten-related disorders. *United European Gastroenterology Journal*, *7* (5), 583–613. Available from https://doi.org/10.1177/2050640619844125.

Amlashi, F. I., Norouzi, Z., Sohrabi, A., Shirzad-Aski, H., Norouzi, A., Ashkbari, A., Gilani, N., Fatemi, S. A., & Besharat, S. (2021). A systematic review and meta-analysis for association of Helicobacter pylori colonization and celiac disease. *PloS One*, *16*(3), e0241156. Available from https://doi.org/10.1371/journal.pone.0241156.

Andersen, V., Möller, S., Jensen, P. B., Møller, F. T., & Green, A. (2020). Caesarean delivery and risk of chronic inflammatory diseases (inflammatory bowel disease, rheumatoid arthritis, coeliac disease, and diabetes mellitus): A population based registry study of 2,699,479 births in Denmark during 1973–2016. *Clinical Epidemiology*, *12*, 287–293. Available from https://doi.org/10.2147/CLEP.S229056.

Andrén Aronsson, C., Lee, H.-S., Hård Af Segerstad, E. M., Uusitalo, U., Yang, J., Koletzko, S., Liu, E., Kurppa, K., Bingley, P. J., Toppari, J., Ziegler, A. G., She, J.-X., Hagopian, W. A., Rewers, M., Akolkar, B., Krischer, J. P., Virtanen, S. M., Norris, J. M., & Agardh, D. (2019). Association of gluten intake during the first 5 years of life with incidence of celiac disease autoimmunity and celiac disease among children at increased risk. *JAMA*, *322*(6), 514–523. Available from https://doi.org/10.1001/jama.2019.10329.

Aronsson, C. A., Lee, H.-S., Liu, E., Uusitalo, U., Hummel, S., Yang, J., Hummel, M., Rewers, M., She, J.-X., Simell, O., Toppari, J., Ziegler, A.-G., Krischer, J., Virtanen, S. M., Norris, J. M., & Agardh, D. (2015). Age at gluten introduction and risk of celiac disease. *Pediatrics*, *135*(2), 239–245. Available from https://doi.org/10.1542/peds.2014-1787.

Assa, A., Waisbourd-Zinman, O., Daher, S., & Shamir, R. (2018). Birth month as a risk factor for the diagnosis of celiac disease later in life: A population-based study. *Journal of Pediatric Gastroenterology and Nutrition*, *67*(3), 367–370. Available from https://doi.org/10.1097/MPG.0000000000002001.

Barroso, M., Beth, S. A., Voortman, T., Jaddoe, V. W. V., van Zelm, M. C., Moll, H. A., & Kiefte-de Jong, J. C. (2018). Dietary patterns after the weaning and lactation period are associated with celiac disease autoimmunity in children. *Gastroenterology*, *154*(8), 2087–2096. Available from https://doi.org/10.1053/j.gastro.2018.02.024, e7.

Beyerlein, A., Chmiel, R., Hummel, S., Winkler, C., Bonifacio, E., & Ziegler, A.-G. (2014). Timing of gluten introduction and islet autoimmunity in young children: updated results from the BABYDIET study. *Diabetes Care*, *37*(9), e194–5. Available from https://doi.org/10.2337/dc14-1208.

Biagi, F., Bianchi, P. I., Vattiato, C., Marchese, A., Trotta, L., Badulli, C., De Silvestri, A., Martinetti, M., & Corazza, G. R. (2012). Influence of HLA-DQ2 and DQ8 on severity in celiac disease. *Journal of Clinical Gastroenterology*, *46*(1), 46–50. Available from https://doi.org/10.1097/MCG.0b013e318221077e.

Bittker, S. S. (2020). Elevated levels of 1,25-dihydroxyvitamin D in Plasma as a missing risk factor for celiac disease. *Clinical and Experimental Gastroenterology*, *13*, 1–15. Available from https://doi.org/10.2147/CEG.S222353.

Bittker, S. S., & Bell, K. R. (2019). Potential risk factors for celiac disease in childhood: A case-control epidemiological survey. *Clinical and Experimental Gastroenterology*, *12*, 303–319. Available from https://doi.org/10.2147/CEG.S210060.

Blutt, S. E., & Conner, M. E. (2007). Rotavirus: to the gut and beyond!. *Current Opinion in Gastroenterology*, *23*(1), 39–43.

Boechler, M., Susi, A., Hisle-Gorman, E., Rogers, P. L., & Nylund, C. M. (2023). Acid suppression and antibiotics administered during infancy are associated with celiac disease. *The Journal of Pediatrics*, *254*, 61–67. Available from https://doi.org/10.1016/j.jpeds.2022.10.013, e1.

Bouziat, R., Hinterleitner, R., Brown, J. J., Stencel-Baerenwald, J. E., Ikizler, M., Mayassi, T., Meisel, M., Kim, S. M., Discepolo, V., Pruijssers, A. J., Ernest, J. D., Iskarpatyoti, J. A., Costes, L. M. M., Lawrence, I., Palanski, B. A., Varma, M., Zurenski, M. A., Khomandiak, S., McAllister, N., & Jabri, B. (2017). Reovirus infection triggers inflammatory responses to dietary antigens and development of celiac disease. *Science (New York, N.Y.)*, *356*(6333), 44–50. Available from https://doi.org/10.1126/science.aah5298.

Caio, G., Volta, U., Sapone, A., Leffler, D. A., De Giorgio, R., Catassi, C., & Fasano, A. (2019). Celiac disease: a comprehensive current review. *BMC Medicine*, *17*(1), 142. Available from https://doi.org/10.1186/s12916-019-1380-z.

Canova, C., Zabeo, V., Pitter, G., Romor, P., Baldovin, T., Zanotti, R., & Simonato, L. (2014). Association of maternal education, early infections, and antibiotic use with celiac disease: A population-based birth cohort study in northeastern Italy. *American Journal of Epidemiology*, *180*(1), 76–85. Available from https://doi.org/10.1093/aje/kwu101.

Capriati, T., Francavilla, R., Castellaneta, S., Ferretti, F., & Diamanti, A. (2015). Impact of the birth's season on the development of celiac disease in Italy. *European Journal of Pediatrics*, *174*(12), 1657–1663. Available from https://doi.org/10.1007/s00431-015-2589-2.

Celdir, M. G., Jansson-Knodell, C. L., Hujoel, I. A., Prokop, L. J., Wang, Z., Murad, M. H., & Murray, J. A. (2022). Latitude and celiac disease prevalence: A meta-analysis and meta-regression. *Clinical Gastroenterology and Hepatology : The Official Clinical Practice Journal of the American Gastroenterological Association*, *20*(6), e1231–e1239. Available from https://doi.org/10.1016/j.cgh.2020.09.052.

Challacombe, D. N., Mecrow, I. K., Elliott, K., Clarke, F. J., & Wheeler, E. E. (1997). Changing infant feeding practices and declining incidence of coeliac disease in West Somerset. *Archives of Disease in Childhood*, *77*(3), 206–209.

Choi, Y. J., Seelbach, M. J., Pu, H., Eum, S. Y., Chen, L., Zhang, B., Hennig, B., & Toborek, M. (2010). Polychlorinated biphenyls disrupt intestinal integrity via NADPH oxidase-induced alterations of tight junction protein expression. *Environmental Health Perspectives*, *118*(7), 976–981. Available from https://doi.org/10.1289/ehp.0901751.

Christen, U., & von Herrath, M. G. (2005). Infections and autoimmunity–Good or bad? *Journal of Immunology (Baltimore, Md. : 1950)*, *174*(12), 7481–7486.

Crespo-Escobar, P., Mearin, M. L., Hervás, D., Auricchio, R., Castillejo, G., Gyimesi, J., Martinez-Ojinaga, E., Werkstetter, K., Vriezinga, S. L., Korponay-Szabo, I. R., Polanco, I., Troncone, R., Stoopman, E., Kolaček, S., Shamir, R., Szajewska, H., Koletzko, S., & Ribes-Koninckx, C. (2017). The role of gluten consumption at an early age in celiac disease development: A further analysis of the prospective PreventCD cohort study. *The American Journal of Clinical Nutrition*, *105*(4), 890–896. Available from https://doi.org/10.3945/ajcn.116.144352.

Cristofori, F., Indrio, F., Miniello, V. L., De Angelis, M., & Francavilla, R. (2018). Probiotics in celiac disease. *Nutrients*, *10*(12). Available from https://doi.org/10.3390/nu10121824.

Daniel, S., Kalansky, A., Tsur, A., Pinsk, V., Ling, G., Rannan, R., & Yerushalmi, B. (2019). Seasonality of birth affects paediatric coeliac disease. *Acta Paediatrica (Oslo, Norway : 1992)*, *108*(3), 529–534. Available from https://doi.org/10.1111/apa.14441.

Decker, E., Engelmann, G., Findeisen, A., Gerner, P., Laass, M., Ney, D., Posovszky, C., Hoy, L., & Hornef, M. W. (2010). Cesarean delivery is associated with celiac disease but not inflammatory bowel disease in children. *Pediatrics*, *125*(6), e1433–40. Available from https://doi.org/10.1542/peds.2009-2260.

Decker, E., Hornef, M., & Stockinger, S. (2011). Cesarean delivery is associated with celiac disease but not inflammatory bowel disease in children. *Gut Microbes*, *2*(2), 91–98.

Duong, Q. A., Pittet, L. F., Curtis, N., & Zimmermann, P. (2022). Antibiotic exposure and adverse long-term health outcomes in children: A systematic review and meta-analysis. *The Journal of Infection*, *85*(3), 213–300. Available from https://doi.org/10.1016/j.jinf.2022.01.005.

Dydensborg Sander, S., Hansen, A. V., Størdal, K., Andersen, A.-M. N., Murray, J. A., & Husby, S. (2018). Mode of delivery is not associated with celiac disease. *Clinical Epidemiology*, *10*, 323–332. Available from https://doi.org/10.2147/CLEP.S152168.

Emilsson, L., Magnus, M. C., & Størdal, K. (2015). Perinatal risk factors for development of celiac disease in children, based on the prospective Norwegian Mother and Child Cohort Study. *Clinical Gastroenterology and Hepatology : The Official Clinical Practice Journal of the American Gastroenterological Association*, *13*(5), 921–927. Available from https://doi.org/10.1016/j.cgh.2014.10.012.

Espino, L., & Núñez, C. (2021). The HLA complex and coeliac disease. *International Review of Cell and Molecular Biology*, *358*, 47–83. Available from https://doi.org/10.1016/bs.ircmb.2020.09.009.

Fallani, M., Young, D., Scott, J., Norin, E., Amarri, S., Adam, R., Aguilera, M., Khanna, S., Gil, A., Edwards, C. A., & Doré, J. (2010). Intestinal microbiota of 6-week-old infants across Europe: Geographic influence beyond delivery mode, breast-feeding, and antibiotics. *Journal of Pediatric Gastroenterology and Nutrition*, *51*(1), 77–84. Available from https://doi.org/10.1097/MPG.0b013e3181d1b11e.

Fasano, A., Berti, I., Gerarduzzi, T., Not, T., Colletti, R. B., Drago, S., Elitsur, Y., Green, P. H. R., Guandalini, S., Hill, I. D., Pietzak, M., Ventura, A., Thorpe, M., Kryszak, D., Fornaroli, F., Wasserman, S. S., Murray, J. A., & Horvath, K. (2003). Prevalence of celiac disease in at-risk and not-at-risk groups in the United States: A large multicenter study. *Archives of Internal Medicine*, *163*(3), 286–292.

Feng, L., Chen, S., Zhang, L., Qu, W., & Chen, Z. (2019). Bisphenol A increases intestinal permeability through disrupting intestinal barrier function in mice. *Environmental Pollution (Barking, Essex : 1987)*, *254*(Pt A), 112960. Available from https://doi.org/10.1016/j.envpol.2019.112960.

Fewtrell, M., Bronsky, J., Campoy, C., Domellöf, M., Embleton, N., Fidler Mis, N., Hojsak, I., Hulst, J. M., Indrio, F., Lapillonne, A., & Molgaard, C. (2017). Complementary feeding: A position paper by the European Society for Paediatric Gastroenterology, Hepatology, and Nutrition (ESPGHAN) Committee on Nutrition. *Journal of Pediatric Gastroenterology and Nutrition*, *64*(1), 119–132. Available from https://doi.org/10.1097/MPG.0000000000001454.

Fouhy, F., Guinane, C. M., Hussey, S., Wall, R., Ryan, C. A., Dempsey, E. M., Murphy, B., Ross, R. P., Fitzgerald, G. F., Stanton, C., & Cotter, P. D. (2012). High-throughput sequencing reveals the incomplete, short-term recovery of infant gut microbiota following parenteral antibiotic treatment with ampicillin and gentamicin. *Antimicrobial Agents and Chemotherapy*, *56*(11), 5811–5820. Available from https://doi.org/10.1128/AAC.00789-12.

Fujinami, R. S., von Herrath, M. G., Christen, U., & Whitton, J. L. (2006). Molecular mimicry, bystander activation, or viral persistence: Infections and autoimmune disease. *Clinical Microbiology Reviews*, *19*(1), 80–94.

García-Montero, C., Fraile-Martínez, O., Gómez-Lahoz, A. M., Pekarek, L., Castellanos, A. J., Noguerales-Fraguas, F., Coca, S., Guijarro, L. G., García-Honduvilla, N., Asúnsolo, A., Sanchez-Trujillo, L., Lahera, G., Bujan, J., Monserrat, J., Álvarez-Mon, M., Álvarez-Mon, M. A., & Ortega, M. A. (2021). Nutritional components in western diet versus mediterranean diet at the gut microbiota-immune system interplay. Implications for health and disease. *Nutrients*, *13*(2). Available from https://doi.org/10.3390/nu13020699.

Gebert, B., Fischer, W., Weiss, E., Hoffmann, R., & Haas, R. (2003). Helicobacter pylori vacuolating cytotoxin inhibits T lymphocyte activation. *Science (New York, N.Y.)*, *301*(5636), 1099–1102.

Getts, D. R., Chastain, E. M. L., Terry, R. L., & Miller, S. D. (2013). Virus infection, antiviral immunity, and autoimmunity. *Immunological Reviews*, *255*(1), 197–209. Available from https://doi.org/10.1111/imr.12091.

Gómez-Rial, J., Sánchez-Batán, S., Rivero-Calle, I., Pardo-Seco, J., Martinón-Martínez, J. M., Salas, A., & Martinón-Torres, F. (2019). Rotavirus infection beyond the gut. *Infection and Drug Resistance*, *12*, 55–64. Available from https://doi.org/10.2147/IDR.S186404.

Hemming-Harlo, M., Lähdeaho, M.-L., Mäki, M., & Vesikari, T. (2019). Rotavirus vaccination does not increase Type 1 diabetes and may decrease celiac disease in children and adolescents. *The Pediatric Infectious Disease Journal*, *38*(5), 539–541. Available from https://doi.org/10.1097/INF.0000000000002281.

Hemming-Harlo, M., Vesikari, T., Uhari, M., Renko, M., Salminen, M., Torcel-Pagnon, L., Hartwig, S., Simondon, F., & Bricout, H. (2017). Sustained high effectiveness of rotateq on hospitalizations attributable to rotavirus-associated gastroenteritis during 4 years in Finland. *Journal of the Pediatric Infectious Diseases Society*, *6*(4), 317–323. Available from https://doi.org/10.1093/jpids/piw061.

Hummel, S., Hummel, M., Banholzer, J., Hanak, D., Mollenhauer, U., Bonifacio, E., & Ziegler, A. G. (2007). Development of autoimmunity to transglutaminase C in children of patients with type 1 diabetes: Relationship to islet autoantibodies and infant feeding. *Diabetologia*, *50*(2), 390–394.

Hummel, S., Pflüger, M., Hummel, M., Bonifacio, E., & Ziegler, A.-G. (2011). Primary dietary intervention study to reduce the risk of islet autoimmunity in children at increased risk for type 1 diabetes: The BABYDIET study. *Diabetes Care*, *34*(6), 1301–1305. Available from https://doi.org/10.2337/dc10-2456.

Hummel, S., Weiß, A., Bonifacio, E., Agardh, D., Akolkar, B., Aronsson, C. A., Hagopian, W. A., Koletzko, S., Krischer, J. P., Lernmark, Å., Lynch, K., Norris, J. M., Rewers, M. J., She, J.-X., Toppari, J., Uusitalo, U., Vehik, K., Virtanen, S. M., Beyerlein, A., & Ziegler, A.-G. (2021). Associations of breastfeeding with childhood autoimmunity, allergies, and overweight: The Environmental Determinants of Diabetes in the Young (TEDDY) study. *The American Journal of Clinical Nutrition*, *114*(1), 134–142. Available from https://doi.org/10.1093/ajcn/nqab065.

Hussein, H. M., & Rahal, E. A. (2019). The role of viral infections in the development of autoimmune diseases. *Critical Reviews in Microbiology*, *45*(4), 394–412. Available from https://doi.org/10.1080/1040841X.2019.1614904.

Inns, T., Fleming, K. M., Iturriza-Gomara, M., & Hungerford, D. (2021). Paediatric rotavirus vaccination, coeliac disease and type 1 diabetes in children: a population-based cohort study. *BMC Medicine*, *19*(1), 147. Available from https://doi.org/10.1186/s12916-021-02017-1.

Ivarsson, A., Hernell, O., Nyström, L., & Persson, L. A. (2003). Children born in the summer have increased risk for coeliac disease. *Journal of Epidemiology and Community Health*, *57*(1), 36–39.

Jakobsson, H. E., Jernberg, C., Andersson, A. F., Sjölund-Karlsson, M., Jansson, J. K., & Engstrand, L. (2010). Short-term antibiotic treatment has differing long-term impacts on the human throat and gut microbiome. *PloS One*, *5*(3), e9836. Available from https://doi.org/10.1371/journal.pone.0009836.

Jansen, M. A. E., Tromp, I. I. M., Kiefte-de Jong, J. C., Jaddoe, V. W. V., Hofman, A., Escher, J. C., Hooijkaas, H., & Moll, H. A. (2014). Infant feeding and anti-tissue transglutaminase antibody concentrations in the Generation R Study. *The American Journal of Clinical Nutrition*, *100*(4), 1095–1101. Available from https://doi.org/10.3945/ajcn.114.090316.

Jansen, M. A. E., van den Heuvel, D., van der Zwet, K. V. M., Jaddoe, V. W. V., Hofman, A., Escher, J. C., Fraaij, P. L. A., Hooijkaas, H., van Zelm, M. C., & Moll, H. A. (2016). Herpesvirus Infections and Transglutaminase Type 2 Antibody Positivity in Childhood: The Generation R Study. *Journal of Pediatric Gastroenterology and Nutrition*, *63*(4), 423–430. Available from https://doi.org/10.1097/MPG.0000000000001163.

Ji, Y., Sakata, Y., & Tso, P. (2011). Nutrient-induced inflammation in the intestine. *Current Opinion in Clinical Nutrition and Metabolic Care*, *14*(4), 315–321. Available from https://doi.org/10.1097/MCO.0b013e3283476e74.

Jiang, H.-Y., Zhang, X., Zhou, Y.-Y., Jiang, C.-M., & Shi, Y.-D. (2020). Infection, antibiotic exposure, and risk of celiac disease: A systematic review and meta-analysis. *Journal of Gastroenterology and Hepatology*, *35*(4), 557–566. Available from https://doi.org/10.1111/jgh.14928.

Kagnoff, M. F., Paterson, Y. J., Kumar, P. J., Kasarda, D. D., Carbone, F. R., Unsworth, D. J., & Austin, R. K. (1987). Evidence for the role of a human intestinal adenovirus in the pathogenesis of coeliac disease. *Gut*, *28*(8), 995–1001.

Kahrs, C. R., Chuda, K., Tapia, G., Stene, L. C., Mårild, K., Rasmussen, T., Rønningen, K. S., Lundin, K. E. A., Kramna, L., Cinek, O., & Størdal, K. (2019). Enterovirus as trigger of coeliac disease: Nested case-control study within prospective birth cohort. *BMJ (Clinical Research Ed.)*, *364*, l231. Available from https://doi.org/10.1136/bmj.l231.

Kemppainen, K. M., Lynch, K. F., Liu, E., Lönnrot, M., Simell, V., Briese, T., Koletzko, S., Hagopian, W., Rewers, M., She, J.-X., Simell, O., Toppari, J., Ziegler, A.-G., Akolkar, B., Krischer, J. P., Lernmark, Å., Hyöty, H., Triplett, E. W., & Agardh, D. (2017). Factors that increase risk of celiac disease autoimmunity after a gastrointestinal infection in early life. *Clinical Gastroenterology and Hepatology : The Official Clinical Practice Journal of the American Gastroenterological Association*, *15*(5), 694–702. Available from https://doi.org/10.1016/j.cgh.2016.10.033, e5.

Kim, B., Kaistha, S. D., & Rouse, B. T. (2006). Viruses and autoimmunity. *Autoimmunity*, *39*(1), 71–77.

Kim, C., Yin, Z., Kamdar, N., & Lee, G. J. (2022). Rotavirus vaccination is not associated with incident celiac disease or autoimmune thyroid disease in a national cohort of privately insured children. *Scientific Reports*, *12*(1), 12941. Available from https://doi.org/10.1038/s41598-022-17187-y.

Kirkwood, C. D. (2010). Genetic and antigenic diversity of human rotaviruses: potential impact on vaccination programs. *The Journal of Infectious Diseases*, *202*, S43–8. Available from https://doi.org/10.1086/653548, Suppl.

Koletzko, S., Lee, H.-S., Beyerlein, A., Aronsson, C. A., Hummel, M., Liu, E., Simell, V., Kurppa, K., Lernmark, Å., Hagopian, W., Rewers, M., She, J.-X., Simell, O., Toppari, J., Ziegler, A.-G., Krischer, J., & Agardh, D. (2018). Cesarean section on the risk of celiac disease in the offspring: The Teddy Study. *Journal of Pediatric Gastroenterology and Nutrition*, *66*(3), 417–424. Available from https://doi.org/10.1097/MPG.0000000000001682.

Lebwohl, B., Green, P. H. R., Murray, J. A., & Ludvigsson, J. F. (2013). Season of birth in a nationwide cohort of coeliac disease patients. *Archives of Disease in Childhood*, *98*(1), 48–51. Available from https://doi.org/10.1136/archdischild-2012-302360.

Lebwohl, B., Sanders, D. S., & Green, P. H. R. (2018). Coeliac disease. *Lancet (London, England)*, *391*(10115), 70–81. Available from https://doi.org/10.1016/S0140-6736(17)31796-8.

Lebwohl, B., Spechler, S. J., Wang, T. C., Green, P. H. R., & Ludvigsson, J. F. (2014). Use of proton pump inhibitors and subsequent risk of celiac disease. *Digestive and Liver Disease : Official Journal of the Italian Society of Gastroenterology and the Italian Association for the Study of the Liver*, *46*(1), 36–40. Available from https://doi.org/10.1016/j.dld.2013.08.128.

Leonard, J., Marshall, J. K., & Moayyedi, P. (2007). Systematic review of the risk of enteric infection in patients taking acid suppression. *The American Journal of Gastroenterology*, *102*(9), 2047–2056, quiz 2057.

Leonard, M. M., Serena, G., Sturgeon, C., & Fasano, A. (2015). Genetics and celiac disease: The importance of screening. *Expert Review of Gastroenterology & Hepatology*, *9*(2), 209–215. Available from https://doi.org/10.1586/17474124.2014.945915.

Lerner, A., Arleevskaya, M., Schmiedl, A., & Matthias, T. (2017). Microbes and viruses are bugging the gut in celiac disease. Are they friends or foes? *Frontiers in Microbiology*, *8*, 1392. Available from https://doi.org/10.3389/fmicb.2017.01392.

Lernmark, Å. (2016). Environmental factors in the etiology of type 1 diabetes, celiac disease, and narcolepsy. *Pediatric Diabetes*, *17*(Suppl 22), 65–72. Available from https://doi.org/10.1111/pedi.12390, Suppl 22.

Levescot, A., Malamut, G., & Cerf-Bensussan, N. (2022). Immunopathogenesis and environmental triggers in coeliac disease. *Gut*, *71*(11), 2337–2349. Available from https://doi.org/10.1136/gutjnl-2021-326257.

Lewy, H., Meirson, H., & Laron, Z. (2009). Seasonality of birth month of children with celiac disease differs from that in the general population and between sexes and is linked to family history and environmental factors. *Journal of Pediatric Gastroenterology and Nutrition*, *48*(2), 181–185. Available from https://doi.org/10.1097/MPG.0b013e3181709530.

Lindehammer, S. R., Björck, S., Lynch, K., Brundin, C., Marsal, K., Agardh, D., & Fex, M. (2011). Early human pregnancy serum cytokine levels predict autoimmunity in offspring. *Autoimmunity*, *44*(6), 445–452. Available from https://doi.org/10.3109/08916934.2010.549530.

Lindfors, K., Lin, J., Lee, H.-S., Hyöty, H., Nykter, M., Kurppa, K., Liu, E., Koletzko, S., Rewers, M., Hagopian, W., Toppari, J., Ziegler, A.-G., Akolkar, B., Krischer, J. P., Petrosino, J. F., Lloyd, R. E., & Agardh, D. (2020). Metagenomics of the faecal virome indicate a cumulative effect of enterovirus and gluten amount on the risk of coeliac disease autoimmunity in genetically at risk children: The TEDDY study. *Gut*, *69*(8), 1416–1422. Available from https://doi.org/10.1136/gutjnl-2019-319809.

Lionetti, E., Castellaneta, S., Francavilla, R., Pulvirenti, A., & Catassi, C. (2017). Mode of delivery and risk of celiac disease: Risk of celiac disease and age at gluten introduction cohort study. *The Journal of Pediatrics*, *184*, 81–86. Available from https://doi.org/10.1016/j.jpeds.2017.01.023, e2.

Lionetti, E., Castellaneta, S., Francavilla, R., Pulvirenti, A., Tonutti, E., Amarri, S., Barbato, M., Barbera, C., Barera, G., Bellantoni, A., Castellano, E., Guariso, G., Limongelli, M. G., Pellegrino, S., Polloni, C., Ughi, C., Zuin, G., Fasano, A., & Catassi, C. (2014). Introduction of gluten, HLA status, and the risk of celiac disease in children. *The New England Journal of Medicine*, *371*(14), 1295–1303. Available from https://doi.org/10.1056/NEJMoa1400697.

Lionetti, E., Galeazzi, T., Dominijanni, V., Acquaviva, I., Catassi, G. N., Iasevoli, M., Malamisura, B., & Catassi, C. (2021). Lower level of Plasma 25-Hydroxyvitamin D in children at diagnosis of celiac disease compared with healthy subjects: A case-control study. *The Journal of Pediatrics*, *228*, 132–137. Available from https://doi.org/10.1016/j.jpeds.2020.08.089, e1.

Logan, K., Perkin, M. R., Marrs, T., Radulovic, S., Craven, J., Flohr, C., Bahnson, H. T., & Lack, G. (2020). Early gluten introduction and celiac disease in the EAT study: A prespecified analysis of the EAT Randomized clinical trial. *JAMA Pediatrics*, *174*(11), 1041–1047. Available from https://doi.org/10.1001/jamapediatrics.2020.2893.

Lu, C., Zhou, W., He, X., Zhou, X., & Yu, C. (2021). Vitamin D status and vitamin D receptor genotypes in celiac disease: a meta-analysis. *Critical Reviews in Food Science and Nutrition*, *61*(12), 2098–2106. Available from https://doi.org/10.1080/10408398.2020.1772716.

Ludvigsson, J. F., & Ludvigsson, J. (2005). Parental smoking and risk of coeliac disease in offspring. *Scandinavian Journal of Gastroenterology*, *40*(3), 336–342.

Lund-Blix, N. A., Mårild, K., Tapia, G., Norris, J. M., Stene, L. C., & Størdal, K. (2019). Gluten intake in early childhood and risk of celiac disease in childhood: A nationwide cohort study. *The American Journal of Gastroenterology*, *114*(8), 1299–1306. Available from https://doi.org/10.14309/ajg.0000000000000331.

Lund-Blix, N. A., Tapia, G., Mårild, K., Brantsæter, A. L., Eggesbø, M., Mandal, S., Stene, L. C., & Størdal, K. (2020). Maternal fibre and gluten intake during pregnancy and risk of childhood celiac disease: the MoBa study. *Scientific Reports, 10*(1), 16439. Available from https://doi.org/10.1038/s41598-020-73244-4.

Mårild, K., Dong, F., Lund-Blix, N. A., Seifert, J., Barón, A. E., Waugh, K. C., Taki, I., Størdal, K., Tapia, G., Stene, L. C., Johnson, R. K., Liu, E., Rewers, M. J., & Norris, J. M. (2019). Gluten intake and risk of celiac disease: Long-term follow-up of an at-risk birth cohort. *The American Journal of Gastroenterology, 114*(8), 1307–1314. Available from https://doi.org/10.14309/ajg.0000000000000255.

Mårild, K., Kahrs, C. R., Tapia, G., Stene, L. C., & Størdal, K. (2015). Infections and risk of celiac disease in childhood: a prospective nationwide cohort study. *The American Journal of Gastroenterology, 110*(10), 1475–1484. Available from https://doi.org/10.1038/ajg.2015.287.

Mårild, K., Kahrs, C. R., Tapia, G., Stene, L. C., & Størdal, K. (2017). Maternal infections, antibiotics, and paracetamol in pregnancy and offspring celiac disease: A cohort study. *Journal of Pediatric Gastroenterology and Nutrition, 64*(5), 730–736. Available from https://doi.org/10.1097/MPG.0000000000001339.

Mårild, K., Ludvigsson, J., Sanz, Y., & Ludvigsson, J. F. (2014). Antibiotic exposure in pregnancy and risk of coeliac disease in offspring: a cohort study. *BMC Gastroenterology, 14*, 75. Available from https://doi.org/10.1186/1471-230X-14-75.

Mårild, K., Stephansson, O., Montgomery, S., Murray, J. A., & Ludvigsson, J. F. (2012). Pregnancy outcome and risk of celiac disease in offspring: a nationwide case-control study. *Gastroenterology, 142*(1), 39–45. Available from https://doi.org/10.1053/j.gastro.2011.09.047, e3.

Mårild, K., Tapia, G., Midttun, Ø., Ueland, P. M., Magnus, M. C., Rewers, M., Stene, L. C., & Størdal, K. (2019). Smoking in pregnancy, cord blood cotinine and risk of celiac disease diagnosis in offspring. *European Journal of Epidemiology, 34*(7), 637–649. Available from https://doi.org/10.1007/s10654-019-00522-5.

Mariné, M., Farre, C., Alsina, M., Vilar, P., Cortijo, M., Salas, A., Fernández-Bañares, F., Rosinach, M., Santaolalla, R., Loras, C., Marquès, T., Cusí, V., Hernández, M. I., Carrasco, A., Ribes, J., Viver, J. M., & Esteve, M. (2011). The prevalence of coeliac disease is significantly higher in children compared with adults. *Alimentary Pharmacology & Therapeutics, 33*(4), 477–486. Available from https://doi.org/10.1111/j.1365-2036.2010.04543.x.

Meijer, C. R., Auricchio, R., Putter, H., Castillejo, G., Crespo, P., Gyimesi, J., Hartman, C., Kolacek, S., Koletzko, S., Korponay-Szabo, I., Ojinaga, E. M., Polanco, I., Ribes-Koninckx, C., Shamir, R., Szajewska, H., Troncone, R., Villanacci, V., Werkstetter, K., & Mearin, M. L. (2022). Prediction models for celiac disease development in children from high-risk families: Data from the PreventCD Cohort. *Gastroenterology, 163*(2), 426–436. Available from https://doi.org/10.1053/j.gastro.2022.04.030.

Monack, D. M., Mueller, A., & Falkow, S. (2004). Persistent bacterial infections: The interface of the pathogen and the host immune system. *Nature Reviews. Microbiology, 2*(9), 747–765.

Moos, C., Duus, K. S., Frederiksen, P., Heitmann, B. L., & Andersen, V. (2020). Exposure to the Danish Mandatory Vitamin D fortification policy in prenatal life and the risk of developing coeliac disease-The importance of season: A semi ecological study. *Nutrients, 12*(5). Available from https://doi.org/10.3390/nu12051243.

Namatovu, F., Lindkvist, M., Olsson, C., Ivarsson, A., & Sandström, O. (2016). Season and region of birth as risk factors for coeliac disease a key to the aetiology? *Archives of Disease in Childhood, 101*(12), 1114–1118. Available from https://doi.org/10.1136/archdischild-2015-310122.

Norris, J. M., Barriga, K., Hoffenberg, E. J., Taki, I., Miao, D., Haas, J. E., Emery, L. M., Sokol, R. J., Erlich, H. A., Eisenbarth, G. S., & Rewers, M. (2005). Risk of celiac disease autoimmunity and timing of gluten introduction in the diet of infants at increased risk of disease. *JAMA, 293*(19), 2343–2351.

Oikarinen, M., Puustinen, L., Lehtonen, J., Hakola, L., Simell, S., Toppari, J., Ilonen, J., Veijola, R., Virtanen, S. M., Knip, M., & Hyöty, H. (2020). Enterovirus infections are associated with the development of celiac disease in a birth cohort study. *Frontiers in Immunology, 11*, 604529. Available from https://doi.org/10.3389/fimmu.2020.604529.

Pecora, F., Persico, F., Argentiero, A., Neglia, C., & Esposito, S. (2020). The role of micronutrients in support of the immune response against viral infections. *Nutrients, 12*(10). Available from https://doi.org/10.3390/nu12103198.

Plot, L., & Amital, H. (2009). Infectious associations of celiac disease. *Autoimmunity Reviews, 8*(4), 316–319. Available from https://doi.org/10.1016/j.autrev.2008.10.001.

Plot, L., Amital, H., Barzilai, O., Ram, M., Bizzaro, N., & Shoenfeld, Y. (2009). Infections may have a protective role in the etiopathogenesis of celiac disease. *Annals of the New York Academy of Sciences, 1173*, 670–674. Available from https://doi.org/10.1111/j.1749-6632.2009.04814.x.

Porpora, M., Conte, M., Lania, G., Bellomo, C., Rapacciuolo, L., Chirdo, F. G., Auricchio, R., Troncone, R., Auricchio, S., Barone, M. V., & Nanayakkara, M. (2022). Inflammation is present, persistent and more sensitive to proinflammatory triggers in celiac disease enterocytes. *International Journal of Molecular Sciences, 23*(4). Available from https://doi.org/10.3390/ijms23041973.

Riddle, M. S., Murray, J. A., Cash, B. D., Pimentel, M., & Porter, C. K. (2013). Pathogen-specific risk of celiac disease following bacterial causes of foodborne illness: a retrospective cohort study. *Digestive Diseases and Sciences, 58*(11), 3242–3245. Available from https://doi.org/10.1007/s10620-013-2733-7.

Roberts, S. E., Williams, J. G., Meddings, D., Davidson, R., & Goldacre, M. J. (2009). Perinatal risk factors and coeliac disease in children and young adults: a record linkage study. *Alimentary Pharmacology & Therapeutics, 29*(2), 222–231. Available from https://doi.org/10.1111/j.1365-2036.2008.03871.x.

Romanos, J., Rosén, A., Kumar, V., Trynka, G., Franke, L., Szperl, A., Gutierrez-Achury, J., van Diemen, C. C., Kanninga, R., Jankipersadsing, S. A., Steck, A., Eisenbarth, G., van Heel, D. A., Cukrowska, B., Bruno, V., Mazzilli, M. C., Núñez, C., Bilbao, J. R., Mearin, M. L., . . . Wijmenga, C. (2014). Improving coeliac disease risk prediction by testing non-HLA variants additional to HLA variants. *Gut, 63*(3), 415–422. Available from https://doi.org/10.1136/gutjnl-2012-304110.

Rostami-Nejad, M., Rostami, K., Sanaei, M., Mohebbi, S. R., Al-Dulaimi, D., Nazemalhosseini-Mojarad, E., Collin, P., Mulder, C. J., & Zali, M. R. (2010). Rotavirus and coeliac autoimmunity among adults with non-specific gastrointestinal symptoms. *Saudi Medical Journal*, *31*(8), 891–894.

Sadiq, A., Khan, J., Ullah, I., Basharat, N., Ali, S., Din, A. U., Ali, I., Farid, A., Shah, M., Abdel-Daim, M. M., Albadrani, G. M., & Khan, I. A. (2022). Seroprevalence of Anti-tTg-IgA among symptomized celiac disease patients and their correlation with rotavirus infection. *BioMed Research International*, *2022*, 6972624. Available from https://doi.org/10.1155/2022/6972624.

Samsel, A., & Seneff, S. (2013). Glyphosate, pathways to modern diseases II: Celiac sprue and gluten intolerance. *Interdisciplinary Toxicology*, *6*(4), 159–184. Available from https://doi.org/10.2478/intox-2013-0026.

Sandberg-Bennich, S., Dahlquist, G., & Källén, B. (2002). Coeliac disease is associated with intrauterine growth and neonatal infections. *Acta Paediatrica (Oslo, Norway : 1992)*, *91*(1), 30–33.

Sarno, M., Discepolo, V., Troncone, R., & Auricchio, R. (2015). Risk factors for celiac disease. *Italian Journal of Pediatrics*, *41*, 57. Available from https://doi.org/10.1186/s13052-015-0166-y.

Sellitto, M., Bai, G., Serena, G., Fricke, W. F., Sturgeon, C., Gajer, P., White, J. R., Koenig, S. S. K., Sakamoto, J., Boothe, D., Gicquelais, R., Kryszak, D., Puppa, E., Catassi, C., Ravel, J., & Fasano, A. (2012). Proof of concept of microbiome-metabolome analysis and delayed gluten exposure on celiac disease autoimmunity in genetically at-risk infants. *PloS One*, *7*(3), e33387. Available from https://doi.org/10.1371/journal.pone.0033387.

Sevelsted, A., Stokholm, J., Bønnelykke, K., & Bisgaard, H. (2015). Cesarean section and chronic immune disorders. *Pediatrics*, *135*(1), e92–8. Available from https://doi.org/10.1542/peds.2014-0596.

Shree, T., Banerjee, P., & Senapati, S. (2022). A meta-analysis suggests the association of reduced serum level of vitamin D and T-allele of Fok1 (rs2228570) polymorphism in the vitamin D receptor gene with celiac disease. *Frontiers in Nutrition*, *9*, 996450. Available from https://doi.org/10.3389/fnut.2022.996450.

Simre, K., Uibo, O., Peet, A., Tillmann, V., Kool, P., Hämäläinen, A.-M., Härkönen, T., Siljander, H., Virtanen, S., Ilonen, J., Knip, M., & Uibo, R. (2016). Exploring the risk factors for differences in the cumulative incidence of coeliac disease in two neighboring countries: The prospective DIABIMMUNE study. *Digestive and Liver Disease : Official Journal of the Italian Society of Gastroenterology and the Italian Association for the Study of the Liver*, *48*(11), 1296–1301. Available from https://doi.org/10.1016/j.dld.2016.06.029.

Singh, P., Arora, A., Strand, T. A., Leffler, D. A., Catassi, C., Green, P. H., Kelly, C. P., Ahuja, V., & Makharia, G. K. (2018). Global prevalence of celiac disease: Systematic review and meta-analysis. *Clinical Gastroenterology and Hepatology : The Official Clinical Practice Journal of the American Gastroenterological Association*, *16*(6), 823–836. Available from https://doi.org/10.1016/j.cgh.2017.06.037, e2.

Smith, L. B., Lynch, K. F., Baxter, J., Lernmark, B., Roth, R., Simell, T., & Johnson, S. B. (2014). Factors associated with maternal-reported actions to prevent type 1 diabetes in the first year of the TEDDY study. *Diabetes Care*, *37*(2), 325–331. Available from https://doi.org/10.2337/dc13-0449.

Soullane, S., Henderson, M., Kang, H., Luu, T. M., Lee, G. E., & Auger, N. (2021). Cesarean delivery and risk of hospitalization for autoimmune disorders before 14 years of age. *European Journal of Pediatrics*, *180*(11), 3359–3366. Available from https://doi.org/10.1007/s00431-021-04132-w.

Steiner, N., Wainstock, T., Sheiner, E., Segal, I., Landau, D., & Walfisch, A. (2019). Small for gestational age as an independent risk factor for long-term pediatric gastrointestinal morbidity of the offspring. *The Journal of Maternal-Fetal & Neonatal Medicine : The Official Journal of the European Association of Perinatal Medicine, the Federation of Asia and Oceania Perinatal Societies, the International Society of Perinatal Obstetricians*, *32*(9), 1407–1411. Available from https://doi.org/10.1080/14767058.2017.1406473.

Stene, L. C., Honeyman, M. C., Hoffenberg, E. J., Haas, J. E., Sokol, R. J., Emery, L., Taki, I., Norris, J. M., Erlich, H. A., Eisenbarth, G. S., & Rewers, M. (2006). Rotavirus infection frequency and risk of celiac disease autoimmunity in early childhood: A longitudinal study. *The American Journal of Gastroenterology*, *101*(10), 2333–2340.

Stephansson, O., Granath, F., Svensson, T., Haglund, B., Ekbom, A., & Kieler, H. (2011). Drug use during pregnancy in Sweden - assessed by the Prescribed Drug Register and the Medical Birth Register. *Clinical Epidemiology*, *3*, 43–50. Available from https://doi.org/10.2147/CLEP.S16305.

Stewart, C. J., Ajami, N. J., O'Brien, J. L., Hutchinson, D. S., Smith, D. P., Wong, M. C., Ross, M. C., Lloyd, R. E., Doddapaneni, H., Metcalf, G. A., Muzny, D., Gibbs, R. A., Vatanen, T., Huttenhower, C., Xavier, R. J., Rewers, M., Hagopian, W., Toppari, J., Ziegler, A.-G., & Petrosino, J. F. (2018). Temporal development of the gut microbiome in early childhood from the TEDDY study. *Nature*, *562*(7728), 583–588. Available from https://doi.org/10.1038/s41586-018-0617-x.

Størdal, K., White, R. A., & Eggesbø, M. (2013). Early feeding and risk of celiac disease in a prospective birth cohort. *Pediatrics*, *132*(5), e1202–9. Available from https://doi.org/10.1542/peds.2013-1752.

Szajewska, H., Shamir, R., Stróżyk, A., Chmielewska, A., Zalewski, B. M., Auricchio, R., Koletzko, S., Korponay-Szabo, I. R., Mearin, M. L., Meijer, C., Ribes-Koninckx, C., & Troncone, R. (2023). Systematic review: early feeding practices and the risk of coeliac disease. A 2022 update and revision. *Alimentary Pharmacology & Therapeutics*, *57*(1), 8–22. Available from https://doi.org/10.1111/apt.17290.

Tanpowpong, P., Li, S., Espinola, J. A., Santos, L. C., James, K. E., Powe, C. E., & Camargo, C. A. (2023). Pregnancy- and birth-related risk factors for the development of childhood celiac disease. *Acta Paediatrica (Oslo, Norway : 1992)*, *112*(5), 1029–1034. Available from https://doi.org/10.1111/apa.16686.

Tanpowpong, P., Obuch, J. C., Jiang, H., McCarty, C. E., Katz, A. J., Leffler, D. A., Kelly, C. P., Weir, D. C., Leichtner, A. M., & Camargo, C. A. (2013). Multicenter study on season of birth and celiac disease: evidence for a new theoretical model of pathogenesis. *The Journal of Pediatrics*, *162*(3), 501–504. Available from https://doi.org/10.1016/j.jpeds.2012.08.056.

Tilg, H., & Moschen, A. R. (2015). Food, immunity, and the microbiome. *Gastroenterology*, *148*(6), 1107–1119. Available from https://doi.org/10.1053/j.gastro.2014.12.036.

Unalp-Arida, A., Ruhl, C. E., Choung, R. S., Brantner, T. L., & Murray, J. A. (2017). Lower prevalence of celiac disease and gluten-related disorders in persons living in Southern vs Northern Latitudes of the United States. *Gastroenterology, 152*(8), 1922–1932. Available from https://doi.org/10.1053/j.gastro.2017.02.012, e2.

Vaarala, O., Jokinen, J., Lahdenkari, M., & Leino, T. (2017). Rotavirus vaccination and the risk of celiac disease or Type 1 Diabetes in Finnish Children at Early Life. *The Pediatric Infectious Disease Journal, 36*(7), 674–675. Available from https://doi.org/10.1097/INF.0000000000001600.

Vriezinga, S. L., Auricchio, R., Bravi, E., Castillejo, G., Chmielewska, A., Crespo Escobar, P., Kolaček, S., Koletzko, S., Korponay-Szabo, I. R., Mummert, E., Polanco, I., Putter, H., Ribes-Koninckx, C., Shamir, R., Szajewska, H., Werkstetter, K., Greco, L., Gyimesi, J., Hartman, C., & Mearin, M. L. (2014). Randomized feeding intervention in infants at high risk for celiac disease. *The New England Journal of Medicine, 371*(14), 1304–1315. Available from https://doi.org/10.1056/NEJMoa1404172.

Welander, A., Tjernberg, A. R., Montgomery, S. M., Ludvigsson, J., & Ludvigsson, J. F. (2010). Infectious disease and risk of later celiac disease in childhood. *Pediatrics, 125*(3), e530–6. Available from https://doi.org/10.1542/peds.2009-1200.

West, J., Fleming, K. M., Tata, L. J., Card, T. R., & Crooks, C. J. (2014). Incidence and prevalence of celiac disease and dermatitis herpetiformis in the UK over two decades: population-based study. *The American Journal of Gastroenterology, 109*(5), 757–768. Available from https://doi.org/10.1038/ajg.2014.55.

Wingren, C. J., Björck, S., Lynch, K. F., Ohlsson, H., Agardh, D., & Merlo, J. (2012). Coeliac disease in children: a social epidemiological study in Sweden. *Acta Paediatrica (Oslo, Norway : 1992), 101*(2), 185–191. Available from https://doi.org/10.1111/j.1651-2227.2011.02434.x.

Yang, X.-Y., Liu, Y.-H., Jiang, H.-Y., & Ying, X.-H. (2022). Cesarean section is not associated with increased risk of celiac disease in the offspring: a meta-analysis. *The Journal of Maternal-Fetal & Neonatal Medicine : The Official Journal of the European Association of Perinatal Medicine, the Federation of Asia and Oceania Perinatal Societies, the International Society of Perinatal Obstetricians, 35*(25), 9570–9577. Available from https://doi.org/10.1080/14767058.2022.2048813.

Yuan, J., Li, P., Tao, J., Shi, X., Hu, B., Chen, H., & Guo, X. (2009). H. pylori escape host immunoreaction through inhibiting ILK expression by VacA. *Cellular & Molecular Immunology, 6*(3), 191–197. Available from https://doi.org/10.1038/cmi.2009.26.

Yue, M., Chen, Q., Zhou, X., Li, L., & Lu, C. (2022). Is Helicobacter pylori Infection Associated with Celiac Disease? A Meta-analysis. *The Turkish Journal of Gastroenterology : The Official Journal of Turkish Society of Gastroenterology, 33*(3), 205–212. Available from https://doi.org/10.5152/tjg.2022.21360.

Zandman-Goddard, G., & Shoenfeld, Y. (2009). Parasitic infection and autoimmunity. *Lupus, 18*(13), 1144–1148. Available from https://doi.org/10.1177/0961203309345735.

Ziegler, A.-G., Schmid, S., Huber, D., Hummel, M., & Bonifacio, E. (2003). Early infant feeding and risk of developing type 1 diabetes-associated autoantibodies. *JAMA, 290*(13), 1721–1728.

Chapter 5

Immunological mechanisms of lesions in celiac disease

Giovanni Monteleone[1,2], Irene Marafini[1,2], Edoardo Troncone[1,2], Emma Calabrese[1,2] and Ivan Monteleone[3]
[1]*Department of Systems Medicine, University of Rome "Tor Vergata", Rome, Italy,* [2]*Gastroenterology Unit, Fondazione Policlinico "Tor Vergata", Rome, Italy,* [3]*Department of Biomedicine and Prevention, University of Rome "Tor Vergata", Rome, Italy*

5.1 Mechanisms controlling the immune homeostasis in the gut

In normal conditions, the gut contains much more inflammatory cells than the rest of the body. This state of "*low-grade inflammation*," which is due to the continuous exposure of the intestinal immune system to a variety of food and microbial antigens, occurs without affecting the absorptive and digestive functions (MacDonald & Monteleone, 2005; Mowat, 2003, 2018). Various factors act in concert to keep under control the "*physiologic*" inflammation. The intestinal epithelium represents one of the main interfaces between our organism and the environment (Berg, 1995). It includes absorptive enterocytes, Paneth cells, goblet cells, tuft and cup cells, microfold (M) cells, and enterochromaffin cells (Neutra et al., 2001). Epithelial cells originate from transit-amplifying cells present in the crypts of Lieberkuhn located at the base of intestinal villi. Enterocytes are tightly associated with each other through the presence of tight junctions and are protected from direct interaction with the microbiota by the mucus secreted by goblet cells, and antimicrobial peptides (i.e., defensins, lysozymes, C-type lectins, and cathelicidins) secreted in the mucus by Paneth cells via a mechanism that is dependent on the Toll-like receptor (TLR) adapter myeloid differentiation factor 88 (MyD88) (Allaire et al., 2018). Intestinal epithelial cells express several pattern recognition receptors (PRR) (i.e., TLRs, retinoic acid-inducible gene-I-like receptors, nucleotide-binding site, and leucine-rich repeat-containing receptors), and studies in germ-free mice have convincingly shown that expression of PRR relies on the presence of the microbiota (Akira & Hemmi, 2003). Therefore the gut epithelium constitutes a physical and chemical barrier, which contributes to providing resistance to potentially dangerous compounds (e.g., allergens, toxic substances), microorganisms, and parasites, often ingested with food (Allaire et al., 2018). Nevertheless, luminal antigens can cross the epithelial layer. This occurs, for example, in the Peyer's patches or follicles, the epithelium of which (the so-called follicle-associated epithelium, FAE) contains specialized epithelial cells (M cells) that transport antigens from the lumen to the gut-associated lymphoid tissue (Neutra et al., 2001). Moreover, macrophages and dendritic cells (DCs) can sample luminal antigens via the extension of protrusions between epithelial cells, particularly in the upper part of the small intestine without disturbing the integrity of the epithelium (Rescigno et al., 2001) (Fig. 5.1).

Intestinal macrophages express high levels of CX3CR1, the receptor of fractalkine, and can capture luminal bacteria, fungi, and soluble proteins (Bain & Mowat, 2011) (Fig. 5.1). These cells display antiinflammatory properties, as they produce interleukin-10 (IL-10), a regulatory cytokine that promotes the expansion of regulatory T cells (Tregs) and inhibits T-cell activation (Bain & Mowat, 2014). DCs express high levels of CD103, the αE integrin, and may or may not express CD11b. CD103 + DCs capture primarily bacteria but not directly soluble antigens, which can be, however, received from macrophages (Farache et al., 2013; Satpathy et al., 2013). CD11b-expressing CD103 + DCs populate the intestinal lamina propria and can migrate to the mesenteric lymph nodes, where they promote differentiation of Tregs, through a mechanism dependent on the activity of transforming growth factor (TGF)-β1, retinoic acid (RA) (a metabolite of vitamin A), and indoleamine 2,3-dioxygenase, an enzyme involved in tryptophan catabolism and Treg differentiation (Coombes et al., 2007; Mucida et al., 2007). Epithelial cells play an important role in the DCs function and induction of immune tolerance, as they produce RA and TGF-β1. Moreover, intestinal epithelial cells are a major source of thymic stromal lymphopoietin, which drives the differentiation of tolerogenic DCs that promote the development of

Pediatric and Adult Celiac Disease. DOI: https://doi.org/10.1016/B978-0-443-13359-6.00021-2

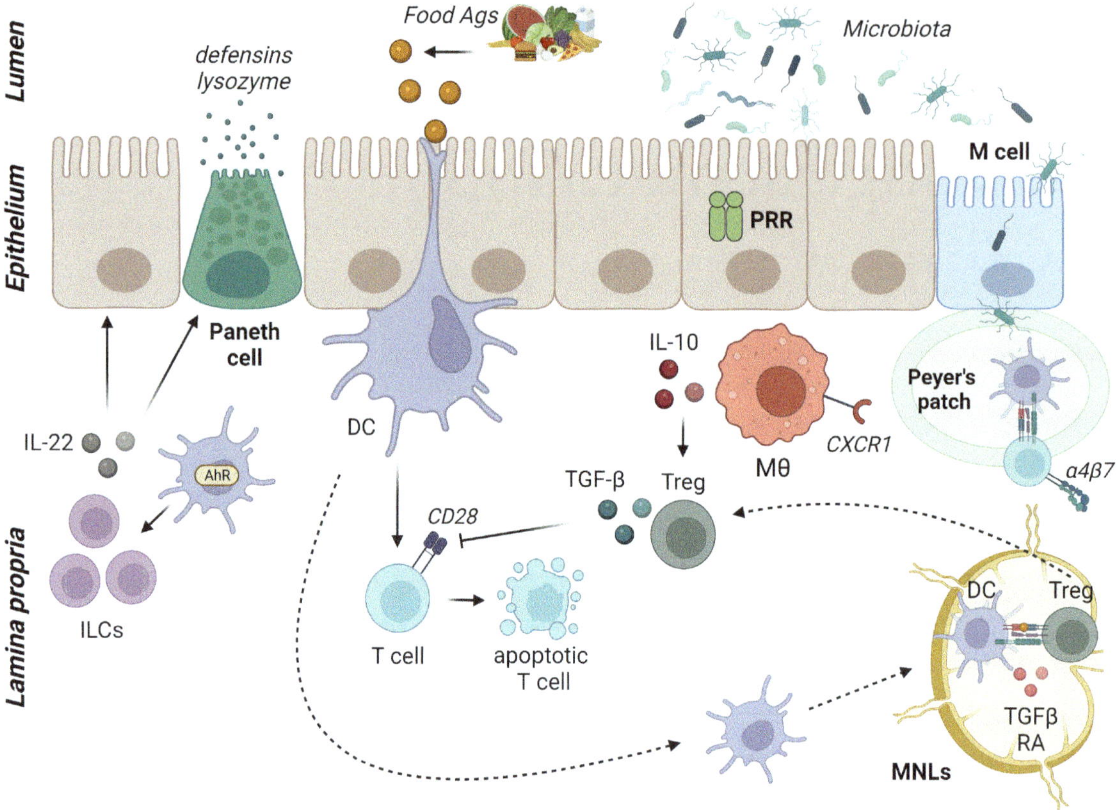

FIGURE 5.1 Schematic representation of some of the mechanisms controlling immune homeostasis in the gut. Although the gut epithelium represents a valid physical (due to the presence of enterocytes tightly associated by tight junctions) and chemical (mainly dependent on the action of Paneth cells secreting antimicrobial peptides) barrier that limit the passage of dietary and microbial antigens (Ags) from the lumen to the lamina propria, the intestinal immune systems is continuously exposed to such antigens. Indeed, Ags can interact with antigen presenting cells in the Peyer's patches, where specialized epithelial cells (M cells) transport the antigens from the lumen to the gut-associated lymphoid tissue thereby promoting activation of a4b7-expressing lymphocytes. Luminal antigens can be also sampled by dendritic cells (DCs) and CXCR1-expressing macrophages (Mθ), which extend protrusions between epithelial cells. Following interaction with antigens, DCs migrate to the mesenteric lymph nodes, where they promote differentiation of regulatory T cells (Tregs) through a mechanism requiring the action of transforming growth factor (TGF)-β1 and retinoic acid (RA). Both interleukin-10 (IL-10), mainly produced by macrophages, and TGF-β1 sustain further the antiinflammatory action of Tregs, thereby suppressing T cell activation and eventually promoting T cell apoptosis. Innate lymphoid cells (ILCs) produce IL-22, a cytokine that promotes epithelial cells functions and repair as well as production of defensins by Paneth cells.

Foxp3-expressing Treg cells, and IL-25, a cytokine that promotes the polarization of regulatory type 2 macrophages (Jiang et al., 2006; Stolfi et al., 2011; Tamachi et al., 2006; Watanabe et al., 2005; Ziegler et al., 2013). However, following interaction with microbial components, epithelial cells can deliver also immunogenic signals to mucosal immune cells, but the exact mechanism by which epithelial cells take such a decision remains unclear.

Epithelial cells interact with intraepithelial αβ and γδ lymphocytes, and innate lymphoid cells (ILCs) (Edelblum et al., 2012; Fan et al., 2019; Witherden & Havran, 2013). Some subsets of ILCs produce IL-22, a cytokine that targets epithelial cells and promotes cell repair, and is required for the formation of Paneth cells, the prime producers of intestinal antimicrobial peptides (Spits & Cupedo, 2012; Walker et al., 2013). IL-22 can be also produced by DCs in response to ligands of the aryl hydrocarbon receptor (AhR) (e.g., products of tryptophan contained in foods and some bacteria) (Vogel et al., 2013) (Fig. 5.1).

Finally, in the normal gut, activated T cells exhibit low expression of costimulatory molecules, have a reduced capacity for cellular expansion, and undergo apoptosis, a phenomenon that selectively reduces cell populations thus limiting unwanted immune responses (Boirivant et al., 1996; Bu et al., 2001; Montufar-Solis et al., 2007).

Consistent with the above observations, defects in the epithelial barrier and/or lack/impairment of regulatory mechanisms can break the immune tolerance toward the luminal antigens thereby resulting in chronic gut inflammation, such as that seen in celiac disease (CD) (MacDonald & Monteleone, 2005).

5.2 HLA and tissue transglutaminase 2 in the initiation of CD-associated immune response

5.2.1 Role of HLA

CD is characterized by an increase in the number of intraepithelial lymphocytes, crypt hyperplasia, and various degrees of villous atrophy (Marsh, 1989). Moreover, in CD patients, there is the production of antibodies against deamidated peptides and tissue transglutaminase 2 (TG2) (Dieterich et al., 1997, 1998), which represent a hallmark of the disease and an important tool in the diagnostic approach to CD.

CD is triggered by the ingestion of gluten in genetically susceptible individuals carrying specific major histocompatibility complex (MHC) class II human leukocyte antigen (HLA) variants. The large majority of CD patients (90%–95%) carry the HLA-DQ2.5 variant (DQA1*05:01, DQB1*02:01), while the remaining patients carry the HLA-DQ2.2 (DQA1*02:01, DQB1*02:02) or HLA-DQ8 (DQA1*03, DQB1*03:02) variant (Abadie et al., 2011; Ludvigsson et al., 2013; Sollid et al., 1989). Nearly 40% of the Western population expresses the CD-associated HLA-DQ2 and HLA-DQ8 molecules, yet CD occurs in less than 2% of these individuals, clearly indicating that HLA variants contribute but are not sufficient for the development of CD. This concept fits with the demonstration that the HLA locus accounts for nearly one-third of the genetic variance of the disease. Non-MHC susceptibility loci, accounting for nearly 15% of the disease risk, are supposed to contribute to CD development (Dubois et al., 2010; Hunt et al., 2008).

HLA-DQ2/DQ8 homozygous individuals are at higher risk of developing CD compared to patients with the predisposing genotype in heterozygosity (Karell et al., 2003; Ploski et al., 1993). This reflects the increased expression of HLA-DQ2 molecules on the surface of antigen-presenting cells and increased antigen presentation seen in homozygosity as compared to heterozygosity (Koning, 2012). The haplotype HLADQ2.5, which binds and presents the full repertoire of gluten peptides that are protected from degradation by gastrointestinal enzymes, confers a higher risk of developing CD as compared to HLA-DQ8, which presents a smaller repertoire of peptides more prone to proteolytic degradation, and HLA-DQ2.2 that can only present a few peptides 35–37 (Shan et al., 2002; Vader et al., 2003). Moreover, HLA-DQ2.5-positive antigen-presenting cells have greater stability of bound peptides and protract gluten presentation as compared to HLA-DQ2.2-positive cells (Fallang et al., 2009). The improved ability of HLA-DQ2.5 to retain its peptide cargo has been ascribed to a polymorphism of DQalpha22 whereby DQ2.5 (tyrosine) can establish a hydrogen bond to the peptide main chain while DQ2.2 (phenylalanine) cannot (Ting et al., 2020). Overall these observations suggest a model in which the development of CD relies on the amplitude of the gluten-specific CD4+ T cell response.

5.2.2 Tissue transglutaminase 2

Gluten has a high content of proline that makes it resistant to gut proteases, and undigested gluten-derived proteins are a good substrate for the action of TG2 (Shan et al., 2002). TG2 catalyzes deamidation, that is the conversion of glutamine residues present in gluten to glutamate, a process that promotes the generation of immunogenic peptides with negatively charged carboxylate residues having a high affinity for positively charged pockets of HLA-DQ2 or HLA-DQ8 proteins (Arentz-Hansen et al., 2000; Molberg et al., 1998; Nilsen et al., 1995). This facilitates the formation of HLA-DQ-gluten complexes at the surface of antigen-presenting cells and consequently an efficient induction of gluten-specific CD4+ T cell responses (Nilsen et al., 1998; Tollefsen et al., 2006). In the normal gut, TG2 is usually found in an inactive state, but inflammation and stress response can up-regulate TG2 expression and enhance the catalytic activity (Ientile et al., 2007; Siegel et al., 2008).

5.3 Role of gluten-reactive T cells

5.3.1 Gluten triggers a Th1-type immune response

The presentation of TG2-deamidated gluten peptides to CD4-positive T cells by antigen-presenting cells bearing HLA-DQ2 and DQ8 triggers a Th1 immune response, with a massive production of IFN-γ and IL-21 (Ientile et al., 2007; Nilsen et al., 1998; Tollefsen et al., 2006) (Fig. 5.2). Consistently, the intestinal mucosal cells of CD patients express the transcription factor signal transducer and activator of transcription (Stat) 1, a downstream target of IFN-γ, and pharmacological inhibition of Stat1 in *ex vivo* organ cultures of CD explants prevents the gliadin-mediated induction of co-stimulatory molecules, such as ICAM-1 and B7–2, involved in the local immune activation (Mazzarella et al., 2003). IFN-γ enhances also the expression of T-bet, a transcription factor that regulates the polarization and function of Th1

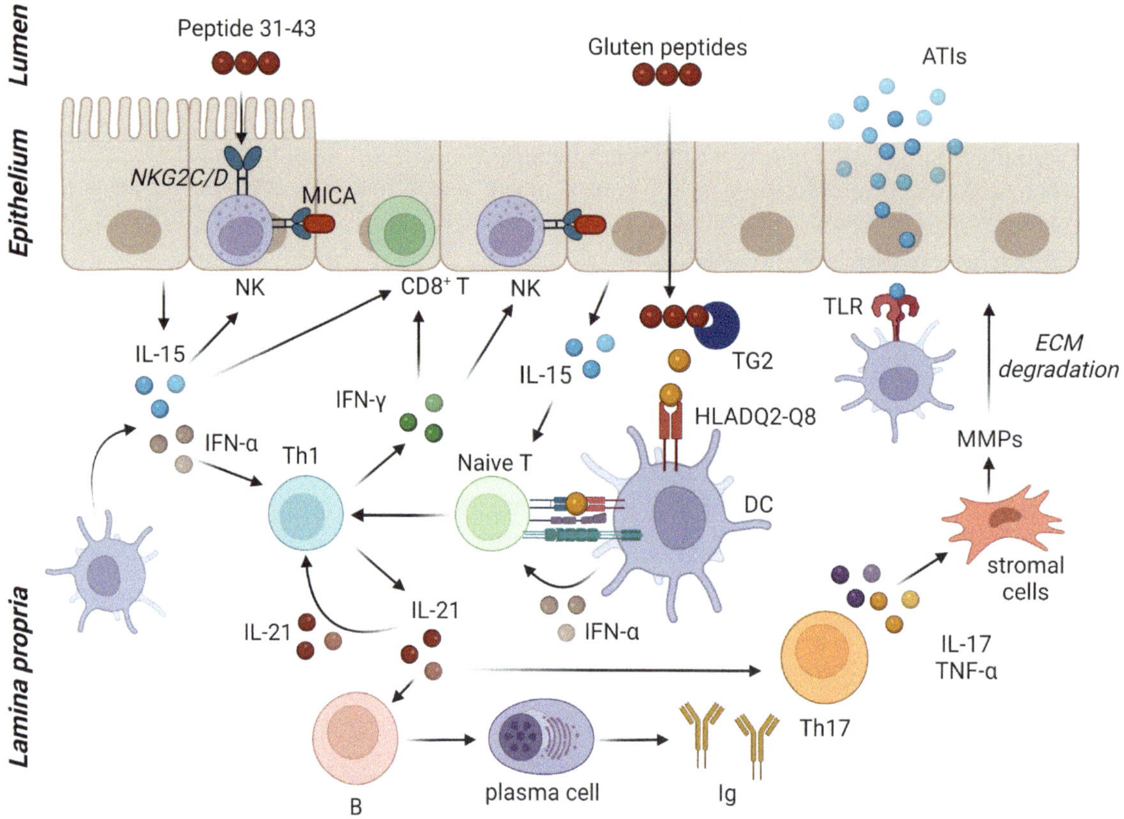

FIGURE 5.2 **Schematic representation of the main pathogenic steps involved in the celiac disease-associated tissue damage.** In celiac disease (CD), gluten peptides entering the lamina propria, are deamidated by transglutaminase 2 (TG2) and presented by human leukocyte antigen (HLA)-DQ2/HLA-DQ8-expressing antigen-presenting cells to naïve T cells. This interaction, in the presence of interleukin (IL)-15 produced mainly by epithelial cells and interferon (IFN)-α secreted by dendritic cells (DC), boosts a Th1 immune response characterized by the production of IFN-γ and IL-21. IFN-γ promotes the function of CD8 + T cells and natural killer (NK) cells, while IL-21 expands the humoral response and sustains the ongoing Th17-type immunity. IL-17A and tumor necrosis factor-(TNF) stimulate stromal cells to secrete matrix metalloproteinases (MMP) thus promoting extracellular matrix (ECM) degradation and epithelial damage. IL-15 enhances also the expression of activating NK receptors on intraepithelial CD8 + T cells and NK cells with the downstream effect of killing MICA-expressing enterocytes. The expression of MICA on epithelial cells is favored by the gliadin-derived peptide P31–49, while the nongluten α-amylase/trypsin inhibitor (ATI) stimulates Toll-like receptors (TLR) on innate immune cells (e.g., DC) thus contributing amplifying the local inflammatory response.

cells (Monteleone et al., 2004). High levels of both IFN-γ and IL-21 are also detectable in the peripheral blood of CD patients on a gluten-containing diet (Table 5.1).

5.3.2 Mechanisms involved in the control of the Th1-immune response

The transcriptional mechanisms and factors that control the commitment of Th1 cells in this condition are not fully understood. Biopsy samples taken from the duodenum of CD patients on gluten exposure express high levels of IL-18, a cytokine capable of expanding IFN- γ production (Salvati et al., 2002). Moreover, in CD mucosa, there is an abundance of IFN-α, a cytokine produced mainly by antigen-presenting cells and able to promote the development of T-cell responses to autoantigens (Nilsen et al., 1998). There are cases of CD developing in patients receiving IFN-α treatment for neoplastic diseases and studies in *ex vivo* organ cultures of fetal gut explants showed the ability of IFN-α to trigger a Th1 cell response, highlighting a possible role for IFN-α in promoting the CD-associated Th1 response to gluten (Bardella et al., 1999; Monteleone et al., 2001). IFN-α is over-produced during viral infections and circumstantial evidence indicates that viruses can act as a driver for loss of oral tolerance in CD (Bardella et al., 1999; McNab et al., 2015). Studies by Bouziat and collaborators have recently shown that the Type-I Lang (T1L) reovirus strain and murine norovirus (MNV), which both infect the gut, are capable of breaking intestinal homeostasis by triggering a transcriptional program in which interferon regulatory factor (IRF)-1 (Bouziat et al., 2017, 2018), a transcription factor overexpressed in the gut of patients with active CD (Salvati et al., 2003), is required for the induction of a Th1 immunity

TABLE 5.1 Expression of cytokines in the duodenum and peripheral blood of celiac disease patients. + and − indicate an increased or decreased expression as compared to the expression of the cytokine in normal subjects.

Cytokine	Duodenum	Peripheral Blood	References
IL-2	DD	+	Lahat et al. (1999), Manavalan et al. (2010), and Penedo-Pita & Peteiro-Cartelle (1991)
IFN-γ	+	+	Bodd et al. (2010) and Nilsen et al. (1998)
IL-21	+	+	Fina et al. (2008) and Iervasi et al. (2020)
IL-18	+	+	Lettesjö et al. (2005) and Salvati et al. (2002)
IFN-α	+	NE	Monteleone et al. (2001)
IL-15	+	NE	Maiuri et al. (2000) and Mention et al. (2003)
IL-17A	+	NE	Monteleone et al. (2010)
IL-4	DD	+	Lahat et al. (1999) and Manavalan et al. (2010)
IL-10	+	+	Lahat et al. (1999) and Manavalan et al. (2010)
TNF-α	DD	+	Manavalan et al. (2010)
IL-6	DD	+	Manavalan et al. (2010)
IL-1β	Not changed	+	Manavalan et al. (2010)

IL, Interleukin; NE = not evaluated. Discordant data (DD) about some cytokines are also indicated.

characterized by the differentiation of IL-12p40-producing dendritic cells, the production of gluten-specific IgG2c antibodies in the serum, TG2 activation in the proximal small intestine, and a delayed-type hypersensitivity reaction to gluten (Bouziat et al., 2017).

The inflamed gut of CD contains high levels of IL-15, the production of which is, at least in part dependent on the function of the mammalian target of Rapamycin (mTOR) (Sarra et al., 2013), a pathway that combines intra- and extracellular signals and acts as a central regulator for the metabolism, growth, and function of immune and nonimmune cells. IL-15 can contribute either directly or indirectly to the ongoing Th1 cell response in CD. For instance, stimulation of human gut lamina propria lymphocytes with IL-15 activates Akt thus enhancing IL-21 synthesis, and blockade of IL-15 with a neutralizing antibody in cultures of mucosal samples of active CD patients reduces IL-21. Blockade of IL-21 activity in cultures of biopsy samples of untreated CD patients reduces T-bet and IFN- γ secretion (Fina et al., 2008). Additionally, stimulation of treated CD biopsies with gluten peptide enhances IL-21 expression, and neutralization of IL-21 largely prevents Th1 cell expansion. Studies in animal models of CD confirmed the ability of IL-15 to increase the Th1 cell response. HLA-DQ8 transgenic mice overexpressing IL-15 in the lamina propria and mesenteric lymph nodes, but not in the intestinal epithelium, display an altered DC phenotype that is shifted from being tolerogenic to inducing the development of IFN-γ-producing Th1 cells. In this model, in conjunction with IL-15, RA activates the Janus kinase MAP kinase pathway in DCs thus promoting secretion of IL-12p70 and IL-23, the former being a master regulator of IFN-γ production and Th1 immunity (Depaolo et al., 2011). In the same model, in presence of IL-6, RA enhances IL-17A, a signature cytokine of Th17 cell responses, which is abundantly produced in active CD by gluten-nonreactive CD4 + T cells and CD8 + T cells. The expansion of Th17 cell response is partly dependent on IL-21, as blockade of IL-21 activity with a neutralizing IL-21 antibody reduces IL-17A expression in cultures of active CD and peptic-tryptic digest of gliadin-treated CD biopsy specimens (Monteleone et al., 2010). These observations suggest that Th1 cytokines, such as IL-21, contribute not only to expanding the ongoing gluten-dependent Th1 cell response but also to amplifying gluten-independent immunity (Fig. 5.2). Both IL-17A and IL-6 are also over-produced in patients with refractory CD (RCD) type I (Caruso et al., 2014) (Table 5.2).

5.3.3 The role of CD4 + T cells in the induction of intestinal damage

Pioneering studies assessing the pathogenic role of IFN-γ in the gut showed the ability of the cytokine to kill human colonic epithelial cell lines in ex-vivo cytotoxic assays (Deem et al., 1991). In addition, it was shown that treatment of

TABLE 5.2 Comparison of duodenal expression of inflammatory cytokines between active celiac disease and refractory celiac disease. + and NT indicate an increased or unchanged expression as compared to the expression of the cytokine in normal subjects.

Cytokine	Celiac disease	Refractory celiac disease	References
IFN-γ	+	NT	Bodd et al. (2010) and Caruso et al. (2014)
IL-21	+	NT	Bodd et al. (2010) and Caruso et al. (2014)
IL-13	NT	+	Gross et al. (2013)
TNF-α	DD	+	Caruso et al. (2014) and Przemioslo et al. (1994)
IL-6	DD	+	Caruso et al. (2014) and Przemioslo et al. (1994)
IL-17A	+	+	Caruso et al. (2014)
IL-15	+	+	Mention et al. (2003)

IL, Interleukin; Discordant data (DD) about some cytokines are also indicated.

intestinal tissue specimens with supernatants from gluten-stimulated T cell clones led to epithelial cell death, a phenomenon that could be counteracted by the addition of a neutralizing IFN-γ antibody to the cultures (Przemioslo et al., 1995). IFN- γ can upregulate the expression of the nonclassical MHC class Ib molecules HLA-E on epithelial cells, the ligands for the activating NK receptor CD94/NKG2C expressed by cytotoxic intraepithelial lymphocytes (Jabri et al., 2000). IFN-γ and IL-21 cooperate in the generation of antideamidated gluten peptide antibodies in a mouse model of CD, characterized by overproduction of IL-15 in the gut epithelium and lamina propria, expression of HLA-DQ8 molecules, and villous atrophy after ingestion of gluten (Abadie et al., 2020). Moreover, IFN-γ alone or in combination with IL-21 promotes the expansion of granzyme B-expressing intra-epithelial lymphocytes (Korneychuk et al., 2014). Altogether these observations underline the role of Th1 cytokines in the development of CD-associated epithelial injury. Nonetheless, accumulating evidence indicates that gluten-specific Th1 immune response is necessary but not sufficient to promote damage to the distressed intestinal epithelium. For example, no villous atrophy is documentable in patients with potential CD, who carry the HLA-DQ2 and HLA-DQ8 variants and exhibit an active adaptive immune response against gluten, but lack activated intraepithelial cytotoxic T lymphocytes (Jabri & Sollid, 2009, 2017). Moreover, HLA-DQ2 or HLA-DQ8 humanized mice with established restricted CD4 + Th1 cell immunity to gluten do not exhibit gut pathology following gluten exposure (De Kauwe et al., 2009). Similarly, the HLA-DQ8 transgenic mice overexpressing IL-15 in the lamina propria and mesenteric lymph nodes but not in the gut epithelium have no villous atrophy despite the abundance of IFN-γ-producing Th1 cells (Abadie et al., 2020).

The profile of mucosal T cell-derived effector cytokines differs between active CD and RCD as both IFN-γ and IL-21 are not over-expressed in RCD patients as compared with normal controls (Caruso et al., 2014), further supporting the concept that the induction of Th1 immunity in CD is strictly dependent on gluten exposure (Table 5.2).

5.3.4 Defective counter-regulatory mechanisms contribute to amplifying the celiac disease-associated immune response

Defects in counterregulatory mechanisms make a further contribution to the amplification of the dominant immune response seen in the active phases of CD and RCD. For example, in both intraepithelial and lamina propria T lymphocytes of patients with active CD there is reduced expression of AhR, the activation of which triggers suppressive signals in T cells (Dinallo et al., 2019). Consistently, treatment of intraepithelial and lamina propria T cells of patients with active CD with a synthetic ligand of AhR reduces the synthesis of inflammatory cytokines and expression of both granzyme B and perforin. The duodenal mucosa of RCD patients contains high Smad7, an intracellular inhibitor of TGF-β1 activity, and this is associated with defective TGF-β1 signaling (Sedda et al., 2017). Knockdown of Smad7 in RCD biopsy samples with a specific antisense oligonucleotide reduces inflammatory cytokine production, supporting the pathogenic role of the Smad7-mediated TGF-β1 activity impairment in RCD. Studies by Benahmed and collaborators documented an IL-15-induced impairment of TGF-β-Smad3-dependent transcriptional responses in T lymphocytes of active CD, which relies on the ability of IL-15 to activate c-jun-N-terminal kinase (Benahmed et al., 2007).

5.4 Innate immune pathways in celiac disease

5.4.1 Innate immune cell activation

Genome-wide association studies have shown that non-HLA genomic regions associated with CD harbor genes involved in innate immunity and stress pathways (Abadie et al., 2011). The fact that patients with potential CD, who exhibit a gluten-specific adaptive immunity but lack the innate epithelial stress response, do not exhibit mucosal damage, underlines the importance of innate immunity in determining villous atrophy. Gliadin peptides can activate monocytes, promote DC maturation, regulate cross-talk between NK cells and DCs, and interfere with proliferative signals for epithelial cells, thus triggering a complex multitude of signals that can eventually contribute to the villous atrophy (Barone et al., 2007; Cinova et al., 2007; Nikulina et al., 2004; Terrazzano et al., 2007). Elegant studies by Maiuri and collaborators showed that treatment of duodenal biopsy samples with α-gliadin peptide P31−43, which does not activate a specific T cell response, up-regulated production of IL-15, and expression of CD83, cyclo-oxygenase (COX)-2, and CD25 in CD3-negative cells, and increased enterocyte apoptosis (Maiuri et al., 2003) (Fig. 5.2). It was also shown that the P31−43 peptide impairs endocytic vesicular trafficking at the epithelial level, due to some similarity with hepatocyte growth factor regulated substrate (HRS) kinase, a protein involved in endocytic maturation (Barone et al., 2010). The P31−43 peptide-induced delay of the vesicular trafficking reduced degradation of receptor tyrosine kinases, and eventually sustained trans-presentation of IL-15/IL-15Rα at the epithelial level (Barone et al., 2011, 2014), contributing to explaining the proliferation of crypt enterocytes, mucosa remodeling, and innate immune activation seen in CD.

Intestinal epithelial cells in CD patients express high levels of stress-inducible MHC class I polypeptide-related molecules and nonclassical MHC class I molecule HLA-E, and in vitro studies showed that gliadin peptides (i.e., peptic-tryptic digest of gliadin or the P31−49 derived peptide) upregulated the expression of such molecules, and increased the synthesis of IL-15 (Hüe et al., 2004). Interestingly, the pro-inflammatory effects of P31−43 occur in a MyD88- and type 1 IFN-dependent manner, and are enhanced by poly I:C (Araya et al., 2016), a TLR3 ligand.

More recently, it has been demonstrated that members of the nongluten α-amylase/trypsin inhibitor (ATI) family contained in wheat and related cereals are powerful inducers of innate immune responses. ATIs stimulate macrophages, monocytes, and DCs via the TLR4-MD2-CD14 complex (Junker et al., 2012). Several processed and unprocessed food products contain high levels of TLR4-activating ATIs, which induce DC activation in mesenteric lymph nodes and the synthesis of inflammatory molecules in the gut (Zevallos et al., 2017). Moreover, ATIs mediate wheat-induced intestinal dysfunction and exacerbate gluten-driven inflammation in susceptible mice, while bacteria with ATI-degrading capacity, such as Lactobacillus, reduce these pro-inflammatory effects (Caminero et al., 2019). These findings suggest that additional wheat proteins other than gluten can contribute to triggering innate immune responses and the multiple pathogenetic events that eventually lead to CD.

5.4.2 Role of intraepithelial lymphocytes in intestinal villous atrophy

The epithelium of untreated CD patients is massively infiltrated by lymphocytes (Ferguson & Murray, 1971), which are supposed to play a crucial role in epithelial cell destruction and subsequent villous atrophy. In untreated CD, CD8 + T intraepithelial lymphocytes with an αβ T cell receptor express high levels of activating NK receptors and show cytotoxic activity. These cells have reduced levels of the inhibitory CD94/NKG2A receptor and increased expression of the activating CD94/NKG2C and NKG2D receptors (Jabri et al., 2000, 2002). Furthermore, they produce large amounts of IFNγ and granzyme B (Jabri et al., 2000; Meresse et al., 2006; Oberhuber et al., 1996). As mentioned above, the cytolytic activity against enterocytes is mediated by the interaction between the receptors and the stress-inducible, nonclassical MHC-like molecule HLA-E (Meresse et al., 2006). Therefore the pathogenic activity of such cell subset in intestinal epithelium is performed in a TCR-independent manner.

A small subset of CD8 + T intraepithelial lymphocytes with a γδ T cell receptor has also been described in CD. These cells represent 10%−15% of all intraepithelial lymphocytes in the normal intestine, while they expand in CD patients (Kutlu et al., 1993). Their role in CD pathogenesis is still not completely understood, with some studies suggesting a regulatory function of these cells (Bhagat et al., 2008). Mayassi and coworkers showed that the phenotype of CD8 + T γδ + intraepithelial lymphocytes can be permanently modified during chronic inflammation, supporting their contribution to CD-related tissue damage (Mayassi et al., 2019).

The mechanisms that trigger intraepithelial lymphocyte activation are not fully understood but locally-produced cytokines are supposed to contribute. IL-15 stimulates intraepithelial CD8 + T lymphocyte expansion and promotes the expression of granzyme B and NKG2D, thus enhancing the cytotoxicity against epithelial cells (Meresse et al., 2004).

Strikingly, IL-15 interferes with the TGF-β1- and Treg-mediated immunosuppression on CD8 + T lymphocytes further contributing to maintaining the cytotoxic properties of intraepithelial lymphocytes (Ahmed et al., 2009; Benahmed et al., 2007). There is also evidence that both IFN-γ and IL-21 can cooperate with IL-15 in promoting CD8 + T lymphocyte expansion/activation. This cooperative role emerges clearly in the above-mentioned mouse model characterized by overexpression of IL-15 in the gut epithelium and lamina propria, which expresses the predisposing HLA-DQ8 molecule and develops villous atrophy after ingestion of gluten (Abadie et al., 2020). Notably, in the model, epithelial damage is reduced by antibodies depleting either CD8 + or CD4 + T cells and by antibodies blocking IFNγ. In another mouse model using ovalbumin as a model dietary antigen, the cooperative interactions between ovalbumin-specific CD4 + T cells and IL-15 overexpressed in the gut were necessary to activate cytotoxic CD8 + T intraepithelial lymphocytes and cause epithelial injury following ovalbumin (Korneychuk et al., 2014).

IL-15 also synergizes with NKG2D signaling, acting as a co-stimulus for the TCR-independent killing of target cells expressing MICA molecules. MICA molecules are stress-induced NKG2D ligands highly expressed in active CD mucosa as a result of the action of gliadin or gliadin-derived peptide P31–49 through an IL-15-dependent mechanism (Hüe et al., 2004).

5.4.3 Innate lymphoid cells in celiac disease

Another class of innate immune cells is the so-called ILCs. ILCs have lymphoid morphology, lack antigen-specific receptors, need IL-7 for their survival, and play important roles in tissue homeostasis and remodeling, inflammation, and protective immune responses against pathogens (Sedda et al., 2014). ILCs constitute a relatively small population of lymphoid cells in the gastrointestinal tract and account for approximately 10% of total intraepithelial cells (Walker et al., 2013). Nonetheless, these cells play an important role in orchestrating intestinal inflammation by rapid cytokine production in response to environmental signals. There are at least six different ILC subsets: NK cells, lymphoid tissue-inducer (LTi) cells, ILC1s, ILC2s, ILC3s, and regulatory ILCs (ILCregs) (Spits & Cupedo, 2012). ILC1s express T-bet and secrete IFN-γ, ILC2s express GATA3 and produce Th2-related cytokines, and ILC3s express the transcription factor ROR-γt and synthesize predominantly IL-17, IL-22, and TNF-α. Intraepithelial ILC1s (ieILC1s) expressing NKp44 and CD103 reside in the gut epithelium, ILC2s are the predominant population in the lung, and ILC3s are mainly present in the small and large intestinal lamina propria (Vivier et al., 2018). We and others have previously shown that NKp44 and NKp46-double positive NK cells are decreased in the intestinal epithelium of active CD patients (Marafini et al., 2016; Uhde et al., 2020). Since NK cells are a first line of defense against viruses and down-regulation of NK cell cytotoxic receptors represents one of the strategies by which viruses escape the host's immune system, these findings support further the hypothesis that viral infections can contribute to the development of CD. In contrast to such a down-regulation of NKp44 in functionally exhausted NK cells, the expression of NKp44 on the surface of other immune cells, such as ILC1s, is increased in CD, and the presence of NKp44-expressing ILC1s correlates with the severity of villous atrophy and enterocyte damage (Marafini et al., 2015). The active CD mucosa contains also high numbers of ILCs producing TNF-α and IFN-γ likely representing ILC3s and possibly a proportion of helper ILC1s (Marafini et al., 2015). Gut ILCs express various TLRs (i.e. TLR2, TLR3, and TLR9) and are functionally capable of responding to poly I:C with an increased TNF-α and IFN-γ production. Interestingly, RAG1-deficient mice given intraperitoneally poly I:C exhibit an increased frequency of TNF-α-producing ILCs in the gut, and depletion of ILCs prevents the poly I:C-driven villous atrophy, thus delineating a further pathway of intestinal damage depending on the action of ILCs.

5.4.4 Defective clearing of apoptotic bodies as a mechanism of amplification of the local inflammation

In the duodenal mucosa of CD patients, there is increased apoptosis of immune cells and enterocytes. In physiologic conditions, the elimination of apoptotic bodies by scavengers (e.g., macrophages/DCs) is a crucial step for preventing secondary necrosis and promoting tolerogenic/antiinflammatory responses. Phagocytosis of apoptotic bodies is mediated by a variety of macrophage membrane-associated proteins (e.g., CD36, CD61) and thrombospondin-1 (TSP-1), an extracellular matrix glycoprotein that bridges apoptotic cells, CD36 and CD61, thus creating an active ternary complex (Canton et al., 2013; Plüddemann et al., 2007). In cases in which there is massive apoptosis, which overwhelms the available scavenging capacity, or when the scavenger mechanism is altered, secondary necrosis predominates thus leading to a deregulated inflammatory response and tissue injury. In the epithelial and lamina propria compartments of active CD patients, accumulation of apoptotic cells and bodies is associated with a defective expression of TSP-1, CD36, and CD61, and diminished capability of CD mucosal cells to phagocyte apoptotic cells (Cupi et al., 2014). In normal mucosal mononuclear cells, IL-15, IL-21, and IFN-γ are all able to inhibit the expression of TSP-1, CD36, and CD61.

5.5 Crosstalk between immune cells and stromal cells

Lamina propria stromal cells, such as fibroblasts and myofibroblasts, play a fundamental role in extracellular matrix building and are involved in intestinal stem cell homeostasis (Beumer & Clevers, 2021). In addition, stromal cells influence immune functions in the gut by interacting with resident immune cells. Indeed, similar to what happens in the bone marrow, intestinal stromal cells may support long-lived plasma cells in the lamina propria, and this effect seems to depend on IL-6 and APRIL, together with the support of molecules involved in cell contacts, such as CXCL12 (Mesin et al., 2011). Ciccocioppo and collaborators demonstrated that mesenchymal stromal cells (MSC) cocultured with gliadin-specific T-cell lines inhibited both the proliferative response of T-cell lines and IFN-γ production (Ciccocioppo et al., 2014). MSCs were also able to modulate different subsets of T cells, determining a reduction of CD4 + cells and expansion of Foxp3 + T cells. On the other hand, studies conducted by MacDonald and collaborators showed that activation of Th1 cells in ex vivo cultures of fetal gut explants stimulates gut myofibroblasts to produce the synthesis of huge amounts of matrix metalloproteinases, a class of proteases that can cleave components of the extracellular matrix thereby causing epithelial damage (Fig. 5.2). Altogether, these findings raise the possibility that active cross-talk between immune cells and stromal cells can contribute to the CD-associated villous atrophy (Salmela et al., 2002, 2004).

5.6 Conclusions and future perspectives

The findings illustrated in this chapter indicate that the induction of a gluten-specific CD4 + T cell immune response requires molecular interactions between gluten, HLA-DQ2/8, and TG2 and that such a response is necessary but not sufficient to induce epithelial damage. The activation of cytotoxic CD8 + intraepithelial lymphocytes is a crucial step in the sequence of events culminating in epithelial damage, and accumulating evidence suggests that IL-15 and cytokines produced by gluten-reactive Th1 cells can cooperate in promoting cytotoxicity against the gut epithelium. If this phenomenon is regulated by additional molecules produced within the inflamed gut remains unknown. Similarly, it is not yet known the exact intracellular pathway involved in the expansion of cytotoxic cells, even though indirect evidence suggests the involvement of the Janus kinase (Jak)/Stat signaling pathway, as both IL-15 and Th1 cytokines activate this pathway in T cells. In this context, it is also noteworthy that malignant cells seen in patients with lymphoma complicating RCD type 2 are marked by somatic mutations in the Jak1/Stat3, which increase their selective growth, as well as CD8 + T intraepithelial lymphocytes in CD show phenotypic similarities (e.g., expression of NKG2D receptors and genes for inflammatory/cytotoxic granzymes, perforin, IFN-γ) with CD8 + T cells seen in patients and mice with autoimmune pathology due to constitutive Stat3 gain-of-function mutations (Masle-Farquhar et al., 2022).

We do not yet know the exact way by which gluten peptides cross the intestinal epithelium barrier. At present, the only evidence is that gluten peptides use the transcellular passage through the endocytic pathway (Barone & Zimmer, 2016). Similarly, it is unclear the site where TG2 encounters gluten and engages the immune system. For its function, TG2 needs Ca2 + concentrations found in the extracellular milieu. Therefore it is plausible that the interaction between gluten peptides and TG2 occurs in the extracellular matrix beneath the gut epithelium. Recent observations underline the possibility that TG2 is released by enterocyte shedding in the intestinal lumen where it could interact with gluten peptides, and the TG2-gluten complexes may eventually be endocytosed by M cells in the follicles and Peyer's patches and activate local immune responses (Iversen et al., 2020).

The exact contribution of B cells in the induction of pathogenic immune responses remains unclear. Data from in vitro experiments indicate that B cells expressing a TG2-specific BCR and HLA-DQ2.5 efficiently present deamidated gluten peptides to T cells, raising the possibility that autoreactive anti-TG2 B cells contribute to the activation of gluten-specific CD4 + T cells and, perhaps, to the epithelial damage (Moens & Tangye, 2014). This hypothesis is supported by the demonstration that villous atrophy induced via a gluten, TG2-dependent manner and dependent on Th1 cell response, in HLA-DQ8 transgenic mice overexpressing IL-15 in the intestinal epithelium and lamina propria, is preventable by depletion of B cells (Lejeune et al., 2021).

Intestinal microbiota alterations have been documented in CD, but it is not fully understood whether such changes can influence the gluten-dependent and/or independent immune response. Activated gliadin reactive HLA-DQ2.5-restricted T cells derived from CD patients cross-react with ubiquitous bacterial proteases, a class of enzymes that cleave peptide bonds (Petersen et al., 2020). Duodenal biopsy samples of patients with active CD have increased proteolytic activity against gluten substrates that correlates with increased Proteobacteria abundance, including Pseudomonas. The Pseudomonas aeruginosa-derived elastase induces gluten-independent, PAR-2-mediated upregulation of inflammatory pathways in C57BL/6 mice without causing villus destruction, while in mice expressing HLA-DQ8 gene, the same protease synergizes with gluten to induce more severe inflammation and moderate villus destruction (Caminero et al., 2019). In active CD, there is decreased duodenal expression of elafin, a human serine protease inhibitor, and supplementation of elafin by a recombinant Lactococcus lactis strain prevents gliadin-induced immunopathology in the NOD/DQ8 mouse model of gluten sensitivity (Galipeau et al., 2014; McCarville et al., 2017). Consistently, in the same mouse model, a serine protease inhibitor produced by Bifidobacterium longum NCC2705 attenuates gluten-induced immunopathology. Regional epithelial expression of the transcription factor GATA4 controls bacterial colonization and immune responses in the proximal small intestine by regulating retinol metabolism and luminal IgA. Notably, in mice lacking jejunal GATA4 expression, the commensal segmented filamentous bacteria promotes pathogenic inflammation and epithelial barrier dysfunction upon Citrobacter rodentium infection (Earley et al., 2023). In CD, GATA4 expression is reduced and such a defect is associated with metabolic alterations, mucosal Actinobacillus, and increased Th17 immunity (Earley et al., 2023). Overall, these findings demonstrate that bacteria can use several tools to impact host immune responses and predispose individuals to the development of CD as well as amplify gluten-dependent detrimental responses.

Although several models of altered immune response to gluten have been developed and used to assess the contribution of specific cell subsets/cytokines in the pathogenesis of gut damage (Table 5.3), no model recapitulates all the genetic, immune, and clinical features of CD. Further studies are needed to generate CD-like animal models to better dissect the pathogenic steps of the disease and to assess the efficacy of novel therapeutic compounds.

TABLE 5.3 Schematic description of the main mouse models of altered immune response to dietary antigens.

Strain and genetic background	Induction	Mucosal changes	Serology	Other phenotypes	References
Germ-free AVN Wistar Rats, WT	Gliadin IG	Villous blunting, crypt hyperplasia, expansion of CD8 + TCRab + IELs	Antigliadin IgG/IgA	–	Stepankova et al. (1996)
AVN Wistar Rats, WT	Gliadin IG + artificial milk + IFN-γ IP	Subtotal villous atrophy, expansion of CD4 + T cells in LP	NA	Diarrhea, vomiting, body loss	Stepankova et al. (2003)
C57BL/6 to RAG-/- transfer	Transfer of memory CD4 + $CD45RB^{low}CD25$ + T cells from immunized C57BL/6 to RAG mice à Gluten-rich diet	Villous atrophy, crypt hyperplasia	NA	–	Freitag et al. (2009, 2014)
NOD.DQ8. Ab	Pepsin–trypsin digests of gliadin + cholera toxin	Barrier dysfunction, altered villus/crypt ratios	Antigliadin IgG, anti-TG2 IgA	Insulitis	Galipeau et al. (2011)
$T3^{b-}$hIL-15Tge	–	Altered villus/crypt ratio, expansion of CD8 + TCRab T cell	Anti-TG2 IgA	–	Ohta et al. (2002) and Yokoyama et al. (2011)
OTII-hIL-15Tge	OVA- containing diet	Villous atrophy, crypt hyperplasia, IELs expansion	NA	–	Korneychuk et al. (2014)
DQ8-Dd-villin-IL-15tg	Gluten-containing diet	Villous atrophy, LP plasmacytosis, expansion of cytotoxic IELs	Antigluten IgG/IgA, anti-DGP IgG	–	Abadie et al. (2020)

WT, wild-type; IG, intra-gastric; IEL, intraepithelial lymphocytes; LP, lamina propria; OVA, ovalbumin; NA, not available. NOD.DQ8.Ab: humanized DQ8 mice crossed with Nonobese Diabetic mice; T3b-hIL-15Tge: IL-15 transgenic mice with IL-15 transgene of human origin placed under the control of the enterocyte-specific promoter T3b; OTII-hIL-15Tge: OTII mice, which possess CD4 + T cells expressing a transgenic TCR specific for a peptide of ovalbumin, crossed with hIL-15Tge mice that overexpress IL-15 in the small intestinal epithelium; DQ8-Dd-villin-IL-15tg: humanized DQ8 mice that overexpress IL-15 in the gut epithelium and lamina propria.

References

Abadie, V., Kim, S. M., Lejeune, T., Palanski, B. A., Ernest, J. D., Tastet, O., Voisine, J., Discepolo, V., Marietta, E. V., Hawash, M. B. F., Ciszewski, C., Bouziat, R., Panigrahi, K., Horwath, I., Zurenski, M. A., Lawrence, I., Dumaine, A., Yotova, V., Grenier, J. C., ... Jabri, B. (2020). IL-15, gluten and HLA-DQ8 drive tissue destruction in coeliac disease. *Nature*, *578*(7796), 600–604. Available from https://doi.org/10.1038/s41586-020-2003-8.

Abadie, V., Sollid, L. M., Barreiro, L. B., & Jabri, B. (2011). Integration of genetic and immunological insights into a model of celiac disease pathogenesis. *Annual Review of Immunology*, *29*, 493–525. Available from https://doi.org/10.1146/annurev-immunol-040210-092915.

Ahmed, M. B., Hmida, N. B., Moes, N., Buyse, S., Abdeladhim, M., Louzir, H., & Cerf-Bensussan, N. (2009). IL-15 renders conventional lymphocytes resistant to suppressive functions of regulatory T cells through activation of the phosphatidylinositol 3-kinase pathway. *Journal of Immunology*, *182*(11), 6763–6770. Available from https://doi.org/10.4049/jimmunol.0801792.

Akira, S., & Hemmi, H. (2003). Recognition of pathogen-associated molecular patterns by TLR family. *Immunology Letters*, *85*(2), 85–95. Available from https://doi.org/10.1016/S0165-2478(02)00228-6.

Allaire, J. M., Crowley, S. M., Law, H. T., Chang, S. Y., Ko, H. J., & Vallance, B. A. (2018). The intestinal epithelium: Central coordinator of mucosal immunity. *Trends in Immunology*, *39*(9), 677–696. Available from https://doi.org/10.1016/j.it.2018.04.002.

Araya, R. E., Castro, M. F. G., Carasi, P., McCarville, J. L., Jury, J., Mowat, A. M., Verdu, E. F., & Chirdo, F. G. (2016). Mechanisms of innate immune activation by gluten peptide p31–43 in mice. *American Journal of Physiology - Gastrointestinal and Liver Physiology*, *311*(1), G40–G49. Available from https://doi.org/10.1152/ajpgi.00435.2015.

Arentz-Hansen, H., Körner, R., Molberg, Ø., Quarsten, H., Vader, W., Kooy, Y. M. C., Lundin, K. E. A., Koning, F., Roepstorff, P., Sollid, L. M., & McAdam, S. N. (2000). The intestinal T cell response to α-gliadin in adult celiac disease is focused on a single deamidated glutamine targeted by tissue transglutaminase. *Journal of Experimental Medicine*, *191*(4), 603–612. Available from https://doi.org/10.1084/jem.191.4.603.

Bain, C. C., & Mowat, A. M. (2011). Intestinal macrophages - specialised adaptation to a unique environment. *European Journal of Immunology*, *41*(9), 2494–2498. Available from https://doi.org/10.1002/eji.201141714.

Bain, C. C., & Mowat, A. M. (2014). Macrophages in intestinal homeostasis and inflammation. *Immunological Reviews*, *260*(1), 102–117. Available from https://doi.org/10.1111/imr.12192.

Bardella, M. T., Marino, R., & Meroni, P. L. (1999). Celiac disease during interferon treatment [5]. *Annals of Internal Medicine*, *131*(2), 157–158. Available from https://doi.org/10.7326/0003-4819-131-2-199907200-00024.

Barone, M. V., Gimigliano, A., Castoria, G., Paolella, G., Maurano, F., Paparo, F., Maglio, M., Mineo, A., Miele, E., Nanayakkara, M., Troncone, R., & Auricchio, S. (2007). Growth factor-like activity of gliadin, an alimentary protein: Implications for coeliac disease. *Gut*, *56*(4), 480–488. Available from https://doi.org/10.1136/gut.2005.086637.

Barone, M. V., Nanayakkara, M., Paolella, G., Maglio, M., Vitale, V., Troiano, R., Ribecco, M. T. S., Lania, G., Zanzi, D., Santagata, S., Auricchio, R., Troncone, R., Auricchio, S., & Bassham, D. (2010). Gliadin peptide P31–43 localises to endocytic vesicles and interferes with their maturation. *PLoS One*, *5*(8), e12246. Available from https://doi.org/10.1371/journal.pone.0012246.

Barone, M. V., Troncone, R., & Auricchio, S. (2014). Gliadin peptides as triggers of the proliferative and stress/innate immune response of the celiac small intestinal mucosa. *International Journal of Molecular Sciences*, *15*(11), 20518–20537. Available from https://doi.org/10.3390/ijms151120518.

Barone, M. V., Zanzi, D., Maglio, M., Nanayakkara, M., Santagata, S., Lania, G., Miele, E., Ribecco, M. T. S., Maurano, F., Auricchio, R., Gianfrani, C., Ferrini, S., Troncone, R., Auricchio, S., & Holowka, D. (2011). Gliadin-mediated proliferation and innate immune activation in celiac disease are due to alterations in vesicular trafficking. *PLoS One*, *6*(2), e17039. Available from https://doi.org/10.1371/journal.pone.0017039.

Barone, M. V., & Zimmer, K. P. (2016). Endocytosis and transcytosis of gliadin peptides. *Molecular and Cellular Pediatrics*, *3*(1). Available from https://doi.org/10.1186/s40348-015-0029-z.

Benahmed, M., Meresse, B., Arnulf, B., Barbe, U., Mention, J. J., Verkarre, V., Allez, M., Cellier, C., Hermine, O., & Cerf-Bensussan, N. (2007). Inhibition of TGF-β signaling by IL-15: A new role for IL-15 in the loss of immune homeostasis in celiac disease. *Gastroenterology*, *132*(3), 994–1008. Available from https://doi.org/10.1053/j.gastro.2006.12.025.

Berg, R. D. (1995). Bacterial translocation from the gastrointestinal tract. *Trends in Microbiology*, *3*(4), 149–154. Available from https://doi.org/10.1016/S0966-842X(00)88906-4.

Beumer, J., & Clevers, H. (2021). Cell fate specification and differentiation in the adult mammalian intestine. *Nature Reviews. Molecular Cell Biology*, *22*(1), 39–53. Available from https://doi.org/10.1038/s41580-020-0278-0.

Bhagat, G., Naiyer, A. J., Shah, J. G., Harper, J., Jabri, B., Wang, T. C., Green, P. H. R., & Manavalan, J. S. (2008). Small intestinal CD8 + TCRγδ + NKG2A + intraepithelial lymphocytes have attributes of regulatory cells in patients with celiac disease. *Journal of Clinical Investigation*, *118*(1), 281–293. Available from https://doi.org/10.1172/JCI30989.

Bodd, M., Ráki, M., Tollefsen, S., Fallang, L. E., Bergseng, E., Lundin, K. E. A., & Sollid, L. M. (2010). HLA-DQ2-restricted gluten-reactive T cells produce IL-21 but not IL-17 or IL-22. *Mucosal Immunology*, *3*(6), 594–601. Available from https://doi.org/10.1038/mi.2010.36.

Boirivant, M., Pica, R., DeMaria, R., Testi, R., Pallone, F., & Strober, W. (1996). Stimulated human lamina propria T cells manifest enhanced Fas-mediated apoptosis. *Journal of Clinical Investigation*, *98*(11), 2616–2622. Available from https://doi.org/10.1172/JCI119082.

Bouziat, R., Biering, S. B., Kouame, E., Sangani, K. A., Kang, S., Ernest, J. D., Varma, M., Brown, J. J., Urbanek, K., Dermody, T. S., Ng, A., Hinterleitner, R., Hwang, S., & Jabri, B. (2018). Murine norovirus infection induces TH1 inflammatory responses to dietary antigens. *Cell Host and Microbe*, *24*(5), 677–688. Available from https://doi.org/10.1016/j.chom.2018.10.004, e5.

Bouziat, R., Hinterleitner, R., Brown, J. J., Stencel-Baerenwald, J. E., Ikizler, M., Mayassi, T., Meisel, M., Kim, S. M., Discepolo, V., Pruijssers, A. J., Ernest, J. D., Iskarpatyoti, J. A., Costes, L. M. M., Lawrence, I., Palanski, B. A., Varma, M., Zurenski, M. A., Khomandiak, S., McAllister, N., ... Jabri, B. (2017). Reovirus infection triggers inflammatory responses to dietary antigens and development of celiac disease. *Science (New York, N.Y.)*, *356*(6333), 44–50. Available from https://doi.org/10.1126/science.aah5298.

Bu, P., Keshavarzian, A., Stone, D. D., Liu, J., Le, P. T., Fisher, S., & Qiao, L. (2001). Apoptosis: One of the mechanisms that maintains unresponsiveness of the intestinal mucosal immune system. *Journal of Immunology*, *166*(10), 6399–6403. Available from https://doi.org/10.4049/jimmunol.166.10.6399.

Caminero, A., McCarville, J. L., Galipeau, H. J., Deraison, C., Bernier, S. P., Constante, M., Rolland, C., Meisel, M., Murray, J. A., Yu, X. B., Alaedini, A., Coombes, B. K., Bercik, P., Southward, C. M., Ruf, W., Jabri, B., Chirdo, F. G., Casqueiro, J., Surette, M. G., ... Verdu, E. F. (2019). Duodenal bacterial proteolytic activity determines sensitivity to dietary antigen through protease-activated receptor-2. *Nature Communications*, *10*(1). Available from https://doi.org/10.1038/s41467-019-09037-9.

Caminero, A., McCarville, J. L., Zevallos, V. F., Pigrau, M., Yu, X. B., Jury, J., Galipeau, H. J., Clarizio, A. V., Casqueiro, J., Murray, J. A., Collins, S. M., Alaedini, A., Bercik, P., Schuppan, D., & Verdu, E. F. (2019). Lactobacilli degrade wheat amylase trypsin inhibitors to reduce intestinal dysfunction induced by immunogenic wheat proteins. *Gastroenterology*, *156*(8), 2266–2280. Available from https://doi.org/10.1053/j.gastro.2019.02.028.

Canton, J., Neculai, D., & Grinstein, S. (2013). Scavenger receptors in homeostasis and immunity. *Nature Reviews. Immunology*, *13*(9), 621–634. Available from https://doi.org/10.1038/nri3515.

Caruso, R., Marafini, I., Sedda, S., Del Vecchio Blanco, G., Giuffrida, P., Macdonald, T. T., Corazza, G. R., Pallone, F., Di Sabatino, A., & Monteleone, G. (2014). Analysis of the cytokine profile in the duodenal mucosa of refractory coeliac disease patients. *Clinical Science*, *126*(6), 451–458. Available from https://doi.org/10.1042/CS20130478.

Ciccocioppo, R., Camarca, A., Cangemi, G. C., Radano, G., Vitale, S., Betti, E., Ferrari, D., Visai, L., Strada, E., Badulli, C., Locatelli, F., Klersy, C., Gianfrani, C., & Corazza, G. R. (2014). Tolerogenic effect of mesenchymal stromal cells on gliadin-specific Tlymphocytes in celiac disease. *Cytotherapy*, *16*(8), 1080–1091. Available from https://doi.org/10.1016/j.jcyt.2014.03.002.

Cinova, J., Palová-Jelínková, L., Smythies, L. E., Černá, M., Pecharová, B., Dvořák, M., Fruhauf, P., Tlaskalová-Hogenová, H., Smith, P. D., & Tučková, L. (2007). Gliadin peptides activate blood monocytes from patients with celiac disease. *Journal of Clinical Immunology*, *27*(2), 201–209. Available from https://doi.org/10.1007/s10875-006-9061-z.

Coombes, J. L., Siddiqui, K. R. R., Arancibia-Cárcamo, C. V., Hall, J., Sun, C. M., Belkaid, Y., & Powrie, F. (2007). A functionally specialized population of mucosal CD103 + DCs induces Foxp3 + regulatory T cells via a TGF-β -and retinoic acid-dependent mechanism. *Journal of Experimental Medicine*, *204*(8), 1757–1764. Available from https://doi.org/10.1084/jem.20070590.

Cupi, M. L., Sarra, M., De Nitto, D., Franzè, E., Marafini, I., Monteleone, I., Del Vecchio Blanco, G., Paoluzi, O. A., Di Fusco, D., Gentileschi, P., Ortenzi, A., Colantoni, A., Pallone, F., Monteleone, G., & Khoury, J. E. (2014). Defective expression of scavenger receptors in celiac disease mucosa. *PLoS One*, *9*(6), e100980. Available from https://doi.org/10.1371/journal.pone.0100980.

De Kauwe, A. L., Chen, Z., Anderson, R. P., Keech, C. L., Price, J. D., Wijburg, O., Jackson, D. C., Ladhams, J., Allison, J., & McCluskey, J. (2009). Resistance to celiac disease in humanized HLA-DR3-DQ2-transgenic mice expressing specific anti-gliadin CD4 + T cells. *Journal of Immunology*, *182*(12), 7440–7450. Available from https://doi.org/10.4049/jimmunol.0900233.

Deem, R. L., Shanahan, F., & Targan, S. R. (1991). Triggered human mucosal T cells release tumour necrosis factor-alpha and interferon-gamma which kill human colonic epithelial cells. *Clinical and Experimental Immunology*, *83*(1), 79–84.

Depaolo, R. W., Abadie, V., Tang, F., Fehlner-Peach, H., Hall, J. A., Wang, W., Marietta, E. V., Kasarda, D. D., Waldmann, T. A., Murray, J. A., Semrad, C., Kupfer, S. S., Belkaid, Y., Guandalini, S., & Jabri, B. (2011). Co-adjuvant effects of retinoic acid and IL-15 induce inflammatory immunity to dietary antigens. *Nature*, *471*(7337), 220–225. Available from https://doi.org/10.1038/nature09849.

Dieterich, W., Ehnis, T., Bauer, M., Donner, P., Volta, U., Riecken, E. O., & Schuppan, D. (1997). Identification of tissue transglutaminase as the autoantigen of celiac disease. *Nature Medicine*, *3*(7), 797–801. Available from https://doi.org/10.1038/nm0797-797.

Dieterich, W., Laag, E., Schopper, H., Volta, U., Ferguson, A., Gillett, H., Riecken, E. O., & Schuppan, D. (1998). Autoantibodies to tissue transglutaminase as predictors of celiac disease. *Gastroenterology*, *115*(6), 1317–1321. Available from https://doi.org/10.1016/S0016-5085(98)70007-1.

Dinallo, V., Marafini, I., Di Fusco, D., Di Grazia, A., Laudisi, F., Dwairi, R., Paoluzi, O. A., Monteleone, G., & Monteleone, I. (2019). Protective effects of aryl hydrocarbon receptor signaling in celiac disease mucosa and in poly I:C-induced small intestinal atrophy mouse model. *Frontiers in Immunology*, *10*. Available from https://doi.org/10.3389/fimmu.2019.00091.

Dubois, P. C. A., Trynka, G., Franke, L., Hunt, K. A., Romanos, J., Curtotti, A., Zhernakova, A., Heap, G. A. R., Ádány, R., Aromaa, A., Bardella, M. T., Van Den Berg, L. H., Bockett, N. A., De La Concha, E. G., Dema, B., Fehrmann, R. S. N., Fernández-Arquero, M., Fiatal, S., Grandone, E., ... Van Heel, D. A. (2010). Multiple common variants for celiac disease influencing immune gene expression. *Nature Genetics*, *42*(4), 295–302. Available from https://doi.org/10.1038/ng.543.

Earley, Z. M., Lisicka, W., Sifakis, J. J., Aguirre-Gamboa, R., Kowalczyk, A., Barlow, J. T., Shaw, D. G., Discepolo, V., Tan, I. L., Gona, S., Ernest, J. D., Matzinger, P., Barreiro, L. B., Morgun, A., Bendelac, A., Ismagilov, R. F., Shulzhenko, N., Riesenfeld, S. J., & Jabri, B. (2023). GATA4 controls regionalization of tissue immunity and commensal-driven immunopathology. *Immunity*, *56*(1), 43–57. Available from https://doi.org/10.1016/j.immuni.2022.12.009, e10.

Edelblum, K. L., Shen, L., Weber, C. R., Marchiando, A. M., Clay, B. S., Wang, Y., Prinz, I., Malissen, B., Sperling, A. I., & Turner, J. R. (2012). Dynamic migration of γδ intraepithelial lymphocytes requires occludin. *Proceedings of the National Academy of Sciences of the United States of America*, *109*(18), 7097–7102. Available from https://doi.org/10.1073/pnas.1112519109.

Fallang, L. E., Bergseng, E., Hotta, K., Berg-Larsen, A., Kim, C. Y., & Sollid, L. M. (2009). Differences in the risk of celiac disease associated with HLA-DQ2.5 or HLA-DQ2.2 are related to sustained gluten antigen presentation. *Nature Immunology*, *10*(10), 1096–1101. Available from https://doi.org/10.1038/ni.1780.

Fan, H., Wang, A., Wang, Y., Sun, Y., Han, J., Chen, W., Wang, S., Wu, Y., & Lu, Y. (2019). Innate lymphoid cells: Regulators of gut barrier function and immune homeostasis. *Journal of Immunology Research*, *2019*. Available from https://doi.org/10.1155/2019/2525984.

Farache, J., Koren, I., Milo, I., Gurevich, I., Kim, K. W., Zigmond, E., Furtado, G. C., Lira, S. A., & Shakhar, G. (2013). Luminal bacteria recruit CD103 + dendritic cells into the intestinal epithelium to sample bacterial antigens for presentation. *Immunity*, *38*(3), 581–595. Available from https://doi.org/10.1016/j.immuni.2013.01.009.

Ferguson, A., & Murray, D. (1971). Quantitation of intraepithelial lymphocytes in human jejunum. *Gut*, *12*(12), 988–994. Available from https://doi.org/10.1136/gut.12.12.988.

Fina, D., Sarra, M., Caruso, R., Del Vecchio Blanco, G., Pallone, F., MacDonald, T. T., & Monteleone, G. (2008). Interleukin 21 contributes to the mucosal T helper cell type 1 response in coeliac disease. *Gut*, *57*(7), 887–892. Available from https://doi.org/10.1136/gut.2007.129882.

Freitag, T. L., Loponen, J., Messing, M., Zevallos, V., Andersson, L. C., Sontag-Strohm, T., Saavalainen, P., Schuppan, D., Salovaara, H., & Meri, S. (2014). Testing safety of germinated rye sourdough in a celiac disease model based on the adoptive transfer of prolamin-primed memory T cells into lymphopenic mice. *American Journal of Physiology - Gastrointestinal and Liver Physiology*, *306*(6), G526–G534. Available from https://doi.org/10.1152/ajpgi.00136.2013.

Freitag, T. L., Rietdijk, S., Junker, Y., Popov, Y., Bhan, A. K., Kelly, C. P., Terhorst, C., & Schuppan, D. (2009). Gliadin-primed CD4 + CD45RBlowCD25- T cells drive gluten-dependent small intestinal damage after adoptive transfer into lymphopenic mice. *Gut*, *58*(12), 1597–1605. Available from https://doi.org/10.1136/gut.2009.186361.

Galipeau, H. J., Rulli, N. E., Jury, J., Huang, X., Araya, R., Murray, J. A., David, C. S., Chirdo, F. G., McCoy, K. D., & Verdu, E. F. (2011). Sensitization to gliadin induces moderate enteropathy and insulitis in nonobese diabetic-DQ8 mice. *Journal of Immunology*, *187*(8), 4338–4346. Available from https://doi.org/10.4049/jimmunol.1100854.

Galipeau, H. J., Wiepjes, M., Motta, J. P., Schulz, J. D., Jury, J., Natividad, J. M., Pinto-Sanchez, I., Sinclair, D., Rousset, P., Martin-Rosique, R., Bermudez-Humaran, L., Leroux, J. C., Murray, J. A., Smecuol, E., Bai, J. C., Vergnolle, N., Langella, P., & Verdu, E. F. (2014). Novel role of the serine protease inhibitor elafin in gluten-related disorders. *American Journal of Gastroenterology*, *109*(5), 748–756. Available from https://doi.org/10.1038/ajg.2014.48.

Gross, S., Van Wanrooij, R. L., Nijeboer, P., Gelderman, K. A., Cillessen, S. A. G. M., Meijer, G. A., Mulder, C. J. J., Bouma, G., Von Blomberg, B. M. E., & Bontkes, H. J. (2013). Differential IL-13 production by small intestinal leukocytes in active coeliac disease versus refractory coeliac disease. *Mediators of Inflammation*, *2013*. Available from https://doi.org/10.1155/2013/939047.

Hüe, S., Mention, J. J., Monteiro, R. C., Zhang, S. L., Cellier, C., Schmitz, J., Verkarre, V., Fodil, N., Bahram, S., Cerf-Bensussan, N., & Caillat-Zucman, S. (2004). A direct role for NKG2D/MICA interaction in villous atrophy during celiac disease. *Immunity*, *21*(3), 367–377. Available from https://doi.org/10.1016/j.immuni.2004.06.018.

Hunt, K. A., Zhernakova, A., Turner, G., Heap, G. A. R., Franke, L., Bruinenberg, M., Romanos, J., Dinesen, L. C., Ryan, A. W., Panesar, D., Gwilliam, R., Takeuchi, F., McLaren, W. M., Holmes, G. K. T., Howdle, P. D., Walters, J. R. F., Sanders, D. S., Playford, R. J., Trynka, G., ... Van Heel, D. A. (2008). Newly identified genetic risk variants for celiac disease related to the immune response. *Nature Genetics*, *40*(4), 395–402. Available from https://doi.org/10.1038/ng.102.

Ientile, R., Caccamo, D., & Griffin, M. (2007). Tissue transglutaminase and the stress response. *Amino Acids*, *33*(2), 385–394. Available from https://doi.org/10.1007/s00726-007-0517-0.

Iervasi, E., Auricchio, R., Strangio, A., Greco, L., & Saverino, D. (2020). Serum IL-21 levels from celiac disease patients correlates with anti-tTG IgA autoantibodies and mucosal damage. *Autoimmunity*, *53*(4), 225–230. Available from https://doi.org/10.1080/08916934.2020.1736047.

Iversen, R., Amundsen, S. F., Kleppa, L., du Pré, M. F., Stamnaes, J., & Sollid, L. M. (2020). Evidence That Pathogenic transglutaminase 2 in celiac disease derives from enterocytes. *Gastroenterology*, *159*(2), 788–790. Available from https://doi.org/10.1053/j.gastro.2020.04.018.

Jabri, B., & Sollid, L. M. (2009). Tissue-mediated control of immunopathology in coeliac disease. *Nature Reviews. Immunology*, *9*(12), 858–870. Available from https://doi.org/10.1038/nri2670.

Jabri, B., & Sollid, L. M. (2017). T cells in celiac disease. *Journal of Immunology*, *198*(8), 3005–3014. Available from https://doi.org/10.4049/jimmunol.1601693.

Jabri, B., De Serre, N. P. M., Cellier, C., Evans, K., Gache, C., Carvalho, C., Mougenot, J. F., Allez, M., Jian, R., Desreumaux, P., Colombel, J. F., Matuchansky, C., Cugnenc, H., LopezBotet, M., Vivier, E., Moretta, A., Roberts, A. I., Ebert, E. C., GuyGrand, D., ... CerfBensussan, N. (2000). Selective expansion of intraepithelial lymphocytes expressing the HLA-E- specific natural killer receptor CD94 in celiac disease. *Gastroenterology*, *118*(5), 867–879. Available from https://doi.org/10.1016/S0016-5085(00)70173-9.

Jabri, B., Selby, J. M., Negulescu, H., Lee, L., Roberts, A. I., Beavis, A., Lopez-Botet, M., Ebert, E. C., & Winchester, R. J. (2002). TCR specificity dictates CD94/NKG2A expression by human CTL. *Immunity*, *17*(4), 487–499. Available from https://doi.org/10.1016/S1074-7613(02)00427-2.

Jiang, Q., Su, H., Knudsen, G., Helms, W., & Su, L. (2006). Delayed functional maturation of natural regulatory T cells in the medulla of postnatal thymus: Role of TSLP. *BMC Immunology*, *7*. Available from https://doi.org/10.1186/1471-2172-7-6.

Junker, Y., Zeissig, S., Kim, S. J., Barisani, D., Wieser, H., Leffler, D. A., Zevallos, V., Libermann, T. A., Dillon, S., Freitag, T. L., Kelly, C. P., & Schuppan, D. (2012). Wheat amylase trypsin inhibitors drive intestinal inflammation via activation of toll-like receptor 4. *Journal of Experimental Medicine*, *209*(13), 2395–2408. Available from https://doi.org/10.1084/jem.20102660.

Karell, K., Louka, A. S., Moodie, S. J., Ascher, H., Clot, F., Greco, L., Ciclitira, P. J., Sollid, L. M., & Partanen, J. (2003). HLA types in celiac disease patients not carrying the DQA1 *05-DQB1 *02 (DQ2) heterodimer: Results from the European genetics cluster on celiac disease. *Human Immunology*, *64*(4), 469–477. Available from https://doi.org/10.1016/S0198-8859(03)00027-2.

Koning, F. (2012). Celiac disease: Quantity matters. *Seminars in Immunopathology*, *34*(4), 541–549. Available from https://doi.org/10.1007/s00281-012-0321-0.

Korneychuk, N., Ramiro-Puig, E., Ettersperger, J., Schulthess, J., Montcuquet, N., Kiyono, H., Meresse, B., & Cerf-Bensussan, N. (2014). Interleukin 15 and CD4 + T cells cooperate to promote small intestinal enteropathy in response to dietary antigen. *Gastroenterology*, *146*(4), 1017–1027. Available from https://doi.org/10.1053/j.gastro.2013.12.023.

Kutlu, T., Brousse, N., Rambaud, C., Le Deist, F., Schmitz, J., & Cerf-Bensussan, N. (1993). Numbers of T cell receptor (TCR) alpha beta + but not of TcR gamma delta + intraepithelial lymphocytes correlate with the grade of villous atrophy in coeliac patients on a long term normal diet. *Gut*, *34*(2), 208–214. Available from https://doi.org/10.1136/gut.34.2.208.

Lahat, N., Shapiro, S., Karban, A., Gerstein, R., Kinarty, A., & Lerner, A. (1999). Cytokine profile in coeliac disease. *Scandinavian Journal of Immunology*, *49*(4), 441–447. Available from https://doi.org/10.1046/j.1365-3083.1999.00523.x.

Lejeune, T., Meyer, C., & Abadie, V. (2021). B lymphocytes contribute to celiac disease pathogenesis. *Gastroenterology*, *160*(7), 2608–2610. Available from https://doi.org/10.1053/j.gastro.2021.02.063, e4.

Lettesjö, H., Hansson, T., Bergqvist, A., Grönlund, J., & Dannaeus, A. (2005). Enhanced interleukin-18 levels in the peripheral blood of children with coeliac disease. *Clinical and Experimental Immunology*, *139*(1), 138–143. Available from https://doi.org/10.1111/j.1365-2249.2005.02661.x.

Ludvigsson, J. F., Leffler, D. A., Bai, J. C., Biagi, F., Fasano, A., Green, P. H. R., Hadjivassiliou, M., Kaukinen, K., Kelly, C. P., Leonard, J. N., Lundin, K. E. A., Murray, J. A., Sanders, D. S., Walker, M. M., Zingone, F., & Ciacci, C. (2013). The Oslo definitions for coeliac disease and related terms. *Gut*, *62*(1), 43–52. Available from https://doi.org/10.1136/gutjnl-2011-301346.

MacDonald, T. T., & Monteleone, G. (2005). Immunity, inflammation, and allergy in the gut. *Science (New York, N.Y.)*, *307*(5717), 1920–1925. Available from https://doi.org/10.1126/science.1106442.

Maiuri, L., Ciacci, C., Auricchio, S., Brown, V., Quaratino, S., & Londei, M. (2000). Interleukin 15 mediates epithelial changes in celiac disease. *Gastroenterology*, *119*(4), 996–1006. Available from https://doi.org/10.1053/gast.2000.18149.

Maiuri, L., Ciacci, C., Ricciardelli, I., Vacca, L., Raia, V., Auricchio, S., Picard, J., Osman, M., Quaratino, S., & Londei, M. (2003). Association between innate response to gliadin and activation of pathogenic T cells in coeliac disease. *Lancet*, *362*(9377), 30–37. Available from https://doi.org/10.1016/S0140-6736(03)13803-2.

Manavalan, J. S., Hernandez, L., Shah, J. G., Konikkara, J., Naiyer, A. J., Lee, A. R., Ciaccio, E., Minaya, M. T., Green, P. H. R., & Bhagat, G. (2010). Serum cytokine elevations in celiac disease: Association with disease presentation. *Human Immunology*, *71*(1), 50–57. Available from https://doi.org/10.1016/j.humimm.2009.09.351.

Marafini, I., Monteleone, I., Di Fusco, D., Sedda, S., Cupi, M. L., Fina, D., Paoluzi, A. O., Pallone, F., Monteleone, G., & Reeves, R. K. (2016). Celiac Disease-Related Inflammation Is Marked by Reduction of Nkp44/Nkp46-Double Positive Natural Killer Cells. *PLoS One*, *11*(5), e0155103. Available from https://doi.org/10.1371/journal.pone.0155103.

Marafini, I., Monteleone, I., Fusco, D., Cupi, D., Paoluzi, M., Colantoni, O., Ortenzi, A., Izzo, A., Vita, R., Luca, D., Sica., Pallone, G., & Monteleone, F. (2015). TNF-alpha producing innate lymphoid cells (ILCs) are increased in active celiac disease and contribute to promote intestinal atrophy in mice. *PLoS One*, *10*(5).

Marsh, M. N. (1989). Immunology and immunopathology of the intestines: The immunopathology of the small intestinal reaction in gluten-sensitivity. *Immunological Investigations*, *18*(1–4), 509–531. Available from https://doi.org/10.3109/08820138909112260.

Masle-Farquhar, E., Jackson, K. J. L., Peters, T. J., Al-Eryani, G., Singh, M., Payne, K. J., Rao, G., Avery, D. T., Apps, G., Kingham, J., Jara, C. J., Skvortsova, K., Swarbrick, A., Ma, C. S., Suan, D., Uzel, G., Chua, I., Leiding, J. W., Heiskanen, K., … Goodnow, C. C. (2022). STAT3 gain-of-function mutations connect leukemia with autoimmune disease by pathological NKG2Dhi CD8 + T cell dysregulation and accumulation. *Immunity*, *55*(12), 2386–2404. Available from https://doi.org/10.1016/j.immuni.2022.11.001, e8.

Mayassi, T., Ladell, K., Gudjonson, H., McLaren, J. E., Shaw, D. G., Tran, M. T., Rokicka, J. J., Lawrence, I., Grenier, J. C., van Unen, V., Ciszewski, C., Dimaano, M., Sayegh, H. E., Kumar, V., Wijmenga, C., Green, P. H. R., Gokhale, R., Jericho, H., Semrad, C. E., … Jabri, B. (2019). Chronic inflammation permanently reshapes tissue-resident immunity in celiac disease. *Cell*, *176*(5), 967–981. Available from https://doi.org/10.1016/j.cell.2018.12.039, e19.

Mazzarella, G., MacDonald, T. T., Salvati, V. M., Mulligan, P., Pasquale, L., Stefanile, R., Lionetti, P., Auricchio, S., Pallone, F., Troncone, R., & Monteleone, G. (2003). Constitutive activation of the signal transducer and activator of transcription pathway in celiac disease lesions. *American Journal of Pathology*, *162*(6), 1845–1855. Available from https://doi.org/10.1016/S0002-9440(10)64319-2.

McCarville, J. L., Dong, J., Caminero, A., Bermudez-Brito, M., Jury, J., Murray, J. A., Duboux, S., Steinmann, M., Delley, M., Tangyu, M., Langella, P., Mercenier, A., Bergonzelli, G., & Verdu, E. F. (2017). A commensal Bifidobacterium longum strain prevents gluten-related immunopathology in mice through expression of a serine protease inhibitor. *Applied and Environmental Microbiology*, *83*(19). Available from https://doi.org/10.1128/AEM.01323-17.

McNab, F., Mayer-Barber, K., Sher, A., Wack, A., & O'Garra, A. (2015). Type I interferons in infectious disease. *Nature Reviews. Immunology*, *15* (2), 87–103. Available from https://doi.org/10.1038/nri3787.

Mention, J. J., Ahmed, M. B., Bègue, B., Barbe, U., Verkarre, V., Asnafi, V., Colombel, J. F., Cugnenc, P. H., Ruemmele, F. M., McIntyre, E., Brousse, N., Cellier, C., & Cerf-Bensussan, N. (2003). Interleukin 15: A key to disrupted intraepithelial lymphocyte homeostasis and lymphomagenesis in celiac disease. *Gastroenterology*, *125*(3), 730–745. Available from https://doi.org/10.1016/S0016-5085(03)01047-3.

Meresse, B., Chen, Z., Ciszewski, C., Tretiakova, M., Bhagat, G., Krausz, T. N., Raulet, D. H., Lanier, L. L., Groh, V., Spies, T., Ebert, E. C., Green, P. H., & Jabri, B. (2004). Coordinated induction by IL15 of a TCR-independent NKG2D signaling pathway converts CTL into lymphokine-activated killer cells in celiac disease. *Immunity*, *21*(3), 357–366. Available from https://doi.org/10.1016/j.immuni.2004.06.020.

Meresse, B., Curran, S. A., Ciszewski, C., Orbelyan, G., Setty, M., Bhagat, G., Lee, L., Tretiakova, M., Semrad, C., Kistner, E., Winchester, R. J., Braud, V., Lanier, L. L., Geraghty, D. E., Green, P. H., Guandalini, S., & Jabri, B. (2006). Reprogramming of CTLs into natural killer-like cells in celiac disease. *Journal of Experimental Medicine*, *203*(5), 1343–1355. Available from https://doi.org/10.1084/jem.20060028.

Mesin, L., Di Niro, R., Thompson, K. M., Lundin, K. E. A., & Sollid, L. M. (2011). Long-lived plasma cells from human small intestine biopsies secrete immunoglobulins for many weeks in vitro. *Journal of Immunology*, *187*(6), 2867–2874. Available from https://doi.org/10.4049/jimmunol.1003181.

Moens, L., & Tangye, S. G. (2014). Cytokine-mediated regulation of plasma cell generation: IL-21 takes center stage. *Frontiers in Immunology*, *5*. Available from https://doi.org/10.3389/fimmu.2014.00065.

Molberg, Ø., Mcadam, S. N., Körner, R., Quarsten, H., Kristiansen, C., Madsen, L., Fugger, L., Scott, H., Norën, O., Roepstorff, P., Lundin, K. E. A., Sjöström, H., & Sollid, L. M. (1998). Tissue transglutaminase selectively modifies gliadin peptides that are recognized by gut-derived T cells in celiac disease. *Nature Medicine*, *4*(6), 713–717. Available from https://doi.org/10.1038/nm0698-713.

Monteleone, G., Pender, S. L. F., Alstead, E., Hauer, A. C., Lionetti, P., & MacDonald, T. T. (2001). Role of interferon α in promoting T helper cell type 1 responses in the small intestine in coeliac disease. *Gut*, *48*(3), 425–429. Available from https://doi.org/10.1136/gut.48.3.425.

Monteleone, G., Pender, S. L. F., Wathen, N. C., & MacDonald, T. T. (2001). Interferon-α drives T cell-mediated immunopathology in the intestine. *European Journal of Immunology*, *31*(8), 2247–2255. Available from https://doi.org/10.1002/1521-4141. (200108)31:8 < 2247::AID-IMMU2247 > 3.0.CO;2-4.

Monteleone, I., Monteleone, G., Del Vecchio Blanco, G., Vavassori, P., Cucchiara, S., MacDonald, T. T., & Pallone, F. (2004). Regulation of the T helper cell type 1 transcription factor T-bet in coeliac disease mucosa. *Gut*, *53*(8), 1090–1095. Available from https://doi.org/10.1136/gut.2003.030551.

Monteleone, I., Sarra, M., Blanco, G. D. V., Paoluzi, O. A., Franzè, E., Fina, D., Fabrizi, A., MacDonald, T. T., Pallone, F., & Monteleone, G. (2010). Characterization of IL-17A-producing cells in celiac disease mucosa. *Journal of Immunology*, *184*(4), 2211–2218. Available from https://doi.org/10.4049/jimmunol.0901919.

Montufar-Solis, D., Garza, T., & Klein, J. R. (2007). T-cell activation in the intestinal mucosa. *Immunological Reviews*, *215*(1), 189–201. Available from https://doi.org/10.1111/j.1600-065X.2006.00471.x.

Mowat, A. M. I. (2003). Anatomical basis of tolerance and immunity to intestinal antigens. *Nature Reviews. Immunology*, *3*(4), 331–341. Available from https://doi.org/10.1038/nri1057.

Mowat, A. M. I. (2018). To respond or not to respond - A personal perspective of intestinal tolerance. *Nature Reviews. Immunology*, *18*(6), 405–415. Available from https://doi.org/10.1038/s41577-018-0002-x.

Mucida, D., Park, Y., Kim, G., Turovskaya, O., Scott, I., Kronenberg, M., & Cheroutre, H. (2007). Reciprocal TH17 and regulatory T cell differentiation mediated by retinoic acid. *Science (New York, N.Y.)*, *317*(5835), 256–260. Available from https://doi.org/10.1126/science.1145697.

Neutra, M. R., Mantis, N. J., & Kraehenbuhl, J. P. (2001). Collaboration of epithelial cells with organized mucosal lymphoid tissues. *Nature Immunology*, *2*(11), 1004–1009. Available from https://doi.org/10.1038/ni1101-1004.

Nikulina, M., Habich, C., Flohé, S. B., Scott, F. W., & Kolb, H. (2004). Wheat gluten causes dendritic cell maturation and chemokine secretion. *Journal of Immunology*, *173*(3), 1925–1933. Available from https://doi.org/10.4049/jimmunol.173.3.1925.

Nilsen, E. M., Jahnsen, F. L., Lundin, K. E. A., Johansen, F. E., Fausa, O., Sollid, L. M., Jahnsen, J., Scott, H., & Brandtzaeg, P. (1998). Gluten induces an intestinal cytokine response strongly dominated by interferon gamma in patients with celiac disease. *Gastroenterology*, *115*(3), 551–563. Available from https://doi.org/10.1016/S0016-5085(98)70134-9.

Nilsen, E. M., Lundin, K. E., Krajci, P., Scott, H., Sollid, L. M., & Brandtzaeg, P. (1995). Gluten specific, HLA-DQ restricted T cells from coeliac mucosa produce cytokines with Th1 or Th0 profile dominated by interferon gamma. *Gut*, *37*(6), 766–776. Available from https://doi.org/10.1136/gut.37.6.766.

Oberhuber, G., Vogelsang, H., Stolte, M., Muthenthaler, S., Kummer, A. J., & Radaszkiewicz, T. (1996). Evidence that intestinal intraepithelial lymphocytes are activated cytotoxic T cells in celiac disease but not in giardiasis. *American Journal of Pathology*, *148*(5), 1351–1357.

Ohta, N., Hiroi, T., Kweon, M. N., Kinoshita, N., Jang, M. H., Mashimo, T., Miyazaki, J. I., & Kiyono, H. (2002). IL-15-dependent activation-induced cell death-resistant Th1 type CD8αβ + NK1.1 + T cells for the development of small intestinal inflammation. *Journal of Immunology*, *169*(1), 460–468. Available from https://doi.org/10.4049/jimmunol.169.1.460.

Penedo-Pita, M., & Peteiro-Cartelle, J. (1991). Increased serum levels of interleukin-2 and soluble interleukin-2 receptor in celiac disease. *Journal of Pediatric Gastroenterology and Nutrition*, *12*(1), 56–60. Available from https://doi.org/10.1097/00005176-199101000-00012.

Petersen, J., Ciacchi, L., Tran, M. T., Loh, K. L., Kooy-Winkelaar, Y., Croft, N. P., Hardy, M. Y., Chen, Z., McCluskey, J., Anderson, R. P., Purcell, A. W., Tye-Din, J. A., Koning, F., Reid, H. H., & Rossjohn, J. (2020). T cell receptor cross-reactivity between gliadin and bacterial peptides in celiac disease. *Nature Structural and Molecular Biology*, *27*(1), 49–61. Available from https://doi.org/10.1038/s41594-019-0353-4.

Ploski, R., Ek, J., Thorsby, E., & Sollid, L. M. (1993). On the HLA-DQ(α1*0501, β1*0201)-associated susceptibility in celiac disease: A possible gene dosage effect of DQB1*0201. *Tissue Antigens*, *41*(4), 173–177. Available from https://doi.org/10.1111/j.1399-0039.1993.tb01998.x.

Plüddemann, A., Neyen, C., & Gordon, S. (2007). Macrophage scavenger receptors and host-derived ligands. *Methods (San Diego, Calif.)*, *43*(3), 207–217. Available from https://doi.org/10.1016/j.ymeth.2007.06.004.

Przemioslo, R. T., Kontakou, M., Nobili, V., & Ciclitira, P. J. (1994). Raised pro-inflammatory cytokines interleukin 6 and tumour necrosis factor α in coeliac disease mucosa detected by immunohistochemistry. *Gut*, *35*(10), 1398–1403. Available from https://doi.org/10.1136/gut.35.10.1398.

Przemioslo, R. T., Lundin, K., Sollid, L. M., Nelufer, J., & Ciclitira, P. J. (1995). Histological changes in small bowel mucosa induced by gliadin sensitive T lymphocytes can be blocked by anti-interferon γ antibody. *Gut, 36*(6), 874–879. Available from https://doi.org/10.1136/gut.36.6.874.

Rescigno, M., Urbano, M., Valzasina, B., Francolini, M., Rotta, G., Bonasio, R., Granucci, F., Kraehenbuhl, J. P., & Ricciardi-Castagnoli, P. (2001). Dendritic cells express tight junction proteins and penetrate gut epithelial monolayers to sample bacteria. *Nature Immunology, 2*(4), 361–367. Available from https://doi.org/10.1038/86373.

Salmela, M. T., MacDonald, T. T., Black, D., Irvine, B., Zhuma, T., Saarialho-Kere, U., & Pender, S. L. F. (2002). Upregulation of matrix metalloproteinases in a model of T cell mediated tissue injury in the gut: Analysis by gene array and in situ hybridisation. *Gut, 51*(4), 540–547. Available from https://doi.org/10.1136/gut.51.4.540.

Salmela, M. T., Pender, S. L. F., Karjalainen-Lindsberg, M. L., Puolakkainen, P., MacDonald, T. T., & Saarialho-Kere, U. (2004). Collagenase-1 (MMP-1), matrilysin-1 (MMP-7), and stromelysin-2 (MMP-10) are expressed by migrating enterocytes during intestinal wound healing. *Scandinavian Journal of Gastroenterology, 39*(11), 1095–1104. Available from https://doi.org/10.1080/00365520410003470.

Salvati, V. M., MacDonald, T. T., Bajaj-Elliott, M., Borrelli, M., Staiano, A., Auricchio, S., Troncone, R., & Monteleone, G. (2002). Interleukin 18 and associated markers of T helper cell type 1 activity in coeliac disease. *Gut, 50*(2), 186–190. Available from https://doi.org/10.1136/gut.50.2.186.

Salvati, V. M., MacDonald, T. T., Del Vecchio Blanco, G., Mazzarella, G., Monteleone, I., Vavassori, P., Auricchio, S., Pallone, F., Troncone, R., & Monteleone, G. (2003). Enhanced expression of interferon regulatory factor-1 in the mucosa of children with celiac disease. *Pediatric Research, 54*(3), 312–318. Available from https://doi.org/10.1203/01.PDR.0000079184.70237.9C.

Sarra, M., Cupi, M. L., Monteleone, I., Franzè, E., Ronchetti, G., Di Sabatino, A., Gentileschi, P., Franceschilli, L., Sileri, P., Sica, G., Del Vecchio Blanco, G., Cretella, M., Paoluzi, O. A., Corazza, G. R., Pallone, F., & Monteleone, G. (2013). IL-15 positively regulates IL-21 production in celiac disease mucosa. *Mucosal Immunology, 6*(2), 244–255. Available from https://doi.org/10.1038/mi.2012.65.

Satpathy, A. T., Briseño, C. G., Lee, J. S., Ng, D., Manieri, N. A., Kc, W., Wu, X., Thomas, S. R., Lee, W. L., Turkoz, M., McDonald, K. G., Meredith, M. M., Song, C., Guidos, C. J., Newberry, R. D., Ouyang, W., Murphy, T. L., Stappenbeck, T. S., Gommerman, J. L., ... Murphy, K. M. (2013). Notch2-dependent classical dendritic cells orchestrate intestinal immunity to attaching-and-effacing bacterial pathogens. *Nature Immunology, 14*(9), 937–948. Available from https://doi.org/10.1038/ni.2679.

Sedda, S., De Simone, V., Marafini, I., Bevivino, G., Izzo, R., Paoluzi, O. A., Colantoni, A., Ortenzi, A., Giuffrida, P., Corazza, G. R., Vanoli, A., Di Sabatino, A., Pallone, F., & Monteleone, G. (2017). High Smad7 sustains inflammatory cytokine response in refractory coeliac disease. *Immunology, 150*(3), 356–363. Available from https://doi.org/10.1111/imm.12690.

Sedda, S., Marafini, I., Figliuzzi, M. M., Pallone, F., & Monteleone, G. (2014). An overview of the role of innate lymphoid cells in gut infections and inflammation. *Mediators of Inflammation, 2014*. Available from https://doi.org/10.1155/2014/235460.

Shan, L., Molberg, Ø., Parrot, I., Hausch, F., Filiz, F., Gray, G. M., Sollid, L. M., & Khosla, C. (2002). Structural basis for gluten intolerance in Celiac Sprue. *Science (New York, N.Y.), 297*(5590), 2275–2279. Available from https://doi.org/10.1126/science.1074129.

Siegel, M., Strnad, P., Watts, R. E., Choi, K., Jabri, B., Omary, M. B., Khosla, C., & El-Shemy, H. A. (2008). Extracellular transglutaminase 2 Is catalytically inactive, but is transiently activated upon tissue injury. *PLoS One, 3*(3), e1861. Available from https://doi.org/10.1371/journal.pone.0001861.

Sollid, L. M., Markussen, G., Ek, J., Gjerde, H., Vartdal, F., & Thorsby, E. (1989). Evidence for a primary association of celiac disease to a particular HLA-DQ α/β heterodimer. *Journal of Experimental Medicine, 169*(1), 345–350. Available from https://doi.org/10.1084/jem.169.1.345.

Spits, H., & Cupedo, T. (2012). Innate lymphoid cells: Emerging insights in development, lineage relationships, and function. *Annual Review of Immunology, 30*, 647–675. Available from https://doi.org/10.1146/annurev-immunol-020711-075053.

Stepankova, R., Kofroňová, O., Tučková, L., Kozáková, H., Cebra, J. J., & Tlaskalová-Hogenová, H. (2003). Experimentally induced gluten enteropathy and protective effect of epidermal growth factor in artificially fed neonatal rats. *Journal of Pediatric Gastroenterology and Nutrition, 36*(1), 96–104. Available from https://doi.org/10.1097/00005176-200301000-00018.

Stepankova, R., Tlaskalová-Hogenová, H., Šinkora, J., Jodl, J., & Frič, P. (1996). Changes in jejunal mucosa after long-term feeding of germfree rats with gluten. *Scandinavian Journal of Gastroenterology, 31*(6), 551–557. Available from https://doi.org/10.3109/00365529609009127.

Stolfi, C., Caruso, R., Franzè, E., Sarra, M., De Nitto, D., Rizzo, A., Pallone, F., & Monteleone, G. (2011). Interleukin-25 fails to activate STAT6 and induce alternatively activated macrophages. *Immunology, 132*(1), 66–77. Available from https://doi.org/10.1111/j.1365-2567.2010.03340.x.

Tamachi, T., Maezawa, Y., Ikeda, K., Iwamoto, I., & Nakajima, H. (2006). Interleukin 25 in allergic airway inflammation. *International Archives of Allergy and Immunology, 140*(1), 59–62. Available from https://doi.org/10.1159/000092713.

Terrazzano, G., Sica, M., Gianfrani, C., Mazzarella, G., Maurano, F., De Giulio, B., De Saint-Mezard, S., Zanzi, D., Maiuri, L., Londei, M., Jabri, B., Troncone, R., Auricchio, S., Zappacosta, S., & Carbone, E. (2007). Gliadin regulates the NK-dendritic cell cross-talk by HLA-E surface stabilization. *Journal of Immunology, 179*(1), 372–381. Available from https://doi.org/10.4049/jimmunol.179.1.372.

Ting, Y. T., Dahal-Koirala, S., Kim, H. S. K., Qiao, S. W., Neumann, R. S., Lundin, K. E. A., Petersen, J., Reid, H. H., Sollid, L. M., & Rossjohn, J. (2020). A molecular basis for the T cell response in HLA-DQ2.2 mediated celiac disease. *Proceedings of the National Academy of Sciences of the United States of America, 117*(6), 3063–3073. Available from https://doi.org/10.1073/pnas.1914308117.

Tollefsen, S., Arentz-Hansen, H., Fleckenstein, B., Molberg, Ø., Ráki, M., Kwok, W. W., Jung, G., Lundin, K. E. A., & Sollid, L. M. (2006). HLA-DQ2 and -DQ8 signatures of gluten T cell epitopes in celiac disease. *Journal of Clinical Investigation, 116*(8), 2226–2236. Available from https://doi.org/10.1172/JCI27620.

Uhde, M., Yu, X., Bunin, A., Brauner, C., Lewis, S. K., Lebwohl, B., Krishnareddy, S., Alaedini, A., Reizis, B., Ghosh, S., Green, P. H., & Bhagat, G. (2020). Phenotypic shift of small intestinal intra-epithelial type 1 innate lymphoid cells in celiac disease is associated with enhanced cytotoxic potential. *Clinical and Experimental Immunology, 200*(2), 163–175. Available from https://doi.org/10.1111/cei.13414.

Vader, W., Stepniak, D., Kooy, Y., Mearin, L., Thompson, A., Van Rood, J. J., Spaenij, L., & Koning, F. (2003). The HLA-DQ2 gene dose effect in celiac disease is directly related to the magnitude and breadth of gluten-specific T cell responses. *Proceedings of the National Academy of Sciences of the United States of America*, *100*(21), 12390–12395. Available from https://doi.org/10.1073/pnas.2135229100.

Vivier, E., Artis, D., Colonna, M., Diefenbach, A., Di Santo, J. P., Eberl, G., Koyasu, S., Locksley, R. M., McKenzie, A. N. J., Mebius, R. E., Powrie, F., & Spits, H. (2018). Innate Lymphoid Cells: 10 Years On. *Cell*, *174*(5), 1054–1066. Available from https://doi.org/10.1016/j.cell.2018.07.017.

Vogel, C. F. A., Wu, D., Goth, S. R., Baek, J., Lollies, A., Domhardt, R., Grindel, A., & Pessah, I. N. (2013). Aryl hydrocarbon receptor signaling regulates NF-κB RelB activation during dendritic-cell differentiation. *Immunology and Cell Biology*, *91*(9), 568–575. Available from https://doi.org/10.1038/icb.2013.43.

Walker, J. A., Barlow, J. L., & McKenzie, A. N. J. (2013). Innate lymphoid cells-how did we miss them? *Nature Reviews. Immunology*, *13*(2), 75–87. Available from https://doi.org/10.1038/nri3349.

Watanabe, N., Wang, Y. H., Lee, H. K., Ito, T., Wang, Y. H., Cao, W., & Liu, Y. J. (2005). Hassall's corpuscles instruct dendritic cells to induce CD4 + CD25 + regulatory T cells in human thymus. *Nature*, *436*(7054), 1181–1185. Available from https://doi.org/10.1038/nature03886.

Witherden, D. A., & Havran, W. L. (2013). Cross-talk between intraepithelial γδ T cells and epithelial cells. *Journal of Leukocyte Biology*, *94*(1), 69–76. Available from https://doi.org/10.1189/jlb.0213101.

Yokoyama, S., Takada, K., Hirasawa, M., Perera, L. P., & Hiroi, T. (2011). Transgenic mice that overexpress human IL-15 in enterocytes recapitulate both B and T cell-mediated pathologic manifestations of celiac disease. *Journal of Clinical Immunology*, *31*(6), 1038–1044. Available from https://doi.org/10.1007/s10875-011-9586-7.

Zevallos, V. F., Raker, V., Tenzer, S., Jimenez-Calvente, C., Ashfaq-Khan, M., Rüssel, N., Pickert, G., Schild, H., Steinbrink, K., & Schuppan, D. (2017). Nutritional Wheat Amylase-Trypsin Inhibitors Promote Intestinal Inflammation via Activation of Myeloid Cells. *Gastroenterology*, *152*(5), 1100–1113. Available from https://doi.org/10.1053/j.gastro.2016.12.006, e12.

Ziegler, S. F., Roan, F., Bell, B. D., Stoklasek, T. A., Kitajima, M., & Han, H. (2013). *The biology of thymic stromal lymphopoietin (TSLP), Advances in pharmacology* (Vol. 66, pp. 129–155). Academic Press Inc. Available from https://doi.org/10.1016/B978-0-12-404717-4.00004-4.

Chapter 6

Celiac disease in pediatric patients

Stefano Guandalini[1] and Naire Sansotta[2]
[1]*Department of Pediatrics, Section of Gastroenterology, Hepatology and Nutrition, University of Chicago Medicine, Chicago, IL, United States,*
[2]*Pediatric Hepatology Gastroenterology and Transplantation Hospital Papa Giovanni XXIII, Bergamo, Italy*

6.1 Introduction

As illustrated in the previous 5 chapters, celiac disease (CD) is a complex autoimmune disorder elicited by the ingestion of gluten in genetically predisposed individuals expressing specific human leukocyte antigen (HLA)-class II haplotypes (called DQ2 and DQ8) and causing an inflammatory disorder of the small intestine. CD is characterized, both in children as well as in adults, by a combination of elevated titers of celiac-specific autoantibodies, an inflammatory enteropathy with various degrees of severity, and gastrointestinal (GI) as well as extra intestinal complaints whose severity varies from asymptomatic to life-threatening. Furthermore, CD is well known to be associated in children with some congenital disorders such as Down, Turner, and Williams-Beuren syndrome; as well as with other autoimmune conditions such as type 1 diabetes and autoimmune thyroiditis.

Over the past few years, considerable advances have been reached in several areas of this condition that are of particular relevance in pediatric age, such as the identification of factors favoring or hampering its onset in genetically predisposed children (see Chapter 4), and the improvement of its diagnostic process.

6.2 Epidemiology

As described in detail in Chapter 1, CD has rates of incidence and prevalence that differ in various parts of the world, and that have been overall increasing over the past decades. This pattern is seen also in children, as previously described in the United States (Catassi et al., 2010) and confirmed in Italy across 25 years in addressing school populations similar for age and geographic location, where a screening detected a rise from 0.88% to 1.58% (Gatti et al., 2020). A recent multicenter mass screening study on a larger school population (Lionetti et al., 2023) found an even higher prevalence (1.65%). Similarly, in North America as well the rates of CD-specific autoimmunity appear to be on the rise (King et al., 2023). Additionally, recently even more striking geographical variations in prevalence have been described in the United States, underlying the importance of differential environmental, genetic, and epigenetic influences (Stahl et al., 2023). Thus, while the availability of reliable serological screening tests (and an increased awareness) are in part responsible for higher detection rates, these data—and other similarly obtained—provide compelling evidence of a true increase. Many environmental factors have been pointed out as plausible causes of such word-wide increase in genetically susceptible individuals. These include, among others, modalities of birth (Tanpowpong et al., 2023), viral and bacterial infections (especially respiratory ones) suffered in the first months of life, modalities of infant and toddler feeding, composition of the intestinal microbiota. Some of the preceding chapters deal in detail with this topic, at present to be considered, however, still a moving target.

Of note, despite improved and easily available serological testing and increasing diagnostic rates, it appears that the majority of children with CD remain undiagnosed: a recent investigation in the United Kingdom showed that about 90% of children remain undiagnosed (Whitburn et al., 2021), based on the expected average of about 1%. In addition, the same study disclosed that while the incidence of diagnosed CD has increased over time in all social classes, the rise has been significantly less evident in lower socioeconomic groups, something that pediatricians and primary care providers should be alerted of.

Pediatric and Adult Celiac Disease. DOI: https://doi.org/10.1016/B978-0-443-13359-6.00010-8

6.3 Clinical manifestations

Celiac disease in children is definitely not the same as CD in adults, just as children are not miniature adults. In fact, the impact on growth and development that CD threatens to cause in children is obviously unique. The presentation of CD in children has substantially changed over time (Pedretti et al., 2021; Almallouhi et al., 2017; Krauthammer et al., 2020; Popp & Maki, 2019; Tapsas et al., 2016), going from a malabsorptive disorder causing mostly obvious gastrointestinal symptoms and malnutrition leading often to failure to thrive, to a more subtle condition causing a variety of extra-intestinal manifestations affecting essentially all organs and systems and occurring either with or without a concomitant gastrointestinal involvement. Clinical presentations are, at least in part, dependent on age of onset. In fact, there is evidence that the occurrence of gastrointestinal manifestations is higher in younger children (Krauthammer et al., 2021) and then tends to decline as age at diagnosis progresses, when extra-intestinal manifestations such as short stature, fatigue, headache occur more frequently and in the majority of cases without concomitant obvious gastrointestinal symptoms (Jericho et al., 2017). Also the levels of anti-tissue transglutaminase (TTG), the autoantibodies specific for celiac disease, at time of diagnosis follow an age-related pattern, being more often higher than 10 times the upper limit of normal in younger children than in older ones (Krauthammer et al., 2021). Of interest, while a recent large multicenter study conducted in Spain confirmed that in children levels of TTG-IgA do correlate with the severity of intestinal damage (Donat et al., 2022), the latter does not correlate with symptoms (Brar et al., 2007) and furthermore they do not predict the long-term outcome of CD (Kröger et al., 2020).

Gastrointestinal and extra-intestinal symptoms and signs presented by celiac children (separately listed in Table 6.1) can also coexist in the same patient, in variable combinations. Furthermore, celiac disease in children, like in adults, can be associated with a number of autoimmune conditions and with certain genetic disorders.

Conditions in *Italics* occur more frequently in younger children. Abbreviation: Attention Deficit Hyperactivity Disorder, ADHD

However, CD may be entirely asymptomatic in children too (Krauthammer et al., 2021), as we have come to know from screening unselected populations or first-degree relatives of known patients (Gould et al., 2023). Incidentally, even if completely asymptomatic, children detected by screening can present intestinal mucosal lesions just as severe as those diagnosed for other reasons (Gould et al., 2023).

The rather high prevalence of asymptomatic patients explains why diagnostic rates around the globe remain low and why diagnosis can be delayed: in fact, physicians are more likely to miss the asymptomatic individuals as well as those whose symptoms do not include overt gastrointestinal involvement (Paez et al., 2017).

TABLE 6.1 Main presenting signs and symptoms of celiac disease in children and teenagers.

GI manifestations	Systemic and extra-GI manifestations
Abdominal distention	Fatigue
Abdominal pain	Weight loss/*failure to thrive*
Diarrhea	*Malnutrition*
Vomiting	Anemia
Constipation	Vitamin D deficiency
Anorexia	Delayed puberty
Intussusception	Short stature
"Celiac crisis"	Dermatitis herpetiformis
Aphthous ulcers/ geographical tongue	Reduced bone mineral density
Dental enamel hypoplasia	Arthritis, arthralgia
	Abnormal liver enzymes
	Neurologic
	Headaches, migraine
	Peripheral neuropathy
	Epilepsy (with or without occipital calcifications)
	Idiopathic seizures
	Psychiatric
	Irritability
	Anxiety/depression
	ADHD symptoms

6.3.1 Gastrointestinal manifestations

Abdominal pain and bloating (Khatib et al., 2016), as well as chronic or intermittent diarrhea, characterized by bulky, foul-smelling, greasy stool, are typical and very common symptoms in children with CD. Counterintuitively, long-standing (and less commonly even severe) constipation can be the presenting manifestation in a significant portion of both pediatric and adult patients (Akman et al., 2018; Ehsani-Ardakani et al., 2013; Khatib et al., 2016). Constipation appears to be related to a well-documented delay in oro-cecal transit time (Benini et al., 2012) possibly caused in part by disturbed upper gastrointestinal motor function (Cucchiara et al., 1995). Because these symptoms can be easily mistaken for irritable bowel syndrome (IBS), both in children and in adults, a recent document by the Italian Societies of Gastroenterology and Endoscopy (SIGE), Neurogastroenterology and Motility (SINGEM), Hospital Gastroenterologists and Endoscopists (AIGO), Digestive Endoscopy (SIED), General Medicine (SIMG), Gastroenterology, Hepatology and Pediatric Nutrition (SIGENP) and Pediatrics (SIP) (Barbara et al., 2022) recommends inclusion of celiac specific serology in screening patients evaluated for IBS.

Of note, the prevalence of functional abdominal pain in celiac children already on a gluten-free diet (GFD) has been reported as similar (Saps et al., 2017) or higher (Cristofori, Tripaldi, et al., 2021) than in the general population.

Infants and toddlers with CD often present with vomiting, weight loss, or failure to thrive, classically associated with distended abdomen and irritability. Poor appetite, occasionally severe like a true anorexia nervosa (Marild et al., 2017), can be an accompanying symptom, leading -particularly in cases of delayed diagnosis- to severe malnutrition and cachexia. However, it should be noted that children with CD can also be overweight or obese, as well documented in the literature and reviewed recently (De Giuseppe et al., 2023), so that the absence of malnutrition should by no means rule out the possibility of CD.

Very young children with CD may rarely also present with intussusception, an uncommon but important gastrointestinal sign. This occurrence was first reported in 1997 (Germann et al., 1997) and is now a well described occurrence in CD children, known to be more frequent in undiagnosed CD children than in control populations (Borkar et al., 2018).

In previous decades, and now exceedingly rare, a dramatic, sudden presentation, referred to as "celiac crisis" may occur. This term describes an acute onset episode of explosive diarrhea associated with protein-losing enteropathy, hypoalbuminemia and profound electrolyte abnormalities with metabolic acidosis and hypokalemia associated to dehydration and lethargy. While this dramatic onset occurs especially in young children (Mones et al., 2007), it has been documented in adults as well (Balaban et al., 2019).

Disorders involving the oral cavity and often ignored are recurrent aphthous stomatitis, dental enamel defects (Alsadat et al., 2021; Nieri et al., 2017) and geographical tongue (Cigic et al., 2016). Factors associated to the etiology of recurrent aphthous stomatitis include family history, stress, local trauma, endocrine disorders, food hypersensitivity and nutritional deficiencies. The specific causes in children with CD, where is found with a prevalence between 20% and 30% (Bucci et al., 2006; Campisi et al., 2008), are, however, unclear and possibly multifactorial. Of note, GFD has been found moderately effective in improving the annoying symptoms caused by them (Campisi et al., 2008).

Dental enamel defects (mostly hypoplasia) have been reported with a widely differing prevalence from 10% to 97% (Acar et al., 2012; Avsar & Kalayci, 2008; Pastore et al., 2008; Souto-Souza et al., 2018; Wierink et al., 2007). They are thought to be secondary to nutritional deficiencies and immune disturbances during the period of enamel formation, mostly the first 6–7 years of life (Cheng, Malahias et al., The). Of note, like recurrent aphthous ulcers, also dental enamel defects can be found in children in the absence of other symptoms, as documented in a large epidemiological Italian study (Martelossi et al., 1996).

As for geographic tongue, this is considered a benign chronic relapsing inflammatory condition typically involving the dorsal surface and lateral aspects of the tongue, caused by a variety of etiological factors including immunological, genetic, atopic, nutritional. Its prevalence is ranging from 1% to 5%, with no sex predilection (Prasanth & Singh, 2021): in children with geographic tongue, CD has been found in about 15% of them, often with no other obvious signs or symptoms suggestive of CD (Cigic et al., 2016), hence representing an often neglected red flag that should call for CD screening.

6.3.2 Extra-intestinal manifestations

While some of the extra-intestinal manifestations such as weight loss, fatigue, short stature and delayed puberty can be attributed to nutritional deficiencies and their metabolic consequences, other manifestations arise from poorly understood pathways.

Extra-intestinal symptoms (Table 6.1) appear to occur at similar rates in children and in adults (Jericho et al., 2017) and, as mentioned, are more common as child age at diagnosis advances (Jericho et al., 2017, Krauthammer et al., 2021). In children, short stature, fatigue and headache appear to be the most common.

6.3.2.1 Short stature

A frequent extra-intestinal manifestation in children, short stature occurred in our series in 33% of the patients (Sansotta et al., 2018). Failure of proper height gain can be an isolated initial presentation of CD (Jericho et al., 2017; Troncone & Kosova, 2010) and can be found in up to 10% of children being investigated for short stature.

Within 2 years of starting a strict GFD, pre-pubertal celiac children should attain appropriate catch up growth and return to their expected trajectory for height. However, if a child is diagnosed post-puberty, their chances for catch up growth are much decreased as the child has likely missed their window.

In such cases, a bone age determination will best predict the child's capacity for further height growth (Jericho et al., 2017). Lack of response to a strict GFD requires careful investigation to the possible presence of another missed comorbidity (inflammatory bowel disease, food aversion, Turner syndrome, or growth hormone deficiency) requiring alternate treatments.

6.3.2.2 Delayed puberty

Defined as a lack of physical or hormonal signs of puberty at the age of usual onset, delayed puberty is another common manifestation of CD affecting roughly 10% of our series of newly diagnosed pediatric celiac patients (Jericho et al., 2017). Visible secondary sexual development usually begins when girls achieve a bone age of 11 years and boys achieve a bone age of 12 years. In CD, this phenomenon is directly related to malabsorption and malnutrition, and should resolve on a GFD, which should prevent any long-term complications and restore normal maturation. If the delay in puberty fails to improve within 12–24 months of starting a GFD, the child should be referred to the evaluation of a pediatric endocrinologist.

6.3.2.3 Anemia

The end result of several different, and sometimes combined, causes, anemia is a frequent finding in untreated celiac children (Harper et al., 2007); however, the single most common type of anemia is due to iron deficiency (IDA). IDA may be the only clinical abnormality identified in many patients and can be the presenting feature of CD, especially in older children (Rajalahti et al., 2017). Anemia at CD diagnosis is associated with more severe histological and serological presentation in children (Rajalahti et al., 2017; Zanini et al., 2013). However, even when asymptomatic, CD can lead to IDA; in a large series of patients with asymptomatic CD, IDA was indeed found in almost half of the patients, with adults having a higher incidence than children: 46% versus 35% (Bottaro et al., 1999).

6.3.2.4 Vitamin D deficiency

Vitamin D deficiency (serum 25-hydroxyvitamin D level less than 20 ng per mL), or insufficiency (serum 25-hydroxyvitamin D level of 20–30 ng per mL) are the most common vitamin deficiencies found in children with CD at diagnosis (Deora et al., 2017; Villanueva et al., 2012), and appear likely to contribute to the low bone mineral density (Mager et al., 2012). Assessment of vitamin D status is therefore recommended (Mearin et al., 2022; Snyder et al., 2016) at diagnosis and, in case of abnormal levels, correction of its deficiency should be carried out (Mearin et al., 2022) to promote immune status and bone health.

6.3.2.5 Dermatitis herpetiformis

Considered the skin manifestation typical of CD (Reunala et al., 2018), DH affects mostly teenagers and adults who present symmetrical, pruritic blisters followed by erosions, excoriations, and hyperpigmentation. This entity is treated specifically in Chapter 16. The most involved sites are mainly the extensor surfaces of elbows and knees, followed by shoulders, buttocks, sacral region, and face. This condition is characterized by the presence of autoantibodies against epidermal transglutaminase (TG3), the supposed autoantigen of DH. Villous atrophy is not constant, occurring in about 2/3 of these patients, more commonly in the younger ones and does not have an impact on the clinical recovery or long-term general health of DH patients (Mansikka et al., 2018). The diagnosis of DH is made by a direct immunofluorescence biopsy of unaffected skin in close proximity to an active lesion (Zone, Meyer et al., 1996), showing pathognomonic granular IgA deposits at the dermo-epidermal junction. Its treatment is based on a strict GFD, that is typically very effective. Dapsone and/or other drugs can be used during the early phase of treatment, until the GFD becomes effective (Bolotin & Petronic-Rosic, 2011).

6.3.2.6 Low bone mineral density

Only a small percentage of newly diagnosed children (Webster et al., 2019) are found to have low bone mineral mass. The etiology of such bone alteration is multifactorial, thought to be mostly secondary to the combination of intestinal malabsorption and chronic inflammation. Low bone mineral density apparently responds well and typically within 1 year to the GFD in CD children (Mager et al., 2012; Snyder et al., 2016; Tuna Kirsaclioglu et al., 2016; Webster et al., 2019), and current recommendations by the ESPGHAN (European Society Pediatric Gastroenterology, Hepatology and Nutrition) (Mearin et al., 2022) do not recommend routine bone mineral density screening, also in consideration of the fact that in celiac children the reduced bone mineral density does not seem to be associated with an increased fracture risk (Canova et al., 2018), even though multiple fractures associated with vitamin D-deficient rickets, have been reported since 1935 (Schlesinger & Keele, 1935).

6.3.2.7 Joint involvement

Arthritis and arthralgia have been described as presenting manifestations of CD both in children, especially teenagers, and more commonly in adults (Dos Santos & Liote, 2017). It is therefore recommended that patients presenting with unexplained articular manifestations be tested for CD (Ghozzi et al., 2013; Sherman et al., 2015).

Some type of joint involvement is seen in fact in up to 10% of celiac pediatric patients at diagnosis (Jericho et al., 2017). When studied with the use of musculoskeletal ultrasound, a technique especially suited to detect early inflammatory and structural abnormalities in joints (Garg et al., 2017), signs of joint involvement in children with CD were found in 32% of newly diagnosed patients as compared to only 3% of those on a GFD for at least six months, suggesting that in CD children a GFD may improve these joint abnormalities.

Indeed, based on these results, treatment with GFD has been also proposed for adults with rheumatoid arthritis and idiopathic arthritis in the absence of CD, with conflicting results (Lidon et al., 2022).

6.3.2.8 Liver involvement

Elevation of liver transaminases is well described in pediatric CD, often idiopathic and thought to be present in about one third of all children with CD (Vajro et al., 2013). This so-called "transaminitis" normalizes in most patients with CD within a few months of GFD.

6.3.2.9 Neurologic manifestations

Several forms of neurologic involvement can be part of the presenting signs and symptoms in celiac children. Among them, headache (Dimitrova et al., 2013; Nenna et al., 2016; Sabino et al., 2020; Zis et al., 2018), is definitely the most common, appearing in 18%–38% of patients (Hom et al., 2020; Lionetti et al., 2009; Niknam et al., 2021; Zis et al., 2018). The prevalence of headache in celiac children follows the same age-related pattern reported for the general population, increasing from 4% in 4–5-year-old to 25% in adolescents. Of interest, within 6–12 months, roughly half of the celiac children on a GFD report a substantial improvement (Jericho et al., 2017; Sansotta et al., 2018).

Peripheral neuropathy (Mearns et al., 2019), and seizures (Bashiri et al., 2016) are well described in pediatric patients, again more commonly in older children and teenagers. Ataxia on the other hand is described almost exclusively in adult cases (Ruggieri et al., 2008).

Epilepsy in CD children appears to occur more frequently than expected, as reported in a recent large epidemiologic study (Canova et al., 2020). Seizures are often generalized tonic-clonic, but partial and occasionally absence seizures are also reported. A "special" form of epilepsy is that associated with symmetrical occipital calcifications (Gobbi, 2005). Patients described are few and mostly from Italy, Spain, and Argentina, suggesting a geographically restricted condition. This type of epilepsy may be benign, but occasionally can evolve into epileptic encephalopathy. The efficacy of GFD seems to be inversely related to the duration of epilepsy and the age of the patient, being more effective in younger ones. In some patients with CD and epilepsy refractory to antiepileptic drugs, seizures have been controlled with a GFD (Canales et al., 2006; Pascotto et al., 1994). Hence, it is recommended that patients with epilepsy without a clear etiology should be screened for CD, as an early diagnosis and treatment might be beneficial.

6.3.2.10 *Psychiatric issues*

Relatively common in children and especially in adolescents and young adults (Lebwohl, Haggard, et al., 2020) psychiatric issues can definitely impact quality of life (Jericho et al., 2021). These disorders include anxiety, often with recurrent panic attacks, hallucinations, depression, cognitive-behavioral disorders and ADHD (Attention Deficit Hyperactivity Disorder), as well documented by a large epidemiological investigation in Sweden (Butwicka et al., 2017).

A moderately increased risk of suicide in CD patients has also been reported (Ludvigsson, Sellgren, et al., 2011). Interestingly, there is some evidence that the GFD may help in alleviating depression in celiac adolescents (Simsek et al., 2015).

A recent epidemiological investigation confirmed that rates of child-reported anxiety and depression symptoms are high and, interestingly, significantly more often reported by those without medical comorbidities than by those with (Germone et al., 2022).

Despite some inconsistencies due to the heterogeneous methodology used in many of the studies (Gaur, 2022), celiac children seem to have a higher incidence of cognitive and behavioral issues that can be grouped under the ADHD (Efe & Tok, 2023; Gaur, 2022), and their response to the GFD can be variable.

6.4 Potential and seronegative celiac children

Two groups of celiac patients observed in children as well as in adults are represented by potential and seronegative patients.

The term potential describes patients who do have positive celiac-specific serology and yet have an intact intestinal mucosa, without the hallmarks of the celiac lesion. These patients may or may not have symptoms; the issue of which patients have a higher likelihood of developing full blown CD and thus should undergo a GFD is of great interest and is being studied with innovative approaches (Picciali et al., 2021). This condition is treated in depth in Chapter 10.

Seronegative celiac disease (SNCD) refers to patients "with or without gastrointestinal signs and symptoms of CD in the presence of villous atrophy and compatible genetics and without TTG-IgA, DGP-IgA, and EMA- IgA, who show clinical and histologic responses to the GFD and for whom other etiologies have been examined" (Leonard et al., 2021). This intriguing entity is dealt with in details in Chapter 11. The prevalence of this condition that at least in adults appears to have a more aggressive disease phenotype than seropositive CD (Schiepatti, Rej, et al., 2021) seems to be around 5% in adults, where is the most frequent cause among the heterogeneous group of seronegative enteropathies (Schiepatti, Cincotta, et al., 2021), but is much rarer in children (Gustafsson et al., 2020), where other causes of seronegative villous atrophy such as Crohn's disease or food allergy are much more frequent (Mandile et al., 2021). Hence, before concluding that a pediatric patient with duodenal villous atrophy can be diagnosed with seronegative CD, it is imperative that an accurate work-up be carried out to look for other conditions. Once this is done with negative results, a GFD for a period of 6–12 months with careful clinical monitoring should be implemented, followed by repeat intestinal biopsies showing normalization of the intestinal lesion.

6.5 Associated conditions

In addition to present with various combinations of gastrointestinal and extra-gastrointestinal manifestations (or being completely asymptomatic) CD is known to be more frequently associated with a variety of disorders. They can be schematically grouped under Autoimmune, genetic/chromosomal and other disorders (see Table 6.2)

6.5.1 Autoimmune disorders

Celiac disease has been known for a long time to carry a strong association with other autoimmune diseases, as recently confirmed in a large-scale study in Israel (Assa et al., 2017). Genome-wide association studies have in fact shown that genetic risk factors are shared by about half of all autoimmune diseases (Jonkers & Wijmenga, 2017), the most relevant group of genes in common between CD and other autoimmune conditions being in the family of Human Leukocyte Antigens (HLA) (Bodis et al., 2018).

A common question is that of the possible role of the GFD in hindering the development of associated autoimmune disorders in CD patients. While the issue is still debated, a GFD has been proposed as a form of treatment in patients with other autoimmune conditions such as Multiple Sclerosis and Psoriasis (Passali et al., 2020).

TABLE 6.2 Main conditions associated with celiac disease in children.

Autoimmune conditions	Genetic/chromosomal disorders	Other disorders
Type 1 Diabetes Mellitus Thyroid disease: Hashimoto Thyroiditis Graves-Basedow disease Autoimmune Hepatitis Idiopathic Juvenile Arthritis Autoimmune skin disorders Psoriasis Vitiligo Alopecia areata	Selective IgA Deficiency Down syndrome Turner syndrome Williams-Beuren syndrome	Eosinophilic Esophagitis Asthma Gastroesophageal reflux Pancreatitis

However, the issue of whether the onset of autoimmune disorders in CD patients is favored by the ingestion of gluten (either before or after diagnosis) remains controversial. An increased prevalence of autoimmune disorders was found in parallel with the increasing age at diagnosis of CD (Ventura et al., 1999), suggesting that prolonged exposure to gluten may indeed favor the onset of autoimmune conditions. Further studies on this issue have revealed conflicting results, so that the issue is still unclear (Elli et al., 2014).

6.5.1.1 *Type 1 diabetes (T1D)*

It has been known for a long time (Walker-Smith & Grigor, 1969) that patients with T1D have a high prevalence not only of CD, but of a number of other autoimmune diseases, including thyroid disease, gastric autoimmunity leading to pernicious anemia, vitiligo, and adrenal gland insufficiency. The prevalence of T1D in celiac patients varies between 1.6% and 16.4%, with T1D onset typically preceding that of CD. however, a large prospective study on a cohort of at risk children followed from 3 until 48 months of age with repeated measurements of specific autoantibodies (Islet cell Antibodies—IAs and TTG-IgA) showed that IAs preceded TTG-IgA development more often (67%) than vice versa (27%) (Hagopian et al., 2017). While the majority of children diagnosed with CD after T1D onset have elevated levels of TTG-IgA at the time of T1D diagnosis, more patients develop CD a few years after T1D onset (Barera et al., 2002), even in later years (Scaramuzza et al., 2013). It is therefore recommended that diabetic children be repeatedly tested for CD even after proven negative at initial screening.

The presence or absence of GI symptoms in children with T1D has no predictable value for CD (Bybrant et al., 2014). On the contrary, it is customary for diabetic children who come to be diagnosed with CD to have minimal or no symptoms at all. All the major medical societies recommend testing for CD children with T1D (Couper et al., 2014; Hill et al., 2005; Hill et al., 2016; Husby et al., 2012; Husby et al., 2020; Murch et al., 2013). To complicate matters however, patients with T1D are known to have low to moderate positive titers of TTG-IgA independently of CD, and such antibodies do fluctuate and even normalize spontaneously in about one-third of patients with T1D who remain on gluten (Castellaneta et al., 2015; Waisbourd-Zinman et al., 2012). These quite common cases end up—once they are shown to have an intact duodenal mucosa - by being considered either a "false positive" TTG-IgA or a potential celiac. For these reasons, it is recommended to proceed with the confirmatory biopsy only when their titer is >3 times the upper limit of normal. Clearly, once diagnosed with CD, children with T1D need to follow a strict GFD regardless of the presence of symptoms. Such diet, in spite of being super-imposed to the restrictions of T1D, does not seem to negatively influence the quality of life and actually has recently been shown to be beneficial to patients with T1D and celiac disease, regardless of symptoms (Nunes-Silva et al., 2017). Data from several recent clinical trials in this regard concluded that a good compliance to the GFD allowed a decrease in hypoglycemic episodes, better glycemic control and fewer episodes of ketoacidosis in these patients (Kaur et al., 2020; Nagl et al., 2019; Soderstrom et al., 2021; Soderstrom et al., 2022).

6.5.1.2 *Autoimmune thyroid diseases*

Hashimoto's thyroiditis and, to a lesser extent, Graves-Basedow disease are autoimmune thyroid diseases occurring more frequently in CD than in the general population (Ghozzi et al., 2014; Minelli et al., 2018) with a prevalence varying between 5%–13% (Freeman, 2016) and up to 30% (Carta et al., 2002), increasing with advancing age. Like many autoimmune conditions, Hashimoto thyroiditis is more common in females and is characterized by antithyroglobulin and/or antithyroid peroxidase antibodies. Of note, the presence of antithyroid antibodies in children with CD has a low predictive value for the development of thyroid hypofunction (Cassio et al., 2010) over at least the ensuing 3 years. Symptoms of autoimmune thyroiditis and CD often overlap: fatigue, weight loss, constipation (or diarrhea) and arthralgia or seronegative arthritis can be found in both. The latter indeed has recently been found to occur significantly more often in patients with autoimmune thyroiditis who are also celiac (Lupoli et al., 2020). In spite of a recently reported apparent decrease of the prevalence of autoimmune thyroiditis in celiac patients over the past 50 years (Castro et al., 2020), it is evident that screening for CD all children and teenagers with autoimmune thyroiditis currently remains recommended (Husby et al., 2020).

6.5.1.3 Autoimmune hepatitis

Autoimmune hepatitis occurs in less than 1% of individuals (mostly adults) (Tunio et al., 2020) but shows a much higher prevalence in CD, where it affects between 2%–6% of pediatric patients (Di Biase et al., 2010, Vajro et al., 2013). The reverse is also true: in fact, about 15% of children with autoimmune hepatitis have been reported to be celiac (Caprai et al., 2008). The response to the GFD has been reported sporadically, and with conflicting outcomes (Di Biase et al., 2010; Mounajjed et al., 2011; Nastasio et al., 2013).

6.5.1.4 Idiopathic juvenile arthritis

The association of this autoimmune chronic inflammation with CD, though previously not found to be strong in Sweden (Oman et al., 2019), has been recently documented in a large series of IJA patients in Italy (Naddei et al., 2022), where CD occurred in 2.4% of the children with IJA, supporting the need for CD screening in all JIA children, and especially those with a family history of autoimmunity, as this has been found in a high proportion of patients presenting both conditions.

6.5.1.5 Autoimmune skin disorders

As confirmed by a large epidemiological investigation (Lebwohl, Soderling, et al., 2020), celiac children are also at increased risks for autoimmune skin disorders such as psoriasis (Acharya & Mathur, 2020; De Bastiani et al., 2015; Ungprasert et al., 2017), vitiligo (Rodrigo et al., 2018) and alopecia areata (Rodrigo et al., 2018). Although the benefit of following a GFD on these skin disorders is still not completely clear, there is some evidence that this diet may be beneficial in treating psoriasis (De Bastiani et al., 2015; Sansotta et al., 2018).

Vitiligo has a worldwide prevalence around 1%, with about half of the cases appearing in childhood. while the association with CD in adults is well known (Zhang et al., 2021). However, its prevalence in childhood CD, as well as the role of GFD in its treatment, is less clear (Persechino et al., 2021).

Alopecia areata is not rare in childhood, affecting up to 2% of the population and characterized by a sudden onset of patchy hair loss on the scalp. Its association with CD was first described in 1995 by Corazza and collaborators (Corazza et al., 1995), and recently confirmed in a metaanalysis (Lee et al., 2022), once again occurring often in the complete absence of GI manifestations. This condition has been shown to be significantly associated with depression and anxiety (Lauron et al., 2023). The efficacy of GFD in ameliorating the alopecia appears to be at best doubtful.

6.5.2 Genetic/chromosomal disorders

Approximately 2% of all celiac children are IgA deficient (Cataldo et al., 1997). On the other hand, about 14% of subjects with selective IgA deficiency are celiac (Lougaris et al., 2019). The presence of IgA deficiency in a celiac child does not appear to be of relevant clinical significance, but of course bears an important meaning in the screening strategy, as the universally recommended screening test is based on IgA antibodies (see later in this chapter).

And in terms of screening, the three well known associations with Down syndrome, Turner syndrome and the quite rare William-Beuren syndrome must be considered, as these patients have a much higher prevalence of CD (Liu et al., 2020; Ostermaier et al., 2020; Marild et al., 2016; Pangallo et al., 2020) than the general population; hence, they definitely need to be screened for CD.

6.5.3 Other associated disorders

In addition to autoimmune disorders and the genetic conditions reviewed above, large-scale epidemiological investigations have shown a growing number of disorders occurring more frequently in celiac patients than in general population, although the evidence for many of these is either weak or controversial. Many of these conditions however are reported to occur mostly in adults, thus they won't be dealt with in this chapter. Among them: asthma (Ludvigsson, Hemminki, et al., 2011), chronic obstructive pulmonary disease (Ludvigsson, Inghammar, et al., 2012), cardiac disorders and strokes (Emilsson et al., 2011; Emilsson et al., 2012; Ludvigsson, West, et al., 2012), gastroesophageal reflux disease(Ludvigsson et al., 2013), pancreatitis (Sadr-Azodi et al., 2012), ocular complications (Mollazadegan et al., 2011; Mollazadegan et al., 2012), idiopathic dilated cardiomyopathy (Emilsson et al., 2012), renal disease (Wijarnpreecha et al., 2016).

However, a higher occurrence of eosinophilic esophagitis has been described with variable prevalence in children, as well as in adults, (Cristofori, D'Abramo, et al., 2021; Hommeida et al., 2017) and the effect of GFD in children has been assessed (Patton et al., 2019). Thus, even when the indication for endoscopy appears to be solely to confirm or rule out celiac disease, esophageal biopsies should be obtained in every case, regardless of the visual appearance of the esophageal mucosa, in order not to miss this association. In fact, there is evidence that adherence to established endoscopic biopsies guidelines is poor, thus lowering detection rates for CD and EoE (Wallach et al., 2017).

6.6 Complications

Non-responsive celiac disease (NRCD) after at least 12 months on a GFD, has been reported in about 15% of celiac children. While in most cases this condition has been found to be secondary to inadvertent (or voluntary) ongoing gluten ingestion (Ali Ibrahim et al., 2022; Veeraraghavan et al., 2021), other causes might be at play. For this reason, it is recommended that when facing a child with an apparent NRCD, proper investigation be conducted, including first and foremost search for any ongoing ingestion of gluten. This can be assisted by measurements of the Gliadin Immunogenic Peptides (GIP) in stools and/or urine (Stefanolo et al., 2021), a useful tool to detect even very small amounts of ingested gluten. Once no evidence for gluten ingestion is found, then other likely causes should be investigated such as disorders of gut/brain axis, chronic constipation, as well as other less frequent causes such as concomitant enteropathies like Crohn's disease, autoimmune enteropathy, cow's milk protein allergy (Capone et al., 2020), and pancreatic insufficiency (Mearin et al., 2022; Rubio-Tapia et al., 2023). In cases where causes of NRCD remain unclear, the institution of a so-called "naked diet" (or Gluten Containing Elimination Diet, that is, a very restricted diet conducted under strict survelillance utilizing only naturally gluten-free foods) and/or the short term use of budesonide might be useful (Ali Ibrahim et al., 2022)

It should in fact be noted that true *refractory* CD (RCD), a serious condition defined by persistent or recurrent villous atrophy and malabsorptive symptoms in CD patients despite adherence to a rigorous GFD and subdivided into two subgroups according to the normal (RCDI) or abnormal phenotype of intraepithelial lymphocytes (RCDII) abnormal IELs, and high risk of overt lymphoma, is considered in children an extremely rare if not non-existing entity. In fact, a recent official position paper by ESPGHAN concluded after an extensive literature search that the evidence for the existence of refractory CD in children is "very poor" (Mearin et al., 2022).

Likewise, while the development of intestinal lymphoma, and especially the so-called EATL (Enteropathy Associated T-Cell Lymphoma) is a rare but well described complication occurring in adult celiac patients, this entity has so far never been reported in children.

6.7 Diagnosing a pediatric patient

6.7.1 The role of serology

Currently, active case finding is the favored strategy to increase detection of CD (Virta et al., 2009). This involves a screening with serological testing individuals presenting overt symptoms (GI or extra-GI) and associated conditions (see Tables 6.1 and 6.2), as well as first degrees of celiac patients like siblings or parents. Serological assays include antibodies directed against small intestinal-enzymatically processed gluten fragments: antideamidated gliadin peptides (DGP) or autoantibodies like antiendomysial (EMA) and TTG antibodies, of both IgA and IgG classes (Fasano & Catassi, 2012).

Universally recommended serological screening for CD both in children and in adults consists of measuring TTG IgA and concurrent measurement of total IgA while on a regular (gluten containing) diet (Husby et al., 2020; Rubio-

Tapia et al., 2023). A positive serological test, especially in the high pretest probability groups described, is supportive of the diagnosis although no single test is 100% sensitive and specific for CD and, furthermore, diagnostic accuracy can vary considerably between laboratories (Leffler & Schuppan, 2010). As mentioned, the diagnostic guidelines from major international societies like ESPGHAN (Husby et al., 2020), NASPGHAN (North American Society for Pediatric Gastroenterology, Hepatology and Nutrition) (Hill et al., 2016) and ACG (American College of Gastroenterology) (Rubio-Tapia et al., 2023) recommend TTG IgA as the most cost effective and reliable screening test to identify CD.

It is now accepted that TTG-IgA are reliable indicators of CD even in young children (where previously an additional DGP testing was recommended), so that the 2020 revised ESPGHAN guidelines recommend that in subjects with normal serum IgA values for their age, only TTG IgA should be used as the initial serological test, regardless of age (Husby et al., 2020). In seven studies involving a combined number of 639 children younger than 2 years with biopsy confirmed CD, the TTG IgA performed as well or better than DGP-IgG in identifying children with disease (Amarri et al., 2013; Basso et al., 2009; Frulio et al., 2015; Mubarak et al., 2011; Olen et al., 2012; Richter et al., 2012; Wolf et al., 2014). A recent systematic review evaluating the diagnostic accuracy of DGP IgG in comparison with TTG IgA in children under two years of age found that the sensitivity of DGP IgG was not statistically superior to TTG IgA (Catassi et al., 2021).

6.7.2 Diagnosis according to the ESPGHAN criteria

Over the last decades, the diagnosis of CD has evolved. The fundamental diagnostic criterium had been from the beginning the documentation of the typical changes in the duodenal mucosa, to which in 1990 the normalization of the CD-specific antibodies on a GFD added weight to diagnosis to the point of eliminating the need for a second or even third biopsy (Walker-Smith et al., 1990) like recommended in the very initial diagnostic guidelines. The discovery of the tissue transglutaminase as the auto-antigen of CD, and the subsequent development and widespread testing of the extremely specific TTG-IgA, allowed for a further diagnostic simplification published in 2012 by ESPGHAN (Husby et al., 2012). These guidelines allowed the diagnosis of celiac disease without the need for a confirmatory biopsy in children with classical clinical presentation, TTG IgA more than 10 times the ULN, positive EMA IgA on a separate blood sample and a consistent HLA DQ2/DQ8 haplotype. The subsequent resolution of symptoms with normalization of the TTG IgA titer after GFD would serve to confirm the diagnosis.

After these recommendations were implemented widely, mostly in Europe, it became further clear not only that this approach was indeed effective with a positive predictive value basically reaching 100%, but also that documenting the compatibility of the HLA status was unnecessary. Thus, in 2020 ESPGHAN published new evidence-based guidelines with the following changes: (1) HLA testing is no longer needed and (2) diagnosis without biopsy is allowed (with TTG IgA more than 10 times the ULN and positive EMA IgA) even in asymptomatic children (Husby et al., 2020). It is estimated that this approach actually applies to about 50% of children presenting to a celiac center with a suspected CD. In addition, a recent study in the United States (Badizadegan et al., 2020) estimated that using this algorithm "between 4891 and 7738 pediatric endoscopies per year could be avoided in the United States," thus resulting in considerable cost savings. The algorithm presented in Figs. 6.1–6.3 (Husby et al., 2020) summarizes accordingly the diagnostic steps for a child/teenager suspected of CD.

In cases where biopsies are obtained, as mentioned in previous sections and fully discussed in other chapters, patients with positive celiac serology and normal histology are considered cases of potential CD; however, it should be noted that before concluding for such diagnosis, it is imperative to confirm that at the time of biopsy the patient was consuming a sufficient amount of gluten and that the biopsies were adequate in number, position and well oriented.

6.8 Follow-up

There is general agreement that follow-up of newly diagnosed celiac children is mandatory for proper disease control and prevention of complications later in life, as lack of proper follow-up has been shown to be associated with poor outcome (Barnea et al., 2014). Indeed, a recent consensus document was published by the ESPGHAN Special Interest Group on Celiac Disease (Mearin et al., 2022), highlighting the crucial points of this process.

Essentially, the follow-up is aimed at two major goals: (1) verify that children and their families have adapted (and are adequately compliant) to an age-appropriate and nutritionally balanced GFD and: (2) monitor the regression of symptoms and the restoration of full health. It is therefore evident that such follow-up should be performed by a pediatric gastroenterologist with specific expertise in CD and—especially in the first 1–2 visits and then whenever needed—a dietitian equally expert in the subtleties of GFD. In this regard, the added availability of electronically mediated

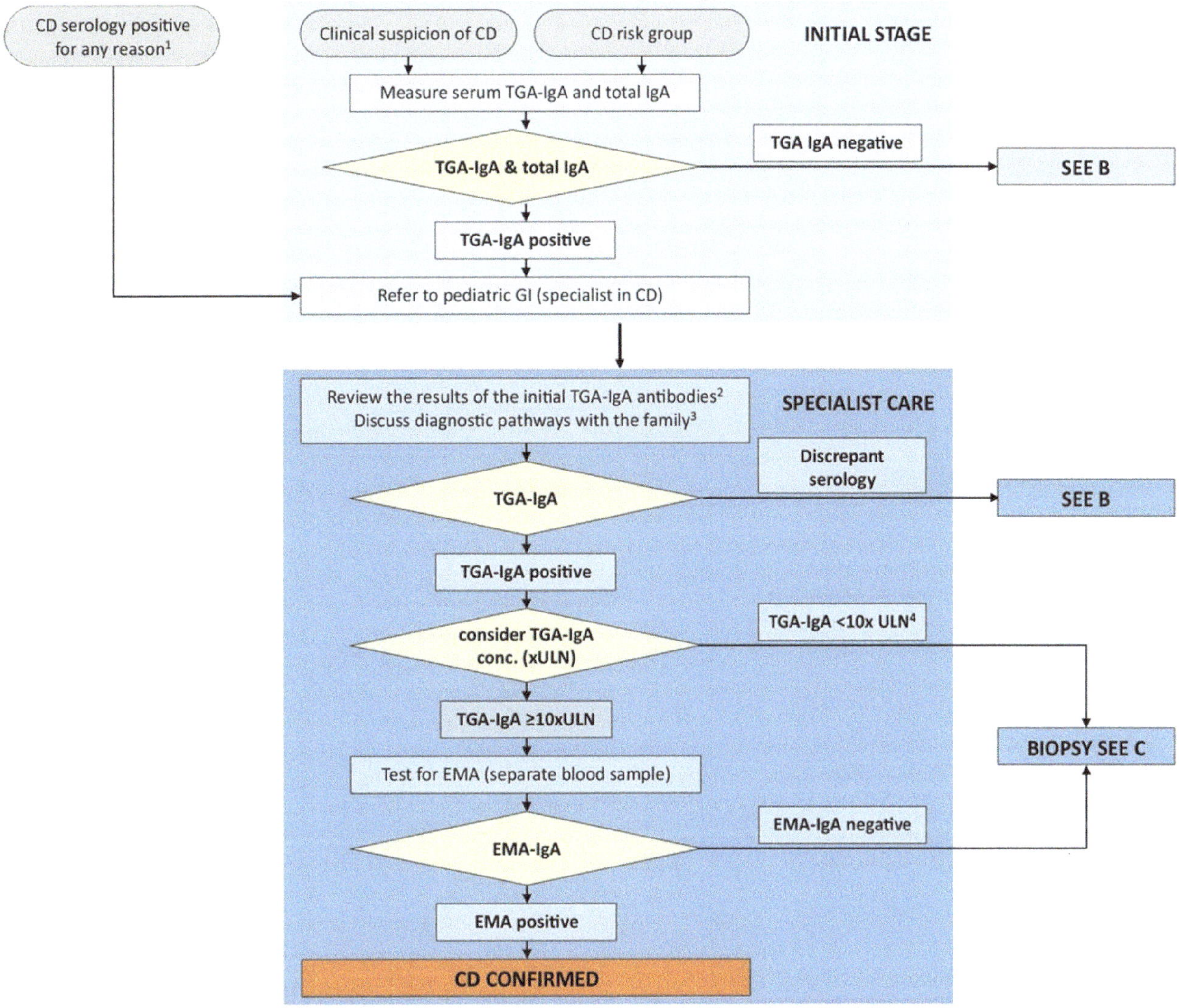

FIGURE 6.1 Diagnostic algorithm in IgA-competent subject. [1] Other than TTG-IgA, including point-of-care tests and DGP. [2] Check the value also in relation to the cut-off and repeat the test if questionable or borderline. No need to retest if done with validated assay with calibration curve. Test with conventional TTG-IgA test if positive POCT and TTG-IgA has not been measured quantitatively. [3] Convey the message that the diagnosis of celiac disease with or without biopsy confirms the need for a lifelong gluten-free diet and that re-evaluation after introduction of the diet would need prolonged re-exposure to gluten with a series of further investigations. [4] If TTG-IgA is only borderline positive confirm sufficient gluten intake and consider retesting of TTG-IgA and EMA.

teaching is showing promising results (Connan et al., 2019; Vriezinga et al., 2018). The frequency of follow-up visits appears to be variable among centers with CD expertise, but generally a first visit should be made after 3–6 months postdiagnosis. Subsequent encounters will depend on several factors, including patient's compliance, speed of normalization of symptoms and serology, considered the best (though imperfect, Silvester et al., 2017) proxy estimate of mucosal healing. The ESPGHAN position paper recommends that subsequent visits "should be every 6 months until normalization of the TTG-IgA levels, and every 12–24 months thereafter" (Mearin et al., 2022).

During follow-up, patients should be evaluated for evolution of their gastrointestinal and/or extraintestinal signs and symptoms, and anthropometric measurements and growth parameters should be monitored. It is important that TTG-IgA antibodies be measured using the same technique as for diagnosis. It should be noted that there is typically a high expectation from families that within a short time TTG-IgA would disappear or at least plunge below the upper limit of normal. This expectation is however unrealistic, as it is well known that such titers decline slowly and reaching a negative (i.e., within normal) value depends not only on full adherence to the GFD, but actually largely on their initial, baseline levels (Gidrewicz et al., 2017; Petroff et al., 2018). Therefore it is not uncommon for a fully compliant child to have still elevated titers 18–24 months after beginning the GFD.

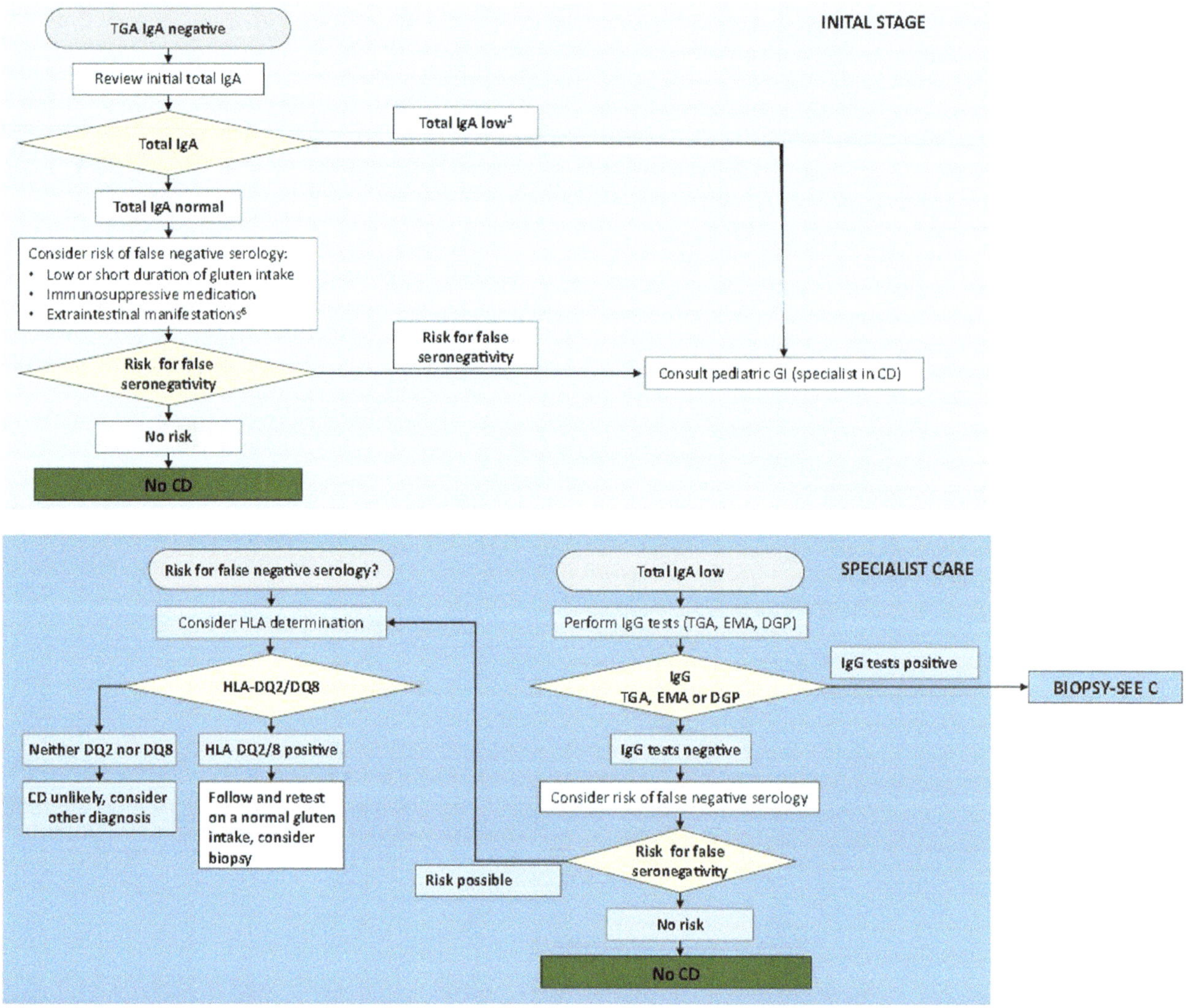

FIGURE 6.2 Diagnostic algorithm in IgA-deficient subjects. [5] Low for age or <0.2 g/L above the age of 3 years [6] For example, Dermatitis Herpetiformis, in which serology can be negative.

Since at diagnosis usually a complete blood count, assessment of micronutrients (e.g., iron, vitamin B12, and vitamin D levels), as well as ALT measurements are often done, any abnormalities must be followed up and appropriately addressed. Assessment of bone mineral density is not routinely recommended (Mearin et al., 2022; Snyder et al., 2016), as there is evidence that even if some degrees of osteopenia are present, they are expected to fully recover within a year on GFD (Mager et al., 2012; Snyder et al., 2016; Wessels et al., 2016).

Symptomatic improvement on the other hand appears in general to be fast in children, and definitely more complete and faster than in adults (Sansotta et al., 2018), with gastrointestinal symptoms being generally quicker to subside or at least improve than extraintestinal ones. In this regard, in prepuberal children who are lagging behind in height, if no catch-up growth is observed within about a year, we would strongly recommend further investigations on other potential causes of the delay and a consultation with endocrinology.

Should follow-up biopsies be performed? Routine assessment of mucosal healing with duodenal biopsies is not recommended in children (Koletzko et al., 2017; Mearin et al., 2022). However, repeating a biopsy should be considered in selected cases of CD, such as in the presence of persistently positive TTG-IgA, lack of clinical improvement, or suspicion of a previously unrecognized additional disorder.

Additionally, when a child with genetic compatibility (i.e., HLA-DQ2 or DQ8 positive) is seen after having already started a GFD before the diagnostic process was completed, a gluten challenge may be necessary to conclusively prove or disprove a CD diagnosis. This is better avoided however in periods of accelerated growth, such as in toddlers and

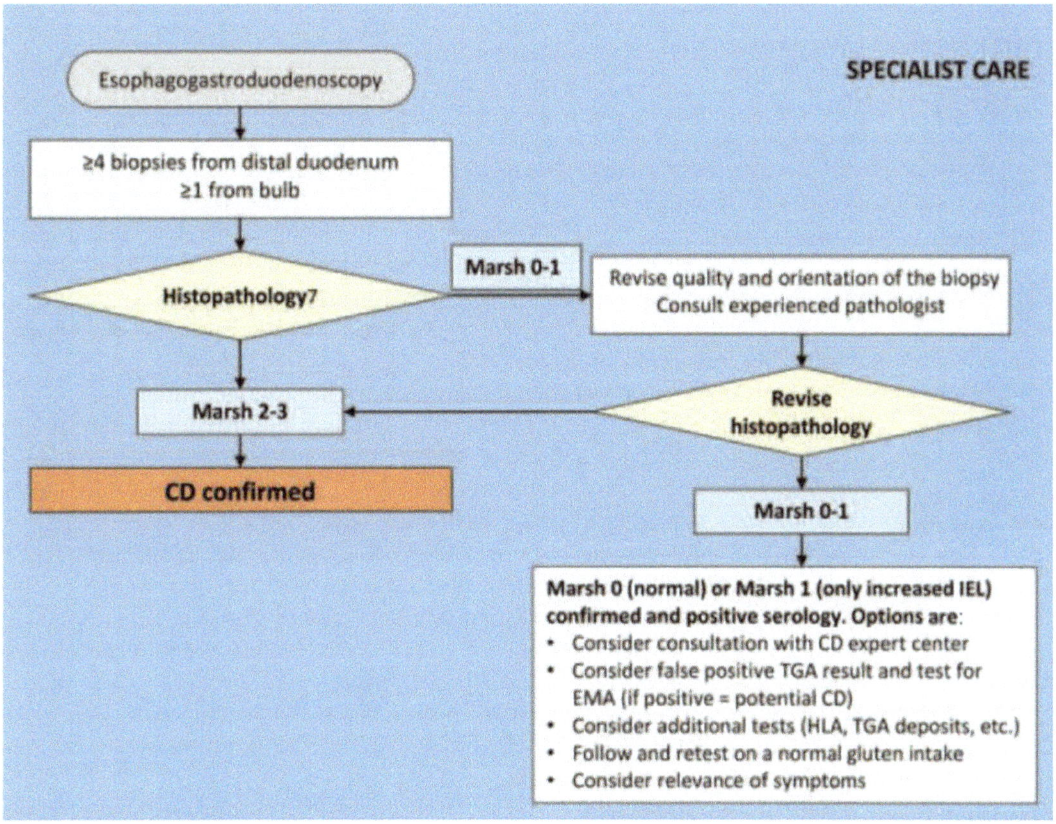

FIGURE 6.3 Diagnostic algorithm in cases undergoing biopsy. [7] The cut-off for normal numbers of IEL is >25 cells/100 enterocytes. The action done at the initial stage in primary care (light blue) is separated from specialist care (dark blue).

teenagers. There is no agreement on the amount of gluten to be ingested or the length of time before assessing relapse with serology and/or biopsy, as several studies have been performed with various methodology. We would endorse the ESPGHAN recommendations (Mearin et al., 2022) of ingesting daily 10–15 g of gluten (as bread slices) for up to 12 months, but with an earlier evaluation in case of appearance of compatible symptoms. In fact, TTG-IgA should be determined 1 month after starting the challenge and repeated every 3 months. In presence of elevation of specific serology and/or symptoms, a duodenal biopsy should be performed. In the absence of both, the challenge is continued for 12 months, and at that point a completely normal diet allowed, with ongoing follow-up and monitoring of TTG-IgA once a year or so.

6.8.1 Issues facing the adolescent patient

6.8.1.1 Dietetic compliance

A strict GFD is the only effective and available therapy for CD but little is known regarding long term dietary patterns of people with CD (Hill et al., 2016; Husby et al., 2020). Patients with CD should be educated to avoid cereals and food products derived from wheat, barley or rye and foods made from gluten-contaminated cereals that are naturally gluten-free like maize, oats (Murray et al., 2004). It is well known that a strict adherence to the GFD is a prerequisite to control symptoms (Jericho et al., 2017; Long et al., 2010; Murray et al., 2004; Sansotta et al., 2018), improve quality of life and decrease the risk of complications. In children, adherence to GFD is very effective in controlling a wide variety of symptoms and reducing health care utilization (Long et al., 2010). Although several societies have recommended routine assessment of GFD adherence, no standard method of assessment is available. A combination of serological markers, self and parent-reported adherence, assessment of symptomatic response and evaluation by an expert dietician are currently considered as the best global proxy indication of dietetic adherence of lack thereof (Husby et al., 2020; Mearin et al., 2022), as described in the above section on Follow-up. The availability of dietitians with specific expertise in CD is highly desirable to evaluate for potential dietary nutrient deficiencies and educate on how to maintain a strict GFD with provision of healthy alternatives to gluten (Madden et al., 2016; Suarez-Gonzalez et al., 2021). Nutritional status, height and weight need to be assessed, as different habits in selecting a GFD may have an impact on the child's growth (Sansotta et al., 2020).

Adherence to GFD in children and teenagers seem to vary between 23% and 98% depending on the definition of adherence and method of assessment (Myleus et al., 2020). The DOGGIE BAG study showed that unintentional gluten intake is common and noninvasive measures of gluten intake such as TGA levels do not necessarily correlate well with gluten exposure in individuals with self-reported strict adherence (Silvester et al., 2020). A recent tool to detect inadvertent gluten ingestion has been developed: in fact, immunologically active fragments of gluten (gliadin peptides) can be detected in samples of urine or stools for up to 48 hours after gluten ingestion, with a quite high degree of accuracy (Cebolla et al., 2018; Moreno et al., 2017). Although this sensitive tool is undoubtedly useful, it must be stressed that only provides a snapshot of recent dietetic transgressions, not a long-term indication of ongoing gluten ingestion. Whether its use enhances diet adherence remains to be seen.

Lifelong adherence to GFD depends upon several factors. For instance, earlier childhood diagnosis seems to lead to improved GFD adherence (Charalampopoulos et al., 2013; Mehta et al., 2023), while individuals that are diagnosed in adolescence may be less adherent than their younger and older comparators (Kurppa et al., 2012). In addition, it has been noticed that children with celiac disease and their families become less interested in dietary education over time, indicating that the greatest opportunity for imparting a correct and healthy diet occurs at the time of diagnosis (Runde et al., 2020). In Europe, accidental transgressions are common (MacCulloch & Rashid, 2014; Tapsas et al., 2014). Adherence to a GFD can be quite difficult (van Doorn et al., 2008), particularly as the children and young teenagers face new challenges like peer pressure and the stigma of "being different," school trips and increasing independence from their parents. Hence, adolescence is a period when adherence becomes unreliable and thus potential risk factors for poor adherence should be recognized (Errichiello et al., 2010). To further aggravate the problem of good dietetic adherence, it should be noted that adolescents and young adults with CD are frequently lost to follow up (Herman et al., 2012) and thus more likely to be not adherent to GFD (Barnea et al., 2014; Schiepatti et al., 2022).

6.8.1.2 Quality of life

Chapter 12 addresses specifically issue affecting quality of life in celiac patients. CD is a chronic, currently incurable disease; and since lifelong adherence to a GFD is demanding and costly, the psychological, social and emotional aspects of these patients are affected (Canova et al., 2021; Nachman et al., 2009; Zingone et al., 2015). Overall, quality of life (QoL) generally seems to improve on a gluten-free diet (GFD) (Zingone et al., 2015) especially in patients with symptoms at diagnosis (Johnston et al., 2004; Mustalahti et al., 2002; Nachman et al., 2009; Ukkola et al., 2011). Nachman et al. described the most significant QoL improvement in the first three months after starting a GFD, with some additional improvement up to one year later (Nachman et al., 2009). However, the continuation of the latter study described a significant QoL deterioration at 4 years postdiagnosis compared with 1 year. Long-term impairment of the QoL was observed in patients who did not strictly comply with GFD (Nachman et al., 2010), a result in part of the perceived degree of difficulty of adhering to a GFD, influenced by social and educational background (Barratt et al., 2011).

Clearly, activity limitations for children and adolescents with CD result from the need to avoid gluten-containing foods while participating in life situations; while for some adherence to GFD may be relatively simple, for others, maintaining the diet in daily life can be challenging and may lead to substantial limitations for them as well as for their families (White et al., 2016). Thus it is not surprising that emotional support and stronger organizational skills are associated with better adherence (White et al., 2016). Understanding the complexity of each activity, its demands, limitations, challenges, and its variable importance to each child and adolescent with CD is challenging, so that important information concerning the actual daily quality of life of this population can be overlooked (Meyer & Rosenblum, 2017).

Of interest, parents of celiac children tend to rate their children's Health-related quality of life (HRQOL) lower than the children's own ratings (Coster & Khetani, 2008), underlying the need to incorporate children's and adolescents' self-reports into the assessment of their HRQOL (Sisk, 2016; Varni et al., 2009).

6.8.1.3 Transitioning care to adult environment

Children with CD grow to be adults with CD and carry with them health and social issues that evolve during adolescence and require answers from health care providers that are subject to age-related changes. It follows that the process of transition from pediatric to adult care is an important one, upon which many factors impinge and that may well carry long-term consequences. Unfortunately, this delicate aspect of care upon which much of the future dietetic adherence is based has not received proper attention, and as a result the process is inconsistent and not agreed upon structured consensus. Additionally, published recommendations have been largely based on expert opinion and evidence from other chronic disorders, thus lacking disease specific data (Italian Society of Paediatric Gastroenterology, Nutrition et al., 2015; Ludvigsson et al., 2016).

At what age should the transition to expert adult care occur? This has not been specifically indicated for CD, although the American Academy of Pediatrics had suggested that the transition for chronic conditions be started at age 12–13 years, developing a transition plan at age 14–15 years, with the actual transfer taking place at ≥18 years of age (American Academy of, P., P. American Academy of Family, P. American College of, G. Transitions Clinical Report Authoring et al., 2011). In reality, the exact timing and modalities of the transition should be flexible, and be the result of a proper, consensual plan fully discussed with family, taking into consideration the adolescent's physical, mental, psychosocial development, and other factors, such as the level of disease activity, dietary adherence, and the patient's autonomy in disease management (Mearin et al., 2022).

Several issues must be discussed during the transition period in CD (Nagra et al., 2015). First of all, patients need to realize—in plain, honest terms but without being alarmed beyond reason—that they have a long-term condition that requires monitoring throughout adult life, as untreated CD may be linked, in adult life, to increased comorbidities (Elfstrom et al., 2008; Jafri et al., 2008; Ludvigsson et al., 2009; Viljamaa et al., 2006). Unfortunately, it has been reported that adolescents with CD transitioning from pediatric care tend to show in general poor follow up (O'Leary et al., 2004) and those lost to follow up to have poor dietary adherence, resulting in elevated CD serologies (Barnea et al., 2014), increased disease burden with symptoms and poorer QoL (Barratt et al., 2011). A good GFD adherence at time of adult referral and classical presentation of CD appear to be independent predictors of good long-term GFD adherence, while, not surprisingly, being lost to follow-up was predictive of poorer long-term GFD adherence (Schiepatti et al., 2022).

A formal transfer of care of an adolescent with CD to adult care, possibly in the context of the same Medical Center, should be implemented. The transfer should be structured and, at minimum, include a transition letter or "celiac passport" providing data on the basis of diagnosis, follow-up, anthropometric data, possible comorbidities and dietary adherence level as also recommended by the recent ESPGHAN position paper on follow-up (Mearin et al., 2022) and by the European Society for the Study of Celiac Disease (ESsCD)(Al-Toma et al., 2019).

References

Acar, S., Yetkiner, A. A., Ersin, N., Oncag, O., Aydogdu, S., & Arikan, C. (2012). Oral findings and salivary parameters in children with celiac disease: A preliminary study. *Medical Principles and Practice: International Journal of the Kuwait University, Health Science Centre*, *21*(2), 129–133.

Acharya, P., & Mathur, M. (2020). Association between psoriasis and celiac disease: A systematic review and meta-analysis. *Journal of the American Academy of Dermatology*, *82*(6), 1376–1385.

Akman, S., Sahaloglu, O., Dalkan, C., Bahceciler, N. N., & Arikan, C. (2018). Is celiac disease misdiagnosed in children with functional constipation? *The Turkish Journal of Gastroenterology: the Official Journal of Turkish Society of Gastroenterology*, *29*(2), 210–214.

Ali Ibrahim, A., Kenyon, V., Fasano, A., & Leonard, M. M. (2022). Budesonide and the gluten containing elimination diet as treatments for non-responsive celiac disease in children. *Journal of Pediatric Gastroenterology and Nutrition*, *75*(5), 616–622.

Almallouhi, E., King, K. S., Patel, B., Wi, C., Juhn, Y. J., Murray, J. A., & Absah, I. (2017). Increasing incidence and altered presentation in a population-based study of pediatric celiac disease in North America. *Journal of Pediatric Gastroenterology and Nutrition*, *65*(4), 432–437.

Alsadat, F. A., Alamoudi, N. M., El-Housseiny, A. A., Felemban, O. M., Dardeer, F. M., & Saadah, O. I. (2021). Oral and dental manifestations of celiac disease in children: A case-control study. *BMC Oral Health*, *21*(1), 669.

Al-Toma, A., Volta, U., Auricchio, R., Castillejo, G., Sanders, D. S., Cellier, C., Mulder, C. J., & Lundin, K. E. A. (2019). European Society for the Study of Coeliac Disease (ESsCD) guideline for coeliac disease and other gluten-related disorders. *United European Gastroenterologisches Journal: Organ der Gesellschaft fur Gastroenterologie der DDR*, *7*(5), 583–613.

Amarri, S., Alvisi, P., De Giorgio, R., Gelli, M. C., Cicola, R., Tovoli, F., Sassatelli, R., Caio, G., & Volta, U. (2013). Antibodies to deamidated gliadin peptides: An accurate predictor of coeliac disease in infancy. *Journal of Clinical Immunology*, *33*(5), 1027–1030.

American Academy of, P., P. American Academy of Family, P. American College of, G. Transitions Clinical Report AuthoringCooley, W. C., & Sagerman, P. J. (2011). Supporting the health care transition from adolescence to adulthood in the medical home. *Pediatrics*, *128*(1), 182–200.

Assa, A., Frenkel-Nir, Y., Tzur, D., Katz, L. H., & Shamir, R. (2017). Large population study shows that adolescents with celiac disease have an increased risk of multiple autoimmune and nonautoimmune comorbidities. *Acta Paediatrica*, *106*(6), 967–972.

Avsar, A., & Kalayci, A. G. (2008). The presence and distribution of dental enamel defects and caries in children with celiac disease. *The Turkish Journal of Pediatrics*, *50*(1), 45–50.

Badizadegan, K., Vanlandingham, D. M., Hampton, W., & Thompson, K. M. (2020). Value of biopsy in a cohort of children with high-titer celiac serologies: observation of dynamic policy differences between Europe and North America. *BMC Health Services Research*, *20*(1), 962.

Balaban, D. V., Dima, A., Jurcut, C., Popp, A., & Jinga, M. (2019). Celiac crisis, a rare occurrence in adult celiac disease: A systematic review. *World Journal of Clinical Cases*, *7*(3), 311–319.

Barbara, G., C. Cremon, M. Bellini, M. Corsetti, G. Di Nardo, F. Falangone, L. Fuccio, F. Galeazzi, P. Iovino, G. Sarnelli, E.V. Savarino, V. Stanghellini, A. Staiano, C. Stasi, C. Tosetti, R. Turco, E. Ubaldi, R.M. Zagari, L. Zenzeri, G. Marasco (2022). Italian guidelines for the management of irritable bowel syndrome: Joint Consensus from the Italian Societies of: Gastroenterology and Endoscopy (SIGE), Neurogastroenterology and Motility (SINGEM), Hospital Gastroenterologists and Endoscopists (AIGO), Digestive Endoscopy (SIED), General Medicine (SIMG), Gastroenterology, Hepatology and Pediatric Nutrition (SIGENP) and Pediatrics (SIP). Dig Liver Dis.

Barera, G., Bonfanti, R., Viscardi, M., Bazzigaluppi, E., Calori, G., Meschi, F., Bianchi, C., & Chiumello, G. (2002). Occurrence of celiac disease after onset of type 1 diabetes: A 6-year prospective longitudinal study. *Pediatrics*, *109*(5), 833–838.

Barnea, L., Mozer-Glassberg, Y., Hojsak, I., Hartman, C., & Shamir, R. (2014). Pediatric celiac disease patients who are lost to follow-up have a poorly controlled disease. *Digestion*, *90*(4), 248–253.

Barratt, S. M., Leeds, J. S., & Sanders, D. S. (2011). Quality of life in Coeliac Disease is determined by perceived degree of difficulty adhering to a gluten-free diet, not the level of dietary adherence ultimately achieved. *Journal of Gastrointestinal and Liver Diseases: JGLD*, *20*(3), 241–245.

Bashiri, H., Afshari, D., Babaei, N., & Ghadami, M. R. (2016). Celiac disease and epilepsy: The effect of gluten-free diet on seizure control. *Advances in Clinical and Experimental Medicine*, *25*(4), 751–754.

Basso, D., Guariso, G., Fogar, P., Meneghel, A., Zambon, C. F., Navaglia, F., Greco, E., Schiavon, S., Rugge, M., & Plebani, M. (2009). Antibodies against synthetic deamidated gliadin peptides for celiac disease diagnosis and follow-up in children. *Clinical Chemistry*, *55*(1), 150–157.

Benini, F., Mora, A., Turini, D., Bertolazzi, S., Lanzarotto, F., Ricci, C., Villanacci, V., Barbara, G., Stanghellini, V., & Lanzini, A. (2012). Slow gallbladder emptying reverts to normal but small intestinal transit of a physiological meal remains slow in celiac patients during gluten-free diet. *Neurogastroenterology and Motility: the Official Journal of the European Gastrointestinal Motility Society*, *24*(2), 100–107, e179–180.

Bodis, G., Toth, V., & Schwarting, A. (2018). Role of Human Leukocyte Antigens (HLA) in autoimmune diseases. *Rheumatology and Therapy*, *5*(1), 5–20.

Bolotin, D., & Petronic-Rosic, V. (2011). Dermatitis herpetiformis. Part II. Diagnosis, management, and prognosis. *Journal of the American Academy of Dermatology*, *64*(6), 1027–1033, quiz 1033-1024.

Borkar, V. V., Poddar, U., Thakral, A., Agarwal, J., Srivastava, A., Yachha, S. K., & Kumar, S. (2018). Intussusception in celiac disease: Is it a common feature in children? *Journal of Gastroenterology and Hepatology*, *33*(2), 380–384.

Bottaro, G., Cataldo, F., Rotolo, N., Spina, M., & Corazza, G. R. (1999). The clinical pattern of subclinical/silent celiac disease: An analysis on 1026 consecutive cases. *The American Journal of Gastroenterology*, *94*(3), 691–696.

Brar, P., Kwon, G. Y., Egbuna, I. I., Holleran, S., Ramakrishnan, R., Bhagat, G., & Green, P. H. (2007). Lack of correlation of degree of villous atrophy with severity of clinical presentation of coeliac disease. *Digestive and Liver Disease: Official Journal of the Italian Society of Gastroenterology and the Italian Association for the Study of the Liver*, *39*(1), 26–29, discussion 30-22.

Bucci, P., Carile, F., Sangianantoni, A., D'Angio, F., Santarelli, A., & Lo Muzio, L. (2006). Oral aphthous ulcers and dental enamel defects in children with coeliac disease. *Acta Paediatr*, *95*(2), 203–207.

Butwicka, A., Lichtenstein, P., Frisen, L., Almqvist, C., Larsson, H., & Ludvigsson, J. F. (2017). Celiac disease is associated with childhood psychiatric disorders: A population-based study. *The Journal of Pediatrics*, *184*(87–93), e81.

Bybrant, M. C., Ortqvist, E., Lantz, S., & Grahnquist, L. (2014). High prevalence of celiac disease in Swedish children and adolescents with type 1 diabetes and the relation to the Swedish epidemic of celiac disease: a cohort study. *Scandinavian Journal of Gastroenterology*, *49*(1), 52–58.

Campisi, G., Di Liberto, C., Carroccio, A., Compilato, D., Iacono, G., Procaccini, M., Di Fede, G., Lo Muzio, L., Craxi, A., Catassi, C., & Scully, C. (2008). Coeliac disease: oral ulcer prevalence, assessment of risk and association with gluten-free diet in children. *Digestive and Liver Disease: Official Journal of the Italian Society of Gastroenterology and the Italian Association for the Study of the Liver*, *40*(2), 104–107.

Canales, P., Mery, V. P., Larrondo, F. J., Bravo, F. L., & Godoy, J. (2006). Epilepsy and celiac disease: Favorable outcome with a gluten-free diet in a patient refractory to antiepileptic drugs. *The Neurologist*, *12*(6), 318–321.

Canova, C., Pitter, G., Zanier, L., Simonato, L., Michaelsson, K., & Ludvigsson, J. F. (2018). Risk of fractures in youths with celiac disease-A population-based study. *The Journal of Pediatrics*, *198*, 117–120.

Canova, C., Rosato, I., Marsilio, I., Valiante, F., Zorzetto, V., Cataudella, G., D'Odorico, A., & Zingone, F. (2021). Quality of life and psychological disorders in coeliac disease: A prospective multicentre study. *Nutrients*, *13*(9).

Canova, C., Ludvigsson, J., Barbiellini Amidei, C., Zanier, L., & Zingone, F. (2020). The risk of epilepsy in children with celiac disease: A population-based cohort study. *European Journal of Neurology: the Official Journal of the European Federation of Neurological Societies*, *27*(6), 1089–1095.

Capone, K., Sansotta, N., Vohra, P., Jericho, H., & Guandalini, S. (2020). Milk protein-induced villous atrophy and elevated serologies in four children with celiac disease on a gluten-free diet. *Annals of Pediatrics*, *3*(1), 1028–1031.

Caprai, S., Vajro, P., Ventura, A., Sciveres, M., & Maggiore, G.S. S. G. f. A. L. D. i. C. Disease. (2008). Autoimmune liver disease associated with celiac disease in childhood: a multicenter study. *Clinical Gastroenterology and Hepatology: the Official Clinical Practice Journal of the American Gastroenterological Association*, *6*(7), 803–806.

Carta, M. G., Hardoy, M. C., Boi, M. F., Mariotti, S., Carpiniello, B., & Usai, P. (2002). Association between panic disorder, major depressive disorder and celiac disease: A possible role of thyroid autoimmunity. *Journal of Psychosomatic Research*, *53*(3), 789–793.

Cassio, A., Ricci, G., Baronio, F., Miniaci, A., Bal, M., Bigucci, B., Conti, V., & Cicognani, A. (2010). Long-term clinical significance of thyroid autoimmunity in children with celiac disease. *The Journal of Pediatrics*, *156*(2), 292–295.

Castellaneta, S., Piccinno, E., Oliva, M., Cristofori, F., Vendemiale, M., Ortolani, F., Papadia, F., Catassi, C., Cavallo, L., & Francavilla, R. (2015). High rate of spontaneous normalization of celiac serology in a cohort of 446 children with type 1 diabetes: a prospective study. *Diabetes Care, 38*(5), 760–766.

Castro, P. D., Harkin, G., Hussey, M., Christopher, B., Kiat, C., Chin, J. L., Trimble, V., McNamara, D., MacMathuna, P., Egan, B., Ryan, B., Kevans, D., Abuzakouk, M., Farrell, R., Feighery, C., Byrnes, V., Mahmud, N., & McManus, R. (2020). Prevalence of coexisting autoimmune thyroidal diseases in coeliac disease is decreasing. *United European Gastroenterology Journal, 8*(2), 148–156.

Cataldo, F., Marino, V., Bottaro, G., Greco, P., & Ventura, A. (1997). Celiac disease and selective immunoglobulin A deficiency. *The Journal of Pediatrics, 131*(2), 306–308.

Catassi, C., Kryszak, D., Bhatti, B., Sturgeon, C., Helzlsouer, K., Clipp, S. L., Gelfond, D., Puppa, E., Sferruzza, A., & Fasano, A. (2010). Natural history of celiac disease autoimmunity in a USA cohort followed since 1974. *Annals of Medicine, 42*(7), 530–538.

Catassi, G. N., Pulvirenti, A., Monachesi, C., Catassi, C., & Lionetti, E. (2021). Diagnostic accuracy of IgA anti-transglutaminase and IgG anti-deamidated gliadin for diagnosis of celiac disease in children under two years of age: A systematic review and meta-analysis. *Nutrients, 14*(1).

Cebolla, A., Moreno, M. L., Coto, L., & Sousa, C. (2018). Gluten immunogenic peptides as standard for the evaluation of potential harmful prolamin content in food and human specimen. *Nutrients, 10*(12).

Charalampopoulos, D., Panayiotou, J., Chouliaras, G., Zellos, A., Kyritsi, E., & Roma, E. (2013). Determinants of adherence to gluten-free diet in Greek children with coeliac disease: A cross-sectional study. *European Journal of Clinical Nutrition, 67*(6), 615–619.

Cheng, J., T. Malahias, P. Brar, M.T. Minaya and P.H. Green (2010). The association between celiac disease, dental enamel defects, and aphthous ulcers in a United States cohort. J Clin Gastroenterol 44(3): 191–194.

Cigic, L., Galic, T., Kero, D., Simunic, M., Medvedec Mikic, I., Kalibovic Govorko, D., & Biocina Lukenda, D. (2016). The prevalence of celiac disease in patients with geographic tongue. *Journal of Oral Pathology & Medicine: Official Publication of the International Association of Oral Pathologists and the American Academy of Oral Pathology, 45*(10), 791–796.

Connan, V., Marcon, M. A., Mahmud, F. H., Assor, E., Martincevic, I., Bandsma, R. H., Vresk, L., & Walsh, C. M. (2019). Online education for gluten-free diet teaching: Development and usability testing of an e-learning module for children with concurrent celiac disease and type 1 diabetes. *Pediatric Diabetes, 20*(3), 293–303.

Corazza, G. R., Andreani, M. L., Venturo, N., Bernardi, M., Tosti, A., & Gasbarrini, G. (1995). Celiac disease and alopecia areata: report of a new association. *Gastroenterology, 109*(4), 1333–1337.

Coster, W., & Khetani, M. A. (2008). Measuring participation of children with disabilities: issues and challenges. *Disability and Rehabilitation, 30*(8), 639–648.

Couper, J. J., Haller, M. J., Ziegler, A. G., Knip, M., Ludvigsson, J., & Craig, M. E.P. International Society for and D. Adolescent. (2014). ISPAD Clinical Practice Consensus Guidelines 2014. Phases of type 1 diabetes in children and adolescents. *Pediatric Diabetes, 15*(20), 18–25.

Cristofori, F., D'Abramo, F. S., Rutigliano, V., Dargenio, V. N., Castellaneta, S., Piscitelli, D., De Benedittis, D., Indrio, F., Raguseo, L. C., Barone, M., & Francavilla, R. (2021). Esophageal eosinophilia and eosinophilic esophagitis in celiac children: A ten year prospective observational study. *Nutrients, 13*(11).

Cristofori, F., Tripaldi, M., Lorusso, G., Indrio, F., Rutigliano, V., Piscitelli, D., Castellaneta, S., Bentivoglio, V., & Francavilla, R. (2021). Functional abdominal pain disorders and constipation in children on gluten-free diet. *Clinical Gastroenterology and Hepatology: the Official Clinical Practice Journal of the American Gastroenterological Association, 19*(12), 2551–2558.

Cucchiara, S., Bassotti, G., Castellucci, G., Minella, R., Betti, C., Fusaro, C., Morelli, A., Bertotto, A., & Auricchio, S. (1995). Upper gastrointestinal motor abnormalities in children with active celiac disease. *Journal of Pediatric Gastroenterology and Nutrition, 21*(4), 435–442.

De Bastiani, R., Gabrielli, M., Lora, L., Napoli, L., Tosetti, C., Pirrotta, E., Ubaldi, E., Bertolusso, L., Zamparella, M., De Polo, M., Nebiacolombo, C., Bortot, M., Mancuso, M., Bacchin, P., Marsala, V., Pinna, R., Tursi, A., Benedetto, E., Cuffari, A., … Gasbarrini, A. (2015). Association between coeliac disease and psoriasis: Italian primary care multicentre study. *Dermatology (Basel, Switzerland), 230*(2), 156–160.

De Giuseppe, R., F. Bergomas, F. Loperfido, F. Giampieri, G. Preatoni, V. Calcaterra and H. Cena (2023). Could celiac disease and overweight/obesity coexist in school-aged children and adolescents? A Systematic Review. Child Obesity.

Deora, V., Aylward, N., Sokoro, A., & El-Matary, W. (2017). Serum vitamins and minerals at diagnosis and follow-up in children with celiac disease. *Journal of Pediatric Gastroenterology and Nutrition, 65*(2), 185–189.

Di Biase, A. R., Colecchia, A., Scaioli, E., Berri, R., Viola, L., Vestito, A., Balli, F., & Festi, D. (2010). Autoimmune liver diseases in a paediatric population with coeliac disease - A 10-year single-centre experience. *Alimentary Pharmacology & Therapeutics, 31*(2), 253–260.

Dimitrova, A. K., Ungaro, R. C., Lebwohl, B., Lewis, S. K., Tennyson, C. A., Green, M. W., Babyatsky, M. W., & Green, P. H. (2013). Prevalence of migraine in patients with celiac disease and inflammatory bowel disease. *Headache, 53*(2), 344–355.

Donat, E., Roca, M., Castillejo, G., Sanchez-Valverde, F., Garcia-Burriel, J. I., Martinez-Ojinaga, E., Eizaguirre, F. J., Barrio, J., Cilleruelo, M. L., Perez-Solis, D., Ochoa-Sangrador, C., Vecino-Lopez, R., Miranda-Cid, M. D. C., Garcia-Calatayud, S., Torres-Peral, R., Juste, M., Armas, H., Barros-Garcia, P., Leis, R., … Ribes-Koninckx, C.H. P. N. S. Coeliac Disease Working Group of the Spanish Gastroenterology. (2022). Correlation of anti-tissue transglutaminase antibodies with the mucosal changes and IgA status of children with celiac disease. *Journal of Pediatric Gastroenterology and Nutrition, 75*(6), 743–748.

Dos Santos, S., & Liote, F. (2017). Osteoarticular manifestations of celiac disease and non-celiac gluten hypersensitivity. *Joint, Bone, Spine: Revue du Rhumatisme, 84*(3), 263–266.

Efe, A., & Tok, A. (2023). A Clinical Investigation on ADHD-traits in childhood celiac disease. *Journal of Attention Disorders*, 10870547221149195.

Ehsani-Ardakani, M. J., Rostami Nejad, M., Villanacci, V., Volta, U., Manenti, S., Caio, G., Giovenali, P., Becheanu, G., Diculescu, M., Pellegrino, S., Magazzu, G., Casella, G., Di Bella, C., Decarli, N., Biancalani, M., Bassotti, G., Hogg-Kollars, S., Zali, M. R., & Rostami, K. (2013). Gastrointestinal and non-gastrointestinal presentation in patients with celiac disease. *Archives of Iranian Medicine*, *16*(2), 78–82.

Elfstrom, P., Montgomery, S. M., Kampe, O., Ekbom, A., & Ludvigsson, J. F. (2008). Risk of thyroid disease in individuals with celiac disease. *The Journal of Clinical Endocrinology and Metabolism*, *93*(10), 3915–3921.

Elli, L., Discepolo, V., Bardella, M. T., & Guandalini, S. (2014). Does gluten intake influence the development of celiac disease-associated complications? *Journal of Clinical Gastroenterology*, *48*(1), 13–20.

Emilsson, L., Andersson, B., Elfstrom, P., Green, P. H., & Ludvigsson, J. F. (2012). Risk of idiopathic dilated cardiomyopathy in 29 000 patients with celiac disease. *J Am Heart Assoc*, *1*(3), e001594.

Emilsson, L., Smith, J. G., West, J., Melander, O., & Ludvigsson, J. F. (2011). Increased risk of atrial fibrillation in patients with coeliac disease: A nationwide cohort study. *European Heart Journal*, *32*(19), 2430–2437.

Errichiello, S., Esposito, O., Di Mase, R., Camarca, M. E., Natale, C., Limongelli, M. G., Marano, C., Coruzzo, A., Lombardo, M., Strisciuglio, P., & Greco, L. (2010). Celiac disease: Predictors of compliance with a gluten-free diet in adolescents and young adults. *Journal of Pediatric Gastroenterology and Nutrition*, *50*(1), 54–60.

Fasano, A., & Catassi, C. (2012). Clinical practice. Celiac disease. *The New England Journal of Medicine*, *367*(25), 2419–2426.

Freeman, H. J. (2016). Endocrine manifestations in celiac disease. *World Journal of Gastroenterology: WJG*, *22*(38), 8472–8479.

Frulio, G., Polimeno, A., Palmieri, D., Fumi, M., Auricchio, R., Piccolo, E., & Carandente Giarrusso, P. (2015). Evaluating diagnostic accuracy of anti-tissue Transglutaminase IgA antibodies as first screening for Celiac Disease in very young children. *Clinica Chimica Acta; International Journal of Clinical Chemistry*, *446*, 237–240.

Garg, K., Agarwal, P., Gupta, R. K., & Sitaraman, S. (2017). Joint involvement in children with celiac disease. *Indian Pediatrics*, *54*(11), 946–948.

Gatti, S., Lionetti, E., Balanzoni, L., Verma, A. K., Galeazzi, T., Gesuita, R., Scattolo, N., Cinquetti, M., Fasano, A., Catassi, C., & Celiac Screening, T. (2020). Increased prevalence of celiac disease in school-age children in Italy. *Clinical Gastroenterology and Hepatology: the Official Clinical Practice Journal of the American Gastroenterological Association*, *18*(3), 596–603.

Gaur, S. (2022). The association between ADHD and celiac disease in children. *Children (Basel)*, *9*(6).

Germann, R., Kuch, M., Prinz, K., Ebbing, A., & Schindera, F. (1997). Celiac disease: An uncommon cause of recurrent intussusception. *Journal of Pediatric Gastroenterology and Nutrition*, *25*(4), 415–416.

Germone, M., Phu, T., Slosky, C., Pan, Z., Jones, A., Stahl, M., Mehta, P., Shull, M., Ariefdjohan, M., & Liu, E. (2022). Anxiety and depression in pediatric patients with celiac disease: A large cross-sectional study. *Journal of Pediatric Gastroenterology and Nutrition*, *75*(2), 181–185.

Ghozzi, M., W. Sakly, A. Mankai, E. Bouajina, F. Bahri, R. Nouira, C. Kechrid and I. Ghedira (2013). Screening for celiac disease, by endomysial antibodies, in patients with unexplained articular manifestations. Rheumatology International.

Ghozzi, M., Sakly, W., Mankai, A., Bouajina, E., Bahri, F., Nouira, R., Kechrid, C., & Ghedira, I. (2014). Screening for celiac disease, by endomysial antibodies, in patients with unexplained articular manifestations. *Rheumatology International*, *34*(5), 637–642.

Gidrewicz, D., Trevenen, C. L., Lyon, M., & Butzner, J. D. (2017). Normalization time of celiac serology in children on a gluten-free diet. *Journal of Pediatric Gastroenterology and Nutrition*, *64*(3), 362–367.

Gobbi, G. (2005). Coeliac disease, epilepsy and cerebral calcifications. *Brain & Development*, *27*(3), 189–200.

Gould, M. J., Dowhaniuk, J., Arredondo, J., Azzopardi, P., Hu, T., Mileski, H., Carpenter, A., Pai, N., & Brill, H. (2023). Characteristics of pediatric patients with celiac disease identified due to an affected first-degree family member. *Journal of Pediatric Gastroenterology and Nutrition*, *76*(1), 49–52.

Gustafsson, I., Repo, M., Popp, A., Kaukinen, K., Hiltunen, P., Arvola, T., Taavela, J., Vornanen, M., Kivela, L., & Kurppa, K. (2020). Prevalence and diagnostic outcomes of children with duodenal lesions and negative celiac serology. *Digestive and Liver Disease: Official Journal of the Italian Society of Gastroenterology and the Italian Association for the Study of the Liver*, *52*(3), 289–295.

Hagopian, W., Lee, H. S., Liu, E., Rewers, M., She, J. X., Ziegler, A. G., Lernmark, A., Toppari, J., Rich, S. S., Krischer, J. P., Erlich, H., Akolkar, B., Agardh, D., & Group, T. S. (2017). Co-occurrence of Type 1 Diabetes and celiac disease autoimmunity. *Pediatrics*, *140*(5).

Harper, J. W., Holleran, S. F., Ramakrishnan, R., Bhagat, G., & Green, P. H. (2007). Anemia in celiac disease is multifactorial in etiology. *American Journal of Hematology*, *82*(11), 996–1000.

Herman, M. L., Rubio-Tapia, A., Lahr, B. D., Larson, J. J., Van Dyke, C. T., & Murray, J. A. (2012). Patients with celiac disease are not followed up adequately. *Clinical Gastroenterology and Hepatology: the Official Clinical Practice Journal of the American Gastroenterological Association*, *10*(8), 893–899, e891.

Hill, I. D., Fasano, A., Guandalini, S., Hoffenberg, E., Levy, J., Reilly, N., & Verma, R. (2016). NASPGHAN clinical report on the diagnosis and treatment of gluten-related disorders. *Journal of Pediatric Gastroenterology and Nutrition*, *63*(1), 156–165.

Hill, I. D., Dirks, M. H., Liptak, G. S., Colletti, R. B., Fasano, A., Guandalini, S., Hoffenberg, E. J., Horvath, K., Murray, J. A., Pivor, M., & Seidman, E. G.H. North American Society for Pediatric Gastroenterology and Nutrition. (2005). Guideline for the diagnosis and treatment of celiac disease in children: recommendations of the North American Society for Pediatric Gastroenterology, Hepatology and Nutrition. *Journal of Pediatric Gastroenterology and Nutrition*, *40*(1), 1–19.

Hom, G. L., Hom, B. L., Kaplan, B., & Rothner, A. D. (2020). A single institution's experience of primary headache in children with celiac disease. *Journal of Child Neurology*, *35*(1), 37–41.

Hommeida, S., Alsawas, M., Murad, M. H., Katzka, D. A., Grothe, R. M., & Absah, I. (2017). The Association Between Celiac Disease and Eosinophilic Esophagitis: Mayo Experience and Meta-analysis of the Literature. *Journal of Pediatric Gastroenterology and Nutrition*, *65*(1), 58–63.

Husby, S., Koletzko, S., Korponay-Szabo, I., Kurppa, K., Mearin, M. L., Ribes-Koninckx, C., Shamir, R., Troncone, R., Auricchio, R., Castillejo, G., Christensen, R., Dolinsek, J., Gillett, P., Hrobjartsson, A., Koltai, T., Maki, M., Nielsen, S. M., Popp, A., Stordal, K., ... Wessels, M. (2020). European Society Paediatric Gastroenterology, Hepatology and Nutrition Guidelines for Diagnosing Coeliac Disease 2020. *Journal of Pediatric Gastroenterology and Nutrition*, *70*(1), 141–156.

Husby, S., Koletzko, S., Korponay-Szabo, I. R., Mearin, M. L., Phillips, A., Shamir, R., Troncone, R., Giersiepen, K., Branski, D., Catassi, C., Lelgeman, M., Maki, M., Ribes-Koninckx, C., Ventura, A., & Zimmer, K. P.E. W. G. O. C. D. Diagnosis, E. G. Committee, H. European Society for Pediatric Gastroenterology and Nutrition. (2012). European Society for Pediatric Gastroenterology, Hepatology, and Nutrition guidelines for the diagnosis of coeliac disease. *Journal of Pediatric Gastroenterology and Nutrition*, *54*(1), 136–160.

Italian Society of Paediatric Gastroenterology, H., I. A. O. H. G. Nutrition, I. S. o. E. I. S. O. G. EndoscopistsElli, L., Maieron, R., Martelossi, S., Guariso, G., Buscarini, E., Conte, D., Di Giulio, E., Staiano, A., Barp, J., Bassotti, G., Bianco, M. A., Buri, L., Carrara, M., Ghidini, B., Giannini, O., Knafelz, D., Miele, E., Peralta, S., ... Guadagnini, T. (2015). Transition of gastroenterological patients from paediatric to adult care: A position statement by the Italian Societies of Gastroenterology. *Digestive and Liver Disease: Official Journal of the Italian Society of Gastroenterology and the Italian Association for the Study of the Liver*, *47*(9), 734–740.

Jafri, M. R., Nordstrom, C. W., Murray, J. A., Van Dyke, C. T., Dierkhising, R. A., Zinsmeister, A. R., & Melton, L. J., 3rd (2008). Long-term fracture risk in patients with celiac disease: A population-based study in Olmsted County, Minnesota. *Digestive Diseases and Sciences*, *53*(4), 964–971.

Jericho, H., Khan, N., Cordova, J., Sansotta, N., Guandalini, S., & Keenan, K. (2021). Call for action: High rates of depression in the pediatric celiac disease population impacts quality of life. *JPGN Reports*, *2*(3), e074.

Jericho, H., Sansotta, N., & Guandalini, S. (2017). Extraintestinal manifestations of celiac disease: Effectiveness of the gluten-free diet. *Journal of Pediatric Gastroenterology and Nutrition*, *65*(1), 75–79.

Johnston, S. D., Rodgers, C., & Watson, R. G. (2004). Quality of life in screen-detected and typical coeliac disease and the effect of excluding dietary gluten. *European Journal of Gastroenterology & Hepatology*, *16*(12), 1281–1286.

Jonkers, I. H., & Wijmenga, C. (2017). Context-specific effects of genetic variants associated with autoimmune disease. *Human Molecular Genetics*, *26*(R2), R185–R192.

Kaur, P., Agarwala, A., Makharia, G., Bhatnagar, S., & Tandon, N. (2020). Effect of gluten-free diet on metabolic control and anthropometric parameters in Type 1 diabetes with subclinical celiac disease: A randomized controlled trial. *Endocrine Practice: Official Journal of the American College of Endocrinology and the American Association of Clinical Endocrinologists*, *26*(6), 660–667.

Khatib, M., Baker, R. D., Ly, E. K., Kozielski, R., & Baker, S. S. (2016). Presenting pattern of pediatric celiac disease. *Journal of Pediatric Gastroenterology and Nutrition*, *62*(1), 60–63.

King, J.A., J.A. Bakal, B. Li, T.A. Whitten, D. Gidrewicz, J.M. Turner, S.V. Van Zanten, H.Q. Huynh, S.J. Heitman, A.A. Shaheen, H. Quan, J. Godley, F.E. Underwood, L. Hracs, D. Bergman, J.F. Ludvigsson, B. Lebwohl, E.I. Benchimol, T. Williamson and G.G. Kaplan (2023). Variation in testing for and incidence of celiac autoimmunity in Canada: A population-based study. Gastroenterology.

Koletzko, S., Auricchio, R., Dolinsek, J., Gillett, P., Korponay-Szabo, I., Kurppa, K., Mearin, M. L., Maki, M., Popp, A., Ribes, C., Shamir, R., Stordal, K., Troncone, R., Werkstetter, K., Wessels, M., Zimmer, K. P., & Husby, S. (2017). No need for routine endoscopy in children with celiac disease on a gluten-free diet. *Journal of Pediatric Gastroenterology and Nutrition*, *65*(3), 267–269.

Krauthammer, A., Guz-Mark, A., Zevit, N., Marderfeld, L., Waisbourd-Zinman, O., Silbermintz, A., Mozer-Glassberg, Y., Nachmias Friedler, V., Rozenfeld Bar Lev, M., Matar, M., Assa, A., & Shamir, R. (2020). Two decades of pediatric celiac disease in a tertiary referral center: What has changed? *Digestive and Liver Disease: Official Journal of the Italian Society of Gastroenterology and the Italian Association for the Study of the Liver*, *52*(4), 457–461.

Krauthammer, A., Guz-Mark, A., Zevit, N., Marderfeld, L., Waisbourd-Zinman, O., Silbermintz, A., Mozer-Glassberg, Y., Nachmias Friedler, V., Rozenfeld Bar Lev, M., Matar, M., Assa, A., & Shamir, R. (2021). Age-dependent trends in the celiac disease: A tertiary center experience. *Journal of Pediatric Gastroenterology and Nutrition*, *72*(6), 894–899.

Kröger, S., Kurppa, K., Repo, M., Huhtala, H., Kaukinen, K., Lindfors, K., Arvola, T., & Kivela, L. (2020). Severity of villous atrophy at diagnosis in childhood does not predict long-term outcomes in celiac disease. *Journal of Pediatric Gastroenterology and Nutrition*, *71*(1), 71–77.

Kurppa, K., Lauronen, O., Collin, P., Ukkola, A., Laurila, K., Huhtala, H., Maki, M., & Kaukinen, K. (2012). Factors associated with dietary adherence in celiac disease: a nationwide study. *Digestion*, *86*(4), 309–314.

Lauron, S., C. Plasse, M. Vaysset, B. Pereira, M. D'Incan, F. Rondepierre and I. Jalenques (2023). Prevalence and odds of depressive and anxiety disorders and symptoms in children and adults with Alopecia Areata: A systematic review and meta-analysis. JAMA Dermatology, 159(3): 281-288.

Lebwohl, B., J. Soderling, B. Roelstraete, M.G. Lebwohl, P.H. Green and J.F. Ludvigsson (2020). Risk of skin disorders in patients with celiac disease: A population-based cohort study. Journal of the American Academy of Dermatology, 85.

Lebwohl, B., L. Haggard, L. Emilsson, J. Soderling, B. Roelstraete, A. Butwicka, P.H. Green and J.F. Ludvigsson (2020). Psychiatric disorders in patients with a diagnosis of celiac disease during childhood from 1973 to 2016. Clinical Gastroenterology and Hepatology: the Official Clinical Practice Journal of the American Gastroenterological Association.

Lee, H., Song, W. J., Koh, S. J., & Park, H. (2022). Epidemiological association between alopecia areata and celiac disease: A meta-analysis. *The Journal of Dermatology*, *49*(11), e399–e400.

Leffler, D. A., & Schuppan, D. (2010). Update on serologic testing in celiac disease. *The American Journal of Gastroenterology*, *105*(12), 2520–2524.

Leonard, M. M., Lebwohl, B., Rubio-Tapia, A., & Biagi, F. (2021). AGA clinical practice update on the evaluation and management of seronegative enteropathies: Expert review. *Gastroenterology, 160*(1), 437–444.

Lidon, A. C., Patricia, M. L., Vinesh, D., & Marta, M. S. (2022). Evaluation of gluten exclusion for the improvement of rheumatoid arthritis in adults. *Nutrients, 14*(24).

Lionetti, E., Pjetraj, D., Gatti, S., Catassi, G., Bellantoni, A., Boffardi, M., Cananzi, M., Cinquetti, M., Francavilla, R., Malamisura, B., Montuori, M., Zuccotti, G., Cristofori, F., Gaio, P., Passaro, T., Penagini, F., Testa, A., Trovato, C. M., & Catassi, C. (2023). Prevalence and detection rate of celiac disease in Italy: Results of a SIGENP multicenter screening in school-age children. *Digestive and Liver Disease: Official Journal of the Italian Society of Gastroenterology and the Italian Association for the Study of the Liver.*

Lionetti, E., Francavilla, R., Maiuri, L., Ruggieri, M., Spina, M., Pavone, P., Francavilla, T., Magista, A. M., & Pavone, L. (2009). Headache in pediatric patients with celiac disease and its prevalence as a diagnostic clue. *Journal of Pediatric Gastroenterology and Nutrition, 49*(2), 202–207.

Liu, E., Wolter-Warmerdam, K., Marmolejo, J., Daniels, D., Prince, G., & Hickey, F. (2020). Routine screening for celiac disease in children with down syndrome improves case finding. *Journal of Pediatric Gastroenterology and Nutrition, 71*(2), 252–256.

Long, K. H., Rubio-Tapia, A., Wagie, A. E., Melton, L. J., 3rd, Lahr, B. D., Van Dyke, C. T., & Murray, J. A. (2010). The economics of coeliac disease: A population-based study. *Alimentary Pharmacology & Therapeutics, 32*(2), 261–269.

Lougaris, V., Sorlini, A., Monfredini, C., Ingrasciotta, G., Caravaggio, A., Lorenzini, T., Baronio, M., Cattalini, M., Meini, A., Ruggeri, L., Salpietro, A., Pilotta, A., Grazzani, L., Prandi, E., Felappi, B., Gualdi, G., Fabiano, A., Fuoti, M., Ravelli, A., ... Plebani, A. (2019). Clinical and Laboratory Features of 184 Italian pediatric patients affected with selective IgA deficiency (SIgAD): A longitudinal single-center study. *Journal of Clinical Immunology, 39*(5), 470–475.

Ludvigsson, J. F., Sellgren, C., Runeson, B., Langstrom, N., & Lichtenstein, P. (2011). Increased suicide risk in coeliac disease—A Swedish nationwide cohort study. *Digestive and Liver Disease: Official Journal of the Italian Society of Gastroenterology and the Italian Association for the Study of the Liver, 43*(8), 616–622.

Ludvigsson, J. F., West, J., Card, T., & Appelros, P. (2012). Risk of stroke in 28,000 patients with celiac disease: A nationwide cohort study in Sweden. *Journal of Stroke and Cerebrovascular Diseases: the Official Journal of National Stroke Association, 21*(8), 860–867.

Ludvigsson, J. F., Hemminki, K., Wahlstrom, J., & Almqvist, C. (2011). Celiac disease confers a 1.sixfold increased risk of asthma: A nationwide population-based cohort study. *The Journal of Allergy and Clinical Immunology, 127*(4), 1071–1073.

Ludvigsson, J. F., Agreus, L., Ciacci, C., Crowe, S. E., Geller, M. G., Green, P. H., Hill, I., Hungin, A. P., Koletzko, S., Koltai, T., Lundin, K. E., Mearin, M. L., Murray, J. A., Reilly, N., Walker, M. M., Sanders, D. S., Shamir, R., Troncone, R., & Husby, S. (2016). Transition from childhood to adulthood in coeliac disease: The Prague consensus report. *Gut, 65*(8), 1242–1251.

Ludvigsson, J. F., Inghammar, M., Ekberg, M., & Egesten, A. (2012). A nationwide cohort study of the risk of chronic obstructive pulmonary disease in coeliac disease. *Journal of Internal Medicine, 271*(5), 481–489.

Ludvigsson, J. F., Aro, P., Walker, M. M., Vieth, M., Agreus, L., Talley, N. J., Murray, J. A., & Ronkainen, J. (2013). Celiac disease, eosinophilic esophagitis and gastroesophageal reflux disease, an adult population-based study. *Scandinavian Journal of Gastroenterology, 48*(7), 808–814.

Ludvigsson, J. F., Montgomery, S. M., Ekbom, A., Brandt, L., & Granath, F. (2009). Small-intestinal histopathology and mortality risk in celiac disease. *JAMA: The Journal of the American Medical Association, 302*(11), 1171–1178.

Lupoli, G.A., M. Tasso, L. Costa, F. Caso, R. Scarpa, A. Del Puente and R. Peluso (2020). Coeliac disease is a risk factor for the development of seronegative arthritis in patients with autoimmune thyroid disease. Rheumatology (Oxford).

MacCulloch, K., & Rashid, M. (2014). Factors affecting adherence to a gluten-free diet in children with celiac disease. *Paediatrics & Child Health, 19*(6), 305–309.

Madden, A. M., Riordan, A. M., & Knowles, L. (2016). Outcomes in coeliac disease: A qualitative exploration of patients' views on what they want to achieve when seeing a dietitian. *Journal of Human Nutrition and Dietetics: the Official Journal of the British Dietetic Association, 29*(5), 607–616.

Mager, D. R., Qiao, J., & Turner, J. (2012). Vitamin D and K status influences bone mineral density and bone accrual in children and adolescents with celiac disease. *European Journal of Clinical Nutrition, 66*(4), 488–495.

Mandile, R., Maglio, M., Pellino, N., Russo, M., Miele, E., Spagnuolo, M. I., Troncone, R., & Auricchio, R. (2021). Seronegative villous atrophy in children: Clinical and immunohistochemical features. *Journal of Pediatric Gastroenterology and Nutrition, 72*(2), 282–287.

Mansikka, E., Hervonen, K., Kaukinen, K., Collin, P., Huhtala, H., Reunala, T., & Salmi, T. (2018). Prognosis of dermatitis herpetiformis patients with and without villous atrophy at diagnosis. *Nutrients, 10*(5).

Marild, K., Stordal, K., Hagman, A., & Ludvigsson, J. F. (2016). Turner syndrome and celiac disease: A case-control study. *Pediatrics, 137*(2), e20152232.

Marild, K., Stordal, K., Bulik, C. M., Rewers, M., Ekbom, A., Liu, E., & Ludvigsson, J. F. (2017). Celiac Disease and anorexia nervosa: A nationwide study. *Pediatrics, 139*(5).

Martelossi, S., Zanatta, E., Del Santo, E., Clarich, P., Radovich, P., & Ventura, A. (1996). Dental enamel defects and screening for coeliac disease. *Acta Paediatrica - Supplement, 412*, 47–48.

Mearin, M. L., Agardh, D., Antunes, H., Al-Toma, A., Auricchio, R., Castillejo, G., Catassi, C., Ciacci, C., Discepolo, V., Dolinsek, J., Donat, E., Gillett, P., Guandalini, S., Husby Md, D. S., Koletzko Md, S., Koltai, T., Korponay-Szabo, I. R., Kurppa, K., Lionetti, E., ... E. S. I. G. o. C. Disease. (2022). ESPGHAN position paper on management and follow-up of children and adolescents with celiac disease. *Journal of Pediatric Gastroenterology and Nutrition, 75*(3), 369–386.

Mearns, E. S., Taylor, A., Thomas Craig, K. J., Puglielli, S., Leffler, D. A., Sanders, D. S., Lebwohl, B., & Hadjivassiliou, M. (2019). Neurological manifestations of neuropathy and ataxia in celiac disease: A systematic review. *Nutrients*, *11*(2).

Mehta, P., Li, Q., Stahl, M., Uusitalo, U., Lindfors, K., Butterworth, M. D., Kurppa, K., Virtanen, S., Koletzko, S., Aronsson, C., Hagopian, W. A., Rewers, M. J., Toppari, J., Ziegler, A. G., Akolkar, B., Krischer, J. P., Agardh, D., Liu, E., & Group, T. S. (2023). Gluten-free diet adherence in children with screening-detected celiac disease using a prospective birth cohort study. *PLoS One*, *18*(2), e0275123.

Meyer, S., & Rosenblum, S. (2017). Activities, participation and quality of life concepts in children and adolescents with celiac disease: A scoping review. *Nutrients*, *9*(9).

Minelli, R., Gaiani, F., Kayali, S., Di Mario, F., Fornaroli, F., Leandro, G., Nouvenne, A., Vincenzi, F., & De' Angelis, G. L. (2018). Thyroid and celiac disease in pediatric age: A literature review. *Acta Bio-Medica: Atenei Parmensis*, *89*(9-S), 11–16.

Mollazadegan, K., Kugelberg, M., Lindblad, B. E., & Ludvigsson, J. F. (2011). Increased risk of cataract among 28,000 patients with celiac disease. *American Journal of Epidemiology*, *174*(2), 195–202.

Mollazadegan, K., Kugelberg, M., Tallstedt, L., & Ludvigsson, J. F. (2012). Increased risk of uveitis in coeliac disease: A nationwide cohort study. *The British Journal of Ophthalmology*, *96*(6), 857–861.

Mones, R. L., Atienza, K. V., Youssef, N. N., Verga, B., Mercer, G. O., & Rosh, J. R. (2007). Celiac crisis in the modern era. *Journal of Pediatric Gastroenterology and Nutrition*, *45*(4), 480–483.

Moreno, M. L., Cebolla, A., Munoz-Suano, A., Carrillo-Carrion, C., Comino, I., Pizarro, A., Leon, F., Rodriguez-Herrera, A., & Sousa, C. (2017). Detection of gluten immunogenic peptides in the urine of patients with coeliac disease reveals transgressions in the gluten-free diet and incomplete mucosal healing. *Gut*, *66*(2), 250–257.

Mounajjed, T., Oxentenko, A., Shmidt, E., & Smyrk, T. (2011). The liver in celiac disease: Clinical manifestations, histologic features, and response to gluten-free diet in 30 patients. *American Journal of Clinical Pathology*, *136*(1), 128–137.

Mubarak, A., Gmelig-Meyling, F. H., Wolters, V. M., Ten Kate, F. J., & Houwen, R. H. (2011). Immunoglobulin G antibodies against deamidated-gliadin-peptides outperform anti-endomysium and tissue transglutaminase antibodies in children <2 years age. *APMIS: Acta Pathologica, Microbiologica, et Immunologica Scandinavica*, *119*(12), 894–900.

Murch, S., Jenkins, H., Auth, M., Bremner, R., Butt, A., France, S., Furman, M., Gillett, P., Kiparissi, F., Lawson, M., McLain, B., Morris, M. A., Sleet, S., Thorpe, M., & Bspghan. (2013). Joint BSPGHAN and Coeliac UK guidelines for the diagnosis and management of coeliac disease in children. *Archives of Disease in Childhood*, *98*(10), 806–811.

Murray, J. A., Watson, T., Clearman, B., & Mitros, F. (2004). Effect of a gluten-free diet on gastrointestinal symptoms in celiac disease. *The American Journal of Clinical Nutrition*, *79*(4), 669–673.

Mustalahti, K., Lohiniemi, S., Collin, P., Vuolteenaho, N., Laippala, P., & Maki, M. (2002). Gluten-free diet and quality of life in patients with screen-detected celiac disease. *Effective Clinical Practice: ECP*, *5*(3), 105–113.

Myleus, A., Reilly, N. R., & Green, P. H. R. (2020). Rate, risk factors, and outcomes of nonadherence in pediatric patients with celiac disease: A systematic review. *Clinical Gastroenterology and Hepatology: the Official Clinical Practice Journal of the American Gastroenterological Association*, *18*(3), 562–573.

Nachman, F., Maurino, E., Vazquez, H., Sfoggia, C., Gonzalez, A., Gonzalez, V., Plancer del Campo, M., Smecuol, E., Niveloni, S., Sugai, E., Mazure, R., Cabanne, A., & Bai, J. C. (2009). Quality of life in celiac disease patients: prospective analysis on the importance of clinical severity at diagnosis and the impact of treatment. *Digestive and Liver Disease: Official Journal of the Italian Society of Gastroenterology and the Italian Association for the Study of the Liver*, *41*(1), 15–25.

Nachman, F., del Campo, M. P., Gonzalez, A., Corzo, L., Vazquez, H., Sfoggia, C., Smecuol, E., Sanchez, M. I., Niveloni, S., Sugai, E., Maurino, E., & Bai, J. C. (2010). Long-term deterioration of quality of life in adult patients with celiac disease is associated with treatment noncompliance. *Digestive and Liver Disease: Official Journal of the Italian Society of Gastroenterology and the Italian Association for the Study of the Liver*, *42*(10), 685–691.

Naddei, R., Di Gennaro, S., Guarino, A., Troncone, R., Alessio, M., & Discepolo, V. (2022). In a large Juvenile Idiopathic Arthritis (JIA) cohort, concomitant celiac disease is associated with family history of autoimmunity and a more severe JIA course: A retrospective study. *Pediatric Rheumatology Online Journal*, *20*(1), 31.

Nagl, K., Bollow, E., Liptay, S., Rosenbauer, J., Koletzko, S., Pappa, A., Nake, A., Frohlich-Reiterer, E., Doring, C., Wolf, J., Salfeld, P., & Prinz, N. (2019). Lower HbA1c in patients with type 1 diabetes and celiac disease who reached celiac-specific antibody-negativity-A multicenter DPV analysis. *Pediatric Diabetes*, *20*(8), 1100–1109.

Nagra, A., McGinnity, P. M., Davis, N., & Salmon, A. P. (2015). Implementing transition: Ready steady go. *Archives of Disease in Childhood. Education and Practice Edition*, *100*(6), 313–320.

Nastasio, S., Sciveres, M., Riva, S., Filippeschi, I. P., Vajro, P., & Maggiore, G. (2013). Celiac disease-associated autoimmune hepatitis in childhood: Long-term response to treatment. *Journal of Pediatric Gastroenterology and Nutrition*, *56*(6), 671–674.

Nenna, R., Petrarca, L., Verdecchia, P., Florio, M., Pietropaoli, N., Mastrogiorgio, G., Bavastrelli, M., Bonamico, M., & Cucchiara, S. (2016). Celiac disease in a large cohort of children and adolescents with recurrent headache: A retrospective study. *Digestive and Liver Disease: Official Journal of the Italian Society of Gastroenterology and the Italian Association for the Study of the Liver*, *48*(5), 495–498.

Nieri, M., Tofani, E., Defraia, E., Giuntini, V., & Franchi, L. (2017). Enamel defects and aphthous stomatitis in celiac and healthy subjects: Systematic review and meta-analysis of controlled studies. *Journal of Dentistry*, *65*, 1–10.

Niknam, R., Seraj, S. R., Fattahi, M. R., Nejati, M., Dehghani, S. M., & Mahmoudi, L. (2021). Impact of serological and histological factors on neurological manifestations in children and adults with celiac disease. *Pediatric Gastroenterology, Hepatology & Nutrition*, *24*(2), 197–206.

Nunes-Silva, J. G., Nunes, V. S., Schwartz, R. P., Mlss Trecco, S., Evazian, D., Correa-Giannella, M. L., Nery, M., & Queiroz, M. S. (2017). Impact of type 1 diabetes mellitus and celiac disease on nutrition and quality of life. *Nutrition & Diabetes*, *7*(1), e239.

O'Leary, C., Wieneke, P., Healy, M., Cronin, C., O'Regan, P., & Shanahan, F. (2004). Celiac disease and the transition from childhood to adulthood: A 28-year follow-up. *The American Journal of Gastroenterology*, *99*(12), 2437–2441.

Olen, O., Gudjonsdottir, A. H., Browaldh, L., Hessami, M., Elvin, K., Liedberg, A. S., Neovius, M., & Grahnquist, L. (2012). Antibodies against deamidated gliadin peptides and tissue transglutaminase for diagnosis of pediatric celiac disease. *Journal of Pediatric Gastroenterology and Nutrition*, *55*(6), 695–700.

Oman, A., Hansson, T., Carlsson, M., & Berntson, L. (2019). Evaluation of screening for coeliac disease in children with juvenile idiopathic arthritis. *Acta Paediatrics*, *108*(4), 688–693.

Ostermaier, K. K., Weaver, A. L., Myers, S. M., Stoeckel, R. E., Katusic, S. K., & Voigt, R. G. (2020). Incidence of celiac disease in down syndrome: A longitudinal, population-based birth cohort study. *Clinical Pediatrics*, *59*(12), 1086–1091.

Paez, M. A., Gramelspacher, A. M., Sinacore, J., Winterfield, L., & Venu, M. (2017). Delay in diagnosis of celiac disease in patients without gastrointestinal complaints. *The American Journal of Medicine*, *130*(11), 1318–1323.

Pangallo, E., Parma, B., Mariani, M., Cianci, P., De Paoli, A., Maitz, S., Fossati, C., Panceri, R., Agosti, M., & Selicorni, A. (2020). Williams-Beuren Syndrome and celiac disease: A real association? *European Journal of Medical Genetics*, *63*(9), 103999.

Pascotto, A., Coppola, G., Ecuba, P., Liguori, G., & Guandalini, S. (1994). Epilepsy and occipital calcifications with or without celiac disease: Report of four cases. *Journal of Epilepsy*, *7*, 130–136.

Passali, M., Josefsen, K., Frederiksen, J. L., & Antvorskov, J. C. (2020). Current evidence on the efficacy of gluten-free diets in multiple sclerosis, psoriasis, Type 1 diabetes and autoimmune thyroid diseases. *Nutrients*, *12*(8).

Pastore, L., Carroccio, A., Compilato, D., Panzarella, V., Serpico, R., & Lo Muzio, L. (2008). Oral manifestations of celiac disease. *Journal of Clinical Gastroenterology*, *42*(3), 224–232.

Patton, T., Chugh, A., Padhye, L., DeGeeter, C., & Guandalini, S. (2019). Pediatric celiac disease and eosinophilic esophagitis: Outcome of dietary therapy. *Journal of Pediatric Gastroenterology and Nutrition*, *69*(2), e43–e48.

Pedretti, M., Sbravati, F., Allegri, D., Labriola, F., Lombardo, V., Spisni, E., Zarbo, C., & Alvisi, P. (2021). Is the clinical pattern of pediatric celiac disease changing? A thirty-years real-life experience of an Italian center. *Italian Journal of Pediatrics*, *47*(1), 235.

Persechino, F., Galli, G., Persechino, S., Valitutti, F., Zenzeri, L., Mauro, A., Corleto, V. D., Parisi, P., Ziparo, C., Evangelisti, M., Quatrale, G., & Di Nardo, G. (2021). Skin manifestations and coeliac disease in paediatric population. *Nutrients*, *13*(10).

Petroff, D., Wolf, J., Richter, T., Auth, M. K. H., Uhlig, H. H., Laass, M. W., Lauenstein, P., Krahl, A., Handel, N., de Laffolie, J., Hauer, A. C., Heiduk, M., Flemming, G., Schmidt, A., Hasenclever, D., & Mothes, T. (2018). Antibody concentrations decrease 14-fold in children with celiac disease on a gluten-free diet but remain high at 3 months. *Clinical Gastroenterology and Hepatology: The Official Clinical Practice Journal of the American Gastroenterological Association*, *16*(9), 1442–1449, e1445.

Piccialli, F., Calabro, F., Crisci, D., Cuomo, S., Prezioso, E., Mandile, R., Troncone, R., Greco, L., & Auricchio, R. (2021). Precision medicine and machine learning towards the prediction of the outcome of potential celiac disease. *Scientific Reports*, *11*(1), 5683.

Popp, A., & Maki, M. (2019). Changing pattern of childhood celiac disease epidemiology: Contributing factors. *Frontiers in Pediatrics*, *7*, 357.

Prasanth, V. J., & Singh, A. (2021). Geographic tongue. *CMAJ: Canadian Medical Association Journal = Journal de l'Association Medicale Canadienne*, *193*(36), E1424.

Rajalahti, T., Repo, M., Kivela, L., Huhtala, H., Maki, M., Kaukinen, K., Lindfors, K., & Kurppa, K. (2017). Anemia in pediatric celiac disease: Association with clinical and histological features and response to gluten-free diet. *Journal of Pediatric Gastroenterology and Nutrition*, *64*(1), e1–e6.

Reunala, T., Salmi, T. T., Hervonen, K., Kaukinen, K., & Collin, P. (2018). Dermatitis herpetiformis: A common extraintestinal manifestation of coeliac disease. *Nutrients*, *10*(5).

Richter, T., Bossuyt, X., Vermeersch, P., Uhlig, H. H., Stern, M., Hauer, A., Zimmer, K. P., Mearin, L., de Roo, J. H., Dahnrich, C., & Mothes, T. (2012). Determination of IgG and IgA antibodies against native gliadin is not helpful for the diagnosis of coeliac disease in children up to 2 years old. *Journal of Pediatric Gastroenterology and Nutrition*, *55*(1), 21–25.

Rodrigo, L., Beteta-Gorriti, V., Alvarez, N., Gomez de Castro, C., de Dios, A., Palacios, L., & Santos-Juanes, J. (2018). Cutaneous and mucosal manifestations associated with celiac disease. *Nutrients*, *10*(7).

Rubio-Tapia, A., Hill, I. D., Semrad, C., Kelly, C. P., & Lebwohl, B. (2023). American College of Gastroenterology Guidelines Update: Diagnosis and management of celiac disease. *The American Journal of Gastroenterology*, *118*(1), 59–76.

Ruggieri, M., Incorpora, G., Polizzi, A., Parano, E., Spina, M., & Pavone, P. (2008). Low prevalence of neurologic and psychiatric manifestations in children with gluten sensitivity. *The Journal of Pediatrics*, *152*(2), 244–249.

Runde, J., Mears, M., Guandalini, S., & Jericho, H. (2020). A narrow window: Booming gluten-free market and fostering healthy dietary habits in children with celiac disease. *Journal of Pediatric Gastroenterology and Nutrition*, *71*(4), 533–535.

Sabino, L., Marino, S., Falsaperla, R., Pisani, F., Massimino, C., & Pavone, P. (2020). Celiac disease and headache in children: A narrative state of the art. *Acta bio-medica: Atenei Parmensis*, *91*(3), e2020056.

Sadr-Azodi, O., Sanders, D. S., Murray, J. A., & Ludvigsson, J. F. (2012). Patients with celiac disease have an increased risk for pancreatitis. *Clinical Gastroenterology and Hepatology: the Official Clinical Practice Journal of the American Gastroenterological Association*, *10*(10), 1136–1142, e1133.

Sansotta, N., Amirikian, K., Guandalini, S., & Jericho, H. (2018). Celiac disease symptom resolution: Effectiveness of the gluten-free diet. *Journal of Pediatric Gastroenterology and Nutrition*, *66*(1), 48–52.

Sansotta, N., Guandalini, S., Romano, S., Amirikian, K., Cipolli, M., Tridello, G., Barzaghi, S., & Jericho, H. (2020). The gluten free diet's impact on growth in children with celiac disease in two different countries. *Nutrients*, *12*(6).

Saps, M., Sansotta, N., Bingham, S., Magazzu, G., Grosso, C., Romano, S., Pusatcioglu, C., & Guandalini, S. (2017). Abdominal pain-associated functional gastrointestinal disorder prevalence in children and adolescents with celiac disease on gluten-free diet: A multinational study. *The Journal of Pediatrics*, *182*, 150–154.

Scaramuzza, A. E., Mantegazza, C., Bosetti, A., & Zuccotti, G. V. (2013). Type 1 diabetes and celiac disease: The effects of gluten free diet on metabolic control. *World J Diabetes*, *4*(4), 130–134.

Schiepatti, A., Rej, A., Maimaris, S., Cross, S. S., Porta, P., Aziz, I., Key, T., Goodwin, J., Therrien, A., Yoosuf, S., Leffler, D. A., Silvester, J. A., Klersy, C., Biagi, F., & Sanders, D. S. (2021). Clinical classification and long-term outcomes of seronegative coeliac disease: A 20-year multicentre follow-up study. *Alimentary Pharmacology & Therapeutics*, *54*(10), 1278–1289.

Schiepatti, A., Cincotta, M., Biagi, F., & Sanders, D. S. (2021). Enteropathies with villous atrophy but negative coeliac serology in adults: Current issues. *BMJ Open Gastroenterol*, *8*(1).

Schiepatti, A., Maimaris, S., de Queiros Mattoso Archela Dos Santos, C., Rusca, G., Costa, S., & Biagi, F. (2022). Long-term adherence to a gluten-free diet and quality of life of celiac patients after transition to an adult referral center. *Digestive Diseases and Sciences*, *67*(8), 3955–3963.

Schlesinger, B., & Keele, K. D. (1935). A case of coeliac disease showing unusual features successfully treated with insulin and glucose. *Archives of Disease in Childhood*, *10*(57), 149–156.

Sherman, Y., Karanicolas, R., DiMarco, B., Pan, N., Adams, A. B., Barinstein, L. V., Moorthy, L. N., & Lehman, T. J. (2015). Unrecognized celiac disease in children presenting for rheumatology evaluation. *Pediatrics*, *136*(1), e68–e75.

Silvester, J. A., Comino, I., Kelly, C. P., Sousa, C., & Duerksen, D. R.D. B. S. Group. (2020). Most patients with celiac disease on gluten-free diets consume measurable amounts of gluten. *Gastroenterology*, *158*(5), 1497–1499, e1491.

Silvester, J. A., Kurada, S., Szwajcer, A., Kelly, C. P., Leffler, D. A., & Duerksen, D. R. (2017). Tests for serum transglutaminase and endomysial antibodies do not detect most patients with celiac disease and persistent villous atrophy on gluten-free diets: A meta-analysis. *Gastroenterology*, *153*(3), 689–701, e681.

Simsek, S., Baysoy, G., Gencoglan, S., & Uluca, U. (2015). Effects of gluten-free diet on quality of life and depression in children with celiac disease. *Journal of Pediatric Gastroenterology and Nutrition*, *61*(3), 303–306.

Sisk, C. (2016). Promoting children's health-related quality of life. *Pediatric Nursing*, *42*(2), 86–88.

Snyder, J., Butzner, J. D., DeFelice, A. R., Fasano, A., Guandalini, S., Liu, E., & Newton, K. P. (2016). Evidence-informed expert recommendations for the management of celiac disease in children. *Pediatrics*, *138*(3).

Soderstrom, H., Rehn, J., Cervin, M., Ahlstermark, C., Bybrant, M. C., & Carlsson, A. (2021). Compliance to a gluten-free diet in swedish children with type 1 diabetes and celiac disease. *Nutrients*, *13*(12).

Soderstrom, H., Cervin, M., Dereke, J., Hillman, M., Tiberg, I., Norstrom, F., & Carlsson, A. (2022). Does a gluten-free diet lead to better glycemic control in children with type 1 diabetes? Results from a feasibility study and recommendations for future trials. *Contemporary Clinical Trials Communications*, *26*, 100893.

Souto-Souza, D., da Consolacao Soares, M. E., Rezende, V. S., de Lacerda Dantas, P. C., Galvao, E. L., & Falci, S. G. M. (2018). Association between developmental defects of enamel and celiac disease: A meta-analysis. *Archives of Oral Biology*, *87*, 180–190.

Stahl, M., Li, Q., Lynch, K., Koletzko, S., Mehta, P., Gragert, L., Norris, J. M., Andren Aronsson, C., Lindfors, K., Kurppa, K., Ilonen, J., Krischer, J., Alkolkar, B., Ziegler, A. G., Toppari, J., Rewers, M. J., Agardh, D., Hagopian, W., Liu, E., & Group, T. S. (2023). Incidence of pediatric celiac disease varies by region. *The American Journal of Gastroenterology*, *118*(3), 539–545.

Stefanolo, J. P., Talamo, M., Dodds, S., de la Paz Temprano, M., Costa, A. F., Moreno, M. L., Pinto-Sanchez, M. I., Smecuol, E., Vazquez, H., Gonzalez, A., Niveloni, S. I., Maurino, E., Verdu, E. F., & Bai, J. C. (2021). Real-world gluten exposure in patients with celiac disease on gluten-free diets, determined from gliadin immunogenic peptides in urine and fecal samples. *Clinical Gastroenterology and Hepatology: The Official Clinical Practice Journal of the American Gastroenterological Association*, *19*(3), 484–491, e481.

Suarez-Gonzalez, M., Bousono-Garcia, C., Jimenez-Trevino, S., & Diaz-Martin, J. J. (2021). Gluten-free diet: Nutritional strategies to improve eating habits in children with celiac disease: A prospective, single-arm intervention study. *Nutrients*, *13*, 4.

Tanpowpong, P., Li, S., Espinola, J. A., Santos, L. C., James, K. E., Powe, C. E., & Camargo, C. A., Jr. (2023). Pregnancy- and birth-related risk factors for the development of childhood celiac disease. *Acta Paediatrics*.

Tapsas, D., Hollen, E., Stenhammar, L., & Falth-Magnusson, K. (2016). The clinical presentation of coeliac disease in 1030 Swedish children: Changing features over the past four decades. *Digestive and Liver Disease: Official Journal of the Italian Society of Gastroenterology and the Italian Association for the Study of the Liver*, *48*(1), 16–22.

Tapsas, D., Falth-Magnusson, K., Hogberg, L., Hammersjo, J. A., & Hollen, E. (2014). Swedish children with celiac disease comply well with a gluten-free diet, and most include oats without reporting any adverse effects: a long-term follow-up study. *Nutrition Research (New York, N.Y.)*, *34*(5), 436–441.

Troncone, R., & Kosova, R. (2010). Short stature and catch-up growth in celiac disease. *Journal of Pediatric Gastroenterology and Nutrition*, *51*, S137–S138.

Tuna Kirsaclioglu, C., Kuloglu, Z., Tanca, A., Kucuk, N. O., Aycan, Z., Ocal, G., Ensari, A., Kalayci, A. G., & Girgin, N. (2016). Bone mineral density and growth in children with coeliac disease on a gluten free-diet. *Turkish Journal of Medical Sciences*, *46*(6), 1816–1821.

Tunio, N.A., E. Mansoor, M.Z. Sheriff, G.S. Cooper, S.N. Sclair and S.M. Cohen (2020). Epidemiology of autoimmune hepatitis (AIH) in the United States between 2014 and 2019: A population-based national study. Journal of Clinical Gastroenterology.

Ukkola, A., Maki, M., Kurppa, K., Collin, P., Huhtala, H., Kekkonen, L., & Kaukinen, K. (2011). Diet improves perception of health and well-being in symptomatic, but not asymptomatic, patients with celiac disease. *Clinical Gastroenterology and Hepatology: the Official Clinical Practice Journal of the American Gastroenterological Association*, *9*(2), 118–123.

Ungprasert, P., Wijarnpreecha, K., & Kittanamongkolchai, W. (2017). Psoriasis and risk of celiac disease: A systematic review and meta-analysis. *Indian Journal of Dermatology*, *62*(1), 41–46.

Vajro, P., Paolella, G., Maggiore, G., & Giordano, G. (2013). Pediatric celiac disease, cryptogenic hypertransaminasemia, and autoimmune hepatitis. *Journal of Pediatric Gastroenterology and Nutrition*, *56*(6), 663–670.

van Doorn, R. K., Winkler, L. M., Zwinderman, K. H., Mearin, M. L., & Koopman, H. M. (2008). CDDUX: A disease-specific health-related quality-of-life questionnaire for children with celiac disease. *Journal of Pediatric Gastroenterology and Nutrition*, *47*(2), 147–152.

Varni, J. W., Limbers, C. A., & Newman, D. A. (2009). Using factor analysis to confirm the validity of children's self-reported health-related quality of life across different modes of administration. *Clinical Trials (London, England)*, *6*(2), 185–195.

Veeraraghavan, G., Therrien, A., Degroote, M., McKeown, A., Mitchell, P. D., Silvester, J. A., Leffler, D. A., Leichtner, A. M., Kelly, C. P., & Weir, D. C. (2021). Non-responsive celiac disease in children on a gluten free diet. *World Journal of Gastroenterology: WJG*, *27*(13), 1311–1320.

Ventura, A., Magazzu, G., & Greco, L. (1999). Duration of exposure to gluten and risk for autoimmune disorders in patients with celiac disease. SIGEP Study Group for Autoimmune Disorders in Celiac Disease. *Gastroenterology*, *117*(2), 297–303.

Viljamaa, M., Kaukinen, K., Pukkala, E., Hervonen, K., Reunala, T., & Collin, P. (2006). Malignancies and mortality in patients with coeliac disease and dermatitis herpetiformis: 30-year population-based study. *Digestive and Liver Disease: Official Journal of the Italian Society of Gastroenterology and the Italian Association for the Study of the Liver*, *38*(6), 374–380.

Villanueva, J., Maranda, L., & Nwosu, B. U. (2012). Is vitamin D deficiency a feature of pediatric celiac disease? *Journal of Pediatric Endocrinology & Metabolism: JPEM*, *25*(5–6), 607–610.

Virta, L. J., Kaukinen, K., & Collin, P. (2009). Incidence and prevalence of diagnosed coeliac disease in Finland: Results of effective case finding in adults. *Scandinavian Journal of Gastroenterology*, *44*(8), 933–938.

Vriezinga, S., Borghorst, A., Van den Akker-Van Marle, E., Benninga, M., George, E., Hendriks, D., Hopman, E., de Meij, T., van der Meulen-de Jong, A., Putter, H., Rings, E., Schaart, M., Schweizer, J., Smit, M., Tabbers, M., Weijerman, M., Wessels, M., & Mearin, M. L. (2018). E-healthcare for celiac disease-A multicenter randomized controlled trial. *The Journal of Pediatrics*, *195*, 154–160, e157.

Waisbourd-Zinman, O., Hojsak, I., Rosenbach, Y., Mozer-Glassberg, Y., Shalitin, S., Phillip, M., & Shamir, R. (2012). Spontaneous normalization of anti-tissue transglutaminase antibody levels is common in children with type 1 diabetes mellitus. *Digestive Diseases and Sciences*, *57*(5), 1314–1320.

Walker-Smith, J. A., & Grigor, W. (1969). Coeliac disease in a diabetic child. *Lancet*, *1*(7603), 1021.

Walker-Smith, J. A., Guandalini, S., Schmitz, J., & Shmerling, D. H.Report of Working Group of European Society of Paediatric Gastroenterology and Nutrition. (1990). Revised criteria for diagnosis of coeliac disease. *Archives of Disease in Childhood*, *65*, 909–911.

Wallach, T., Genta, R. M., Lebwohl, B., Green, P. H. R., & Reilly, N. R. (2017). Adherence to celiac disease and eosinophilic esophagitis biopsy guidelines is poor in children. *Journal of Pediatric Gastroenterology and Nutrition*, *65*(1), 64–68.

Webster, J., Vajravelu, M. E., Choi, C., Zemel, B., & Verma, R. (2019). Prevalence of and risk factors for low bone mineral density in children with celiac disease. *Clinical Gastroenterology and Hepatology: the Official Clinical Practice Journal of the American Gastroenterological Association*, *17*(8), 1509–1514.

Wessels, M. M., Van, V., II, Vriezinga, S. L., Putter, H., Rings, E. H., & Mearin, M. L. (2016). Complementary serologic investigations in children with celiac disease is unnecessary during follow-up. *The Journal of Pediatrics*, *169*, 55–60.

Whitburn, J., Rao, S. R., Paul, S. P., & Sandhu, B. K. (2021). Diagnosis of celiac disease is being missed in over 80% of children particularly in those from socioeconomically deprived backgrounds. *European Journal of Pediatrics*, *180*(6), 1941–1946.

White, L. E., Bannerman, E., & Gillett, P. M. (2016). Coeliac disease and the gluten-free diet: A review of the burdens; factors associated with adherence and impact on health-related quality of life, with specific focus on adolescence. *Journal of Human Nutrition and Dietetics: The Official Journal of the British Dietetic Association*, *29*(5), 593–606.

Wierink, C. D., van Diermen, D. E., Aartman, I. H., & Heymans, H. S. (2007). Dental enamel defects in children with coeliac disease. *International Journal of Paediatric Dentistry / the British Paedodontic Society [and] the International Association of Dentistry for Children*, *17* (3), 163–168.

Wijarnpreecha, K., Thongprayoon, C., Panjawatanan, P., Thamcharoen, N., Pachariyanon, P., Nakkala, K., & Cheungpasitporn, W. (2016). Celiac disease and the risk of kidney diseases: A systematic review and meta-analysis. *Digestive and Liver Disease: Official Journal of the Italian Society of Gastroenterology and the Italian Association for the Study of the Liver*, *48*(12), 1418–1424.

Wolf, J., Hasenclever, D., Petroff, D., Richter, T., Uhlig, H. H., Laass, M. W., Hauer, A., Stern, M., Bossuyt, X., de Laffolie, J., Flemming, G., Villalta, D., Schlumberger, W., & Mothes, T. (2014). Antibodies in the diagnosis of coeliac disease: A biopsy-controlled, international, multicentre study of 376 children with coeliac disease and 695 controls. *PLoS One*, *9*(5), e97853.

Zanini, B., Caselani, F., Magni, A., Turini, D., Ferraresi, A., Lanzarotto, F., Villanacci, V., Carabellese, N., Ricci, C., & Lanzini, A. (2013). Celiac disease with mild enteropathy is not mild disease. *Clinical Gastroenterology and Hepatology: the Official Clinical Practice Journal of the American Gastroenterological Association*, *11*(3), 253–258.

Zhang, J. Z., Abudoureyimu, D., Wang, M., Yu, S. R., & Kang, X. J. (2021). Association between celiac disease and vitiligo: A review of the literature. *World Journal of Clinical Cases*, *9*(34), 0430–10437.

Zingone, F., Swift, G. L., Card, T. R., Sanders, D. S., Ludvigsson, J. F., & Bai, J. C. (2015). Psychological morbidity of celiac disease: A review of the literature. *United European Gastroenterologisches Journal: Organ der Gesellschaft fur Gastroenterologie der DDR*, *3*(2), 136–145.

Zis, P., Julian, T., & Hadjivassiliou, M. (2018). Headache associated with coeliac disease: A systematic review and meta-analysis. *Nutrients*, *10*(10).

Zone, J. J., Meyer, L. J., & Petersen, M. J. (1996). Deposition of granular IgA relative to clinical lesions in dermatitis herpetiformis. *Archives of Dermatology*, *132*(8), 912–918.

Chapter 7

Celiac disease in adult patients

Carolina Ciacci[1], Ida De Micco[1], Michele Di Stefano[2] and Caterina Mengoli[2]
[1]*Department of Medicine Surgery and Dentistry, Scuola Medica Salernitana University of Salerno, Baronissi, Salerno, Italy,* [2]*Department of Internal Medicine, University of Pavia, Fondazione IRCCS Policlinico San Matteo, Pavia, Italy*

7.1 Introduction

Celiac disease (CD) may present at any age, with different signs, symptoms, and levels of severity (Green & Cellier, 2007). There is significant variability in clinical presentation and country-specific reported prevalence, and this can be explained by factors intrinsically related to CD, such as the prevalence of HLA predisposing genotypes, or environmental causes, as the amount of consumed gluten, the type of study, or the physicians' awareness of the disease itself. For example, in Japan, CD has a significantly lower prevalence than in the rest of the world (0.05%) (Fukunaga et al., 2018), as well as diagnostic criteria may vary worldwide.

A series of studies suggested that the multifaceted clinical presentation in adults can explain the diagnostic delay in CD. A case-control study involving UK general practitioners showed that undiagnosed CD patients had a threefold increased risk of misdiagnosis, even 10 years earlier than their final diagnosis (Card et al., 2013). Similarly, in a Finnish study, 32% of the CD patients reported a diagnostic delay of >10 years (Fuchs et al., 2014). There is no consensus on whether gender differences account for this diagnostic delay: a Danish study indicated that women had a diagnostic delay longer than men (Kårhus et al., 2022), but a USA study did not confirm this (Jansson-Knodell et al., 2018).

Overall, this delay is decreasing due to the greater awareness of doctors and the higher accuracy of diagnostic tests (Makharia et al., 2011). A recent Italian multicenter study found that in a cohort observed from 2011 to 2021, the median diagnostic delay was 8 months, while patient- and physician-dependent delays were 3 and 4 months, respectively (Lenti et al., 2022). Further improving CD awareness may warrant fewer complications and better quality of life.

Thirty years ago, the prevalence of CD was represented as an iceberg (Catassi et al., 1994), whose submerged base represented still undiagnosed patients. Nowadays, in Europe and North America, most undiagnosed CD is either people with silent or subclinical CD. Conversely, the symptomatic disease remains underdiagnosed in Asia and Africa due to poor local awareness of the clinical presentation and scarce screening activity (Makharia & Catassi, 2019; Poddighe & Abdukhakimova, 2021). Otherwise, when actively searched, CD disease prevalence and clinical presentation appear similar to that of the Western world. A recent retrospective study from Northern China confirmed that among 2800 patients with gastrointestinal symptoms investigated by serology, the prevalence of CD was 1.4%. Notably, 17% of the pregnant women included in the study were tested for severe gastrointestinal symptoms (Wang et al., 2022).

7.2 Clinical disease presentation in adults

Clinical presentation of adult CD is highly variable, and the patient may present with a combination of digestive and nondigestive symptoms and signs, ranging from silent/subclinical to a less frequent overt malabsorption syndrome. There have been some attempts to classify CD clinical manifestations according to their presentation. The most common way to define CD presentation at diagnosis is to describe the presence/absence of gastrointestinal and extraintestinal symptoms and their severity.

In 2013 a team of experts attempted to define clinical presentation *in silent, classical*, and *subclinical CD* (Ludvigsson et al., 2013). Patients with *silent* or *asymptomatic* CD do not report symptoms at diagnosis, even on direct questioning. Usually, asymptomatic patients are found through family/mass screenings. Mass screening studies showed

Pediatric and Adult Celiac Disease. DOI: https://doi.org/10.1016/B978-0-443-13359-6.00001-7

that more than half of the undiagnosed population likely have an asymptomatic or subclinical form of CD (Kvamme et al., 2022). It has been shown that patients who self-report as asymptomatic but show characteristic CD intestinal lesions have improved quality of life after starting a gluten-free diet, which is often reported as a "new normality" (Kurppa et al., 2014).

The *classical* CD is characterized by signs and symptoms of malabsorption, such as diarrhea, steatorrhea, weight loss, or growth failure. The term recalls the "old" CD diagnoses, which were exclusively made in patients with gastrointestinal symptoms (Collin et al., 1997).

Finally, there is the *subclinical* CD, characterized by few or less severe gastrointestinal symptoms or by the presence of minor, transient, or apparently unrelated symptoms, that is, a patient suffering from a mild functional gut disease, fatigue, or anemia, or diagnosed because of the presence of an associated condition.

It is common knowledge that the severity of symptoms is unrelated to the degree of mucosal damage or the serum level of celiac antibodies. A possible explanation is that biopsy explores an extremely small part of the intestine, whereas the extension of the damaged mucosa may play a role in determining symptoms. In the future, videocapsule endoscopy, facilitated by computer-assisted artificial intelligence, may become a quantitative and reproducible measurement tool (Chetcuti Zammit et al., 2023).

As previously mentioned, CD patients may present with both intestinal and extraintestinal symptoms and signs. In patients with only hypochromic anemia, for example, plasma cholesterol in the high-to-normal range could be used to exclude the presence of CD (Ciacci et al., 1999). The most recent systematic review of serological screening strategies showed that individuals with dermatitis herpetiformis, a family history of CD, migraine, anemia, type 1 diabetes, osteoporosis, or chronic hypertransaminasemia are 1.5–2 times more likely to have CD than the general population. Otherwise, the presence of gastrointestinal symptoms did not help identifying CD (Elwenspoek et al., 2022). Table 7.1 summarizes the signs and symptoms of CD in adults divided by gastrointestinal and extraintestinal.

7.2.1 Gastrointestinal symptoms

7.2.1.1 Diarrhea and constipation

Diarrhea was once the most frequent symptom of CD. The advent of sensitive and specific serology tests and their use in family screening and case-finding has changed the face of this condition. More recent data show that chronic diarrhea is now present in less than 40% of adults with characteristics similar to irritable bowel syndrome (IBS) (Tursi et al., 2010).

Conversely, the so-called "celiac crisis" presents as an acute onset, with a rapid worsening of gastrointestinal symptoms, and may require hospitalization and/or parenteral nutrition. The patients present with severe dehydration, metabolic acidosis, hypoproteinemia, weight loss, and neurologic dysfunction. The prevalence of such a condition is unknown. The rare cases described in the literature are mostly pediatric (Mones et al., 2007); however, in a series of about 1200 adult patients in the United States, 12 cases met the criteria of the celiac crisis at diagnosis (Balaban et al., 2019).

TABLE 7.1 Presentation of celiac disease in adults: signs and symptoms.

Gastrointestinal signs or symptoms	Extraintestinal signs or symptoms
Chronic diarrhea	Iron deficiency and anemia, vitamin B12 and folic acid deficiency
Acute diarrhea (celiac crisis)	Late menarche, early menopause, miscarriage
Weight loss	Skin diseases: dermatitis herpetiformis, urticaria, psoriasis, alopecia
Anorexia, vomiting	Malaise and fatigue
Bloating or abdominal distension	Reduced bone density
Dyspepsia	Neurological and psychiatric disorders
Constipation	Dental enamel defects, recurrent aphthous stomatitis
Recurrent abdominal pain mimicking irritable bowel syndrome	Cryptic chronic hypertransaminasemia

A UK study suggested the concurrence of CD with inflammatory bowel disease (IBD), reporting that the prevalence of IBD in CD was 10-fold increased compared to the general population. In contrast, the prevalence of CD in IBD was comparable with that of the controls (Leeds et al., 2007). A more recent study investigating the histology records of IBD, microscopic colitis, and CD showed that the prevalence of CD in microscopic colitis is 4.4% (Sonnenberg & Genta, 2023).

Adults, as well as children, may present with constipation. A recent study on the clinical presentation of patients with biopsy-proven CD seen at a community-based adult gastroenterology practice showed that constipation at diagnosis was relatively frequent (17%) and not severe enough to ask for a gastroenterologist consultation (Roldan et al., 2023).

7.2.1.2 Irritable bowel syndrome-like symptoms

According to the Rome IV criteria, IBS is a disease defined by recurrent abdominal pain at least 1 day/week in the last 3 months, associated with changes in the frequency/form of stools (Drossman & Hasler, 2016). IBS is highly prevalent, affecting 1 in 10 individuals worldwide (Lovell & Ford, 2012). Some CD patients present with clinical manifestations similar to IBS, including abdominal pain, bloating, and bowel habit changes. Particularly in these cases, the diagnostic delay is due to misclassification (Aziz & Simrén, 2021). A meta-analysis showed a higher prevalence of positive celiac serology and biopsy-proven CD in people with IBS symptoms than in healthy controls. However, it is still unclear whether testing for CD in people with IBS-like symptoms is useful (Irvine et al., 2017).

7.2.1.3 Symptoms of gastroesophageal reflux

Gastroesophageal reflux disease (GERD) causes heartburn, regurgitation, and extraesophageal manifestations (Vaezi et al., 2018). These symptoms are often present in CD patients at diagnosis. One of the earliest studies showed that CD patients with steatorrhea presented a higher prevalence of esophageal symptoms and reduced lower esophageal sphincter pressure than CD patients without steatorrhea and controls (Iovino et al., 1998). Further studies have consistently shown the efficacy of a GFD in rapidly controlling esophageal symptoms, preventing their recurrence (Cuomo et al., 2003; Nachman et al., 2011; Barratt et al., 2011). Conversely, upper GI endoscopy of patients with GERD symptoms may result in the diagnosis of previously unsuspected CD (Collin et al., 2004; Zádori et al., 2022).

The putative mechanisms for GERD symptoms in untreated CD range from the arrival of intact gluten to the distal gastrointestinal segment inducing esophageal dysmotility to delayed gastric emptying, and small intestine dysmotility, and/or altered neuroendocrine control of upper GI functions (Leffler & Kelly, 2011). However, the intimate mechanisms of GERD in untreated CD have yet to be elucidated, and available evidence still shows inconsistent results.

7.2.1.4 Weight loss, overweight and obesity

A recent retrospective study showed that weight loss, extremely frequent in the past, was seen in only 25% of patients at diagnosis (Bledsoe et al., 2019). Patients with CD may be overweight or even obese at diagnosis. The literature indicates that the prevalence of overweight patients with CD at diagnosis is increasing: a recent USA study comparing patients diagnosed from 2000 to 2007 and from 2013 to 2016 in the community versus patients seen at a tertiary center showed that the most recent cohort had a higher proportion of overweight/obese patients compared with the previous and compared with that of the tertiary center (Roldan et al., 2023).

7.2.2 Extraintestinal symptoms

7.2.2.1 Dermatitis herpetiformis

Dermatitis herpetiformis is an extraintestinal manifestation of CD presenting with itchy papules and vesicles symmetrically present mainly on the extensor surfaces of the body (elbows, knees, and buttocks). Rarely, patients with dermatitis herpetiformis present overt malabsorption or gastrointestinal symptoms, and intestinal lesions are often minor. For a more extensive description of dermatitis herpetiformis, see Chapter 16.

7.2.2.2 Other skin and oral diseases

The prevalence and the pathophysiology of the mucocutaneous manifestations associated with CD other than DH are poorly known. Skin diseases associated with CD are divided into four categories: autoimmune, allergic, inflammatory, and miscellaneous (Humbert et al., 2006). Among the autoimmune diseases, Corazza et al. (1995), described the association between alopecia areata, characterized by areas of hair loss in the scalp with no apparent inflammation, and CD (Corazza et al., 1995). Alopecia can be the only sign of CD, also in children. Both CD and alopecia areata show the presence of organ-specific autoantibodies, T-lymphocyte infiltration at the site of the lesion, and association with HLA genes. Alopecia areata in adult CD patients responds poorly to a GFD (Bardella et al., 2000; Lebwohl et al., 2021). Psoriasis is also an autoimmune disease, about two times more frequent in undiagnosed CD than in controls (Ludvigsson et al., 2011). Atopic dermatitis and chronic urticaria have a weaker association with adult CD. Urticaria has been shown to have a genetic association with the human leukocyte antigen HLA-DQ8 alleles (O'Donnell et al., 1999). Data from 43,300 CD patients from the Swedish Registry of Health showed that, after 11 years of follow-up, the incidence of skin diseases other than dermatitis herpetiformis in CD patients was 50% more than in controls. Increased risks were found for eczema (HR 1.67), psoriasis (HR 1.55), urticaria (HR 1.52), vitiligo (HR 1.90), acne (HR 1.39), and alopecia areata (HR 1.78). Risks from multiple common skin disorders persist long-term (Lebwohl et al., 2021).

A recent comprehensive review showed that CD patients might suffer from several oral and teeth diseases at diagnosis, including dental enamel defects, oral aphthous stomatitis, periodontitis, dental enamel lesions, and cavities (Wieser et al., 2023). Dental enamel defects appear to be a peculiar sign of gluten-related immune activity before the eruption of definitive teeth. Patients with dental enamel defects show an odds ratio of CD of 5,69 in 30 studies pooled in a met-analysis (Nieri et al., 2017).

7.2.2.3 Alterations of bone mass and mineral metabolism

Seventy-five % of untreated CD patients with overt malabsorption at presentation and half of those with minor symptoms, or asymptomatic harbor a derangement of bone mineralization (Corazza et al., 2005).

The pathophysiology encompasses two overlapping mechanisms in untreated or noncompliant patients. Calcium malabsorption causes a negative calcium balance worsened by reduced calcium intake due to secondary lactose intolerance, intraluminal precipitation as soaps, and increased intestinal secretion. Consequently, hypersecretion of parathyroid hormone directly activates bone resorption pathways and increases the activity of renal 1-α-hydroxylase, inducing the conversion of 25-hydroxy-vitamin D into 1,25-dihydroxy-vitamin D to improve calcium active transport. In CD, negligible amounts of vitamin D-dependent calcium-binding protein, calbindin, in immature enterocytes make this effort ineffective. Moreover, high levels of 1,25-dihydroxy-vitamin D may negatively affect bone levels, causing bone resorption (Di Stefano et al., 2013).

The overproduction of proinflammatory cytokines IL-6, TNF-a, and IL-1beta enhances cell-to-cell communications stimulating bone resorption (Di Stefano et al., 1997, 2019; Okamoto & Takayanagi, 2023; Zhou et al., 2022).

RANKL-RANK-Osteoprotegerin (OPG) pathway regulates bone cell activity (Khosla, 2001). RANKL links to its receptor RANK stimulating osteoclast activity and differentiation. OPG prevents this proresorptive pathway. In untreated CD patients, RANKL/OPG ratio is increased (Taranta et al., 2004), supporting the pathogenic role of an imbalance between osteoblast and osteoclast activity. The alteration of RANKL-RANK-OPG seems the main pathophysiological mechanism for persistent bone loss when a strict GFD improves intestinal absorption (Di Stefano et al., 2019). Metabolic osteopathy of CD is characterized by changes in serum markers of both bone synthesis and bone resorption. The unavailability of calcium and the faster bone resorption process, supported by PTH, result in a net bone loss, depicting a high turnover osteoporosis (Corazza et al., 2005). PTH increase is associated with the reduction of 25-hydroxy-vitamin D and the increase of 1,25-dihydroxy-vitamin D, making both inappropriate and useless vitamin D3 supplementation. Moreover, if markers of bone resorption are generally high in untreated patients, markers of bone synthesis are increased only in a subgroup of them, and low levels of PICP at diagnosis proved to be predictive of an insufficient bone mass gain after two years of strict GFD (Corazza et al., 1997; Di Stefano et al., 2019).

At present, guidelines for managing CD do not include specific indications on the need for determining vitamin D status and serum markers of bone metabolism either at diagnosis or after a GFD (Al-Toma et al., 2019; Ludvigsson et al., 2014; Rubio-Tapia et al., 2023; Zingone et al., 2022).

Studies on the prevalence of fractures in CD patients agree on a significantly increased risk (Olmos et al., 2008). However, the odds ratio or the hazard ratio calculated in studies dealing with large samples of CD patients (Ludvigsson, Osby et al., 2007; West et al., 2003) were generally higher than one, but always lower than two, raising some doubts on the clinical significance of these results. However, in large population-based studies, enrolling CD patient cohorts with a wide age range, a mild increase in fracture risk is a meaningful result. An increased risk of fractures is confirmed in women before diagnosis, when overt malabsorption and/or a low BMI value is present, in patients with a long diagnostic delay, and patients not fully compliant with GFD (Jafri et al., 2008; Sánchez et al., 2011; West et al., 2003).

In women with hypothyroidism, the risk of fractures is increased during the first years of hormonal treatment. Similarly, other condition such type I diabetes, rheumatologic disorders and other autoimmune comorbidities, lactase deficiency, are responsible for a failure to achieve an adequate peak bone (Di Stefano et al., 2002).

Dual-energy X-ray absorptiometry (DEXA) is currently considered the gold standard for measuring BMD. However, in adult CD, the timing and frequency of BMD measurement is still a matter of debate. As a strict GFD from childhood is able to normalize BMD in children, the measurement of BMD in this subset of patients is considered useless (Fouda et al., 2012). Guidelines agree on the need for BMD measurement at diagnosis in adults, in the postmenopausal period for women and after the age of 55 years in men (Fouda et al., 2012; Scott et al., 2000) and it seems appropriate to stratify CD patients according to the presence of overt malabsorption at diagnosis.

Even if metabolic bone disease in CD is known since 60s, managing this complication in real life is not yet routine as only one out of three newly diagnosed CD patients undergoes DEXA (Singh & Garber, 2016). Programs directed to a widen diffusion of guidelines and to increase awareness of this complication are needed. Treatment of metabolic osteopathy in CD will be addressed in Chapter 13.

7.2.2.4 Kidney diseases

Enteric hyperoxaluria due to malabsorption and steatorrhea causes renal calculi and chronic tubulointerstitial renal damage (Lindsjö, 1989). Few studies assessed the prevalence of renal diseases in CD. CD prevalence is increased in people with chronic renal diseases (Sahin et al., 2012). Adult untreated CD patients have 4 times the risk of nephrolithiasis compared to controls due to a 120% higher oxalate and 43% lower calcium urinary excretion. GFD reduces the risk of nephrolithiasis (Ciacci et al., 2008).

IgA nephropathy (IgAN) is a glomerulonephritis characterized by the deposition of under galactosylated polymeric IgA1 antibodies in the glomerular mesangium. Studies have described the co-occurrence of IgAN with CD (Collin et al., 2002). Recently, it has been shown that tTG-targeted IgA deposits can be found in the kidney biopsies of gluten-consuming patients (Dutta et al., 2023; Nurmi et al., 2021).

7.2.2.5 Psychological and neurological manifestations

Epidemiological studies indicate an association between undiagnosed CD and mood and psychiatric disorders. Mood and psychiatric disorders may also appear after diagnosis. At diagnosis, adults with CD frequently perceive anxiety and depression (Zingone et al., 2015), unexplained fatigue (Siniscalchi et al., 2005), sleep disorders (Zingone et al., 2010), and lack of concentration (Kristensen et al., 2019). Mood disorders seem not to be associated with the intensity of CD symptoms. It has also been described a significant bidirectional association between CD and eating disorders (Nikniaz et al., 2021; Peters et al., 2022).

Adults with CD may carry a higher risk of depression, anxiety, headache, epilepsy, panic disorder, and dysthymia compared with non-CD controls, but these data need to be confirmed (Sharma et al., 2021). Among psychiatric disorders, schizophrenia has been associated with CD in some observational studies (Eaton et al., 2004), but not in others (Ludvigsson et al., 2007).

Regarding neurological diseases, a systematic review indicates that epilepsy is 1.8 times more prevalent in CD patients than the general population, and patients with epilepsy have 2 times more prevalence of CD than the general population (Julian et al., 2019). Patients with CD frequently claim to suffer from "brain fog," a mild impairment of cognition present in other chronic diseases characterized by lack of attention and reduced verbal fluidity, which eventually improves with GFD. This has been confirmed by comparing CD patients with controls of UK Biobank, showing an increased frequency of cognitive deficits (slow reaction time), anxiety and depression, thoughts of self-harm, and health-related unhappiness in CD (Lichtwark et al., 2014). Furthermore, MRI scan demonstrated white matter brain alterations in CD patients compared with controls (Croall et al., 2020).

Ataxia, a disorder characterized by loss of coordination, clumsiness, and gait instability, is one of the neurological manifestations of CD (Mearns et al., 2019). However, idiopathic sporadic ataxia may be associated with positive serum antigliadin antibodies, even without duodenal enteropathy. The few cases that underwent autopsy revealed cerebellar atrophy and loss of Purkinje cells (Hadjivassiliou et al., 2016).

Indeed, more research is needed to shed light on the pathophysiology of both mood and neurological manifestations of CD at diagnosis and the effect of the GFD (Clappison et al., 2020).

7.2.2.6 Hematological alterations

It is not uncommon for patients with undiagnosed CD and hematological manifestations to be seen by a hematologist before receiving a CD diagnosis (Halfdanarson et al., 2007). CD should always be considered in the presence of an unexplained or isolated hematological finding.

Iron deficiency is the most common hematological manifestation in newly diagnosed patients with CD (Harper et al., 2007). Iron deficiency and related anemia may represent the only signs of an underlying subclinical CD (Corazza et al., 1995; Repo et al., 2022).

A recent retrospective study evaluating more than 500 USA CD patients showed that anemia due to iron deficiency was present in 25% of them at diagnosis, more frequently in women (84.4%) than men (58.3%) (Roldan et al., 2022). Most of the longitudinal studies demonstrate a prompt recovery of iron deficiency following a GFD. Besides iron malabsorption, folate and vitamin B12 malabsorption, nutritional deficiencies, inflammation, blood loss, development of refractory CD, concomitant *Helicobacter pylori* infection or autoimmune gastritis, are other causes of anemia in CD patients. Anemia management at diagnosis should target the patient's specific needs (Montoro-Huguet et al., 2021). This topic will also be treated in Chapter 13.

Leukopenia/neutropenia is an uncommon hematological finding in untreated CD. It can be autoimmune or secondary to deficiencies of folate, vitamin B12 or copper (Jameson et al., 1985). Conversely, thrombocytosis is relatively common (Nelson et al., 1976). The etiology of thrombocytosis is unknown, but it may be secondary to inflammation or, less commonly, to iron-deficiency anemia and functional hyposplenism. Thrombocytosis as well may resolve after the institution of a GFD.

Coagulation may be altered in untreated CD. Coagulation indices values can be elevated due to concomitant vitamin K deficiency secondary to malabsorption. Prothrombin time can be elevated in about 18% of untreated adult patients (Cavallaro et al., 2004; Rubio-Tapia & Murray, 2008).

Folic acid deficiency usually causes macrocytic and megaloblastic anemia. CD patients who have iron deficiency along with folate and vitamin B_{12} deficiency may not show the characteristic macrocytosis (Brusco et al., 2000).

Although early investigations and case reports showed that micronutrient deficiencies are common in CD patients (Ince et al., 2008), little is known about the true prevalence of micronutrient alterations. In a retrospective study conducted in the USA CD patients from a tertiary center were compared with a matching NANHES cohort. The analysis showed that zinc was deficient in 59.4% of CD patients versus 33.2% of controls ($p < 0.001$). Copper was low in 6.4% of CD patients versus 2.1% of controls ($P = 0.003$) while vitamin B12 was low in 5.3% CD patients versus 1.8% of controls ($P = 0.004$); folate was low in 3.6% CD patients versus 0.3% of controls ($p = 0.002$) (Bledsoe et al., 2019).

7.2.2.7 Liver abnormalities

Among the extraintestinal manifestation of CD, liver and biliary tract alterations have been widely described. Untreated CD patients may suffer from cryptogenic hypertransaminasemia with nonspecific histologic hepatic changes ranging from mild to severe liver dysfunction. According to the most recent meta-analysis of 20 studies, altered liver chemistry in untreated CD patients has an 18.7% pooled prevalence. It often leads to CD diagnosis in asymptomatic patients and disappears in a few weeks, at maximum within 1 year of GFD in most cases (Aggarwal et al., 2023). The reasons for altered liver chemistry in CD are unknown.

Certain autoimmune liver diseases, such as autoimmune hepatitis and autoimmune cholangitis, are more frequent in CD patients than in the general population and seem to be unrelated to gluten toxicity (Volta, 2009). Future research is needed to understand the pathophysiology of these alterations, however, several hypotheses have been done, such as the involvement of gut- liver axis, circulating antitissue transglutaminase antibodies, and disruption of the intestinal barrier (Hoffmanová et al., 2018).

What we know and what we do not know yet regarding the clinical presentation of CD is reported in Table 7.2.

7.3 Gender and clinical presentation

The prevalence of CD is higher in adult women than men, although early and recent studies showed a similar gender prevalence of CD in children (Green & Cellier, 2007; Lionetti et al., 2023). A recent review of the existing studies showed that the pooled incidence rates of CD in women and men are 17.4 and 7.8 per 100.000 person-years, respectively (Makharia et al., 2022). Otherwise, the prevalence of the types of clinical presentation—silent, subclinical, and overtly symptomatic—seems similar between men and women according to most studies (Jansson-Knodell et al., 2018).

TABLE 7.2 Clinical presentation and follow-up of celiac disease: what we know and what we should know.

What we know	What we should know and do not know yet
Proactive search for CD by case-finding will increase its frequency	Which are the determinants for such a heterogeneity of symptoms at onset of CD in adults?
The clinical presentation varies from overt malabsorption to asymptomatic cases	Will subclinical CD eventually, over time, evolve into overt CD?
Symptoms are not related to the severity of mucosal damage	What are the risks for the subclinical/asymptomatic CD cases of being persistently undiagnosed while on a gluten-containing diet?
Severity of symptoms may be related to the diagnostic delay, but also to the quantity of gluten intake	What are the risks for subclinical or asymptomatic cases of CD with low adherence to a GFD?
Older patients at diagnosis and a clinical presentation with gastrointestinal symptoms are risk factors for malignancy	Why does the level of circulating gluten-related antibodies not correlate with histology and the type of CD clinical manifestation?

Old and new studies indicate that women affected by CD have more severe presentations at diagnosis, such as severe signs of malabsorption, longer duration of symptoms and/or signs before the diagnosis, more dyspepsia, nausea/vomiting, constipation, and more frequently than men have a low body mass index (Ciacci et al., 1995; Lenti & Di Sabatino, 2021; Lima et al., 2019). It should be underlined that, at diagnosis, the vast majority of women show iron deficiency, with or without anemia, likely due to the sum of iron malabsorption, iron loss for gynecological causes, and malnutrition (Seidita et al., 2022).

Studies on small series of patients demonstrate that untreated CD interfere with women's reproductive life, causing delayed puberty/menarche and miscarriage (Santonicola et al., 2011).

A recent meta-analysis including 14 cohorts and 4 case-control studies showed that the risk for spontaneous abortion, fetal growth restriction, stillbirth, preterm delivery, cesarean delivery, and lower mean birthweight was significantly higher in pregnant women with CD. Early diagnosis of CD was not linked to any adverse pregnancy outcome (Arvanitakis et al., 2023). On the other side, a systematic review with meta-analysis showed that the prevalence of CD in women with infertility is not increased. Thus the screening for CD in infertile women may have a low yield (Glimberg et al., 2021).

Regarding the male gender, young adult males are likely to be less frequently diagnosed than women, which may be partially related to women's more frequent interactions with the healthcare system (Dixit et al., 2014). However, late in life, the difference disappears (Holmes & Muirhead, 2017).

Male infertility has not been associated with undiagnosed CD. A Swedish nationwide population-based cohort of 7121 adult men with CD showed that the overall fertility hazard ratio in men with biopsy-verified CD was 1.02 (Zugna et al., 2011).

A recent study on 7.49 million individuals in Germany followed by general practitioners suggests that nonalcoholic fatty liver disease is associated with a higher incidence of CD, more frequently in men (HR 2.02). The authors hypothesize that NAFLD promotes the development of CD, particularly in young men (Roderburg et al., 2022).

7.4 Clinical presentation in the elderly

Data from the United Kingdom indicate that approximately 25% of all CD diagnoses are now in patients over 60 and 20% are in those over 65 (Holmes & Muirhead, 2017). The number of diagnoses in older people has increased over time, as they were 0% in 1990–91 and 18.7% in 2016–17 (Shiha et al., 2020). Elderly patients with CD often present extraintestinal symptoms, such as anemia. The extraintestinal clinical presentation and the mild intensity likely explain the diagnostic delay in older patients (Collin et al., 2018; Laurikka et al., 2018). Noteworthy, it should be remembered that diagnosis in older age is a risk factor for complications (Biagi, Gobbi, et al., 2014).

7.5 Celiac disease in type 1 diabetes mellitus

A recent meta-analysis, including more than 26,000 children and adults, indicates that the pooled prevalence of CD in type 1 diabetes mellitus is 6.0% (95% CI = 5.0%–6.9%). The prevalence was similar in different parts of the world

(Europe 6.1%; North America 4.8%, and other countries 4.8%). The proportion of women among the screened type 1 diabetes mellitus individuals varies between 42% and 59% (Elfström et al., 2014).

7.6 The optimal strategy for identifying adults with celiac disease

The serological strategy of mass screening or case finding in children and adults are also treated in Chapters 1, 6, and 8. Presently, it is felt that the appropriate approach is proactive case-finding in high-risk groups. Table 7.3 summarizes the relevant diseases that warrant the active search of CD.

All first-degree relatives should be screened, as 7%–10% of them may be affected or develop the disease. The risk of having CD is highest in siblings (8.9%), followed by offspring (7.9%) and parents (3.0%), and second-degree relatives of patients (2.3%) (Singh et al., 2015). It is recommended to perform serology and HLA typing in all first-degree relatives, and those with predisposing HLA typing should follow up with serology further down the line. First-line relatives of CD patients with negative HLA DQ2/8 do not require further follow-up.

Identifying adults and children with CD outside the high-risk groups is more challenging, and the goal remains avoiding diagnostic delay and waste of economic and human resources. Recent guidelines recommend that adults with symptoms, signs, or laboratory evidence suggestive of malabsorption, IgA deficiency, or an unexplained elevation of liver chemistry should be tested for CD (Al-Toma et al., 2019).

7.7 Follow-up of adult celiac disease patients

CD is a lifelong disease. Follow-up aims to accomplish CD patients' goals, which is mainly to have a better quality of life, resolve symptoms, and reduce the risk of complications. However, in the meantime, clinicians aspire to reach disease remission and healing of enteropathy. Due to the complexity of life-long gluten avoidance, all patients with a childhood or adulthood diagnosis of CD should undergo structured follow-up according to all the international guidelines. Another issue far to be cleared is who should follow up CD patients. There is a concern that when patients transition to adulthood, they might be lost at follow-up (Ulnes et al., 2023). Regarding adults, some guidelines recommend follow-up in the care of gastroenterology tertiary centers (Al-Toma et al., 2019). However, due to the high prevalence, in some countries, the follow-up is often handed over to primary care, and only in case of complications is the patient referred back to a more specialized team (Rouvroye et al., 2021).

The timing of follow-up visits after diagnosis is not clearly defined. As a general rule, consultation, including a dietitian, should be scheduled after 3–6 months from the diagnosis, as the patients might have questions and uncertainties to discuss with the doctor or dietitian. A close follow-up is necessary in case of older patients with gastrointestinal

TABLE 7.3 Case-finding. Clinical conditions in which celiac disease should be searched in adults.

First-degree family history of CD
Type 1 diabetes mellitus
Iron-deficiency anemia
Dermatitis herpetiformis-like rash
Selective IgA deficiency
Autoimmune thyroid disorders
IBS-diarrhea
Short stature, significant nonvoluntary weight loss
Delayed puberty, amenorrhea, early menopause, recurrent miscarriage/infertility
Unexplained increase in liver enzymes and autoimmune liver disease
Persistent fatigue
Osteopenia/osteoporosis especially in men or young women; rheumatoid arthritis
Turner Syndrome, Down's syndrome, Williams' syndrome

symptoms, as they are at greater risk of complications compared to asymptomatic and younger patients (Biagi et al., 2014).

When patients are well informed and compliant with the GFD, the follow-up might be scheduled yearly or even more distanced. Telemedicine has proven very useful in CD follow up, as shown during the COVID-19 pandemic (Siniscalchi et al., 2020).

The main reasons to follow up symptomatic and asymptomatic CD patients at diagnosis are the possible nutritional imbalance or dietary mistakes (voluntary or nonvoluntary) during GFD, onset of new diseases or complications, and the perceived low quality of life.

7.7.1 Nutrition during follow-up

Even during a GFD, nutritional imbalances that should be monitored may occur. Since avoiding gluten may reduce the consumption of beneficial whole grains, the extent of the cardiovascular risk may increase and a recent study from Norway, has shown that CD women had an unbalanced diet with a higher intake of total and saturated fatty acids and a lower fiber intake than the general population (van Megen et al., 2023).

A prospective USA cohort study showed that long-term dietary gluten intake was not associated with risk of coronary heart disease (Lebwohl et al., 2017), whereas a recent Italian retrospective study on more than 8000 CD patients showed that CD generally reduces the risk of cardiovascular diseases, particularly of carotid lesions, especially in those patients on a long-term GFD (Dore et al., 2023). On the contrary a study from the UK National Biobank on about 2000 CD patients without cardiovascular disease at diagnosis showed an increased incidence rate of them than in controls (9.0 compared with 7.4 per 1000 person-years) (Conroy et al., 2023).

Regarding the need for combining to a GFD the supplementation of iron, calcium and other oligoelements, vitamins, and bone-active drugs, refer to Chapter 13.

7.7.2 Monitoring the gluten free diet

Adherence to a strict GFD is currently the only effective treatment for CD. Although good adherence to a GFD can adequately control CD-related symptoms, disease relapse may occur even with a very low amount of gluten, considering that the duration of exposure to gluten is also important. Moreover, gluten contamination of gluten-free food is an issue for patients following a strict GFD. A recent review has shown the limit of the current literature on the topic and the need for further clinical trials to fully understand the relationship among gluten exposure, symptoms, and histological damage (Rostami-Nejad et al., 2023).

Questionnaires have been developed and validated to assess the adherence of GFD (Biagi et al., 2009, 2012), but this issue is treated more in depth in Chapter 12.

7.7.3 Follow-up biopsies

It is still being uncertain whether a follow-up biopsy is necessary and if so, when. Currently, there is no data on the degree of mucosal healing that constitutes sufficient remission in CD patients, and therefore there is no guideline-recommended follow-up. However, discrepancies exist in the available data, likely due to differences in population and study designs (Ciacci et al., 1995; Hære et al., 2016; Rubio-Tapia et al., 2010).

The current practice is that of reserving follow-up biopsies for patients feeling unwell. Therefore the evidence derives from retrospective studies which are intrinsically biased by the "need for biopsy" because asymptomatic patients will not usually undergo follow-up endoscopy. Moreover, the clinical significance of mucosal abnormality in asymptomatic patients is yet to be clarified. It must be noted that a follow-up biopsy showing mucosal damage also might help to convince patients reluctant to adhere to a GFD or otherwise disclose other gastrointestinal diseases (Biagi, Vattiato, et al., 2014; Harder et al., 2020).

Based on available evidence, CD patients with persistent villous atrophy have a 3.78-fold increase in malignancies (mostly lymphomas) versus CD patients with mucosal healing, who have a 1.5-fold increased risk (Lebwohl et al., 2013). In contrast, normal architecture with increased intraepithelial lymphocytes (Marsh 1) is linked to a modest mortality increase in CD (Ludvigsson et al., 2009).

7.7.4 Monitoring bone density

No doubt that the most critical factor is ensuring that adult CD patients receive the appropriate screening and follow-up care to maintain optimal bone health (Khurana et al., 2023; Ludvigsson et al., 2014; Rubio-Tapia et al., 2023; Tye-Din, 2022).

The American College of Gastroenterology recommends that follow-up BMD measurement is suggested after one year, if bone loss is detected at diagnosis, and after two years, in presence of normal values (Rubio-Tapia et al., 2023). However, considering the efficacy of a GFD on the improvement of BMD during the first two years of treatment (Corazza et al., 1996, 1997) and the lack of information on the usefulness of the association of mineral-active drugs to GFD (Kemppainen et al., 1999), the most important variable determining DEXA follow-up timing is essentially represented by the adherence to a strict GFD: in nonadherent patients, yearly evaluation could be indicated until a strict GFD is adopted, on the contrary, in patients with normal BMD at the first evaluation and following a strict GFD, the next BMD measurement could be scheduled at perimenopausal period in females and after 50 years in males (Cosman et al., 2014).

Patients with severe bone loss or multiple risk factors for osteoporosis may require more frequent monitoring, while those with stable bone density may be monitored less frequently.

In addition to monitoring bone density, patients with CD should be advised to consume adequate amounts of calcium after calcium: and ,if a clear deficiency is present, also vitamin D supplementation (Ciacci et al., 2020), engage in weight-bearing exercise, and avoid smoking and excessive alcohol consumption (Passananti et al., 2012; Singh & Whelan, 2011) (see also Chapter 13).

7.7.5 Follow up on IBS-celiac patients

Up to 50% of CD patients continue to experience symptoms compatible with IBS despite optimal adherence to a gluten-free diet (Laurikka et al., 2016; Schiepatti et al., 2022; Silvester et al., 2017; van Megen et al., 2021). These patients also show higher rates of anxiety and depression and generally have a lower quality of life (Parker et al., 2022). Moreover, a recent retrospective study showed that CD patients from Israel have an odd ratio of 2.08 for the occurrence of fibromyalgia (Savin et al., 2023).

As dietary treatment and lifestyle interventions are commonly recommended for managing IBS (McKenzie et al., 2016), the same can be done for CD patients. Additional dietary modifications or medications may be necessary to manage their IBS symptoms. The use of probiotics, prebiotics, and low FODMAP diets have been shown to improve these symptoms (Muir et al., 2019; Tye-Din, 2022; van Megen et al., 2022).

If the patient continues to suffer from symptoms or signs, the first step is to recheck the quality of the diagnosis with a careful revision of the records at the time of diagnosis. If doubt of misdiagnosis of CD persists, a long-term gluten challenge followed by an assessment of serology and histology or, in the future, immuno-markers of gluten intolerance might exclude o confirm the presence of CD (Anderson, 2022).

The follow-up of IBS-celiac patients is crucial to optimize their management and improve their quality of life. In patients with IBS-celiac, follow-up visits should also include an assessment of their IBS symptoms, including stool frequency/consistency, bloating, and abdominal pain. The follow-up frequency may vary depending on the patient's clinical status, with more frequent visits for patients with ongoing symptoms or those struggling to adhere to a GFD.

7.7.6 Psychological, psychiatric symptoms and quality of life

It is essential for healthcare providers to be aware of the potential psychiatric symptoms associated with CD and to include psychiatric evaluations in their follow-up care plans for these patients (Tye-Din, 2022) (Raiteri et al., 2022). Anxiety is a psychiatric symptom frequently reported in patients with gastrointestinal symptoms (Addolorato et al., 2008). In CD, anxiety seems to improve after diagnosis and GFD (Addolorato et al., 2001). Depression is also commonly reported in untreated and treated conditions, with variable prevalence rates (Slim et al., 2018; Zylberberg et al., 2017). A recent systematic review and meta-analysis found that a long-term GFD may significantly reduce and normalize the severity of depressive symptoms for subjects with CD, with an effect for both symptomatic and extraintestinal CD patients, but no effect for asymptomatic/silent cases (Busby et al., 2018).

Other psychiatric symptoms reported in CD patients at follow-up include irritability, mood swings, cognitive impairment, and sleep disturbances (Liester & Liester, n.d.). These symptoms may be related to the effects of long-term undiagnosed disease and/or the psychosocial impact of living with a chronic illness. Psychiatric treatment options may include psychotherapy and medication, although studies on the outcome of these treatments are scanty. A cross-sectional study of more of 1000 USA CD patients at moment of endoscopy, compared to non-CD patients showed the use of psychotropic medications was not different in the two groups, but CD patients used more antidepressants (Zylberberg et al., 2018). However, systematic reviews of literature underlined the scarcity of clinical trials on psychological disturbances at diagnosis and follow-up of CD patients (Busby et al., 2018; Smith & Gerdes, 2012).

The impact of a GFD on adult patients' quality of life will be specifically addressed in Chapter 12.

7.8 Mortality in celiac disease

Although CD has long been considered a condition with a favorable outcome, mortality rates were proven to be higher than in the general population, contrary to what has been documented for inflammatory bowel disease (GBD, 2017).

After diagnosis, the risk of death was found to be elevated during the initial three years, but gradually decreases over time following the implementation of a GFD, eventually reaching mortality levels comparable to those of the general population (Corrao et al., 2001; West et al., 2004).

In a multicentric, retrospective cohort study conducted in Italy in patients diagnosed between 1962 and 1994, standardized mortality ratio (SMR) resulted as high as 2.0. No evidence was found for any influence of sex or year of presentation. On the contrary, the SMR even doubles in patients with a diagnostic delay longer than 10 years, and is anyway increased in patients with malabsorption at diagnosis, and in those who did not adhere to a GFD; patients diagnosed with mild symptoms or by antibody screening did not show any relevant excess of mortality (Corrao et al., 2001). In line with the latter result, in most studies performed on biobank of frozen blood samples collected several years before testing, the positivity of antitransglutaminase and/or antiendomysial antibodies, suggesting a lasting condition of "unrecognized CD," was not associated to a higher mortality risk (Canavan et al., 2011; Godfrey et al., 2010; Lohi et al., 2009).

In conclusion, heterogeneity in clinical presentation, degree of adherence to a GDF and age at diagnosis, all crucial factors in determining mortality risk, can explain the wide differences in literature data, that nowadays globally show contradictory results (Biagi & Corazza, 2015). In fact, reported mortality rates span from 0.52 (notably, a reduced mortality rate), and 3.9 (Biagi & Corazza, 2010). Some recent population studies, performed on large, national databases did not evidence long-term higher mortality in patients with a diagnosis of CD (Abdul Sultan et al., 2015; Koskinen et al., 2020; Quarpong et al., 2019), even confirming a higher risk for lymphoproliferative disease compared with the general population. A large meta-analysis evidenced in celiac population a mortality ratio at the limit for statistical significancy (odds ratio 1.24) (Tio et al., 2012), and, similarly, a recent wide population-based, retrospective cohort study demonstrated only a small increased mortality risk (hazard ratio 1.21) (Lebwohl, Larsson et al., 2021). Besides differences in methodology and possible selection bias derived by registry-based studies, it could be postulated that these results reflect a progressive change in the natural history over time, thanks to the increasing awareness of clinicians and to larger availability of tests that allow earlier diagnosis. Moreover, wider diffusion of gluten free-products and of educational information on GFD could have increased patient adherence to diet, so reducing the rate of complications. There is evidence that the risk of mortality in individuals with CD who were diagnosed after 2000 is lower compared to those who were diagnosed before (respectively, SMR 1.6 vs SMR 2.23) (Holmes & Muirhead, 2018).

Surprisingly enough, long-term mortality of celiac patients diagnosed in childhood was reported to be persistently higher than in the general population. As shown in two papers on the same cohort followed-up for 45 years (Quarpong et al., 2019; Solaymani-Dodaran et al., 2007), long-term mortality excess (SMR 3.32) in children was not related to classic complications of CD, but rather to external causes, such as accidents, suicides, and violence. Authors postulated a role for psychiatric morbidity related to the status of having a chronic illness.

Two recent population-based studies (Koskinen et al., 2020; Lebwohl et al., 2020) addressed causes of mortality in adults CD. These studies showed that excess mortality was mainly due to gastrointestinal tract malignancies (HR 2.18) and, above all, to enteropathy-associated T cell lymphoma (HR 5.07), as discussed elsewhere in this book (Chapter 14). A slight risk excess, of questionable significance, was found for death due to cardiovascular (HR 1.08) and respiratory diseases (HR 1.21) (Lebwohl et al., 2020), the latter due to infection by encapsulated bacteria (Simons et al., 2018) (see Chapter 15). Indeed, putative causes of increased mortality risk related to cardiovascular disease are less clear, and available data are conflicting.

7.9 Celiac disease during the Covid-19 pandemic

The recent pandemic has been a challenge also for adult CD patients. It has been shown that the SARS-CoV-2 targets intestinal enterocytes because of their high expression of ACE2 and TMPRSS receptors. Moreover, the Covid-19-

related systemic cytokine storm involves the intestinal mucosa, disrupting the intestinal barrier and increasing permeability. One hypothesis suggests that during a Covid-19 infection, gliadin fragments may pass the mucosa and elicit a gluten-related specific response in the lamina propria of HLA DQ2/DQ8 patients (Trovato et al., 2021). If this is confirmed, we should face an outbreak of Covid-19-related CD in the following years. Due to the protective measures adopted during the Covid-19 pandemic, as expected, there was a decrease in CD diagnoses, as they were limited to the most severe and symptomatic cases (Crocco et al., 2023; Valitutti et al., 2021).

During the pandemic, and for the first time, the British Society of Gastroenterology suggested omitting the biopsy in the case of adults with high suspicion of CD presenting a ten-fold increase of serum anti-transglutaminase IgA, so adopting the same diagnostic criteria as in children (BSG Interim Guidance: Covid-19 specific nonbiopsy protocol for those with suspected celiac disease, 2022 https://www.bsg.org.uk/covid-19-advice/covid-19-specific-non-biopsy-protocol-guidance-for-those-with-suspected-celiac-disease/).

Regarding the susceptibility of CD patients to the SARS-CoV-2 infection, international prospective studies showed that patients with CD had similar odds of contracting Covid-19, did not have a higher frequency of complications, and may not need to take additional precautions to prevent exposure aside from that recommended to the general population (Lebwohl, Larsson et al., 2021; Zhen et al., 2021). CD patients showed an anti-SARS-CoV-2 Ig profile similar to non-CD controls, except for the anti-N IgA subtype. No risk factors were identified among CD parameters and GFD adherence (Elli et al., 2022).

A study looking at complicated CD, examined 21 Italian patients affected by refractory CD (RCD) during Covid. None of the 21 RCD patients reported symptoms suggestive of Covid-19 or changes in their symptoms in the index period; no cohabitants developed Covid-19; consequently, none was hospitalized. All the RCD patients referred to have strictly applied social distancing due to their consciousness of their fragility and risk of developing severe forms of Covid-19 (Elli et al., 2020).

During the pandemic, a study found that 2/3 of patients preferred online over face-to-face appointments. The most common difficulties faced during the lockdown period were the availability of gluten-free food delivery at home, increased prices of gluten-free food, and traveling long distances to arrange gluten-free food supplies (Mehtab et al., 2021).

Despite the countless problems caused by the event, the silver lining of the Covid-19 pandemic has been the opportunity to successfully implement telemedicine in the follow-up of many diseases, including CD (Månsson et al., 2022; Mehtab et al., 2021; Siniscalchi et al., 2020).

References

Abdul Sultan, A., Crooks, C. J., Card, T., Tata, L. J., Fleming, K. M., & West, J. (2015). Causes of death in people with coeliac disease in England compared with the general population: A competing risk analysis. *Gut*, *64*(8), 1220–1226. Available from https://doi.org/10.1136/gutjnl-2014-308285.

Addolorato, G., Capristo, E., Ghittoni, G., Valeri, C., Mascianà, R., Ancona, C., & Gasbarrini, G. (2001). Anxiety but not depression decreases in coeliac patients after one-year gluten-free diet: A longitudinal study. *Scandinavian Journal of Gastroenterology*, *36*(5), 502–506. Available from https://doi.org/10.1080/00365520119754.

Addolorato, G., Mirijello, A., D'Angelo, C., Leggio, L., Ferrulli, A., Abenavoli, L., Vonghia, L., Cardone, S., Leso, V., Cossari, A., Capristo, E., & Gasbarrini, G. (2008). State and trait anxiety and depression in patients affected by gastrointestinal diseases: Psychometric evaluation of 1641 patients referred to an internal medicine outpatient setting. *International Journal of Clinical Practice*, *62*(7), 1063–1069. Available from https://doi.org/10.1111/j.1742-1241.2008.01763.x.

Aggarwal, M., Garg, R., Kumar, P., Lindenmeyer, C. C., Wakim-Fleming, J., Jansson-Knodell, C., & Rubio-Tapia, A. (2023). Bi-directional relationship between celiac disease and liver chemistries: A systematic review and meta-analysis. *Digestive Diseases and Sciences*, *68*(4), 1369–1380. Available from https://doi.org/10.1007/s10620-022-07663-w.

Al-Toma, A., Volta, U., Auricchio, R., Castillejo, G., Sanders, D. S., Cellier, C., Mulder, C. J., & Lundin, K. E. A. (2019). European Society for the Study of Coeliac Disease (ESsCD) guideline for coeliac disease and other gluten-related disorders. *United European Gastroenterology Journal*, *7*(5), 583–613. Available from https://doi.org/10.1177/2050640619844125.

Anderson, R. P. (2022). Review article: Diagnosis of coeliac disease: a perspective on current and future approaches. *Alimentary Pharmacology & Therapeutics*, *56*(Suppl 1), S18–S37. Available from https://doi.org/10.1111/apt.16840.

Arvanitakis, K., Siargkas, A., Germanidis, G., Dagklis, T., & Tsakiridis, I. (2023). Adverse pregnancy outcomes in women with celiac disease: A systematic review and meta-analysis. *Annals of Gastroenterology*, *36*(1), 12–24. Available from https://doi.org/10.20524/aog.2022.0764.

Aziz, I., & Simrén, M. (2021). The overlap between irritable bowel syndrome and organic gastrointestinal diseases. *The Lancet. Gastroenterology & Hepatology*, *6*(2), 139–148. Available from https://doi.org/10.1016/S2468-1253(20)30212-0.

Balaban, D. V., Dima, A., Jurcut, C., Popp, A., & Jinga, M. (2019). Celiac crisis, a rare occurrence in adult celiac disease: A systematic review. *World Journal of Clinical Cases*, *7*(3), 311–319. Available from https://doi.org/10.12998/wjcc.v7.i3.311.

Bardella, M. T., Marino, R., Barbareschi, M., Bianchi, F., Faglia, G., & Bianchi, P. (2000). Alopecia areata and coeliac disease: No effect of a gluten-free diet on hair growth. *Dermatology (Basel, Switzerland), 200*(2), 108–110. Available from https://doi.org/10.1159/000018340.

Barratt, S. M., Leeds, J. S., Robinson, K., Shah, P. J., Lobo, A. J., McAlindon, M. E., & Sanders, D. S. (2011). Reflux and irritable bowel syndrome are negative predictors of quality of life in coeliac disease and inflammatory bowel disease. *European Journal of Gastroenterology & Hepatology, 23*(2), 159–165. Available from https://doi.org/10.1097/MEG.0b013e328342a547.

Biagi, F., Andrealli, A., Bianchi, P. I., Marchese, A., Klersy, C., & Corazza, G. R. (2009). A gluten-free diet score to evaluate dietary compliance in patients with coeliac disease. *The British Journal of Nutrition, 102*(6), 882–887. Available from https://doi.org/10.1017/S0007114509301579.

Biagi, F., Bianchi, P. I., Marchese, A., Trotta, L., Vattiato, C., Balduzzi, D., Brusco, G., Andrealli, A., Cisarò, F., Astegiano, M., Pellegrino, S., Magazzù, G., Klersy, C., & Corazza, G. R. (2012). A score that verifies adherence to a gluten-free diet: A cross-sectional, multicentre validation in real clinical life. *The British Journal of Nutrition, 108*(10), 1884–1888. Available from https://doi.org/10.1017/S0007114511007367.

Biagi, F., & Corazza, G. R. (2010). Mortality in celiac disease. *Nature Reviews. Gastroenterology & Hepatology, 7*(3), 158–162. Available from https://doi.org/10.1038/nrgastro.2010.2.

Biagi, F., & Corazza, G. R. (2015). Do different patients with coeliac disease have different mortality rates. *Gut, 64*(8), 1187–1188. Available from https://doi.org/10.1136/gutjnl-2014-308567.

Biagi, F., Gobbi, P., Marchese, A., Borsotti, E., Zingone, F., Ciacci, C., Volta, U., Caio, G., Carroccio, A., Ambrosiano, G., Mansueto, P., & Corazza, G. R. (2014). Low incidence but poor prognosis of complicated coeliac disease: A retrospective multicentre study. *Digestive and Liver Disease: Official Journal of the Italian Society of Gastroenterology and the Italian Association for the Study of the Liver, 46*(3), 227–230. Available from https://doi.org/10.1016/j.dld.2013.10.010.

Biagi, F., Vattiato, C., Agazzi, S., Balduzzi, D., Schiepatti, A., Gobbi, P., & Corazza, G. R. (2014). A second duodenal biopsy is necessary in the follow-up of adult coeliac patients. *Annals of Medicine, 46*(6), 430–433. Available from https://doi.org/10.3109/07853890.2014.913378.

Bledsoe, A. C., King, K. S., Larson, J. J., Snyder, M., Absah, I., Choung, R. S., & Murray, J. A. (2019). Micronutrient deficiencies are common in contemporary celiac disease despite lack of overt malabsorption symptoms. *Mayo Clinic Proceedings, 94*(7), 1253–1260. Available from https://doi.org/10.1016/j.mayocp.2018.11.036.

Brusco, G., Di Stefano, M., & Corazza, G. R. (2000). Increased red cell distribution width and coeliac disease. *Digestive and Liver Disease, 32*(2), 128–130. Available from https://doi.org/10.1016/s1590-8658(00)80399-0, In this issue.

Busby, E., Bold, J., Fellows, L., & Rostami, K. (2018). Mood disorders and gluten: It's not all in your mind! A systematic review with meta-analysis. *Nutrients, 10*(11), 1708. Available from https://doi.org/10.3390/nu10111708.

Canavan, C., Logan, R. F., Khaw, K.-T., & West, J. (2011). No difference in mortality in undetected coeliac disease compared with the general population: A UK cohort study. *Alimentary Pharmacology & Therapeutics, 34*(8), 1012–1019. Available from https://doi.org/10.1111/j.1365-2036.2011.04811.x.

Card, T. R., Siffledeen, J., West, J., & Fleming, K. M. (2013). An excess of prior irritable bowel syndrome diagnoses or treatments in Celiac disease: Evidence of diagnostic delay. *Scandinavian Journal of Gastroenterology, 48*(7), 801–807. Available from https://doi.org/10.3109/00365521.2013.786130.

Catassi, C., Rätsch, I. M., Fabiani, E., Rossini, M., Bordicchia, F., Candela, F., Coppa, G. V., & Giorgi, P. L. (1994). Coeliac disease in the year 2000: Exploring the iceberg. *Lancet (London, England), 343*(8891), 200–203. Available from https://doi.org/10.1016/s0140-6736(94)90989-x.

Cavallaro, R., Iovino, P., Castiglione, F., Palumbo, A., Marino, M., Di Bella, S., Sabbatini, F., Labanca, F., Tortora, R., Mazzacca, G., & Ciacci, C. (2004). Prevalence and clinical associations of prolonged prothrombin time in adult untreated coeliac disease. *European Journal of Gastroenterology & Hepatology, 16*(2), 219–223. Available from https://doi.org/10.1097/00042737-200402000-00016.

Chetcuti Zammit, S., McAlindon, M. E., Greenblatt, E., Maker, M., Siegelman, J., Leffler, D. A., Yardibi, O., Raunig, D., Brown, T., & Sidhu, R. (2023). Quantification of celiac disease severity using video capsule endoscopy: A comparison of human experts and machine learning algorithms. *Current Medical Imaging*. Available from https://doi.org/10.2174/1573405619666230123110957.

Ciacci, C., Bilancio, G., Russo, I., Iovino, P., Cavallo, P., Santonicola, A., Bucci, C., Cirillo, M., & Zingone, F. (2020). 25-Hydroxyvitamin D, 1,25-dihydroxyvitamin D, and peripheral bone densitometry in adults with celiac disease. *Nutrients, 12*(4), 929. Available from https://doi.org/10.3390/nu12040929.

Ciacci, C., Cirillo, M., Giorgetti, G., Alfinito, F., Franchi, A., Mazzetti di Pietralata, M., & Mazzacca, G. (1999). Low plasma cholesterol: A correlate of nondiagnosed celiac disease in adults with hypochromic anemia. *The American Journal of Gastroenterology, 94*(7), 1888–1891. Available from https://doi.org/10.1111/j.1572-0241.1999.01225.x.

Ciacci, C., Cirillo, M., Sollazzo, R., Savino, G., Sabbatini, F., & Mazzacca, G. (1995). Gender and clinical presentation in adult celiac disease. *Scandinavian Journal of Gastroenterology, 30*(11), 1077–1081. Available from https://doi.org/10.3109/00365529509101610.

Ciacci, C., Spagnuolo, G., Tortora, R., Bucci, C., Franzese, D., Zingone, F., & Cirillo, M. (2008). Urinary stone disease in adults with celiac disease: Prevalence, incidence and urinary determinants. *The Journal of Urology, 180*(3), 974–979. Available from https://doi.org/10.1016/j.juro.2008.05.007.

Clappison, E., Hadjivassiliou, M., & Zis, P. (2020). Psychiatric manifestations of coeliac disease, a systematic review and meta-analysis. *Nutrients, 12*(1), 142. Available from https://doi.org/10.3390/nu12010142.

Collin, P., Mustalahti, K., Kyrönpalo, S., Rasmussen, M., Pehkonen, E., & Kaukinen, K. (2004). Should we screen reflux oesophagitis patients for coeliac disease? *European Journal of Gastroenterology & Hepatology, 16*(9), 917–920. Available from https://doi.org/10.1097/00042737-200409000-00017.

Collin, P., Reunala, T., Rasmussen, M., Kyrönpalo, S., Pehkonen, E., Laippala, P., & Mäki, M. (1997). High incidence and prevalence of adult coeliac disease. *Augmented diagnostic approach. Scandinavian Journal of Gastroenterology*, *32*(11), 1129–1133. Available from https://doi.org/10.3109/00365529709002992.

Collin, P., Syrjänen, J., Partanen, J., Pasternack, A., Kaukinen, K., & Mustonen, J. (2002). Celiac disease and HLA DQ in patients with IgA nephropathy. *The American Journal of Gastroenterology*, *97*(10), 2572–2576. Available from https://doi.org/10.1111/j.1572-0241.2002.06025.x.

Collin, P., Vilppula, A., Luostarinen, L., Holmes, G. K. T., & Kaukinen, K. (2018). Review article: Coeliac disease in later life must not be missed. *Alimentary Pharmacology & Therapeutics*, *47*(5), 563–572. Available from https://doi.org/10.1111/apt.14490.

Conroy, M., Allen, N., Lacey, B., Soilleux, E., & Littlejohns, T. (2023). Association between coeliac disease and cardiovascular disease: Prospective analysis of UK Biobank data. *BMJ Medicine*, *2*(1), e000371. Available from https://doi.org/10.1136/bmjmed-2022-000371.

Corazza, G. R., Andreani, M. L., Venturo, N., Bernardi, M., Tosti, A., & Gasbarrini, G. (1995). Celiac disease and alopecia areata: Report of a new association. *Gastroenterology*, *109*(4), 1333–1337. Available from https://doi.org/10.1016/0016-5085(95)90597-9.

Corazza, G. R., Di Sario, A., Cecchetti, L., Jorizzo, R. A., Di Stefano, M., Minguzzi, L., Brusco, G., Bernardi, M., & Gasbarrini, G. (1996). Influence of pattern of clinical presentation and of gluten-free diet on bone mass and metabolism in adult coeliac disease. *Bone*, *18*(6), 525–530. Available from https://doi.org/10.1016/8756-3282(96)00071-3.

Corazza, G. R., Di Stefano, M., Jorizzo, R. A., Cecchetti, L., Minguzzi, L., & Gasbarrini, G. (1997). Propeptide of type I procollagen is predictive of posttreatment bone mass gain in adult celiac disease. *Gastroenterology*, *113*(1), 67–71. Available from https://doi.org/10.1016/s0016-5085(97)70081-7.

Corazza, G. R., Di Stefano, M., Mauriño, E., & Bai, J. C. (2005). Bones in coeliac disease: Diagnosis and treatment. Best practice & research. *Clinical Gastroenterology*, *19*(3), 453–465. Available from https://doi.org/10.1016/j.bpg.2005.01.002.

Corazza, G. R., Valentini, R. A., Andreani, M. L., D'Anchino, M., Leva, M. T., Ginaldi, L., De Feudis, L., Quaglino, D., & Gasbarrini, G. (1995). Subclinical coeliac disease is a frequent cause of iron-deficiency anaemia. *Scandinavian Journal of Gastroenterology*, *30*(2), 153–156. Available from https://doi.org/10.3109/00365529509093254.

Corrao, G., Corazza, G. R., Bagnardi, V., Brusco, G., Ciacci, C., Cottone, M., Sategna Guidetti, C., Usai, P., Cesari, P., Pelli, M. A., Loperfido, S., Volta, U., Calabró, A., & Certo, M. (2001). Mortality in patients with coeliac disease and their relatives: A cohort study, & Club del Tenue Study Group*Lancet (London, England)*, *358*(9279), 356–361. Available from https://doi.org/10.1016/s0140-6736(01)05554-4.

Cosman, F., de Beur, S. J., LeBoff, M. S., Lewiecki, E. M., Tanner, B., Randall, S., & Lindsay, R. (2014). Clinician's guide to prevention and treatment of osteoporosis. *Osteoporosis International*, *25*(10), 2359–2381. Available from https://doi.org/10.1007/s00198-014-2794-2.

Croall, I. D., Sanders, D. S., Hadjivassiliou, M., & Hoggard, N. (2020). Cognitive deficit and white matter changes in persons with celiac disease: A population-based study. *Gastroenterology*, *158*(8), 2112–2122. Available from https://doi.org/10.1053/j.gastro.2020.02.028.

Crocco, M., Calvi, A., Canzoneri, F., Malerba, F., Zampatti, N., Chiaro, A., Arrigo, S., Gandullia, P., Proietti, S., & Bonassi, S. (2023). The Influence of SARS-CoV-2 pandemic on the diagnosis of celiac disease and clinical practice in pediatric gastroenterology. *Nutrients*, *15*(3), 559. Available from https://doi.org/10.3390/nu15030559.

Cuomo, A., Romano, M., Rocco, A., Budillon, G., Del Vecchio Blanco, C., & Nardone, G. (2003). Reflux oesophagitis in adult coeliac disease: Beneficial effect of a gluten free diet. *Gut*, *52*(4), 514–517. Available from https://doi.org/10.1136/gut.52.4.514.

Di Stefano, M., Bergonzi, M., Benedetti, I., De Amici, M., Torre, C., Brondino, N., Miceli, E., Pagani, E., Marseglia, G. L., Corazza, G. R., & Di Sabatino, A. (2019). Alterations of inflammatory and matrix production indices in celiac disease with low bone mass on long-term gluten-free diet. *Journal of Clinical Gastroenterology*, *53*(6), e221–e226. Available from https://doi.org/10.1097/MCG.0000000000001032.

Di Stefano, M., Mengoli, C., Bergonzi, M., & Corazza, G. R. (2013). Bone mass and mineral metabolism alterations in adult celiac disease: Pathophysiology and clinical approach. *Nutrients*, *5*(11), 4786–4799. Available from https://doi.org/10.3390/nu5114786.

Di Stefano, M., Sciarra, G., Jorizzo, R. A., Grillo, R. L., Cecchetti, L., Speziale, D., Pichini, L., Gasbarrini, G., & Corazza, G. R. (1997). Local and gonadal factors in the pathogenesis of celiac bone loss. *Italian Journal of Gastroenterology and Hepatology*, *29*(suppl 2), 31. Available from https://pubmed.ncbi.nlm.nih.gov/?term = Di + Stefano + M%2C + Sciarra + G%2C + Jorizzo + RA%2C + Grillo + RL%2C + Cecchetti + L%2C + Speziale + D%2C + Pichini + L%2C + Gasbarrini + G%2C + Corazza + GR. + Local + and + gonadal + factors + in + the + pathogenesis + of + celiac + bone + loss. + Ital + J + Gastroenterol + Hepatol + 1997%3B29 + %28suppl + 2%29%3A31.&sort = date&size = 200.

Di Stefano, M., Veneto, G., Malservisi, S., Cecchetti, L., Minguzzi, L., Strocchi, A., & Corazza, G. R. (2002). Lactose malabsorption and intolerance and peak bone mass. *Gastroenterology*, *122*(7), 1793–1799. Available from https://doi.org/10.1053/gast.2002.33600.

Dixit, R., Lebwohl, B., Ludvigsson, J. F., Lewis, S. K., Rizkalla-Reilly, N., & Green, P. H. R. (2014). Celiac disease is diagnosed less frequently in young adult males. *Digestive Diseases and Sciences*, *59*(7), 1509–1512. Available from https://doi.org/10.1007/s10620-014-3025-6.

Dore, M. P., Mereu, S., Saba, P. S., Portoghese, M., & Pes, G. M. (2023). Celiac disease and cardiovascular risk: A retrospective case-control study. *Journal of Clinical Medicine*, *12*(6), 2087. Available from https://doi.org/10.3390/jcm12062087.

Drossman, D. A., & Hasler, W. L. (2016). Rome IV-functional GI disorders: Disorders of gut-brain interaction. *Gastroenterology*, *150*(6), 1257–1261. Available from https://doi.org/10.1053/j.gastro.2016.03.035.

Dutta, R., Rawat, R., Das, P., Singh, G., Kumari, A., Ahmad, M., Chauhan, A., Ahuja, V., Agrawal, S. K., & Makharia, G. K. (2023). Identification of celiac disease associated IgA nephropathy by IgA anti-tissue transglutaminase2 antibody deposits in archived formalin-fixed tissues. *Saudi Journal of Gastroenterology: Official Journal of the Saudi Gastroenterology Association*, *29*(1), 59–65. Available from https://doi.org/10.4103/sjg.sjg_326_22.

Eaton, W., Mortensen, P. B., Agerbo, E., Byrne, M., Mors, O., & Ewald, H. (2004). Coeliac disease and schizophrenia: population based case control study with linkage of Danish national registers. *BMJ (Clinical Research ed.)*, *328*(7437), 438–439. Available from https://doi.org/10.1136/bmj.328.7437.438.

Elfström, P., Sundström, J., & Ludvigsson, J. F. (2014). Systematic review with meta-analysis: Associations between coeliac disease and type 1 diabetes. *Alimentary Pharmacology & Therapeutics*, *40*(10), 1123–1132. Available from https://doi.org/10.1111/apt.12973.

Elli, L., Facciotti, F., Lombardo, V., Scricciolo, A., Sanders, D. S., Vaira, V., Barisani, D., Vecchi, M., Costantino, A., Scaramella, L., dell'Osso, B., Doneda, L., & Roncoroni, L. (2022). Anti-SARS-CoV-2 immunoglobulin profile in patients with celiac disease living in a high incidence area. *Digestive and Liver Disease: Official Journal of the Italian Society of Gastroenterology and the Italian Association for the Study of the Liver*, *54*(1), 3–9. Available from https://doi.org/10.1016/j.dld.2021.08.027.

Elli, L., Scaramella, L., Lombardo, V., Scricciolo, A., Doneda, L., Roncoroni, L., & Vecchi, M. (2020). Refractory celiac disease and COVID-19 outbreak: Findings from a high incidence scenario in Northern Italy. *Clinics and Research in Hepatology and Gastroenterology*, *44*(5), e115–e120. Available from https://doi.org/10.1016/j.clinre.2020.07.026.

Elwenspoek, M. M., Thom, H., Sheppard, A. L., Keeney, E., O'Donnell, R., Jackson, J., Roadevin, C., Dawson, S., Lane, D., Stubbs, J., Everitt, H., Watson, J. C., Hay, A. D., Gillett, P., Robins, G., Jones, H. E., Mallett, S., & Whiting, P. F. (2022). Defining the optimum strategy for identifying adults and children with coeliac disease: Systematic review and economic modelling. *Health Technology Assessment (Winchester, England)*, *26*(44), 1–310. Available from https://doi.org/10.3310/ZUCE8371.

Fouda, M. A., Khan, A. A., Sultan, M. S., Rios, L. P., McAssey, K., & Armstrong, D. (2012). Evaluation and management of skeletal health in celiac disease: position statement. *Canadian Journal of Gastroenterology = Journal Canadien de Gastroenterologie*, *26*(11), 819–829. Available from https://doi.org/10.1155/2012/823648.

Fuchs, V., Kurppa, K., Huhtala, H., Collin, P., Mäki, M., & Kaukinen, K. (2014). Factors associated with long diagnostic delay in celiac disease. *Scandinavian Journal of Gastroenterology*, *49*(11), 1304–1310. Available from https://doi.org/10.3109/00365521.2014.923502.

Fukunaga, M., Ishimura, N., Fukuyama, C., Izumi, D., Ishikawa, N., Araki, A., Oka, A., Mishiro, T., Ishihara, S., Maruyama, R., Adachi, K., & Kinoshita, Y. (2018). Celiac disease in non-clinical populations of Japan. *Journal of Gastroenterology*, *53*(2), 208–214. Available from https://doi.org/10.1007/s00535-017-1339-9.

GBD (2017). Inflammatory Bowel Disease Collaborators. The global, regional, and national burden of inflammatory bowel disease in 195 countries and territories, 1990-2017: a systematic analysis for the Global Burden of Disease Study 2017. The Lancet. *Gastroenterology & Hepatology*, 5(1), 17–30. Available from https://doi.org/10.1016/S2468-1253(19)30333-4.

Glimberg, I., Haggård, L., Lebwohl, B., Green, P. H. R., & Ludvigsson, J. F. (2021). The prevalence of celiac disease in women with infertility-A systematic review with meta-analysis. *Reproductive Medicine and Biology*, *20*(2), 224–233. Available from https://doi.org/10.1002/rmb2.12374.

Godfrey, J. D., Brantner, T. L., Brinjikji, W., Christensen, K. N., Brogan, D. L., Van Dyke, C. T., Lahr, B. D., Larson, J. J., Rubio-Tapia, A., Melton, L. J., Zinsmeister, A. R., Kyle, R. A., & Murray, J. A. (2010). Morbidity and mortality among older individuals with undiagnosed celiac disease. *Gastroenterology*, *139*(3), 763–769. Available from https://doi.org/10.1053/j.gastro.2010.05.041.

Green, P. H. R., & Cellier, C. (2007). Celiac disease. *The New England Journal of Medicine*, *357*(17), 1731–1743. Available from https://doi.org/10.1056/NEJMra071600.

Hadjivassiliou, M., Rao, D. G., Grìnewald, R. A., Aeschlimann, D. P., Sarrigiannis, P. G., Hoggard, N., Aeschlimann, P., Mooney, P. D., & Sanders, D. S. (2016). Neurological dysfunction in coeliac disease and non-coeliac gluten sensitivity. *Official Journal of the American College of Gastroenterology | ACG*, *111*(4), 561. Available from https://doi.org/10.1038/ajg.2015.434.

Halfdanarson, T. R., Litzow, M. R., & Murray, J. A. (2007). Hematologic manifestations of celiac disease. *Blood*, *109*(2), 412–421. Available from https://doi.org/10.1182/blood-2006-07-031104.

Harder, G., Schiepatti, A., Biagi, F., Borrelli de Andreis, F., Agazzi, S., Gabrielli, G. M., & Klersy, C. (2020). Optimising the follow-up of adult coeliac disease with a clinical-based score to identify patients in need of a histological reassessment: A retrospective single centre study. *The British Journal of Nutrition*, *123*(10), 1159–1164. Available from https://doi.org/10.1017/S0007114520000537.

Harper, J. W., Holleran, S. F., Ramakrishnan, R., Bhagat, G., & Green, P. H. R. (2007). Anemia in celiac disease is multifactorial in etiology. *American Journal of Hematology*, *82*(11), 996–1000. Available from https://doi.org/10.1002/ajh.20996.

Hoffmanová, I., Sánchez, D., Tučková, L., & Tlaskalová-Hogenová, H. (2018). Celiac disease and liver disorders: From putative pathogenesis to clinical implications. *Nutrients*, *10*(7), 892. Available from https://doi.org/10.3390/nu10070892.

Holmes, G. K. T., & Muirhead, A. (2017). Epidemiology of coeliac disease in a single centre in Southern Derbyshire 1958-2014. *BMJ Open Gastroenterology*, *4*(1), e000137. Available from https://doi.org/10.1136/bmjgast-2017-000137.

Holmes, G. K. T., & Muirhead, A. (2018). Mortality in coeliac disease: A population-based cohort study from a single centre in Southern Derbyshire, UK. *BMJ Open Gastroenterology*, *5*(1), e000201. Available from https://doi.org/10.1136/bmjgast-2018-000201.

Humbert, P., Pelletier, F., Dreno, B., Puzenat, E., & Aubin, F. (2006). Gluten intolerance and skin diseases. *European Journal of Dermatology: EJD*, *16*(1), 4–11.

Hære, P., Høie, O., Schulz, T., Schönhardt, I., Raki, M., & Lundin, K. E. A. (2016). Long-term mucosal recovery and healing in celiac disease is the rule - not the exception. *Scandinavian Journal of Gastroenterology*, *51*(12), 1439–1446. Available from https://doi.org/10.1080/00365521.2016.1218540.

Ince, A. T., Kayadibi, H., Soylu, A., Ovunç, O., Gültepe, M., Toros, A. B., Yaşar, B., Kendir, T., & Abut, E. (2008). Serum copper, ceruloplasmin and 24-h urine copper evaluations in celiac patients. *Digestive Diseases and Sciences*, *53*(6), 1564–1572. Available from https://doi.org/10.1007/s10620-007-0043-7.

Iovino, P., Ciacci, C., Sabbatini, F., Acioli, D. M., D'Argenio, G., & Mazzacca, G. (1998). Esophageal impairment in adult celiac disease with steatorrhea. *The American Journal of Gastroenterology*, *93*(8), 1243–1249. Available from https://doi.org/10.1111/j.1572-0241.1998.00403.x.

Irvine, A. J., Chey, W. D., & Ford, A. C. (2017). Screening for celiac disease in irritable bowel syndrome: An updated systematic review and meta-analysis. *The American Journal of Gastroenterology*, *112*(1), 65–76. Available from https://doi.org/10.1038/ajg.2016.466.

Jafri, M. R., Nordstrom, C. W., Murray, J. A., Van Dyke, C. T., Dierkhising, R. A., Zinsmeister, A. R., & Melton, L. J. (2008). Long-term fracture risk in patients with celiac disease: A population-based study in Olmsted County, Minnesota. *Digestive Diseases and Sciences*, *53*(4), 964–971. Available from https://doi.org/10.1007/s10620-007-9976-0.

Jameson, S., Hellsing, K., & Magnusson, S. (1985). Copper malabsorption in coeliac disease. *The Science of the Total Environment*, *42*(1–2), 29–36. Available from https://doi.org/10.1016/0048-9697(85)90004-x.

Jansson-Knodell, C. L., King, K. S., Larson, J. J., Van Dyke, C. T., Murray, J. A., & Rubio-Tapia, A. (2018). Gender-based differences in a population-based cohort with celiac disease: More alike than unalike. *Digestive Diseases and Sciences*, *63*(1), 184–192. Available from https://doi.org/10.1007/s10620-017-4835-0.

Julian, T., Hadjivassiliou, M., & Zis, P. (2019). Gluten sensitivity and epilepsy: A systematic review. *Journal of Neurology*, *266*(7), 1557–1565. Available from https://doi.org/10.1007/s00415-018-9025-2.

Kemppainen, T., Kröger, H., Janatuinen, E., Arnala, I., Lamberg-Allardt, C., Kärkkäinen, M., Kosma, V. M., Julkunen, R., Jurvelin, J., Alhava, E., & Uusitupa, M. (1999). Bone recovery after a gluten-free diet: A 5-year follow-up study. *Bone*, *25*(3), 355–360. Available from https://doi.org/10.1016/s8756-3282(99)00171-4.

Khosla, S. (2001). Minireview: the OPG/RANKL/RANK system. *Endocrinology*, *142*(12), 5050–5055. Available from https://doi.org/10.1210/endo.142.12.8536.

Khurana, A., Leffler, D. A., Gomez, K., & Thukral, C. (2023). Short and long-term follow-up and clinical outcomes in patients with celiac disease in a large private practice setting. *BMC Gastroenterology*, *23*(1), 9. Available from https://doi.org/10.1186/s12876-023-02643-4.

Koskinen, I., Virta, L. J., Huhtala, H., Ilus, T., Kaukinen, K., & Collin, P. (2020). Overall and cause-specific mortality in adult celiac disease and dermatitis herpetiformis diagnosed in the 21st century. *The American Journal of Gastroenterology*, *115*(7), 1117–1124. Available from https://doi.org/10.14309/ajg.0000000000000665.

Kristensen, V. A., Valeur, J., Brackmann, S., Jahnsen, J., Brunborg, C., & Tveito, K. (2019). Attention deficit and hyperactivity disorder symptoms respond to gluten-free diet in patients with coeliac disease. *Scandinavian Journal of Gastroenterology*, *54*(5), 571–576. Available from https://doi.org/10.1080/00365521.2019.1608467.

Kurppa, K., Paavola, A., Collin, P., Sievänen, H., Laurila, K., Huhtala, H., Saavalainen, P., Mäki, M., & Kaukinen, K. (2014). Benefits of a gluten-free diet for asymptomatic patients with serologic markers of celiac disease. *Gastroenterology*, *147*(3), 610–617. Available from https://doi.org/10.1053/j.gastro.2014.05.003, e1.

Kvamme, J.-M., Sørbye, S., Florholmen, J., & Halstensen, T. S. (2022). Population-based screening for celiac disease reveals that the majority of patients are undiagnosed and improve on a gluten-free diet. *Scientific Reports*, *12*(1), 12647. Available from https://doi.org/10.1038/s41598-022-16705-2.

Kårhus, L. L., Hansen, S., Rumessen, J. J., & Linneberg, A. (2022). Diagnostic delay in coeliac disease: A survey among Danish patients. *Canadian Journal of Gastroenterology & Hepatology*, *2022*, 5997624. Available from https://doi.org/10.1155/2022/5997624.

Laurikka, P., Nurminen, S., Kivelä, L., & Kurppa, K. (2018). Extraintestinal manifestations of celiac disease: Early detection for better long-term outcomes. *Nutrients*, *10*(8), 1015. Available from https://doi.org/10.3390/nu10081015.

Laurikka, P., Salmi, T., Collin, P., Huhtala, H., Mäki, M., Kaukinen, K., & Kurppa, K. (2016). Gastrointestinal symptoms in celiac disease patients on a long-term gluten-free diet. *Nutrients*, *8*(7), 429. Available from https://doi.org/10.3390/nu8070429.

Lebwohl, B., Cao, Y., Zong, G., Hu, F. B., Green, P. H. R., Neugut, A. I., Rimm, E. B., Sampson, L., Dougherty, L. W., Giovannucci, E., Willett, W. C., Sun, Q., & Chan, A. T. (2017). Long term gluten consumption in adults without celiac disease and risk of coronary heart disease: Prospective cohort study. *BMJ (Clinical Research ed.)*, *357*, j1892. Available from https://doi.org/10.1136/bmj.j1892.

Lebwohl, B., Granath, F., Ekbom, A., Smedby, K. E., Murray, J. A., Neugut, A. I., Green, P. H. R., & Ludvigsson, J. F. (2013). Mucosal healing and risk for lymphoproliferative malignancy in celiac disease: A population-based cohort study. *Annals of Internal Medicine*, *159*(3), 169–175. Available from https://doi.org/10.7326/0003-4819-159-3-201308060-00006.

Lebwohl, B., Green, P. H. R., Söderling, J., Roelstraete, B., & Ludvigsson, J. F. (2020). Association between celiac disease and mortality risk in a Swedish population. *JAMA: the Journal of the American Medical Association*, *323*(13), 1277–1285. Available from https://doi.org/10.1001/jama.2020.1943.

Lebwohl, B., Larsson, E., Söderling, J., Roelstraete, B., Murray, J. A., Green, P. H. R., & Ludvigsson, J. F. (2021). Risk of severe Covid-19 in patients with celiac disease: A population-based cohort study. *Clinical Epidemiology*, *13*, 121–130. Available from https://doi.org/10.2147/CLEP.S294391.

Lebwohl, B., Söderling, J., Roelstraete, B., Lebwohl, M. G., Green, P. H. R., & Ludvigsson, J. F. (2021). Risk of skin disorders in patients with celiac disease: A population-based cohort study. *Journal of the American Academy of Dermatology*, *85*(6), 1456–1464. Available from https://doi.org/10.1016/j.jaad.2020.10.079.

Leeds, J. S., Höroldt, B. S., Sidhu, R., Hopper, A. D., Robinson, K., Toulson, B., Dixon, L., Lobo, A. J., McAlindon, M. E., Hurlstone, D. P., & Sanders, D. S. (2007). Is there an association between coeliac disease and inflammatory bowel diseases? A study of relative prevalence in comparison with population controls. *Scandinavian Journal of Gastroenterology*, *42*(10), 1214–1220.

Leffler, D. A., & Kelly, C. P. (2011). Celiac disease and gastroesophageal reflux disease: Yet another presentation for a clinical chameleon. *Clinical Gastroenterology and Hepatology: The Official Clinical Practice Journal of the American Gastroenterological Association*, *9*(3), 192–193. Available from https://doi.org/10.1016/j.cgh.2010.11.039.

Lenti, M. V., Aronico, N., Bianchi, P. I., D'Agate, C. C., Neri, M., Volta, U., Mumolo, M. G., Astegiano, M., Calabrò, A. S., Zingone, F., Latella, G., Di Sario, A., Carroccio, A., Ciacci, C., Luzza, F., Bagnato, C., Fantini, M. C., Elli, L., Cammarota, G., … Di Sabatino, A. (2022). Diagnostic delay in adult coeliac disease: An Italian multicentre study. *Digestive and Liver Disease: Official Journal of the Italian Society of Gastroenterology and the Italian Association for the Study of the Liver*, *S1590–8658*(22), 00819. Available from https://doi.org/10.1016/j.dld.2022.11.021, -2.

Lenti, M. V., & Di Sabatino, A. (2021). Disease- and gender-related characteristics of coeliac disease influence diagnostic delay. *European Journal of Internal Medicine*, *83*, 12–13. Available from https://doi.org/10.1016/j.ejim.2020.09.023.

Lichtwark, I. T., Newnham, E. D., Robinson, S. R., Shepherd, S. J., Hosking, P., Gibson, P. R., & Yelland, G. W. (2014). Cognitive impairment in coeliac disease improves on a gluten-free diet and correlates with histological and serological indices of disease severity. *Alimentary Pharmacology & Therapeutics*, *40*(2), 160–170. Available from https://doi.org/10.1111/apt.12809.

Liester, M.B., & Liester, M.G. (n.d.). A Review of Psychiatric Disorders Associated with Celiac Disease. Dual Diagnosis: Open Access, 3(1), 0–0. https://doi.org/10.21767/2472-5048.100035.

Lima, R. F., Maria da Silva Kotze, L., Kotze, L. R., Chrisostomo, K. R., & Nisihara, R. (2019). Gender-related differences in celiac patients at diagnosis. *Archives of Medical Research*, *50*(7), 437–441. Available from https://doi.org/10.1016/j.arcmed.2019.11.007.

Lindsjö, M. (1989). Oxalate metabolism in renal stone disease with special reference to calcium metabolism and intestinal absorption. *Scandinavian Journal of Urology and Nephrology. Supplementum*, *119*, 1–53.

Lionetti, E., Pjetraj, D., Gatti, S., Catassi, G., Bellantoni, A., Boffardi, M., Cananzi, M., Cinquetti, M., Francavilla, R., Malamisura, B., Montuori, M., Zuccotti, G., Cristofori, F., Gaio, P., Passaro, T., Penagini, F., Testa, A., Trovato, C. M., & Catassi, C. (2023). Prevalence and detection rate of celiac disease in Italy: Results of a SIGENP multicenter screening in school-age children. *Digestive and Liver Disease: Official Journal of the Italian Society of Gastroenterology and the Italian Association for the Study of the Liver*, *S1590–8658*(22), 00871–00874. Available from https://doi.org/10.1016/j.dld.2022.12.023.

Lohi, S., Mäki, M., Rissanen, H., Knekt, P., Reunanen, A., & Kaukinen, K. (2009). Prognosis of unrecognized coeliac disease as regards mortality: A population-based cohort study. *Annals of Medicine*, *41*(7), 508–515. Available from https://doi.org/10.1080/07853890903036199.

Lovell, R. M., & Ford, A. C. (2012). Global prevalence of and risk factors for irritable bowel syndrome: A meta-analysis. *Clinical Gastroenterology and Hepatology : The Official Clinical Practice Journal of the American Gastroenterological Association*, *10*(7), 712–721. Available from https://doi.org/10.1016/j.cgh.2012.02.029, e4.

Ludvigsson, J. F., Bai, J. C., Biagi, F., Card, T. R., Ciacci, C., Ciclitira, P. J., Green, P. H. R., Hadjivassiliou, M., Holdoway, A., van Heel, D. A., Kaukinen, K., Leffler, D. A., Leonard, J. N., Lundin, K. E. A., McGough, N., Davidson, M., Murray, J. A., Swift, G. L., & Walker, M. M. (2014). Diagnosis and management of adult coeliac disease: Guidelines from the British Society of Gastroenterology, British Society of Gastroenterology. *Gut*, *63*(8), 1210–1228. Available from https://doi.org/10.1136/gutjnl-2013-306578.

Ludvigsson, J. F., Leffler, D. A., Bai, J. C., Biagi, F., Fasano, A., Green, P. H. R., Hadjivassiliou, M., Kaukinen, K., Kelly, C. P., Leonard, J. N., Lundin, K. E. A., Murray, J. A., Sanders, D. S., Walker, M. M., Zingone, F., & Ciacci, C. (2013). The Oslo definitions for coeliac disease and related terms. *Gut*, *62*(1), 43–52. Available from https://doi.org/10.1136/gutjnl-2011-301346.

Ludvigsson, J. F., Lindelöf, B., Zingone, F., & Ciacci, C. (2011). Psoriasis in a nationwide cohort study of patients with celiac disease. *The Journal of Investigative Dermatology*, *131*(10), 2010–2016. Available from https://doi.org/10.1038/jid.2011.162.

Ludvigsson, J. F., Michaelsson, K., Ekbom, A., & Montgomery, S. M. (2007). Coeliac disease and the risk of fractures - A general population-based cohort study. *Alimentary Pharmacology & Therapeutics*, *25*(3), 273–285. Available from https://doi.org/10.1111/j.1365-2036.2006.03203.x.

Ludvigsson, J. F., Montgomery, S. M., Ekbom, A., Brandt, L., & Granath, F. (2009). Small-intestinal histopathology and mortality risk in celiac disease. *JAMA: the Journal of the American Medical Association*, *302*(11), 1171–1178. Available from https://doi.org/10.1001/jama.2009.1320.

Ludvigsson, J. F., Osby, U., Ekbom, A., & Montgomery, S. M. (2007). Coeliac disease and risk of schizophrenia and other psychosis: A general population cohort study. *Scandinavian Journal of Gastroenterology*, *42*(2), 179–185. Available from https://doi.org/10.1080/00365520600863472.

Makharia, G. K., & Catassi, C. (2019). Celiac disease in Asia. *Gastroenterology Clinics of North America*, *48*(1), 101–113. Available from https://doi.org/10.1016/j.gtc.2018.09.007.

Makharia, G. K., Chauhan, A., Singh, P., & Ahuja, V. (2022). Review article: Epidemiology of coeliac disease. *Alimentary Pharmacology & Therapeutics*, *56*(Suppl 1), S3–S17. Available from https://doi.org/10.1111/apt.16787.

Makharia, G. K., Verma, A. K., Amarchand, R., Bhatnagar, S., Das, P., Goswami, A., Bhatia, V., Ahuja, V., Datta Gupta, S., & Anand, K. (2011). Prevalence of celiac disease in the northern part of India: A community based study. *Journal of Gastroenterology and Hepatology*, *26*(5), 894–900. Available from https://doi.org/10.1111/j.1440-1746.2010.06606.x.

McKenzie, Y. A., Bowyer, R. K., Leach, H., Gulia, P., Horobin, J., O'Sullivan, N. A., Pettitt, C., Reeves, L. B., Seamark, L., Williams, M., Thompson, J., & Lomer, M. C. E. (2016). British Dietetic Association systematic review and evidence-based practice guidelines for the dietary management of irritable bowel syndrome in adults (2016 update. (IBS Dietetic Guideline Review Group on behalf of Gastroenterology Specialist Group of the British Dietetic Association)*Journal of Human Nutrition and Dietetics: The Official Journal of the British Dietetic Association*, *29* (5), 549–575. Available from https://doi.org/10.1111/jhn.12385.

Mearns, E. S., Taylor, A., Thomas Craig, K. J., Puglielli, S., Leffler, D. A., Sanders, D. S., Lebwohl, B., & Hadjivassiliou, M. (2019). Neurological manifestations of neuropathy and ataxia in celiac disease: A systematic review. *Nutrients*, *11*(2), 380. Available from https://doi.org/10.3390/nu11020380.

van Megen, F., Fossli, M., Skodje, G. I., Carlsen, M. H., Andersen, L. F., Veierød, M. B., Lundin, K. E. A., & Henriksen, C. (2023). Nutritional assessment of women with celiac disease compared to the general population. *Clinical Nutrition ESPEN*, *54*, 251–257. Available from https://doi.org/10.1016/j.clnesp.2023.01.031.

van Megen, F., Skodje, G. I., Lergenmuller, S., Zühlke, S., Aabakken, L., Veierød, M. B., Henriksen, C., & Lundin, K. E. A. (2022). A low FODMAP diet reduces symptoms in treated celiac patients with ongoing symptoms-A randomized controlled trial. *Clinical Gastroenterology and Hepatology: The Official Clinical Practice Journal of the American Gastroenterological Association*, *20*(10), 2258–2266. Available from https://doi.org/10.1016/j.cgh.2022.01.011, e3.

van Megen, F., Skodje, G. I., Stendahl, M., Veierød, M. B., Lundin, K. E. A., & Henriksen, C. (2021). High disease burden in treated celiac patients - A web-based survey. *Scandinavian Journal of Gastroenterology*, *56*(8), 882–888. Available from https://doi.org/10.1080/00365521.2021.1930146.

Mehtab, W., Chauhan, A., Agarwal, A., Singh, A., Rajput, M. S., Mohta, S., Jindal, V., Banyal, V., Ahmed, A., Pramanik, A., Vij, N., Miyan, A., Singh, N., Malhotra, A., & Makharia, G. K. (2021). Impact of Corona Virus Disease 2019 pandemic on adherence to gluten-free diet in Indian patients with celiac disease. *Indian Journal of Gastroenterology*, *40*(6), 613–620. Available from https://doi.org/10.1007/s12664-021-01213-4.

Mones, R. L., Atienza, K. V., Youssef, N. N., Verga, B., Mercer, G. O., & Rosh, J. R. (2007). Celiac crisis in the modern era. *Journal of Pediatric Gastroenterology and Nutrition*, *45*(4), 480–483. Available from https://doi.org/10.1097/MPG.0b013e318032c8e7.

Montoro-Huguet, M. A., Santolaria-Piedrafita, S., Cañamares-Orbis, P., & García-Erce, J. A. (2021). Iron deficiency in celiac disease: Prevalence, health impact, and clinical management. *Nutrients*, *13*(10), 3437. Available from https://doi.org/10.3390/nu13103437.

Muir, J. G., Varney, J. E., Ajamian, M., & Gibson, P. R. (2019). Gluten-free and low-FODMAP sourdoughs for patients with coeliac disease and irritable bowel syndrome: A clinical perspective. *International Journal of Food Microbiology*, *290*, 237–246. Available from https://doi.org/10.1016/j.ijfoodmicro.2018.10.016.

Månsson, A. L., Meijer-Boekel, C., & Mårild, K. (2022). Utilization and effectiveness of ehealth technology in the follow-up of celiac disease: A systematic review. *Journal of Pediatric Gastroenterology and Nutrition*, *74*(6), 812–818. Available from https://doi.org/10.1097/MPG.0000000000003423.

Nachman, F., Vázquez, H., González, A., Andrenacci, P., Compagni, L., Reyes, H., Sugai, E., Moreno, M. L., Smecuol, E., Hwang, H. J., Sánchez, I. P., Mauriño, E., & Bai, J. C. (2011). Gastroesophageal reflux symptoms in patients with celiac disease and the effects of a gluten-free diet. *Clinical Gastroenterology and Hepatology: The Official Clinical Practice Journal of the American Gastroenterological Association*, *9*(3), 214–219. Available from https://doi.org/10.1016/j.cgh.2010.06.017.

Nelson, E. W., Ertan, A., Brooks, F. P., & Cerda, J. J. (1976). Thrombocytosis in patients with celiac sprue. *Gastroenterology*, *70*(6), 1042–1044.

Nieri, M., Tofani, E., Defraia, E., Giuntini, V., & Franchi, L. (2017). Enamel defects and aphthous stomatitis in celiac and healthy subjects: Systematic review and meta-analysis of controlled studies. *Journal of Dentistry*, *65*, 1–10. Available from https://doi.org/10.1016/j.jdent.2017.07.001.

Nikniaz, Z., Beheshti, S., Abbasalizad Farhangi, M., & Nikniaz, L. (2021). A systematic review and meta-analysis of the prevalence and odds of eating disorders in patients with celiac disease and vice-versa. *The International Journal of Eating Disorders*, *54*(9), 1563–1574. Available from https://doi.org/10.1002/eat.23561.

Nurmi, R., Korponay-Szabó, I., Laurila, K., Huhtala, H., Niemelä, O., Mustonen, J., Mäkelä, S., Kaukinen, K., & Lindfors, K. (2021). Celiac disease-type tissue transglutaminase autoantibody deposits in kidney biopsies of patients with IgA nephropathy. *Nutrients*, *13*(5), 1594. Available from https://doi.org/10.3390/nu13051594.

Okamoto, K., & Takayanagi, H. (2023). Effect of T cells on bone. *Bone*, *168*, 116675. Available from https://doi.org/10.1016/j.bone.2023.116675.

Olmos, M., Antelo, M., Vazquez, H., Smecuol, E., Mauriño, E., & Bai, J. C. (2008). Systematic review and meta-analysis of observational studies on the prevalence of fractures in coeliac disease. *Digestive and Liver Disease: Official Journal of the Italian Society of Gastroenterology and the Italian Association for the Study of the Liver*, *40*(1), 46–53. Available from https://doi.org/10.1016/j.dld.2007.09.006.

O'Donnell, B. F., O'Neill, C. M., Francis, D. M., Niimi, N., Barr, R. M., Barlow, R. J., Kobza Black, A., Welsh, K. I., & Greaves, M. W. (1999). Human leucocyte antigen class II associations in chronic idiopathic urticaria. *The British Journal of Dermatology*, *140*(5), 853–858. Available from https://doi.org/10.1046/j.1365-2133.1999.02815.x.

Parker, S., Palsson, O., Sanders, D. S., Simren, M., Sperber, A. D., Törnblom, H., Urwin, H., Whitehead, W., & Aziz, I. (2022). Functional gastrointestinal disorders and associated health impairment in individuals with celiac disease. *Clinical Gastroenterology and Hepatology: The Official Clinical Practice Journal of the American Gastroenterological Association*, *20*(6), 1315–1325. Available from https://doi.org/10.1016/j.cgh.2021.07.026, e4.

Passananti, V., Santonicola, A., Bucci, C., Andreozzi, P., Ranaudo, A., Di Giacomo, D. V., & Ciacci, C. (2012). Bone mass in women with celiac disease: Role of exercise and gluten-free diet. *Digestive and Liver Disease: Official Journal of the Italian Society of Gastroenterology and the Italian Association for the Study of the Liver*, *44*(5), 379–383. Available from https://doi.org/10.1016/j.dld.2011.12.012.

Peters, J. E., Basnayake, C., Hebbard, G. S., Salzberg, M. R., & Kamm, M. A. (2022). Prevalence of disordered eating in adults with gastrointestinal disorders: A systematic review. *Neurogastroenterology and Motility: The Official Journal of the European Gastrointestinal Motility Society*, *34*(8), e14278. Available from https://doi.org/10.1111/nmo.14278.

Poddighe, D., & Abdukhakimova, D. (2021). Celiac disease in Asia beyond the Middle East and Indian subcontinent: Epidemiological burden and diagnostic barriers. *World Journal of Gastroenterology*, *27*(19), 2251–2256. Available from https://doi.org/10.3748/wjg.v27.i19.2251.

Quarpong, W., Card, T. R., West, J., Solaymani-Dodaran, M., Logan, R. F., & Grainge, M. J. (2019). Mortality in people with coeliac disease: Long-term follow-up from a Scottish cohort. *United European Gastroenterology Journal*, *7*(3), 377–387. Available from https://doi.org/10.1177/2050640618814662.

Raiteri, A., Granito, A., Giamperoli, A., Catenaro, T., Negrini, G., & Tovoli, F. (2022). Current guidelines for the management of celiac disease: A systematic review with comparative analysis. *World Journal of Gastroenterology*, *28*(1), 154–175. Available from https://doi.org/10.3748/wjg.v28.i1.154.

Repo, M., Kurppa, K., Huhtala, H., Luostarinen, L., Kaukinen, K., & Kivelä, L. (2022). Significance of low ferritin without anaemia in screen-detected, adult coeliac disease patients. *Journal of Internal Medicine*, *292*(6), 904–914. Available from https://doi.org/10.1111/joim.13548.

Roderburg, C., Loosen, S., Kostev, K., Demir, M., Joerdens, M. S., & Luedde, T. (2022). Nonalcoholic fatty liver disease is associated with a higher incidence of coeliac disease. *European Journal of Gastroenterology & Hepatology*, *34*(3), 328–331. Available from https://doi.org/10.1097/MEG.0000000000002234.

Roldan, G. A., Goyes, D., Villafuerte-Gálvez, J. A., Urquiaga, M., Dennis, M., Murray, J. A., Leffler, D. A., & Kelly, C. P. (2022). Anemia etiology and the response to a gluten-free diet in untreated patients with celiac disease: A 2-year follow-up. *The American Journal of Gastroenterology*, *117*(10), 1684–1692. Available from https://doi.org/10.14309/ajg.0000000000001875.

Roldan, G. A., Jamot, S., Kopec, K., Charoen, A., Leffler, D., Feller, E. R., & Shah, S. A. (2023). Celiac disease presenting in a community-based gastroenterology practice: Obesity and bone disease are common. *Digestive Diseases and Sciences*, *68*(3), 860–866. Available from https://doi.org/10.1007/s10620-022-07521-9.

Rostami-Nejad, M., Asri, N., Olfatifar, M., Khorsand, B., Houri, H., & Rostami, K. (2023). Systematic review and dose-response meta-analysis on the relationship between different gluten doses and risk of coeliac disease relapse. *Nutrients*, *15*(6), 1390. Available from https://doi.org/10.3390/nu15061390.

Rouvroye, M. D., Slottje, P., van Gils, T., Mulder, C. J., Muris, J. W., Walstock, D., Reinders, M., & Bouma, G. (2021). Insight in the diagnosis and treatment of coeliac disease in general practice: A survey and case vignette study among 106 general practitioners. *The European Journal of General Practice*, *27*(1), 313–319. Available from https://doi.org/10.1080/13814788.2021.1985455.

Rubio-Tapia, A., Hill, I. D., Semrad, C., Kelly, C. P., Greer, K. B., Limketkai, B. N., & Lebwohl, B. (2023). American College of Gastroenterology Guidelines Update: Diagnosis and management of celiac disease. *The American Journal of Gastroenterology*, *118*(1), 59–76. Available from https://doi.org/10.14309/ajg.0000000000002075.

Rubio-Tapia, A., & Murray, J. A. (2008). Liver involvement in celiac disease. *Minerva Medica*, *99*(6), 595–604.

Rubio-Tapia, A., Rahim, M. W., See, J. A., Lahr, B. D., Wu, T.-T., & Murray, J. A. (2010). Mucosal recovery and mortality in adults with celiac disease after treatment with a gluten-free diet. *The American Journal of Gastroenterology*, *105*(6), 1412–1420. Available from https://doi.org/10.1038/ajg.2010.10.

Sahin, I., Eminbeyli, L., Andic, S., Tuncer, I., & Koz, S. (2012). Screening for celiac disease among patients with chronic kidney disease. *Renal Failure*, *34*(5), 545–549. Available from https://doi.org/10.3109/0886022X.2012.669299.

Santonicola, A., Iovino, P., Cappello, C., Capone, P., Andreozzi, P., & Ciacci, C. (2011). From menarche to menopause: The fertile life span of celiac women. *Menopause (New York, N.Y.)*, *18*(10), 1125–1130. Available from https://doi.org/10.1097/gme.0b013e3182188421.

Savin, E., Tsur, A. M., Watad, A., Gendelman, O., Kopylov, U., Cohen, A. D., & Amital, H. (2023). Association of fibromyalgia with cancerous and non-cancerous gastrointestinal comorbidities: A cross-sectional study. *Clinical and Experimental Rheumatology*. Available from https://doi.org/10.55563/clinexprheumatol/rp5lkn.

Schiepatti, A., Maimaris, S., Lusetti, F., Scalvini, D., Minerba, P., Cincotta, M., Fazzino, E., & Biagi, F. (2022). High prevalence of functional gastrointestinal disorders in celiac patients with persistent symptoms on a gluten-free diet: A 20-year follow-up study. *Digestive Diseases and Sciences*. Available from https://doi.org/10.1007/s10620-022-07727-x.

Scott, E. M., Gaywood, I., & Scott, B. B. (2000). Guidelines for osteoporosis in coeliac disease and inflammatory bowel disease. *British Society of Gastroenterology. Gut*, *46*(Suppl 1), i1–i8. Available from https://doi.org/10.1136/gut.46.suppl_1.i1, Suppl 1.

Seidita, A., Mansueto, P., Compagnoni, S., Castellucci, D., Soresi, M., Chiarello, G., Cavallo, G., De Carlo, G., Nigro, A., Chiavetta, M., Mandreucci, F., Giuliano, A., Disclafani, R., & Carroccio, A. (2022). Anemia in celiac disease: Prevalence, associated clinical and laboratory features, and persistence after gluten-free diet. *Journal of Personalized Medicine*, *12*(10), 1582. Available from https://doi.org/10.3390/jpm12101582.

Sharma, N., Singh, K., & Senapati, S. (2021). Celiac disease poses significant risk in developing depression, anxiety, headache, epilepsy, panic disorder, dysthymia: A meta-analysis. *Indian Journal of Gastroenterology: Official Journal of the Indian Society of Gastroenterology*, *40*(5), 453–462. Available from https://doi.org/10.1007/s12664-021-01215-2.

Shiha, M. G., Marks, L. J., & Sanders, D. S. (2020). Diagnosing coeliac disease in the elderly: A United Kingdom cohort study. *Gastroenterology and Hepatology from Bed to Bench*, *13*(1), 37–43.

Silvester, J. A., Graff, L. A., Rigaux, L., Bernstein, C. N., Leffler, D. A., Kelly, C. P., Walker, J. R., & Duerksen, D. R. (2017). Symptoms of functional intestinal disorders are common in patients with celiac disease following transition to a gluten-free diet. *Digestive Diseases and Sciences*, *62*(9), 2449–2454. Available from https://doi.org/10.1007/s10620-017-4666-z.

Simons, M., Scott-Sheldon, L. A. J., Risech-Neyman, Y., Moss, S. F., Ludvigsson, J. F., & Green, P. H. R. (2018). Celiac Disease and Increased Risk of Pneumococcal Infection: A Systematic Review and Meta-Analysis. *The American Journal of Medicine*, *131*(1), 83–89. Available from https://doi.org/10.1016/j.amjmed.2017.07.021.

Singh, P., Arora, S., Lal, S., Strand, T. A., & Makharia, G. K. (2015). Risk of celiac disease in the first- and second-degree relatives of patients with celiac disease: A systematic review and meta-analysis. *The American Journal of Gastroenterology*, *110*(11), 1539–1548. Available from https://doi.org/10.1038/ajg.2015.296.

Singh, P., & Garber, J. J. (2016). Implementation and adherence to osteoporosis screening guidelines among coeliac disease patients. *Digestive and Liver Disease: Official Journal of the Italian Society of Gastroenterology and the Italian Association for the Study of the Liver*, *48*(12), 1451–1456. Available from https://doi.org/10.1016/j.dld.2016.08.121.

Singh, J., & Whelan, K. (2011). Limited availability and higher cost of gluten-free foods. *Journal of Human Nutrition and Dietetics: The Official Journal of the British Dietetic Association*, *24*(5), 479–486. Available from https://doi.org/10.1111/j.1365-277X.2011.01160.x.

Siniscalchi, M., Iovino, P., Tortora, R., Forestiero, S., Somma, A., Capuano, L., Franzese, M. D., Sabbatini, F., & Ciacci, C. (2005). Fatigue in adult coeliac disease. *Alimentary Pharmacology & Therapeutics*, *22*(5), 489–494. Available from https://doi.org/10.1111/j.1365-2036.2005.02619.x.

Siniscalchi, M., Zingone, F., Savarino, E. V., D'Odorico, A., & Ciacci, C. (2020). COVID-19 pandemic perception in adults with celiac disease: An impulse to implement the use of telemedicine. *Digestive and Liver Disease: Official Journal of the Italian Society of Gastroenterology and the Italian Association for the Study of the Liver*, *52*(10), 1071–1075. Available from https://doi.org/10.1016/j.dld.2020.05.014.

Slim, M., Rico-Villademoros, F., & Calandre, E. P. (2018). Psychiatric comorbidity in children and adults with gluten-related disorders: A narrative review. *Nutrients*, *10*(7), 875. Available from https://doi.org/10.3390/nu10070875.

Smith, D. F., & Gerdes, L. U. (2012). Meta-analysis on anxiety and depression in adult celiac disease. *Acta Psychiatrica Scandinavica*, *125*(3), 189–193. Available from https://doi.org/10.1111/j.1600-0447.2011.01795.x.

Solaymani-Dodaran, M., West, J., & Logan, R. F. A. (2007). Long-term mortality in people with celiac disease diagnosed in childhood compared with adulthood: A population-based cohort study. *The American Journal of Gastroenterology*, *102*(4), 864–870. Available from https://doi.org/10.1111/j.1572-0241.2007.01111.x.

Sonnenberg, A., & Genta, R. M. (2023). Case-control study of the concurrence of coeliac disease with inflammatory bowel disease. *Alimentary Pharmacology & Therapeutics*. Available from https://doi.org/10.1111/apt.17408.

Sánchez, M. I. P., Mohaidle, A., Baistrocchi, A., Matoso, D., Vázquez, H., González, A., Mazure, R., Maffei, E., Ferrari, G., Smecuol, E., Crivelli, A., de Paula, J. A., Gómez, J. C., Pedreira, S., Mauriño, E., & Bai, J. C. (2011). Risk of fracture in celiac disease: Gender, dietary compliance, or both. *World Journal of Gastroenterology*, *17*(25), 3035–3042. Available from https://doi.org/10.3748/wjg.v17.i25.3035.

Taranta, A., Fortunati, D., Longo, M., Rucci, N., Iacomino, E., Aliberti, F., Facciuto, E., Migliaccio, S., Bardella, M. T., Dubini, A., Borghi, M. O., Saraifoger, S., Teti, A., & Bianchi, M. L. (2004). Imbalance of osteoclastogenesis-regulating factors in patients with celiac disease. *Journal of Bone and Mineral Research: The Official Journal of the American Society for Bone and Mineral Research*, *19*(7), 1112–1121. Available from https://doi.org/10.1359/JBMR.040319.

Tio, M., Cox, M. R., & Eslick, G. D. (2012). Meta-analysis: Coeliac disease and the risk of all-cause mortality, any malignancy and lymphoid malignancy. *Alimentary Pharmacology & Therapeutics*, *35*(5), 540–551. Available from https://doi.org/10.1111/j.1365-2036.2011.04972.x.

Trovato, C. M., Montuori, M., Pietropaoli, N., & Oliva, S. (2021). COVID-19 and celiac disease: A pathogenetic hypothesis for a celiac outbreak. *International Journal of Clinical Practice*, *75*(9), e14452. Available from https://doi.org/10.1111/ijcp.14452.

Tursi, A., Elisei, W., Giorgetti, G. M., Gaspardone, A., Lecca, P. G., Di Cesare, L., & Brandimarte, G. (2010). Prevalence of celiac disease and symptoms in relatives of patients with celiac disease. *European Review for Medical and Pharmacological Sciences*, *14*(6), 567–572.

Tye-Din, J. A. (2022). Review article: Follow-up of coeliac disease. *Alimentary Pharmacology & Therapeutics*, *56*(Suppl 1), S49–S63. Available from https://doi.org/10.1111/apt.16847.

Ulnes, M., Albrektsson, H., Størdal, K., Saalman, R., Ludvigsson, J. F., & Mårild, K. (2023). Regional Swedish study found that one in seven coeliac patients experienced loss of follow up during childhood. *Acta Paediatrica (Oslo, Norway)*, *112*(3), 510–521. Available from https://doi.org/10.1111/apa.16633.

Vaezi, M. F., Katzka, D., & Zerbib, F. (2018). Extraesophageal symptoms and diseases attributed to GERD: Where is the pendulum swinging now? *Clinical Gastroenterology and Hepatology: The Official Clinical Practice Journal of the American Gastroenterological Association*, *16*(7), 1018–1029. Available from https://doi.org/10.1016/j.cgh.2018.02.001.

Valitutti, F., Troncone, R., Pisano, P., & Ciacci, C. (2021). Where have all the other coeliacs gone in 2020? Road for a 2021 catch-up with missed diagnoses. *Digestive and Liver Disease*, *53*(4), 504–505. Available from https://doi.org/10.1016/j.dld.2021.01.008.

Volta, U. (2009). Pathogenesis and clinical significance of liver injury in celiac disease. *Clinical Reviews in Allergy & Immunology*, *36*(1), 62–70. Available from https://doi.org/10.1007/s12016-008-8086-x.

Wang, M., Kong, W.-J., Feng, Y., Lu, J.-J., Hui, W.-J., Liu, W.-D., Li, Z.-Q., Shi, T., Cui, M., Sun, Z.-Z., & Gao, F. (2022). Epidemiological, clinical, and histological presentation of celiac disease in Northwest China. *World Journal of Gastroenterology*, *28*(12), 1272–1283. Available from https://doi.org/10.3748/wjg.v28.i12.1272.

West, J., Logan, R. F. A., Card, T. R., Smith, C., & Hubbard, R. (2003). Fracture risk in people with celiac disease: A population-based cohort study. *Gastroenterology*, *125*(2), 429–436. Available from https://doi.org/10.1016/s0016-5085(03)00891-6.

West, J., Logan, R. F. A., Smith, C. J., Hubbard, R. B., & Card, T. R. (2004). Malignancy and mortality in people with coeliac disease: Population based cohort study. *BMJ (Clinical Research ed.)*, *329*(7468), 716–719. Available from https://doi.org/10.1136/bmj.38169.486701.7C.

Wieser, H., Amato, M., Caggiano, M., & Ciacci, C. (2023). Dental manifestations and celiac disease-An overview. *Journal of Clinical Medicine*, *12*(8). Available from https://doi.org/10.3390/jcm12082801.

Zhen, J., Stefanolo, J. P., Temprano, M., de la, P., Tedesco, S., Seiler, C., Caminero, A. F., de-Madaria, E., Huguet, M. M., Vivas, S., Niveloni, S. I., Bercik, P., Smecuol, E., Uscanga, L., Trucco, E., Lopez, V., Olano, C., Mansueto, P., Carroccio, A., Green, P. H. R., ... Pinto-Sanchez, M. I. (2021). The risk of contracting COVID-19 is not increased in patients with celiac disease. *Clinical Gastroenterology and Hepatology*, *19*(2), 391–393. Available from https://doi.org/10.1016/j.cgh.2020.10.009.

Zhou, P., Zheng, T., & Zhao, B. (2022). Cytokine-mediated immunomodulation of osteoclastogenesis. *Bone*, *164*, 116540. Available from https://doi.org/10.1016/j.bone.2022.116540.

Zingone, F., Maimaris, S., Auricchio, R., Caio, G. P. I., Carroccio, A., Elli, L., Galliani, E., Montagnani, M., Valiante, F., & Biagi, F. (2022). Guidelines of the Italian societies of gastroenterology on the diagnosis and management of coeliac disease and dermatitis herpetiformis. *Digestive and Liver Disease: Official Journal of the Italian Society of Gastroenterology and the Italian Association for the Study of the Liver*, *54*(10), 1304–1319. Available from https://doi.org/10.1016/j.dld.2022.06.023.

Zingone, F., Siniscalchi, M., Capone, P., Tortora, R., Andreozzi, P., Capone, E., & Ciacci, C. (2010). The quality of sleep in patients with coeliac disease. *Alimentary Pharmacology & Therapeutics*, *32*(8), 1031–1036. Available from https://doi.org/10.1111/j.1365-2036.2010.04432.x.

Zingone, F., Swift, G. L., Card, T. R., Sanders, D. S., Ludvigsson, J. F., & Bai, J. C. (2015). Psychological morbidity of celiac disease: A review of the literature. *United European Gastroenterology Journal*, *3*(2), 136–145. Available from https://doi.org/10.1177/2050640614560786.

Zugna, D., Richiardi, L., Akre, O., Stephansson, O., & Ludvigsson, J. F. (2011). Celiac disease is not a risk factor for infertility in men. *Fertility and Sterility, 95(5), 1709-1713.e1-3*. Available from https://doi.org/10.1016/j.fertnstert.2011.01.132.

Zylberberg, H. M., Demmer, R. T., Murray, J. A., Green, P. H. R., & Lebwohl, B. (2017). Depression and insomnia among individuals with celiac disease or on a gluten-free diet in the USA: Results from a national survey. *European Journal of Gastroenterology & Hepatology*, *29*(9), 1091–1096. Available from https://doi.org/10.1097/MEG.0000000000000932.

Zylberberg, H. M., Ludvigsson, J. F., Green, P. H. R., & Lebwohl, B. (2018). Psychotropic medication use among patients with celiac disease. *BMC Psychiatry*, *18*(1), 76. Available from https://doi.org/10.1186/s12888-018-1668-0.

Zádori, N., Németh, D., Frim, L., Vörhendi, N., Szakó, L., Váncsa, S., Hegyi, P., & Czimmer, J. (2022). Dyspepsia-like symptoms in Helicobacter pylori-negative chronic gastritis are associated with ASCA-, ANCA-, and celiac seropositivity but not with other autoimmune parameters: A single-centre, retrospective cross-sectional study. *International Journal of General Medicine*, *15*, 7789–7796. Available from https://doi.org/10.2147/IJGM.S380419.

Chapter 8

Serology and screening in celiac disease

Umberto Volta[1], Giacomo Caio[2] and Roberto De Giorgio[2]

[1]*Department of Medical and Surgical Sciences (DIMEC), University of Bologna, Bologna, Italy,* [2]*Department of Translational Medicine, University of Ferrara, Ferrara, Italy*

8.1 Introduction

In the latest decades, serology has revolutionized the knowledge of celiac disease (CD) (Di Sabatino & Corazza, 2009; Lebwohl & Rubio-Tapia, 2021). In fact, before the availability of immunological markers, CD could only be diagnosed by clinical suspicion and duodenal biopsy. Although an intestinal flat mucosa is the typical histopathological picture still regarded as the "gold standard" for CD diagnosis, antibody markers have introduced a radical change in the celiac world by identifying unsuspected at-risk groups of patients with a wide range of clinical manifestations (Volta et al., 2014).

8.2 Sensitivity and specificity of serological markers: review of the literature

An amazingly high number of immunological markers for CD has been identified in the last 40 years (Fig. 8.1)

The beginning of serological era dates back to the early 1970s and 1980s with the detection of R1 antireticulin (R1-ARA) and antigliadin antibodies (AGA), respectively (Rizzetto & Doniach, 1973; Volta et al., 1985). Both R1-ARA and AGA of IgA class performed better than R1-ARA and AGA of IgG class. IgA R1-ARA, detected by indirect immunofluorescence, have been observed in untreated CD with a prevalence varying from 36% to 78%. Because of the low sensitivity, R1-ARA IgA antibodies have proved of limited value in the diagnosis of CD. Conversely, their specificity has been reported up as high, although only rare "false positive" cases have been described in other gastrointestinal and autoimmune disorders (Volta et al., 1985). The analysis of the existing evidence on the sensitivity and specificity of AGA showed considerable heterogeneity, which reflects different homemade or commercially available enzyme-linked immunosorbent assays (ELISA) with a high variability in the cutoffs. Indeed, sensitivity of AGA ranges from 80% to 90% with a low specificity (up to 29% of false positive cases) (Hill, 2005). Due to their low diagnostic accuracy both R1-ARA and AGA are no longer used in the diagnostic approach of CD being replaced by new tests developed between mid 80s and 90s. The modern era of antibody testing includes the screening of three key markers, that is, antiendomysial (EmA) IgA, antitissue transglutaminase (anti-tTG) IgA, and antideamidated gliadin peptides (DGP) IgG and IgA (Aleanzi et al., 2001; Chorzelski et al., 1983; Dieterich et al., 1998). The identification of EmA IgA was developed in the mid 80 s and represented a significant step forward for CD serology. Unlike anti-tTG and DGP, which depend on standardized and reproducible ELISA, EmA test relies on the subjective interpretation of the immunofluorescent pattern on monkey esophagus or human umbilical cord tissues with a high inter- and intraobserver variability (Volta et al., 1996). Nevertheless, in centers with significant experience with EmA, this is the test with the highest positive predictive value for CD diagnosis. In 1997, tTG was identified as the celiac autoantigen (Dieterich et al., 1997). This milestone discovery allowed for the development of an ELISA test that overcame the difficulties related to the indirect immunofluorescence required for EmA detection. The early assays used tTG from guinea pig liver as a substrate leading to false-positive results due to impurities linked to this antigen. ELISAs based on human recombinant tTG were eventually developed to improve the diagnostic accuracy of this test. However, physicians should be aware that among many commercially available ELISA kits for anti-tTG not all of them perform well, as demonstrated by a number of studies documenting a high variability in accuracy (Singh et al., 2020). At the beginning of the new millennium DGP have been introduced in the serological celiac scenario (Aleanzi et al., 2001). Deamidation of gliadin peptides is elicited by tTG (or transglutaminase 2, TG2, based on molecular classification) by replacing

Pediatric and Adult Celiac Disease. DOI: https://doi.org/10.1016/B978-0-443-13359-6.00012-1

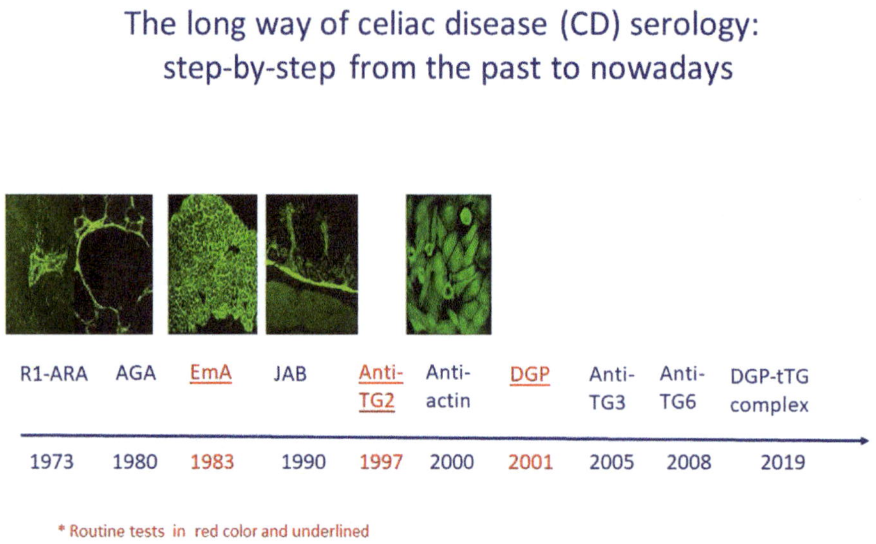

FIGURE 8.1 The long way of celiac disease serology: from the beginning to nowadays. The figure illustrates the various steps of celiac disease (CD) serology which started in early 70 s with the identification of R1-antireticulin (R1-ARA) and antigliadin antibodies (AGA), now not any longer used because of their poor predictive ability, to the discovery of the three key markers, that is, antiendomysial (EmA), antitissue transglutaminase (anti-tTG), otherwise known as antitransglutaminase 2 (TG2), and the deamidated gliadin peptide antibodies (DGP), commonly applied in testing patients at-risk for CD. Both EmA and anti-tTG show best accuracy in the IgA immunoglobulin class, whereas DGP exhibit higher performance in the IgG class. Other abandoned tests for CD screening included antijejunal antibodies (JAB) and antiactin antibodies. Anti-TG3 are indicated in dermatitis herpetiformis, while anti-TG6 underlie patients with neurologically-related clinical manifestations, specifically gluten ataxia. The anti-DGP-tTG complexes represent the frontiers of CD serology although their detection is still mainly in the research area or pertains to sophisticated diagnostic approaches available in some tertiary referral centers. Note the readily visualizable immunofluorescent patterns for R1-ARA, EmA, JAB and antiactin antibodies.

glutamine with glutamic acid (Molberg et al., 1998). This process plays a relevant role in CD pathogenesis, since deamidated gliadin peptides are specifically recognized by gliadin T lymphocytes, thereby enhancing the antibody response in celiac patients. The result is the production of antibodies to DGP of both IgG and IgA class. These antibodies display a higher diagnostic accuracy for CD than those to native gliadin, but lower than anti-tTG and EmA of IgA class (Di Sabatino & Corazza, 2009; Lebwohl & Rubio-Tapia, 2021). DGP IgG can be useful for detecting CD in the first infancy since at this age they may precede the appearance of the other serological markers (Amarri et al., 2013). Data from the literature show that children exhibit higher antibody titers, which tend to decrease by age (Vivas et al., 2009). This can explain why seronegative CD is virtually exclusive of adulthood, being very rare in childhood (Gustafsson et al., 2020; Volta et al., 2016). To assess the real diagnostic accuracy of the three routinely tested CD markers, that is, anti tTG, EmA and DGP, a literature search was carried out by using PubMed in the last 25 years (from 1997 to date). Studies satisfying the following criteria were considered in our analysis: (1) Inclusion of both untreated celiac patients and controls; (2) Investigations including only children and adults as well as studies with both children and adults; (3) All CD patients and controls tested for at least one of the three antibodies and all examined by small intestinal biopsy; (4) Exclusion of selective IgA deficiency. A high number of studies were ruled out since only a subset of CD cases and controls tested for serology were actually biopsied. Twenty-six studies fulfilled the required criteria for a total of 2910 CD patients and 3315 controls (Agardh, 2007; Bardella et al., 2001; Basso et al., 2009; Bazzigaluppi et al., 2006; Biagi et al., 1999, 2001; Bonamico et al., 2001; Bürgin-Wolff et al., 2002; Carroccio et al., 2002; Collin et al., 2005; Dahele et al., 2001; Dickey et al., 2001; Kaukinen et al., 2007; Llorente et al., 2004; Lock et al., 1999; Niveloni et al., 2007; Prause et al., 2009; Rashtak et al., 2008; Srinivas et al., 2014; Sugai et al., 2006; Sulkanen et al., 1998; Tesei et al., 2003; Troncone et al., 1999; Volta et al., 2008, Volta, Granito et al., 2010; Wolf et al., 2017). Sensitivity and specificity of serological markers were determined in the combined group of children and adults as well as separately in these two age subsets (Table 8.1).

In the whole group of patients (children and adults) anti-tTG IgA exhibited the highest sensitivity for CD, but the gap with EmA was very small (92.8 vs 92.4%). Interestingly, in children anti-tTG displayed a more pronounced sensitivity than EmA (95.4 vs 92.5%), whereas in adults EmA IgA sensitivity was superior than that of anti-tTG IgA (94.9 vs 91.1%). With regard to specificity EmA IgA showed better results than anti-tTG IgA in the three groups under consideration in this chapter (i.e., adults + children 98.6 vs 95.9%; children 97.8 vs 96.0%; adults 99.4 vs 94.7%). Both DGP IgG and IgA confirmed to

TABLE 8.1 Sensitivity and specificity of serological tests for celiac disease: a review of the literature.

Study	UCD	Controls	Age groups	Anti-tTG IgA		EmA IgA		DGP IgG		DGP IgA	
	all biopsied	all biopsied		sensitivity	specificity	sensitivity	specificity	sensitivity	specificity	sensitivity	specificity
Sulkanen (1998)a	136	207	Children	94.9	93.7	92.6	99.5	NT	NT	NT	NT
Biagi (1999)a	39	61	Adults	94.9	91.8	100	100	NT	NT	NT	NT
Troncone (1999)a	48	63	Children	92.0	98.0	87.5	98.4	NT	NT	NT	NT
Lock (1999)a	27	65	Adults	85.2	96.9	100	100	NT	NT	NT	NT
Biagi et al. (2001)a	56	52	Adults	98.2	84.6	94.6	100	NT	NT	NT	NT
Bardella (2001)a	40	110	Adults	100	98.2	100	97.3	NT	NT	NT	NT
Dahele (2001)a	114	65	Adults	80.7	96.9	86.8	100	NT	NT	NT	NT
Dickey (2001)a	73	58	Children/ adults	75.3	98.3	80.6	96.6	NT	NT	NT	NT
Bonamico (2001)b	62	56	Children	100	100	95.2	98.2	NT	NT	NT	NT
Carroccio (2002)a	24	183	Adults	100	91.8	100	100	NT	NT	NT	NT
Burgin-Wolff (2002) b	208	157	Children/ adults	96.2	99.4	96.6	100	NT	NT	NT	NT
Tesei (2003)b	250	176	Children	91.0	96.0	86.0	100	NT	NT	NT	NT
Llorente (2004)b	61	64	Children/ adults	98.4	97.0	97.7	95.0	NT	NT	NT	NT
Collin (2005)b	126	105	Children/ adults	94.0	99.0	89.0	98.0	NT	NT	NT	NT
Sugai (2006)b	92	84	Adults	95.0	95.6	NT	NT	96.7	100	94.6	93.8
Bazzigaluppi (2006) b	143	64	Children	99.3	95.3	97.9	96.9	NT	NT	NT	NT
Agardh (2007)b	119	57	Children	97.0	96.0	NT	NT	95.0	86.0	91.0	91.0
Kaukinen (2007)b	44	46	Children/ adults	88.6	97.8	79.5	100	90.9	97.8	NT	NT

(Continued)

TABLE 8.1 (Continued)

Study	UCD	Controls	Age groups	Anti-tTG IgA		EmA IgA		DGP IgG		DGP IgA	
	all biopsied	all biopsied		sensitivity	specificity	sensitivity	specificity	sensitivity	specificity	sensitivity	specificity
Niveloni (2007)b	60	81	Adults	95.0	97.5	NT	NT	96.7	100	98.3	93.8
Volta (2008)b	128	43	Children/ adults	96.8	95.4	93.7	100	84.4	100	83.6	86.0
Rashtak (2008)b	92	124	Adults	78.0	98.0	NT	NT	65.0	98.0	74.0	95.0
Prause (2009)b	142	160	Children	95.1	98.1	NT	NT	98.6	93.1	95.1	86.3
Basso (2009)b	161	129	Children	92.5	97.6	NT	NT	80.1	96.9	80.7	92.9
Volta, Granito et al. (2010)	48	96	Children/ adults	93.7	96.6	91.6	100	82.3	98.9	84.3	79.8
Srinivas (2014)b	88	664	Adults	84.0	96.0	83.0	99.0	NT	NT	NT	NT
Wolf (2017)b	529	345	Children	97.1	89.3	95.8	94.0	76.9	94.2	NT	NT
Total cases (children + adults)											
	2910	3315		92.8	95.9	92.4	98,6	86.7	96.5	87.7	89.8
Only Childrenc											
	1590	1257		95.4	96.0	92.5	97.8	87.6	92.5	88.9	90.1
Only Adultsc											
	632	1489		91.1	94.7	94.9	99.4	86.1	99.3	88,9	94.2

Abbreviations: UCD, untreated celiac disease; anti-tTG: antitissue transglutaminase.
[a]*anti-tTG against guinea-pig liver.*
[b]*anti-tTG against human recombinant tissue transglutaminase; EmA: antiendomysial antibodies; DGP: deamidated gliadin peptide antibodies; NT: not tested.*
[c]*7 studies reported a mixed series of patients without specifying the number of children and adults.*

have a lower diagnostic accuracy than anti-tTG and EmA IgA in the three groups. Sensitivity of DGP IgG ranged from 86.1 for adulthood CD to 87.6% for pediatric CD, whereas DGP IgA sensitivity resulted to be 88.9% for celiac children and adults. In terms of specificity, DGP IgG performed better than DGP IgA with a peak very close to that of EmA IgA, particularly in adults (99.3 vs 99.4%).

In addition to standardized ELISA tests for anti-tTG IgA, a tTG-based point of care, that is, fingerstick test, has been introduced in the clinical practice. Advantages of this test include the easy interpretation (positive or negative), the simple processing technique and the results available in a few minutes (Nemec et al., 2006; Raivio et al., 2008). Although some studies reported a very high diagnostic accuracy for these tests, their sensitivity and specificity are lower than those obtained with ELISA (Mooney et al., 2014; Volta, Fabbri et al., 2010). Moreover, the major limitation of the fingerstick test is the lack of a "titer" indicating the entity of the positivity. Recent data (Choung et al., 2019) demonstrated a very high sensitivity and specificity of a new assay directed to identify the serum immune response to epitopes of the tTG-DGP complex. These promising results need confirmatory evidence from other studies.

The high levels of sensitivity and specificity reported in the literature for serological CD markers (EmA, anti-tTG and DGP) can be partly influenced by the fact that in many studies only patients with positivity for serological markers underwent duodenal biopsy to confirm the diagnosis of CD (Lebwohl et al., 2012; Leffler & Schuppan, 2010). This is a relevant bias since 2%–8% of CD patients are seronegative and the preselection by means of serology for intestinal biopsy overestimates the predictive value of serology in CD diagnosis.

8.3 Different protocols in various clinical settings

Different protocols for detecting CD serological markers have been proposed in various clinical settings (Di Sabatino & Corazza, 2009; Lebwohl & Rubio-Tapia, 2021; Lebwohl et al., 2012; Leffler & Schuppan, 2010). In the general population with a low CD expected prevalence, after having ruled out IgA deficiency (via total serum IgA assay), the test of choice is anti-tTG IgA, which is the most sensitive marker for CD. Should anti-tTG IgA antibodies test negative, then CD can be directly excluded; conversely, in case of anti-tTG positivity the candidate must always undergo duodenal biopsy. The finding of villous atrophy, regardless of its severity, is confirmatory of CD, whereas a normal villous/crypt ratio rules out CD (Fig. 8.2).

In patients with malabsorption (diarrhea/steatorrhea, weight loss and marked impairment of the general conditions) in whom a high CD prevalence is expected and provided that the total IgA pool is normal, EmA IgA should be assayed since in this clinical setting EmA prevalence is comparable with that of anti-tTG IgA, thus being superfluous the search for antibodies directed against tTG. Because of seronegative CD, all these patients should undergo duodenal biopsy regardless of EmA result. There are three scenarios that we would like to tackle, namely (1) cases with positive EmA IgA: the finding of more or less severe villous atrophy will confirm CD diagnosis; (2) cases with negative EmA IgA: the finding of normal villi will rule out CD; (3) cases with negative EmA displaying villous atrophy: HLA typing should always be performed. More specifically, a subject positive to DQ2 and/or DQ8 should be considered a suspected CD, thus requiring further clinical and histological evaluation after one year of a gluten free diet (GFD). Conversely, in those cases testing negative for DQ2 and DQ8, CD can be ruled out and other conditions with non-gluten dependent villous atrophy should be sought (Fig. 8.3).

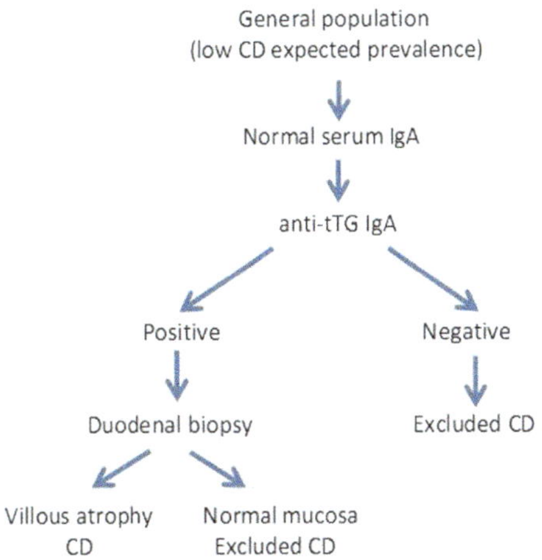

FIGURE 8.2 Algorithm for celiac disease (CD) screening in the general population Provided that IgA serum level is normal, the subject should be tested for anti-tTG IgA. If positive, a duodenal biopsy is mandatory. The underlying histopathology may unravel various degree villous atrophy, confirming CD diagnosis or a normal mucosa, which excludes CD. Conversely, in the right wing of the present scheme, with negative anti-tTG IgA CD is ruled out.

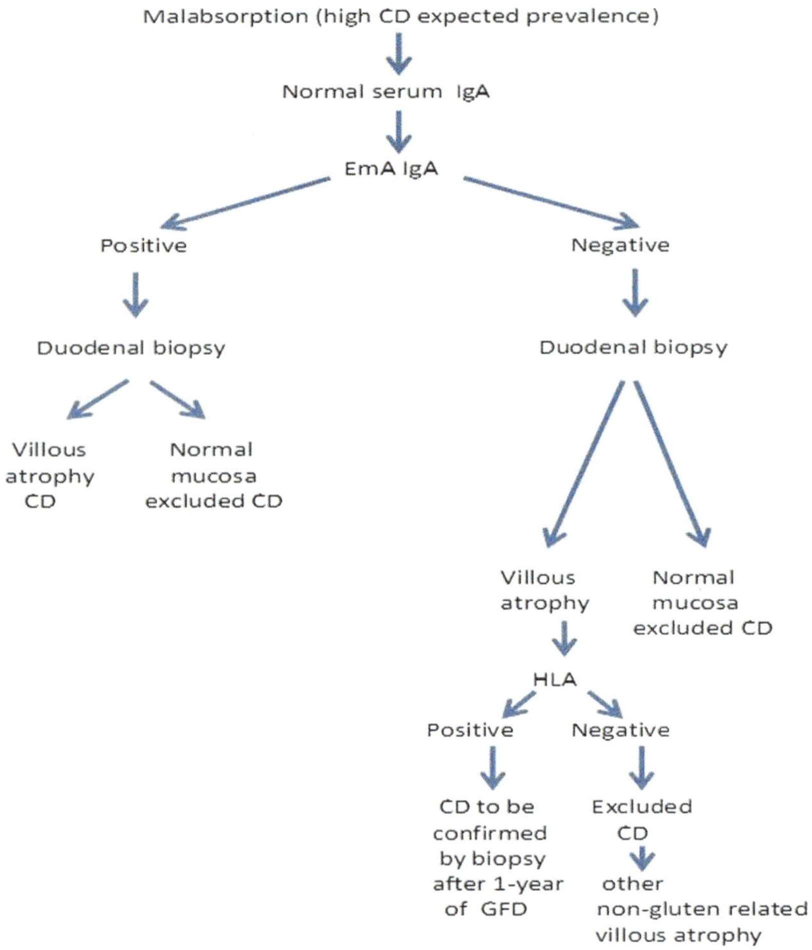

FIGURE 8.3 Algorithm for celiac disease (CD) in patients with malabsorption. Once ruled out IgA deficiency with normal serum IgA levels, the subject should be tested for EmA IgA. Note that a positive EmA IgA leads to a similar approach to what has been indicated in Fig. 8.2 (left wing of the scheme). In patients testing negative for EmA IgA the duodenal biopsy may result positive, that is, villous atrophy or negative, hence CD is ruled out. In the former case, HLA typing is key: if positive, the patient should be followed-up for at least one year of gluten free diet (GFD) and then retested by biopsy. If HLA is negative, CD is excluded and other causes of villous atrophy should be investigated.

Since a selective IgA deficiency (i.e., serum IgA <7 mg/dl) is a clinical condition closely related to CD (Pallav et al., 2016; Villalta et al., 2010), such patients should be tested for anti-tTG IgG or DGP IgG or EmA IgG. A positive result indicates that these patients require duodenal biopsy to confirm or exclude the diagnosis of CD. In contrast, in cases testing negative for IgG antibodies, CD is excluded; however, patients with a clinical suspect of CD and IgA deficiency should undergo duodenal biopsy (Fig. 8.4).

8.4 High antibody titers: a different diagnostic strategy in children and adults

Current data based on both prospective and retrospective studies have clearly shown that high anti-tTG and EmA IgA titers strongly correlate with a severe CD-related small intestinal damage. Based on these findings, the ESPGHAN criteria for CD diagnosis have established that CD can be diagnosed in children and adolescents with high anti-tTG IgA titers (over ten times the cut-off) and detectable EmA (confirmed in a second serum sample) without the need for duodenal biopsy (Husby et al., 2012; Husby et al., 2020). These criteria are extensively defined in Chapter 6. Although large multicenter European studies have shown that the ESPGHAN criteria display a high diagnostic accuracy in identifying CD, it is important to point out that they are not followed worldwide. In some countries, such as the United States, these criteria are not recommended because of the poor reproducibility of most anti-tTG assays (Egner et al., 2012). The ESPGHAN criteria showed a good diagnostic accuracy also in detecting adult CD patients (Fuchs et al., 2019; Penny et al., 2021); however, several lines of evidence indicate that duodenal biopsy still represents a pillar to diagnose CD in adult patients (Corazza & Lenti, 2017). First, endoscopy at the time of diagnosis is useful not only for duodenal biopsy, but also to identify comorbidities frequently detectable in adult CD patients, such as lymphocytic or autoimmune gastritis, peptic ulcer and Helicobacter pylori infection. Second, although rarely, there might be the possibility to diagnose a simultaneous occurrence of CD and small bowel adenocarcinoma or enteropathy-associated T-cell lymphoma (EATL). Third, the

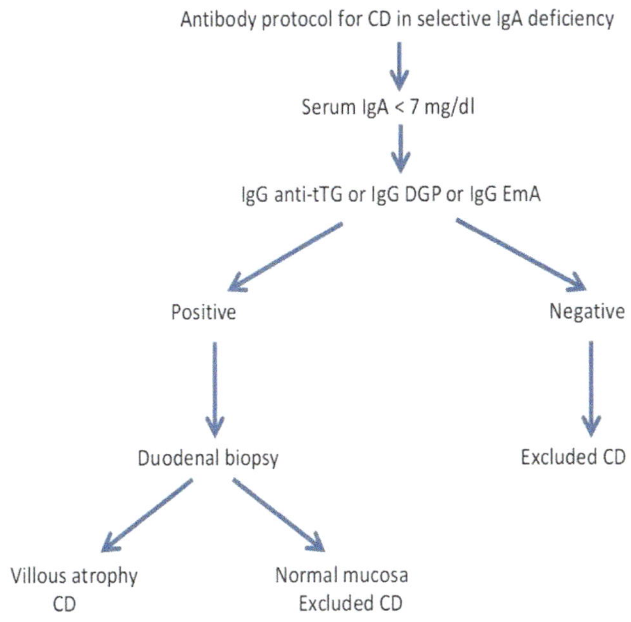

FIGURE 8.4 Algorithm for celiac disease (CD) in patients with selective IgA deficiency. This occurs when the IgA level is <7 mg/dl. IgA deficiency is a closely CD-related disorder. In cases with IgA deficiency, antitissue transglutaminase (anti-tTG) or antideamidated gliadin peptide (DGP) or anti endomysial antibodies (EmA) of IgG class should be assayed. If positive (left wing of the scheme), then the patient should undergo duodenal biopsy/ies attesting either a normal mucosa (thus ruling out CD) or villous atrophy which is conclusive for the disease.

histological picture at diagnosis is extremely useful for the management of CD follow-up especially in those patients with a poor compliance with GFD and in nonresponsive CD (see also Chapter 14). In these cases, it is crucial to compare the biopsy performed after GFD with that at diagnosis in order to verify whether the histological picture is improved, unchanged or worsened. Fourth, the first biopsy is essential for patients developing refractory CD (RCD). Fifth, duodenal biopsy in adults is particularly important to persuade asymptomatic/paucisymptomatic CD patients that the diagnosis is true and that it is mandatory to adhere to GFD strictly. Sixth, another reason supporting the need of a confirmatory duodenal biopsy in adults is that anti-tTG (even at very high titer) may yield false positive results particularly in patients with autoimmune disorders such as type 1 diabetes mellitus, Hashimoto thyroiditis and connective tissue disorders—that is, systemic lupus erythematosus, rheumatoid arthritis and Sjögren syndrome (personal observation). Finally, seronegative CD, which is virtually confined to adult patients, should not be disregarded. Relying only on serology for CD diagnosis, while skipping duodenal biopsy, is a strategy risking to miss seronegative CD cases that occur in 2%–8% of all CD diagnoses.

8.5 Case finding or mass screening: which strategy should be adopted?

In latest years, there has been an open debate on the strategy for the identification of CD (Collin, 2005; Lebwohl & Rubio-Tapia, 2021). The scientific community questioned whether to perform a general population screening or, alternatively, set up a case finding approach in at-risk groups for CD. CD fulfills the WHO guidelines for mass screening: (1) a very common disorder (>1% in the general population); (2) available sensitive and specific serological tests; and (3) high effective treatment, that is, GFD. On the other hand, mass screening would lead to the identification of a large number of asymptomatic people with mild intestinal lesions for whom GFD not only does not add any advantage, but can be even harmful in terms of quality of life and psychological problems (Ukkola et al., 2011). In a recent mass screening children aged 5–11 years were screened for CD by means of a two step approach, based on HLA-DQ2 and -DQ8 testing by a simplified genetic assay followed by serology only in cases with genetic CD predisposition (Lionetti et al., 2023). The overall prevalence of CD (established by ESPGHAN criteria) resulted to be one of the highest in the world (1.65%) with 60% of newly diagnosed celiac patients (only 40% of CD cases had been previously diagnosed). However, it is noteworthy to underline that approximately half (43.7%) of the newly diagnosed CD did not report any symptom and some of them had minimal lesions of intestinal mucosa without atrophy (potential CD). Some aspects of CD mass screening remain to be clarified, possibly by large prospective studies with long-term follow-up, such as the best target age and the high cost for the Health Care System. Nowadays, the majority of the experts in the field does not support mass screening for CD (Collin, 2005; Lebwohl & Rubio-Tapia, 2021). Cost-effectiveness and cost-utility studies are necessary before the implementation of mass screening for CD. Currently, the best strategy for the identification of CD is based on the case finding approach by means of serological testing in at-risk groups for CD. A number of conditions carry an increased risk for CD, including first-degree relatives of CD patients, iron- or folic acid-deficiency anemia, aphthous stomatitis, dental enamel defects, short stature, cryptogenic hypertransaminasaemia, unexplained osteoporosis, infertility, recurrent miscarriages and irritable bowel syndrome. Many autoimmune disorders, such as dermatitis herpetiformis, Hashimoto thyroiditis, Graves disease, type 1 diabetes mellitus, Sjögren syndrome, autoimmune liver disease, some neurological impairment (cerebellar ataxia, cryptogenic epilepsy and peripheral neuropathy) as well as some chromosomal disorders (Down's, Turner's and Williams' syndrome) should be regarded as at-risk groups for CD (Collin, 2005; Di Sabatino & Corazza, 2009; Lebwohl & Rubio-Tapia, 2021; Volta et al., 2014). The case finding approach allows for the identification of a high number of symptomatic patients who will significantly benefit from GFD (Table 8.2). Patients with CD identified by means of case finding usually display a better compliance with GFD in comparison with those identified by mass screening (Roma et al., 2010). This topic will be also discussed in chapter 1. A proposal for serological testing in at-risk groups for CD is reported in Fig. 8.5.

8.6 Antibodies in assessing adherence and response to gluten free diet: strengths and weaknesses of serology

Existing data clearly show that the gold-standard for assessing the adherence and response to GFD in CD patients is the evaluation of small intestinal histopathology (Dipper et al., 2009; Leonard et al., 2017; Silvester et al., 2017). Although

TABLE 8.2 Case finding strategy vs mass screening.

Case finding strategy is preferred to mass screening for celiac disease (CD) detection		
	Mass screening	**Case finding**
Pros	Fulfills the WHO criteria (high CD prevalence, availability of good markers, optimal diet-therapy)	Identifies a high number of symptomatic patients with an underlying villous atrophy who greatly benefit from a GFD
Cons	Identification of a high number of asymptomatic patients with mild intestinal lesions for whom a GFD is not indicated being even potentially harmful	No argument against this approach
	Not recommended	Highly recommended

Abbreviations: gluten-free diet, GFD; World Health Organization, WHO.

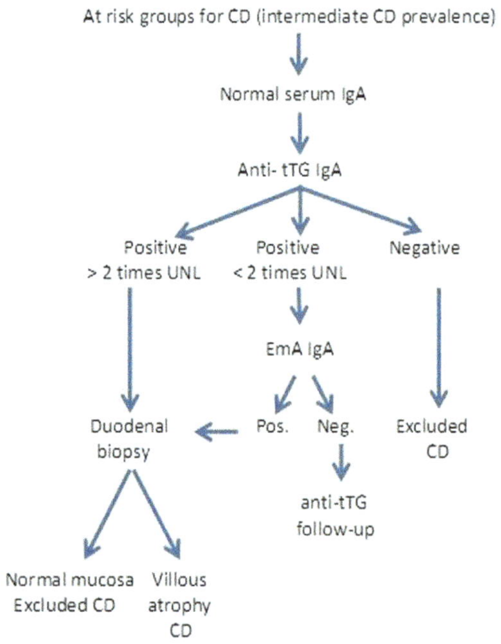

FIGURE 8.5 Algorithm for celiac disease (CD) in at-risk groups for CD. Provided that total IgA level is normal, anti-tTG IgA should be determined. The following scenarios can be identified: anti-tTG IgA are positive >2 times the upper normal limit (UNL) (left wing), therefore duodenal biopsy is mandatory yielding either to a normal mucosa result (CD is excluded) or villous atrophy indicative of CD. In the central part of the scheme, anti-tTG IgA <2 times UNL and in this case EmA IgA should be assessed with two possibilities: if positive, a duodenal biopsy is required (thus with the two options previously illustrated); if negative, a follow-up with anti-tTG is recommended. Finally, the right wing of this scheme indicates the third possibility, that is, a negative assay that rules out CD.

TABLE 8.3 Sensitivity and specificity of antitissue transglutaminase (anti-tTG) and antiendomysial antibodies (EmA) for the detection of persistent villous atrophy in celiac disease after GFD.

	Anti-tTG IgA	EmA IgA
Sensitivity (%)	50 (95% CI, 41–60)	45 (95% CI,34–57)
Specificity (%)	83 (95% CI,79–87)	91 (95% CI, 87–94)

Meta-analysis on 26 studies: 20 with adult CD and 6 with childhood CD for a total of 2277 patients, all biopsied at diagnosis and after gluten-free diet (GFD, range 2–600 months). Of the 26 studies, 6 with anti-tTG IgA, 15 with EmA IgA, 5 with both anti-tTG IgA and EmA IgA. Anti-tTG and EmA had similar performance in both children and adults.

Source: From Silvester AJ et al. Gastroenterology 2017, 153:689–701.

widely used to follow-up CD patients, serologic tests do not exhibit a similar accuracy as in CD diagnosis. An ideal serological test able to substitute the small intestinal biopsy in CD follow-up should disclose the vast majority of CD patients with persistent villous atrophy after gluten withdrawal. In this view, false-positive results would not be a relevant limitation (as negative ones would be), since positive cases could be checked by biopsy if clinically deemed appropriate. In contrast, since a substantial number of celiac patients with negative antibodies after GFD still exhibit mucosal villous atrophy, it is quite obvious that serology cannot replace the histological evaluation. This has been confirmed by a meta-analysis on the usefulness of anti-tTG IgA and EmA IgA in CD follow-up (Silvester et al., 2017). In this study 442 articles were reviewed, but only 26 of them (20 with adult and six with pediatric CD for a total number of 2277 patients) met the inclusion criteria (i.e., duodenal biopsy in both negative and positive antibody testing after GFD). While anti-tTG IgA and EmA IgA showed a relatively high specificity for villous atrophy (83% and 91%, respectively), their sensitivities were very low (50% for anti-tTG and 45% for EmA) (Table 8.3). The tests had similar levels of performance in children and adults.

In the same meta-analysis DGP were evaluated only in three studies (two with pediatric and one with adult CD for a total of 256 patients) showing a slightly higher sensitivity (62%), but a lower specificity (80%) than anti-tTG and EmA IgA for detecting villous atrophy after GFD. Based on these results, the normalization of anti-tTG, EmA and DGP in treated CD should be interpreted neither as an indicator of mucosal healing nor as a sign of strict adherence to GFD. A variable degree of villous atrophy in CD under GFD may be explained by intentional or inadvertent exposure to minimal amount of gluten in such patients. Indeed, trace amounts of gluten may be sufficient to cause persistent mucosal damage, while the antibodies became negative (Caio et al., 2019). Notably, since serology normalizes within 12 months of GFD in about 50%–67% of CD patients, it is recommended waiting for this time interval before new duodenal biopsies are obtained (Ashton et al., 2022; Sbravati et al., 2022). The persistence of antibody positivity at a high titer after 12 months of gluten withdrawal is regarded as an indicator of lack of compliance with GFD and should immediately prompt physicians to check for the status of small intestinal mucosa. However, it has been demonstrated that CD patients with high titer anti-tTG IgA at diagnosis (≥10 times the upper normal limit) may display a slow antibody normalization despite a strict adherence to GFD (Sansotta et al., 2020; Sbravati et al., 2022). In these patients, histological re-evaluation may be postponed in the absence of signs and symptoms suggestive of disease activity. Other factors significantly associated with a longer time of anti-tTG remission are female gender and the association with autoimmune comorbidities such as type 1 diabetes mellitus and autoimmune thyroiditis (Sbravati et al., 2022).

Recommendation to monitor routinely the serology in CD follow-up is still ongoing because there are no other validated noninvasive measures that predict with high reliability the adherence to GFD and mucosal healing. Novel tools to evaluate the excretion of gluten immunogenic peptides (GIP) in the urine or stools appeared promising as a marker of both GFD adherence and mucosal recovery in clinical trials (Moreno et al., 2017), but their usefulness has been recently questioned in the clinical practice (Monachesi et al., 2021). Their routine implementation needs to be confirmed by further studies on a large series of CD patients. In conclusion, serology casts more doubts than certainties on the adherence to the diet and mucosal healing in CD follow-up. Both positive and negative antibody results in CD patients on GFD must be always integrated by clinical and dietary assessment. Although the use of CD antibodies has some indication in evaluating the adherence to GFD, they are not sensitive enough to reveal dietary transgression and poor small bowel recovery. Therefore in the absence of validated biomarkers for monitoring CD patients on GFD, follow-up duodenal biopsy is still the only appropriate test to assess mucosal recovery in CD patients (Farina et al., 2021).

References

Agardh, D. (2007). Antibodies against synthetic deamidated gliadin peptides and tissue transglutaminase for the identification of childhood celiac disease. *Clinical Gastroenterology and Hepatology*, *5*(11), 1276–1281. Available from https://doi.org/10.1016/j.cgh.2007.05.024.

Aleanzi, M., Demonte, A. M., Esper, C., Garcilazo, S., & Waggener, M. (2001). Celiac disease: Antibody recognition against native and selectively deamidated gliadin peptides. *Clinical Chemistry*, *47*(11), 2023–2028. Available from https://doi.org/10.1093/clinchem/47.11.2023.

Amarri, S., Alvisi, P., De Giorgio, R., Gelli, M. C., Cicola, R., Tovoli, F., Sassatelli, R., Caio, G., & Volta, U. (2013). Antibodies to deamidated gliadin peptides: An accurate predictor of coeliac disease in infancy. *Journal of Clinical Immunology*, *33*(5), 1027–1030. Available from https://doi.org/10.1007/s10875-013-9888-z.

Ashton, J. J., Driscoll, K., Himsworth, J., Beattie, R. M., & Batra, A. (2022). Time to normalisation of tissue transglutaminase in paediatric coeliac disease is dependent on initial titre and half of patients will normalise within 12 months. *Archives of Disease in Childhood*, *107*(7), 660–664. Available from https://doi.org/10.1136/archdischild-2021-323250.

Bardella, M. T., Trovato, C., Cesana, B. M., Pagliari, C., Gebbia, C., & Peracchi, M. (2001). Serological markers for coeliac disease: Is it time to change. *Digestive and Liver Disease*, *33*(5), 426–431. Available from https://doi.org/10.1016/S1590-8658(01)80015-3.

Basso, D., Guariso, G., Fogar, P., Meneghel, A., Zambon, C. F., Navaglia, F., Greco, E., Schiavon, S., Rugge, M., & Plebani, M. (2009). Antibodies against synthetic deamidated gliadin peptides for celiac disease diagnosis and follow-up in children. *Clinical Chemistry*, *55*(1), 150–157. Available from https://doi.org/10.1373/clinchem.2008.110395.

Bazzigaluppi, E., Roggero, P., Parma, B., Brambillasca, M. F., Meroni, F., Mora, S., Bosi, E., & Barera, G. (2006). Antibodies to recombinant human tissue-transglutaminase in coeliac disease: Diagnostic effectiveness and decline pattern after gluten-free diet. *Digestive and Liver Disease*, *38*(2), 98–102. Available from https://doi.org/10.1016/j.dld.2005.10.020.

Biagi, F., Ellis, H. J., Yiannakou, J. Y., Brusco, G., Swift, G. L., Smith, P. M., Corazza, G. R., & Ciclitira, P. J. (1999). Tissue transglutaminase antibodies in celiac disease. *American Journal of Gastroenterology*, *94*(8), 2187–2192. Available from https://doi.org/10.1111/j.1572-0241.1999.01292.x.

Biagi, F., Pezzimenti, D., Campanella, J., Vadacca, G. B., & Corazza, G. R. (2001). Endomysial and tissue transglutaminase antibodies in coeliac sera: A comparison not influenced by previous serological testing. *Scandinavian Journal of Gastroenterology*, *36*(9), 955–958. Available from https://doi.org/10.1080/003655201750305477.

Bonamico, M., Tiberti, C., Picarelli, A., Mariani, P., Rossi, D., Cipolletta, E., Greco, M., Di Tola, M., Sabbatella, L., Carabba, B., Magliocca, F. M., Strisciuglio, P., & Di Mario, U. (2001). Radioimmunoassay to detect antitransglutaminase autoantibodies is the most sensitive and specific screening method for celiac disease. *American Journal of Gastroenterology*, *96*(5), 1536–1540. Available from https://doi.org/10.1016/S0002-9270(01)02317-6.

Bürgin-Wolff, A., Dahlbom, I., Hadziselimovic, F., & Petersson, C. J. (2002). Antibodies against human tissue transglutaminase and endomysium in diagnosing and monitoring coeliac disease. *Scandinavian Journal of Gastroenterology*, *37*(6), 685–691. Available from https://doi.org/10.1080/00365520212496.

Caio, G., Volta, U., Sapone, A., Leffler, D. A., De Giorgio, R., Catassi, C., & Fasano, A. (2019). Celiac disease: A comprehensive current review. *BMC Medicine*, *17*(1), 142. Available from https://doi.org/10.1186/s12916-019-1380-z.

Carroccio, A., Vitale, G., Di Prima, L., Chifari, N., Napoli, S., La Russa, C., Gulotta, G., Averna, M. R., Montalto, G., Mansueto, S., & Notarbartolo, A. (2002). Comparison of anti-transglutaminase ELISAs and an anti-endomysial antibody assay in the diagnosis of celiac disease: A prospective study. *Clinical Chemistry*, *48*(9), 1546–1550. Available from https://doi.org/10.1093/clinchem/48.9.1546.

Chorzelski, T. P., Sulej, J., Tchorzewska, H., Jablonska, S., Beutner, E. H., & Kumar, V. (1983). IgA class endomysium antibodies in dermatitis herpetiformis and coeliac disease. *Annals of the New York Academy of Sciences*, *420*, 325–334. Available from https://doi.org/10.1111/j.1749-6632.1983.tb22220.x.

Choung, R. S., Rostamkolaei, K., Ju, J. M., Marietta, E. V., Dyke., Rajasekaran, J. J., Jayaraman, V., Wang, T., Bei, K., Rajasekaran, K. E., Krishna, K., Krishnamurthy, H. K., & Murray, A., J. (2019). Synthetic neoepitopes of the transglutaminase-deamidated gliadin complex as biomarkers for diagnosing and monitoring celiac disease. *Gastroenterology*, *156*(3), 582–591. Available from https://doi.org/10.1053/j.gastro.2018.10.025.

Collin, P. (2005). Should adults be screened for celiac disease? What are the benefits and harms of screening. *Gastroenterology*, *128*(4), S104–S108. Available from https://doi.org/10.1053/j.gastro.2005.02.021.

Collin, P., Kaukinen, K., Vogelsang, H., Korponay-Szabó, I., Sommer, R., Schreier, E., Volta, U., Granito, A., Veronesi, L., Mascart, F., Ocmant, A., Ivarsson, A., Lagerqvist, C., Bürgin-Wolff, A., Hadziselimovic, F., Furlano, R. I., Sidler, M. A., Mulder, C. J. J., Goerres, M. S., & Mäki, M. (2005). Antiendomysial and antihuman recombinant tissue transglutaminase antibodies in the diagnosis of coeliac disease: A biopsy-proven European multicentre study. *European Journal of Gastroenterology and Hepatology*, *17*(1), 85–91. Available from https://doi.org/10.1097/00042737-200501000-00017.

Corazza, G. R., & Lenti, M. V. (2017). Does biopsy still have a role for adult coeliac disease. *The Lancet. Gastroenterology & Hepatology*, *2*(11), 773–774. Available from https://doi.org/10.1016/S2468-1253(17)30251-0.

Dahele, A. V. M., Aldhous, M. C., Humphreys, K., & Ghosh, S. (2001). Serum IgA tissue transglutaminase antibodies in coeliac disease and other gastrointestinal diseases. *QJM - Monthly Journal of the Association of Physicians*, *94*(4), 195–205. Available from https://doi.org/10.1093/qjmed/94.4.195.

Di Sabatino, A., & Corazza, G. R. (2009). Coeliac disease. *The Lancet*, *373*(9673), 1480–1493. Available from https://doi.org/10.1016/S0140-6736(09)60254-3.

Dickey, W., McMillan, S. A., & Hughes, D. F. (2001). Sensitivity of serum tissue transglutaminase antibodies for endomysial antibody positive and negative coeliac disease. *Scandinavian Journal of Gastroenterology*, *36*(5), 511–514. Available from https://doi.org/10.1080/003655201750153359.

Dieterich, W., Ehnis, T., Bauer, M., Donner, P., Volta, U., Riecken, E. O., & Schuppan, D. (1997). Identification of tissue transglutaminase as the autoantigen of celiac disease. *Nature Medicine*, *3*(7), 797–801. Available from https://doi.org/10.1038/nm0797-797.

Dieterich, W., Laag, E., Schopper, H., Volta, U., Ferguson, A., Gillett, H., Riecken, E. O., & Schuppan, D. (1998). Autoantibodies to tissue transglutaminase as predictors of celiac disease. *Gastroenterology*, *115*(6), 1317–1321. Available from https://doi.org/10.1016/S0016-5085(98)70007-1.

Dipper, C. R., Maitra, S., Thomas, R., Lamb, C. A., McLean-Tooke, A. P. C., Ward, R., Smith, D., Spickett, G., & Mansfield, J. C. (2009). Anti-tissue transglutaminase antibodies in the follow-up of adult coeliac disease. *Alimentary Pharmacology and Therapeutics*, *30*(3), 236–244. Available from https://doi.org/10.1111/j.1365-2036.2009.04039.x.

Egner, W., Shrimpton, A., Sargur, R., Patel, D., & Swallow, K. (2012). ESPGHAN guidance on coeliac disease 2012: Multiples of ULN for decision making do not harmonise assay performance across centres. *Journal of Pediatric Gastroenterology and Nutrition*, *55*(6), 733–735. Available from https://doi.org/10.1097/MPG.0b013e31826531f6.

Farina, E., Roncoroni, L., Lombardo, V., Scricciolo, A., Vecchi, M., Doneda, L., & Elli, L. (2021). Clinical value of tissue transglutaminase antibodies in celiac patients over a long term follow-up. *Nutrients*, *13*(9), 3057. Available from https://doi.org/10.3390/nu13093057.

Fuchs, V., Kurppa, K., Huhtala, H., Laurila, K., Mäki, M., Collin, P., Salmi, T., Luostarinen, L., Saavalainen, P., & Kaukinen, K. (2019). Serology-based criteria for adult coeliac disease have excellent accuracy across the range of pre-test probabilities. *Alimentary Pharmacology and Therapeutics*, *49*(3), 277–284. Available from https://doi.org/10.1111/apt.15109.

Gustafsson, I., Repo, M., Popp, A., Kaukinen, K., Hiltunen, P., Arvola, T., Taavela, J., Vornanen, M., Kivelä, L., & Kurppa, K. (2020). Prevalence and diagnostic outcomes of children with duodenal lesions and negative celiac serology. *Digestive and Liver Disease*, *52*(3), 289–295. Available from https://doi.org/10.1016/j.dld.2019.11.011.

Hill, I. D. (2005). What are the sensitivity and specificity of serologic tests for celiac disease? Do sensitivity and specificity vary in different populations. *Gastroenterology*, *128*(4), S25–S32. Available from https://doi.org/10.1053/j.gastro.2005.02.012.

Husby, S., Koletzko, S., Korponay-Szabó, I. R., Mearin, M. L., Phillips, A., Shamir, R., Troncone, R., Giersiepen, K., Branski, D., Catassi, C., Lelgeman, M., Mäki, M., Ribes-Koninckx, C., Ventura, A., & Zimmer, K. P. (2012). European society for pediatric gastroenterology, hepatology, and nutrition guidelines for the diagnosis of coeliac disease. *Journal of Pediatric Gastroenterology and Nutrition*, *54*(1), 136–160. Available from https://doi.org/10.1097/MPG.0b013e31821a23d0.

Husby, S., Koletzko, S., Korponay-Szabó, I., Kurppa, K., Mearin, M. L., Ribes-Koninckx, C., Shamir, R., Troncone, R., Auricchio, R., Castillejo, G., Christensen, R., Dolinsek, J., Gillett, P., Hróbjartsson, A., Koltai, T., Maki, M., Nielsen, S. M., Popp, A., Størdal, K., & Wessels, M. (2020). European society paediatric gastroenterology, hepatology and nutrition guidelines for diagnosing coeliac disease 2020. *Journal of Pediatric Gastroenterology and Nutrition*, *70*(1), 141–156. Available from https://doi.org/10.1097/MPG.0000000000002497.

Kaukinen, K., Collin, P., Laurila, K., Kaartinen, T., Partanen, J., & Mäki, M. (2007). Resurrection of gliadin antibodies in coeliac disease. Deamidated gliadin peptide antibody test provides additional diagnostic benefit. *Scandinavian Journal of Gastroenterology*, *42*(12), 1428–1433. Available from https://doi.org/10.1080/00365520701452217.

Lebwohl, B., & Rubio-Tapia, A. (2021). Epidemiology, presentation, and diagnosis of celiac disease. *Gastroenterology*, *160*(1), 63–75. Available from https://doi.org/10.1053/j.gastro.2020.06.098.

Lebwohl, B., Rubio-Tapia, A., Assiri, A., Newland, C., & Guandalini, S. (2012). Diagnosis of celiac disease. *Gastrointestinal Endoscopy Clinics of North America*, *22*(4), 661–677. Available from https://doi.org/10.1016/j.giec.2012.07.004.

Leffler, D. A., & Schuppan, D. (2010). Update on serologic testing in celiac disease. *American Journal of Gastroenterology*, *105*(12), 2520–2524. Available from https://doi.org/10.1038/ajg.2010.276.

Leonard, M. M., Weir, D. C., Degroote, M., Mitchell, P. D., Singh, P., Silvester, J. A., Leichtner, A. M., & Fasano, A. (2017). Value of IgA tTG in predicting mucosal recovery in children with celiac disease on a gluten-free diet. *Journal of Pediatric Gastroenterology and Nutrition*, *64*(2), 286–291. Available from https://doi.org/10.1097/MPG.0000000000001460.

Lionetti, E., Pjetraj, D., Gatti, S., Catassi, G., Bellantoni, A., Boffardi, M., Cananzi, M., Cinquetti, M., Francavilla, R., Malamisura, B., Montuori, M., Zuccotti, G., Cristofori, F., Gaio, P., Passaro, T., Penagini, F., Testa, A., Trovato, C.M., & Catassi, C. (2023). Prevalence and detection rate of celiac disease in Italy: Results of a SIGENP multicenter screening in school-age children. Digestive and Liver Disease : Official Journal of the Italian Society of Gastroenterology and the Italian Association for the Study of the Liver. https://doi.org/10.1016/j.dld.2022.12.023

Llorente, M. J., Sebastián, M., Fernández-Aceñero, M. J., Prieto, G., & Villanueva, S. (2004). IgA antibodies against tissue transglutaminase in the diagnosis of celiac disease: Concordance with intestinal biopsy in children and adults. *Clinical Chemistry*, *50*(2), 451–453. Available from https://doi.org/10.1373/clinchem.2003.024976.

Lock, R. J., Pitcher, M. C., & Unsworth, D. J. (1999). IgA anti-tissue transglutaminase as a diagnostic marker of gluten sensitive enteropathy. *Journal of Clinical Pathology*, *52*(4), 274–277. Available from https://doi.org/10.1136/jcp.52.4.274.

Molberg, Ø., Mcadam, S. N., Körner, R., Quarsten, H., Kristiansen, C., Madsen, L., Fugger, L., Scott, H., Norën, O., Roepstorff, P., Lundin, K. E. A., Sjöström, H., & Sollid, L. M. (1998). Tissue transglutaminase selectively modifies gliadin peptides that are recognized by gut-derived T cells in celiac disease. *Nature Medicine*, *4*(6), 713–717. Available from https://doi.org/10.1038/nm0698-713.

Monachesi, C., Verma, A. K., Catassi, G. N., Franceschini, E., Gatti, S., Gesuita, R., Lionetti, E., & Catassi, C. (2021). Determination of urinary gluten immunogenic peptides to assess adherence to the gluten-free diet: A randomized, double-blind, controlled study. *Clinical and Translational Gastroenterology*, *12*(10), e00411. Available from https://doi.org/10.14309/ctg.0000000000000411.

Mooney, P. D., Kurien, M., Evans, K. E., Chalkiadakis, I., Hale, M. F., Kannan, M. Z., Courtice, V., Johnston, A. J., Irvine, A. J., Hadjivassiliou, M., & Sanders, D. S. (2014). Point-of-care testing for celiac disease has a low sensitivity in endoscopy. *Gastrointestinal Endoscopy*, *80*(3), 456–462. Available from https://doi.org/10.1016/j.gie.2014.02.009.

Moreno, M., de, L., Cebolla, Á., Muñoz-Suano, A., Carrillo-Carrion, C., Comino, I., Pizarro, Á., León, F., Rodríguez-Herrera, A., & Sousa, C. (2017). Detection of gluten immunogenic peptides in the urine of patients with coeliac disease reveals transgressions in the gluten-free diet and incomplete mucosal healing. *Gut*, *66*(2), 250–257. Available from https://doi.org/10.1136/gutjnl-2015-310148.

Nemec, G., Ventura, A., Stefano, M., Di Leo, G., Baldas, V., Tommasini, A., Ferrara, F., Taddio, A., Città, A., Sblattero, D., Marzari, R., & Not, T. (2006). Looking for celiac disease: Diagnostic accuracy of two rapid commercial assays. *American Journal of Gastroenterology*, *101*(7), 1597–1600. Available from https://doi.org/10.1111/j.1572-0241.2006.00597.x.

Niveloni, S., Sugai, E., Cabanne, A., Vazquez, H., Argonz, J., Smecuol, E., Moreno, M. L., Nachman, F., Mazure, R., Kogan, Z., Gomez, J. C., Mauriño, E., & Bai, J. C. (2007). Antibodies against synthetic deamidated gliadin peptides as predictors of celiac disease: Prospective assessment in an adult population with a high pretest probability of disease. *Clinical Chemistry*, *53*(12), 2186–2192. Available from https://doi.org/10.1373/clinchem.2006.081364.

Pallav, K., Xu, H., Leffler, D. A., Kabbani, T., & Kelly, C. P. (2016). Immunoglobulin A deficiency in celiac disease in the United States. *Journal of Gastroenterology and Hepatology*, *31*(1), 133–137. Available from https://doi.org/10.1111/jgh.13176.

Penny, H. A., Raju, S. A., Lau, M. S., Marks, L. J. S., Baggus, E. M. R., Bai, J. C., Bassotti, G., Bontkes, H. J., Carroccio, A., Danciu, M., Derakhshan, M. H., Ensari, A., Ganji, A., Green, P. H. R., Johnson, M. W., Ishaq, S., Lebwohl, B., Levene, A., Maxim, R., & Sanders, D. S. (2021). Accuracy of a no-biopsy approach for the diagnosis of coeliac disease across different adult cohorts. *Gut*, *70*(5), 876–883. Available from https://doi.org/10.1136/gutjnl-2020-320913.

Prause, C., Ritter, M., Probst, C., Daehnrich, C., Schlumberger, W., Komorowski, L., Lieske, R., Richter, T., Hauer, A. C., Stern, M., Uhlig, H. H., Laass, M. W., Zimmer, K. P., & Mothes, T. (2009). Antibodies against deamidated gliadin as new and accurate biomarkers of childhood coeliac disease. *Journal of Pediatric Gastroenterology and Nutrition*, *49*(1), 52–58. Available from https://doi.org/10.1097/MPG.0b013e318195dae3.

Raivio, T., Korponay-Szabó, I. R., Paajanen, T., Ashorn, M., Iltanen, S., Collin, P., Laurila, K., Nemes, E., Kovács, J. B., Carrard, G., Saramäki, M., Mäki, M., & Kaukinen, K. (2008). Comparison of a novel whole blood transglutaminase-based ELISAWith a whole blood rapid antibody test and established conventional serological celiac disease assays. *Journal of Pediatric Gastroenterology and Nutrition*, *47*(5), 562–567. Available from https://doi.org/10.1097/MPG.0b013e3181615cde.

Rashtak, S., Ettore, M. W., Homburger, H. A., & Murray, J. A. (2008). Comparative usefulness of deamidated gliadin antibodies in the diagnosis of celiac disease. *Clinical Gastroenterology and Hepatology*, *6*(4), 426–432. Available from https://doi.org/10.1016/j.cgh.2007.12.030.

Rizzetto, M., & Doniach, D. (1973). Types of "reticulin" antibodies detected in human sera by immunofluorescence. *Journal of Clinical Pathology*, *26*(11), 841–851. Available from https://doi.org/10.1136/jcp.26.11.841.

Roma, E., Roubani, A., Kolia, E., Panayiotou, J., Zellos, Aglaia, & Syriopoulou, V. P. (2010). Dietary compliance and life style of children with coeliac disease. *Journal of Human Nutrition and Dietetics*. Available from https://doi.org/10.1111/j.1365-277X.2009.01036.x.

Sansotta, N., Alessio, M. G., Norsa, L., Previtali, G., Ferrari, A., Guerra, G., & D'Antiga, L. (2020). Trend of antitissue transglutaminase antibody normalization in children with celiac disease started on gluten-free diet: A comparative study between chemiluminescence and ELISA serum assays. *Journal of Pediatric Gastroenterology and Nutrition*, *70*(1), 37–41. Available from https://doi.org/10.1097/MPG.0000000000002519.

Sbravati, F., Cosentino, A., Lenzi, J., Fiorentino, M., Ambrosi, F., Salerno, A., Di Biase, A., Righi, B., Brusa, S., Valin, P. S., Bruni, L., Battistini, B., Pagano, S., Grondona, A. G., Labriola, F., & Alvisi, P. (2022). Antitissue transglutaminase antibodies' normalization after starting a gluten-free diet in a large population of celiac children-a real-life experience. *Digestive and Liver Disease*, *54*(3), 336–342. Available from https://doi.org/10.1016/j.dld.2021.06.026.

Silvester, J. A., Kurada, S., Szwajcer, A., Kelly, C. P., Leffler, D. A., & Duerksen, D. R. (2017). Tests for serum transglutaminase and endomysial antibodies do not detect most patients with celiac disease and persistent villous atrophy on gluten-free diets: A meta-analysis. *Gastroenterology*, *153*(3), 689–701. Available from https://doi.org/10.1053/j.gastro.2017.05.015, e1.

Singh, P., Singh, A., Silvester, J. A., Sachdeva, V., Chen, X., Xu, H., Leffler, D. A., Ahuja, V., Duerksen, D. R., Kelly, C. P., & Makharia, G. K. (2020). Inter- and intra-assay variation in the diagnostic performance of assays for anti-tissue transglutaminase in 2 populations. *Clinical Gastroenterology and Hepatology*, *18*(11), 2628–2630. Available from https://doi.org/10.1016/j.cgh.2019.09.018.

Srinivas, M., Basumani, P., Podmore, G., Shrimpton, A., & Bardhan, K. D. (2014). Utility of testing patients, on presentation, for serologic features of celiac disease. *Clinical Gastroenterology and Hepatology*, *12*(6), 946–952. Available from https://doi.org/10.1016/j.cgh.2013.10.037.

Sugai, E., Vázquez, H., Nachman, F., Moreno, M. L., Mazure, R., Smecuol, E., Niveloni, S., Cabanne, A., Kogan, Z., Gómez, J. C., Mauriño, E., & Bai, J. C. (2006). Accuracy of testing for antibodies to synthetic gliadin-related peptides in celiac disease. *Clinical Gastroenterology and Hepatology*, *4*(9), 1112–1117. Available from https://doi.org/10.1016/j.cgh.2006.05.004.

Sulkanen, S., Halttunen, T., Laurila, K., Kolho, K. L., Korponay-Szabo, I. R., Sarnesto, A., Savilahti, E., Collin, P., & Maki, M. (1998). Tissue transglutaminase autoantibody enzyme-linked immunosorbent assay in detecting celiac disease. *Gastroenterology*, *115*(6), 1322–1328. Available from https://doi.org/10.1016/S0016-5085(98)70008-3.

Tesei, N., Sugai, E., Vázquez, H., Smecuol, E., Niveloni, S., Mazure, R., Moreno, M. L., Gomez, J. C., Mauriño, E., & Bai, J. C. (2003). Antibodies to human recombinant tissue transglutaminase may detect coeliac disease patients undiagnosed by endomysial antibodies. *Alimentary Pharmacology and Therapeutics*, *17*(11), 1415–1423. Available from https://doi.org/10.1046/j.1365-2036.2003.01595.x.

Troncone, R., Maurano, F., Rossi, M., Micillo, M., Greco, L., Auricchio, R., Salerno, G., Salvatore, F., & Sacchetti, L. (1999). IgA antibodies to tissue transglutaminase: An effective diagnostic test for celiac disease. *Journal of Pediatrics*, *134*(2), 166–171. Available from https://doi.org/10.1016/S0022-3476(99)70410-5.

Ukkola, A., Mäki, M., Kurppa, K., Collin, P., Huhtala, H., Kekkonen, L., & Kaukinen, K. (2011). Diet improves perception of heatlth and well-being in symptomatic, but not asymptomatic patients, with celiac disease. *Clinical Gastroenterology and Hepatology*. Available from https://doi.org/10.1016/j.cgh.2010.10.011.

Villalta, D., Tonutti, E., Prause, C., Koletzko, S., Uhlig, H. H., Vermeersch, P., Bossuyt, X., Stern, M., Laass, M. W., Ellis, J. H., Ciclitira, P. J., Richter, T., Daehnrich, C., Schlumberger, W., & Mothes, T. (2010). IgG antibodies against deamidated gliadin peptides for diagnosis of celiac disease in patients with IgA deficiency. *Clinical Chemistry*, *56*(3), 464–468. Available from https://doi.org/10.1373/clinchem.2009.128132.

Vivas, S., Ruiz De Morales, J. G., Riestra, S., Arias, L., Fuentes, D., Alvarez, N., Calleja, S., Hernando, M., Herrero, B., Casqueiro, J., & Rodrigo, L. (2009). Duodenal biopsy may be avoided when high transglutaminase antibody titers are present. *World Journal of Gastroenterology*, *15*(38), 4775–4780. Available from https://doi.org/10.3748/wjg.15.4775.

Volta, U., Lenzi, M., Lazzari, R., Cassani, F., Collina, A., Bianchi, F. B., & Pisi, E. (1985). Antibodies to gliadin detected by immunofluorescence and a micro-ELISA method: markers of active childhood and adult coeliac disease. *Gut*, *26*(7), 667–671. Available from https://doi.org/10.1136/gut.26.7.667.

Volta, U., Caio, G., Boschetti, E., Giancola, F., Rhoden, K. J., Ruggeri, E., Paterini, P., & De Giorgio, R. (2016). Seronegative celiac disease: Shedding light on an obscure clinical entity. *Digestive and Liver Disease*, *48*(9), 1018–1022. Available from https://doi.org/10.1016/j.dld.2016.05.024.

Volta, U., Caio, G., Stanghellini, V., & Giorgio, D. (2014). The changing clinical profile of celiac disease: A 15-year experience (1998-2012) in an Italian Referral center. *BMC Gastroenterology*, *14*, 194. Available from https://doi.org/10.1186/s12876-014-0194-x.

Volta, U., Granito, A., Fiorini, E., Parisi, C., Piscaglia, M., Pappas, G., Muratori, P., & Bianchi, F. B. (2008). Usefulness of antibodies to deamidated gliadin peptides in celiac disease diagnosis and follow-up. *Digestive Diseases and Sciences*, *53*(6), 1582–1588. Available from https://doi.org/10.1007/s10620-007-0058-0.

Volta, U., Granito, A., Parisi, C., Fabbri, A., Fiorini, E., Piscaglia, M., Tovoli, F., Grasso, V., Muratori, P., Pappas, G., & De Giorgio, R. (2010). Deamidated gliadin peptide antibodies as a routine test for celiac disease: A prospective analysis. *Journal of Clinical Gastroenterology*, *44*(3), 186–190. Available from https://doi.org/10.1097/MCG.0b013e3181c378f6.

Volta, U., Molinaro, N., De Franceschi, L., & Bianchi, F. B. (1996). Human umbilical cord as substrate for IgA antiendomysial antibodies allows large scale screening for celiac sprue. *Journal of Clinical Gastroenterology*, *23*(1), 18–20. Available from https://doi.org/10.1097/00004836-199607000-00006.

Volta, U., Fabbri, A., Parisi, C., Piscaglia, M., Caio, G., Tovoli, F., & Fiorini, E. (2010). Old and new serological tests for celiac disease screening. *Expert Review of Gastroenterology & Hepatology*, *4*(1), 31–35. Available from https://doi.org/10.1586/egh.09.66.

Wolf, J., Petroff, D., Richter, T., Auth, M. K. H., Uhlig, H. H., Laass, M. W., Lauenstein, P., Krahl, A., Händel, N., de Laffolie, J., Hauer, A. C., Kehler, T., Flemming, G., Schmidt, F., Rodrigues, A., Hasenclever, D., & Mothes, T. (2017). Validation of antibody-based strategies for diagnosis of pediatric celiac disease without biopsy. *Gastroenterology*, *153*(2), 410–419. Available from https://doi.org/10.1053/j.gastro.2017.04.023, e17.

Chapter 9

Role of intestinal biopsy in childhood and adult celiac disease

Alessandro Vanoli[1], Marco Vincenzo Lenti[2], Giuseppe De Lisi[1] and Gino Roberto Corazza[2]
[1]*Pathology Unit, Department of Molecular Medicine, University of Pavia, Fondazione IRCCS Policlinico San Matteo, Pavia, Italy,*
[2]*Department of Internal Medicine, University of Pavia, Fondazione IRCCS Policlinico San Matteo, Pavia, Italy*

9.1 Indications to intestinal biopsy for celiac disease diagnosis

The decision-making algorithm for establishing the indication of the intestinal biopsy depends on the patient's age, the clinical features, and the laboratory data.

In *children*, according to the latest ESPGHAN guidelines, a biopsy-free approach for celiac disease (CD) diagnosis is advised in case of IgA class antibodies against transglutaminase 2 (TGA-IgA) ≥ 10 times the upper limit of normal, with accurate, validated tests, and positive IgA class antiendomysial antibodies (EMA-IgA) in a second serum sample (Husby et al., 2020). Children with positive TGA-IgA but lower titers (<10 times upper limit of normal) should undergo biopsies. Conversely, the previous ESPGHAN guidelines supporting the biopsy-free approach for the first time were more stringent, and included that the patients were symptomatic, had positive serology (TGA-IgA values ≥ 10 times the upper limit of normal and EMA-IgA), as well as HLA DQ2/DQ8 alleles (Husby et al., 2012). This is an important point, as the validation study of the ESPGHAN guidelines (Werkstetter et al., 2017) was only performed for the first version, in which both clinical and laboratory criteria must be fulfilled. Indeed, following the latest pediatric guidelines, the separation between potential and *bona fide* (active) CD is not possible, as well as histology comparison before and after a gluten-free diet, when clinically required (Donat et al., 2021) (Chapter 6).

In *adults*, according to the latest guidelines (Al-Toma et al., 2019; Husby et al., 2019; Ludvigsson et al., 2014; Rubio-Tapia et al., 2023; Zingone et al., 2022), the bioptic approach remains the gold standard for diagnosing CD. Exceptions to perform a biopsy in case of positive CD serology include major contraindications to the endoscopic examination (e.g., severe coagulation disorders) and pregnant women, in whom the biopsy should be postponed until postpartum (Ludvigsson et al., 2014). Patients with a strong clinical suspicion for CD should undergo duodenal biopsy regardless of serology to exclude/confirm seronegative CD or other rare forms of seronegative enteropathy (Zingone et al., 2022) (Chapter 11).

Despite the current guidelines, some Authors argue, by analogy to the pediatric population, that a biopsy-free approach could be feasible in the adult setting, given the high positive predictive value of serology in predicting villous atrophy (Efthymakis et al., 2017; Fuchs et al., 2019; Holmes & Hill, 2017; Penny et al., 2021; Wakim-Fleming et al., 2013). This perspective, despite following a common trend of less invasive and cheaper medical practices, comes with its flaws (Corazza & Lenti, 2017; Corazza & Lenti, 2021). First, a biopsy-free approach would increase the rate of missed or delayed diagnosis of seronegative, potential, and refractory CD. Second, the lack of an "index biopsy," which represents the histological reference to compare with follow-up biopsies, precludes the accurate assessment of response to a gluten-free diet. Third, as villous atrophy has been associated with a greater risk of developing lymphoproliferative malignancies (Elfström et al., 2011; Lebwohl et al., 2013), the risk stratification of patients would not be possible, thus precluding a proper follow-up. Furthermore, other upper gastrointestinal conditions potentially associated with CD would be missed.

In conclusion, in the adult setting, moving to a biopsy-free approach seems to be premature and should be supported by stronger evidence. In children, even though there is still a slight risk of overdiagnosis and overtreatment, diagnosis based on serology only is endorsed by the current guidelines.

Pediatric and Adult Celiac Disease. DOI: https://doi.org/10.1016/B978-0-443-13359-6.00015-7

9.2 Sampling and handling of small intestinal biopsies

9.2.1 Role of endoscopy

A reliable histological diagnosis of CD needs an adequate number of specimens, a standardized sampling of the small intestinal mucosa, and a proper handling of the biopsies. Indeed, the patient must be on a gluten-containing diet at the time of endoscopy, and the endoscopist should always provide a gross description of the sampled duodenal areas.

To note, even a macroscopically normal-appearing mucosa can hide villous atrophy, as demonstrated by previous studies (Dickey & Hughes, 2001; Robson et al., 2014). The demonstration that the diagnostic adequacy of endoscopic biopsies is better than by suction capsule biopsy (Mee et al., 1985) has fueled further research aimed at finding endoscopic signs of CD. Over the years, the following signs have been reported: the mosaic appearance of the mucosa (Stevens & McCarthy, 1976), the reduction or loss of Kerckring folds seen at maximal inflation in the descending duodenum (Brocchi et al., 1988), and the scalloped configuration of Kerckring folds (Jabbari et al., 1988) that turned out to be the prominent sign in the childhood form (Corazza et al., 1993). Some of these results have been promptly confirmed (Mauriño et al., 1993; Mclntyre et al., 1992), but the accuracy reported in the former studies has been later doubted due to a selection bias consisting in a high pretest probability of CD in such study populations (Bardella et al., 2000; Oxentenko et al., 2002). No doubt that in suspected patients, biopsies should be taken irrespective of the appearance of the mucosa, but the attention to these signs has allowed the diagnosis of several patients who were not previously suspected and who performed endoscopy for other reasons (Corazza et al., 1990; Green et al., 2000; Vjero et al., 2003). Thus this is the proper setting for looking at endoscopic signs, not that of patients with positive CD serology or suspected on clinical grounds.

Over the last few years, novel and much more sensitive endoscopic techniques have allowed a better characterization of duodenal lesions in CD, such as water-immersion endoscopy, magnification, with or without electronic or dye-spray chromoendoscopy (Cammarota et al., 2009), and confocal laser endomicroscopy (Zambelli et al., 2007). However, in unselected patients in the setting of routine endoscopy, the simple white-light, high-definition endoscopy was found to be comparable to these more advanced techniques (i.e., electronic chromoendoscopy) (Penny et al., 2016). Hence, expensive, non-routinary, time-consuming, advanced endoscopic imaging techniques are not needed in this setting.

Despite some research purposes, videocapsule endoscopy, which does not allow to take biopsies, has no role in the routinary diagnosis of CD, while its major role is in that of documenting CD complications more distally localized (see Chapters 14 and 15).

9.2.2 Biopsy sampling and handling

Since lesions of CD may be patchy in distribution, especially in children (Ravelli et al., 2010), the number of biopsy specimens correlates with the likelihood of detecting histologic evidence of CD (Lebwohl et al., 2011; Pais et al., 2008). Current pediatric guidelines for CD diagnosis state that at least four specimens from the distal (D2 or D3) part of the duodenum and at least one from the duodenal bulb (D1) should be taken (Husby et al., 2020). Most guidelines for adult CD diagnosis recommend taking multiple specimens (again, at least four) from the distal duodenum (Allen et al., 2015; Al-Toma et al., 2019; Ludvigsson et al., 2014; Robert et al., 2018; Rubio-Tapia et al., 2023). In addition, these guidelines also suggest that taking samples from the duodenal bulb (D1) may increase the diagnostic accuracy. Such an approach is particularly relevant considering that around 5% of both pediatric and adult CD can manifest itself in the form of "ultrashort CD," that involves the duodenal bulb only (Kurien et al., 2012; McCarty et al., 2018; Mooney et al., 2016).

A proper handling of intestinal biopsies is key for preventing detrimental misinterpretations (Dickson et al., 2006; Lagana & Bhagat, 2019; Perera et al., 1975; Serra & Jani, 2006). First, biopsy samples from the small intestine are extremely fragile and are easily disrupted and crushed during handling. As a rule, special care is needed whilst taking, transferring, and placing the specimen. The samples should be handled gently with gloved hands or with smooth or rubber-tipped forceps. Obtaining a single biopsy per pass of the forceps improves the quality of the subsequent histological examination (Al-Toma et al., 2019; Latorre et al., 2015; Robert et al., 2018).

Contentious but important points are whether the endoscopist or other practitioner should orient the specimens before formalin fixation, by mounting them on a support with villi upward, and whether orientation could improve diagnostic reliability (Ludvigsson et al., 2014). Dissecting microscopes may help in orienting the sample in the endoscopic room and can give an early hint about villous architecture. A study by Biagi et al. showed a substantial agreement in detecting villous atrophy between dissecting microscopy and histologic evaluations (κ statistics: 0.78), providing an immediate tool for the detection of severe villous atrophy (Biagi et al., 2016). Nevertheless, this method cannot replace traditional histology, particularly in the cases of infiltrative or milder lesions or in the diagnosis of the complicated forms of CD (Lenti & Corazza, 2019). In keeping with the indications provided by Perera et al. (Perera et al., 1975), we think that proper orientation of small bowel biopsies on an adequate support (preferably on millipore cellulose acetate filters) is pivotal for a proper interpretation.

To date, there is no evidence-based data comparing diagnostic accuracy between well oriented and randomly embedded small intestinal biopsies. However, some authors believe that attempting to orient biopsies is not helpful (Asmussen et al., 2009; Robert et al., 2018) and that poor orientation can be mitigated by obtaining the recommended number of samples, by entrusting a specialized technician with handling the samples in the pathology lab and by performing serial sections through tissue blocks (Robert et al., 2018).

Common artifacts that may lead to misinterpretation by the histopathologist during evaluation of duodenal biopsies for mucosal architecture are summarized in Table 9.1. If these artifacts are not correctly recognized, villi may erroneously appear shorter than normal or distorted, giving a false impression of villous atrophy (Haggitt et al., 1982;

TABLE 9.1 Common artifacts found in duodenal biopsies which may lead to misinterpretation.

Type of artifacts	What causes them	Clues to recognize them	What to do to mitigate them
Tangential artifacts	Tangential or oblique sections linked to inappropriate orientation of the specimens	Presence of several crypts with elliptical or round shapes Multilayered arrangement of the crypts Pseudostratification of villous surface epithelium with a prevalence of roundish rather than elongated enterocyte nuclei which can mimic IELs	Proper orientation of the specimens Multiple histological sections
Brunner's gland-related artifacts	Presence of Brunner's gland, which may cause shortening of overlying villi	Look at Brunner's glands in the specimens. They are lined by mucinous cells and are more common in D1 biopsies	Avoid evaluation of VA in areas including Brunner's glands The presence of Brunner's glands should be recorded in the pathology report
Lymphoid aggregate-related artifacts	Presence of mucosal lymphoid aggregates. They can distort villous morphology. An increased number of IELs is physiologically present over these aggregates	Look at nodular aggregates of lymphocytes in the lamina propria, with or without a germinal center.	Areas showing lymphoid nodules should be avoided in histological examination of VA and IELs
Biopsy trauma-related artifacts	They are derived by poorly handled specimens and include crushing artifacts at the margins of the biopsy, separation of villous epithelium from the lamina propria and artifactual hemorrhages	These artifacts are more pronounced at the periphery of the biopsy and can be easily spotted by looking at the margin of the specimen	Sampling and handling of the specimens by specialized professionals and by following guidelines recommendations (e.g., obtaining a single biopsy per pass of the forceps, obtaining the recommended number of samples, etc.)
Superficial sampling-related artifacts	These are seen in specimens with absence of *muscularis mucosae* with consequentlateral spread of the villous bases and apparent shortening of the villi.	The absence of *muscularis mucosae* helps to recognize them.	Proper sampling and handling of the specimens
Fixation-related artifacts	Prolonged cold ischemia time and dehydration due to exposure to air may lead to suboptimal cytomorphology.	The presence of signs of autolytic phenomena (e.g., poorly distinguishable nuclei) and a wrinkled appearance of the specimens.	Putting the biopsies quickly in the standard 10% neutral-buffered formalin solution (other fixative solutions should be avoided).

Abbreviations: IEL: intraepithelial lymphocyte; VA: villous atrophy.

Perera et al., 1975). In conclusion, these potential pitfalls stress the importance of adequate sampling, handling, and proper orientation of the specimens, particularly for specimens taken from D1.

9.3 Histological lesions associated with celiac disease

The elementary histologic lesions in the small intestine of active CD, both in children and adults, are intraepithelial lymphocytosis [i.e., an increase in the density of intraepithelial lymphocytes (IELs)] and mucosal architectural changes including villous atrophy and crypt hyperplasia (Marsh, 1992; Rubin et al., 1960). Indeed, this feature is characteristic, but not pathognomonic, for CD and thus cannot grant a CD diagnosis if not properly related with gluten toxicity (Chen et al., 2023; Dickson et al., 2006). The severity gradient of histological lesions in CD is reported to be cranio-caudal, that is, they are usually more severe in the duodenum and decrease toward the distal small intestine. By capsule endoscopy, the majority of untreated CD patients display an extensive small intestinal enteropathy, that is, involvement of both the duodenum and the jejunum, with the latter usually having a patchier involvement (Murray et al., 2008). Although villous atrophy has been reported to be very rare in the terminal ileum, CD ileal mucosa may show intraepithelial lymphocytosis (Dickey & Hughes, 2004; Hopper et al., 2006; Thompson, 1974). Contrary to what is commonly believed, no clear correlation between disease extension, as assessed by wireless capsule endoscopy, and clinical manifestations has been reported (Murray et al., 2008).

9.3.1 Intraepithelial lymphocytosis

Intraepithelial lymphocytosis is the most frequent histological feature of CD. For a more detailed characterization of IELs, their origin, and their functions in the normal and CD small bowel (see Chapters 5 and 14). Duodenal biopsies from all untreated CD patients show an increase in surface epithelial infiltration by mature T lymphocytes (Ferguson & Murray, 1971; Marsh & Crowe, 1995; Marsh, 1990). The upper limit of normal IELs has progressively decreased over time, being 40 per 100 epithelial cells in its first description (Ferguson & Murray, 1971) to 20–25 per 100 epithelial cells (Hayat et al., 2002; Robert et al., 2018). A cutoff set to 40 IELs per 100 enterocytes has a 80% sensitivity rate for CD diagnosis whereas a cutoff of 25 intraepithelial lymphocytes per 100 enterocytes has a 99% sensitivity rate (Rostami et al., 2017). In most untreated CD cases the proportion of IELs is higher than 40%. In evidence-based practice and in recent classifications, intraepithelial lymphocytosis is defined as >25 IELs per 100 enterocytes, or, better, per 100 enterocyte nuclei (Husby et al., 2019; Ludvigsson et al., 2014; Walker et al., 2010; Zingone et al., 2022). IEL count should be performed in H&E-stained sections, where the optimal cutoff of 25% has been established (Rostami et al., 2017).

IELs should be counted along the whole, well oriented, villi, excluding crypts and areas over lymphoid aggregates, where falsely elevated IEL counts could be observed. At least 500 epithelial cells must be counted, as IEL distribution may be heterogeneous (Hayat et al., 2002). In normal mucosa, IELs are usually more numerous at the bases of villi and along their lateral aspects compared to the villous tips, the so-called "*decrescendo pattern*" (Goldstein & Underhill, 2001). In those cases of CD with milder lesions, a loss of "decrescendo pattern" is seen, with a more uniform distribution of IELs along the villus length. Therefore if some villi are still detectable, a more rapid method consisting in counting IELs per 20 enterocytes just at the tips of at least five villi (the so called "villous tip method") has been suggested (Biagi et al., 2004; Goldstein & Underhill, 2001; Järvinen et al., 2004; Ludvigsson & Murray, 2019). Using the latter method, the cutoff for intraepithelial lymphocytosis is considered to be 6 IELs per 20 enterocytes.

The usefulness of CD3 immunohistostaining to better visualize IELs is still debated. CD3 immunohistostaining has some shortcomings, including its inherent costs, the potential pitfall of highlighting lamina propria T lymphocytes near the enterocyte basement membrane leading to a potential overestimation of IELs, and the need for redefining the optimal threshold for intraepithelial lymphocytosis on CD3-stained sections (Hudacko et al., 2013; Marsh & Heal, 2017; Robert et al., 2018). However, in agreement with other authors, it is accepted that performing CD3 immunohistochemistry may be useful in cases with borderline IEL counts on H&E-stained sections and to highlight the distribution pattern of IELs within the villus epithelium (Ensari, 2010; Ludvigsson et al., 2014; Mino & Lauwers, 2003). In addition, it may have a role in the differential diagnosis between CD and other atrophic nongluten-sensitive enteropathies, which are inconstantly characterized by intraepithelial lymphocytosis (see Chapter 11).

Finally, preliminary findings about the use of immunohistochemistry for γδ T cell markers to support CD diagnosis are reported (Spencer et al., 1991). Although T cell receptor (TCR) γδ-positive IELs are expanded in untreated CD duodenal mucosa, as they increase from 5%–10% to 15%–40% of total IELs (Malamut et al., 2009), there is no definitive evidence that adding TCR-γδ immunochemistry improves CD diagnosis when compared to H&E and CD3 evaluations (Robert et al., 2018).

9.3.2 Architectural changes

Architectural changes, represented by villous atrophy/blunting and crypt hyperplasia, are key alterations of active CD. Normal duodenal mucosa has a ratio between villous height and crypt depth that varies from 3:1 to 5:1 in D2 and D3, as well as in the jejunum. In D1, duodenal villi in healthy individuals are usually shorter, may have a leaf-like morphology, with a broad flattened base and a tapering apex, and may show unusual forms, with fused or branching tips. As a consequence, the normal villous-to-crypt ratio in D1 is lower (about 2:1) (Al-Toma et al., 2019; Ensari, 2010; Villanacci et al., 2011). Therefore villous atrophy is defined as a villous-to-crypt ratio less than 3:1 in D2-D3 and less than 2:1 in D1 specimens.

Histologically, crypt hyperplasia is defined as an elongation of the intestinal crypts (Marsh & Crowe, 1995). It is associated with a reduced number of goblet cells within the crypt and an increased number of crypt mitotic figures (i.e., presence of more than 1 mitosis per crypt). To correctly evaluate these architectural changes, at least 3–4 consecutive well-oriented villus-crypt units should be present in the specimen to avoid misinterpretation of "short" villi in poorly oriented samples with tangential artifacts as villous atrophy (Fig. 9.1). Additional care should be given when analyzing samples for the presence of trauma and superficial artifacts, lymphoid aggregates, and nodules of Brunner glands (the so-called "Brunner artifact"), especially in D1, which can distort overlying villous morphology giving the erroneous impression of atrophic villi (Fig. 9.1) (Robert et al., 2018).

9.3.3 Other histological abnormalities

Enterocyte cytological changes (e.g., cuboidal or flattened shapes and cytoplasmic vacuolization, nuclear pyknosis, and loss of basal nuclear orientation), focal gastric metaplasia and a slight increase of crypt cell apoptosis can be detected in CD patients (Shalimar et al., 2013), while a marked increase of crypt apoptosis is not a typical sign of CD and may

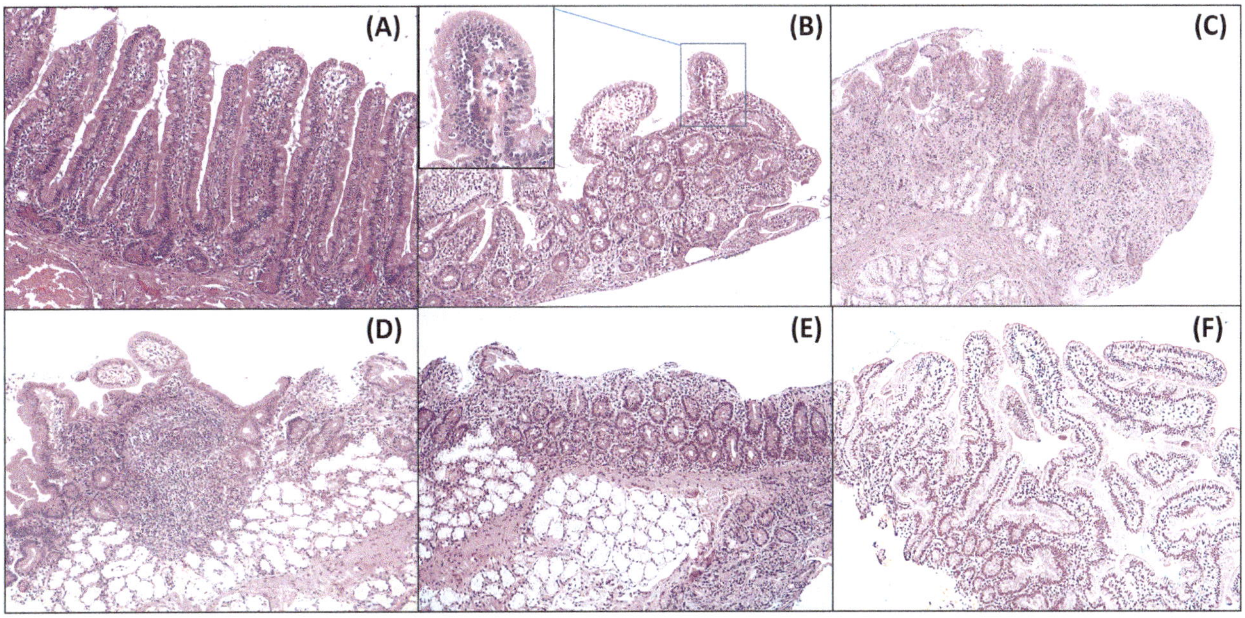

FIGURE 9.1 Representative images of duodenal biopsy artifacts commonly encountered in routine practice, to be compared to an optimal duodenal biopsy. (A) A representative histologic image of a well-oriented duodenal biopsy with preserved villous architecture (A, ×100, H&E). (B) Tangential artifacts. Note, in the inlet, a roundish and pseudostratified appearance of enterocyte nuclei on the left lateral aspect of a villus opposed to the normal elongated nuclei on the right lateral aspect of the same villus (B, ×100, H&E; inlet, ×200, H&E). (C) Brunner's glands artifacts with a group of Brunner's glands that determine an alteration of the normal villous architecture above them, mimicking a partial villous atrophy (C, ×100, H&E). (D) Lymphoid aggregate-related artifacts with a lymphoid aggregate that distorts the villi above it giving them an atrophic appearance (D, ×100, H&E). (E) Mixed artifacts (tangential and Brunner's gland-related, and even some trauma-related artifacts) with zonal loss of superficial epithelium (E, ×100, H&E). (F) Superficial sampling-related artifacts. Note the absence of the *muscolaris mucosae* (F, ×100, H&E).

be related to other conditions such as drug-related injuries or graft-versus-host disease (Matsuda et al., 2020). Although goblet cells may be reduced in number in CD duodenal mucosa (Capuano et al., 2011), the lack of Paneth cells or the increase of neuroendocrine cells has been reported in complicated CD (Di Sabatino et al., 2008), and in other nongluten sensitive enteropathies (Chapters 11 and 14).

CD duodenal biopsies may also show variably thickened subepithelial collagen, which may be highlighted with Masson's trichrome staining, but to a lesser extent than in collagenous sprue (Chapter 14) (Bossart et al., 1975).

Other changes that can be noticed are the increased infiltration of T and B lymphocytes, eosinophils, and neutrophils in the inflammatory milieu of the lamina propria. Significant duodenal neutrophilic infiltrates have been reported to be more frequent in the pediatric population (Moran et al., 2012). Even neutrophilic cryptitis can be focally seen when mucosal flattening is widespread, but CD is seldom associated with full-blown crypt destruction, which is more common in other inflammatory diseases (Brown et al., 2021; Chen et al., 2023). Plasma cells are usually numerous within the lamina propria, and their absence is indicative of a concomitant common variable immunodeficiency (Agarwal & Mayer, 2013; Biagi et al., 2012).

The histologic features of special forms of CD, including potential CD, refractory CD, enteropathy-associated T-cell lymphoma, and dermatitis herpetiformis, as well as those of other atrophic nongluten-sensitive enteropathies, are addressed in other Chapters (Chapters 10, 14, 16, and 11, respectively).

9.4 Assessing severity of celiac disease mucosal lesions

In clinical practice, qualitative histology, which uses descriptive and categorical elements to assess characteristics of samples, is the most widely used method to define histological properties and injuries of specimens. However, this method suffers from a variable rate of interobserver variation over the degree of damage and cannot accurately evaluate changes over time, especially when assessing the response to the gluten-free diet (Corazza et al., 1982; Corazza et al., 2007; Mubarak et al., 2011; Rostom et al., 2006). Some semiquantitive systems can mitigate some of these problems by having less variability and greater agreement among pathologists.

Thus even though evidence has shown that grading of mucosal damage has no association with the severity of symptoms nor can predict patient's prognosis (Biagi et al., 2016; Brar et al., 2007; Rostami et al., 2015), it may have clinical implications, mainly in the case of serial comparisons.

9.4.1 Semiquantitative grading of lesions

Several classification grading systems have been proposed. It is however necessary to highlight that the application of each of them, despite being useful for introducing a shared nomenclature, inevitably implies a certain degree of subjective evaluation by the observer. Hence, even these systems may lead to some degrees of intra- or interobserver variations.

The first and the most widely recognized classification was proposed by Marsh in 1992 (Marsh, 1992). This classification is based on four histopathologically recognizable stages of mucosal remodeling, including Marsh 0 (normal mucosa), Marsh 1 (i.e., infiltrative lesion, characterized by normal villous architecture with increased intraepithelial lymphocytes), Marsh 2 (i.e., hyperplastic lesion, characterized by Marsh 1 lesions + crypt hyperplasia), and Marsh 3 (i.e., destructive lesion with mucosal flattening) (Fig. 9.2). In the original Marsh classification, a stage 4, characterized by villous atrophy and hypoplastic crypts was also included at the end of the spectrum; however, it was subsequently removed, as it is not a part of the pathological spectrum of CD.

A more detailed semiquantitative classification was proposed by Oberhuber in 1999 (Oberhuber et al., 1999), that differs from Marsh classification, as Marsh 3 pattern was split into 3 A (mild villous atrophy), 3B (moderate or subtotal villous atrophy), and 3 C (complete loss of villi or severe villous atrophy). However, Marsh himself highlighted that Marsh 3 subclassification is untenable and open to misinterpretation (Marsh & Heal, 2017; Marsh et al., 2015). Additionally, interobserver agreement and reproducibility of the Oberhuber's system are low (Arguelles-Grande et al., 2012; Corazza & Villanacci, 2005; Corazza et al., 2007).

A more recently proposed grading system for CD-related histologic injuries in duodenal biopsies is the Corazza-Villanacci classification scheme (Corazza & Villanacci, 2005; Corazza et al., 2007) which condenses the histologic changes into three categories, namely normal villous architectures with increased IELs (grade A), shortened but still detectable villi (grade B1), and completely flattened mucosa with no detectable villi (grade B2). This scheme has been proved to have a higher interobserver agreement when compared with the Oberhuber system and has an increased

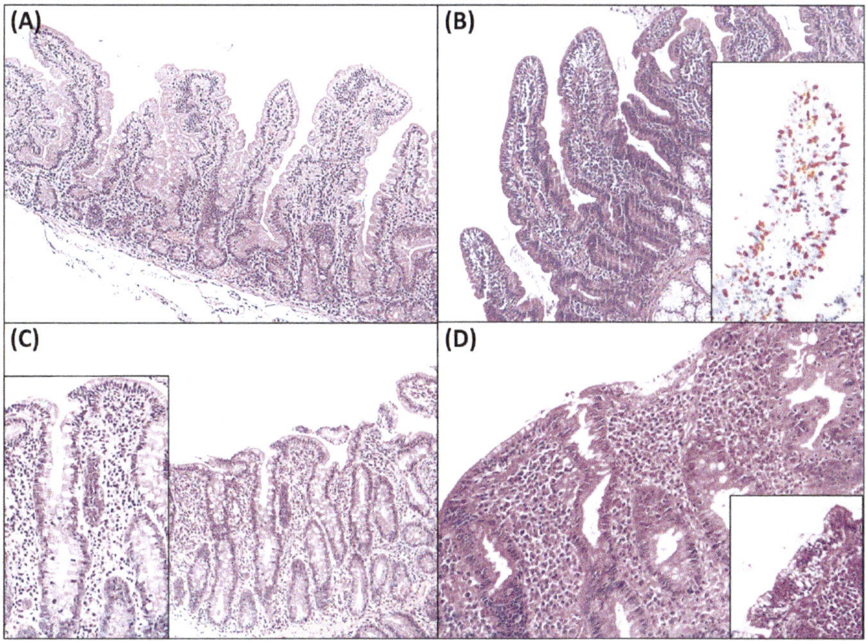

FIGURE 9.2 **Representative images of histologic spectrum of duodenal mucosa in celiac disease and corresponding histologic grades.** (A) A duodenal biopsy with preserved villous architecture and a normal number of intraepithelial lymphocytes (i.e., Marsh 0) (A, × 100, H&E). (B) A bioptic specimen with marked intraepithelial lymphocytosis and preserved mucosal architecture (i.e., Marsh 1/ Corazza A). Note, in the inlet, a higher magnification of a villus of the same specimen depicted in (B) immunostained for CD3, to highlight the high number of CD3 + intraepithelial T lymphocytes (B, × 100, H&E; inlet, × 200). (C) A bioptic sample showing marked, albeit not complete, villous atrophy (i.e., Marsh 3/ Corazza B1). Note, in the inlet, marked crypt hyperplasia, with numerous mitotic figures with an elongated crypt (C, × 100, H&E; inlet, × 200, H&E). (D) A mucosal sample with completely flat mucosa without recognizable villi (i.e., Marsh 3/ Corazza B2). Note, in the inlet, signs of enterocyte vacuolization, a relatively common ancillary cytologic finding in the atrophic mucosa of untreated celiac patients (D, × 100, H&E; inlet, × 200, H&E).

TABLE 9.2 Comparison of the most common classification grading systems for CD histologic lesions.

Intraepithelial lymphocytes	Architectural alterations	Marsh	Marsh—Oberhuber	Corazza—Villanacci	Ensari
Normal	Absent	Stage 0	Stage 0	Normal	Normal
Increased	Absent	Stage 1	Stage 1	Grade A	Type 1
Increased	Crypt hyperplasia without villous atrophy	Stage 2	Stage 2	Grade A	Type 1
Increased	Mild villous atrophy	Stage 3	Stage 3 A	Grade B1	Type 2
Increased	Marked/subtotal villous atrophy	Stage 3	Stage 3B	Grade B1	Type 2
Increased	Total villous atrophy (flat mucosa)	Stage 3	Stage 3 C	Grade B2	Type 3

reproducibility, thus facilitating the comparison of biopsies obtained throughout follow-up (Corazza et al., 2007; Kaur Bilkhoo et al., 2016).

In addition, Ensari proposed a three-tiered classification scheme similar to that of Corazza-Villanacci, but instead of using grade A, B1, and B2 categories, she provided numerical labels (type 1, 2, and 3) (Ensari, 2010). A comparison of the different grading systems is provided in Table 9.2. In general, it is the specimen with the greater villous atrophy that determines the grade.

Regardless of the underlying etiology, the patterns Marsh 0 coupled with ultrastructural enterocyte alterations, Marsh 1 and Marsh 2 leading to gastrointestinal symptoms or micronutrient deficiencies have also been collectively termed "microscopic enteritis" by Rostami et al. (Rostami et al., 2015). However, this term is not currently used in the histopathologic reports and its clinical usefulness is still to be validated.

TABLE 9.3 Conditions that can cause duodenal intraepithelial lymphocytosis without villous atrophy (the so-called Marsh 1 pattern).

Potential CD	
Treated CD	
Nonceliac gluten sensitivity	
Infections	Viral enteritis Parasitic infections (e.g., cryptosporidiosis and giardiasis) Small bowel bacterial overgrowth *Helicobacter pylori* gastritis and/or duodenitis HIV-related enteritis
Iatrogenic injury	NSAID PPI
Immune-related conditions	IBDs Food hypersensitivity (e.g., milk, soy, eggs, and fish) Primary and acquired immunodeficiencies (e.g., CVID, IgA deficiency, and AIDS) Autoimmune diseases (e.g., SLE, AG) Graft versus host disease
Other	Family members of CD patients Blind loop syndrome
Idiopathic	

AG: autoimmune gastritis; AIDS: acquired immune deficiency syndrome; CD: celiac disease; CVID: common variable immunodeficiency; HIV: human immunodeficiency virus; IBD: inflammatory bowel disease; NSAID: nonsteroidal antiinflammatory drugs; PPI: proton-pump inhibitors; SLE: systemic lupus erythematosus.

Finally, it should be stressed that Marsh 1 or Corazza A patterns may also be seen in some celiac-related and celiac-unrelated conditions, as summarized in Table 9.3. Among the latter, *Helicobacter pylori*-related gastritis is one of the most commonly encountered (Memeo et al., 2005; Smyrk, 2017).

9.4.2 Morphometric measurements

Quantitative histology, that is, morphometric measurements, is a method that uses accurate measurements, taken with a light microscope with a calibrated micrometer or a graticule, or with computer-assisted methods, to quantify architectural changes as well as counting of IELs.

Morphometric measurements had been used in the setting of clinical research for decades (Corazza et al., 1982; Corazza et al., 1985) to provide a more precise and reproducible tool compared to qualitative and semiquantitive observations. Early morphometric studies in CD were performed using the stereological method proposed by Dunnill and Whitehead, using a Weibel graticule inserted in the microscope eyepiece, which is superimposed on a biopsy specimen, allowing to calculate the surface area-to-volume ratio of the small bowel mucosa (Dunnill & Whitehead, 1972).

More recently standardized morphometric procedures, like measuring villous height and crypt depth in 5–12 villous-crypt units from at least four different samples, were implemented (Murray et al., 2016; Taavela et al., 2013), such as the Quantitative-Mucosal Algorithmic Rules for Scoring Histology (Q-MARSH) system. This linear system allows a morphometric quantitative analysis of the average villous height-to-crypt depth ratio (Adelman et al., 2018). A ratio ≥ 3.0 is considered normal, while a ratio <2 represents a villous atrophy. A composite quantitative histology scale based on both villous height to crypt depth ratio and IELs has been proposed, with a better accuracy compared to villous height/crypt depth ratio alone (Syage et al., 2022). It should be stressed that these linear morphometric measurements based on villous to crypt ratio suffers greatly from the subjectivity in the assessment of crypt-villous border and from even minor orientation artifacts (Isola et al., 2022; Lenti & Corazza, 2019). Indeed, an early investigation comparing linear, stereological and computer-aided methods showed that only stereological techniques can overcome these limitations (G R Corazza et al., 1985).

To summarize, although morphometric analyses are time-consuming and lack firm evidence of clinical benefit to be fully integrated into routine diagnostic CD protocols, they may be important in the research setting, especially to precisely monitor subtle histologic changes in clinical trials.

TABLE 9.4 Core items to be covered in the histopathology report.

- the number of biopsies assessed
- whether biopsies are correctly oriented or not
- whether Brunner's glands are seen
- the presence or absence of villous atrophy
- the presence or absence crypt hyperplasia
- the presence or absence of intraepithelial lymphocytosis
- the presence of lamina propria inflammation (specify the presence of increase of plasma cells and neutrophils)
- the use of a semiquantitative grading system in the case of serial comparison

9.5 Conclusions

As a practical key-point for all pathologists, a list of the main items to be covered in the histopathology reports are summarized in Table 9.4. These practical points, if applied, could be helpful in harmonizing the reports across different pathology Units, hence reducing the risk of misinterpretation and misdiagnosis.

References

Adelman, D. C., Murray, J., Wu, T. T., Mäki, M., Green, P. H., & Kelly, C. P. (2018). Measuring change in small intestinal histology in patients with celiac disease. *American Journal of Gastroenterology*, *113*(3), 339–347. Available from https://doi.org/10.1038/ajg.2017.480.

Agarwal, S., & Mayer, L. (2013). Diagnosis and treatment of gastrointestinal disorders in patients with primary immunodeficiency. *Clinical Gastroenterology and Hepatology*, *11*(9), 1050–1063. Available from https://doi.org/10.1016/j.cgh.2013.02.024.

Allen, J. I., Katzka, D., Robert, M., & Leontiadis, G. I. (2015). American gastroenterological association institute technical review on the role of upper gastrointestinal biopsy to evaluate dyspepsia in the adult patient in the absence of visible mucosal lesions. *Gastroenterology*, *149*(4), 1088–1118. Available from https://doi.org/10.1053/j.gastro.2015.07.040.

Al-Toma, A., Volta, U., Auricchio, R., Castillejo, G., Sanders, D. S., Cellier, C., Mulder, C. J., & Lundin, K. E. A. (2019). European Society for the Study of Coeliac Disease (ESsCD) guideline for coeliac disease and other gluten-related disorders. *United European Gastroenterology Journal*, *7*(5), 583–613. Available from https://doi.org/10.1177/2050640619844125.

Arguelles-Grande, C., Tennyson, C. A., Lewis, S. K., Green, P. H. R., & Bhagat, G. (2012). Variability in small bowel histopathology reporting between different pathology practice settings: Impact on the diagnosis of coeliac disease. *Journal of Clinical Pathology*, *65*(3), 242–247. Available from https://doi.org/10.1136/jclinpath-2011-200372.

Asmussen, L., Bernstein, I., Matzen, P., & Holck, S. (2009). Bidrager montering af gastrointestinale biopsier på milliporefilter til en forbedret snitkvalitet? *Ugeskrift for Laeger*, *171*(37), 2646–2650. Available from http://www.ugeskriftet.dk/portal/page/portal/LAEGERDK/UGESKRIFT_FOR_LAEGER/TIDLIGERE_NUMRE/2009/UFL_2009_37/UFL_2009_171_37_2646.

Bardella, M. T., Minoli, G., Radaelli, F., Quatrini, M., Bianchi, P. A., & Conte, D. (2000). Reevaluation of duodenal endoscopic markers in the diagnosis of celiac disease. *Gastrointestinal Endoscopy*, *51*(6), 714–716. Available from https://doi.org/10.1067/mge.2000.104653.

Biagi, F., Bianchi, P. I., Zilli, A., Marchese, A., Luinetti, O., Lougaris, V., Plebani, A., Villanacci, V., & Corazza, G. R. (2012). The significance of duodenal mucosal atrophy in patients with common variable immunodeficiency: A clinical and histopathologic study. *American Journal of Clinical Pathology*, *138*(2), 185–189. Available from https://doi.org/10.1309/AJCPEIILH2C0WFYE.

Biagi, F., Luinetti, O., Campanella, J., Klersy, C., Zambelli, C., Villanacci, V., Lanzini, A., & Corazza, G. R. (2004). Intraepithelial lymphocytes in the villous tip: Do they indicate potential coeliac disease? *Journal of Clinical Pathology*, *57*(8), 835–839. Available from https://doi.org/10.1136/jcp.2003.013607.

Biagi, F., Vattiato, C., Burrone, M., Schiepatti, A., Agazzi, S., Maiorano, G., Luinetti, O., Alvisi, C., Klersy, C., & Corazza, G. R. (2016). Is a detailed grading of villous atrophy necessary for the diagnosis of enteropathy. *Journal of Clinical Pathology*, *69*(12), 1051–1054. Available from https://doi.org/10.1136/jclinpath-2016-203711.

Bossart, R., Henry, K., Booth, C. C., & Doe, W. F. (1975). Subepithelial collagen in intestinal malabsorption. *Gut*, *16*(1), 18–22. Available from https://doi.org/10.1136/gut.16.1.18.

Brar, P., Kwon, G. Y., Egbuna, I. I., Holleran, S., Ramakrishnan, R., Bhagat, G., & Green, P. H. R. (2007). Lack of correlation of degree of villous atrophy with severity of clinical presentation of coeliac disease. *Digestive and Liver Disease*, *39*(1), 26–29. Available from https://doi.org/10.1016/j.dld.2006.07.014.

Brocchi, E., Corazza, G. R., Caletti, G., Treggiari, E. A., Barbara, L., & Gasbarrini, G. (1988). Endoscopic demonstration of loss of duodenal folds in the diagnosis of celiac disease. *New England Journal of Medicine*, *319*(12), 741–744. Available from https://doi.org/10.1056/NEJM198809223191202.

Brown, I., Bettington, M., & Rosty, C. (2021). The role of histopathology in the diagnosis and management of coeliac disease and other malabsorptive conditions. *Histopathology*, *78*(1), 88–105. Available from https://doi.org/10.1111/his.14262.

Cammarota, G., Fedeli, P., & Gasbarrini, A. (2009). Emerging technologies in upper gastrointestinal endoscopy and celiac disease. *Nature Clinical Practice Gastroenterology and Hepatology*, *6*(1), 47–56. Available from https://doi.org/10.1038/ncpgasthep1298.

Capuano, M., Iaffaldano, L., Tinto, N., Montanaro, D., Capobianco, V., Izzo, V., Tucci, F., Troncone, G., Greco, L., & Sacchetti, L. (2011). MicroRNA-449a overexpression, reduced NOTCH1 signals and scarce goblet cells characterize the small intestine of celiac patients. *PLoS One*, *6*(12). Available from https://doi.org/10.1371/journal.pone.0029094.

Chen, Z. E., Lee, H. E., & Wu, T.-T. (2023). Histologic evaluation in the diagnosis and management of celiac disease: practical challenges, current best practice recommendations and beyond. *Human Pathology*, *132*, 20–30. Available from https://doi.org/10.1016/j.humpath.2022.07.017.

Corazza, G. R., Brocchi, E., Caletti, G., & Gasbarrini, G. (1990). Loss of duodenal folds allows diagnosis of unsuspected coeliac disease. *Gut*, *31*(9), 1080–1081. Available from https://doi.org/10.1136/gut.31.9.1080.

Corazza, G. R., Frazzoni, M., Dixon, M. F., & Gasbarrini, G. (1985). Quantitative assessment of the mucosal architecture of jejunal biopsy specimens: A comparison between linear measurement, stereology, and computer aided microscopy. *Journal of Clinical Pathology*, *38*(7), 765–770.

Corazza, G. R., & Lenti, M. V. (2021). Intestinal biopsy, coeliac disease and resistance to change. *European Journal of Gastroenterology and Hepatology*, *33*(1), E31–E32. Available from https://doi.org/10.1097/MEG.0000000000002171.

Corazza, G. R., & Villanacci, V. (2005). Coeliac disease. *Journal of Clinical Pathology*, *58*(6), 573–574. Available from https://doi.org/10.1136/jcp.2004.023978.

Corazza, G. R., Bonvicini, F., Frazzoni, M., Gatto, M., & Gasbarrini, G. (1982). Observer Variation in assessment of jejunal biopsy specimens: A comparison between subjective criteria and morphometric measurement. *Gastroenterology*, *83*(6), 1217–1222. Available from https://doi.org/10.1016/S0016-5085(82)80131-5.

Corazza, G. R., Caletti, G. C., Lazzari, R., Collina, A., Brocchi, E., Di Sario, A., Ferrari, A., & Gasbarrini, G. (1993). Scalloped duodenal folds in childhood celiac disease. *Gastrointestinal Endoscopy*, *39*(4), 543–545. Available from https://doi.org/10.1016/S0016-5107(93)70167-4.

Corazza, G. R., & Lenti, M. V. (2017). Does biopsy still have a role for adult coeliac disease. *Lancet Gastroenterology and Hepatology*, *2*(11), 773–774. Available from https://doi.org/10.1016/S2468-1253(17)30251-0.

Corazza, G. R., Villanacci, V., Zambelli, C., Milione, M., Luinetti, O., Vindigni, C., Chioda, C., Albarello, L., Bartolini, D., & Donato, F. (2007). Comparison of the interobserver reproducibility with different histologic criteria used in celiac disease. *Clinical Gastroenterology and Hepatology*, *5*(7), 838–843. Available from https://doi.org/10.1016/j.cgh.2007.03.019.

Di Sabatino, A., Miceli, E., Dhaliwal, W., Biancheri, P., Salerno, R., Cantoro, L., Vanoli, A., De Vicenzi, M., Del Vecchio Blanco, C., MacDonald, T. T., & Corazza, G. R. (2008). Distribution, proliferation, and function of Paneth cells in uncomplicated and complicated adult celiac disease. *American Journal of Clinical Pathology*, *130*(1), 34–42. Available from https://doi.org/10.1309/5ADNAR4VN11TTKQ6.

Dickey, W., & Hughes, D. (2001). Disappointing sensitivity of endoscopic markers for villous atrophy in a high-risk population: Implications for celiac disease diagnosis during routine endoscopy. *American Journal of Gastroenterology*, *96*(7), 2126–2128. Available from https://doi.org/10.1111/j.1572-0241.2001.03947.x.

Dickey, W., & Hughes, D. F. (2004). Histology of the terminal ileum in coeliac disease. *Scandinavian Journal of Gastroenterology*, *39*(7), 665–667. Available from https://doi.org/10.1080/00365520410004901.

Dickson, B. C., Streutker, C. J., & Chetty, R. (2006). Coeliac disease: An update for pathologists. *Journal of Clinical Pathology*, *59*(10), 1008–1016. Available from https://doi.org/10.1136/jcp.2005.035345.

Donat, E., Roca, M., Masip, E., Polo, B., Ramos, D., & Ribes-Koninckx, C. (2021). Screening-detected positive serology for celiac disease: A real clinical challenge. *European Journal of Gastroenterology and Hepatology*, *33*(1), E999–E1002. Available from https://doi.org/10.1097/MEG.0000000000002093.

Dunnill, M. S., & Whitehead, R. (1972). A method for the quantitation of small intestinal biopsy specimens. *Journal of Clinical Pathology*, *25*(3), 243–246. Available from https://doi.org/10.1136/jcp.25.3.243.

Efthymakis, K., Serio, M., Milano, A., Laterza, F., Bonitatibus, A., Di Nicola, M., & Neri, M. (2017). Application of the biopsy-sparing ESPGHAN guidelines for celiac disease diagnosis in adults: A real-life study. *Digestive Diseases and Sciences*, *62*(9), 2433–2439. Available from https://doi.org/10.1007/s10620-017-4672-1.

Elfström, P., Granath, F., Ekström Smedby, K., Montgomery, S. M., Askling, J., Ekbom, A., & Ludvigsson, J. F. (2011). Risk of lymphoproliferative malignancy in relation to small intestinal histopathology among patients with celiac disease. *Journal of the National Cancer Institute*, *103*(5), 436–444. Available from https://doi.org/10.1093/jnci/djq564.

Ensari, A. (2010). Gluten-sensitive enteropathy (celiac disease): Controversies in diagnosis and classification. *Archives of Pathology & Laboratory Medicine*, *134*(6), 826–836. Available from https://doi.org/10.5858/134.6.826.

Ferguson, A., & Murray, D. (1971). Quantitation of intraepithelial lymphocytes in human jejunum. *Gut*, *12*(12), 988–994. Available from https://doi.org/10.1136/gut.12.12.988.

Fuchs, V., Kurppa, K., Huhtala, H., Laurila, K., Mäki, M., Collin, P., Salmi, T., Luostarinen, L., Saavalainen, P., & Kaukinen, K. (2019). Serology-based criteria for adult coeliac disease have excellent accuracy across the range of pre-test probabilities. *Alimentary Pharmacology and Therapeutics*, *49*(3), 277–284. Available from https://doi.org/10.1111/apt.15109.

Goldstein, N. S., & Underhill, J. (2001). Morphologic features suggestive of gluten sensitivity in architecturally normal duodenal biopsy specimens. *American Journal of Clinical Pathology*, *116*(1), 63–71. Available from https://doi.org/10.1309/5PRJ-CM0U-6KLD-6KCM.

Green, P. H. R., Shane, E., Rotterdam, H., Forde, K. A., & Grossbard, L. (2000). Significance of unsuspected celiac disease detected at endoscopy. *Gastrointestinal Endoscopy*, *51*(1), 60–65. Available from https://doi.org/10.1016/S0016-5107(00)70389-0.

Haggitt, R. C., Appelman, H. D., Correa, P., Fenoglio, C. M., Goldman, H., Hamilton, S. R., Morson, B. C., Ransohoff, D. F., Riddell, R. H., Sommers, S. C., & Yardley, J. H. (1982). Carcinoma or dysplasia in Crohn's disease. *Archives of Pathology and Laboratory Medicine*, *106*(6), 308–309.

Hayat, M., Cairns, A., Dixon, M. F., & O'Mahony, S. (2002). Quantitation of intraepithelial lymphocytes in human duodenum: What is normal. *Journal of Clinical Pathology*, *55*(5), 393–394. Available from https://doi.org/10.1136/jcp.55.5.393.

Holmes, G. K. T., & Hill, P. G. (2017). Coeliac disease: Further evidence that biopsy is not always necessary for diagnosis. *European Journal of Gastroenterology and Hepatology*, *29*(10), 1189–1190. Available from https://doi.org/10.1097/MEG.0000000000000937.

Hopper, A. D., Hurlstone, D. P., Leeds, J. S., McAlindon, M. E., Dube, A. K., Stephenson, T. J., & Sanders, D. S. (2006). The occurrence of terminal ileal histological abnormalities in patients with coeliac disease. *Digestive and Liver Disease*, *38*(11), 815–819. Available from https://doi.org/10.1016/j.dld.2006.04.003.

Hudacko, R., Kathy Zhou, X., & Yantiss, R. K. (2013). Immunohistochemical stains for CD3 and CD8 do not improve detection of gluten-sensitive enteropathy in duodenal biopsies. *Modern Pathology*, *26*(9), 1241–1245. Available from https://doi.org/10.1038/modpathol.2013.57.

Husby, S., Koletzko, S., Korponay-Szabó, I., Kurppa, K., Mearin, M. L., Ribes-Koninckx, C., Shamir, R., Troncone, R., Auricchio, R., Castillejo, G., Christensen, R., Dolinsek, J., Gillett, P., Hróbjartsson, A., Koltai, T., Maki, M., Nielsen, S. M., Popp, A., Størdal, K., ... Wessels, M. (2020). European Society Paediatric Gastroenterology, Hepatology and Nutrition Guidelines for Diagnosing Coeliac Disease 2020. *Journal of Pediatric Gastroenterology and Nutrition*, *70*(1), 141–156. Available from https://doi.org/10.1097/MPG.0000000000002497.

Husby, S., Koletzko, S., Korponay-Szabó, I., Mearin, M., Phillips, A., Shamir, R., Troncone, R., Giersiepen, K., Branski, D., Catassi, C., Lelgeman, M., Mäki, M., Ribes-Koninckx, C., Ventura, A., & Zimmer, K. P. (2012). ESPGHAN working group on coeliac disease diagnosis; ESPGHAN gastroenterology committee; European Society for Pediatric Gastroenterology, Hepatology, and Nutrition. European Society for Pediatric Gastroenterology, Hepatology, and Nutrition guidelines for the diagnosis of coeliac disease. *Journal of Pediatric Gastroenterology and Nutrition*, *54*(1). Available from https://doi.org/10.1097/MPG.0b013e31821a23d0, Erratum.

Husby, S., Murray, J. A., & Katzka, D. A. (2019). AGA clinical practice update on diagnosis and monitoring of celiac disease—Changing utility of serology and histologic measures: Expert review. *Gastroenterology*, *156*(4), 885–889. Available from https://doi.org/10.1053/j.gastro.2018.12.010.

Isola, J., Maxwell, J., Välimäki, Leffler, Raunig, D., Condover, D., Murray, J., Robert, M., Chen, Z., Lei, Q., Wu, & Smithson, G. (2022). Strategies for improving inter-reader variability of villus height and crypt depth measurements across different institutions. In: Proceedinngs of the 19th international celiac disease symposium, Sorrento, 123.

Jabbari, M., Wild, G., Goresky, C. A., Daly, D. S., Lough, J. O., Cleland, D. P., & Kinnear, D. G. (1988). Scalloped valvulae conniventes: An endoscopic marker of celiac sprue. *Gastroenterology*, *95*(6), 1518–1522. Available from https://doi.org/10.1016/S0016-5085(88)80071-4.

Järvinen, T. T., Collin, P., Rasmussen, M., Kyrönpalo, S., Mäki, M., Partanen, J., Reunala, T., & Kaukinen, K. (2004). Villous tip intraepithelial lymphocytes as markers of early-stage coeliac disease. *Scandinavian Journal of Gastroenterology*, *39*(5), 428–433. Available from https://doi.org/10.1080/00365520310008773.

Kaur Bilkhoo, H., Ducruet, T., Marchand, V., Deslandres, C., Djemli, A., Dal Soglio, D., & Patey, N. (2016). Revisiting pathological criteria for earlier diagnosis of coeliac disease. *Journal of Pediatric Gastroenterology & Nutrition*, *62*(5), 734–738. Available from https://doi.org/10.1097/mpg.0000000000001026.

Kurien, M., Evans, K. E., Hopper, A. D., Hale, M. F., Cross, S. S., & Sanders, D. S. (2012). Duodenal bulb biopsies for diagnosing adult celiac disease: Is there an optimal biopsy site? *Gastrointestinal Endoscopy*, *75*(6), 1190–1196. Available from https://doi.org/10.1016/j.gie.2012.02.025.

Lagana, S. M., & Bhagat, G. (2019). Biopsy diagnosis of celiac disease: The pathologist's perspective in light of recent advances. *Gastroenterology Clinics of North America*, *48*(1), 39–51. Available from https://doi.org/10.1016/j.gtc.2018.09.003.

Latorre, M., Lagana, S. M., Freedberg, D. E., Lewis, S. K., Lebwohl, B., Bhagat, G., & Green, P. H. R. (2015). Endoscopic biopsy technique in the diagnosis of celiac disease: One bite or two? *Gastrointestinal Endoscopy*, *81*(5), 1228–1233. Available from https://doi.org/10.1016/j.gie.2014.10.024.

Lebwohl, B., Granath, F., Ekbom, A., Smedby, K. E., Murray, J. A., Neugut, A. I., Green, P. H. R., & Ludvigsson, J. F. (2013). Mucosal healing and risk for lymphoproliferative malignancy in celiac disease: A population-based cohort study. *Annals of Internal Medicine*, *159*(3), 169–175. Available from https://doi.org/10.7326/0003-4819-159-3-201308060-00006.

Lebwohl, B., Kapel, R. C., Neugut, A. I., Green, P. H. R., & Genta, R. M. (2011). Adherence to biopsy guidelines increases celiac disease diagnosis. *Gastrointestinal Endoscopy*, *74*(1), 103–109. Available from https://doi.org/10.1016/j.gie.2011.03.1236.

Lenti, M. V., & Corazza, G. R. (2019). Measuring too much or too little in adult coeliac disease. *Journal of Clinical Pathology*, *72*(5), 341–342. Available from https://doi.org/10.1136/jclinpath-2018-205667.

Ludvigsson, J. F., & Murray, J. A. (2019). Epidemiology of celiac disease. *Gastroenterology Clinics of North America*, *48*(1), 1–18. Available from https://doi.org/10.1016/j.gtc.2018.09.004.

Ludvigsson, J. F., Bai, J. C., Biagi, F., Card, T. R., Ciacci, C., Ciclitira, P. J., Green, P. H. R., Hadjivassiliou, M., Holdoway, A., van Heel, D. A., Kaukinen, K., Leffler, D. A., Leonard, J. N., Lundin, K. E. A., McGough, N., Davidson, M., Murray, J. A., Swift, G. L., Walker, M. M., ... Sanders, D. S. (2014). Diagnosis and management of adult coeliac disease: guidelines from the British Society of Gastroenterology. *Gut*, *63*(8), 1210–1228. Available from https://doi.org/10.1136/gutjnl-2013-306578.

Malamut, M., Cellier, B., & Cerf-Bensussan, N. (2009). La maladie coeliaque en 2009: Un futur sans régime? Celiac disease in 2009: A future without gluten-free diet? *Gastroenterologie Clinique et Biologique*, *33*(8–9). Available from https://doi.org/10.1016/j.gcb.2009.07.007, Epub.

Marsh, M. N. (1990). Grains of truth: Evolutionary changes in small intestinal mucosa in response to environmental antigen challenge. *Gut*, *31*(1), 111–114. Available from https://doi.org/10.1136/gut.31.1.111.

Marsh, M. N. (1992). Gluten, major histocompatibility complex, and the small intestine. A molecular and immunobiologic approach to the spectrum of gluten sensitivity ('celiac sprue'). *Gastroenterology*, *102*(1), 330–354. Available from https://doi.org/10.1016/0016-5085(92)91819-P.

Marsh, M. N., & Crowe, P. T. (1995). 5 Morphology of the mucosal lesion in gluten sensitivity. *Bailliere's Clinical Gastroenterology*, *9*(2), 273–293. Available from https://doi.org/10.1016/0950-3528(95)90032-2.

Marsh, M. N., & Heal, C. J. (2017). Evolutionary developments in interpreting the gluten-induced mucosal celiac lesion: An archimedian heuristic. *Nutrients*, *9*(3). Available from https://doi.org/10.3390/nu9030213.

Marsh, M. N., Johnson, M. W., & Rostami, K. (2015). Mucosal histopathology in celiac disease: A rebuttal of Oberhuber's sub-division of Marsh III. *Gastroenterology and Hepatology from Bed to Bench*, *8*(2), 99–109. Available from http://journals.sbmu.ac.ir/ghfbb/index.php/ghfbb/article/download/713/487.

Matsuda, K., Ono, S., Tanaka, I., Inoue, M., Kinowaki, S., Ishikawa, M., Tsuda, M., Yamamoto, K., Shimizu, Y., Takahashi, S., Hayase, E., Hashimoto, D., Teshima, T., & Sakamoto, N. (2020). Histological and magnified endoscopic evaluation of villous atrophy in gastrointestinal graft-versus-host disease. *Annals of Hematology*, *99*(5), 1121–1128. Available from https://doi.org/10.1007/s00277-020-03966-y.

Mauriño, E., Capizzano, H., Niveloni, S., Kogan, Z., Valero, J., Boerr, L., & Cesar Bai, J. (1993). Value of endoscopic markers in celiac disease. *Digestive Diseases and Sciences*, *38*(11), 2028–2033. Available from https://doi.org/10.1007/BF01297080.

McCarty, T. R., O'Brien, C. R., Gremida, A., Ling, C., & Rustagi, T. (2018). Efficacy of duodenal bulb biopsy for diagnosis of celiac disease: A systematic review and meta-analysis. *Endoscopy International Open*, *06*(11), E1369–E1378. Available from https://doi.org/10.1055/a-0732-5060.

McIntyre, A. S., Ng, D. P. K., Smith, J. A., Amoah, J., & Long, R. G. (1992). The endoscopic appearance of duodenal folds is predictive of untreated adult celiac disease. *Gastrointestinal Endoscopy*, *38*(2), 148–151. Available from https://doi.org/10.1016/S0016-5107(92)70380-0.

Mee, A. S., Burke, M., Vallon, A. G., Newman, J., & Cotton, P. B. (1985). Small bowel biopsy for malabsorption: Comparison of the diagnostic adequacy of endoscopic forceps and capsule biopsy specimens. *British Medical Journal (Clinical Research ed.)*, *291*(6498), 769–772. Available from https://doi.org/10.1136/bmj.291.6498.769.

Memeo, L., Jhang, J., Hibshoosh, H., Green, P. H., Rotterdam, H., & Bhagat, G. (2005). Duodenal intraepithelial lymphocytosis with normal villous architecture: Common occurrence in H. pylori gastritis. *Modern Pathology*, *18*(8), 1134–1144. Available from https://doi.org/10.1038/modpathol.3800404.

Mino, M., & Lauwers, G. Y. (2003). Role of lymphocytic immunophenotyping in the diagnosis of gluten-sensitive enteropathy with preserved villous architecture. *American Journal of Surgical Pathology*, *27*(9), 1237–1242. Available from https://doi.org/10.1097/00000478-200309000-00007.

Mooney, P. D., Kurien, M., Evans, K. E., Rosario, E., Cross, S. S., Vergani, P., Hadjivassiliou, M., Murray, J. A., & Sanders, D. S. (2016). Clinical and immunologic features of ultra-short celiac disease. *Gastroenterology*, *150*(5), 1125–1134. Available from https://doi.org/10.1053/j.gastro.2016.01.029.

Moran, C. J., Kolman, O. K., Russell, G. J., Brown, I. S., & Mino-Kenudson, M. (2012). Neutrophilic infiltration in gluten-sensitive enteropathy is neither uncommon nor insignificant: Assessment of duodenal biopsies from 267 pediatric and adult patients. *American Journal of Surgical Pathology*, *36*(9), 1339–1345. Available from https://doi.org/10.1097/PAS.0b013e318254f413.

Mubarak, A., Nikkels, P., Houwen, R., & Ten Kate, F. (2011). Reproducibility of the histological diagnosis of celiac disease. *Scandinavian Journal of Gastroenterology*, *46*(9), 1065–1073. Available from https://doi.org/10.3109/00365521.2011.589471.

Murray, J. A., Rubio-Tapia, A., Van Dyke, C. T., Brogan, D. L., Knipschield, M. A., Lahr, B., Rumalla, A., Zinsmeister, A. R., & Gostout, C. J. (2008). Mucosal atrophy in celiac disease: Extent of involvement, correlation with clinical presentation, and response to treatment. *Clinical Gastroenterology and Hepatology : The Official Clinical Practice Journal of the American Gastroenterological Association*, *6*(2), 186–193.

Murray, J., Kelly., Green, P., Marcantonio., Wu, T., Mäki., & Adelman, D. (2016). CeliAction study group of investigators. No difference between latiglutenase and placebo in reducing villous atrophy or improving symptoms in patients with symptomatic celiac disease. *Gastroenterology*, *152*(4). Available from https://doi.org/10.1053/j.gastro.2016.11.004, Epub.

Oberhuber, G., Granditsch, G., & Vogelsang, H. (1999). The histopathology of coeliac disease: Time for a standardized report scheme for pathologists. *European Journal of Gastroenterology and Hepatology*, *11*(10), 1185–1194. Available from https://doi.org/10.1097/00042737-199910000-00019.

Oxentenko, A. S., Grisolano, S. W., Murray, J. A., Burgart, L. J., Dierkhising, R. A., & Alexander, J. A. (2002). The insensitivity of endoscopic markers in celiac disease. *The American Journal of Gastroenterology*, *97*(4), 933–938. Available from https://doi.org/10.1111/j.1572-0241.2002.05612.x.

Pais, W. P., Duerksen, D. R., Pettigrew, N. M., & Bernstein, C. N. (2008). How many duodenal biopsy specimens are required to make a diagnosis of celiac disease? *Gastrointestinal Endoscopy*, *67*(7), 1082–1087. Available from https://doi.org/10.1016/j.gie.2007.10.015.

Penny, H. A., Mooney, P. D., Burden, M., Patel, N., Johnston, A. J., Wong, S. H., Teare, J., & Sanders, D. S. (2016). High definition endoscopy with or without I-Scan increases the detection of celiac disease during routine endoscopy. *Digestive and Liver Disease*, *48*(6), 644–649. Available from https://doi.org/10.1016/j.dld.2016.02.009.

Penny, H. A., Raju, S. A., Lau, M. S., Marks, L. J. S., Baggus, E. M. R., Bai, J. C., Bassotti, G., Bontkes, H. J., Carroccio, A., Danciu, M., Derakhshan, M. H., Ensari, A., Ganji, A., Green, P. H. R., Johnson, M. W., Ishaq, S., Lebwohl, B., Levene, A., Maxim, R., ... Sanders, D. S. (2021). Accuracy of a no-biopsy approach for the diagnosis of coeliac disease across different adult cohorts. *Gut*, *70*(5), 876–883. Available from https://doi.org/10.1136/gutjnl-2020-320913.

Perera, D. R., Weinstein, W. M., & Rubin, C. E. (1975). Small intestinal biopsy. *Human Pathology*, *6*(2), 157–217. Available from https://doi.org/10.1016/S0046-8177(75)80176-6.

Ravelli, A., Villanacci, V., Monfredini, C., Martinazzi, S., Grassi, V., & Manenti, S. (2010). How patchy is patchy villous atrophy?: Distribution pattern of histological lesions in the duodenum of children with celiac disease. *American Journal of Gastroenterology*, *105*(9), 2103–2110. Available from https://doi.org/10.1038/ajg.2010.153.

Robert, M. E., Crowe, S. E., Burgart, L., Yantiss, R. K., Lebwohl, B., Greenson, J. K., Guandalini, S., & Murray, J. A. (2018). Statement on best practices in the use of pathology as a diagnostic tool for celiac disease: A guide for clinicians and pathologists. *American Journal of Surgical Pathology*, *42*(9), e44–e58. Available from https://doi.org/10.1097/PAS.0000000000001107.

Robson, K., Alizart, M., Martin, J., Nagel, R., & Green, J. (2014). Coeliac Patients Are Undiagnosed at Routine Upper Endoscopy. *PLoS One*, *9*(3), e90552. Available from https://doi.org/10.1371/journal.pone.0090552.

Rostami, K., Aldulaimi, D., Holmes, G., Johnson, M. W., Robert, M., Srivastava, A., Fléjou, J. F., Sanders, D. S., Volta, U., Derakhshan, M. H., Going, J. J., Becheanu, G., Catassi, C., Danciu, M., Materacki, L., Ghafarzadegan, K., Ishaq, S., Rostami-Nejad, M., Peña, A. S., ... Villanacci, V. (2015). Microscopic enteritis: Bucharest consensus. *World Journal of Gastroenterology*, *21*(9), 2593–2604. Available from https://doi.org/10.3748/wjg.v21.i9.2593.

Rostami, K., Marsh, M. N., Johnson, M. W., Mohaghegh, H., Heal, C., Holmes, G., Ensari, A., Aldulaimi, D., Bancel, B., Bassotti, G., Bateman, A., Becheanu, G., Bozzola, A., Carroccio, A., Catassi, C., Ciacci, C., Ciobanu, A., Danciu, M., Derakhshan, M. H., ... Srivastava, A. (2017). Rocking onwards: Intraepithelial lymphocyte counts, distribution & role in coeliac disease mucosal interpretation. *Gut*, *66*(12), 2080–2086. Available from https://doi.org/10.1136/gutjnl-2017-314297.

Rostom, A., Murray, J. A., & Kagnoff, M. F. (2006). American Gastroenterological Association (AGA) Institute Technical Review on the Diagnosis and Management of Celiac Disease. *Gastroenterology*, *131*(6), 1981–2002. Available from https://doi.org/10.1053/j.gastro.2006.10.004.

Rubin, C. E., Brandborg, L. L., Phelps, P. C., & Taylor, H. C. (1960). Studies of celiac disease. I. The apparent identical and specific nature of the duodenal and proximal jejunal lesion in celiac disease and idiopathic sprue. *Gastroenterology*, *38*, 28–49.

Rubio-Tapia, A., Hill, I. D., Semrad, C., Kelly, C. P., & Lebwohl, B. (2023). American College of Gastroenterology Guidelines Update: Diagnosis and Management of Celiac Disease. *American Journal of Gastroenterology*, *118*(1), 59–76. Available from https://doi.org/10.14309/ajg.0000000000002075.

Serra, S., & Jani, P. A. (2006). An approach to duodenal biopsies. *Journal of Clinical Pathology*, *59*(11), 1133–1150. Available from https://doi.org/10.1136/jcp.2005.031260.

Shalimar, D. M., Das, P., Sreenivas, V., Gupta, S. D., Panda, S. K., & Makharia, G. K. (2013). Mechanism of villous atrophy in celiac disease: Role of apoptosis and epithelial regeneration. *Archives of Pathology and Laboratory Medicine*, *137*(9), 1262–1269. Available from https://doi.org/10.5858/arpa.2012-0354-OA.

Smyrk, T. C. (2017). Practical approach to the flattened duodenal biopsy. *Surgical Pathology Clinics*, *10*(4), 823–839. Available from https://doi.org/10.1016/j.path.2017.07.004.

Spencer, J., Isaacson, P. G., MacDonald, T. T., Thomas, A. J., & Walker-Smith, J. A. (1991). Gamma/delta T cells and the diagnosis of coeliac disease. *Clinical and Experimental Immunology*, *85*(1), 109–113.

Stevens, F., & McCarthy, C. (1976). The endoscopic demonstration of coeliac disease. *Endoscopy*, *08*(04), 177–180. Available from https://doi.org/10.1055/s-0028-1098406.

Syage, J., Sealey-Voyksner, Wu, & Murray. (2022). Consideration for a composite quantitative histology scale based on VH:CD and IEL. In: Proceedings of the 19th international celiac disease symposium, Sorrento, 136.

Taavela, J., Koskinen, O., Huhtala, H., Lähdeaho, M.-L., Popp, A., Laurila, K., Collin, P., Kaukinen, K., Kurppa, K., Mäki, M., & Sestak, K. (2013). Validation of morphometric analyses of small-intestinal biopsy readouts in celiac disease. *PLoS One*, *8*(10), e76163. Available from https://doi.org/10.1371/journal.pone.0076163.

Thompson, H. (1974). Necropsy studies on adult coeliac disease. *Journal of Clinical Pathology*, *27*(9), 710–721. Available from https://doi.org/10.1136/jcp.27.9.710.

Villanacci, V., Ceppa, P., Tavani, E., Vindigni, C., & Volta, U. (2011). Coeliac disease: The histology report. *Digestive and Liver Disease : Official Journal of the Italian Society of Gastroenterology and the Italian Association for the Study of the Liver*, *43*(Suppl 4), S385–95. Available from https://doi.org/10.1016/S1590-8658(11)60594-X.

Vjero, K., Martucci, S., Alvisi, C., Broglia, F., Viera, F. T., Perego, M., & Corazza, G. R. (2003). Defining a proper setting for endoscopy in coeliac disease. *European Journal of Gastroenterology and Hepatology*, *15*(6), 675–678. Available from https://doi.org/10.1097/00042737-200306000-00015.

Wakim-Fleming, J., Pagadala, M. R., Lemyre, M. S., Lopez, R., Kumaravel, A., Carey, W. D., & Zein, N. N. (2013). Diagnosis of celiac disease in adults based on serology test results, without small-bowel biopsy. *Clinical Gastroenterology and Hepatology*, *11*(5), 511–516. Available from https://doi.org/10.1016/j.cgh.2012.12.015.

Walker, M. M., Murray, J. A., Ronkainen, J., Aro, P., Storskrubb, T., D'Amato, M., Lahr, B., Talley, N. J., & Agreus, L. (2010). Detection of celiac disease and lymphocytic enteropathy by parallel serology and histopathology in a population-based study. *Gastroenterology*, *139*(1), 112–119. Available from https://doi.org/10.1053/j.gastro.2010.04.007.

Werkstetter, K. J., Korponay-Szabó, I. R., Popp, A., Villanacci, V., Salemme, M., Heilig, G., Lillevang, S. T., Mearin, M. L., Ribes-Koninckx, C., Thomas, A., Troncone, R., Filipiak, B., Mäki, M., Gyimesi, J., Najafi, M., Dolinšek, J., Dydensborg Sander, S., Auricchio, R., Papadopoulou, A., ... Eftekhar Sadat, A. T. (2017). Accuracy in diagnosis of celiac disease without biopsies in clinical practice. *Gastroenterology*, *153*(4), 924–935. Available from https://doi.org/10.1053/j.gastro.2017.06.002.

Zambelli, A., Villanacci, V., Buscarini, E., Lupinacci, G., De Grazia, F., Brambilla, G., Menozzi, F., La Mantia, L., & Bassotti, G. (2007). Confocal laser endomicroscopy in celiac disease: Description of findings in two cases. *Endoscopy*, *39*(11), 1018–1020. Available from https://doi.org/10.1055/s-2007-966783.

Zingone, F., Maimaris, S., Auricchio, R., Caio, G. P. I., Carroccio, A., Elli, L., Galliani, E., Montagnani, M., Valiante, F., & Biagi, F. (2022). Guidelines of the Italian societies of gastroenterology on the diagnosis and management of coeliac disease and dermatitis herpetiformis. *Digestive and Liver Disease*, *54*(10), 1304–1319. Available from https://doi.org/10.1016/j.dld.2022.06.023.

Chapter 10

Potential celiac disease

Roberta Mandile, Renata Auricchio, Valentina Discepolo and Riccardo Troncone
Department of Translational Medical Science and European Laboratory for the Investigation of Food-Induced Diseases, University of Naples Federico II, Naples, Italy

10.1 Introduction

Celiac disease (CD) is an immune-mediated disorder elicited by gluten in genetically susceptible individuals, characterized by CD-specific antibodies, small bowel lesions, and variable clinical manifestations (Husby et al., 2012). Despite the enteropathy has historically been described as the most prominent feature, it is now widely accepted that the degree of intestinal damage can range from total villous atrophy to minor or absent histological alterations (Marsh, 1988). Quantitative histology and computerized image analysis have shown that these features occur in a continuum, with flat mucosa at one end of the spectrum and normal villous architecture, with or without an abnormally higher count of intraepithelial lymphocytes, at the other end (Marsh, 1988).

The term latent CD was introduced in the 90s for patients who display a normal jejunal biopsy while consuming a normal diet and, at some other time, before or since, have had a flat mucosa recovering on a gluten free diet (GFD) (Ferguson et al., 1993). These subjects, despite a normal mucosa, may still express subtle immunological alterations similar to those present in untreated CD; however, this definition is rarely fulfilled, only by subjects who underwent two duodenal biopsies, one with normal intestinal architecture and the other with villous atrophy (Troncone, 1995). At that same time, the term potential CD (PCD) was proposed for subjects who showed a normal mucosa with no or mild infiltration of intraepithelial lymphocytes and immunological alterations specific of active CD, such as high TCR-γδ intraepithelial lymphocytes (Savilahti et al., 1990) or celiac-like intestinal antibody pattern (O'Mahony et al., 1991). Subsequently, with the Oslo definitions (Ludvigsson et al., 2013), while the use of the term latent CD was discouraged, the term PCD was attributed to a subset of patients with normal intestinal architecture, with or without high number of intraepithelial lymphocytes, who are at risk of developing CD as indicated by positive celiac serology [increased serum levels of antitissue transglutaminase (anti-TG2) and antiendomysium antibodies (EMA)]. The European Society of Paediatric Gastroenterology Hepatology and Nutrition (ESPGHAN) gave a similar definition in its 2012 guidelines (Husby et al., 2012), adding that PCD patients may present or not gluten-related clinical symptoms, and may develop or not villous atrophy over time. Anti-TG2 IgA antibodies titers are usually lower in these subjects compared to untreated CD (Setty et al., 2015). Interestingly, during follow-up, fluctuating titers or disappearance of antibodies over time are noted (Auricchio et al., 2019). From a genetic standpoint, patients with PCD more frequently show HLA haplotypes related to low or moderate risk of CD (Biagi et al., 2012; Sperandeo et al., 2011) and a different asset of non-HLA polymorphisms (Sperandeo et al., 2011). This clinical, genetic and immunological evidence, altogether suggests that PCD patients show different features compared to active CD, and also among PCD patients, a certain degree of heterogeneity.

Clinical management of PCD is still widely debated since short and long-term risks of this condition have not been sufficiently explored. Is PCD a pathological condition or a predisease state? Do all patients require a GFD regardless of the presence of symptoms? Despite an increasing interest in the scientific community in recent years, many of these issues remain unanswered. Certainly, PCD represents an incomparably useful model for the study of early molecular and immunological mechanisms involved in CD, with relevant information for the implementation of prevention strategies.

In the present chapter we aim at summarizing the most recent evidence about this emerging condition both from a pathogenetic and a clinical prospective.

Pediatric and Adult Celiac Disease. DOI: https://doi.org/10.1016/B978-0-443-13359-6.00009-1

10.2 Pathogenesis

10.2.1 T cells and cytokines

The presence of intestinal gluten-specific CD4 T cells plays a key role in the development of CD (Jabri & Sollid, 2017). Pathogenic gluten-specific T cells, HLA DQ2 and/or HLA DQ8 restricted, have a Th1 phenotype characterized by the production of gamma interferon (IFN-γ) (Nilsen et al., 1998). The pattern of cytokines produced by gluten-specific CD4 T cells includes IFN-γ and IL-21 (Fina et al., 2008; Iervasi et al., 2020), but not IL-17 (Bodd et al., 2010). IL-21 is thought to play an important role in the development of tissue damage and particularly in the crosstalk between the gluten-specific adaptive response in the lamina propria and the intestinal epithelium. Synergistically with IL-15, IL-21 has been shown to increase cytolytic properties in CD8 cytotoxic T cells expressing natural killer like receptors (NKRs) (Ciszewski et al., 2020). IL-15 is a key cytokine in the pathogenesis of CD. It is significantly upregulated in both the small intestinal lamina propria and in the epithelium of CD patients (Maiuri et al., 2000). IL-15 seems to be critical in the development of the gluten-specific Th1 response, but also in enabling effector cytotoxic CD8 T cells to kill stressed tissue. In fact, IL15, promoting the expression of activating NK receptors by intraepithelial lymphocytes, enables them to kill epithelial cells following the recognition of stress signals (Meresse et al., 2004; Meresse et al., 2006). Finally, IL-15 could contribute to the mucosal damage by impairing the suppressive function of regulatory T cell (Tregs) (Benahmed et al., 2007; Zanzi et al., 2011).

In PCD patients a reduced number of inflammatory cells infiltrate lamina propria and the epithelium. However, not only the magnitude, but also the nature of the adaptive gluten-specific T cell response is different in PCD compared to overt CD. As a matter of fact, a different cytokine profile has been identified: an increased density of IL4-producing T cells with a T helper 2 (Th2) phenotype was found in PCD (Vitale et al., 2021), suggesting that Th2 cells could in part contrast the shift toward a IFN-γ producing Th1 response, and eventually contribute to prevention of tissue damage. In line with this concept, IFN-γ transcripts increase in parallel with the degree of mucosal alterations: IFN-γ RNA expression levels are lower in PCD patients with Marsh 0 lesion than in those with Marsh 1 lesion and in the latter significantly lower than in active CD (Abadie & Jabri, 2014; Zanzi et al., 2011). Recent findings indicate that in the process of transition to villous atrophy the almost disappearance of IL4-producing cells is accompanied by an increased expansion of TcR$\gamma\delta$ + intraepithelial lymphocytes (Vitale et al., 2019). Importantly, IL-15, is not hyperexpressed in PCD at the same levels as it is in active CD (Borrelli et al., 2013). Even more strikingly, IL-21 is downregulated in PCD (Iervasi et al., 2020; Sperandeo et al., 2011). As mentioned above, IL-15 and IL-21 synergize to promote cytolytic properties of intraepithelial CD8 T cells responsible for tissue damage (Ciszewski et al., 2020).

10.2.2 T regulatory cells

The possible role of Treg is disputed in CD, including the possibility that in PCD they could contribute to maintain mucosal integrity despite the presence of gluten-specific Th1 cells. The IL-10/IFN-γ ratio was shown to be higher in PCD compared to active CD, and among PCD patients, in Marsh 1 compared to Marsh 0 (Abadie & Jabri, 2014). This suggests that immunoregulatory pathways may act in the intestinal mucosa of PCD patients preventing the progression toward mucosal damage. In line with this hypothesis, both the percentage and the absolute number of Foxp3 + T cells in the lamina propria of PCD patients are higher compared to controls. However, they are lower in PCD compared to active CD patients, implicating that although the immune system attempts to downregulate the ongoing inflammation through a rapid redistribution of Foxp3 + Tregs from the circulation to the tissue, yet, tissue damage occurs when immunoregulatory pathways do not counterbalance inflammation. Importantly, IL-15 seems to play a key role in this process. In fact, while intestinal Tregs from PCD patients are able to suppress T responder cells, regardless of the presence of IL-15 (Borrelli et al., 2013), in patients with villous atrophy IL-15 is able to inhibit T regulatory cells (Meresse et al., 2006).

10.2.3 Mechanisms of mucosal damage

The activation of lamina propria gluten-specific T cell response alone is not sufficient to induce villous atrophy. The effector phase leading to tissue damage requires the activation of cytotoxic T cells that infiltrate and progressively destroy the target organ. Under physiological conditions CD8 + TcRgd cytotoxic T cells express the NKG2A inhibitory receptor and low levels of the activating NKG2D receptor (Jabri et al., 2000; Jabri et al., 2002; Roberts et al., 2001). In active CD, intraepithelial lymphocytes downregulate the expression of the CD94/NKG2A complex and upregulate the activating CD94/NKG2C and NKG2D receptors (Meresse et al., 2004; Meresse et al., 2006). At the same time, intestinal epithelial cells upregulate MIC-A/B and HLA-E (Meresse et al., 2004), ligands for NKG2D and CD94/NKG2C, respectively (Bauer et al., 1999; Braud et al., 1998; Lee et al., 1998). In fact, in overt CD, intraepithelial lymphocytes recognize stress signals on enterocytes, such as MIC-A/B or HLA-E. These molecules are not expressed by healthy small-bowel intestinal epithelial cells, but can be induced in the context of stress, inflammation, and infection. Evidence increasingly suggests that activated CD8 + cytotoxic T cells do not induce tissue damage when they infiltrate a nonstressed/inflamed tissue, even if the appropriate antigen is expressed. So, the event that leads to enterocyte destruction and overt CD is the "licensing" of intraepithelial lymphocytes to kill enterocytes (Fig. 10.1).

In conclusion, PCD patients miss the contextual presence of an antigluten CD4 T cell immunity and an upregulation of NKRs that render CD8 intraepithelial lymphocytes able to kill epithelial cells based on the recognition of stress signals. In line with this, the absence of villous atrophy in PCD is associated with the lack of a fully activated NKR phenotype in CD8 cytotoxic T cells and of increased levels of epithelial stress markers, including IL-15.

10.2.4 Production of antitissue transglutaminase antibodies

The gluten-specific CD 4 T cells is also responsible for the production of the specific autoantibodies found in CD patients sera: antideamidated gliadin antibodies and anti-TG2 antibodies. TG2-specific B cells internalize TG2 complexed with gliadin peptides and present gluten derived peptides to specific T cells that provide help to naïve B cells. This results in the activation of plasma cells producing antideamidated gliadin antibodies and anti-TG2 antibodies

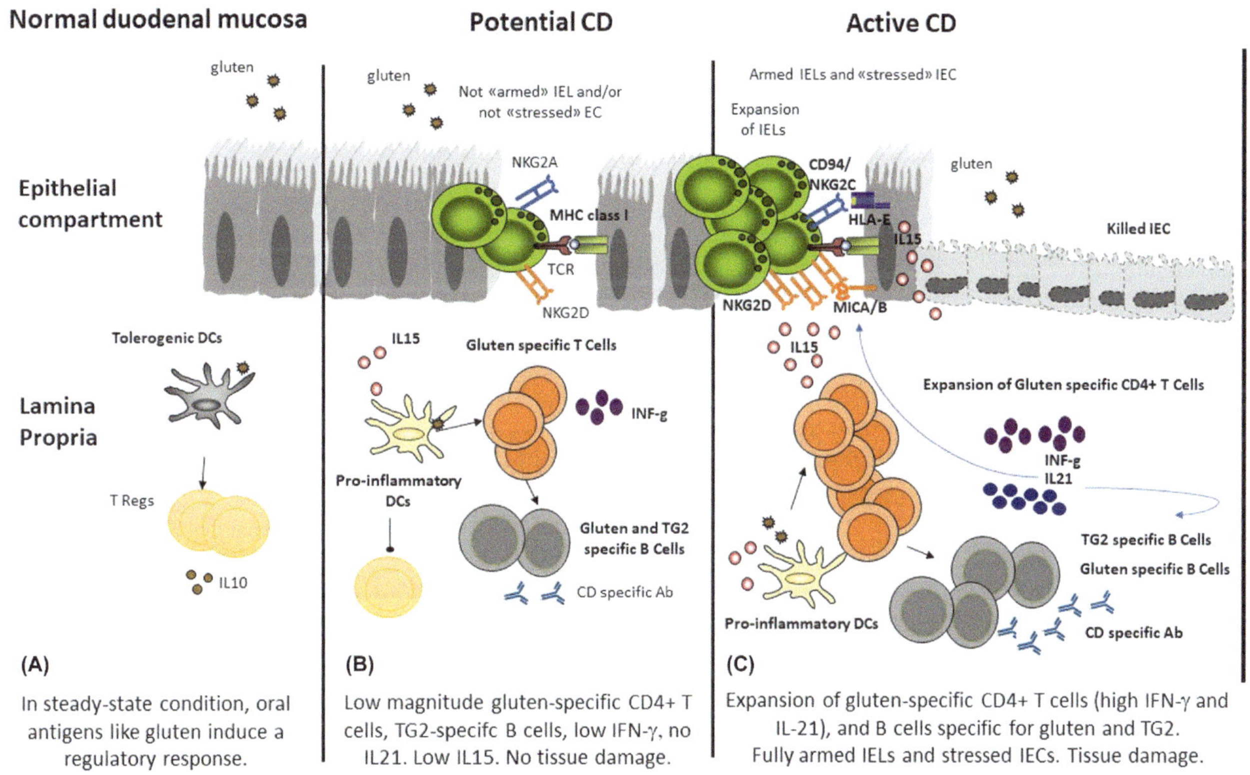

FIGURE 10.1 Schematic representation of the immunopathogenic events leading to villous atrophy. At steady-state, in the absence of a proinflammatory environment in the small intestinal mucosa, gluten prompts a regulatory response as a consequence of its presentation via dendritic cells (DC) with a tolerogenic phenotype (A, *left panel*). In potential celiac disease (CD), a certain degree of IL-15 upregulation in the lamina propria leads to proinflammatory DC, that present deamidated gluten peptides to naïve T cells leading to the rise of a gluten-specific CD4 + T cell response, featuring the secretion of IFN-γ. Those T cells in turns activate a B cell response characterized by the production of antibodies specific to tissue transglutaminase 2 (TG2) and deamidated gliadin peptides. The magnitude of the T cell-INF-γ response is low, and not sufficient to determine a full activation of intraepithelial lymphocytes (IELs). Due to this, and to the lack of epithelial IL-15 and IL-21 in the lamina propria, in potential CD, IELs lack of a fully activated phenotype and thus fail to kill intestinal epithelial cells (IECs). In fact, potential CD patients typically lack tissue damage (B, *middle panel*). In active CD, a higher magnitude of gluten-specific T cell leads to higher IFN-γ and IL-21 levels. High levels of IL-15 are also found both in the epithelium and lamina propria. More cytotoxic IELs are recruited, expressing lower levels of inhibiting natural killer receptors (NKR) such as NKG2A and high levels of activating receptors like NKG2D, that bind stress signal molecules (MICA/B) on IECs, this asset on both IELs and IECs leads to the final development of villous atrophy (C, *right panel*).

(Sollid et al., 1997). Supported by observations in animals (Lindstad et al., 2021), this model explains the gluten dependence of anti-TG2 antibodies production. If this model is true (Lindstad et al., 2021), the presence of anti-TG2 antibodies indicates that gluten specific T cells are present in the gut of PCD patients. The lower levels of IgA antiTG2 antibodies observed in such patients (Setty et al., 2015) likely reflect a smaller amplitude of the CD4 T cell response as compared to active CD. The clinical observation that autoantibodies may disappear in a subset of PCD patients during follow-up (Auricchio et al., 2019) suggests that the loss of tolerance to gluten may still be reversible in some patients. This seems to be the case especially in younger children with PCD followed-up prospectively, a significant part of whom tend to become serologically negative over time (Lionetti et al., 2019; Simell et al., 2007). The comprehension of the mechanisms behind this observation could be particularly relevant for the implementation of preventive strategies.

10.2.5 Epithelium

The epithelial compartment in PCD is not completely normal. The expression of all junction proteins has been found already decreased in early stage CD when compared with nonceliac controls (Rauhavirta et al., 2014), possibly on a constitutive basis. In fact, junction protein expression correlated positively with mucosal villus morphology and negatively with the number of intraepithelial lymphocytes, the intensity of small-intestinal anti-TG2 deposits, and serum anti-TG2 antibodies. These findings show that the mucosal epithelial integrity is disrupted already in early stage CD before the disorder progresses to full-blown enteropathy.

Although the immunological events leading to the villous atrophy have not been fully elucidated, PCD represents an early disease stage and thus a unique opportunity to dissect those events involved in the progression toward disruption of tissue architecture and overt CD.

10.2.6 Evidence from animal models

Animal models are critical to understand the pathogenesis of diseases and to test new therapeutic strategies. The very first experiments conducted to create an animal model of CD showed the lack of intestinal lesions in mice after systemic immunization with gliadin and subsequent gluten feeding (Troncone & Ferguson, 1991). Once the critical role of HLA emerged, mice were engineered to express HLA-DQ8 (Marietta et al., 2004) and subsequently DQ2.5 (De Kauwe et al., 2009) to mimic the antigen presentation occurring in human patients. Transgenic mice expressing HLA-Dr3-DQ2 and human CD4, one with a NOD background and another TCR transgenic having over 90% of CD4(+) T cells specific for the DQ2-alpha-II epitope with a Th1 phenotype, did not show any sign of mucosal damage when consuming gluten. These humanized mouse models indicate that gluten ingestion can be tolerated without intestinal pathology even when HLA-DQ2-restricted CD4(+) T cell immunity to gluten is established, confirming that the presence of an adaptive immune response to gluten is not enough to induce enteropathy.

As mentioned before, the majority of CD patients show a significant upregulation of IL-15 both in the epithelium and in the lamina propria of the duodenal mucosa (Abadie & Jabri, 2014). To test the hypothesis that both an adaptive immune response and innate epithelial stress markers are necessary to induce a full activation of cytotoxic IELs, mice expressing the human HLA-DQ8 were engineered to overexpress IL-15 in both the lamina propria and the intestinal epithelium. Interestingly, neither mice expressing high IL-15 levels in exclusively haematopoietic cells mimicking the lamina propria of CD patients (DQ8-Dd IL-15 transgenic mice), nor those expressing high IL-15 exclusively in the epithelium (DQ8-villin-IL-15 transgenic mice) developed villous atrophy (Abadie et al., 2020). The first model resembles PCD patients, where a Th1 response to gluten is not enough to cause mucosal destruction.

Only triple transgenic mice overexpressing IL-15 in both the epithelial and the lamina propria compartments, developed villous atrophy when fed gluten for several weeks; interestingly, the damage was blocked by anti-TG2 inhibitors and reverted upon gluten withdrawal (Abadie et al., 2020). This model, despite not perfect, significantly helped to dissect the events required for CD development and represents a unique model to test potential therapeutics for CD.

10.2.7 Analogies with type 1 diabetes

The condition featuring a specific CD4 T cell response and circulating autoantibodies without tissue damage is not unique of PCD. In prediabetes, pancreatic beta islets are infiltrated by autoreactive CD4 + T cells and autoantibodies are measurable in the serum even in the absence of beta cells destruction (Ciecko et al., 2023). Indeed, similar pathogenic mechanisms are shared between the two disorders. Like in CD, in NOD model of type 1 diabetes, NKG2D was expressed on autoreactive intrapancreatic CD8 + cells and the use of an anti-NKG2D prevented the progression the overt disease (Trembath et al., 2020). Equally, IL-15 was found to be increased in the beta islet cells of type 1 diabetes, and its inhibition delays the evolution to overt diabetes (Ciecko et al., 2023). Finally, in both diseases, stress molecules like MHC-like molecules are expressed in the target tissue, and its destruction is mediated by CD8 + T cells in an antigen independent manner. Like in PCD, also in prediabetes, IL-21 is downregulated and it becomes overexpressed only when tissue damage occurs (Ciecko et al., 2023). The notion that cytotoxic T cells need to be licensed in tissues to become effective killer cells and that licensing depends on IL-15 is supported by the observation in NOD mice that inhibition of IL-15 signaling at the prediabetic stage delayed diabetes development (Herold et al., 2019).

The presence of a predisease stage offers a unique opportunity to test preventive strategies to block or postpone overt-disease onset. Interestingly, it has been recently demonstrated that a one course administration of teplizumab, an Fc receptor–nonbinding anti-CD3 monoclonal antibody, delays the development from prediabetes to diabetes of about 3 years (Chen et al., 2013). Given the similarities discussed above PCD could be a candidate to such approach.

10.3 Clinical aspects

10.3.1 Diagnosis of potential celiac disease

The diagnosis of PCD is established when a positive CD serology (anti-TG2 and EMA) is associated to a normal duodenal architecture (Marsh 0 or Marsh 1 lesion) (Husby et al., 2020). Some considerations must be made in order to reduce the risk of PCD misdiagnosis.

First, anti-TG2 IgA positivity must be confirmed in a second blood sample, together with EMA. Anti-TG2 IgA measurement is regarded as the most appropriate primary test in the diagnostic work up for CD by the major international diagnostic guidelines from ESPGHAN (Husby et al., 2020), the North American Society for Pediatric Gastroenterology, Hepatology, and Nutrition (NASPGHAN) (Hill et al., 2016), and the American College of Gastroenterology (ACG). Anti-TG2 IgA sensitivity for the diagnosis of CD ranges from 74% to 100%, but their specificity is lower particularly when they are marginally elevated as it is often the case of PCD (Giersiepen et al., 2012). For this reason, before a diagnosis of PCD is established, the presence of EMA antibodies must be confirmed in a second blood sample. In fact EMA show a higher specificity for the diagnosis of CD (Giersiepen et al., 2012; Sheppard et al., 2022). One may fear that, since biopsy is necessary to assess the presence of a normal intestinal architecture and thus diagnose PCD, the "biopsy sparing approach" allowed in children by the latest ESPGHAN guidelines could miss this diagnosis in some cases. This appears to be unlikely as different multicentric studies realized in recent years have demonstrated that for high levels of anti-TG2 (> 10 times the upper limit of normality) the probability to have a concomitant villous atrophy at intestinal biopsy is virtually 100% (Werkstetter et al., 2017; Wolf et al., 2014).

The second necessary contribution is a reliable histological evaluation based on a careful morphometrical analysis and performed by an experienced pathologist. In fact, despite intestinal biopsy remains the gold standard for CD diagnosis, both a "false" villous atrophy or, vice versa, a falsely normal duodenal architecture, can be the consequence of an improper orientation of the biopsy. To make a diagnosis of PCD a normal duodenal architecture must be documented: this is defined by a villous/crypt ratio higher than 2 (Husby et al., 2020).

To overcome pitfalls in the analysis of duodenal biopsies, it has been recently proposed a "molecular morphometry" using PAXgene fixed paraffin-embedded biopsies: the mRNA expression ratio of villous epithelium-specific gene APOA4 to crypt proliferation gene Ki67 showed a similar significant distinction between paired baseline and postgluten challenge biopsies as quantitative histomorphometry (Taavela et al., 2019; Taavela et al., 2021). In the future, the mRNA analyses from duodenal biopsies could serve as a molecular surrogate for the morphometrical villous/crypt ratio, but at present, in clinical settings, the correct histological evaluation of duodenal biopsy samples remains the gold standard for diagnosis.

Finally, before confirming a diagnosis of PCD, it should be made clear that the patient is on a gluten-containing diet.

10.3.2 Epidemiology

Screening strategies of the general population and at risk groups have determined an increase of both CD and PCD diagnoses. In a recent systematic review and metaanalysis, the prevalence of CD based on serology was 1.4% and based on biopsy was 0.7% (Singh et al., 2018). Fewer data are available on PCD prevalence worldwide: globally it has been estimated to be of around 10% of the total CD diagnoses, but the range is quite wide ranging from 3% to 30% (Fig. 10.2). Paparo et al. in 2005 estimated a PCD prevalence of 6.2% in Italian children with CD (Paparo et al., 2005), while Kurppa

	Population	Prevalence (%)	Symptomatic (%)	Response to GFD (%)	Follow-up (years)	Evolution to VA (%)
Volta et., al 2016	Adults	79/735 (10%)	61/79 (79%)	100%	10	1/10 (10%)
Biagi et., al 2013	Adults	47/187 (23%)	23/47 (69%)	100%	7	4/14 (36%)
Imperatore et., al 2017	Adults	56/452 (12%)	43/56 (77%)	p<0,05	6	6/13 (46&)
Kurppa et., al 2009	Adults	23/70 (32%)	20/23 (87%)	100%	1	8/10 (80%)
Newton et., al 2023	Adults	84/2775 (3%)	77/84 (92%)	29/41 (71%)	6	5/15 (30%)
Kondala et., al 2016	Adults and children	n.e.	29/57 (49%)	n.e.	1	5/57 (9%)
Tampowpong et., al 2012	Children	62/320 (19%)	n.e.	n.e.	n.e.	n.e.
Tosco et., al 2011	Children	n.e.	20/106 (19%)	11/20 (55%)	3	10/86 (11%)
Auricchio et., al 2014	Children	n.e.	26/210 (12%)	n.e.	9	23/175 (13%)
Auricchio et., al 2019	Children	n.e.	44/340 (13%)	n.e.	12	42/280 (15%)
Lionetti et., al 2012	Children	24/96 (25%)	1/24 (4%)	n.e.	2	1/21 (5%)
Lionetti et., al 2019	Children	26/553 (5%)	0	n.e.	10	3/23 (13%)
Kurppa et., al 2010	Children	17/59 (29%)	16/17 (94%)	n.e.	2	7/8 (87%)
Mandile et., al 2017	Children	n.e.	44/330 (19%)	19/35 (54%)	3	n.e.

FIGURE 10.2 Summary of the most important evidences about potential celiac disease patients in adults and children; n.e.: not evaluated.

et al. estimated a prevalence up to 28% (Kurppa, Collin, et al., 2010) and 33% (Kurppa et al., 2009), respectively, in Finnish children and adults. Just like CD, PCD is more frequent in first degree relatives of CD patients and in patients with other concomitant autoimmune disorders compared to general population. Although some patients are diagnosed due to clinical symptoms, most of PCD diagnosis are the result of screening of first-degree relatives of CD patients or subjects with associated autoimmune conditions. Interestingly, an increased prevalence of PCD in patients biopsied because of positive CD autoimmunity has been found in children with type 1 diabetes in comparison to nondiabetic subjects (12% vs 7%) (Franzese et al., 2011) suggesting in type 1 diabetes patients with CD autoimmunity a lower intensity of mucosal T cell reaction to gluten, possibly on a genetic basis.

10.3.3 Genetics in potential celiac disease

CD has an important hereditary component: approximately 10% of first-degree relatives are affected by the disease (David & Ajdukiewicz, 1975) with concordance rates of 75%–80% in monozygotic and 10% in dizygotic twins (Greco et al., 2002). The strongest and best characterized genetic susceptibility factors are HLA class II genes known as HLA-DQ2 and HLA-DQ8 (see Chapter 4). As discussed, the presence of DQ2 and/or DQ8 HLA molecules is pathogenetically essential to produce anti TG2 antibodies, thus it is not surprising that PCD patients, just like CD patients, bear one of these molecules. However, patients with PCD more frequently show low-to-moderate HLA-related risk and in particular an increased frequency of DQB1*0302 and a reduced frequency of DQB1*02 homozygosity (Sperandeo et al., 2011). This "gene-dosage" effect could be responsible for a less intense mucosal T cell reaction to gluten.

Also a subset of non HLA genes seems to differ in PCD patients compared to CD. Sperandeo et al. demonstrated that one marker of the KIAA1109/IL-2/IL-21 candidate region differentiated potentials from overt celiac. Overall it seems that at least 6 different non-HLA polymorphisms are differently represented in PCD and CD patients and the presence of some of them can help predicting evolution to villous atrophy in PCD (Sperandeo et al., 2011). The same authors showed that the expression of IL-21 was completely suppressed in potentials compared to overt celiacs and to controls. In contrast IL-2, KIAA1109 and c-REL expression were over-expressed.

10.3.4 Clinical symptoms and response to gluten-free diet

The presence or not of clinical symptoms does not affect the diagnosis of PCD that only requires a positive CD serology with a preserved intestinal architecture, but symptoms are essential to establish the most adequate clinical management. In fact, despite clinical management of asymptomatic patients is still not clear, symptomatic patients with PCD are usually addressed to a GFD trial (Auricchio et al., 2019).

Percentage of symptomatic cases among PCD patients, as well as symptoms described, varies considerably among studies. In general, it seems that in children most of PCD patients are asymptomatic while symptoms increase in adult population (Fig. 10.2). It should be considered however that screening of at risk subjects, such as first degree relatives is most frequently performed by pediatricians and this could partially explain such a discordance. In a cohort of 340 PCD children, only 60 (17%) started a gluten-free diet (GFD) because of the presence of symptoms (Auricchio et al., 2019). Similar proportion were reported in previous studies from the same group ranging from 12% (Auricchio et al., 2014) to

19% (Tosco et al., 2011). Mandile et al. (Mandile et al., 2018) reported 14% of symptomatic PCD patients and the most frequently reported symptoms were failure to thrive, abdominal pain and diarrhea. The CELIPREV study also demonstrated that the majority of PCD children are asymptomatic (Lionetti et al., 2019). However, this result could be underestimated because of the study design: the CELIPREV study in fact investigates the incidence of PCD in a prospective cohort of children with CD-predisposing HLA genes followed-up since birth. Vice versa, the Finnish study group reported that in 96% of the cases PCD patients presented clinical symptoms (mainly abdominal pain and signs of malabsorption) and only in 1% of their cohort patients were asymptomatic and screened because of first-degree family history (Kurppa et al., 2010). More recently the same group also reported that more subtle signs of malabsorption were already present in PCD children. They reported that the development of anemia and iron deficiency in PCD is a continuum, thus suggesting a possible dietary treatment (Repo et al., 2017).

As mentioned before, in adult population, symptoms in PCD seems to be more common and do not differ significantly from those of patients with villous atrophy. Volta et al. (Volta et al., 2016) demonstrated that 79% of PCD patients were symptomatic at diagnosis: among these symptomatic cases, 16% had the classic phenotype characterized by diarrhea and weight loss, whereas the other 51 symptomatic PCD patients showed the occurrence of iron deficiency anemia (more frequently than folic acid deficiency anemia), osteopenia, aphthous stomatitis, irritable bowel syndrome. Similarly, in a study conducted by Biagi et al. (Biagi et al., 2013) 44% presented diarrhea and/or weight loss and 44% anemia or minor symptoms of malabsorption. More recently Newton et al. reported that the vast majority (around 92%) of PCD adult patients were symptomatic at diagnosis. Most of them were screened because of gastrointestinal symptoms while extraintestinal manifestation and hematinic deficiencies were less common in PCD compared to CD patients (Newton et al., 2023).

Volta and Biagi both showed a response rate to a GFD of 100% (Biagi et al., 2013; Volta et al., 2016), while in the study conducted by Newton et al. one third of patients did not improve after a GFD (Newton et al., 2023). In the pediatric study cohort of Mandile et al. the percentage of responding and nonresponding patients to a GFD was similar, even though anti-TG2 serum level always declined upon antigen elimination, confirming the adherence to the diet. In particular, about half of the patients showed a positive clinical response in the first 12 months while the other half showed a partial or absent clinical response. Moreover, not all symptoms showed the same tendency to disappear after GFD. For instance, abdominal pain and diarrhea were more prone to improve than failure to thrive. So, it was clear that not all symptoms were gluten dependent. Authors also analyzed intestinal changes in duodenal biopsies at the time of PCD diagnosis and after at least one year of GFD: the mild intestinal inflammation and infiltration were not significantly affected by GFD. Based on these data, the authors concluded that caution is necessary in pediatric PCD before attributing all abnormalities to gluten, as often symptoms and inflammatory markers of intestinal mucosa do not improve even after a prolonged period of GFD (Mandile et al., 2018). A different experience has been reported in Finland: in a pediatric cohort all the patients put on a GFD resolved their symptoms and, vice-versa, in all symptomatic patients left on a gluten-containing diet (GCD) symptoms persisted (Kurppa, Ashorn, et al., 2010). In adults, Newton et al. investigated more deeply the group of patients with PCD who did not respond to GFD, and the cause of the symptoms other than CD (including inflammatory bowel disease, irritable bowel syndrome, and pancreatic insufficiency) was identified in 50% of the cases (Newton et al., 2023).

Unfortunately, there is a paucity of randomized studies to assess the effect of GFD in PCD. In a study conducted by Kurppa et al. (Kurppa et al., 2009) 23 Marsh I–II patients were randomized either to continue on a GCD diet or start a GFD. After 1 year, clinical, serologic, and histologic evaluations were repeated. A total of 47 participants had small-bowel mucosal lesions compatible with CD (Marsh III), and these served as disease controls. In subjects with Marsh I–II lesion left on GCD the small-bowel mucosal villous architecture deteriorated in all participants, and the symptoms and abnormal antibody titers persisted. In contrast, in the GFD group (Marsh I–II) the symptoms were alleviated, antibody titers decreased, and mucosal inflammation decreased. When the trial was completed, all participants chose to continue a life-long GFD. Quality of life and bone mineral density of subjects with PCD were later assessed (Kurppa, Ashorn, et al., 2010). 27 PCD and 46 CD patients were evaluated at baseline and after 1 year of GFD. The total Gastrointestinal Symptoms Rating Scale (GSRS) score significantly decreased only in CD, but dyspepsia decreased significantly in PCD. Equally the self perceived well being assessed by PWBG total score significantly increased only in CD, while depression significantly improved in PCD patients. Most of PCD patients also had increased bone mineral density after the GFD was commenced. However, this trend did not reach the statistical significance. In conclusion, the authors demonstrated that also asymptomatic adult PCD, who were randomized to a GFD, despite the fact that all subjects prior to the study described themselves as "asymptomatic," improved significantly the GSRS. The authors thus concluded that even apparently asymptomatic patients with positive antibodies benefit from a GFD.

10.3.5 Natural history

Studies conducted in prospective pediatric cohorts have clearly shown that in PCD progression to villous atrophy is not the rule. More than 20 years ago Auricchio et al. started to prospectively follow up PCD children from diagnosis with periodical clinical, serological and histological evaluations. Patients with clinical symptoms were addressed to a GFD trial whereas asymptomatic patients were suggested to continue on a GCD unless different parental choice. The first published study (Tosco et al., 2011) in 2011, highlighted a cumulative incidence of villous atrophy of 30% in three years, and the subsequent two studies on the same cohort reported similar rates (33% at 9 years and 43% at 12 years) (Auricchio et al., 2014, 2019), with most of the "events" (progression to villous atrophy) occurring in the first years of follow up. This study also demonstrated how a relevant proportion of patients, around one third of them, even stopped producing auto-antibodies (anti-TG2 and EMA) in a permanent way, meaning that in some cases the autoimmune process may be reversible. The disappearance of CD autoantibodies was particularly frequent in individuals diagnosed in young age. It is possible that infants, especially those under three years of age, have a more "plastic" immune system, and, in some cases, their oral tolerance to gluten may be regained. Other prospective studies on children showed similar results with even lower progression rates. Results from the CELIPREV study group, designed to prospectively follow up children with CD-predisposing HLA genes since birth, have revealed a progression rate of 13% at 10 years. In this case, up to 83% of the subjects permanently stopped producing antibodies (Lionetti et al., 2019).

When adult patients are monitored on a GCD, progression rates to overt CD are comparable to the pediatric population. Biagi et al. (Biagi et al., 2013) prospectively followed up a small cohort of adult PCD patients for twelve years. Out of 24 patients maintained on a GCD, 14 underwent at least one histological reevaluation that revealed overt CD in 5 of them (time between diagnostic and follow-up biopsy: 12 ± 8 months), while the remaining nine patients still had a normal mucosal architecture after a mean follow-up time of 30 ± 29 months. Volta et al. (Volta et al., 2016) few years later prospectively followed another cohort of adult PCD patients: 16 of the 77 PCD patients (21%) were put on a GCD and remained asymptomatic. All of them were followed up from 1 to 10 years. Only 1 out of the 16 PCD patients became symptomatic after 2-year follow-up and duodenal biopsy confirmed the concomitant development of overt CD. CD positive associated serology disappeared in four patients in the first 6 months. More recently, Newton et al. (Newton et al., 2023) reported that only one third of PCD patients who continue to manifest clinical symptoms on a GCD developed villous atrophy, with a mean progression time of two years. In a Finnish prospective study, both in adult and children, a more consistent incidence of CD in patients previously diagnosed with PCD has been found, respectively of 87% and 96%, prompting the authors to suggest a GFD to all PCD patients (Kurppa et al., 2009; Kurppa, Ashorn, et al., 2010).

10.3.6 Predictive biomarkers

Since evolution to an overt form of CD is anyway uncertain, the appropriateness of a lifelong indication to GFD appears to be questionable, especially where symptoms are absent. It is then important to predict who has more chances to evolve to villous atrophy. Predictive biomarkers associated to the development of villous atrophy have been carefully analyzed in different studies. It has been hypothesized that deposition of anti-TG2 IgA antibodies along the epithelial basement membrane precedes the development of villous atrophy (Borrelli et al., 2018; Koskinen et al., 2010; Maglio & Troncone, 2020), but their role has recently been reconsidered, especially when the impact of anti-TG2 deposits is compared to other risk factors present at the same time. Also, intraepithelial lymphocytosis, the presence of a high risk HLA haplotype, a set of non-HLA genes, and a young age at diagnosis seem to predispose to an increased risk to develop villous atrophy (Auricchio et al., 2019).

A discriminant analysis was conducted to explore the capacity of each variable to discriminate between children who remained potential on follow-up from those who developed flat mucosa. It emerged that IELs with gamma delta TCR were the best discriminators, followed by the age at diagnosis and genetic profile of the individual. For each case, the model obtained by the discriminant equation allowed the computation of an individual discriminant score able to predict the chances of an individual to belong to the group of those who will develop villous atrophy or not. Overall, the discriminant analysis model allows to correctly predict 80% of children (Auricchio et al., 2019).

10.3.7 Risks

There is still a scarce knowledge about possible nutritional and immunological long term risks developing in asymptomatic individuals maintained on a GCD. Imperatore et al. (Imperatore et al., 2017) have demonstrated that patients kept on GCD have a higher risk of developing immune-mediated disorder than those following a GFD (61% vs 18%, $p = 0.03$, OR = 3.3), with no difference between patients with Marsh 0 or Marsh 1 lesion. This result however has never been confirmed in other studies with acceptable sample sizes and the hypothesis that continuous gluten exposure can itself increase the risk of developing other concomitant autoimmune disorders remains an open issue in PCD. Equally, because the lack of long term randomized studies, it is still unknown whether and to what extent subtle signs of malabsorption can appear years after the diagnosis in PCD individual regardless of the presence of clinical symptoms. Interestingly, Dickey et al. (Dickey et al., 2005) measured bone mineral density in 31 PCD patients. They found osteopenia to be present in 30% and osteoporosis in 10% of these patients, and the degree of bone disease did not differ from that found in patients diagnosed with overt CD. No similar studies have been performed in children, where the possible alterations, for example, bone density, are still reversible and malnutrition risk is higher because of the growth process.

10.4 Conclusions

In conclusion, PCD is a very heterogeneous condition. The mechanisms that regulate the eventual progression are not known, nor are environmental factors that may precipitate full blown CD. Risk factors at diagnosis associated to future evolution have been analyzed and include, among others, increased IELs, age at diagnosis, HLA and non-HLA genes. This can guide the clinician in the delicate choice of starting or not a GFD regardless of the presence of symptoms. Even when symptoms are present, however, their relation to gluten must be verified. Further studies are needed to address the issue of whether a continuous gluten exposure can increase long-term immunological and nutrition risks in PCD patients kept on a GCD.

References

Abadie, V., & Jabri, B. (2014). IL-15: A central regulator of celiac disease immunopathology. *Immunological Reviews*, *260*(1), 221–234. Available from https://doi.org/10.1111/imr.12191.

Abadie, V., Kim, S. M., Lejeune, T., Palanski, B. A., Ernest, J. D., Tastet, O., Voisine, J., Discepolo, V., Marietta, E. V., Hawash, M. B. F., Ciszewski, C., Bouziat, R., Panigrahi, K., Horwath, I., Zurenski, M. A., Lawrence, I., Dumaine, A., Yotova, V., Grenier, J. C., ... Jabri, B. (2020). IL-15, gluten and HLA-DQ8 drive tissue destruction in coeliac disease. *Nature*, *578*(7796), 600–604. Available from https://doi.org/10.1038/s41586-020-2003-8.

Auricchio, R., Mandile, R., Del Vecchio, M. R., Scapaticci, S., Galatola, M., Maglio, M., Discepolo, V., Miele, E., Cielo, D., Troncone, R., & Greco, L. (2019). Progression of celiac disease in children with antibodies against tissue transglutaminase and normal duodenal architecture. *Gastroenterology*, *157*(2), 413–420. Available from https://doi.org/10.1053/j.gastro.2019.04.004, e3.

Auricchio, R., Tosco, A., Piccolo, E., Galatola, M., Izzo, V., Maglio, M., Paparo, F., Troncone, R., & Greco, L. (2014). Potential celiac children: 9-year follow-up on a gluten-containing diet. *American Journal of Gastroenterology*, *109*(6), 913–921. Available from https://doi.org/10.1038/ajg.2014.77.

Bauer, S., Groh, V., Wu, J., Steinle, A., Phillips, J. H., Lanier, L. L., & Spies, T. (1999). Activation of NK cells and T cells by NKG2D, a receptor for stress- inducible MICA. *Science (New York, N.Y.)*, *285*(5428), 727–729. Available from https://doi.org/10.1126/science.285.5428.727.

Benahmed, M., Meresse, B., Arnulf, B., Barbe, U., Mention, J. J., Verkarre, V., Allez, M., Cellier, C., Hermine, O., & Cerf-Bensussan, N. (2007). Inhibition of TGF-β Signaling by IL-15: A new role for IL-15 in the loss of immune homeostasis in celiac disease. *Gastroenterology*, *132*(3), 994–1008. Available from https://doi.org/10.1053/j.gastro.2006.12.025.

Biagi, F., Bianchi, P. I., Vattiato, C., Marchese, A., Trotta, L., Badulli, C., De Silvestri, A., Martinetti, M., & Corazza, G. R. (2012). Influence of HLA-DQ2 and DQ8 on severity in celiac disease. *Journal of Clinical Gastroenterology*, *46*(1), 46–50. Available from https://doi.org/10.1097/MCG.0b013e318221077e.

Biagi, F., Trotta, L., Alfano, C., Balduzzi, D., Staffieri, V., Bianchi, P. I., Marchese, A., Vattiato, C., Zilli, A., Luinetti, O., Gobbi, P., & Corazza, G. R. (2013). Prevalence and natural history of potential celiac disease in adult patients. *Scandinavian Journal of Gastroenterology*, *48*(5), 537–542. Available from https://doi.org/10.3109/00365521.2013.777470.

Bodd, M., Ráki, M., Tollefsen, S., Fallang, L. E., Bergseng, E., Lundin, K. E. A., & Sollid, L. M. (2010). HLA-DQ2-restricted gluten-reactive T cells produce IL-21 but not IL-17 or IL-22. *Mucosal Immunology*, *3*(6), 594–601. Available from https://doi.org/10.1038/mi.2010.36.

Borrelli, M., Maglio, M., Korponay-Szabó, I. R., Vass, V., Mearin, M. L., Meijer, C., Niv-Drori, H., Ribes-Koninckx, C., Roca, M., Shamir, R., Troncone, R., & Auricchio, R. (2018). Intestinal anti-transglutaminase 2 immunoglobulin A deposits in children at risk for coeliac disease (CD): Data from the PreventCD study. *Clinical and Experimental Immunology*, *191*(3), 311–317. Available from https://doi.org/10.1111/cei.13078.

Borrelli, M., Salvati, V. M., Maglio, M., Zanzi, D., Ferrara, K., Santagata, S., Ponticelli, D., Aitoro, R., Mazzarella, G., Lania, G., Gianfrani, C., Auricchio, R., & Troncone, R. (2013). Immunoregulatory pathways are active in the small intestinal mucosa of patients with potential celiac disease. *American Journal of Gastroenterology*, *108*(11), 1775–1784. Available from https://doi.org/10.1038/ajg.2013.303.

Braud, V. M., Allan, D. S. J., O'Callaghan, C. A., Soderstrom, K., D'Andrea, A., Ogg, G. S., Lazetic, S., Young, N. T., Bell, J. I., Phillips, J. H., Lanler, L. L., & McMichael, A. J. (1998). HLA-E binds to natural killer cell receptors CD94/NKG2A, B and C. *Nature*, *391*(6669), 795–799. Available from https://doi.org/10.1038/35869.

Chen, J., Feigenbaum, L., Awasthi, P., Butcher, D. O., Anver, M. R., Golubeva, Y. G., Bamford, R., Zhang, X., Claire, M. B., St., Thomas, C. J., Discepolo, V., Jabri, B., & Waldmann, T. A. (2013). Insulin-dependent diabetes induced by pancreatic beta cell expression of IL-15 and IL-15R? *Proceedings of the National Academy of Sciences of the United States of America*, *110*(33), 13534–13539. Available from https://doi.org/10.1073/pnas.1312911110.

Ciecko, A. E., Wang, Y., Harleston, S., Drewek, A., Serreze, D. V., Geurts, A. M., Lin, C. W., & Chen, Y. G. (2023). Heterogeneity of islet-infiltrating IL-21 + CD4 T cells in a mouse model of Type 1 diabetes. *Journal of Immunology*, *210*(7), 935–946. Available from https://doi.org/10.4049/jimmunol.2200712.

Ciszewski, C., Discepolo, V., Pacis, A., Doerr, N., Tastet, O., Mayassi, T., Maglio, M., Basheer, A., Al-Mawsawi, L. Q., Green, P. H. R., Auricchio, R., Troncone, R., Waldmann, T. A., Azimi, N., Tagaya, Y., Barreiro, L. B., & Jabri, B. (2020). Identification of a γc receptor antagonist that prevents reprogramming of human tissue-resident cytotoxic T cells by IL15 and IL21. *Gastroenterology*, *158*(3), 625–637. Available from https://doi.org/10.1053/j.gastro.2019.10.006, e13.

David, T. J., & Ajdukiewicz, A. B. (1975). A family study of coeliac disease. *Journal of Medical Genetics*, *12*(1), 79–82. Available from https://doi.org/10.1136/jmg.12.1.79.

De Kauwe, A. L., Chen, Z., Anderson, R. P., Keech, C. L., Price, J. D., Wijburg, O., Jackson, D. C., Ladhams, J., Allison, J., & McCluskey, J. (2009). Resistance to celiac disease in humanized HLA-DR3-DQ2-transgenic mice expressing specific anti-gliadin CD4 + T cells. *Journal of Immunology*, *182*(12), 7440–7450. Available from https://doi.org/10.4049/jimmunol.0900233.

Dickey, W., Hughes, D. F., & McMillan, S. A. (2005). Patients with serum IgA endomysial antibodies and intact duodenal villi: Clinical characteristics and management options. *Scandinavian Journal of Gastroenterology*, *40*(10), 1240–1243. Available from https://doi.org/10.1080/00365520510023747.

Ferguson, A., Arranz, E., & O'Mahony, S. (1993). Clinical and pathological spectrum of coeliac disease - Active, silent, latent, potential. *Gut*, *34*(2), 150–151. Available from https://doi.org/10.1136/gut.34.2.150.

Fina, D., Sarra, M., Caruso, R., Del Vecchio Blanco, G., Pallone, F., MacDonald, T. T., & Monteleone, G. (2008). Interleukin 21 contributes to the mucosal T helper cell type 1 response in coeliac disease. *Gut*, *57*(7), 887–892. Available from https://doi.org/10.1136/gut.2007.129882.

Franzese, A., Iafusco, D., Spadaro, R., Cavaliere, O., Prisco, F., Auricchio, R., Troncone, R., Valerio, G., Lera, R., Fontana, F., Cherubini, V., Biagioni, M., Pasquino, B., Gallo, F., Suprani, T., Tumini, S., Lazzaro, N., Toni, S., D'Annunzio, G., ... Pinelli, L. (2011). Potential celiac disease in type 1 diabetes: A multicenter study. *Diabetes Research and Clinical Practice*, *92*(1), 53–56. Available from https://doi.org/10.1016/j.diabres.2010.12.028.

Giersiepen, K., Lelgemann, M., Stuhldreher, N., Ronfani, L., Husby, S., Koletzko, S., & Korponay-Szabó, I. R. (2012). Accuracy of diagnostic antibody tests for coeliac disease in children: Summary of an evidence report. *Journal of Pediatric Gastroenterology and Nutrition*, *54*(2), 229–241. Available from https://doi.org/10.1097/MPG.0b013e318216f2e5.

Greco, L., Romino, R., Coto, I., Di Cosmo, N., Percopo, S., Maglio, M., Paparo, F., Gasperi, V., Limongelli, M. G., Cotichini, R., D'Agate, C., Tinto, N., Sacchetti, L., Tosi, R., & Stazi, M. A. (2002). The first large population based twin study of coeliac disease. *Gut*, *50*(5), 624–628. Available from https://doi.org/10.1136/gut.50.5.624.

Herold, K. C., Bundy, B. N., Alice Long, S., Bluestone, J. A., DiMeglio, L. A., Dufort, M. J., Gitelman, S. E., Gottlieb, P. A., Krischer, J. P., Linsley, P. S., Marks, J. B., Moore, W., Moran, A., Rodriguez, H., Russell, W. E., Schatz, D., Skyler, J. S., Tsalikian, E., Wherrett, D. K., ... Greenbaum, C. J. (2019). An anti-CD3 antibody, teplizumab, in relatives at risk for type 1 diabetes. *New England Journal of Medicine*, *381*(7), 603–613. Available from https://doi.org/10.1056/NEJMoa1902226.

Hill, I. D., Fasano, A., Guandalini, S., Hoffenberg, E., Levy, J., Reilly, N., & Verma, R. (2016). NASPGHAN clinical report on the diagnosis and treatment of gluten-related disorders. *Journal of Pediatric Gastroenterology and Nutrition*, *63*(1), 156–165. Available from https://doi.org/10.1097/MPG.0000000000001216.

Husby, S., Koletzko, S., Korponay-Szabó, I. R., Mearin, M. L., Phillips, A., Shamir, R., Troncone, R., Giersiepen, K., Branski, D., Catassi, C., Lelgeman, M., Mäki, M., Ribes-Koninckx, C., Ventura, A., & Zimmer, K. P. (2012). European society for pediatric gastroenterology, hepatology, and nutrition guidelines for the diagnosis of coeliac disease. *Journal of Pediatric Gastroenterology and Nutrition*, *54*(1), 136–160. Available from https://doi.org/10.1097/MPG.0b013e31821a23d0.

Husby, S., Koletzko, S., Korponay-Szabó, I., Kurppa, K., Mearin, M. L., Ribes-Koninckx, C., Shamir, R., Troncone, R., Auricchio, R., Castillejo, G., Christensen, R., Dolinsek, J., Gillett, P., Hróbjartsson, A., Koltai, T., Maki, M., Nielsen, S. M., Popp, A., Størdal, K., ... Wessels, M. (2020). European Society Paediatric Gastroenterology, Hepatology and Nutrition Guidelines for Diagnosing Coeliac Disease 2020. *Journal of Pediatric Gastroenterology and Nutrition*, *70*(1), 141–156. Available from https://doi.org/10.1097/MPG.0000000000002497.

Iervasi, E., Auricchio, R., Strangio, A., Greco, L., & Saverino, D. (2020). Serum IL-21 levels from celiac disease patients correlates with anti-tTG IgA autoantibodies and mucosal damage. *Autoimmunity*, *53*(4), 225–230. Available from https://doi.org/10.1080/08916934.2020.1736047.

Imperatore, N., Tortora, R., De Palma, G. D., Capone, P., Gerbino, N., Donetto, S., Testa, A., Caporaso, N., & Rispo, A. (2017). Beneficial effects of gluten free diet in potential coeliac disease in adult population. *Digestive and Liver Disease*, *49*(8), 878–882. Available from https://doi.org/10.1016/j.dld.2017.03.009.

Jabri, B., & Sollid, L. M. (2017). T cells in celiac disease. *Journal of Immunology*, *198*(8), 3005–3014. Available from https://doi.org/10.4049/jimmunol.1601693.

Jabri, B., De Serre, N. P. M., Cellier, C., Evans, K., Gache, C., Carvalho, C., Mougenot, J. F., Allez, M., Jian, R., Desreumaux, P., Colombel, J. F., Matuchansky, C., Cugnenc, H., LopezBotet, M., Vivier, E., Moretta, A., Roberts, A. I., Ebert, E. C., GuyGrand, D., . . . CerfBensussan, N. (2000). Selective expansion of intraepithelial lymphocytes expressing the HLA-E- specific natural killer receptor CD94 in celiac disease. *Gastroenterology*, *118*(5), 867–879. Available from https://doi.org/10.1016/S0016-5085(00)70173-9.

Jabri, B., Selby, J. M., Negulescu, H., Lee, L., Roberts, A. I., Beavis, A., Lopez-Botet, M., Ebert, E. C., & Winchester, R. J. (2002). TCR specificity dictates CD94/NKG2A expression by human CTL. *Immunity*, *17*(4), 487–499. Available from https://doi.org/10.1016/S1074-7613(02)00427-2.

Koskinen, O., Collin, P., Lindfors, K., Laurila, K., Mäki, M., & Kaukinen, K. (2010). Usefulness of small-bowel mucosal transglutaminase-2 specific autoantibody deposits in the diagnosis and follow-up of celiac disease. *Journal of Clinical Gastroenterology*, *44*(7), 483–488. Available from https://doi.org/10.1097/MCG.0b013e3181b64557.

Kurppa, K., Ashorn, M., Iltanen, S., Koskinen, L. L. E., Saavalainen, P., Koskinen, O., Mäki, M., & Kaukinen, K. (2010). Celiac disease without villous atrophy in children: A prospective study. *Journal of Pediatrics*, *157*(3), 373–e1. Available from https://doi.org/10.1016/j.jpeds.2010.02.070.

Kurppa, K., Collin, P., Sievänen, H., Huhtala, H., Mäki, M., & Kaukinen, K. (2010). Gastrointestinal symptoms, quality of life and bone mineral density in mild enteropathic coeliac disease: A prospective clinical trial. *Scandinavian Journal of Gastroenterology*, *45*(3), 305–314. Available from https://doi.org/10.3109/00365520903555879.

Kurppa, K., Collin, P., Viljamaa, M., Haimila, K., Saavalainen, P., Partanen, J., Laurila, K., Huhtala, H., Paasikivi, K., Mäki, M., & Kaukinen, K. (2009). Diagnosing mild enteropathy celiac disease: A randomized, controlled clinical study. *Gastroenterology*, *136*(3), 816–823. Available from https://doi.org/10.1053/j.gastro.2008.11.040.

Lee, N., Llano, M., Carretero, M., Ishitani, A., Navarro, F., López-Botet, M., & Geraghty, D. E. (1998). HLA-E is a major ligand for the natural killer inhibitory receptor CD94/NKG2A. *Proceedings of the National Academy of Sciences*, *95*(9), 5199–5204. Available from https://doi.org/10.1073/pnas.95.9.5199.

Lindstad, C. B., Dewan, A. E., Stamnaes, J., Sollid, L. M., du Pré, M. F., & Kanellopoulos-Langevin, C. (2021). TG2-gluten complexes as antigens for gluten-specific and transglutaminase-2 specific B cells in celiac disease. *PLoS One*, *16*(11), e0259082. Available from https://doi.org/10.1371/journal.pone.0259082.

Lionetti, E., Castellaneta, S., Francavilla, R., Pulvirenti, A., Naspi Catassi, G., & Catassi, C. (2019). Long-term outcome of potential celiac disease in genetically at-risk children: The prospective CELIPREV cohort study. *Journal of Clinical Medicine*, *8*(2), 186. Available from https://doi.org/10.3390/jcm8020186.

Ludvigsson, J. F., Leffler, D. A., Bai, J. C., Biagi, F., Fasano, A., Green, P. H. R., Hadjivassiliou, M., Kaukinen, K., Kelly, C. P., Leonard, J. N., Lundin, K. E. A., Murray, J. A., Sanders, D. S., Walker, M. M., Zingone, F., & Ciacci, C. (2013). The Oslo definitions for coeliac disease and related terms. *Gut*, *62*(1), 43–52. Available from https://doi.org/10.1136/gutjnl-2011-301346.

Maglio, M., & Troncone, R. (2020). Intestinal anti-tissue transglutaminase2 autoantibodies: Pathogenic and clinical implications for celiac disease. *Frontiers in Nutrition*, *7*. Available from https://doi.org/10.3389/fnut.2020.00073.

Maiuri, L., Ciacci, C., Auricchio, S., Brown, V., Quaratino, S., & Londei, M. (2000). Interleukin 15 mediates epithelial changes in celiac disease. *Gastroenterology*, *119*(4), 996–1006. Available from https://doi.org/10.1053/gast.2000.18149.

Mandile, R., Discepolo, V., Scapaticci, S., Del Vecchio, M. R., Maglio, M. A., Greco, L., Troncone, R., & Auricchio, R. (2018). The effect of gluten-free diet on clinical symptoms and the intestinal mucosa of patients with potential celiac disease. *Journal of Pediatric Gastroenterology and Nutrition*, *66*(4), 654–656. Available from https://doi.org/10.1097/MPG.0000000000001745.

Marietta, E., Black, K., Camilleri, M., Krause, P., Rogers, R. S., David, C., Pittelkow, M. R., & Murray, J. A. (2004). A new model for dermatitis herpetiformis that uses HLA-DQ8 transgenic NOD mice. *Journal of Clinical Investigation*, *114*(8), 1090–1097. Available from https://doi.org/10.1172/JCI200421055.

Marsh, M. N. (1988). Studies of intestinal lymphoid tissue. XI - The immunopathology of cell-mediated reactions in gluten sensitivity and other enteropathies. *Scanning Microscopy*, *2*(3), 1663–1684.

Meresse, B., Chen, Z., Ciszewski, C., Tretiakova, M., Bhagat, G., Krausz, T. N., Raulet, D. H., Lanier, L. L., Groh, V., Spies, T., Ebert, E. C., Green, P. H., & Jabri, B. (2004). Coordinated induction by IL15 of a TCR-independent NKG2D signaling pathway converts CTL into lymphokine-activated killer cells in celiac disease. *Immunity*, *21*(3), 357–366. Available from https://doi.org/10.1016/j.immuni.2004.06.020.

Meresse, B., Curran, S. A., Ciszewski, C., Orbelyan, G., Setty, M., Bhagat, G., Lee, L., Tretiakova, M., Semrad, C., Kistner, E., Winchester, R. J., Braud, V., Lanier, L. L., Geraghty, D. E., Green, P. H., Guandalini, S., & Jabri, B. (2006). Reprogramming of CTLs into natural killer-like cells in celiac disease. *Journal of Experimental Medicine*, *203*(5), 1343–1355. Available from https://doi.org/10.1084/jem.20060028.

Newton, M., Greenaway, E. A., Holland, W. J., Raju, S. A., Rej, A., & Sanders, D. S. (2023). What are the clinical consequences of 'potential' coeliac disease? *Digestive and Liver Disease*, *55*(4), 478–484. Available from https://doi.org/10.1016/j.dld.2022.10.019.

Nilsen, E. M., Jahnsen, F. L., Lundin, K. E. A., Johansen, F. E., Fausa, O., Sollid, L. M., Jahnsen, J., Scott, H., & Brandtzaeg, P. (1998). Gluten induces an intestinal cytokine response strongly dominated by interferon gamma in patients with celiac disease. *Gastroenterology*, *115*(3), 551–563. Available from https://doi.org/10.1016/S0016-5085(98)70134-9.

O'Mahony, S., Arranz, E., Barton, J. R., & Ferguson, A. (1991). Dissociation between systemic and mucosal humoral immune responses in coeliac disease. *Gut*, *32*(1), 29–35. Available from https://doi.org/10.1136/gut.32.1.29.

Paparo, F., Petrone, E., Tosco, A., Maglio, M., Borrelli, M., Salvati, V. M., Miele, E., Greco, L., Auricchio, S., & Troncone, R. (2005). Clinical, HLA, and small bowel immunohistochemical features of children with positive serum antiendomysium antibodies and architecturally normal small intestinal mucosa. *American Journal of Gastroenterology*, *100*(10), 2294–2298. Available from https://doi.org/10.1111/j.1572-0241.2005.41134.x.

Rauhavirta, T., Lindfors, K., Koskinen, O., Laurila, K., Kurppa, K., Saavalainen, P., Mäki, M., Collin, P., & Kaukinen, K. (2014). Impaired epithelial integrity in the duodenal mucosa in early stages of celiac disease. *Translational Research*, *164*(3), 223–231. Available from https://doi.org/10.1016/j.trsl.2014.02.006.

Repo, M., Lindfors, K., Mäki, M., Huhtala, H., Laurila, K., Lähdeaho, M. L., Saavalainen, P., Kaukinen, K., & Kurppa, K. (2017). Anemia and iron deficiency in children with potential celiac disease. *Journal of Pediatric Gastroenterology and Nutrition*, *64*(1), 56–62. Available from https://doi.org/10.1097/MPG.0000000000001234.

Roberts, A. I., Lee, L., Schwarz, E., Groh, V., Spies, T., Ebert, E. C., & Jabri, B. (2001). Cutting edge: NKG2D receptors induced by IL-15 costimulate CD28-negative effector CTL in the tissue microenvironment. *Journal of Immunology*, *167*(10), 5527–5530. Available from https://doi.org/10.4049/jimmunol.167.10.5527.

Savilahti, E., Arato, A., & Verkasalo, M. (1990). Intestinal γ/δ receptor-bearing T lymphocytes in celiac disease and inflammatory bowel diseases in children. *Constant increase in celiac disease. Pediatric Research*, *28*(6), 579–581. Available from https://doi.org/10.1203/00006450-199012000-00005.

Setty, M., Discepolo, V., Abadie, V., Kamhawi, S., Mayassi, T., Kent, A., Ciszewski, C., Maglio, M., Kistner, E., Bhagat, G., Semrad, C., Kupfer, S. S., Green, P. H., Guandalini, S., Troncone, R., Murray, J. A., Turner, J. R., & Jabri, B. (2015). Distinct and synergistic contributions of epithelial stress and adaptive immunity to functions of intraepithelial killer cells and active celiac disease. *Gastroenterology*, *149*(3), 681–691. Available from https://doi.org/10.1053/j.gastro.2015.05.013, e10.

Sheppard, A. L., Elwenspoek, M. M. C., Scott, L. J., Corfield, V., Everitt, H., Gillett, P. M., Hay, A. D., Jones, H. E., Mallett, S., Watson, J., & Whiting, P. F. (2022). Systematic review with meta-analysis: The accuracy of serological tests to support the diagnosis of coeliac disease. *Alimentary Pharmacology and Therapeutics*, *55*(5), 514–527. Available from https://doi.org/10.1111/apt.16729.

Simell, S., Hoppu, S., Hekkala, A., Simell, T., Ståhlberg, M. R., Viander, M., Yrjänäinen, H., Grönlund, J., Markula, P., Simell, V., Knip, M., Ilonen, J., Hyöty, H., & Simell, O. (2007). Fate of five celiac disease-associated antibodies during normal diet in genetically at-risk children observed from birth in a natural history study. *American Journal of Gastroenterology*, *102*(9), 2026–2035. Available from https://doi.org/10.1111/j.1572-0241.2007.01360.x.

Singh, P., Arora, A., Strand, T. A., Leffler, D. A., Catassi, C., Green, P. H., Kelly, C. P., Ahuja, V., & Makharia, G. K. (2018). Global prevalence of celiac disease: systematic review and meta-analysis. *Clinical Gastroenterology and Hepatology*, *16*(6), 823–836. Available from https://doi.org/10.1016/j.cgh.2017.06.037, e2.

Sollid, L. M., Molberg., Mcadam, S., & Lundin, K. E. A. (1997). Autoantibodies in coeliac disease: Tissue transglutaminase guilt by association? *Gut*, *41*(6), 851–852. Available from https://doi.org/10.1136/gut.41.6.851.

Sperandeo, M. P., Tosco, A., Izzo, V., Tucci, F., Troncone, R., Auricchio, R., Romanos, J., Trynka, G., Auricchio, S., Jabri, B., Greco, L., & Schönbach, C. (2011). Potential celiac patients: A model of celiac disease pathogenesis. *PLoS One*, *6*(7)e21281. Available from https://doi.org/10.1371/journal.pone.0021281.

Taavela, J., Viiri, K., Popp, A., Oittinen, M., Dotsenko, V., Peräaho, M., Staff, S., Sarin, J., Leon, F., Mäki, M., & Isola, J. (2019). Histological, immunohistochemical and mRNA gene expression responses in coeliac disease patients challenged with gluten using PAXgene fixed paraffin-embedded duodenal biopsies. *BMC Gastroenterology*, *19*(1). Available from https://doi.org/10.1186/s12876-019-1089-7.

Taavela, J., Viiri, K., Välimäki, A., Sarin, J., Salonoja, K., Mäki, M., & Isola, J. (2021). Apolipoprotein A4 defines the villus-crypt border in duodenal specimens for celiac disease morphometry. *Frontiers in Immunology*, *12*. Available from https://doi.org/10.3389/fimmu.2021.713854.

Tosco, A., Salvati, V. M., Auricchio, R., Maglio, M., Borrelli, M., Coruzzo, A., Paparo, F., Boffardi, M., Esposito, A., D'Adamo, G., Malamisura, B., Greco, L., & Troncone, R. (2011). Natural history of potential celiac disease in children. *Clinical Gastroenterology and Hepatology*, *9*(4), 320–325. Available from https://doi.org/10.1016/j.cgh.2010.09.006.

Trembath, A. P., Krausz, K. L., Sharma, N., Gerling, I. C., Mathews, C. E., & Markiewicz, M. A. (2020). Nkg2d signaling within the pancreatic islets reduces nod diabetes and increases protective central memory cd8 + t-cell numbers. *Diabetes*, *69*(8), 1749–1762. Available from https://doi.org/10.2337/db19-0979.

Troncone, R. (1995). Latent coeliac disease in Italy. *Acta Pædiatrica*, *84*(11), 1252–1257. Available from https://doi.org/10.1111/j.1651-2227.1995.tb13543.x.

Troncone, R., & Ferguson, A. (1991). Animal model of gluten induced enteropathy in mice. *Gut*, *32*(8), 871–875. Available from https://doi.org/10.1136/gut.32.8.871.

Vitale, S., Maglio, M., Picascia, S., Mottola, I., Miele, E., Troncone, R., Auricchio, R., & Gianfrani, C. (2021). Intestinal cellular biomarkers of mucosal lesion progression in pediatric celiac disease. *Pharmaceutics*, *13*(11), 1971. Available from https://doi.org/10.3390/pharmaceutics13111971.

Vitale, S., Santarlasci, V., Camarca, A., Picascia, S., Pasquale, A. D., Maglio, M., Maggi, E., Cosmi, L., Annunziato, F., Troncone, R., Auricchio, R., & Gianfrani, C. (2019). The intestinal expansion of TCRγδ + and disappearance of IL4 + T cells suggest their involvement in the evolution from potential to overt celiac disease. *European Journal of Immunology*, *49*(12), 2222–2234. Available from https://doi.org/10.1002/eji.201948098.

Volta, U., Caio, G., Giancola, F., Rhoden, K. J., Ruggeri, E., Boschetti, E., Stanghellini, V., & De Giorgio, R. (2016). Features and progression of potential celiac disease in adults. *Clinical Gastroenterology and Hepatology*, *14*(5), 686–693. Available from https://doi.org/10.1016/j.cgh.2015.10.024, e1.

Werkstetter, K. J., Korponay-Szabó, I. R., Popp, A., Villanacci, V., Salemme, M., Heilig, G., Lillevang, S. T., Mearin, M. L., Ribes-Koninckx, C., Thomas, A., Troncone, R., Filipiak, B., Mäki, M., Gyimesi, J., Najafi, M., Dolinšek, J., Dydensborg Sander, S., Auricchio, R., Papadopoulou, A.,

... Eftekhar Sadat, A. T. (2017). Accuracy in diagnosis of celiac disease without biopsies in clinical practice. *Gastroenterology*, *153*(4), 924–935. Available from https://doi.org/10.1053/j.gastro.2017.06.002.

Wolf, J., Hasenclever, D., Petroff, D., Richter, T., Uhlig, H. H., Laaß, M. W., Hauer, A., Stern, M., Bossuyt, X., De Laffolie, J., Flemming, G., Villalta, D., Schlumberger, W., & Mothes, T. (2014). Antibodies in the diagnosis of coeliac disease: A biopsy-controlled, international, multicentre study of 376 children with coeliac disease and 695 controls. *PLoS One*, *9*(5). Available from https://doi.org/10.1371/journal.pone.0097853.

Zanzi, D., Stefanile, R., Santagata, S., Iaffaldano, L., Iaquinto, G., Giardullo, N., Lania, G., Vigliano, I., Vera, A. R., Ferrara, K., Auricchio, S., Troncone, R., & Mazzarella, G. (2011). IL-15 interferes with suppressive activity of intestinal regulatory T cells expanded in celiac disease. *American Journal of Gastroenterology*, *106*(7), 1308–1317. Available from https://doi.org/10.1038/ajg.2011.80.

Chapter 11

Chronic enteropathies with negative celiac serology in adults

Annalisa Schiepatti, Stiliano Maimaris and Federico Biagi
Department of Internal Medicine and Medical Therapeutics, University of Pavia; Gastroenterology Unit, IRCCS Istituti Clinici Scientifici Maugeri, Pavia, Italy

11.1 Introduction

Villous atrophy (VA) of the small bowel is a histopathological feature found in various chronic enteropathies. These conditions, with diverse underlying causes, present clinically as malabsorption syndromes of differing severity. In most cases, VA is a result of conventional celiac disease (CD), a gluten-dependent chronic enteropathy characterized by a high prevalence and increased mortality compared to the general population (Biagi & Corazza, 2010; Jansson-Knodell et al., 2020; Lebwohl et al., 2018; Lebwohl et al., 2020; Leonard et al., 2021; Schiepatti et al., 2019; Schiepatti et al., 2021; Schiepatti et al., 2022). CD diagnosis in adults relies on positive celiac-specific serology, including IgA tissue transglutaminase antibodies (TTA), endomysial antibodies (EmA), and deamidated gliadin peptide antibodies (DGP), together with VA responsive to a gluten-free diet (GFD) (Lebwohl et al., 2018; Ludvigsson et al., 2014). However, recently pediatric guidelines allow for a biopsy-sparing diagnostic strategy in select children (Husby et al., 2020) and this possibility is also being investigated for adults (Penny et al., 2021). While CD diagnosis is generally straightforward, challenges may arise when patients exhibit VA and negative celiac serology (Jansson-Knodell et al., 2020; Leonard et al., 2021; Schiepatti et al., 2019; Schiepatti et al., 2021). According to a recent international expert consensus, clinical scenarios characterized by VA and negative celiac serology—also called seronegative enteropathies (SNEs)—can be divided into two categories: (1) forms of CD with negative celiac antibodies, including true seronegative CD and CD with IgA deficiency, and (2) chronic enteropathies unrelated to CD and gluten intake, defined as non-celiac enteropathies (NCEs) (Schiepatti et al., 2022). Even though the introduction of international consensus on the terminology and diagnostic criteria for seronegative CD (SNCD) and NCEs in adults represents progress towards a more systematic approach in diagnosing and treating these conditions, distinguishing between SNCD and NCEs remains a challenge (Jansson-Knodell et al., 2020; Leonard et al., 2021; Schiepatti et al., 2019; Schiepatti et al., 2021). Patients with NCEs are often mistakenly diagnosed as having seronegative CD (Leonard et al., 2021; Pallav et al., 2012; Schiepatti et al., 2019; Schiepatti et al., 2021; Schiepatti et al., 2021; Schiepatti et al., 2020), and in one out of four patients with an initial SNCD diagnosis, the original diagnosis was not confirmed, as recently reported (Schiepatti et al., 2021). The rarity of SNEs, overlapping clinical and histopathological traits, lack of biomarkers for some conditions, and methodological errors in obtaining and processing duodenal samples—such as incorrect orientation and insufficient sampling of duodenal lesions—contribute to the difficulty in differentiating SNEs (Jansson-Knodell et al., 2020; Leonard et al., 2021; Schiepatti et al., 2019; Schiepatti et al., 2021; Schiepatti et al., 2021; Schiepatti et al., 2020; Taavela et al., 2013).

Although SNCD is the most common cause of VA with negative celiac serology in adults (Aziz et al., 2017; Degaetani et al., 2013; Fernández-Bañares et al., 2020; Jansson-Knodell et al., 2020; Leonard et al., 2021; Schiepatti et al., 2017, 2019; Schiepatti et al., 2021; Volta et al., 2016), distinguishing between seronegative CD and NCEs is crucial to minimize diagnostic delays and prevent unwarranted lifelong GFD. Properly characterizing NCEs also ensures appropriate treatment for conditions accompanied by long-lasting morbidity and mortality (Aziz et al., 2017; Degaetani et al., 2013; Pensieri et al., 2019; Schiepatti et al., 2021; Schiepatti et al., 2020; Volta et al., 2016).

Pediatric and Adult Celiac Disease. DOI: https://doi.org/10.1016/B978-0-443-13359-6.00018-2

In this chapter, we provide an updated summary of the current literature and a clinically-oriented perspective on the management and differential diagnosis of SNEs in adults.

11.2 Epidemiology of seronegative enteropathies

Few data on the epidemiology of SNEs are available (Schiepatti et al., 2021). Overall, SNEs are rare conditions, although figures vary according to the type of population under investigation (general population vs patients with VA versus patients with VA and negative celiac serology).

11.2.1 Prevalence of seronegative enteropathies in the general population

Firstly, while the prevalence of conventional CD in the general population is usually reported around 1% (Biagi et al., 2010; Lebwohl et al., 2018; Ludvigsson et al., 2014; Singh et al., 2018), the true prevalence of SNEs (considered as all the patients with VA and negative celiac serology, regardless of their underlying etiology) in the general population is currently unknown (Schiepatti et al., 2021). Estimations can only be made regarding the prevalence of various forms of SNEs. For instance, olmesartan-associated enteropathy (OAE), the most common type of drug-induced SNEs, is thought to occur in 1/4000 to 1/5000 cases among olmesartan users (Marietta et al., 2015; Schiepatti et al., 2019). Autoimmune enteropathy has an estimated incidence of 1/million person-years (Akram et al., 2007; Corazza et al., 1997; Masia et al., 2014; Unsworth & Walker-Smith, 1985). Data on the prevalence of gastrointestinal involvement in patients with common variable immunodeficiency (CVID) is even more scarce. Gastrointestinal involvement in CVID patients is undoubtedly a significant clinical issue, as persistent diarrhea has been reported in 20%–60% of CVID patients (Hermaszewski & Webster, 1993; Jørgensen et al., 2016; Kalha & Sellin, 2004; Khodadad et al., 2007; Park et al., 2008). Nonetheless, the true prevalence of VA in CVID patients remains unknown (Schiepatti et al., 2019). A study conducted in Northern Italy estimated the prevalence of Whipple's disease in the general population to be around three cases per million people (Biagi et al., 2015). As for SNCD, there is currently no available data on its prevalence in the general population (Schiepatti et al., 2021).

11.2.2 Prevalence of seronegative enteropathies among all patients with villous atrophy

When all patients affected by VA are considered, the prevalence of SNEs overall (i.e., all patients with VA and negative celiac serology, irrespective of their etiology) has been determined to be 5.1%, based on data from an Italian referral center over a 15-year-period (Schiepatti et al., 2017).

It is important to note the studies examining the prevalence of SNCD among all celiac patients. Initially, research on SNCD indicated a prevalence of 10%–20% among all celiac patients (Abrams et al., 2004; Dickey et al., 2000; Rostami et al., 1998, 1999; Salmi et al., 2006; Schiepatti et al., 2018), while more recently prevalence figures between 2%–6.5% have been reported (Collin et al., 2005; Schiepatti et al., 2017; Schiepatti et al., 2021; Volta, Caio et al., 2016). A recent international multicenter study on SNCD has shown that 4.5% of all celiac patients directly diagnosed over a 20 year-period in the participating centers presented with negative celiac serology at baseline. More precisely, the prevalence of "true" SNCD was 2.7% (95% CI 2.06%–3.47%) and the prevalence of CD associated with IgA deficiency was 1.8% (95% CI 1.27%–2.44%) (Schiepatti et al., 2021). This study has also reported that there were no statistically significant differences among the three centers, and that the prevalence of SNCD among celiac patients remained stable over two decades (2000–20), therefore confirming that the difference between the former and the latter prevalence figures in the literature data are likely to be due to different diagnostic criteria adopted rather than a true change in the prevalence of the disease over time. Although, interestingly, the prevalence figures provided by this paper (Schiepatti et al., 2021) reflect the sensitivity of contemporary celiac-specific serology (Leffler & Schuppan, 2010; Leonard et al., 2021; Schiepatti et al., 2019; Schiepatti et al., 2021), it is undoubtedly difficult to ascertain the true prevalence of SNCD in a population-based setting. Only two papers provided data on the frequency of SNCD in a more general setting, which was that of patients referred to endoscopy (Hopper et al., 2007; Mooney et al., 2015). Of course, also in this scenario prevalence figures are likely to overestimate the real prevalence of SNCD in the general population.

11.2.3 Prevalence of different etiologies among all patients with seronegative enteropathies

Considering the cases of seronegative enteropathies (SNEs) documented in the literature (Aziz et al., 2017; Degaetani et al., 2013; Schiepatti et al., 2017; Volta et al., 2016), it becomes apparent that seronegative celiac disease (SNCD) is

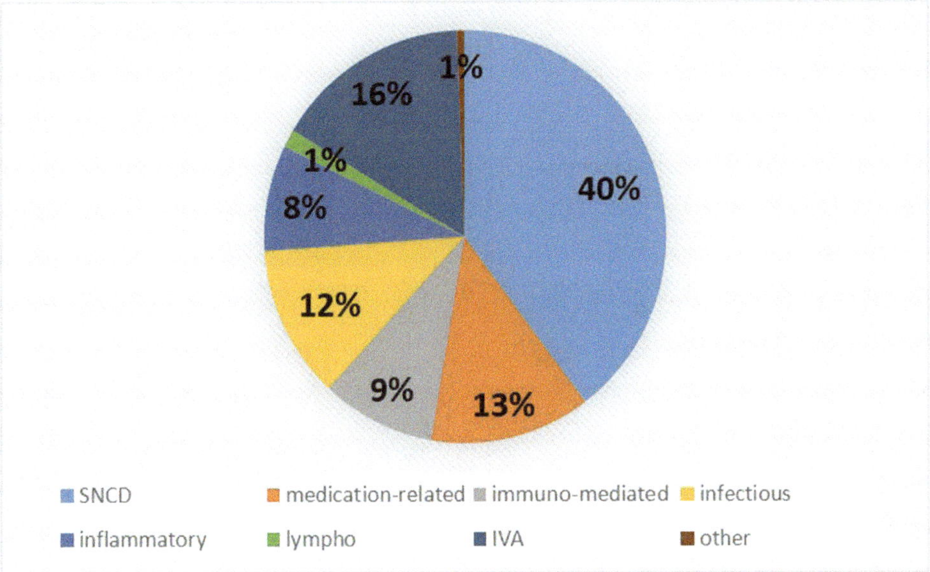

FIGURE 11.1 Percentages of seronegative enteropathies based on 349 cases published by five major international centers (Aziz et al., 2017; Degaetani et al., 2013; Schiepatti et al., 2017; Volta et al., 2016). *SNCD*, seronegative celiac disease; *IVA*, idiopathic villous atrophy; lympho: lymphomas.

the leading cause of SNEs in adults, followed by iatrogenic and infectious origins. Notably, up to 14% of adult patients do not have a definitive cause identified, resulting in a final diagnosis of idiopathic villous atrophy/undefined sprue (Aziz et al., 2017; Degaetani et al., 2013; Schiepatti et al., 2017; Volta et al., 2016). This is illustrated in Fig. 11.1 (Aziz et al., 2017; Degaetani et al., 2013; Schiepatti et al., 2017; Volta et al., 2016).

While SNEs are also reported in children, the primary causes in this group are inflammatory, infectious, and immune-related (Gustafsson et al., 2020; Mandile et al., 2021). Conversely, SNCD and iatrogenic reasons are virtually nonexistent in these cases.

11.3 Pathogenesis

The topic of the immunopathogenesis of SNEs is a very complex one to deal with, and it is still unexplored for the vast majority of these conditions. In this regard, a few considerations are necessary. First, given the clinical heterogeneity of SNEs, it is likely that different underlying and still obscure pathogenetic mechanisms are in place. Therefore to provide a global overview on the immunopathogenetic mechanisms of all the SNEs is hardly feasible. A description of the molecular mechanisms involved in each type of NCEs would be preferable, but this is definitely limited by the lack of studies which has specifically investigated these aspects so far. Hypothetically, as it happens in celiac disease, where VA of the small bowel is the final result of a complex interplay between gluten, the trigger agent, environmental factors and predisposing factors of the host (Di Sabatino & Corazza, 2009), also in NCEs a similar pathogenetic model may be applicable.

For some types of NCEs the etiological agent is known, as is the case for giardiasis (Levinson & Nastro, 1978) and iatrogenic enteropathies, particularly those due to olmesartan and other angiotensin II receptor blockers (Rubio-Tapia et al., 2012) and the more recently described forms of VA due to immune-checkpoint inhibitors (Schiepatti et al., 2022). However, also in these forms of SNEs in which the trigger agent has been identified, the molecular mechanisms leading to mucosal damage are still unknown. Even more complex are the cases of patients affected by CVID and AE. In patients with CVID, VA has been described to occur either secondary to infections such as giardiasis (Biagi et al., 2012; Malamut et al., 2010) and chronic Norovirus infection (Woodward et al., 2015), but also without evidence of any infectious triggers (Biagi et al., 2012; Malamut et al., 2010); AE may be found either in isolation, or in the context of complex immunodeficiency syndromes which are genetically determined and associated with an increased risk of autoimmunity and malignancies, such as the IPEX syndrome (Natacha et al., 2009).

Finally, a genetic predisposition of the host could be relevant in these enteropathies. While CD is invariably associated with HLA-DQ2 and DQ8 molecules (Lebwohl et al., 2018; Ludvigsson et al., 2014), data on the genetics of NCEs are very scarce. Only a recent study has evaluated the HLA genetic profile of patients with NCEs compared with patients affected by CD and healthy controls (bone marrow donors) in an Italian and British cohort (De Silvestri et al., 2023). This study has shown that in both the Italian and in the British group, the DQA1*0102 DQB1*0602 haplotype occurred significantly more frequently in patients with NCEs than those with CD, although no significant difference was found in comparison to healthy controls (De Silvestri et al., 2023). This result could be helpful for addressing the differential diagnosis between NCEs and seronegative and complicated forms of CD and may also be relevant for future developments in the pathogenesis of NCEs.

In conclusion, regardless from the trigger and the underlying mechanisms of each type of SNEs, the resulting phenotype is always characterized by VA and a malabsorption syndrome, meaning that the physiological response of the small bowel to a pathogenetic trigger is always the same (MacDonald & Spencer, 1988). However, it calls for investigation of the molecular mechanisms leading to VA in SNEs, as this could be key for developing targeted approaches and therapies for these patients.

11.4 Long-term outcomes and mortality

Literature data on long-term outcomes in SNEs indicate a poor prognosis (Aziz et al., 2017; Schiepatti et al., 2017). The UK study by Aziz et al. revealed that patients with SNEs, encompassing both SNCD and NCEs, experienced higher mortality rates compared to conventional seropositive CD (Aziz et al., 2017). The Italian study by Schiepatti et al. supported these findings and also offered information on the risks of complications and death (Schiepatti et al., 2017). In their research, complications arose in 4/260 CD patients (type 1 refractory CD, abdominal B-cell lymphoma, small bowel carcinoma, and enteropathy associated T-cell lymphoma), as opposed to 4/14 SNEs patients, resulting in complication rates of 0.2 (95% CI: 0.1–0.6) and 6.3 per 100 person-years (95% CI: 2.4–17.0), respectively (Volta et al., 2016). In the CD group, 4/260 patients died (three due to unrelated causes and one from enteropathy associated T-cell lymphoma), while 4/14 died in the SNEs group, resulting in mortality rates of 0.2 (95% CI: 0.1–0.6) and 6.0 deaths per 100 person-years (95% CI: 2.2–16), respectively (HR = 25.37, 95% CI 6.15–104.63, $P < 0.001$) (Schiepatti et al., 2017).

11.4.1 Prognosis of seronegative celiac disease

It has recently been shown that individuals with true SNCD and CD associated with IgA deficiency exhibit a more severe disease phenotype and less favorable prognosis than conventional seropositive CD patients (Schiepatti et al., 2021). Those with true SNCD and CD + IgA deficiency face an elevated risk of developing premalignant and malignant complications and experience increased mortality, particularly over an extended period, compared to traditional CD. The study identified true SNCD or CD + IgA deficiency, an older age at diagnosis, a classical pattern at diagnosis, and a lack of clinical response to a GFD as major predictors of complications through multivariate analysis.

11.4.2 Prognosis in nonceliac enteropathies

There is currently a lack of comparative data on whether the prognosis of nonceliac enteropathies (NCEs) differs based on their underlying cause. However, an analysis of our cases evaluated over a 20 year-period suggests that long-term outcomes vary depending on the underlying etiology of NCEs (Schiepatti et al., 2022). In fact, patients with NCEs due to identifiable factors, such as medications (particularly olmesartan) and infections, typically have an excellent prognosis once the causative agent is removed. On the other hand, patients with common variable immunodeficiency (CVID), autoimmune enteropathy (AE), primary small bowel lymphomas, and idiopathic villous atrophy (IVA) may experience severe complications that worsen their prognosis (Aziz et al., 2017; Basson et al., 2016; Ciccocioppo et al., 2012, 2018; Costetti et al., 2021; Malamut et al., 2012; Pensieri et al., 2019; Resnick et al., 2012; Schiepatti et al., 2022; Schiepatti et al., 2017; Schiepatti et al., 2020; Sharma et al., 2018).

High mortality rates are seen in patients with CVID enteropathy (Pensieri et al., 2019; Resnick et al., 2012), primarily due to severe infections and malignant complications. Similarly, patients with AE may develop lymphoproliferative complications (Ciccocioppo et al., 2018; Malamut et al., 2012), and an American study found an increased mortality rate in AE patients, comparable to those with refractory CD (Sharma et al., 2018).

A recent Italian and UK study (Schiepatti et al., 2020) has classified patients with idiopathic villous atrophy (IVA) into three main types, each with distinct clinical phenotypes and prognoses (see paragraph 5.4.6). Specifically, type 1 IVA, likely caused by an infectious agent, has a good prognosis (5-year survival of 96%). Type 2 IVA features persistent nonclonal VA and has an excellent survival rate (5-year survival of 100%). Lastly, type 3 IVA comprises a range of enteropathies with lymphoproliferative characteristics and a poor outcome (5-year survival of 27%).

There has not been a systematic investigation into the clinical predictors of poor long-term outcomes in NCEs. Age at diagnosis and low albumin have been identified as mortality predictors in IVA (Schiepatti, Sanders et al., 2020) and a UK study found that more extensive small bowel disease involvement on capsule endoscopy was associated with an unfavorable prognosis ($P = 0.019$) in SNEs (Chetcuti Zammit et al., 2020). Identifying clinical predictors of poor long-term outcomes and mortality in NCEs is essential for optimizing patient follow-up.

11.5 Clinical features of patients with seronegative enteropathies

In the vast majority of cases, the clinical picture of SNEs is dominated by a severe malabsorption syndrome, often requiring hospital admission. Patients with SNEs (both SNCD and NCEs) are older at diagnosis than conventional seropositive CD, while there is no specific association with gender (Aziz et al., 2017; Jansson-Knodell et al., 2020; Leonard et al., 2021; Pallav et al., 2012; Schiepatti et al., 2019; Schiepatti et al., 2021; Schiepatti et al., 2021). In these patients, upper GI endoscopy with duodenal biopsy must be considered as first line investigation despite the negative celiac serology, given the high pretest probability of having VA (Jansson-Knodell et al., 2020; Leonard et al., 2021; Schiepatti et al., 2019; Schiepatti et al., 2021). A detailed description of clinical features is provided in each dedicated paragraph.

11.5.1 Histopathological features of patients with seronegative enteropathies

Although VA is the key histological feature for NCEs and it is mandatory for the diagnosis, there are currently no specific histological markers allowing the differential diagnosis between SNCD and other nonceliac SNEs. As reported in the Paris consensus, some of these histological findings can be supportive for a specific etiology (see Table 11.1), but their significance must always to be considered in context with the whole clinical and biochemical scenario (Schiepatti et al., 2022). A more detailed description on histopathological features which are relevant for the diagnosis of SNEs will be provided in the paragraphs dedicated to each type of enteropathy.

11.5.2 Classification of seronegative enteropathies

A recent Delphi-based international consensus has provided a classification of SNEs in adults into three groups: (1) forms of CD presenting with negative serology; (2) NCEs posing problems of differential diagnosis with seronegative CD; (3) NCEs not posing problems of differential diagnosis with CD. The consensus also identified some conditions that should no longer be considered as causes of VA, as evidence in this regard was insufficient (Schiepatti et al., 2022). This classification is summarized in Table 11.1. Finally, there were some very rare conditions, for which it was not possible to assign to a specific diagnostic category (Al-Saleem & Al-Mondhiry, 2005; Fung et al., 1970; Kaosombatwattana et al., 2017; Korpe & Petri, 2012; Lappinga et al., 2010; Marth et al., 2016; Rubio-Tapia et al., 2010; Weinstein et al., 1970). These conditions will be only briefly described (Table 11.2) (Al-Saleem & Al-Mondhiry, 2005; Basson et al., 2016; Chetcuti Zammit et al., 2020; Fung et al., 1970; Kaosombatwattana et al., 2017; Korpe & Petri, 2012; Lappinga et al., 2010; Malamut et al., 2012; Marth et al., 2016; Resnick et al., 2012; Rubio-Tapia et al., 2010; Schiepatti et al., 2020; Sharma et al., 2018; Weinstein et al., 1970).

TABLE 11.1 Nonceliac enteropathies categorized based on their differential diagnosis with seronegative celiac disease (Schiepatti, Premoli et al., 2022; Schiepatti et al., 2022).

Seronegative celiac disease	Conditions entering differential diagnosis with seronegative celiac disease	Conditions not entering differential diagnosis with seronegative celiac disease	Not causes of villous atrophy
True seronegative CD CD and IgA deficiency GFD already started or on immunosuppressants CD and CVID Dermatitis herpetiformis	Giardiasis Tropical sprue Autoimmune enteropathy CVID Indolent $CD4^+$ T-cell lymphoma Idiopathic villous atrophy	Type 1 and 2 EATL Iatrogenic enteropathies* HIV enteropathy Crohn's disease Eosinophilic enteritis	*Helicobacter pylori* infection Peptic duodenitis NSAIDs enteropathy

CD, celiac disease; *CVID*, common variable immunodeficiency; *GFD*, gluten-free diet; *EATL*, enteropathy associated T-cell lymphoma; *IPSID*, immunoproliferative small intestinal disease; *NSAIDs*, nonsteroidal anti-inflammatory drugs.
**iatrogenic causes: olmesartan & other angiotensin II receptor blockers, mycophenolate mofetil, chemotherapy, radiotherapy, graft-versus-host disease, transplanted small intestine.*

TABLE 11.2 Conditions in which frank villous atrophy is still debated (Schiepatti, Premoli et al., 2022; Schiepatti, Sanders et al., 2022).

Condition	Clinical features	Diagnostic tests	Diagnostic findings
Whipple's disease (Marth et al., 2016)	Middle-aged Caucasians with seronegative migratory arthritis, fever, lymphadenopathy, diarrhea and weight loss, late neurological involvement (fatal)	Duodenal biopsy; PCR for *Tropheryma whipplei*	PAS + macrophages infiltrating the *lamina propria*
Tuberculosis (Fung et al., 1970)	Cough, fever, night sweats	Mantoux tuberculin skin test; interferon-gamma release assay; chest X-ray; CT	Caseating granulomas
Small intestinal bacterial overgrowth (Lappinga et al., 2010)	Diarrhea and weight loss of variable severity in patients with predisposing conditions (e.g., gastric resection and right hemicolectomy)	H2-glucose breath test	Intraepithelial lymphocytosis with or without a variable degree of villous blunting
Collagenous sprue (Rubio-Tapia et al., 2010; Weinstein et al., 1970)	Diarrhea and weight loss, abdominal pain; sometimes suggestive history for CD or TS or ARBs	Duodenal biopsy	Subepithelial collagen band (> 20 μm)
Enteropathy associated with IPSID (Al-Saleem & Al-Mondhiry, 2005)	Diarrhea and weight loss of different severity in young adults living in lower socioeconomic areas and in the Middle East, Far East, and Africa	Full-thickness intestinal biopsy; abdomen CT;	Plasmacytic infiltrate of CD20 + B-cells and plasma cells expressing alpha heavy chain but no light chains
Environmental/tropical enteropathy (Korpe & Petri, 2012)	Poorly understood condition ranging in severity from subclinical (paraphysiological) to overt malabsorption in people living in poor hygienic conditions in rural areas of developing Countries	Duodenal biopsy	Variable degree of villous blunting and T-cell infiltration
Enteropathy associated with mefenamic acid (Kaosombatwattana et al., 2017)	Malabsorption + suggestive pharmacological history	Duodenal biopsy and response to drug withdrawal	Variable degree of villous blunting and mixed inflammatory infiltrate of the *lamina propria*, including eosinophilia

CD, celiac disease; *CT*, computed tomography; *PCR*, polymerase chain reaction; *ARBs*, angiotensin II receptors blockers; *PAS*, Periodic-acid schiff.

11.5.3 Seronegative celiac disease

Seronegative CD, as recently defined, is a broad term referring to various forms of CD characterized by VA responsive to a GFD and negative celiac antibodies at the time of diagnosis (Schiepatti et al., 2022). More specifically, this category includes true SNCD, which should be distinguished from CD associated with selective IgA deficiency. Previously, there was significant debate over whether patients with total IgA deficiency who exhibit positive class IgG TTA or IgG EmA and VA should be considered as having SNCD. Since these patients generate an IgG-based serological response, this factor may suggest a diagnosis of traditional seropositive CD rather than true SNCD (Schiepatti et al., 2018). However, comparative data on the clinical characteristics and long-term outcomes of these patients has revealed that both true SNCD and CD + IgA deficiency exhibit a more severe phenotype and disease progression with an increased risk of complications and mortality compared to conventional seropositive CD (Schiepatti et al., 2021).

Additionally, up to 30% of individuals with biopsy-confirmed dermatitis herpetiformis may have negative serology (Reunala et al., 2021; Schiepatti et al., 2019; Schiepatti et al., 2021) and some cases of seronegative CD associated with CVID have been reported (Biagi et al., 2012; Heneghan et al., 1997; Malamut et al., 2010; Webster et al., 1981). Histological improvement of VA following gluten withdrawal is the primary method for diagnosing CD in patients with CVID. In addition, negative HLA-DQ2 and DQ8 typing is highly effective in ruling out CD (Biagi et al., 2012; Malamut et al., 2010). In contrast, celiac-specific antibodies do not play a role in this diagnostic process. In fact, class

IgG EmA were detected in CVID patients who previously received IV immunoglobulin replacement therapy and in whom CD was ruled out due to the absence of HLA-DQ2 and DQ8 molecules (Biagi et al., 2012; Jørgensen et al., 2016; Malamut et al., 2010). Certain histopathological features, such as a lack of plasma cells, the presence of a polymorphonuclear infiltrate, and graft-versus-host disease-like lesions (Biagi et al., 2012; Daniels et al., 2007; Malamut et al., 2010) have been observed in patients with CVID and duodenal VA but not in untreated CD, making them potentially useful for differential diagnosis. Other criteria, including the extension of intestinal lesions throughout small intestine, crypt hyperplasia, increased IELs, increased inflammation of the *lamina propria*, crypt apoptotic bodies, HLA-DQ2 and/or DQ8 positivity, were considered unable to confirm or exclude a CD diagnosis in CVID (Schiepatti et al., 2022).

Lastly, there are two patient groups frequently encountered in clinical practice that may exhibit negative celiac serology during serological testing, but should be considered as conventional CD rather than seronegative CD. If not correctly identified, these patients often lead to diagnostic errors (Schiepatti et al., 2020). These two clinical situations include: (1) patients with negative serology because they have already initiated a GFD or immunosuppressive therapies at the time of serological testing—these patients can restore their positive serological response if exposed to gluten or if immunosuppressants are discontinued (Schiepatti et al., 2018; Schiepatti et al., 2020); (2) patients with VA but discrepancies between TTA and EmA results while on a regular gluten-containing diet (e.g., borderline/low titer-positive TTA with negative EmA or vice versa).

11.5.3.1 Diagnostic criteria for true seronegative celiac disease and celiac disease associated with IgA deficiency

A clinical picture characterized by severe malabsorption, often occurring in patients aged 45 or older, together with a possible association with autoimmune disorders and family history of CD should prompt suspicion of SNCD. In patients with a high clinical suspicion of SNCD a gastroscopy with four duodenal biopsies taken form the second duodenal portion must be performed as a first line investigation, regardless of the negative result of celiac serology (Jørgensen et al., 2016; Schiepatti, Cincotta, et al., 2021). The following criteria must be satisfied in order to make a diagnosis of both seronegative CD and CD associated with IgA deficiency:

1. VA, crypt hyperplasia and an increased intra-epithelial lymphocyte count, on correctly-oriented duodenal specimens, with subsequent full recovery on a GFD;
2. evidence of HLA typing showing specific celiac haplotypes, that is, HLA-DQ2.5 (DQA1*0501, DQB1*0201), HLA-DQ8 (DQA1*03, DQB1*0302), HLA-DQ2.2 (DQA1*0201, DQB1*0202) or HLA-DQ7.5 (DQA1*05, DQB1*0301);
3. systematic exclusion of all other causes of VA.

It is of paramount importance that all the investigations are performed before starting the patient on a GFD or immunosuppressive therapy, as they may lead to false negative serology (Schiepatti et al., 2021).

In some cases, individuals suspected of having SNCD may need a diagnostic gluten challenge to confirm the diagnosis. Currently, there is no standard method for this diagnostic tool, but a minimum of 10 g of gluten per day for 6–8 weeks has been proposed by some authors (Al-Toma et al., 2019; Aziz et al., 2017). HLA typing should always be conducted in unclear cases of VA with negative celiac serology, since it can help differentiate seronegative CD from NCEs. Despite the fact that 30%–40% of Caucasian individuals may carry HLA-DQ2 or DQ8 haplotypes, negative HLA typing rules out seronegative CD (Jansson-Knodell et al., 2020; Leonard et al., 2021; Schiepatti et al., 2019; Schiepatti et al., 2021). Furthermore, a recent study from Italy and the UK found that NCEs are more frequently associated with HLA- DQA1*0102 DQB1*0602 than CD (De Silvestri et al., 2023), which can be helpful in making a differential diagnosis.

When available, small bowel mucosal transglutaminase 2-specific IgA deposits can support a SNCD diagnosis and help distinguish it from other NCEs in patients with normal serum IgA levels (Salmi et al., 2006).

In some cases, patients with VA who recover on a GFD, exhibit compatible celiac HLA typing, negative IgA celiac antibodies, and show selective IgA deficiency (total serum IgA level $<5-7$ mg/dL) $\pm$ positive IgG celiac serology, which allows for the diagnosis of CD associated with IgA deficiency. In patients with IgA deficiency, the sensitivity of IgG TTA and IgG DGP antibodies is greater than IgG EmA (91% vs 82% vs 76%) and should be the preferred method (Leffler & Schuppan, 2010; Villalta et al., 2010).

A strict lifelong GFD is essential for treating SNCD. It is recommended for these patients to undergo thorough clinical and endoscopic follow-ups due to their higher risk of developing complications and mortality compared to conventional seropositive CD (Schiepatti et al., 2021).

11.5.4 Enteropathies posing problems of differential diagnosis with seronegative celiac disease

This group includes enteropathies characterized by malabsorption, a variable degree of VA with negative celiac serology, which are unresponsive to a GFD. These conditions have been identified as the most challenging diagnoses by a recent international consensus (Schiepatti et al., 2022). This group includes autoimmune enteropathy, enteropathy associated with CVID, tropical sprue, giardiasis, CD4 + indolent T-cell lymphoma, and idiopathic villous atrophy. A detailed description is provided for each one of these conditions.

11.5.4.1 Autoimmune enteropathy

Autoimmune enteropathy (AE) is an extremely rare enteropathy with an estimated prevalence of about one case per million individuals, first identified in children and later also in adults (Akram et al., 2007; Catassi et al., 1999; Marthinsen et al., 2008; Unsworth & Walker-Smith, 1985). AE patients typically exhibit a severe malabsorption syndrome, often necessitating hospitalization and management at tertiary referral centers. It is not unusual for patients with AE to be initially misdiagnosed as having seronegative and refractory forms of CD (Akram et al., 2007; Corazza et al., 1997; Schiepatti et al., 2021; Schiepatti et al., 2020; Volta et al., 2016). The diagnosis of autoimmune enteropathy is established through the combination of severe malabsorption, overt villous atrophy (VA) with negative celiac serology that does not respond to dietary restrictions, and the presence of positive IgA/IgG enterocyte antibodies (indirect immunofluorescence on human/monkey jejunum) (Fig. 11.2). It is also essential to rule out other causes of VA before confirming a diagnosis (Akram et al., 2007; Corazza et al., 1997; Jansson-Knodell et al., 2020; Leonard et al., 2021; Masia et al., 2014; Schiepatti et al., 2019; Schiepatti et al., 2021; Unsworth & Walker-Smith, 1985). There are additional clinical and histopathological features that may support the diagnosis of AE, but if alone they are not sufficient for making the diagnosis. These criteria include:

1. A history of associated autoimmune conditions
2. A clinical response to immunosuppressive therapies
3. Deep crypt lymphocytosis and/or plasma cell infiltration, neutrophilic cryptitis ± crypt microabscesses, reduced or absent Paneth cells on duodenal histology
4. The presence of positive serum anti-AIE 75KD antibodies (ELISA) or nonorgan specific autoantibodies

HLA typing has no relevance for AE diagnosis (Schiepatti et al., 2022): HLA-DQ2 and DQ8 molecules can be found in up to 20% of AE patients (Akram et al., 2007; Corazza et al., 1997; Volta et al., 2016).

There remains a lack of agreement on several aspects of AE, including the absence of serious immunodeficiencies, involvement of other gastrointestinal tract locations, the diagnostic significance of serum antigoblet cell antibodies (Fig. 11.2) and certain duodenal histopathological features.

Regarding the utility of goblet cell antibodies, both literary sources and our experience indicate that they have poor specificity for AE, as they have been found in patients with chronic inflammatory bowel diseases (Hibi et al., 1994), their first-degree relatives (Folwaczny et al., 1997) and control patients with normal villi. Consequently, we believe their use in AE diagnosis should be discouraged (Biagi et al., 2009).

In conclusion, while certain histological characteristics have been proposed to support the diagnosis of AE, as mentioned earlier, histology alone cannot serve as the sole determinant for an AE diagnosis (Greenson, 2015). The range of histopathological features in AE is extensive, and there is currently no broad agreement on the importance of factors such as IEL count, crypt hyperplasia and crypt apoptotic bodies, lack of gamma-delta T-cells, and depletion of goblet

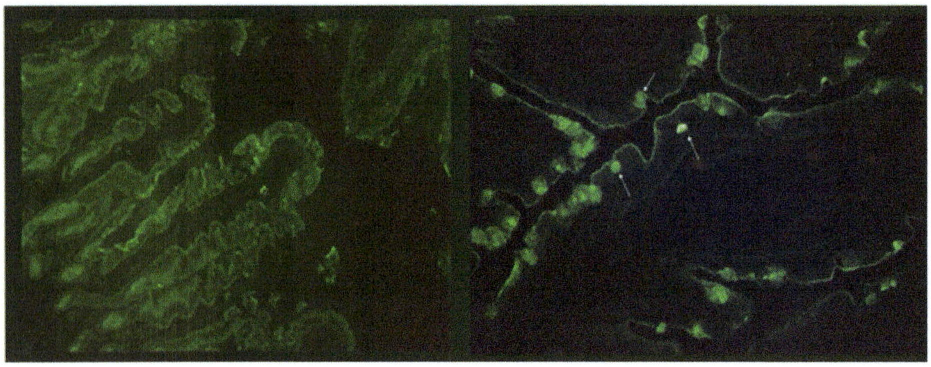

FIGURE 11.2 Positive enterocyte antibodies (image on the left) and positive goblet cell antibodies (image on the right). Positive enterocyte antibodies are the serological markers of autoimmune enteropathy, whereas goblet cells antibodies are totally nonspecific and must not be used for the diagnosis of autoimmune enteropathy.

cells in diagnosing AE (Akram et al., 2007; Corazza et al., 1997; Greenson, 2015; Masia et al., 2014; Sharma et al., 2018). Specifically, besides the required VA for diagnosis (Akram et al., 2007; Catassi et al., 1999; Corazza et al., 1997; Marthinsen et al., 2008; Schiepatti et al., 2019; Unsworth & Walker-Smith, 1985; Volta et al., 2016), cases of AE exhibiting a celiac-like pattern with increased IELs have been reported (Corazza et al., 1997), along with cases showing a low IEL count accompanied by dense lymphoplasmacellular infiltrates in the lamina propria, neutrophilic cryptitis with or without crypt microabscesses, and increased cryptic apoptosis (Akram et al., 2007; Greenson, 2015; Masia et al., 2014). Histological features that overlap between sprue-like patterns and other AE patterns have been documented, as well as the simultaneous occurrence of CD and AE in a single patient (Akram et al., 2007). As a result, given the lack of distinctive characteristics for this enteropathy, making an accurate histological diagnosis of AE remains difficult. A comprehensive integration of clinical and histopathological information continues to be essential when considering or diagnosing AE (Corazza et al., 1997; Schiepatti et al., 2019; Schiepatti et al., 2021; Volta et al., 2016).

11.5.4.2 Enteropathy associated with common variable immune-deficiency

Common variable immune-deficiency (CVID) is the most prevalent primary symptomatic immune-deficiency, with an estimated occurrence of 1 in 10,000 to 250,000 people in the general population (Chapel et al., 2008; Conley et al., 1999; Park et al., 2008). Gastrointestinal complications frequently occur in CVID patients, but they present with considerable clinical and histological variability (Hermaszewski & Webster, 1993; Jørgensen et al., 2016; Kalha & Sellin, 2004; Khodadad et al., 2007; Park et al., 2008). The prevalence and causes of frank VA in CVID patients are not well-defined. Although giardiasis and other gastrointestinal infections, such as chronic Norovirus infection (Biagi et al., 2012; Greenson, 2015; Malamut et al., 2010; Woodward et al., 2015), have been described in association with CVID, intestinal lesions can occur also without an underlying infection. Furthermore, CD has been reported in patients with CVID (Biagi et al., 2012; Heneghan et al., 1997; Malamut et al., 2010; Webster et al., 1981). Refer to the paragraph on seronegative CD for more information.

The diagnosis of chronic noninfectious enteropathy in patients with a confirmed CVID diagnosis (Conley et al., 1999) requires the presence of gastrointestinal symptoms (regardless of severity, ranging from occasional diarrhea to overt malabsorption syndrome), VA, and the exclusion of other VA causes, including *Giardia lamblia* and other chronic gastrointestinal infections (Biagi et al., 2012; Malamut et al., 2010; Woodward et al., 2015). Certain duodenal histopathological characteristics, such as increased intraepithelial lymphocytes in the duodenum, increased inflammation in the *lamina propria*, crypt apoptotic bodies, and graft-versus-host disease-like lesions, support a CVID enteropathy diagnosis. However, it remains uncertain whether other histopathological features, such as plasma cell depletion in the mucosa, follicular/nodular lymphoid hyperplasia, Crohn's-like lesions/granulomas, eosinophilic infiltrate, and crypt abscesses/neutrophilic infiltrate, contribute to the CVID diagnosis (Bästlein et al., 1988; Biagi et al., 2012; Daniels et al., 2007; Greenson, 2015; Heneghan et al., 1997; Lougaris et al., 2015; Malamut et al., 2010; Webster et al., 1981). Additionally, although gastrointestinal manifestations in CVID patients are highly heterogeneous and may involve any part of the gastrointestinal tract, there is no consensus on the relationship between CVID, lymphocytic gastritis, atrophic gastritis, microscopic colitis, and inflammatory bowel disease (Schiepatti et al., 2022). Prospective studies on large cohorts are needed to assess these complex associations and, for now, each case should be individually evaluated for diagnosis and optimal management. Finally, HLA typing can be beneficial for excluding CD in patients with CVID.

11.5.4.3 Tropical sprue

Tropical sprue (TS) is a chronic disorder first identified in 1948 as a severe malabsorption syndrome affecting soldiers in tropical regions (Stefanini, 1948). The condition is characterized by malabsorption, a certain level of VA (usually mild) accompanied by intraepithelial lymphocytosis, and should be considered in patients with a history of residing in, or visiting tropical countries (especially areas in South Asia, Southeast Asia, Africa, and South America within the Tropics) for a minimum of 2 months and under unsanitary conditions. The diagnosis can be confirmed by ruling out giardiasis, CD, and other causes of VA, as well as a prompt clinical and histological response to oral antibiotics. Supportive diagnostic factors include low folate levels that improve with supplementation, low vitamin B12 levels potentially causing megaloblastic anemia, and fat-soluble vitamin deficiencies, while HLA typing has no diagnostic significance (Yadav et al., 2011).

Despite some authors proposing a primary involvement of the distal small bowel and terminal ileum in TS, the question of whether ileal involvement in TS is more prominent remains unresolved and is currently not a diagnostic criterion (Yadav et al., 2011).

11.5.4.4 Giardiasis

Giardiasis is an infection caused by the flagellated intestinal protozoan *Giardia lamblia* (also known as *Giardia duodenalis* or *intestinalis*) (Levinson & Nastro, 1978). This condition can present with a wide range of symptoms, from severe malabsorption syndrome to irritable bowel syndrome-like manifestations, or even no symptoms at all. As a result, giardiasis should be considered not only in patients experiencing villous atrophy (VA) with negative celiac antibodies and significant malabsorption, but also in milder clinical situations. Diagnosis can be made through one or more of the following tests:

1. detection of *Giardia*-specific antigens in stool samples;
2. trophozoites in formalin-fixed paraffin embedded haematoxylin and eosin stained duodenal specimens and/or duodenal aspirate;
3. direct identification of cysts/trophozoites in fresh feces;
4. *Giardia*-specific PCR.

Antibiotic treatment (typically metronidazole) is the first-line treatment, and a positive response to this therapy further supports the diagnosis. While HLA typing is not essential for diagnosis, it may be considered in patients with borderline TTA to rule out CD. Though giardiasis has been identified in patients with CVID, IgA deficiency, and CD, systematic testing for giardiasis in such cases is not necessary (Schiepatti, Sanders et al., 2022). Decisions to investigate other causes of VA should be determined on an individual basis.

11.5.4.5 Small bowel indolent $CD4^+$ T-cell lymphoma

Indolent CD4 + T-cell lymphoma of the small bowel is an uncommon non-Hodgkin lymphoma that primarily affects the small intestine (Malamut et al., 2014). This lymphoma is frequently misdiagnosed as type 2 refractory CD because of persistent VA and malabsorption despite a GFD and the monoclonal nature of intraepithelial lymphocytes (Malamut et al., 2014). Duodenal VA is essential for diagnosis after ruling out other causes of VA.

The diagnosis relies on immunohistochemistry that reveals widespread infiltration of the epithelium and/or increased presence of small/medium CD3 + CD4 + T-cells in the lamina propria, along with monoclonal rearrangement for beta-TCR and/or gamma-TCR in duodenal biopsy specimens. Elevated CD3 + CD4 + intraepithelial/*lamina propria* lymphocytes detected by flow cytometry are also diagnostic.

Determining whether to conduct a bone marrow biopsy, additional endoscopic or radiologic examinations to evaluate the involvement of other gastrointestinal tract areas, or molecular diagnostics for STAT3-JAK2 fusions (Sharma, Oishi, et al., 2018) should be considered on an individual case basis.

11.5.4.6 Idiopathic villous atrophy

Idiopathic villous atrophy (IVA) is a condition defined by VA accompanied by an increased count of duodenal intraepithelial lymphocytes, negative celiac serology, no response to a gluten-free diet, and exclusion of known causes of VA (Aziz et al., 2017; Schiepatti et al., 2020). HLA typing is useful for ruling out CD if celiac haplotypes are absent. IVA has only recently been identified (Aziz et al., 2017; Schiepatti et al., 2020), so various aspects of this condition remain to be fully understood. It remains uncertain whether the severity of malabsorption symptoms at the time of presentation, family history of CD, past medical history of autoimmune conditions (such as dermatitis herpetiformis), potential involvement of other parts of the gastrointestinal tract, and mucosal deposits of IgA TTG play a significant role in diagnosis.

A recent study suggests that IVA is not a singular condition but rather a group of enteropathies with undefined causes and varying prognoses (Schiepatti et al., 2020). IVA can be categorized into three types: type 1 IVA, which involves transient, self-limiting partial VA, resolving spontaneously within 6−12 months and possibly due to infections, having a favorable prognosis; type 2 IVA, characterized by ongoing nonclonal VA, associated with HLA DQB1*0301 and DQB1*06, and an excellent long-term prognosis; and type 3 IVA, characterized by persistent VA and lymphoproliferative features (abnormal T-cell populations or ongoing gamma-TCR monoclonality, or a history of lymphoproliferative disorders) with a poor prognosis (Schiepatti et al., 2020).

11.5.5 Nonceliac enteropathies not posing problems of differential diagnosis with seronegative celiac disease

Chronic enteropathies are defined by varying levels of duodenal VA and malabsorption syndromes of different severities. Suspicion of these conditions usually arises from typical clinical or histopathological features, as well as suggestive

TABLE 11.3 Enteropathies not posing problems of differential diagnosis with seronegative celiac disease because of clues guiding to the correct diagnosis.

Condition	Clinical features	Diagnostic findings	Diagnostic tests	Treatment
EATL (type 1 and 2)	Diarrhea and weight loss, abdominal pain, fever, bleeding, obstruction and/or perforation; type 1 usually associated to CD, unlike type 2 (also known as MEITL)	Inflammatory markers, abdominal CT/PET scan, capsule endoscopy, enteroscopy	Malignant IEL on IHC or aberrant IEL flow cytometry	Hematologic, surgical and chemotherapic treatment
Angiotensin II receptor blockers	Diarrhea and weight loss and suggestive pharmacological history	Duodenal biopsy and clinical response to drug withdrawal	Villous atrophy similar to CD	Drug withdrawal
Immune-checkpoint inhibitors	Diarrhea and weight loss and suggestive pharmacological history	Duodenal biopsy and drug withdrawal	Villous atrophy similar to CD	Drug withdrawal, steroids, rescue therapy (infliximab, vedolizumab, fecal microbiota transplantation)
Chemotherapy	Diarrhea and weight loss and suggestive pharmacological history	Duodenal biopsy	Villous atrophy similar to CD, *lamina propria* fibrosis	Steroids, consult oncologist to evaluate alternative regimens
Radiotherapy	Diarrhea and weight loss and history of radiotherapy	Duodenal biopsy	*Lamina propria* fibrosis	Steroids
GVHD	Diarrhea and weight loss and history of bone marrow transplantation	Duodenal biopsy	Crypt cell necrosis, loss of epithelium	Steroids or budesonide
HIV enteropathy	Known history of AIDS, presence of opportunistic infections	HIV test	Decrease CD4 + T lymphocytes, increase in CD8 + T lymphocytes	Antiretroviral therapy, treatment of opportunistic infections
Eosinophilic gastroenteritis	History of atopy and allergies	Duodenal biopsy and peripheral hyper eosinophilia	Massive eosinophilic infiltration on duodenal biopsy	Steroids and dietary therapy
Crohn's disease	Chronic diarrhea, abdominal pain, fever, weight loss	Duodenal biopsy, colonoscopy + biopsies, abdomen CT, MR enterography	Villous atrophy, granulomas	Steroids, antibiotics, azathioprine, biologic therapy

CD, celiac disease; *EATL*, enteropathy associated T-cell lymphoma; *MEITL*, monomorphic epitheliotropic T-cell lymphoma; *MR*, magnetic resonance; *IHC*, immunohistochemistry; *IEL*, intaepithelial lymphocytes. Special consideration should be given to medication-induced enteropathies, as detailed in the dedicated paragraph.

elements in a patient's past medical and pharmacological history, which dramatically increase the likelihood of the diagnosis. These enteropathies and the factors guiding their differential diagnosis are presented in Table 11.3 (Boscá et al., 2008; Costetti et al., 2021; Culliford et al., 2004; Foukas & de Leval, 2015; Kamar et al., 2004; Kapembwa et al., 1989; Patey-Mariaud de Serre et al., 2002; Rubio-Tapia et al., 2012; Schiepatti, Cincotta, et al., 2021; Stacey & Green, 2014; Touchefeu et al., 2014; Walker et al., 2018; Ziegler et al., 2003).

11.5.5.1 Medication-related and iatrogenic forms of villous atrophy

Some chronic NCEs, histologically indistinguishable from untreated CD and typically characterized by severe malabsorption, may occur in patients taking certain medications such as angiotensin II receptor blockers (ARB2s), particularly olmesartan (Costetti et al., 2021; Rubio-Tapia et al., 2012), mephenamic acid (Kaosombatwattana et al., 2017), azathioprine (Ziegler et al., 2003), methotrexate (Boscá et al., 2008) and mycophenolate mofetil (Kamar et al., 2004). An emerging cause of iatrogenic enteropathy are checkpoint-inhibitors, which are increasingly used for the treatment of many types of cancer (Schiepatti, Premoli et al., 2022). Chemotherapy, radiotherapy, graft-versus-host disease and transplanted small intestine are also listed among the iatrogenic causes of NCEs (Patey-Mariaud de Serre et al., 2002; Stacey & Green, 2014; Touchefeu et al., 2014). Although nonsteroidal anti-inflammatory drugs are known for their potential to cause small bowel damage (Tai & McAlindon, 2018), they should not be considered as a cause of VA with negative celiac serology (Schiepatti et al., 2022).

In these patients, clinical presentation and medical history can help raise diagnostic suspicion and guide towards the correct diagnosis. Despite being the second most common reason for SNEs in adults and their classification within NCEs not causing differential diagnostic issues with SNCD, awareness about iatrogenic enteropathies among clinicians remains low. As a result, these patients can be easily overlooked and often misdiagnosed as having seronegative or refractory forms of CD (Costetti et al., 2021; Jansson-Knodell et al., 2020; Leonard et al., 2021; Lundin & Green, 2022; Rubio-Tapia et al., 2012; Schiepatti et al., 2019; Schiepatti et al., 2021; Schiepatti et al., 2021; Schiepatti et al., 2020). This is an important issue, as in most cases, the prognosis of iatrogenic forms of VA significantly improves upon discontinuation of the offending medication (Boscá et al., 2008; Costetti et al., 2021; Jansson-Knodell et al., 2020; Kamar et al., 2004; Kaosombatwattana et al., 2017; Leonard et al., 2021; Rubio-Tapia et al., 2012; Schiepatti et al., 2019; Schiepatti, Cincotta, et al., 2021; Ziegler et al., 2003). A comprehensive pharmacological history is essential for the timely recognition of these enteropathies. Histological and clinical improvement following drug withdrawal remains the primary diagnostic approach. A brief overview of the most common iatrogenic enteropathies is provided below.

11.5.5.2 Angiotensin II receptor blocker-associated enteropathy

Mayo Clinic investigators first documented olmesartan-associated enteropathy (OAE) in 2012 (Rubio-Tapia et al., 2012). Subsequent reports have indicated additional cases due to other ARB2s, such as irbesartan (Marthey et al., 2014), valsartan (Herman et al., 2013), telmisartan (Cyrany et al., 2014) and possibly candesartan (Aziz et al., 2017; Costetti et al., 2021). Though gastrointestinal side effects may align with a class effect of ARB2s (Cyrany et al., 2014), a large French population study (Basson et al., 2016) and a multidatabase US study (Dong et al., 2018) suggest that users of non-olmesartan ARB2s do not have an increased risk for malabsorption. No apparent correlation exists with ARB2s dosage, while the impact of treatment duration is debated. European papers primarily link enteropathy occurrence to the length of ARB2 therapy (Basson et al., 2016; Costetti et al., 2021; Marietta et al., 2015; Marthey et al., 2014; Rubio-Tapia et al., 2012; Schiepatti et al., 2016), whereas in the United States, a higher rate of enteropathy was found among olmesartan initiators (Dong et al., 2018). Despite estimates indicating 1/4000 to 1/5000 olmesartan users experiencing OAE (Marietta et al., 2015), the true prevalence remains uncertain. Typical ARB2s-associated enteropathy patients have arterial hypertension and, compared to those with CD, are older and more likely to experience severe malabsorption requiring hospitalization (Costetti et al., 2021). There is no gender predominance or specific association with HLA haplotypes (Costetti et al., 2021; Marietta et al., 2015; Marthey et al., 2014; Rubio-Tapia et al., 2012; Schiepatti et al., 2016). Serological markers for ARB2s-associated enteropathy have not been identified, though antinuclear antibodies have been reported (Marthey et al., 2014; Schiepatti et al., 2016). Duodenal histology is nearly indistinguishable from untreated CD. Various authors describe neutrophilic infiltration, increased crypt apoptosis resembling AE (Burbure et al., 2016; Scialom et al., 2015), an increased eosinophil count (Costetti et al., 2021) and increased subepithelial collagen deposition (Burbure et al., 2016; Rubio-Tapia et al., 2010; Scialom et al., 2015) in OAE patients. However, these histopathological findings are nonspecific and insufficient for diagnosing ARB2s-associated enteropathy based on histology alone. Accurate diagnosis relies on the integration of the patient's clinical and pharmacological history (Schiepatti et al., 2022).

In some patients, ARB2s-enteropathy was associated with microscopic colitis (lymphocytic and collagenous) and chronic lymphocytic gastritis (Costetti et al., 2021; Rubio-Tapia et al., 2012).

Upon discontinuing the causative drug, ARB2s-associated enteropathy patients generally experience rapid clinical and histological recovery (Costetti et al., 2021; Rubio-Tapia et al., 2012), although long-term outcome data is lacking (Costetti et al., 2021).

11.5.5.3 Enteropathy with villous atrophy due to immune checkpoint inhibitors

Immune-checkpoint inhibitors (ICIs) are monoclonal antibodies that target immune-checkpoint molecules, which are involved in suppressing cytotoxic T-cells. By targeting these molecules, ICIs enhance the survival and functionality of cytotoxic T-cells, ultimately boosting the innate immune system's tumor surveillance activity. The targeted molecules include cytotoxic T-lymphocyte-associated protein 4 (CTLA-4) and programmed cell death protein 1 (PD-1), which are receptors found on cytotoxic T-cell surfaces that interact with ligands CD80/CD86 and programmed death-ligand 1 for PD-1 on antigen-presenting cells. The first ICI, ipilimumab (an anti-CTLA-4), was approved by the Food and Drug Administration for treating metastatic melanoma in 2011 (Bellaguarda & Hanauer, 2020). Since then, other ICIs targeting PD-1 (i.e., pembrolizumab and nivolumab) and programmed death-ligand 1 (i.e., atezolizumab and durvalumab) have become available. ICIs have revolutionized the standard of care for numerous cancer types, and their usage is anticipated to increase in the future.

Although ICIs are highly effective in enhancing anti-tumor T-cell activity, they can also cause a broader T-cell response, which may result in severe immune-related adverse events in some cases. These toxicities can vary depending on the agent, cancer type, and individual susceptibilities (Bellaguarda & Hanauer, 2020; Marin-Acevedo et al., 2019). The skin and colon are the most commonly affected organs by these autoimmune-like toxicities (Marin-Acevedo et al., 2019). Immune-related adverse events involving the colon while on ICI therapy can manifest as life-threatening colitis resembling inflammatory bowel disease (Powell et al., 2020). However, ICIs can also cause gastrointestinal adverse events in the small bowel (Schiepatti, Premoli et al., 2022) an under-recognized condition that can develop independent of colonic involvement. The prevalence and risk factors for small bowel involvement during ICI therapy remain unknown. Small bowel involvement can present as severe malabsorption, often necessitating hospitalization, and with VA indistinguishable from untreated CD. Cases of severe ulcerative enteritis without VA have also been reported (Schiepatti, Premoli et al., 2022). Treatment involves immediately stopping ICIs and administering high-dose steroids. In some instances, infliximab has been used as rescue therapy; however, these therapies are based on expert opinions and should be evaluated on a case-by-case basis.

11.5.6 Conditions that can cause villous atrophy and should be considered on a case-by-case basis

In certain situations, VA may be just one part of a complex clinical presentation. This occurs in tuberculosis, where a malabsorption syndrome is a rare occurrence, and other signs usually lead to the diagnosis. On the other hand, in conditions like Whipple's disease and small intestinal bacterial overgrowth, although VA is possible, it is an extremely rare histopathological feature, and additional clinical and histopathological factors aid clinicians in making the final diagnosis. Table 11.2 provides a comprehensive list of conditions which should be assessed on an individual basis (Al-Saleem & Al-Mondhiry, 2005; Fung et al., 1970; Kaosombatwattana et al., 2017; Korpe & Petri, 2012; Lappinga et al., 2010; Marth et al., 2016; Rubio-Tapia et al., 2010; Weinstein et al., 1970).

11.5.7 Conditions not clearly associated with villous atrophy

Nonsteroidal anti-inflammatory drug use, *H. Pylori* infection, and peptic duodenitis should not be considered as causes of VA. The evidence supporting their causal role in VA is extremely weak and nearly anecdotal, with their significance in clinical practice remaining unclear (Schiepatti et al., 2022).

11.6 Methodological approach to the differential diagnosis of seronegative enteropathies

The release of an international consensus regarding the nomenclature and diagnostic guidelines for SNEs (Schiepatti et al., 2022) undoubtedly represents a significant milestone in identifying these disorders, yet their treatment predominantly continues to involve tertiary referral centers. Drawing from our clinical expertise and the latest research, we established a set of examinations (Table 11.4) and a diagnostic work-up for the differential diagnosis of SNEs (Fig. 11.3) (Schiepatti et al., 2021).

In individuals with severe malabsorption, a gastroscopy with duodenal biopsy is essential, even if celiac serology is negative. A comprehensive set of laboratory, fecal, molecular, and imaging evaluations is needed to fully understand the patient's condition (Jansson-Knodell et al., 2020; Leonard et al., 2021; Schiepatti et al., 2019; Schiepatti et al., 2021). Proper collection and processing of duodenal biopsies from the second part of the duodenum are critical for assessing VA. Ensuring that the tests are performed while the patient is consuming gluten is required, and if not, additional or repeated testing should be done after a gluten challenge. For those referred with a prior history of VA, an expert gastrointestinal pathologist should carefully examine the duodenal biopsy slides while being informed about the patient's pharmacological history and clinical characteristics. Therefore patients with only an elevated intraepithelial

TABLE 11.4 Useful tests in the evaluation of patients with seronegative enteropathies.

Laboratory tests
IgA and IgG Ema, TTA and DGP
Serum IgA, IgG, IgM
HLA typing
Fecal calprotectin
Flow cytometry for aberrant IELs
Enterocyte antibodies
HIV testing
Quantiferon
Tests for gastrointestinal infections
Giardia lamblia and other parasites
Viruses
Duodenal histology
H&E and PAS staining
PCR for *Giardia*, tuberculosis and Whipple's disease
PCR for beta and gamma-TCR clonality assessment
Endoscopic examinations
Capsule endoscopy
Enteroscopy
Colonoscopy + biopsies
Abdominal imaging (CT/PET)

EMA, endomysial antibodies; *TTA*, tissue transglutaminase antibodies; *DGP*, deamidated gliadin antibodies; *H&E*, hematoxylin and eosin staining; *PAS*, Periodic-acid-Schiff staining; *IELs*, intraepithelial lymphocytes; *CT*, computed tomography; *PET*, positron emission tomography.

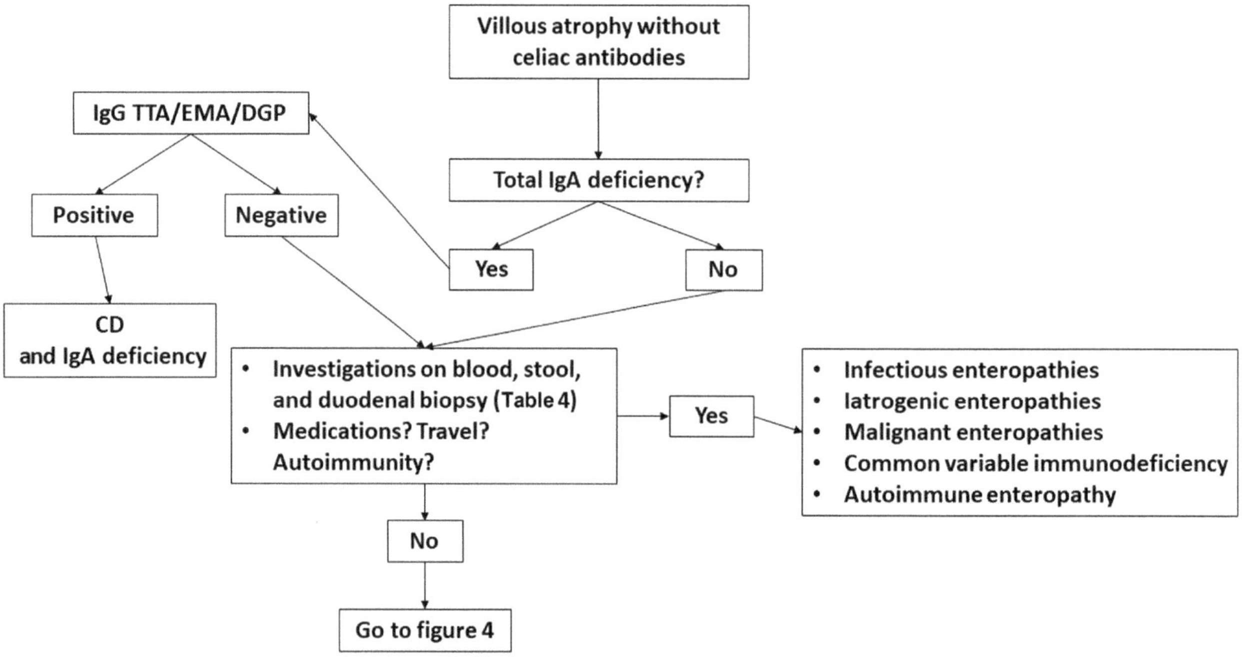

FIGURE 11.3 Flowchart showing the differential diagnosis of villous atrophy with negative celiac serology. *TTA*, tissue transglutaminase antibodies; *EMA*, endomysial antibodies; *DGP*, deamidated gliadin antibodies.

lymphocyte count without VA cannot be included in this algorithm. Regrettably, inadequate sampling during endoscopy and poor orientation of duodenal biopsy samples contribute significantly to diagnostic errors in SNEs, resulting in overdiagnosis of seronegative CD and unwarranted treatment (Pallav et al., 2012; Schiepatti et al., 2021; Schiepatti et al., 2021).

The other crucial point is absence of IgA EmA/TTA/DGP (Jansson-Knodell et al., 2020; Leonard et al., 2021; Schiepatti et al., 2019; Schiepatti et al., 2021). After confirming VA and the lack of celiac-specific antibodies, all potential causes need to be ruled out before considering seronegative CD or IVA (Jansson-Knodell et al., 2020; Leonard et al., 2021; Schiepatti et al., 2019; Schiepatti et al., 2021). In this context, relevant medical and clinical history, along with specific laboratory and molecular analyses can help in reaching the right diagnosis in many cases. Determining HLA types for celiac haplotypes should be considered, and is especially useful when negative to rule out seronegative CD (Jansson-Knodell et al., 2020; Leonard et al., 2021; Schiepatti et al., 2019; Schiepatti et al., 2021).

11.7 Clinical management and follow-up of seronegative enteropathies

The ideal approach to managing SNEs is still unclear. Attention has recently been directed towards creating uniform definitions and diagnostic criteria for SNCD and SNEs. However, there remains a lack of agreement on the most effective treatment options and the optimal timing for clinical and endoscopic follow-up of SNE patients (Schiepatti et al., 2021; Schiepatti et al., 2022).

11.7.1 Treatment

Drawing from our clinical experience, we suggest the following strategies for treating and monitoring these patients. Treatment for patients with SNEs caused by a known factor is dependent on the underlying etiology (see Table 11.1). Patients with seronegative CD should adhere to a rigorous and lifelong GFD, with histological response to the diet being the cornerstone for confirming the diagnosis. The best approach for patients with SNEs of unknown origin remains unclear. When HLA typing does not show celiac haplotypes, it is probable that these patients do not require a GFD. However, questions remain about the proper timing for initiating a GFD in patients with SNEs, negative

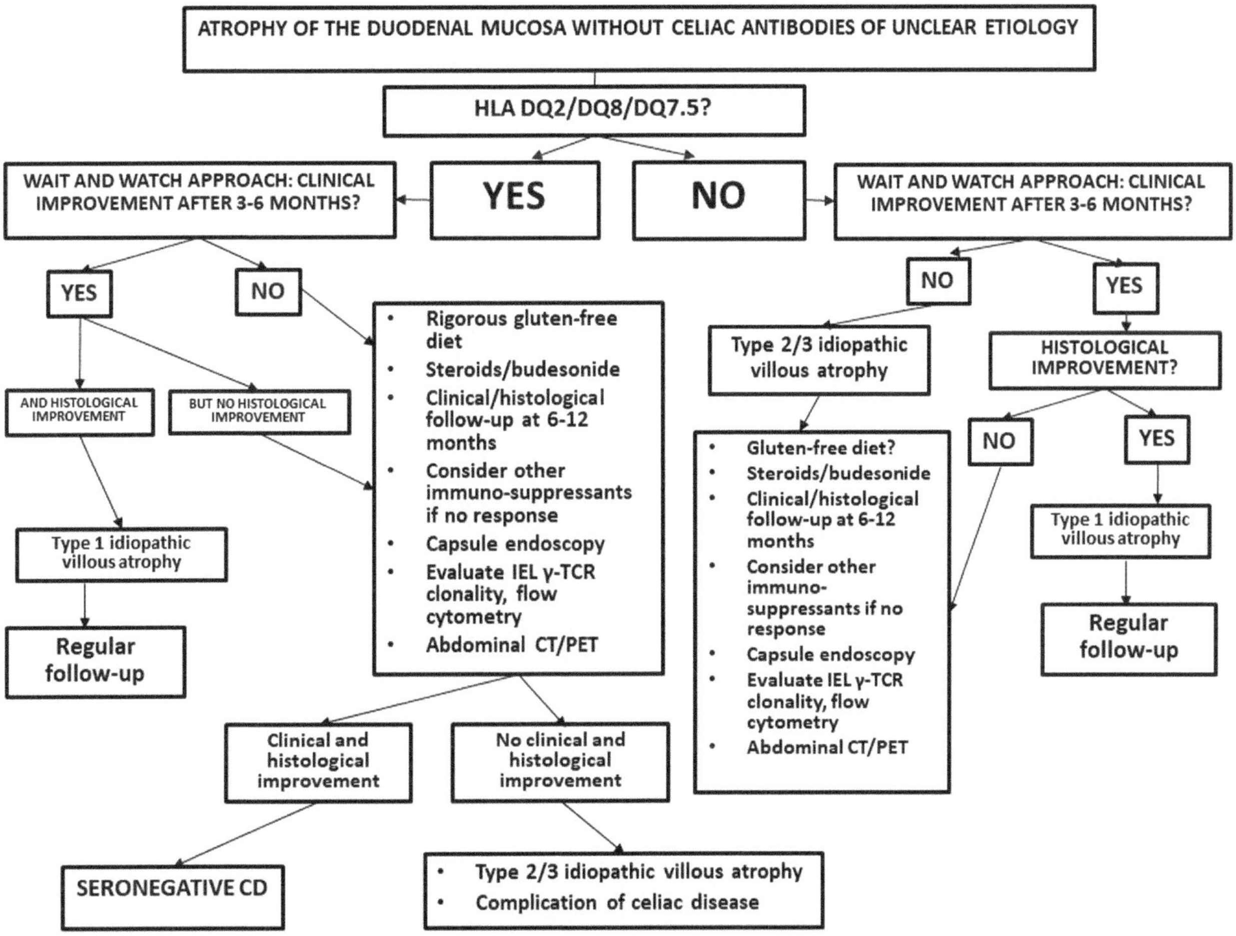

FIGURE 11.4 Evaluation and management of atrophy of the duodenal mucosa with negative celiac antibodies. *TCR*, T-cell receptor; *IEL*, intraepithelial lymphocytes.

investigations, and positive HLA-DQ2/DQ8/DQ7.5. We believe that the decision to start a GFD should depend on variables such as ethnicity, HLA typing, clinical picture severity, and the extent of small bowel involvement on baseline capsule endoscopy (Aziz et al., 2017; Chetcuti Zammit et al., 2020; Schiepatti et al., 2021; Schiepatti et al., 2020). We propose that Caucasian patients carrying HLA-DQ2/DQ8 with no alternative etiology identified could begin a GFD. For patients with these characteristics who exhibit severe malabsorption features and are aged >50 years, corticosteroids followed by oral budesonide may be considered to facilitate mucosal recovery. Parenteral nutrition should be contemplated for patients with severe malabsorption and malnutrition. In patients with no identified cause for VA, mild symptoms, and no laboratory abnormalities at diagnosis (IVA 1), adopting a "watch and wait" strategy with histological follow-up may be reasonable, as spontaneous histological recovery may occur (Aziz et al., 2017; Schiepatti et al., 2019; Schiepatti et al., 2021; Schiepatti et al., 2020). Fig. 11.4 illustrates a flowchart for diagnosing and managing seronegative CD and IVA.

11.7.2 Follow-up

Considering the potential for negative long-term outcomes in SNEs (Aziz et al., 2017; Jansson-Knodell et al., 2020; Leonard et al., 2021; Schiepatti et al., 2017, 2019; Schiepatti et al., 2021; Schiepatti et al., 2021), we strongly advocate for rigorous clinical and endoscopic follow-up for these patients, irrespective of the cause. Although it may be debatable whether to conduct a follow-up duodenal biopsy in patients with SNEs resulting from a discernible and reversible cause (e.g., iatrogenic or giardiasis in patients without predisposing factors) and who have experienced a full and satisfactory response to treatment, we feel it is essential to confirm both histological recovery and the extent of the affected small bowel using capsule endoscopy in patients with seronegative CD, CVID, autoimmune enteropathy, IVA, iatrogenic causes, and lymphomas (Schiepatti et al., 2021).

11.8 Conclusions

Differential diagnosis and management of SNEs in adults is still challenging. A methodical approach is required to promptly identify conditions that, despite their rarity, are associated with unfavorable outcomes. Evaluating histological responses to treatment is crucial for diagnosing seronegative CD and for guiding the clinical management and follow-up of other types of SNE. Future research opportunities might involve exploring the natural history of NCEs, pinpointing molecular mechanisms that lead to VA, and discovering noninvasive biomarkers for diagnosing and monitoring these conditions. This would promote a more uniform approach to their clinical management.

References

Abrams, J. A., Diamond, B., Rotterdam, H., & Green, P. H. R. (2004). Seronegative celiac disease: Increased prevalence with lesser degrees of villous atrophy. *Digestive Diseases and Sciences*, *49*(4), 546–550. Available from https://doi.org/10.1023/B:DDAS.0000026296.02308.00.

Akram, S., Murray, J. A., Pardi, D. S., Alexander, G. L., Schaffner, J. A., Russo, P. A., & Abraham, S. C. (2007). Adult autoimmune enteropathy: Mayo clinic rochester experience. *Clinical Gastroenterology and Hepatology*, *5*(11), 1282–1290. Available from https://doi.org/10.1016/j.cgh.2007.05.013.

Al-Saleem, T., & Al-Mondhiry, H. (2005). Immunoproliferative small intestinal disease (IPSID): A model for mature B-cell neoplasms. *Blood*, *105*(6), 2274–2280. Available from https://doi.org/10.1182/blood-2004-07-2755.

Al-Toma, A., Volta, U., Auricchio, R., Castillejo, G., Sanders, D. S., Cellier, C., Mulder, C. J., & Lundin, K. E. A. (2019). European Society for the Study of Coeliac Disease (ESsCD) guideline for coeliac disease and other gluten-related disorders. *United European Gastroenterology Journal*, *7*(5), 583–613. Available from https://doi.org/10.1177/2050640619844125.

Aziz, I., Peerally, M. F., Barnes, J. H., Kandasamy, V., Whiteley, J. C., Partridge, D., Vergani, P., Cross, S. S., Green, P. H., & Sanders, D. S. (2017). The clinical and phenotypical assessment of seronegative villous atrophy; A prospective UK centre experience evaluating 200 adult cases over a 15-year period (2000–2015). *Gut*, *66*(9), 1563–1572. Available from https://doi.org/10.1136/gutjnl-2016-312271.

Basson, M., Mezzarobba, M., Weill, A., Ricordeau, P., Allemand, H., Alla, F., & Carbonnel, F. (2016). Severe intestinal malabsorption associated with olmesartan: A French nationwide observational cohort study. *Gut*, *65*(10), 1664–1669. Available from https://doi.org/10.1136/gutjnl-2015-309690.

Bästlein, C., Burlefinger, R., Holzberg, E., Voeth, C., Garbrecht, M., & Ottenjann, R. (1988). Common variable immunodeficiency syndrome and nodular lymphoid hyperplasia in the small intestine. *Endoscopy*, *20*(5), 272–275. Available from https://doi.org/10.1055/s-2007-1018192.

Bellaguarda, E., & Hanauer, S. (2020). Checkpoint inhibitor–induced colitis. *American Journal of Gastroenterology*, *115*(2), 202–210. Available from https://doi.org/10.14309/ajg.0000000000000497.

Biagi, F., Balduzzi, D., Delvino, P., Schiepatti, A., Klersy, C., & Corazza, G. R. (2015). Prevalence of Whipple's disease in north-western Italy. *European Journal of Clinical Microbiology & Infectious Diseases*, *34*(7), 1347–1348. Available from https://doi.org/10.1007/s10096-015-2357-2.

Biagi, F., Bianchi, P. I., Trotta, L., & Corazza, G. R. (2009). Anti-goblet cell antibodies for the diagnosis of autoimmune enteropathy. *American Journal of Gastroenterology*, *104*(12). Available from https://doi.org/10.1038/ajg.2009.511.

Biagi, F., Bianchi, P. I., Zilli, A., Marchese, A., Luinetti, O., Lougaris, V., Plebani, A., Villanacci, V., & Corazza, G. R. (2012). The significance of duodenal mucosal atrophy in patients with common variable immunodeficiency: A clinical and histopathologic study. *American Journal of Clinical Pathology*, *138*(2), 185–189. Available from https://doi.org/10.1309/AJCPEIILH2C0WFYE.

Biagi, F., & Corazza, G. R. (2010). Mortality in celiac disease. *Nature Reviews Gastroenterology and Hepatology*, *7*(3), 158–162. Available from https://doi.org/10.1038/nrgastro.2010.2.

Biagi, F., Klersy, C., Balduzzi, D., & Corazza, G. R. (2010). Are we not over-estimating the prevalence of coeliac disease in the general population. *Annals of Medicine*, *42*(8), 557–561. Available from https://doi.org/10.3109/07853890.2010.523229.

Boscá, M. M., Añón, R., Mayordomo, E., Villagrasa, R., Balza, N., Amorós, C., Corts, J. R., & Benages, A. (2008). Methotrexate induced sprue-like syndrome. *World Journal of Gastroenterology*, *14*(45), 7009–7011. Available from https://doi.org/10.3748/wjg.14.7009.

Burbure, N., Lebwohl, B., Arguelles-Grande, C., Green, P. H. R., Bhagat, G., & Lagana, S. (2016). Olmesartan-associated sprue-like enteropathy: A systematic review with emphasis on histopathology. *Human Pathology*, *50*, 127–134. Available from https://doi.org/10.1016/j.humpath.2015.12.001.

Catassi, C., Fabiani, E., Spagnuolo, M. I., Barera, G., & Guarino, A. (1999). Severe and protracted diarrhea: Results of the 3-year SIGEP multicenter survey. *Journal of Pediatric Gastroenterology and Nutrition*, *29*(1), 63–68. Available from https://doi.org/10.1097/00005176-199907000-00016.

Chapel, H., Lucas, M., Lee, M., Bjorkander, J., Webster, D., Grimbacher, B., Fieschi, C., Thon, V., Abedi, M. R., & Hammarstrom, L. (2008). Common variable immunodeficiency disorders: Division into distinct clinical phenotypes. *Blood*, *112*(2), 277–286. Available from https://doi.org/10.1182/blood-2007-11-124545.

Chetcuti Zammit, S., Schiepatti, A., Aziz, I., Kurien, M., Sanders, D. S., & Sidhu, R. (2020). Use of small-bowel capsule endoscopy in cases of equivocal celiac disease. *Gastrointestinal Endoscopy*, *91*(6), 1312–1321. Available from https://doi.org/10.1016/j.gie.2019.12.044, e2.

Ciccocioppo, R., Croci, G. A., Biagi, F., Vanoli, A., Alvisi, C., Cavenaghi, G., Riboni, R., Arra, M., Gobbi, P. G., Paulli, M., & Corazza, G. R. (2018). Intestinal T-cell lymphoma with enteropathy-associated T-cell lymphoma-like features arising in the setting of adult autoimmune enteropathy. *Hematological Oncology*, *36*(2), 481–488. Available from https://doi.org/10.1002/hon.2494.

Ciccocioppo, R., Russo, M. L., Bernardo, M. E., Biagi, F., Catenacci, L., Avanzini, M. A., Alvisi, C., Vanoli, A., Manca, R., Luinetti, O., Locatelli, F., & Corazza, G. R. (2012). Mesenchymal stromal cell infusions as rescue therapy for corticosteroid-refractory adult autoimmune enteropathy. *Mayo Clinic Proceedings*, *87*(9), 909–914. Available from https://doi.org/10.1016/j.mayocp.2012.04.014.

Collin, P., Kaukinen, K., Vogelsang, H., Korponay-Szabó, I., Sommer, R., Schreier, E., Volta, U., Granito, A., Veronesi, L., Mascart, F., Ocmant, A., Ivarsson, A., Lagerqvist, C., Bürgin-Wolff, A., Hadziselimovic, F., Furlano, R. I., Sidler, M. A., Mulder, C. J. J., Goerres, M. S., ... Mäki, M. (2005). Antiendomysial and antihuman recombinant tissue transglutaminase antibodies in the diagnosis of coeliac disease: A biopsy-proven European multicentre study. *European Journal of Gastroenterology and Hepatology*, *17*(1), 85–91. Available from https://doi.org/10.1097/00042737-200501000-00017.

Conley, M. E., Notarangelo, L. D., & Etzioni, A. (1999). Diagnostic criteria for primary immunodeficiencies. *Clinical Immunology*, *93*(3), 190–197. Available from https://doi.org/10.1006/clim.1999.4799.

Corazza, G. R., Biagi, F., Volta, U., Andreani, M. L., De Franceschi, L., & Gasbarrini, G. (1997). Autoimmune enteropathy and villous atrophy in adults. *Lancet*, *350*(9071), 106–109. Available from https://doi.org/10.1016/S0140-6736(97)01042-8.

Costetti, M., Schiepatti, A., Fraticelli, S., Costa, S., Maimaris, S., Lenti, M. V., Villani, L., Bianchi, P. I., Di Sabatino, A., Corazza, G. R., Vanoli, A., & Biagi, F. (2021). Clinical and gastro-duodenal histopathological features of enteropathy due to angiotensin II receptor blockers. *Digestive and Liver Disease*, *53*(10), 1262–1267. Available from https://doi.org/10.1016/j.dld.2021.07.002.

Culliford, A., Markowitz, D., Rotterdam, H., & Green, P. H. R. (2004). Scalloping of Duodenal Mucosa in Crohn's Disease. *Inflammatory Bowel Diseases*, *10*(3), 270–273. Available from https://doi.org/10.1097/00054725-200405000-00015.

Cyrany, J., Vasatko, T., Machac, J., Nova, M., Szanyi, J., & Kopacova, M. (2014). Letter: telmisartan-associated enteropathy - Is there any class effect. *Alimentary Pharmacology & Therapeutics*, *40*(5), 569–570. Available from https://doi.org/10.1111/apt.12850.

Daniels, J. A., Lederman, H. M., Maitra, A., & Montgomery, E. A. (2007). Gastrointestinal tract pathology in patients with common variable immunodeficiency (CVID): A clinicopathologic study and review. *American Journal of Surgical Pathology*, *31*(12), 1800–1812. Available from https://doi.org/10.1097/PAS.0b013e3180cab60c.

De Silvestri, A., Schiepatti, A., Fowler, B. D., Badulli, C., Belvisi, F., Cacciatore, R., Pasi, A., Zingone, F., Goodwin, J., Key, T., Rej, A., Sanders, D. S., & Biagi, F. (2023). DQA1*0102 DQB1*0602 haplotype distinguishes coeliac disease and its complications from gluten unrelated enteropathies. *European Journal of Gastroenterology and Hepatology*, *35*(1), 64–72. Available from https://doi.org/10.1097/MEG.0000000000002480.

Degaetani, M., Tennyson, C. A., Lebwohl, B., Lewis, S. K., Abu Daya, H., Arguelles-Grande, C., Bhagat, G., & Green, P. H. R. (2013). Villous atrophy and negative celiac serology: A diagnostic and therapeutic dilemma. *American Journal of Gastroenterology*, *108*(5), 647–653. Available from https://doi.org/10.1038/ajg.2013.45.

Di Sabatino, A., & Corazza, G. R. (2009). Coeliac disease. *The Lancet*, *373*(9673), 1480–1493. Available from https://doi.org/10.1016/S0140-6736(09)60254-3.

Dickey, W., Hughes, D. F., & McMillan, S. A. (2000). Reliance on serum endomysial antibody testing underestimates the true prevalence of coeliac disease by one fifth. *Scandinavian Journal of Gastroenterology*, *35*(2), 181–183. Available from https://doi.org/10.1080/003655200750024362.

Dong, Y. H., Jin, Y., Tsacogianis, T. N., He, M., Hsieh, P. H., & Gagne, J. J. (2018). Use of olmesartan and enteropathy outcomes: A multi-database study. *Alimentary Pharmacology and Therapeutics*, *47*(6), 792–800. Available from https://doi.org/10.1111/apt.14518.

Fernández-Bañares, F., Crespo, L., Núñez, C., López-Palacios, N., Tristán, E., Vivas, S., Farrais, S., Arau, B., Vidal, J., Roy, G., & Esteve, M. (2020). Gamma delta + intraepithelial lymphocytes and coeliac lymphogram in a diagnostic approach to coeliac disease in patients with seronegative villous atrophy. *Alimentary Pharmacology & Therapeutics*, *51*(7), 699–705. Available from https://doi.org/10.1111/apt.15663.

Folwaczny, C., Noehl, N., Tschop, K., Endres, S. P., Heldwein, W., Loeschke, K., & Fricke, H. (1997). Goblet cell autoantibodies in patients with inflammatory bowel disease and their first-degree relatives. *Gastroenterology*, *113*(1), 101–106. Available from https://doi.org/10.1016/s0016-5085(97)70085-4.

Foukas, P. G., & de Leval, L. (2015). Recent advances in intestinal lymphomas. *Histopathology*, *66*(1), 112–136. Available from https://doi.org/10.1111/his.12596.

Fung, W. P., Tan, K. K., Yu, S. F., & Sho, K. M. (1970). Malabsorption and subtotal villous atrophy secondary to pulmonary and intestinal tuberculosis. *Gut*, *11*(3), 212–216. Available from https://doi.org/10.1136/gut.11.3.212.

Greenson, J. K. (2015). The biopsy pathology of non-coeliac enteropathy. *Histopathology*, *66*(1), 29–36. Available from https://doi.org/10.1111/his.12522.

Gustafsson, I., Repo, M., Popp, A., Kaukinen, K., Hiltunen, P., Arvola, T., Taavela, J., Vornanen, M., Kivelä, L., & Kurppa, K. (2020). Prevalence and diagnostic outcomes of children with duodenal lesions and negative celiac serology. *Digestive and Liver Disease*, *52*(3), 289–295. Available from https://doi.org/10.1016/j.dld.2019.11.011.

Heneghan, M. A., Stevens, F. M., Cryan, E. M., Warner, R. H., & McCarthy, C. F. (1997). Celiac sprue and immunodeficiency states: A 25-year review. *Journal of Clinical Gastroenterology*, *25*(2), 421–425. Available from https://doi.org/10.1097/00004836-199709000-00004.

Herman, M., Rubio-Tapia, A., Marietta, E., Wu, T.-T., & Murray, J. (2013). Severe Enteropathy in a Patient on Valsartan. *American Journal of Gastroenterology*, 108. Available from https://doi.org/10.14309/00000434-201310001-01011.

Hermaszewski, R. A., & Webster, A. D. B. (1993). Primary hypogammaglobulinaemia: A survey of clinical manifestations and complications. *QJM*, *86*(1), 31–42. Available from https://doi.org/10.1093/oxfordjournals.qjmed.a068735.

Hibi, T., Ohara, M., Kobayashi, K., Brown, W. R., Toda, K., Takaishi, H., Hosoda, Y., Hayashi, A., Iwao, Y., & Watanabe, M. (1994). Enzyme linked immunosorbent assay (ELISA) and immunoprecipitation studies on anti-goblet cell antibody using a mucin producing cell line in patients with inflammatory bowel disease. *Gut*, *35*(2), 224–230. Available from https://doi.org/10.1136/gut.35.2.224.

Hopper, A. D., Cross, S. S., Hurlstone, D. P., McAlindon, M. E., Lobo, A. J., Hadjivassiliou, M., Sloan, M. E., Dixon, S., & Sanders, D. S. (2007). Pre-endoscopy serological testing for coeliac disease: Evaluation of a clinical decision tool. *British Medical Journal*, *334*(7596), 729–732. Available from https://doi.org/10.1136/bmj.39133.668681.BE.

Husby, S., Koletzko, S., Korponay-Szabó, I., Kurppa, K., Mearin, M. L., Ribes-Koninckx, C., Shamir, R., Troncone, R., Auricchio, R., Castillejo, G., Christensen, R., Dolinsek, J., Gillett, P., Hróbjartsson, A., Koltai, T., Maki, M., Nielsen, S. M., Popp, A., Størdal, K., … Wessels, M. (2020). European Society Paediatric Gastroenterology, Hepatology and Nutrition Guidelines for Diagnosing Coeliac Disease 2020. *Journal of Pediatric Gastroenterology and Nutrition*, *70*(1), 141–156. Available from https://doi.org/10.1097/MPG.0000000000002497.

Jansson-Knodell, C. L., Murray, J. A., & Rubio-Tapia, A. (2020). Management of small bowel villous atrophy in patients seronegative for celiac disease. *American Journal of Gastroenterology*, *115*(4), 492–497. Available from https://doi.org/10.14309/ajg.0000000000000575.

Jørgensen, S. F., Reims, H. M., Frydenlund, D., Holm, K., Paulsen, V., Michelsen, A. E., Jørgensen, K. K., Osnes, L. T., Bratlie, J., Eide, T. J., Dahl, C. P., Holter, E., Tronstad, R. R., Hanevik, K., Brattbakk, H. R., Kaveh, F., Fiskerstrand, T., Kran, A. M. B., Ueland, T., … Fevang, B. (2016). A cross-sectional study of the prevalence of gastrointestinal symptoms and pathology in patients with common variable immunodeficiency. *American Journal of Gastroenterology*, *111*(10), 1467–1475. Available from https://doi.org/10.1038/ajg.2016.329.

Kalha, I., & Sellin, J. H. (2004). Common variable immunodeficiency and the gastrointestinal tract. *Current Gastroenterology Reports*, *6*(5), 377–383. Available from https://doi.org/10.1007/s11894-004-0053-y.

Kamar, N., Faure, P., Dupuis, E., Cointault, O., Joseph-Hein, K., Durand, D., Moreau, J., & Rostaing, L. (2004). Villous atrophy induced by mycophenolate mofetil in renal-transplant patients. *Transplant International*, *17*(8), 463–467. Available from https://doi.org/10.1007/s00147-004-0745-0.

Kaosombatwattana, U., Limsrivilai, J., Pongpaibul, A., Maneerattanaporn, M., & Charatcharoenwitthaya, P. (2017). Severe enteropathy with villous atrophy in prolonged mefenamic acid users – A currently under-recognized in previously well-recognized complication Case report and review of literature. *Medicine (United States)*, *96*(44). Available from https://doi.org/10.1097/MD.0000000000008445.

Kapembwa, M. S., Batman, P. A., Fleming, S. C., & Griffin, G. E. (1989). HIV enteropathy. *The Lancet*, *334*(8678–8679), 1521–1522. Available from https://doi.org/10.1016/S0140-6736(89)92961-9.

Khodadad, A., Aghamohammadi, A., Parvaneh, N., Rezaei, N., Mahjoob, F., Bashashati, M., Movahedi, M., Fazlollahi, M. R., Zandieh, F., Roohi, Z., Abdollahzade, S., Salavati, A., Kouhi, A., Talebpour, B., & Daryani, N. E. (2007). Gastrointestinal manifestations in patients with common variable immunodeficiency. *Digestive Diseases and Sciences*, *52*(11), 2977–2983. Available from https://doi.org/10.1007/s10620-006-9736-6.

Korpe, P. S., & Petri, W. A. (2012). Environmental enteropathy: Critical implications of a poorly understood condition. *Trends in Molecular Medicine*, *18*(6), 328–336. Available from https://doi.org/10.1016/j.molmed.2012.04.007.

Lappinga, P. J., Abraham, S. C., Murray, J. A., Vetter, E. A., Patel, R., & Wu, T. T. (2010). Small intestinal bacterial overgrowth histopathologic features and clinical correlates in an underrecognized entity. *Archives of Pathology and Laboratory Medicine*, *134*(2), 264–270. Available from https://doi.org/10.1043/1543-2165-134.2.264, http://www.archivesofpathology.org/doi/pdf.

Lebwohl, B., Green, P. H. R., Söderling, J., Roelstraete, B., & Ludvigsson, J. F. (2020). Association between celiac disease and mortality risk in a Swedish population. *JAMA - Journal of the American Medical Association*, *323*(13), 1277–1285. Available from https://doi.org/10.1001/jama.2020.1943.

Lebwohl, B., Sanders, D. S., & Green, P. H. R. (2018). Coeliac disease. *The Lancet*, *391*(10115), 70–81. Available from https://doi.org/10.1016/S0140-6736(17)31796-8.

Leffler, D. A., & Schuppan, D. (2010). Update on serologic testing in celiac disease. *American Journal of Gastroenterology*, *105*(12), 2520–2524. Available from https://doi.org/10.1038/ajg.2010.276.

Leonard, M. M., Lebwohl, B., Rubio-Tapia, A., & Biagi, F. (2021). AGA clinical practice update on the evaluation and management of seronegative enteropathies: Expert review. *Gastroenterology*, *160*(1), 437–444. Available from https://doi.org/10.1053/j.gastro.2020.08.061.

Levinson, J. D., & Nastro, L. J. (1978). Giardiasis with total villous atrophy. *Gastroenterology*, *74*(2), 271–275. Available from https://doi.org/10.1016/0016-5085(78)90809-0.

Lougaris, V., Ravelli, A., Villanacci, V., Salemme, M., Soresina, A., Fuoti, M., Lanzarotto, F., Lanzini, A., Plebani, A., & Bassotti, G. (2015). Gastrointestinal pathologic abnormalities in pediatric- and adult-onset common variable immunodeficiency. *Digestive Diseases and Sciences*, *60* (8), 2384–2389. Available from https://doi.org/10.1007/s10620-015-3638-4.

Ludvigsson, J. F., Bai, J. C., Biagi, F., Card, T. R., Ciacci, C., Ciclitira, P. J., Green, P. H. R., Hadjivassiliou, M., Holdoway, A., Van Heel, D. A., Kaukinen, K., Leffler, D. A., Leonard, J. N., Lundin, K. E. A., McGough, N., Davidson, M., Murray, J. A., Swift, G. L., Walker, M. M., … Sanders, D. S. (2014). Diagnosis and management of adult coeliac disease: Guidelines from the British society of gastroenterology. *Gut*, *63*(8), 1210–1228. Available from https://doi.org/10.1136/gutjnl-2013-306578.

Lundin, K. E. A., & Green, P. H. R. (2022). Seronegative coeliac disease and non-coeliac enteropathies: Precision medicine, precision medicine, where are you. *Gut*, *71*(11), 2148–2149. Available from https://doi.org/10.1136/gutjnl-2022-327768.

MacDonald, T. T., & Spencer, J. (1988). Evidence that activated mucosal T cells play a role in the pathogenesis of enteropathy in human small intestine. *Journal of Experimental Medicine*, *167*(4), 1341–1349. Available from https://doi.org/10.1084/jem.167.4.1341.

Malamut, G., Meresse, B., Kaltenbach, S., Derrieux, C., Verkarre, V., Macintyre, E., Ruskone–Fourmestraux, A., Fabiani, B., Radford–Weiss, I., Brousse, N., Hermine, O., Cerf–Bensussan, N., & Cellier, C. (2014). Small intestinal CD4 + T-cell lymphoma is a heterogenous entity with common pathology features. *Clinical Gastroenterology and Hepatology*, *12*(4), 599–608. Available from https://doi.org/10.1016/j.cgh.2013.11.028, e1.

Malamut, G., Verkarre, V., Callens, C., Colussi, O., Rahmi, G., Macintyre, E., Haïoun, C., Meresse, B., Brousse, N., Romana, S., Hermine, O., Cerf–Bensussan, N., & Cellier, C. (2012). Enteropathy-associated T-cell lymphoma complicating an autoimmune enteropathy. *Gastroenterology*, *142*(4), 726–729. Available from https://doi.org/10.1053/j.gastro.2011.12.040, e3.

Malamut, G., Verkarre, V., Suarez, F., Viallard, J. F., Lascaux, A. S., Cosnes, J., Bouhnik, Y., Lambotte, O., Béchade, D., Ziol, M., Lavergne, A., Hermine, O., Cerf-Bensussan, N., & Cellier, C. (2010). The enteropathy associated with common variable immunodeficiency: The delineated frontiers with celiac disease. *American Journal of Gastroenterology*, *105*(10), 2262–2275. Available from https://doi.org/10.1038/ajg.2010.214.

Mandile, R., Maglio, M., Pellino, N., Russo, M., Miele, E., Spagnuolo, M. I., Troncone, R., & Auricchio, R. (2021). Seronegative villous atrophy in children: Clinical and immunohistochemical features. *Journal of Pediatric Gastroenterology & Nutrition*, *72*(2), 282–287. Available from https://doi.org/10.1097/mpg.0000000000002917.

Marietta, E. V., Cartee, A., Rishi, A., & Murray, J. A. (2015). Drug-induced enteropathy. *Digestive Diseases*, *33*(2), 215–220. Available from https://doi.org/10.1159/000370205.

Marin-Acevedo, J. A., Chirila, R. M., & Dronca, R. S. (2019). Immune checkpoint inhibitor toxicities. *Mayo Clinic Proceedings*, *94*(7), 1321–1329. Available from https://doi.org/10.1016/j.mayocp.2019.03.012.

Marth, t, Moos, V., & Müller, C. (2016). Infection with Tropheryma whipplei and Whipple's disease. *The Lancet Infectious Diseases*, *16*, 13–22.

Marthey, L., Cadiot, G., Seksik, P., Pouderoux, P., Lacroute, J., Skinazi, F., Mesnard, B., Chayvialle, J. A., Savoye, G., Druez, A., Parlier, D., Abitbol, V., Gompel, M., Eoche, M., Poncin, E., Bobichon, R., Colardelle, P., Wils, P., Salloum, H., … Carbonnel, F. (2014). Olmesartan-associated enteropathy: Results of a national survey. *Alimentary Pharmacology & Therapeutics*, *40*(9), 1103–1109. Available from https://doi.org/10.1111/apt.12937.

Marthinsen, L. M., Scott, H., & Ejderhamn, J. (2008). Autoimmune enteropathy in Swedish children, 1985–2002: A call for strict diagnostic criteria. *Scandinavian Journal of Gastroenterology*, *43*(9), 1102–1107. Available from https://doi.org/10.1080/00365520802029864.

Masia, R., Peyton, S., Lauwers, G. Y., & Brown, I. (2014). Gastrointestinal biopsy findings of autoimmune enteropathy: A review of 25 cases. *American Journal of Surgical Pathology*, *38*(10), 1319–1329. Available from https://doi.org/10.1097/PAS.0000000000000317.

Mooney, P. D., Wong, S. H., Johnston, A. J., Kurien, M., Avgerinos, A., & Sanders, D. S. (2015). Increased detection of celiac disease with measurement of deamidated gliadin peptide antibody before endoscopy. *Clinical Gastroenterology and Hepatology*, *13*(7), 1278–1284. Available from https://doi.org/10.1016/j.cgh.2015.01.010, e1.

Natacha, P.-M. d. S., Canioni, D., Ganousse, S., Rieux-Laucat, F., Goulet, O., Ruemmele, F., & Brousse, N. (2009). Digestive histopathological presentation of IPEX syndrome. *Modern Pathology*, *22*(1), 95–102. Available from https://doi.org/10.1038/modpathol.2008.161.

Pallav, K., Leffler, D. A., Tariq, S., Kabbani, T., Hansen, J., Peer, A., Bhansali, A., Najarian, R., & Kelly, C. P. (2012). Noncoeliac enteropathy: The differential diagnosis of villous atrophy in contemporary clinical practice. *Alimentary Pharmacology and Therapeutics*, *35*(3), 380–390. Available from https://doi.org/10.1111/j.1365-2036.2011.04938.x.

Park, M. A., Li, J. T., Hagan, J. B., Maddox, D. E., & Abraham, R. S. (2008). Common variable immunodeficiency: A new look at an old disease. *The Lancet*, *372*(9637), 489–502. Available from https://doi.org/10.1016/S0140-6736(08)61199-X.

Patey-Mariaud de Serre, N., Reijasse, D., Verkarre, V., Canioni, D., Colomb, V., Haddad, E., & Brousse, N. (2002). Chronic intestinal graft-versus-host disease: Clinical, histological and immunohistochemical analysis of 17 children. *Bone Marrow Transplantation*, *29*(3), 223–230. Available from https://doi.org/10.1038/sj.bmt.1703329.

Penny, H. A., Raju, S. A., Lau, M. S., Marks, L. J. S., Baggus, E. M. R., Bai, J. C., Bassotti, G., Bontkes, H. J., Carroccio, A., Danciu, M., Derakhshan, M. H., Ensari, A., Ganji, A., Green, P. H. R., Johnson, M. W., Ishaq, S., Lebwohl, B., Levene, A., Maxim, R., … Sanders, D. S. (2021). Accuracy of a no-biopsy approach for the diagnosis of coeliac disease across different adult cohorts. *Gut*, *70*(5), 876–883. Available from https://doi.org/10.1136/gutjnl-2020-320913.

Pensieri, M. V., Pulvirenti, F., Schiepatti, A., Maimaris, S., Lattanzio, S., Quinti, I., Klersy, C., Corazza, G. R., & Biagi, F. (2019). The high mortality of patients with common variable immunodeficiency and small bowel villous atrophy. *Scandinavian Journal of Gastroenterology*, *54*(2), 164–168. Available from https://doi.org/10.1080/00365521.2019.1568543.

Powell, N., Ibraheim, H., Raine, T., Speight, R. A., Papa, S., Brain, O., Green, M., Samaan, M. A., Spain, L., Yousaf, N., Hunter, N., Eldridge, L., Pavlidis, P., Irving, P., Hayee, B. 'H., Turajlic, S., Larkin, J., Lindsay, J. O., & Gore, M. (2020). British Society of Gastroenterology endorsed guidance for the management of immune checkpoint inhibitor-induced enterocolitis. *The Lancet Gastroenterology & Hepatology*, *5*(7), 679–697. Available from https://doi.org/10.1016/s2468-1253(20)30014-5.

Resnick, E. S., Moshier, E. L., Godbold, J. H., & Cunningham-Rundles, C. (2012). Morbidity and mortality in common variable immune deficiency over 4 decades. *Blood*, *119*(7), 1650–1657. Available from https://doi.org/10.1182/blood-2011-09-377945.

Reunala, T., Hervonen, K., & Salmi, T. (2021). Dermatitis Herpetiformis: An update on diagnosis and management. *American Journal of Clinical Dermatology*, *22*(3), 329–338. Available from https://doi.org/10.1007/s40257-020-00584-2.

Rostami, K., Kerckhaert, J., Tiemessen, R., Von Blomberg, B. M. E., Meijer, J. W. R., & Mulder, C. J. J. (1999). Sensitivity of antiendomysium and antigliadin antibodies in untreated celiac disease: Disappointing in clinical practice. *American Journal of Gastroenterology*, *94*(4), 888–894. Available from https://doi.org/10.1016/S0002-9270(99)00038-6.

Rostami, K., Kerckhaert, J., Von Blomberg, B. M. E., Meijer, J. W. R., Wahab, P., & Mulder, C. J. J. (1998). SAT and serology in adult coeliacs, seronegative coeliac disease seems a reality. *Netherlands Journal of Medicine*, *53*(1), 15–19. Available from https://doi.org/10.1016/S0300-2977(98)00050-3.

Rubio-Tapia, A., Herman, M. L., Ludvigsson, J. F., Kelly, D. G., Mangan, T. F., Wu, T. T., & Murray, J. A. (2012). Severe spruelike enteropathy associated with olmesartan. *Mayo Clinic Proceedings*, *87*(8), 732–738. Available from https://doi.org/10.1016/j.mayocp.2012.06.003.

Rubio-Tapia, A., Talley, N. J., Gurudu, S. R., Wu, T. T., & Murray, J. A. (2010). Gluten-free diet and steroid treatment are effective therapy for most patients with collagenous sprue. *Clinical Gastroenterology and Hepatology*, *8*(4), 344. Available from https://doi.org/10.1016/j.cgh.2009.12.023, e3.

Salmi, T. T., Collin, P., Korponay-Szabo, I. R., Laurila, K., Partanen, J., Huhtala, H., Kiraly, R., Lorand, L., Reunala, T., Maki, M., & Kaukinen, K. (2006). Endomysial antibody-negative coeliac disease: Clinical characteristics and intestinal autoantibody deposits. *Gut*, *55*(12), 1746–1753. Available from https://doi.org/10.1136/gut.2005.071514.

Schiepatti, A., Cincotta, M., Biagi, F., & Sanders, D. S. (2021). Enteropathies with villous atrophy but negative coeliac serology in adults: Current issues. *BMJ Open Gastroenterology*, *8*(1), e000630. Available from https://doi.org/10.1136/bmjgast-2021-000630.

Schiepatti, A., Premoli, A., Maimaris, S., Rizzo, M., Marples, M., Villani, L., Scott, N., Sottotetti, F., Sanders, D. S., Biagi, F., & Donnellan, C. (2022). Small bowel villous atrophy due to immune-checkpoint inhibitors: Report of two cases and literature review. *Drugs in Context*, *11*, 1–12. Available from https://doi.org/10.7573/dic.2022-6-3.

Schiepatti, A., Rej, A., Maimaris, S., Cross, S. S., Porta, P., Aziz, I., Key, T., Goodwin, J., Therrien, A., Yoosuf, S., Leffler, D. A., Silvester, J. A., Klersy, C., Biagi, F., & Sanders, D. S. (2021). Clinical classification and long-term outcomes of seronegative coeliac disease: A 20-year multicentre follow-up study. *Alimentary Pharmacology & Therapeutics*, *54*(10), 1278–1289. Available from https://doi.org/10.1111/apt.16599.

Schiepatti, A., Savioli, J., Vernero, M., Borrelli de Andreis, F., Perfetti, L., Meriggi, A., & Biagi, F. (2020). Pitfalls in the diagnosis of coeliac disease and gluten-related disorders. *Nutrients*, *12*(6). Available from https://doi.org/10.3390/nu12061711.

Schiepatti, A., Biagi, F., Cumetti, D., Luinetti, O., Sonzogni, A., Mugellini, A., & Corazza, G. R. (2016). Olmesartan-associated enteropathy: New insights on the natural history? Report of two cases. *Scandinavian Journal of Gastroenterology*, *51*(2), 152–156. Available from https://doi.org/10.3109/00365521.2015.1074719.

Schiepatti, A., Biagi, F., Fraternale, G., Vattiato, C., Balduzzi, D., Agazzi, S., Alpini, C., Klersy, C., & Corazza, G. R. (2017). Short article: Mortality and differential diagnoses of villous atrophy without coeliac antibodies. *European Journal of Gastroenterology and Hepatology*, *29*(5), 572–576. Available from https://doi.org/10.1097/MEG.0000000000000836.

Schiepatti, A., Sanders, D. S., Aziz, I., De Silvestri, A., Goodwin, J., Key, T., Quaye, L., Giuffrida, P., Vanoli, A., Paulli, M., Cross, S. S., Vergani, P., Betti, E., Maiorano, G., Ellis, R., Snowden, J. A., Di Sabatino, A., Corazza, G. R., & Biagi, F. (2020). Clinical phenotype and mortality in patients with idiopathic small bowel villous atrophy: A dual-centre international study. *European Journal of Gastroenterology and Hepatology*, *32*(8), 938–949. Available from https://doi.org/10.1097/MEG.0000000000001726.

Schiepatti, A., Sanders, D. S., Baiardi, P., Caio, G., Ciacci, C., Kaukinen, K., Lebwohl, B., Leffler, D., Malamut, G., Murray, J. A., Rostami, K., Rubio-Tapia, A., Volta, U., & Biagi, F. (2022). Nomenclature and diagnosis of seronegative coeliac disease and chronic non-coeliac enteropathies in adults: The Paris consensus. *Gut*, *71*(11), 2218–2225. Available from https://doi.org/10.1136/gutjnl-2021-326645.

Schiepatti, A., Sanders, D. S., & Biagi, F. (2018). Seronegative coeliac disease: Clearing the diagnostic dilemma. *Current Opinion in Gastroenterology*, *34*(3), 154–158. Available from https://doi.org/10.1097/MOG.0000000000000436.

Schiepatti, A., Sanders, D. S., Zuffada, M., Luinetti, O., Iraqi, A., & Biagi, F. (2019). Overview in the clinical management of patients with seronegative villous atrophy. *European Journal of Gastroenterology and Hepatology*, *31*(4), 409–417. Available from https://doi.org/10.1097/MEG.0000000000001340.

Scialom, S., Malamut, G., Meresse, B., Guegan, N., Brousse, N., Verkarre, V., Derrieux, C., Macintyre, E., Seksik, P., Savoye, G., Cadiot, G., Vuitton, L., Marthey, L., Carbonnel, F., Cerf-Bensussan, N., Cellier, C., & Sampson, A. P. (2015). Gastrointestinal disorder associated with olmesartan mimics autoimmune enteropathy. *PLOS ONE*, *10*(6). Available from https://doi.org/10.1371/journal.pone.0125024.

Sharma, A., Choung, R. S., Wang, X. J., Russo, P. A., Wu, T. T., Nehra, V., & Murray, J. A. (2018). Features of adult autoimmune enteropathy compared with refractory celiac disease. *Clinical Gastroenterology and Hepatology*, *16*(6), 877–883. Available from https://doi.org/10.1016/j.cgh.2017.12.044, e1.

Sharma, A., Oishi, N., Boddicker, R. L., Hu, G., Benson, H. K., Ketterling, R. P., Greipp, P. T., Knutson, D. L., Kloft-Nelson, S. M., He, R., Eckloff, B. W., Jen, J., Nair, A. A., Davila, J. I., Dasari, S., Lazaridis, K. N., Nora Bennani, N., Wu, T. T., Nowakowski, G. S., ... Feldman, A. L. (2018). Recurrent stat3-jak2 fusions in indolent t-cell lymphoproliferative disorder of the gastrointestinal tract. *Blood*, *131*(20), 2262–2266. Available from https://doi.org/10.1182/blood-2018-01-830968.

Singh, P., Arora, A., Strand., Tor, A., Leffler., Daniel, A., Catassi, C., Green, P. H., Kelly, C. P., Ahuja, V., & Makharia, G. K. (2018). Global prevalence of celiac disease: Systematic review and meta-analysis. *Clinical Gastroenterology and Hepatology*, *16*(6), 823–836. Available from https://doi.org/10.1016/j.cgh.2017.06.037, e2.

Stacey, R., & Green, J. T. (2014). Radiation-induced small bowel disease: Latest developments and clinical guidance. *Therapeutic Advances in Chronic Disease*, *5*(1), 15–29. Available from https://doi.org/10.1177/2040622313510730.

Stefanini, M. (1948). Hypoproteinemia and edema in the course of tropical sprue. *Gastroenterology*, *11*(1), 50–55.

Taavela, J., Koskinen, O., Huhtala, H., Lähdeaho, M.-L., Popp, A., Laurila, K., Collin, P., Kaukinen, K., Kurppa, K., Mäki, M., & Sestak, K. (2013). Validation of morphometric analyses of small-intestinal biopsy readouts in celiac disease. *PLoS ONE*, *8*(10). Available from https://doi.org/10.1371/journal.pone.0076163.

Tai, F. W. D., & McAlindon, M. E. (2018). NSAIDs and the small bowel. *Current Opinion in Gastroenterology*, *34*(3), 175–182. Available from https://doi.org/10.1097/MOG.0000000000000427.

Touchefeu, Y., Montassier, E., Nieman, K., Gastinne, T., Potel, G., Bruley Des Varannes, S., Le Vacon, F., & De La Cochetière, M. F. (2014). Systematic review: The role of the gut microbiota in chemotherapy- or radiation-induced gastrointestinal mucositis - Current evidence and potential clinical applications. *Alimentary Pharmacology and Therapeutics*, *40*(5), 409–421. Available from https://doi.org/10.1111/apt.12878.

Unsworth, D. J., & Walker-Smith, J. A. (1985). Autoimmunity in diarrhoeal disease. *Journal of Pediatric Gastroenterology and Nutrition*, *4*(3), 375–380. Available from https://doi.org/10.1097/00005176-198506000-00009.

Villalta, D., Tonutti, E., Prause, C., Koletzko, S., Uhlig, H. H., Vermeersch, P., Bossuyt, X., Stern, M., Laass, M. W., Ellis, J. H., Ciclitira, P. J., Richter, T., Daehnrich, C., Schlumberger, W., & Mothes, T. (2010). IgG antibodies against deamidated gliadin peptides for diagnosis of celiac disease in patients with IgA deficiency. *Clinical Chemistry*, *56*(3), 464–468. Available from https://doi.org/10.1373/clinchem.2009.128132.

Volta, U., Caio, G., Boschetti, E., Giancola, F., Rhoden, K. J., Ruggeri, E., Paterini, P., & De Giorgio, R. (2016). Seronegative celiac disease: Shedding light on an obscure clinical entity. *Digestive and Liver Disease*, *48*(9), 1018–1022. Available from https://doi.org/10.1016/j.dld.2016.05.024.

Volta, U., Mumolo, M. G., Caio, G., Boschetti, E., Latorre, R., Giancola, F., Paterini, P., & De Giorgio, R. (2016). Autoimmune enteropathy: Not all flat mucosa mean coeliac disease. *Gastroenterology and Hepatology from Bed to Bench*, *9*(2), 140–145. Available from http://journals.sbmu.ac.ir/ghfbb/index.php/ghfbb/article/download/907/572.

Walker, M. M., Potter, M., & Talley, N. J. (2018). Eosinophilic gastroenteritis and other eosinophilic gut diseases distal to the oesophagus. *The Lancet Gastroenterology and Hepatology*, *3*(4), 271–280. Available from https://doi.org/10.1016/S2468-1253(18)30005-0.

Webster, A. D., Slavin, G., Shiner, M., Platts-Mills, T. A., & Asherson, G. L. (1981). Coeliac disease with severe hypogammaglobulinaemia. *Gut*, *22*(2), 153–157. Available from https://doi.org/10.1136/gut.22.2.153.

Weinstein, W. M., Saunders, D. R., Tytgat, G. N., & Rubin, C. E. (1970). Collagenous Sprue—An Unrecognized Type of Malabsorption. *New England Journal of Medicine*, *283*(24), 1297–1301. Available from https://doi.org/10.1056/NEJM197012102832401.

Woodward, J. M., Gkrania-Klotsas, E., Cordero-Ng, A. Y. K., Aravinthan, A., Bandoh, B. N., Liu, H., Davies, S., Zhang, H., Stevenson, P., Curran, M. D., & Kumararatne, D. (2015). The role of chronic norovirus infection in the enteropathy associated with common variable immunodeficiency. *American Journal of Gastroenterology*, *110*(2), 320–327. Available from https://doi.org/10.1038/ajg.2014.432.

Yadav, P., Das, P., Mirdha, B. R., Gupta, S. D., Bhatnagar, S., Pandey, R. M., & Makharia, G. K. (2011). Current spectrum of malabsorption syndrome in adults in India. *Indian Journal of Gastroenterology*, *30*(1), 22–28. Available from https://doi.org/10.1007/s12664-011-0081-0.

Ziegler, T. R., Fernández-Estívariz, C., Gu, L. H., Fried, M. W., & Leader, L. M. (2003). Severe villus atrophy and chronic malabsorption induced by azathioprine. *Gastroenterology*, *124*(7), 1950–1957. Available from https://doi.org/10.1016/S0016-5085(03)00405-0.

Chapter 12

Gluten free diet, assessment of its adherence, and quality of life

Fabiana Zingone and Annalisa Zanini
Department of Surgery, Oncology and Gastroenterology, University of Padua, Padua, Italy

12.1 Introduction

A gluten-free diet (GFD) is the only effective celiac disease (CD) treatment. Celiac patients should not consume foods containing wheat, barley, and rye or their derivates (Hollon et al., 2013). Patients need to learn to select naturally gluten-free food, choose certified gluten-free products, read food labels, and avoid contamination (Concerning the composition and labelling of foodstuffs suitable for people intolerant to gluten (Text with EEA relevance), 2009). It is well known that adherence to a GFD improves in patients who receive better CD instructions. Still, support from family members and healthcare workers also has a favorable impact (Ludvigsson et al., 2015). Following a GFD is a learning process, not only for the patients but also for their families and friends. The GFD determines the disappearance of celiac-related symptoms and serum autoantibodies, intestinal mucosa recovery, and long-term complication prevention.

12.2 Allowed cereals and naturally gluten-free food

GFD combines naturally gluten-free foods and substitutes prepared with gluten-free foods and cereals. The gluten-free food groups include fruits, vegetables, meat, fish, dairy, beans, legumes, and nuts. Gluten-Free Cereals are, for example, teff, rice, maize, and sorghum. Pseudocereals are buckwheat, amaranth, quinoa, and chia. Pseudocereals are gluten-free and are a suitable option to enhance the nutritional value of gluten-free foods. Among cereals, there is a degree of reluctance in some countries to permit the consumption of oats by CD patients due to the possibility that commercially available oats are not pure and contaminated by gluten-containing cereals. However, Pinto-Sanchez et al. (Pinto-Sánchez et al., 2017) have recently shown that they are safe for most CD patients. Excluding oats from the diet may be advisable for at least the first few months after diagnosis (Zingone et al., 2022), and when introduced, patients should be monitored (Al-Toma et al., 2019). Including alternative grain-based products significantly increased the gluten-free dietary pattern's nutrient profile. Conversely, not consuming recommended grain products could have important implications for dietary intakes of B vitamins, iron, and fiber. Enriched grain foods are a significant contributor to a daily intake of iron. However, most gluten-free flour, bread products, pasta, and cold cereals are not enriched (Lee et al., 2009; Lionetti et al., 2020; Thompson et al., 2005). Table 12.1 reports a list of products safely, risky, and forbidden for CD patients.

12.3 Definition of safe gluten threshold

A product is gluten-free if it contains no wheat, rye, or barley or is processed to contain less than 20 ppm of gluten. Codex Alimentarius, which represents a consensus of international standards for food safety, established in 2008 that the cut-off for "gluten-free" products is 20 ppm. Many countries have set local cut-offs, ranging from 20 ppm in European Union, the United Kingdom, Canada, and the United States, to 10 ppm in Japan, 1 ppm in Chile, and 0 in Australia.

Several studies have been conducted trying to evaluate a safe threshold for gluten consumption. In a 2007 study by Catassi et al., CD patients were given a capsule containing 0, 10, or 50 mg of gluten daily for 90 days (13 patients per group). Clinical, serologic, and histologic evaluations were performed at t0 and after 90 days from the gluten microchallenge. After the microchallenge, the IgA anti-tTG did not show any significant change. In contrast, a significant

Pediatric and Adult Celiac Disease. DOI: https://doi.org/10.1016/B978-0-443-13359-6.00011-X

TABLE 12.1 List of products safely, risky, and forbidden for CD patients.

	ALWAYS SAFE	Risky foods (possible contamination)	FORBIDDEN
CEREALS	Rice Corn Buckwheat Amaranth Millet Quinoa Tubers Teff	Mix of permitted grains and legumes Creams of permitted cereals Deglutinated wheat Tapioca flour Chickpeas, soybean, chestnut, almond, hazelnut flour Ready-made risottos Couscous, tacos, tortillas from allowed cereals Oat	Wheat Emmer/spelled Barley Spelled Rye Triticum monococcum Khorasan wheat
MEAT, FISH and EGGS	Meat Fish, shellfish, and crustaceans Eggs	Pork, beef, poultry salami, sausage Ready meals or precooked Homogenized meat/fish/ham	Meat or fish breaded or floured or mixed with breadcrumbs
DAIRY PRODUCTS	Milk: diary long-life delactosed, highly digestible Fresh and aged, even delactosed, cheese Natural yogurt, even delactosed Cream: fresh and long-life, even delactosed	White creamy yogurt, thickeners, flavorings or other substances added Soy, rice yogurt Light, spreadable cheese added with thickeners, flavors or other substances Condensed milk	Convenience cheese-based food, breaded with forbidden flours Malt, cereal, biscuits yogurt Cereal milk, biscuits milk
LEGUMES	All legumes	Legumes mix	
VEGETABLES and FRUITS	All types of vegetables All types of dried fruit with or without nuts	Prepared vegetable soup made Candied, caramelized, frosted fruits	Vegetables with not allowed cereals Covered with breadcrumbs, floured, battered with not allowed ingredients vegetables
DRINKS	Fruit juice Fizzy drinks without sweeteners Coffee, tea, chamomile tea, herbal tea Wine, champagne, distillates	Light drinks with sweeteners Soy, rice, almond milk drinks Drinks with ginseng or ginseng flavor Alcoholic beverages with added flavorings or other substances	Barley and/or wheat beers Instant coffee, coffee substitutes, gluten cereal-based drinks
SWEETS and FATS	Honey, sugar Butter, clarified butter, lard, cocoa butter, vegetable oils Jam and marmalade Chocolate	Sweeteners Light butter, margarine, light margarine	Chocolate, cake, biscuits with prohibited cereals Sourdough, mother dough Wheat gluten (seitan)

difference in villous height/crypt depth was observed between the placebo and the 50-mg groups. Instead, no significant differences in intraepithelial lymphocyte ratio were detected among the three groups. The authors concluded that daily ingestion of contaminating gluten should be kept lower than 50 mg, even if 10 mg may still induce damage to the duodenal mucosa (Catassi et al., 2007). A year later, a systematic review with metaanalysis was published, including 13 studies (Akobeng & Thomas, 2008). The authors found that the daily amount of tolerable gluten varied widely between studies. While some patients tolerated an average of 34–36 mg of gluten daily, others who consumed about 10 mg of gluten daily developed mucosal abnormalities. Therefore the authors concluded that a daily gluten intake of <10 mg is unlikely to cause significant histological abnormalities.

12.4 The role of gluten contamination

Unintentional ongoing gluten exposure is more common than commonly believed, even when patients self-report excellent/good GFD adherence. Traces of gluten contamination could be a contributing factor to persistent villous atrophy and lack of clinical recovery (Fernández-Bañares et al., 2021; Silvester et al., 2020). Hollon et al. (2013) tested a diet to eliminate exposure to any possible sources of gluten cross-contamination from packaged/processed foods, including those labeled gluten-free. In particular, this study aimed to evaluate the effect of the Gluten Contamination Elimination Diet (GCED) on nonresponsive CD patients with persistent symptoms and/or villous atrophy despite strict adherence to a gluten-free diet. This diet was tested on 17 patients (15 women and 2 men with a median age of 42) over 3–6 months. Clinical, serological, and histological evaluations of the small intestine were performed before and after the GCED. The response rate of the GCED was 82%, with 14 of the 17 compliant patients responding to the diet, with a resolution of symptoms and resolution of their prior celiac enteropathy. Five patients, wrongly considered to have a Refractory Celiac Disease type 1 before the GCED, were asymptomatic after the GCED and no longer met the refractory criteria. Furthermore, of the 14 patients who responded to the diet, 11 (79%) returned to their previous traditional GFD without a repeat of symptoms or elevated serology. Therefore a gluten contamination elimination diet for at least three months may be an effective therapeutic option for slow responder patients to a GFD, underlining the possible negative influence of unintentional contamination on CD patients.

Sourcing gluten-free food (which can be particularly challenging when traveling or eating away from home) and avoiding cross-contamination, checking food labels, buying and cooking gluten-free food, and the cost of gluten-free food can trigger some concerns in our patients. However, after initial frustrations, most people adapted to the changed prescription policy (Peters et al., 2020). It is essential to give the message that a gluten-free diet should be followed rigorously, avoiding contamination and maintaining a natural GFD.

12.5 How to monitor gluten-free diet adherence

At each follow-up medical consultation, the patient's clinical condition and blood test results should be reviewed and compared with the clinical picture at diagnosis. As reported by the recent CD guidelines (Al-Toma et al., 2019; Zingone et al., 2022), GFD adherence should be evaluated by dietary interviews and IgA anti-tTG titers, which should gradually reduce and eventually become negative. Fig. 12.1 shows how to monitor gluten-free diet adherence.

12.5.1 Dietitian's role

After the CD diagnosis, we must give our patients all the correct instructions to follow a balanced GFD (Madden et al., 2016). Therefore the role of specialist dietitians in diagnosis becomes essential in our clinical practice. Their role includes nutritional and anthropometric assessment and giving all the information to correctly follow a GFD to provide a nutritionally-balanced diet that avoids nutrient deficiencies and gluten contamination. However, there is a limited publication on the efficacy of dietetic evaluation in detecting GFD compliance in CD adult patients. One study found that dietary adherence evaluation had a sensitivity and specificity of 56% and 85% when detecting persistent villous atrophy (Biagi et al., 2014). A prospective study aimed to measure GFD adherence in CD patients comparing the Standardized Dietician Evaluation (SDE), based on a detailed interview conducted by an experienced dietitian, and a standard questionnaire named the Celiac Dietary Adherence Test (CDAT), which takes into account several aspects from the patients' knowledge to the reported dietary adherence (Gładyś et al., 2020). Ninety-two CD adults were included, and the authors found that 24%–52% of study participants did not adhere well to a GFD, finding that SDE is the most accurate method in assessing compliance with a GFD. Another study (Leffler et al., 2007) comparing five measures of GFD adherence:

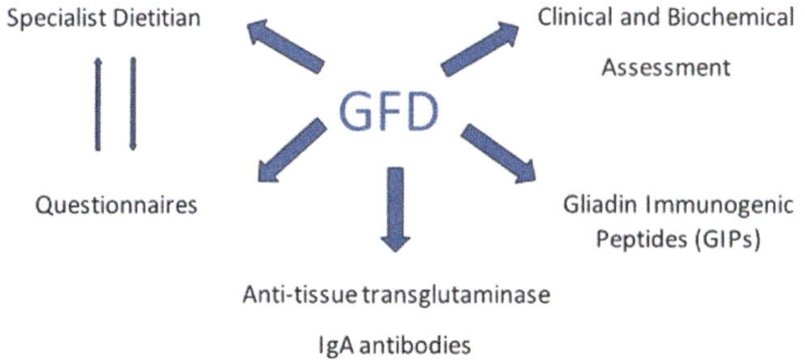

FIGURE 12.1 How to monitor gluten-free diet adherence. Schematic representation of adherence monitoring of a gluten-free diet.

patient self-report of the believed level of adherence dietary history, an evaluation conducted by a professional nutritionist, serologic screening test, and small-bowel biopsy, found that a standardized assessment by an expert nutritionist was the best available standard of GFD adherence, whereas biopsy histology is the gold standard measure of CD. Finally, Fera et al. found that, among 82 CD patients reporting strict compliance to a GFD, only in 49, dietitians confirmed a strict GFD after the direct inquiry by an expert dietitian (Fera et al., 2003).

The dietitian's role becomes more important due to the evidence that CD patients on a GFD may have an imbalanced diet due to higher consumption of fatty and sugary foods and a lower intake of iron, calcium, vitamins, and fiber. Moreover, gluten-free processed foods have a higher glycaemic index than conventional products, which may further affect CD patients' nutritional status (Barone et al., 2016; Caeiro et al., 2022; Lee et al., 2009; Levran et al., 2018; Wild et al., 2010). Several studies also point out the risk of CD patients on a GFD developing metabolic syndrome and fatty liver (Ciccone et al., 2019; Tortora et al., 2015). Tortora et al. (2015) conducted a prospective observational study including 98 patients with CD, and Ciccone et al. (2019) included 185 patients with CD in a retrospective study. They collected waist circumference, BMI, blood pressure, lipid profile (HDL cholesterol, triglycerides), and blood glucose levels at diagnosis of CD and after 12 months of GFD. Both studies confirmed a high risk of metabolic syndrome 1 year after starting a GFD. More recently, an increased risk of metabolic-associated fatty liver disease (MAFLD) in CD patients on a GFD has been described (Rispo et al., 2021). Rispo et al. retrospectively enrolled 221 newly diagnosed CD patients who underwent clinical, laboratory, and ultrasonography investigations at diagnosis and after two years of follow-up. This study explored the prevalence of nonalcoholic fatty liver disease (NAFLD) and MAFLD in CD patients at the time of CD diagnosis and after 2 years of a GFD. At 2 years of follow-up, 46.6% of patients developed NAFLD, while 32.6% had MAFLD. However, other studies have found contrasting results (Ehteshami et al., 2018; García-Manzanares, 2011; Ukkola et al., 2012). Ukkola et al. (2012) prospectively investigated weight and disease-related issues in 698 newly detected adults diagnosed with CD. At diagnosis, 4% of subjects were underweight, 57% normal, 28% overweight, and 11% obese. On a GFD, 69% of underweight patients gained weight, and 18% of overweight and 42% of obese patients lost weight; in the rest, BMI remained stable. Another case report (García-Manzanares, 2011) described the resolution of metabolic syndrome after six months of a GFD in an adult woman diagnosed with CD. After GFD, a slight weight loss, normalization of the transaminase levels, and decreased cholesterol and triglycerides were observed. This possible negative impact of a GFD may be avoided through a collaboration of dietitians and physicians to guide the patients to follow a balanced diet, preferring naturally gluten free foods to those packaged ones.

12.5.2 Questionnaires

In support of an expert dietitian or where access to this expertise is limited, standardized patient-reported adherence questionnaires should be considered. The most common questionnaires are the Biagi Score (Biagi et al., 2009) and the CDAT (Leffler et al., 2009). Coleman et al. (2021) described higher specificity for the Biagi score and higher sensitivity for the CDAT in assessing persistent villous atrophy in adult CD. However, there is only little evidence evaluating the ability of questionnaires to detect gluten contamination in non-responsive CD patients. One, cited above, reported that a dietitian evaluation is the most accurate method in assessing compliance with a GFD compared to CDAT (Gładyś et al., 2020).

12.5.3 Serology

CD-specific antibodies start to decline within months of the introduction of the GFD, dropping quite rapidly during the first year and eventually becoming negative in most patients (Ciacci et al., 2002; Sugai et al., 2010). Persistent positivity strongly predicts poor dietary adherence [34], however, normalizing antibody titers does not predict histological recovery of villous atrophy. A recent metaanalysis has reported that IgA anti-tTG during follow-up has a low sensitivity (<50%) in detecting the persistence of villous atrophy (Vahedi et al., 2003). However, despite that, checking for routine serology with anti-tTG IgA in CD patients on a GFD is still recommended to assess dietary adherence in CD (Al-Toma et al., 2019; Zingone et al., 2022). In patients with IgA deficiency, anti-tTG IgG tends to decline but often remains positive despite a strict GFD and should, therefore not be used to assess adherence (Wang et al., 2014). The need for follow-up duodenal biopsy should be evaluated on a case-by-case basis. It should be performed in patients with a lack of clinical response to a GFD, seronegative CD, and in those who, at diagnosis, had factors predictive of an increased risk of developing complications (Biagi et al., 2018).

12.5.4 Gluten immunogenic peptide

CD patients can report a high GFD adherence to the standard questionnaires and have negative serology despite the persistence of symptoms and/or duodenal histological damage. These situations often increase the diagnosis of refractory celiac disease or the search for alternative diseases. In 2017, Moreno et al. (2017) described a non-invasive method to evaluate the presence of immunogenic gluten peptides (GIP) in urine, distinguishing gluten consumers from nonconsumers. The authors collected the urine samples of 76 healthy subjects and 58 patients with CD subjected to different gluten dietary conditions. The GIP detection in urine was revealed in about 50% of the patients on a GFD, and all patients with quantifiable GIP in urine showed incomplete intestinal mucosa recovery. In 2012, Comino et al. detected gluten peptides in healthy individuals' feces after consent to a regular gluten-containing diet, after consuming a GFD combined with controlled ingestion of a fixed amount of gluten, and after ingestion of <100 mg gluten per day (Comino et al., 2016). A systematic review in 2021 confirmed that GIP detection is a precise approach for determining voluntary or involuntary gluten consumption, and its use can prevent possible complications (Coto et al., 2021). The GIP test in the urine should be performed within 12–24 hours after the possible gluten contamination, while in the stool, within 1–7 days. Moreover, the results can be related to the amount of ingested gluten and the individual gluten metabolism (Coto et al., 2021). A recent prospective double-blinded placebo-controlled study (Burger et al., 2022) enrolled 15 CD adults adhering to a strictly GFD. The study provided ascending doses of 50, 100, 200, and 500 mg alternated with placebo, and urine portions from 2, 5–17 h after the ingestion were collected for the GIP analysis. GIP was detected from urine in 47% of the patients receiving 50 mg gluten and 86% with 500 mg gluten. Using at least two samples for GIP determinations could be combined with dietary and serological assessments in non-response to a GFD to exclude gluten contamination. Recently, Garzón-Benavides et al. evaluated the interval of determination of urinary GIP for the monitoring of GFD adherence in 94 patients with CD on a GFD for at least 24 months and showed that 94% of patients with negative urinary GIP in ≥2 follow-up visits showed the absence of histological lesions (Garzón-Benavides et al., 2023). In their clinical trial, discordant results of urinary GIP were described by Monachesi et al. (2021). In this study, 25 healthy adults (group A) on a standard GFD performed six gluten challenges (0, 10, 50, 100, 500, and 1,000 mg) with quantification of urinary GIP before (T0) and during the following 24 hours and other 12 participants (Group B) on a gluten contamination elimination diet underwent urinary GIP determination at T0 and after challenge with 5 or 10 mg gluten. The authors found that traces of gluten in the standard GFD may cause positivity of urinary GIP determination, whereas false negativity was also common after a gluten intake of 10–1000 mg.

12.6 Living with CD

Several reports have described the difficulties of living with CD, particularly regarding the impact of this condition on physical, social, and emotional factors.

12.6.1 Quality of life

In recent years significant attention has been placed on evaluating the quality of life (QOL) in CD patients, both at diagnosis and while on a GFD (Zingone et al., 2015). Most studies have found a reduction in QOL at diagnosis, which improves once on a GFD. Sex, age, presenting symptoms, and GFD adherence may impact the QOL. Nachman et al. reported that at 4 years postdiagnosis, Qol and Depression scales showed significant deterioration compared with 1 year, even if the scores remained significantly better than those at diagnosis (Nachman et al., 2010). The long-term impairment of QoL was observed in patients who were not strictly compliant with the GFD. A year later, Berratt et al. described that the main contributing factor to a reduced QoL is the perceived degree of difficulty adhering to a GFD, which can be influenced by social and educational background (Barratt et al., 2011). A systematic review by Burger et al. (2017), including studies using different questionnaires [Psychological general well-being (PGWB), Short-Form 36-Item Health Survey (SF-36) analyses of the physical component score (PCS) and mental component score (MCS), aimed to evaluate the effect of GFD on health-related quality of life (HRQoL) in CD. The authors found that a GFD significantly improves but does not normalize HRQoL in CD, finding that better dietary adherence results in higher HRQoL. Marsilio et al. (2020) analyzed 100 patients on a GFD, reporting an overall good QoL in CD patients on a GFD. Patients older than 35 years were more worried about their general health. Patients with good compliance to a GFD tended to have a better QoL, while noncompliant to GFD appeared to suffer from dysphoria (...generalized dissatisfaction with life); however, it was difficult to identify which is the cause and which is the effect. No statistically significant difference was observed according to gender and time from diagnosis.

12.6.2 Depression

The pooled prevalence of depression in CD was 3.5% (Clappison et al., 2020). Depression can start as a consequence of ongoing gastrointestinal symptoms, and more severe symptoms correlate well with the severity of depressive symptoms. Therefore several studies reported that a GFD improved depression as the disease symptoms improved (Corvaglia et al., 1999; Nachman et al., 2010). However, other studies showed the persistence of depressive symptoms, possibly due to dietary restrictions that impair patients' social relationships and reduce their QoL (Addolorato et al., 2001; Siniscalchi et al., 2005; Zingone et al., 2010). Still, noncompliance with a GFD might also be a consequence or a cause of persistent depression (Addolorato et al., 2004). An increased risk of suicide has also been reported in individuals with CD compared with controls (Ludvigsson et al., 2011). Not all studies agree with the increasing prevalence of depression in individuals with CD. Some authors have suggested that depression in CD patients results from having a chronic disease rather than being directly related to having CD (Fera et al., 2003; Garud et al., 2009; Häuser et al., 2010; Roos et al., 2006).

12.6.3 Anxiety

The pooled prevalence of anxiety in CD was 3.7% (Clappison et al., 2020). Addolorato et al. found that state anxiety was present in more individuals with CD than controls, whereas no difference was found in trait anxiety. Anxiety may be a reaction to the disabling symptoms of CD rather than a personality trait of the individuals who suffer from CD. After 1 year on a GFD, state anxiety symptoms improved significantly (Addolorato et al., 1996, 2001). Anxiety was more remarkable in female CD patients than in male patients (Häuser et al., 2010). Social phobia has been reported to be associated with CD, both at diagnosis and on a GFD (Addolorato et al., 2008; Carta et al., 2002). Not all studies confirmed this association (Fera et al., 2003; Roos et al., 2006). Zingone et al. described sleep disorders as standard in CD at diagnosis and during treatment with a GFD. They are related to depression, anxiety, and fatigue and are inversely related to the quality of life scale scores (Zingone et al., 2010).

A prospective study has recently reported an improvement in QOL and psychological disorders after one and two years on a GFD, describing dietary compliance as the leading risk factor (Canova et al., 2021; Zingone et al., 2021). Before CD diagnosis, patients may express concerns about unexplained symptoms and feel frustrated about repeated consultations that do not adequately explain their problems. At the time of diagnosis, there may be concerns about investigations and a diagnosis of a long-term condition. However, this may be accompanied by relief that a diagnosis has finally been made. While on a GFD, some psychological problems may lessen over time as knowledge of the condition improves, and biochemical abnormalities are corrected. Still, some patients may have ongoing concerns about coping with the diet and do not adhere to it, particularly during social interaction. Overall, findings suggest that, in adults with CD, monitoring to ensure GFD adherence may be essential to promote to improve QOL. However, there remains a concern that extreme GFD vigilance may diminish QOL and should also be monitored.

12.7 Conclusion

This chapter provides practical suggestions on a correct gluten-free diet, details on allowed cereals, naturally gluten-free food, contamination risks, and a safe gluten threshold definition. It is important to give our patients all the correct instructions to follow a GFD. This can be achieved through specialist dietitians who can offer nutritional and anthropometric assessments. They can provide all the information needed to correctly follow a GFD, including guidance on a nutritionally-balanced diet. This diet should be based mainly on naturally gluten free foods, to avoid nutrient deficiencies and gluten contamination. A gluten-free diet adherence can be monitored by physicians through the combination of an expert dietitian and/or validated questionnaires, the search for antitissue transglutaminase IgA, and the use of urinary gluten peptides. Finally, before CD diagnosis, patients may express concerns about unexplained symptoms and feel frustrated about repeated consultations that do not adequately explain their problems. At the time of diagnosis, there may be concerns about investigations and a diagnosis of a long-term condition. However, this may be accompanied by relief that a diagnosis has finally been made. While on a GFD, some psychological problems may lessen over time as knowledge of the condition improves, and biochemical abnormalities are corrected. Still, some patients may have ongoing concerns about coping with the diet and may not adhere to it, particularly during social interaction. Findings suggest that monitoring may be essential in adults with CD to ensure GFD adherence to improve QOL. However, there remains a concern that extreme GFD vigilance may diminish QoL and should also be monitored.

References

Addolorato, G., Capristo, E., Ghittoni, G., Valeri, C., Mascianà, R., Ancona, C., & Gasbarrini, G. (2001). Anxiety but not depression decreases in coeliac patients after one-year gluten-free diet: A longitudinal study. *Scandinavian Journal of Gastroenterology*, *36*(5), 502–506. Available from https://doi.org/10.1080/00365520119754.

Addolorato, G., De Lorenzi, G., Abenavoli, L., Leggio, L., Capristo, E., & Gasbarrini, G. (2004). Psychological support counselling improves gluten-free diet compliance in coeliac patients with affective disorders. *Alimentary Pharmacology and Therapeutics*, *20*(7), 777–782. Available from https://doi.org/10.1111/j.1365-2036.2004.02193.x.

Addolorato, G., Mirijello, A., D'Angelo, C., Leggio, L., Ferrulli, A., Abenavoli, L., Vonghia, L., Cardone, S., Leso, V., Cossari, A., Capristo, E., & Gasbarrini, G. (2008). State and trait anxiety and depression in patients affected by gastrointestinal diseases: Psychometric evaluation of 1641 patients referred to an internal medicine outpatient setting. *International Journal of Clinical Practice*, *62*(7), 1063–1069. Available from https://doi.org/10.1111/j.1742-1241.2008.01763.x.

Addolorato, G., Stefanini, G. F., Capristo, E., Caputo, F., Gasbarrini, A., & Gasbarrini, G. (1996). Anxiety and depression in adult untreated celiac subjects and in patients affected by inflammatory bowel disease: A personality \trait\ or a reactive illness? *Hepato-Gastroenterology*, *43*(12), 1513–1517.

Akobeng, A. K., & Thomas, A. G. (2008). Systematic review: Tolerable amount of gluten for people with coeliac disease. *Alimentary Pharmacology and Therapeutics*, *27*(11), 1044–1052. Available from https://doi.org/10.1111/j.1365-2036.2008.03669.x.

Al-Toma, A., Volta, U., Auricchio, R., Castillejo, G., Sanders, D. S., Cellier, C., Mulder, C. J., & Lundin, K. E. A. (2019). European Society for the Study of Coeliac Disease (ESsCD) guideline for coeliac disease and other gluten-related disorders. *United European Gastroenterology Journal*, *7*(5), 583–613. Available from https://doi.org/10.1177/2050640619844125.

Barone, M., Della Valle, N., Rosania, R., Facciorusso, A., Trotta, A., Cantatore, F. P., Falco, S., Pignatiello, S., Viggiani, M. T., Amoruso, A., De Filippis, R., Di Leo, A., & Francavilla, R. (2016). A comparison of the nutritional status between adult celiac patients on a long-term, strictly gluten-free diet and healthy subjects. *European Journal of Clinical Nutrition*, *70*(1), 23–27. Available from https://doi.org/10.1038/ejcn.2015.114.

Barratt, S. M., Leeds, J. S., & Sanders, D. S. (2011). Quality of life in coeliac disease is determined by perceived degree of difficulty adhering to a gluten-free diet, not the level of dietary adherence ultimately achieved. *Journal of Gastrointestinal and Liver Diseases*, *20*(3), 241–245. Available from http://www.jgld.ro/2011/3/5.pdf.

Biagi, F., Andrealli, A., Bianchi, P. I., Marchese, A., Klersy, C., & Corazza, G. R. (2009). A gluten-free diet score to evaluate dietary compliance in patients with coeliac disease. *British Journal of Nutrition*, *102*(6), 882–887. Available from https://doi.org/10.1017/S0007114509301579.

Biagi, F., Schiepatti, A., Maiorano, G., Fraternale, G., Agazzi, S., Zingone, F., Ciacci, C., Volta, U., Caio, G., Tortora, R., Klersy, C., & Corazza, G. R. (2018). Risk of complications in coeliac patients depends on age at diagnosis and type of clinical presentation. *Digestive and Liver Disease*, *50*(6), 549–552. Available from https://doi.org/10.1016/j.dld.2017.12.001.

Biagi, F., Vattiato, C., Agazzi, S., Balduzzi, D., Schiepatti, A., Gobbi, P., & Corazza, G. R. (2014). A second duodenal biopsy is necessary in the follow-up of adult coeliac patients. *Annals of Medicine*, *46*(6), 430–433. Available from https://doi.org/10.3109/07853890.2014.913378.

Burger, J. P. W., de Brouwer, B., IntHout, J., Wahab, P. J., Tummers, M., & Drenth, J. P. H. (2017). Systematic review with meta-analysis: Dietary adherence influences normalization of health-related quality of life in coeliac disease. *Clinical Nutrition*, *36*(2), 399–406. Available from https://doi.org/10.1016/j.clnu.2016.04.021.

Burger, J.P. W., van Lochem, E.G., Roovers, E.A., Drenth, J.P. H., & Wahab, P, J. (2022). Dose-escalating (50-500 mg) gluten administration leads to detectable gluten-immunogenic-peptides in urine of patients with coeliac disease which Is unrelated to symptoms, a placebo controlled trial. *Nutrients, 14*(9), 1771. https://doi.org/10.3390/nu14091771.

Caeiro, C., Pragosa, C., Cruz, M. C., Pereira, C. D., & Pereira, S. G. (2022). The role of pseudocereals in celiac disease: Reducing nutritional deficiencies to improve well-being and health. *Journal of Nutrition and Metabolism*, *2022:8502169*. Available from https://doi.org/10.1155/2022/8502169.

Canova, C., Rosato, I., Marsilio, I., Valiante, F., Zorzetto, V., Cataudella, G., D'odorico, A., & Zingone, F. (2021). Quality of life and psychological disorders in coeliac disease: A prospective multicentre study. *Nutrients*, *13*(9), 3233. Available from https://doi.org/10.3390/nu13093233.

Carta, M. G., Hardoy, M. C., Boi, M. F., Mariotti, S., Carpiniello, B., & Usai, P. (2002). Association between panic disorder, major depressive disorder and celiac disease: A possible role of thyroid autoimmunity. *Journal of Psychosomatic Research*, *53*(3), 789–793. Available from https://doi.org/10.1016/S0022-3999(02)00328-8.

Catassi, C., Fabiani, E., Iacono, G., D'Agate, C., Francavilla, R., Biagi, F., Volta, U., Accomando, S., Picarelli, A., De Vitis, I., Pianelli, G., Gesuita, R., Carle, F., Mandolesi, A., Bearzi, I., & Fasano, A. (2007). A prospective, double-blind, placebo-controlled trial to establish a safe gluten threshold for patients with celiac disease. *The American Journal of Clinical Nutrition*, *85*(1), 160–166.

Ciacci, C., Cavallaro, R., della Valle, N., & d'Argenio, G. (2002). The use of serum tTG-ab assay in patients on gluten-free diet as a measure of dietetic compliance. *Gastroenterology*, *122*(2), 588. Available from https://doi.org/10.1053/gast.2002.31598.

Ciccone, A., Gabrieli, D., Cardinale, R., Di Ruscio, M., Vernia, F., Stefanelli, G., Necozione, S., Melideo, D., Viscido, A., Frieri, G., & Latella, G. (2019). Metabolic alterations in celiac disease occurring after following a gluten-free diet. *Digestion*, *100*(4), 262–268. Available from https://doi.org/10.1159/000495749.

Clappison, E., Hadjivassiliou, M., & Zis, P. (2020). Psychiatric manifestations of coeliac disease, a systematic review and meta-analysis. *Nutrients*, *12*(1). Available from https://doi.org/10.3390/nu12010142.

Coleman, S. H., Rej, A., Baggus, E. M. R., Lau, M. S., Marks, L. J., Hadjivassiliou, M., Cross, S. S., Leffler, D. A., Elli, L., & Sanders, D. S. (2021). What is the optimal method assessing for persistent villous atrophy in adult coeliac disease? *Journal of Gastrointestinal and Liver Diseases*, *30* (2). Available from https://doi.org/10.15403/jgld-3370.

Comino, I., Fernández-Bañares, F., Esteve, M., Ortigosa, L., Castillejo, G., Fambuena, B., Ribes-Koninckx, C., Sierra, C., Rodríguez-Herrera, A., Salazar, J. C., Caunedo, Á., Marugán-Miguelsanz, J. M., Garrote, J. A., Vivas, S., Lo Iacono, O., Nuñez, A., Vaquero, L., Vegas, A. M., Crespo, L., & Sousa, C. (2016). Fecal gluten peptides reveal limitations of serological tests and food questionnaires for monitoring gluten-free diet in celiac disease patients. *American Journal of Gastroenterology*, *111*(10), 1456–1465. Available from https://doi.org/10.1038/ajg.2016.439.

Concerning the composition and labelling of foodstuffs suitable for people intolerant to gluten (Text with EEA relevance). (2009). *Commission Regulation*, *16*, 3–5.

Corvaglia, L., Catamo, R., Pepe, G., Lazzari, R., & Corvaglia, E. (1999). Depression in adult untreated celiac subjects: Diagnosis by the pediatrician. *American Journal of Gastroenterology*, *94*(3), 839–843. Available from https://doi.org/10.1016/S0002-9270(99)00011-8.

Coto, L., Mendia, I., Sousa, C., Bai, J. C., & Cebolla, A. (2021). Determination of gluten immunogenic peptides for the management of the treatment adherence of celiac disease: A systematic review. *World Journal of Gastroenterology*, *27*(37), 6306–6321. Available from https://doi.org/10.3748/wjg.v27.i37.6306.

Coto, L., Sousa, C., & Cebolla, A. (2021). Dynamics and considerations in the determination of the excretion of gluten immunogenic peptides in urine: Individual variability at low gluten intake. *Nutrients*, *13*(8). Available from https://doi.org/10.3390/nu13082624.

Ehteshami, M., Shakerhosseini, R., Sedaghat, F., Hedayati, M., Eini-Zinab, H., & Hekmatdoost, A. (2018). The effect of gluten free diet on components of metabolic syndrome: A randomized clinical trial. *Asian Pacific Journal of Cancer Prevention*, *19*(10), 2979–2984. Available from https://doi.org/10.22034/APJCP.2018.19.10.2979.

Fera, T., Cascio, B., Angelini, G., Martini, S., & Guidetti, C. S. (2003). Affective disorders and quality of life in adult coeliac disease patients on a gluten-free diet. *European Journal of Gastroenterology and Hepatology*, *15*(12), 1287–1292. Available from https://doi.org/10.1097/00042737-200312000-00006.

Fernández-Bañares, F., Beltrán, B., Salas, A., Comino, I., Ballester-Clau, R., Ferrer, C., Molina-Infante, J., Rosinach, M., Modolell, I., Rodríguez-Moranta, F., Arau, B., Segura, V., Fernández-Salazar, L., Santolaria, S., Esteve, M., & Sousa, C. (2021). Persistent villous atrophy in de novo adult patients with celiac disease and strict control of gluten-free diet adherence: A Multicenter Prospective Study (CADER Study). *American Journal of Gastroenterology*, *116*(5), 1036–1043. Available from https://doi.org/10.14309/ajg.0000000000001139.

García-Manzanares, Á. (2011). Resolution of metabolic syndrome after following a gluten free diet in an adult woman diagnosed with celiac disease. *World Journal of Gastrointestinal Pathophysiology*, *2*(3), 49. Available from https://doi.org/10.4291/wjgp.v2.i3.49.

Garud, S., Leffler, D., Dennis, M., Edwards-George, J., Saryan, D., Sheth, S., Schuppan, D., Jamma, S., & Kelly, C. P. (2009). Interaction between psychiatric and autoimmune disorders in coeliac disease patients in the northeastern United States. *Alimentary Pharmacology and Therapeutics*, *29*(8), 898–905. Available from https://doi.org/10.1111/j.1365-2036.2009.03942.x.

Garzón-Benavides, M., Ruiz-Carnicer, Á., Segura, V., Fombuena, B., García-Fernandez, F., Sobrino-Rodriguez, S., Gómez-Izquierdo, L., Montes-Cano, M. A., Millan-Domínguez, R., del Carmen Rico, M., González-Naranjo, C., Bozada-García, J. M., Coronel-Rodríguez, C., Espin, B., Díaz, J., Comino, I., Argüelles-Arias, F., Cebolla, Á., Romero-Gómez, M., … Sousa, C. (2023). Clinical utility of urinary gluten immunogenic peptides in the follow-up of patients with coeliac disease. *Alimentary Pharmacology and Therapeutics*, *57*(9), 993–1003. Available from https://doi.org/10.1111/apt.17417.

Gładyś, K., Dardzińska, J., Guzek, M., Adrych, K., & Małgorzewicz, S. (2020). Celiac dietary adherence test and standardized dietician evaluation in assessment of adherence to a gluten-free diet in patients with celiac disease. *Nutrients*, *12*(8), 1–10. Available from https://doi.org/10.3390/nu12082300.

Häuser, W., Janke, K. H., Klump, B., Gregor, M., & Hinz, A. (2010). Anxiety and depression in adult patients with celiac disease on a gluten-free diet. *World Journal of Gastroenterology*, *16*(22), 2780–2787. Available from https://doi.org/10.3748/wjg.v16.i22.2780.

Hollon, J. R., Cureton, P. A., Martin, M. L., Puppa, E. L. L., & Fasano, A. (2013). Trace gluten contamination may play a role in mucosal and clinical recovery in a subgroup of diet-adherent non-responsive celiac disease patients. *BMC Gastroenterology*, *13*(1). Available from https://doi.org/10.1186/1471-230X-13-40.

Lee, A. R., Ng, D. L., Dave, E., Ciaccio, E. J., & Green, P. H. R. (2009). The effect of substituting alternative grains in the diet on the nutritional profile of the gluten-free diet. *Journal of Human Nutrition and Dietetics*, *22*(4), 359–363. Available from https://doi.org/10.1111/j.1365-277X.2009.00970.x.

Leffler, D. A., Dennis, M., Edwards George, J. B., Jamma, S., Magge, S., Cook, E. F., Schuppan, D., & Kelly, C. P. (2009). A simple validated gluten-free diet adherence survey for adults with celiac disease. *Clinical Gastroenterology and Hepatology*, *7*(5), 530–e2. Available from https://doi.org/10.1016/j.cgh.2008.12.032.

Leffler, D. A., Edwards George, J. B., Dennis, M., Cook, E. F., Schuppan, D., & Kelly, C. P. (2007). A prospective comparative study of five measures of gluten-free diet adherence in adults with coeliac disease. *Alimentary Pharmacology and Therapeutics*, *26*(9), 1227–1235. Available from https://doi.org/10.1111/j.1365-2036.2007.03501.x.

Levran, N., Wilschanski, M., Livovsky, J., Shachar, E., Moskovitz, M., Assaf-Jabrin, L., & Shteyer, E. (2018). Obesogenic habits among children and their families in response to initiation of gluten-free diet. *European Journal of Pediatrics*, *177*(6), 859–866. Available from https://doi.org/10.1007/s00431-018-3128-8.

Lionetti, E., Antonucci, N., Marinelli, M., Bartolomei, B., Franceschini, E., Gatti, S., Catassi, G. N., Verma, A. K., Monachesi, C., & Catassi, C. (2020). Nutritional status, dietary intake, and adherence to the mediterranean diet of children with celiac disease on a gluten-free diet: A case-control prospective study. *Nutrients*, *12*(1). Available from https://doi.org/10.3390/nu12010143.

Ludvigsson, J. F., Card, T., Ciclitira, P. J., Swift, G. L., Nasr, I., Sanders, D. S., & Ciacci, C. (2015). Support for patients with celiac disease: A literature review. *United European Gastroenterology Journal*, *3*(2), 146–159. Available from https://doi.org/10.1177/2050640614562599.

Ludvigsson, J. F., Sellgren, C., Runeson, B., Långström, N., & Lichtenstein, P. (2011). Increased suicide risk in coeliac disease-a swedish nationwide cohort study. *Digestive and Liver Disease*, *43*(8), 616–622. Available from https://doi.org/10.1016/j.dld.2011.02.009.

Madden, A. M., Riordan, A. M., & Knowles, L. (2016). Outcomes in coeliac disease: a qualitative exploration of patients' views on what they want to achieve when seeing a dietitian. *Journal of Human Nutrition and Dietetics*, *29*(5), 607–616. Available from https://doi.org/10.1111/jhn.12378.

Marsilio, I., Canova, C., D'odorico, A., Ghisa, M., Zingone, L., Lorenzon, G., Savarino, E. V., & Zingone, F. (2020). Quality-of-life evaluation in Coeliac patients on a gluten-free diet. *Nutrients*, *12*(10), 1–10. Available from https://doi.org/10.3390/nu12102981.

Monachesi, C., Anil, K., Verma., Giulia, N., Catassi, E., Franceschini, S., Gatti, R., Gesuita, E., Lionetti, E., & Catassi, C. (2021). Determination of urinary gluten immunogenic peptides to assess adherence to the gluten-free diet: A randomized, double-blind, controlled study. *Clinical and Translational Gastroenterology*, *12*(10), e00411. Available from https://doi.org/10.14309/ctg.0000000000000411.

Moreno, M., de, L., Cebolla, Á., Muñoz-Suano, A., Carrillo-Carrion, C., Comino, I., Pizarro, Á., León, F., Rodríguez-Herrera, A., & Sousa, C. (2017). Detection of gluten immunogenic peptides in the urine of patients with coeliac disease reveals transgressions in the gluten-free diet and incomplete mucosal healing. *Gut*, *66*(2), 250–257. Available from https://doi.org/10.1136/gutjnl-2015-310148.

Nachman, F., del Campo, M. P., González, A., Corzo, L., Vázquez, H., Sfoggia, C., Smecuol, E., Sánchez, M. I. P., Niveloni, S., Sugai, E., Mauriño, E., & Bai, J. C. (2010). Long-term deterioration of quality of life in adult patients with celiac disease is associated with treatment noncompliance. *Digestive and Liver Disease*, *42*(10), 685–691. Available from https://doi.org/10.1016/j.dld.2010.03.004.

Peters, M., Crocker, H., Jenkinson, C., & Violato, M. (2020). Withdrawing gluten-free food from prescriptions in England: A mixed-methods study to examine the impact of policy changes on quality of life. *Journal of Human Nutrition and Dietetics*, *33*(4), 453–464. Available from https://doi.org/10.1111/jhn.12728.

Pinto-Sánchez, M. I., Causada-Calo, N., Bercik, P., Ford, A. C., Murray, J. A., Armstrong, D., Semrad, C., Kupfer, S. S., Alaedini, A., Moayyedi, P., Leffler, D. A., Verdú, E. F., & Green, P. (2017). Safety of adding oats to a gluten-free diet for patients with celiac disease: Systematic review and meta-analysis of clinical and observational studies. *Gastroenterology*, *153*(2), 395–409. Available from https://doi.org/10.1053/j.gastro.2017.04.009, e3.

Rispo, A., Imperatore, N., Gurino, M., Tortora, R., Alisi, A., Cossiga, V., Testa, A., Ricciolino, S., Fiorentino, A., & Morisco, F. (2021). Metabolic-associated fatty liver disease (MAFLD) in coeliac disease. *Liver International*, *41*(4), 788–798. Available from https://doi.org/10.1111/liv.14767.

Roos, S., Kärner, A., & Hallert, C. (2006). Psychological well-being of adult coeliac patients treated for 10 years. *Digestive and Liver Disease*, *38*(3), 177–180. Available from https://doi.org/10.1016/j.dld.2006.01.004.

Silvester, J. A., Comino, I., Rigaux, L. N., Segura, V., Green, K. H., Cebolla, A., Weiten, D., Dominguez, R., Leffler, D. A., Leon, F., Bernstein, C. N., Graff, L. A., Kelly, C. P., Sousa, C., & Duerksen, D. R. (2020). Exposure sources, amounts and time course of gluten ingestion and excretion in patients with coeliac disease on a gluten-free diet. *Alimentary Pharmacology and Therapeutics*, *52*(9), 1469–1479. Available from https://doi.org/10.1111/apt.16075.

Siniscalchi, M., Iovino, P., Tortora, R., Forestiero, S., Somma, A., Capuano, L., Franzese, M. D., Sabbatini, F., & Ciacci, C. (2005). Fatigue in adult coeliac disease. *Alimentary Pharmacology and Therapeutics*, *22*(5), 489–494. Available from https://doi.org/10.1111/j.1365-2036.2005.02619.x.

Sugai, E., Nachman, F., Váquez, H., González, A., Andrenacci, P., Czech, A., Niveloni, S., Mazure, R., Smecuol, E., Cabanne, A., Mauriño, E., & Bai, J. C. (2010). Dynamics of celiac disease-specific serology after initiation of a gluten-free diet and use in the assessment of compliance with treatment. *Digestive and Liver Disease*, *42*(5), 352–358. Available from https://doi.org/10.1016/j.dld.2009.07.011.

Thompson, T., Dennis, M., Higgins, L. A., Lee, A. R., & Sharrett, M. K. (2005). Gluten-free diet survey: Are Americans with coeliac disease consuming recommended amounts of fibre, iron, calcium and grain foods. *Journal of Human Nutrition and Dietetics*, *18*(3), 163–169. Available from https://doi.org/10.1111/j.1365-277X.2005.00607.x.

Tortora, R., Capone, P., De Stefano, G., Imperatore, N., Gerbino, N., Donetto, S., Monaco, V., Caporaso, N., & Rispo, A. (2015). Metabolic syndrome in patients with coeliac disease on a gluten-free diet. *Alimentary Pharmacology and Therapeutics*, *41*(4), 352–359. Available from https://doi.org/10.1111/apt.13062.

Ukkola, A., Mäki, M., Kurppa, K., Collin, P., Huhtala, H., Kekkonen, L., & Kaukinen, K. (2012). Changes in body mass index on a gluten-free diet in coeliac disease: A nationwide study. *European Journal of Internal Medicine*, *23*(4), 384–388. Available from https://doi.org/10.1016/j.ejim.2011.12.012.

Vahedi, K., Mascart, F., Mary, J. Y., Laberenne, J. E., Bouhnik, Y., Morin, M. C., Ocmant, A., Velly, C., Colombel, J. F., & Matuchansky, C. (2003). Reliability of antitransglutaminase antibodies as predictors of gluten-free diet compliance in adult celiac disease. *American Journal of Gastroenterology*, *98*(5), 1079–1087. Available from https://doi.org/10.1111/j.1572-0241.2003.07284.x.

Wang, N., Truedsson, L., Elvin, K., Andersson, B. A., Rönnelid, J., Mincheva-Nilsson, L., Lindkvist, A., Ludvigsson, J. F., Hammarström, L., & Dahle, C. (2014). Serological assessment for celiac disease in IgA deficient adults. *PLoS One*, *9*(4), e93180. Available from https://doi.org/10.1371/journal.pone.0093180.

Wild, D., Robins, G. G., Burley, V. J., & Howdle, P. D. (2010). Evidence of high sugar intake, and low fibre and mineral intake, in the gluten-free diet. *Alimentary Pharmacology and Therapeutics*, *32*(4), 573–581. Available from https://doi.org/10.1111/j.1365-2036.2010.04386.x.

Zingone, F., Maimaris, S., Auricchio, R., Caio, G. P. I., Carroccio, A., Elli, L., Galliani, E., Montagnani, M., Valiante, F., & Biagi, F. (2022). Guidelines of the Italian societies of gastroenterology on the diagnosis and management of coeliac disease and dermatitis herpetiformis. *Digestive and Liver Disease*, *54*(10), 1304–1319. Available from https://doi.org/10.1016/j.dld.2022.06.023.

Zingone, F., Secchettin, E., Marsilio, I., Valiante, F., Zorzetto, V., Cataudella, G., D'Odorico, A., & Canova, C. (2021). Clinical features and psychological impact of celiac disease at diagnosis. *Digestive and Liver Disease*, *53*(12), 1565–1570. Available from https://doi.org/10.1016/j.dld.2021.05.016.

Zingone, F., Siniscalchi, M., Capone, P., Tortora, R., Andreozzi, P., Capone, E., & Ciacci, C. (2010). The quality of sleep in patients with coeliac disease. *Alimentary Pharmacology and Therapeutics*, *32*(8), 1031–1036. Available from https://doi.org/10.1111/j.1365-2036.2010.04432.x.

Zingone, F., Swift, G. L., Card, T. R., Sanders, D. S., Ludvigsson, J. F., & Bai, J. C. (2015). Psychological morbidity of celiac disease: A review of the literature. *United European Gastroenterology Journal*, *3*(2), 136–145. Available from https://doi.org/10.1177/2050640614560786.

Chapter 13

Medical treatments and follow-up for clinical conditions associated to celiac disease

Antonio Gasbarrini[1,2], Stefano Bibbò[1,2], Gianluca Ianiro[1,2], Giovanni Cammarota[1,2] and Giovanni Gasbarrini[1,2]

[1]*Digestive Disease Center, Fondazione Policlinico Universitario "A. Gemelli" IRCCS, Rome, Italy,* [2]*Department of Translational Medicine and Surgery, Catholic University of Rome, Rome, Italy*

13.1 Introduction

Celiac disease (CD) is very often associated with many clinical conditions such as anemia, osteoporosis, metabolic disorders, skin disorders, reproductive disorders, polyautoimmunity, and neurological and psychiatric disorders (Table 13.1). The underlying pathophysiology for most of these conditions is malabsorption secondary to villous atrophy induced by gluten-related mechanisms. In addition, the increased autoimmune response can promote the development of inflammation even at a distance from the intestine. For most of these conditions, the gluten-free diet (GFD) represents the cornerstone of treatment, however, for some of them, specific supportive treatments may be appropriate, which will be detailed in this chapter.

TABLE 13.1 Clinical conditions associated to CD.

Anemia	Iron deficiency anemia Vitamin B12 or folate deficiency Blood loss Anemia of chronic disease
Bone diseases	Osteoporosis
Metabolic disorders	MAFLD/NAFLD
Skin disorders	Dermatitis herpetiformis Alopecia areata Deficiency skin diseases
Reproductive disorders	Amenorrhea Infertility Recurrent miscarriages Hypogonadism Low birth weight or preterm deliveries
Autoimmune diseases	Increased occurrence of poliautoimmunity (diabetes, thyroid diseases, arthritis, psoriatic skin disorders)
Neurological diseases	Gluten ataxia Peripheral neuropathies Cognitive impairment Epilepsy
Psychiatric diseases	Depression and anxiety

Abbreviations: metabolic-associated fatty liver disease, MAFLD; nonalcoholic fatty liver disease, NAFLD.

Pediatric and Adult Celiac Disease. DOI: https://doi.org/10.1016/B978-0-443-13359-6.00004-2

13.2 Anemia

Anemia is a frequent manifestation of CD, which involves about 40% of celiac patients, both in children and adults. In most cases, anemia is explained by the deficiency of iron caused by malabsorbition. The majority of patients respond to a GFD, with hemoglobin levels returning to normal in parallel with the healing of the intestinal mucosa and the restoration of the absorptive capacity. In the most severe cases, a supplementation of dietary iron and vitamins can favor a rapid return to normal laboratory values. Unfortunately, there are cases of persistent iron deficiency anemia due to several conditions, such as nonadherence to a GFD, intake of a low-iron GFD, or the persistence of intestinal mucosal atrophy (Seidita et al., 2022; Stefanelli et al., 2020). The refractoriness to oral iron supplementation can be explained by several reasons. For example, oral iron intake may be limited by possible side effects, such as abdominal discomfort, stomach pain, constipation and diarrhea, which may lead to a reduction in therapeutic adherence. However, newer formulations such as sucrosomial iron (a ferric pyrophosphate coated with a phospholipid and sucrester membrane) and feralgine (a solution of ferrous bisglycinate chelate and sodium alginate) may be effective in providing iron supplementation in celiac patients intolerant to other classical oral therapies, ensuring optimal intestinal absorption (Elli et al., 2018; Giancotti et al., 2019). Celiac patients who are refractory to dietary correction and to oral iron supplementation require periodic intravenous iron administration (Lichtenstein & Onken, 2018). Furthermore, anemia in CD has a multifactorial pathogenesis and, although it is frequently a consequence of iron deficiency, it can be caused by deficiencies of folate or vitamin B12, or by blood loss or other associated diseases (Martín-Masot et al., 2019). For this reason it is mandatory to evaluate vitamin B12 and folate levels and, eventually, to supplement deficiencies (Wierdsma et al., 2013). In some cases, nutritional deficiencies alone do not explain anemia, that may instead be explained by in inflammation in the so called anemia of chronic disease (ACD). In these patients, downregulation of intestinal inflammatory induced by GFD improves anemia (Bergamaschi et al., 2008; Repo et al., 2022; Roldan et al., 2022). Finally, the presence of other comorbidities could explain a loss of blood through the gastrointestinal tract. In fact, celiacs with anemia refractory to GFD or to iron and vitamin supplementation should always be investigated with upper/lower endoscopy or video capsule endoscopy to evaluate the presence of associated inflammatory, erosive or neoplastic conditions (Efthymakis et al., 2017).

13.3 Osteopenia and osteoporosis

Bone metabolism disorders, such as osteopenia or osteoporosis, have been reported in several studies on celiac patients, both in adults and children (see Chapters 6 and 7), and this may be due to malabsorption of nutrients essential for bone health, such as calcium and vitamin D (Lungaro et al., 2023; Mora et al., 1999; Zanchetta et al., 2016). Accordingly, the risk of fractures is slightly increased in patients with CD (Mosca et al., 2022; West et al., 2003).

Several mechanisms, such as malnutrition or hormonal alterations, can cause the reduced bone mineralization in celiac patients. Malabsorption of calcium and vitamin D is in fact a common feature of CD. In particular, vitamin D contributes to bone mineralization and regulates intestinal calcium absorption (G et al., 2021; Krupa-Kozak, 2014). Furthermore, calcium deficiency can be caused by multifactorial aspects. In fact, calcium is often malabsorbed (Molteni et al., 1995), but, in addition, a GFD based on industrial products is often deficient in calcium (Krupa-Kozak & Drabińska, 2016). As regards hormonal mechanisms, growth hormone (GH) and prolactin (PRL) have been considered. In particular, a study highlighted the association between CD and a high presence of antipituitary antibodies which could alter bone metabolism by mediating the reduction of IGF-1 (Mirza & Canalis, 2015); moreover, patients with untreated CD would present lower than normal levels of PRL with consequent reduction of bone density (Capozzi et al., 2015; Reifen et al., 1997).

Diagnosis of osteopenia/osteoporosis is made by dual-energy X-ray absorptiometry (DEXA). DEXA may be performed at the diagnosis especially in patients with severe malabsorption or with long delay in diagnosis and in particular it is mandatory in individuals with clinical presentation suggestive of bone disease. Moreover, at least one DEXA should be recommended in adults no later than age 35 and then repeated at regular 3–5 year intervals based on the presence of baseline abnormalities (Al-Toma et al., 2019).

Treatment is based on strict adherence to a GFD that is able to restore normal levels of vitamin D and calcium in some patients. However, supplementation is required in some individuals with increased risk factors for osteoporosis that fails to resolve only with diet (Caruso et al., 2013). In fact, calcium and vitamin D supplementation alone are not sufficient to restore bone loss, since they may promote a rise in vitamin D and calcium levels without inducing a prompt normalization in bone density (Barera et al., 2020; Mautalen et al., 1997). In those cases where dietary measures and supplementation of vitamin D and calcium are not sufficient, and there is an increased risk of fractures due to the presence of comorbidities or personal risk factors, it is advisable to start an appropriate treatment with specific medications, namely bisphosphonates, estrogens, teriparatide, or denosumab (Fouda et al., 2012; Lewiecki et al., 2015; Nuti et al., 2019).

13.4 Metabolic disorders

In recent years, much evidence has been emerging to support the association between CD and metabolic disorders (see also Chapter 12). The most widely accepted hypothesis supports the idea that a metabolic imbalance acts as a trigger for liver fat accumulation and inflammation. In support of this hypothesis, most CD patients who start a GFD tend to gain weight, and in particular have an increased incidence of metabolic-associated fatty liver disease (MAFLD) and nonalcoholic fatty liver disease (NAFLD), both characterized by liver steatosis, and impaired lipid and carbohydrate metabolism. These alterations could be favored by the retrieval of intestinal absorption, by the consumption of easily available industrial high-in-fat and carbohydrate gluten-free products (Raiteri et al., 2022), together with an immune-mediated hepatic damage triggered by antitransglutaminases (Scapaticci et al., 2023; Valvano et al., 2020). Nutritional deficiencies, such as low blood levels of vitamin D, may contribute to the development of a pro-inflammatory state that can trigger metabolic syndrome and associated liver disorders (Carambia & Herkel, 2018). A retrospective study on a population of 221 adult CD patients showed a prevalence of NAFLD higher than MAFLD at diagnosis (29.4% and 14.5%, respectively), whereas the majority of NAFLD patients (81.8%) "progressed" to MAFLD after 2 years of a GFD (Rispo et al., 2021). To monitor the onset and progression of these metabolic disorders, it is mandatory to carry out yearly blood tests, and in selected patients, an abdominal ultrasound. Considering that the GFD is crucial in the management of CD, a proper nutritional education is mandatory for preventing such metabolic disorders. The ideal GFD should be nutrient, rich in naturally gluten-free foods, balanced with macro- and micronutrients. For these reasons, the pseudograin rotation provides a good source of fiber, complex carbohydrates, protein, fatty acids, minerals and vitamins. Therefore dietitians should play a leading role in GFD education and maintenance (Theethira & Dennis, 2015).

13.5 Skin disorders

Dermatitis herpetiformis (DH) is an extraintestinal manifestation of CD (see Chapter 16). The activity of the cutaneous lesions does not always parallel the gastrointestinal symptoms; in fact, dermatitis often occurs in individuals who are otherwise symptomless. DH usually has a good clinical response to a GFD that resolves both skin lesions and gastrointestinal symptoms (Reunala et al., 2018); however, gastrointestinal symptoms usually resolve in a shorter time (Hietikko et al., 2018; Paek et al., 2011). In more severe or very slow responding cases, the use of dapsone, topical steroids, or immunosuppressors may be considered (Al-Toma et al., 2019).

Another dermatological disease that is in some cases associated with CD is alopecia areata. The response of skin lesions to a GFD appears to be age-dependent: several case series have reported an optimal response in pediatric patients, while this appeared partial or inadequate in cases of adults or with long-standing alopecia (Bardella et al., 2000; Corazza et al., 1995; Fessatou et al., 2003). Finally, a series of dermatological alterations related to trace elements deficiencies may occur in CD patients. In particular, zinc deficiency induced by intestinal malabsorption can cause acrodermatitis, brittle nails, and thinning hair. Therefore it appears important to consider a possible malabsorption of trace elements and to supplement promptly this deficiency (Zou et al., 2023).

13.6 Reproductive disorders

Reproductive disorders in CD may include infertility, recurrent miscarriage, low birth weight or preterm delivery, delayed menarche, amenorrhea, premature menopause, impotence, and hypogonadism. Mechanisms, such as autoimmunity mediated by tissue antitransglutaminase antibodies and nutritional deficiencies have been hypothesized (Özgör & Selimoğlu, 2010). More in depth, CD may affect fertility through immune-mediated mechanisms, also affecting angiogenesis, and malabsorption of micronutrients such as iron, zinc, folic acid, vitamin B12, or fat-soluble vitamins, promoting hyperhomocysteinemia and increasing the risk of thrombosis and coagulation disorders (Di Simone et al., 2021; Khizroeva et al., 2019; Simopoulou et al., 2019). A meta-analysis confirmed that the risk of reproductive disorders is significantly higher in CD than in the general population. In particular, the OR for CD was 5.82 in women with recurrent miscarriage, 5.06 in women with unexplained infertility, and 8.73 in patients with intrauterine fetal growth retardation (Tersigni et al., 2014).

As regards the treatment, a strict GFD is mandatory, as it is able both to improve malabsorption deficiencies and to reduce the proinflammatory state, furthermore close monitoring of gluten contamination is essential in women who continue to have problems despite a GFD. Additionally, nutritional support with folate supplementation may improve reproductive outcomes, whereas concomitant vitamin B12 supplementation is mandatory to avoid deleterious effects on the fetal nervous system (Fabozzi et al., 2022).

13.7 Autoimmune diseases associated to CD

CD is often associated with other autoimmune disorders. In particular, findings from a large epidemiological study in CD patients, insulin dependent diabetes (OR 5.5), thyroid diseases (OR = 1.8), arthritis (OR = 2.4) and psoriatic skin disorders (OR = 1.6) were the most common (Assa et al., 2017). Regarding treatment, the role of a GFD as a supportive treatment of the associated autoimmune disease is not yet clear. Some studies have reported that a GFD is able to improve glycemic control, clinical parameters, and BMI in individuals with type 1 diabetes (Amin et al., 2002; Hansen et al., 2006). Instead, there is no evidence supporting a role of a GFD in modulating the thyroid hormonal and immunological profile (Malandrini et al., 2022), nor in the control of symptoms or improvement of quality of life in rheumatological disorders (Lidón et al., 2022).

13.8 Neurological and psychiatric disorders

Several neurological disorders, such as gluten ataxia (GA), peripheral neuropathies, cognitive impairment and epilepsy have been found in celiac patients. This association can be explained through gluten dependent mechanisms or vitamin deficiency (Gala et al., 2022). GA is characterized by gluten-related autoimmunity and it is the most common neurological manifestation of CD. Patients with GA clinically present with gait ataxia or postural deficits. Symptoms are usually subtle and chronic, with possible associations with myoclonus or sensorimotor neuropathy (Hadjivassiliou et al., 2015). The strict GFD is the first line treatment for GA. In particular, the downregulation of serum antibodies induced by a GFD seems to cause a partial reversal of the immune-mediated damage to the central nervous system (Hadjivassiliou et al., 2003). Unfortunately, improvement of ataxia is not always complete, as cerebellar atrophy and Purkinje cell loss are not reversible. In cases where patients have followed a strict GFD for at least 1 year with no improvement in symptoms, immunosuppression and intravenous immunoglobulin can be considered (Hadjivassiliou et al., 2017).

The peripheral neuropathies associated with CD are often secondary to vitamin or trace element deficiencies. In particular, deficiencies of vitamin B12 or copper favor the onset of this neurological impairment. These deficiencies need to be promptly addressed to avoid irreversible damage (Goodman et al., 2009; Stabler, 2013). A form called "gluten neuropathy" is also known, not always associated with active enteropathy. It is an idiopathic neuropathy, in which an alternative etiology cannot be found. In these patients, the GFD slightly improves pain perception and promotes a better quality of life (Hadjivassiliou et al., 2006; Hadjivassiliou et al., 2010; Zis et al., 2018).

Patients with CD sometimes have cognitive impairment that is mild during the early stages of CD and worsens as the disease progresses (Longarini et al., 2019). Pathophysiology remains unclear, but it can be speculated to be secondary to malabsorption of vitamin B12, vitamin E, and folate. Furthermore, it has been proposed that proinflammatory cytokines bind and incorporate into the blood-brain barrier favoring cognitive deficits. Vitamin supplementation does not lead to evident improvement, however, cognitive function recovers completely with a strict GFD and reappears with gluten exposure (Hu et al., 2006; Makhlouf et al., 2018).

Epilepsy may be present in 5% of adult CD patients, with a 1.fourfold increased risk compared to controls (Ludvigsson et al., 2012). Various factors, including vitamin B12 malabsorption, local deposition of tTG antibodies, may favor its development (Casciato et al., 2015). The strict adherence to GFD and vitamin supplementation have shown promising results in reducing the frequency of seizures and therefore represent a first-line treatment (Lea et al., 1995).

Finally, several psychiatric disorders have also been associated with CD, in particular depression and anxiety, as well as manic-depressive illness, schizophrenia, and bipolar disorder. This association is likely multifactorial, and the examination of the mental health status is recommended when a diagnosis of CD is made, in order to establish an adequate management (Alkhiari, 2023).

References

Alkhiari, R. (2023). Psychiatric and neurological manifestations of celiac disease in adults. *Cureus*, *15*, e35712. Available from https://doi.org/10.7759/cureus.35712.

Al-Toma, A., Volta, U., Auricchio, R., Castillejo, G., Sanders, D. S., Cellier, C., Mulder, C. J., & Lundin, K. E. A. (2019). European Society for the Study of Coeliac Disease (ESsCD) guideline for coeliac disease and other gluten-related disorders. *United European Gastroenterology Journal*, *7* (5), 583–613. Available from https://doi.org/10.1177/2050640619844125.

Amin, R., Murphy, N., Edge, J., Ahmed, M. L., Acerini, C. L., & Dunger, D. B. (2002). A longitudinal study of the effects of a gluten-free diet on glycemic control and weight gain in subjects with type 1 diabetes and celiac disease. *Diabetes Care*, *25*(7), 1117–1122. Available from https://doi.org/10.2337/diacare.25.7.1117.

Assa, A., Frenkel-Nir, Y., Tzur, D., Katz, L. H., & Shamir, R. (2017). Large population study shows that adolescents with celiac disease have an increased risk of multiple autoimmune and nonautoimmune comorbidities. *Acta Paediatrica, International Journal of Paediatrics*, *106*(6), 967–972. Available from https://doi.org/10.1111/apa.13808.

Bardella, M. T., Marino, R., Barbareschi, M., Bianchi, F., Faglia, G., & Bianchi, P. (2000). Alopecia areata and coeliac disease: No effect of a gluten-free diet on hair growth. *Dermatology*, *200*(2), 108–110. Available from https://doi.org/10.1159/000018340.

Barera, G., Maruca, K., Sgaramella, P., Di Stefano, M., & Mora, S. (2020). Short-term, low dose Vitamin D supplementation in young patients with celiac disease: A pilot study. *European Journal of Gastroenterology and Hepatology*, *32*(5), 663–664. Available from https://doi.org/10.1097/MEG.0000000000001627.

Bergamaschi, G., Markopoulos, K., Albertini, R., Di Sabatino, A., Biagi, F., Ciccocioppo, R., Arbustini, E., & Corazza, G. R. (2008). Anemia of chronic disease and defective erythropoietin production in patients with celiac disease. *Haematologica*, *93*(12), 1785–1791. Available from https://doi.org/10.3324/haematol.13255.

Capozzi, A., Scambia, G., Pontecorvi, A., & Lello, S. (2015). Hyperprolactinemia: Pathophysiology and therapeutic approach. *Gynecological Endocrinology*, *31*(7), 506–510. Available from https://doi.org/10.3109/09513590.2015.1017810.

Carambia, A., & Herkel, J. (2018). Dietary and metabolic modulators of hepatic immunity. *Seminars in Immunopathology*, *40*(2), 175–188. Available from https://doi.org/10.1007/s00281-017-0659-4.

Caruso, R., Pallone, F., Stasi, E., Romeo, S., & Monteleone, G. (2013). Appropriate nutrient supplementation in celiac disease. *Annals of Medicine*, *45*(8), 522–531. Available from https://doi.org/10.3109/07853890.2013.849383.

Casciato, S., Morano, A., Albini, M., Fanella, M., Lapenta, L., Fattouch, J., Carnì, M., Colonnese, C., Manfredi, M., Teresa Giallonardo, A., & Di Bonaventura, C. (2015). Cryptogenic focal epilepsy and »hidden» celiac disease in adulthood: A causal or accidental link? *International Journal of Neuroscience*, *125*(12), 913–917. Available from https://doi.org/10.3109/00207454.2014.983227.

Corazza, G. R., Andreani, M. L., Venturo, N., Bernardi, M., Tosti, A., & Gasbarrini, G. (1995). Celiac disease and alopecia areata: Report of a new association. *Gastroenterology*, *109*(4), 1333–1337. Available from https://doi.org/10.1016/0016-5085(95)90597-9.

Di Simone, N., Gratta, M., Castellani, R., D'Ippolito, S., Specchia, M., Scambia, G., & Tersigni, C. (2021). Celiac disease and reproductive failures: An update on pathogenic mechanisms. *American Journal of Reproductive Immunology*, *85*(4). Available from https://doi.org/10.1111/aji.13334.

Efthymakis, K., Milano, A., Laterza, F., Serio, M., & Neri, M. (2017). Iron deficiency anemia despite effective gluten-free diet in celiac disease: Diagnostic role of small bowel capsule endoscopy. *Digestive and Liver Disease*, *49*(4), 412–416. Available from https://doi.org/10.1016/j.dld.2016.12.007.

Elli, L., Ferretti, F., Branchi, F., Tomba, C., Lombardo, V., Scricciolo, A., Doneda, L., & Roncoroni, L. (2018). Sucrosomial iron supplementation in anemic patients with celiac disease not tolerating oral ferrous sulfate: A prospective study. *Nutrients*, *10*(3), 330. Available from https://doi.org/10.3390/nu10030330.

Fabozzi, G., Verdone, G., Allori, M., Cimadomo, D., Tatone, C., Stuppia, L., Franzago, M., Ubaldi, N., Vaiarelli, A., Ubaldi, F. M., Rienzi, L., & Gennarelli, G. (2022). Personalized nutrition in the management of female infertility: New insights on chronic low-grade inflammation. *Nutrients*, *14*(9). Available from https://doi.org/10.3390/nu14091918.

Fessatou, S., Kostaki, M., & Karpathios, T. (2003). Coeliac disease and alopecia areata in childhood. *Journal of Paediatrics and Child Health*, *39*(2), 152–154. Available from https://doi.org/10.1046/j.1440-1754.2003.00116.x.

Fouda, M. A., Khan, A. A., Sultan, M., Rios, L. P., McAssey, K., & Armstrong, D. (2012). Evaluation and management of skeletal health in celiac disease: Position statement. *Canadian Journal of Gastroenterology*, *26*(11), 819–829. Available from https://doi.org/10.1155/2012/823648.

G, A., Seth, A., Kumar, P., & Jain, A. (2021). Prevalence and management of vitamin D deficiency in children with newly diagnosed coeliac disease: Cohort study. *Paediatrics and International Child Health*, *41*(4), 247–252. Available from https://doi.org/10.1080/20469047.2021.1996089.

Gala, D., Scharf, S., Kudlak, M., Green, C., Khowaja, F., Shah, M., Kumar, V., & Ullal, G. (2022). A comprehensive review of the neurological manifestations of celiac disease and its treatment. *Diseases*, *10*, 111.

Giancotti, L., Talarico, V., Mazza, G. A., Marrazzo, S., Gangemi, P., Miniero, R., & Bertini, M. (2019). FeralgineTM a new approach for iron deficiency anemia in celiac patients. *Nutrients*, *11*(4). Available from https://doi.org/10.3390/nu11040887.

Goodman, B. P., Mistry, D. H., Pasha, S. F., & Bosch, P. E. (2009). Copper deficiency myeloneuropathy due to occult celiac disease. *Neurologist*, *15*(6), 355–356. Available from https://doi.org/10.1097/NRL.0b013e31819428a8.

Hadjivassiliou, M., Davies-Jones, G. A. B., Sanders, D. S., & Grünewald, R. A. (2003). Dietary treatment of gluten ataxia. *Journal of Neurology, Neurosurgery and Psychiatry*, *74*(9), 1221–1224. Available from https://doi.org/10.1136/jnnp.74.9.1221.

Hadjivassiliou, M., Grünewald, R. A., Sanders, D. S., Shanmugarajah, P., & Hoggard, N. (2017). Effect of gluten-free diet on cerebellar MR spectroscopy in gluten ataxia. *Neurology*, *89*(7), 705–709. Available from https://doi.org/10.1212/WNL.0000000000004237.

Hadjivassiliou, M., Kandler, R. H., Chattopadhyay, A. K., Davies-Jones, A. G. B., Jarratt, J. A., Sanders, D. S., Sharrack, B., & Grünewald, R. A. (2006). Dietary treatment of gluten neuropathy. *Muscle and Nerve*, *34*(6), 762–766. Available from https://doi.org/10.1002/mus.20642.

Hadjivassiliou, M., Rao, D. G., Wharton, S. B., Sanders, D. S., Grünewald, R. A., & Davies-Jones, A. G. B. (2010). Sensory ganglionopathy due to gluten sensitivity. *Neurology*, *75*(11), 1003–1008. Available from https://doi.org/10.1212/WNL.0b013e3181f25ee0.

Hadjivassiliou, M., Sanders, D. D., & Aeschlimann, D. P. (2015). Gluten-related disorders: Gluten ataxia. *Digestive Diseases*, *33*(2), 264–268. Available from https://doi.org/10.1159/000369509.

Hansen, D., Brock-Jacobsen, B., Lund, E., Bjørn, C., Hansen, L. P., Nielsen, C., Fenger, C., Lillevang, S. T., & Husby, S. (2006). Clinical benefit of a gluten-free diet in type 1 diabetic children with screening-detected celiac disease: A population-based screening study with 2 years' follow-up. *Diabetes Care*, *29*(11), 2452–2456. Available from https://doi.org/10.2337/dc06-0990.

Hietikko, M., Hervonen, K., Salmi, T., Ilus, T., Zone, J. J., Kaukinen, K., Reunala, T., & Lindfors, K. (2018). Disappearance of epidermal transglutaminase and IgA deposits from the papillary dermis of patients with dermatitis herpetiformis after a long-term gluten-free diet. *British Journal of Dermatology*, *178*(3), e198–e201. Available from https://doi.org/10.1111/bjd.15995.

Hu, W. T., Murray, J. A., Greenaway, M. C., Parisi, J. E., & Josephs, K. A. (2006). Cognitive impairment and celiac disease. *Archives of Neurology*, *63*(10), 1440–1446. Available from https://doi.org/10.1001/archneur.63.10.1440.

Khizroeva, J., Nalli, C., Bitsadze, V., Lojacono, A., Zatti, S., Andreoli, L., Tincani, A., Shoenfeld, Y., & Makatsariya, A. (2019). Infertility in women with systemic autoimmune diseases. *Best Practice & Research Clinical Endocrinology & Metabolism*, *33*(6), 101369. Available from https://doi.org/10.1016/j.beem.2019.101369.

Krupa-Kozak, U. (2014). Pathologic bone alterations in celiac disease: Etiology, epidemiology, andtreatment. *Nutrition*, *30*(1), 16–24. Available from https://doi.org/10.1016/j.nut.2013.05.027.

Krupa-Kozak, U., & Drabińska, N. (2016). Calcium in gluten-free life: Health-related and nutritional implications. *Foods*, *5*(3), 1–11. Available from https://doi.org/10.3390/foods5030051.

Lea, M. E., Harbord, M., & Sage, M. R. (1995). Bilateral occipital calcification associated with celiac disease, folate deficiency, and epilepsy. *American Journal of Neuroradiology*, *16*(7), 1498–1500.

Lewiecki, E. M., Bilezikian, J. P., Binkley, N., Hans, D., Krueger, D., Miller, P. D., Oates, M., & Shane, E. (2015). Update on osteoporosis from the 2014 Santa Fe Bone symposium. *Endocrine Research*, *40*(2), 106–119. Available from https://doi.org/10.3109/07435800.2015.1005746.

Lichtenstein, G. R., & Onken, J. E. (2018). Improved hemoglobin response with ferric carboxymaltose in patients with gastrointestinal-related iron-deficiency anemia versus oral iron. *Digestive Diseases and Sciences*, *63*(11), 3009–3019. Available from https://doi.org/10.1007/s10620-018-5204-3.

Lidón, A. C., Patricia, M. L., Vinesh, D., & Marta, M. S. (2022). Evaluation of gluten exclusion for the improvement of rheumatoid arthritis in adults. *Nutrients*, *14*(24). Available from https://doi.org/10.3390/nu14245396.

Longarini, G., Richly, P., Temprano, M. D. L. P., Costa, A. F., Vázquez, H., Moreno, M. L., Niveloni, S., López, P., Smecuol, E., Mazure, R., González, A., Mauriño, E., & Bai, J. C. (2019). A prospective study on cognitive impairment in middle-aged adults with newly diagnosed celiac disease. *Journal of Clinical Gastroenterology*, *53*(4), 290–294. Available from https://doi.org/10.1097/MCG.0000000000001018.

Ludvigsson, J. F., Zingone, F., Tomson, T., Ekbom, A., & Ciacci, C. (2012). Increased risk of epilepsy in biopsy-verified celiac disease A population-based cohort study. *Neurology*, *78*(18), 1401–1407. Available from https://doi.org/10.1212/WNL.0b013e3182544728.

Lungaro, L., Manza, F., Costanzini, A., Barbalinardo, M., Gentili, D., Caputo, F., Guarino, M., Zoli, G., Volta, U., De Giorgio, R., & Caio, G. (2023). Osteoporosis and celiac disease: Updates and hidden pitfalls. *Nutrients*, *15*(5), 1089. Available from https://doi.org/10.3390/nu15051089.

Makhlouf, S., Messelmani, M., Zaouali, J., & Mrissa, R. (2018). Cognitive impairment in celiac disease and non-celiac gluten sensitivity: Review of literature on the main cognitive impairments, the imaging and the effect of gluten free diet. *Acta Neurologica Belgica*, *118*(1), 21–27. Available from https://doi.org/10.1007/s13760-017-0870-z.

Malandrini, S., Trimboli, P., Guzzaloni, G., Virili, C., & Lucchini, B. (2022). What about TSH and anti-thyroid antibodies in patients with autoimmune thyroiditis and celiac disease using a gluten-free diet? A systematic review. *Nutrients*, *14*(8), 1681. Available from https://doi.org/10.3390/nu14081681.

Martín-Masot, R., Nestares, M. T., Diaz-Castro, J., López-Aliaga, I., Alférez, M. J. M., Moreno-Fernandez, J., & Maldonado, J. (2019). Multifactorial etiology of anemia in celiac disease and effect of gluten-free diet: A comprehensive review. *Nutrients*, *11*(11). Available from https://doi.org/10.3390/nu11112557.

Mautalen, C., Gonzalez, D., Mazure, R., Vazquez, H., Lorenzetti, M. P., Maurino, E., Niveloni, S., Pedreira, S., Smecuol, E., Boerr, L. A., & Bai, J. C. (1997). Effect of treatment on bone mass, mineral metabolism, and body composition in untreated celiac disease patients. *American Journal of Gastroenterology*, *92*(2), 313–318.

Mirza, F., & Canalis, E. (2015). Secondary osteoporosis: Pathophysiology and management. *European Journal of Endocrinology*, *173*(3), R131–R151. Available from https://doi.org/10.1530/EJE-15-0118.

Molteni, N., Bardella, M. T., Vezzoli, G., Pozzoli, E., & Bianchi, P. (1995). Intestinal calcium absorption as shown by stable strontium test in celiac disease before and after gluten-free diet. *The American Journal of Gastroenterology*, *90*(11), 2025–2028. Available from https://doi.org/10.1111/j.1572-0241.1995.tb10186.x.

Mora, S., Barera, G., Beccio, S., Proverbio, M. C., Weber, G., Bianchi, C., & Chiumello, G. (1999). Bone density and bone metabolism are normal after long-term gluten-free diet in young celiac patients. *American Journal of Gastroenterology*, *94*(2), 398–403. Available from https://doi.org/10.1111/j.1572-0241.1999.867_r.x.

Mosca, C., Thorsteinsdottir, F., Abrahamsen, B., Rumessen, J. J., & Händel, M. N. (2022). Newly diagnosed celiac disease and bone health in young adults: A systematic literature review. *Calcified Tissue International*, *110*(6), 641–648. Available from https://doi.org/10.1007/s00223-021-00938-w.

Nuti, R., Brandi, M. L., Checchia, G., Di Munno, O., Dominguez, L., Falaschi, P., Fiore, C. E., Iolascon, G., Maggi, S., Michieli, R., Migliaccio, S., Minisola, S., Rossini, M., Sessa, G., Tarantino, U., Toselli, A., & Isaia, G. C. (2019). Guidelines for the management of osteoporosis and fragility fractures. *Internal and Emergency Medicine*, *14*(1), 85–102. Available from https://doi.org/10.1007/s11739-018-1874-2.

Özgör, B., & Selimoğlu, M. A. (2010). Coeliac disease and reproductive disorders. *Scandinavian Journal of Gastroenterology*, *45*(4), 395–402. Available from https://doi.org/10.3109/00365520903508902.

Paek, S. Y., Steinberg, S. M., & Katz, S. I. (2011). Remission in dermatitis herpetiformis: A cohort study. *Archives of Dermatology*, *147*(3), 301–305. Available from https://doi.org/10.1001/archdermatol.2010.336.

Raiteri, A., Granito, A., Faggiano, C., Giamperoli, A., Catenaro, T., Negrini, G., & Tovoli, F. (2022). Hepatic steatosis in patients with celiac disease: The role of packaged gluten-free foods. *Nutrients*, *14*(14), 2942. Available from https://doi.org/10.3390/nu14142942.

Reifen, R., Buskila, D., Maislos, M., Press, J., & Lerner, A. (1997). Serum prolactin in coeliac disease: A marker for disease activity. *Archives of Disease in Childhood*, *77*(2), 155–157. Available from https://doi.org/10.1136/adc.77.2.155.

Repo, M., Kurppa, K., Huhtala, H., Luostarinen, L., Kaukinen, K., & Kivelä, L. (2022). Significance of low ferritin without anaemia in screen-detected, adult coeliac disease patients. *Journal of Internal Medicine*, *292*(6), 904–914. Available from https://doi.org/10.1111/joim.13548.

Reunala, T., Salmi, T. T., Hervonen, K., Kaukinen, K., & Collin, P. (2018). Dermatitis herpetiformis: A common extraintestinal manifestation of coeliac disease. *Nutrients*, *10*(5). Available from https://doi.org/10.3390/nu10050602.

Rispo, A., Imperatore, N., Guarino, M., Tortora, R., Alisi, A., Cossiga, V., Testa, A., Ricciolino, S., Fiorentino, A., & Morisco, F. (2021). Metabolic-associated fatty liver disease (MAFLD) in coeliac disease. *Liver International*, *41*(4), 788–798. Available from https://doi.org/10.1111/liv.14767.

Roldan, G. A., Goyes, D., Villafuerte-Gálvez, J. A., Urquiaga, M., Dennis, M., Murray, J. A., Leffler, D. A., & Kelly, C. P. (2022). Anemia etiology and the response to a gluten-free diet in untreated patients with celiac disease: A 2-year follow-up. *The American Journal of Gastroenterology*, *117*(10), 1684–1692. Available from https://doi.org/10.14309/ajg.0000000000001875.

Scapaticci, S., Venanzi, A., Chiarelli, F., & Giannini, C. (2023). MAFLD and Celiac Disease in children. *International Journal of Molecular Sciences*, *24*(2).

Seidita, A., Mansueto, P., Compagnoni, S., Castellucci, D., Soresi, M., Chiarello, G., Cavallo, G., Carlo, De, Nigro, A., Chiavetta, M., Mandreucci, F., Giuliano, A., Disclafani, R., & Carroccio, A. (2022). Anemia in celiac disease: Prevalence, associated clinical and laboratory features, and persistence after gluten-free diet. *Journal of Personalized Medicine*, *12*.

Simopoulou, M., Sfakianoudis, K., Maziotis, E., Grigoriadis, S., Giannelou, P., Rapani, A., Tsioulou, P., Pantou, A., Kalampokas, T., Vlahos, N., Pantos, K., & Koutsilieris, M. (2019). The impact of autoantibodies on IVF treatment and outcome: A systematic review. *International Journal of Molecular Sciences*, *20*(4), 892. Available from https://doi.org/10.3390/ijms20040892.

Stabler, S. P. (2013). Vitamin B12 deficiency. *New England Journal of Medicine*, *368*(2), 149–160. Available from https://doi.org/10.1056/NEJMcp1113996.

Stefanelli, G., Viscido, A., Longo, S., Magistroni, M., & Latella, G. (2020). Persistent Iron deficiency anemia in patients with celiac disease despite a gluten-free diet. *Nutrients*, *12*(8), 2176. Available from https://doi.org/10.3390/nu12082176.

Tersigni, C., Castellani, R., De waure, C., Fattorossi, A., De Spirito, M., Gasbarrini, A., Scambia, G., & Di Simone, N. (2014). Celiac disease and reproductive disorders: Meta-analysis of epidemiologic associations and potential pathogenic mechanisms. *Human Reproduction Update*, *20*(4), 582–593. Available from https://doi.org/10.1093/humupd/dmu007.

Theethira, T. G., & Dennis, M. (2015). Celiac disease and the gluten-free diet: Consequences and recommendations for improvement. *Digestive Diseases*, *33*(2), 175–182. Available from https://doi.org/10.1159/000369504.

Valvano, M., Longo, S., Stefanelli, G., Frieri, G., Viscido, A., Latella, G., & Disease, C. (2020). Gluten-free diet, and metabolic and liver disorders. *Nutrients*, *12*(4).

West, J., Logan, R. F. A., Card, T. R., Smith, C., & Hubbard, R. (2003). Fracture risk in people with celiac disease: A population-based cohort study. *Gastroenterology*, *125*(2), 429–436. Available from https://doi.org/10.1016/S0016-5085(03)00891-6.

Wierdsma, N. J., van Bokhorst-de van der Schueren, M. A. E., Berkenpas, M., Mulder, C. J. J., & van Bodegraven, A. A. (2013). Vitamin and mineral deficiencies are highly prevalent in newly diagnosed celiac disease patients. *Nutrients*, *5*(10), 3975–3992. Available from https://doi.org/10.3390/nu5103975.

Zanchetta, M. B., Longobardi, V., & Bai, J. C. (2016). Bone and celiac disease. *Current Osteoporosis Reports*, *14*(2), 43–48. Available from https://doi.org/10.1007/s11914-016-0304-5.

Zis, P., Sarrigiannis, P. G., Rao, D. G., & Hadjivassiliou, M. (2018). Quality of life in patients with gluten neuropathy: A case-controlled study. *Nutrients*, *10*(6). Available from https://doi.org/10.3390/nu10060662.

Zou, P., Du, Y., Yang, C., & Cao, Y. (2023). Trace element zinc and skin disorders. *Frontiers in Medicine*, *9*. Available from https://doi.org/10.3389/fmed.2022.1093868.

Chapter 14

Refractory celiac disease and lymphomagenesis

Marco Vincenzo Lenti[1], Giacomo Broglio[1], Marco Lucioni[2] and Gino Roberto Corazza[1]
[1]*Department of Internal Medicine, University of Pavia, Fondazione IRCCS Policlinico San Matteo, Pavia, Italy,* [2]*Department of Experimental Medicine, University of Pavia, Fondazione IRCCS Policlinico San Matteo, Pavia, Italy*

14.1 Introduction

Refractory celiac disease (RCD) and enteropathy-associated T-cell lymphoma (EATL) are established complications of adult celiac disease (CD) (Di Sabatino & Corazza, 2009; Green & Cellier, 2007). Most adult patients respond clinically and histologically to a gluten-free diet (GFD) within 12 months, with both the regression of symptoms and improvement of the small bowel mucosal lesions (Al-Toma et al., 2019; Ludvigsson et al., 2014; Rubio-Tapia et al., 2013). However, in up to 30%−40% of patients on a GFD, symptoms and/or villous atrophy may be persisting (Lebwohl et al., 2014; Penny, Rej et al., 2022; Tye-Din, 2022), although this figure is affected by the definitions used, the follow-up time, and differences in the clinical setting. Indeed, there are still uncertainties regarding the correct timing for assessing the response to diet, which is currently based on expert opinion rather than ad hoc studies, reporting a range of 12−24 months (Al-Toma et al., 2019; Lebwohl et al., 2014; Penny, Rej et al., 2022; Rubio-Tapia et al., 2013, 2023).

Importantly, it should be made clear that the lack of response to a GFD has heterogeneous reasons that encompass two main nosological categories, namely "nonresponsive CD" and "RCD" (Fig. 14.1). Further, when coping with RCD, there may be some frequent and relevant causes of false refractoriness that make this diagnosis even more difficult. This categorization has a major clinical relevance, as it implies a critical dissection of the causes of lack of response to a GFD, hence avoiding detrimental misdiagnoses.

14.2 Nonresponsive celiac disease

This condition has been defined by the persistence or recurrence of symptoms (diarrhea, steatorrhea, abdominal pain, weight loss, anemia-related symptoms, others) after at least a 12-month strict GFD, with significant improvement of small bowel mucosal lesions (Abdulkarim et al., 2002; Leffler et al., 2007). The remission of intestinal lesions implies that a follow-up duodenal biopsy is obtained after at least 12 months of symptom persistency. Nonresponsive CD essentially relates to the possible concomitance of other nongluten-sensitive conditions, the symptoms of which do not remit after diet (Fig. 14.1). Overall, its prevalence is poorly known and based on small, retrospective series.

The conditions showed in Fig. 14.1 have been repeatedly reported as the most frequent causes of symptom nonresponse (Abdulkarim et al., 2002; Leffler et al., 2007; Penny et al., 2020). *Irritable bowel syndrome* may clinically overlap CD, and it may be the most common misdiagnose leading to CD diagnostic delay in adults [(Lenti, Aronico, et al., 2022)]. In the largest series from the United States and the United Kingdom, irritable bowel syndrome was responsible for symptom persistency in almost one quarter of the cases (Leffler et al., 2007; Penny, Rej et al., 2022). According to a metaanalysis, the pooled prevalence of irritable bowel syndrome-like symptoms in patients with CD was 38% (Sainsbury et al., 2013). On the other hand, other putative causes of CD nonresponsiveness must be ruled out. Notably, a fermentable oligosaccharides, disaccharides, monosaccharides and polyols *(FODMAP)-rich diet* could mimic irritable bowel syndrome and may perpetrate symptom persistency (Testa et al., 2018).

Paying attention to certain clinical features could be of help in discriminating among other causes of CD nonresponsiveness. A clear relation between diarrhea and intake of dairy products should prompt the work-up for *lactose malabsorption,*

Pediatric and Adult Celiac Disease. DOI: https://doi.org/10.1016/B978-0-443-13359-6.00007-8

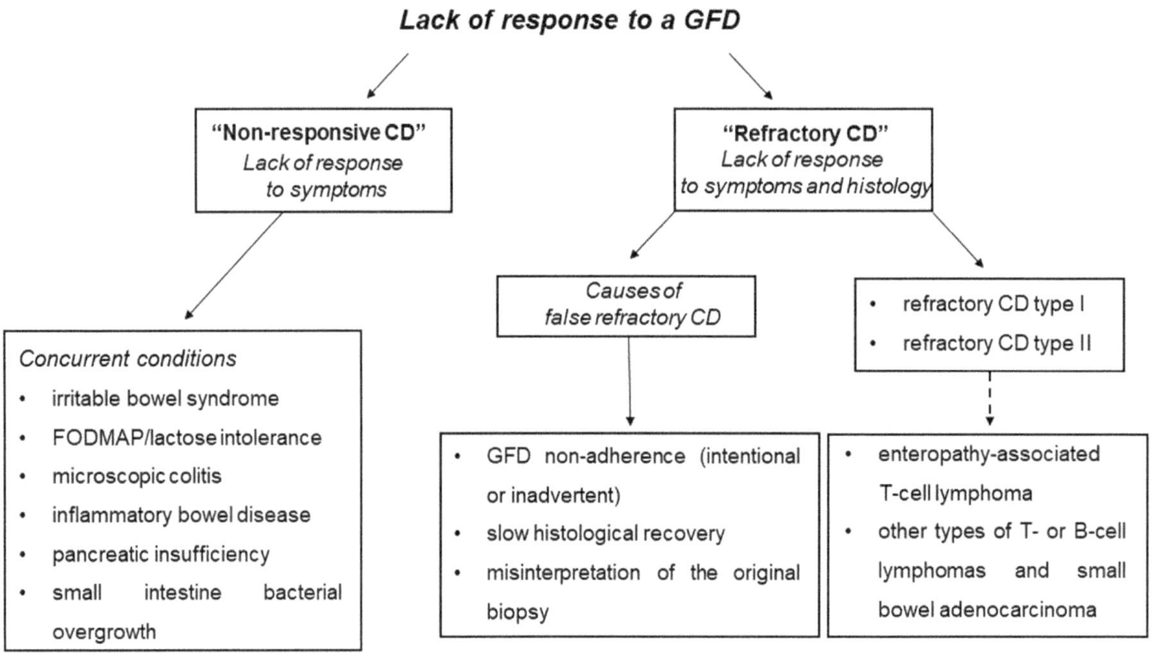

FIGURE 14.1 Flow-chart of the lack of response to a gluten-free diet. In CD, the lack of response to a GFD encompasses heterogeneous causes that can be summarized into two main categories, namely "nonresponsive CD" and "RCD." RCD type II can further complicate (dashed arrow) into EATL and, more rarely, into other lymphomas and small bowel adenocarcinoma.

although its prevalence in nonresponsive CD (2.7%) turned out to be lower than previously thought (Penny, Rej et al., 2022). Watery diarrhea, especially at night, with no correlation with food intake, should raise the suspicion of *microscopic colitis*. The concomitance of microscopic colitis is substantial, as up to 4% of CD patients have been found to have both diseases (Green et al., 2009). Data regarding the co-existence of *inflammatory bowel disease* and CD showed that there may be a 10-fold increased risk of this association (Leeds et al., 2007). In nonresponsive CD patients with suspected microscopic colitis or inflammatory bowel disease a colonoscopy with biopsies should be performed. *Exocrine pancreatic insufficiency* should be suspected in case of steatorrhea and weight loss, and it can be confirmed by decreased fecal elastase concentrations (Rana et al., 2016). In these patients, however, pancreatic enzyme supplementation gave unsatisfactory results (Yoosuf et al., 2023). As for *small intestine bacterial overgrowth*, a recent pooled-data analysis did not support a clear relationship with CD (Losurdo et al., 2017). Nevertheless, the authors concluded that this condition may be more common in CD when symptoms do not improve after a GFD. Many of the results obtained with hydrogen breath tests should be dismissed due to the interpretation difficulties related to the high fasting basal hydrogen levels often present in CD as a consequence of the fermentation of endogenous substrates (Corazza et al., 1987). However, a study conducted through quantitative culture of duodenal aspirates confirms a prevalence of 11% in nonresponsive CD (Rubio-Tapia, Barton et al., 2009).

14.3 Refractory celiac disease

RCD has been defined as the persistence or recurrence of malabsorption symptoms and villous atrophy after 12–24 months of GFD (Biagi & Corazza, 2001; Cellier et al., 1998; Green et al., 2022; Ludvigsson et al., 2013; Malamut & Cellier, 2019). RCD can be either primary, in case of resistance to a GFD since the very first diagnosis of CD, or secondary, when a patient relapses after initial response to diet. In case of primary RCD, an initial diagnosis of uncomplicated CD is usually made, and only after a GFD, when symptoms and villous atrophy persist or worsen, a diagnosis of refractoriness may be postulated. Indeed, primary RCD, due to the lack of response to a GFD and the frequent CD seronegativity, can be misdiagnosed with other disorders much more than secondary RCD. A further distinction is that of RCD type I and II; this latter in many cases evolve into EATL, and in few cases into other types of lymphoma (Malamut & Cellier, 2019) or, anecdotally, into small bowel adenocarcinoma (Vanoli et al., 2017) (Fig. 14.1). For all these reasons, and due to other challenges that pertain to this condition, diagnosis and management of RCD should be reserved to tertiary referral centers. Before dissecting the characteristics of RCD, the causes that may lead to a false diagnosis of refractoriness will be discussed.

14.3.1 Causes of false refractory celiac disease

Three main situations have been advocated to lead to a false diagnosis of RCD (Fig. 14.1): (1) GFD nonadherence, either for intentional or inadvertent gluten intake, (2) slow histological recovery, and (3) misinterpretation of the original biopsy. It is worth noting that false RCD is more common than the true counterpart. In a few cases, a clear and univocal cause of persistent villous atrophy cannot be found, and these patients must be closely followed-up (Kaukinen et al., 2007).

14.3.1.1 Dietary nonadherence or inadvertent gluten intake

Persistent gluten intake is considered a leading cause of enduring lesions in nominally treated CD patients (Dewar et al., 2012). Overall, both intentional and inadvertent gluten intake are very common, the former reported in up to 40% of adult patients (Hall et al., 2013). Intentional gluten intake is multifactorial, and may depend on poor self-efficacy, lower symptom burden, scarce knowledge of CD, and low educational level (Lebwohl et al., 2014).

Inadvertent occasional gluten intake was an important cause of persistence or recurrence of mucosal lesions in a study from the UK (Dewar et al., 2012). Noteworthy, these lesions tended to be milder (e.g., intraepithelial lymphocytosis and crypt hyperplasia) compared to those found in patients with true RCD. The role of very low-dose gluten intake in maintaining symptoms or lesions is still debated. An established safe gluten threshold is that of 10 mg (Catassi et al., 2007). However, it has been anecdotally reported that the sporadic (Scotta et al., 1982) or daily (Biagi et al., 2004) intake of a communion wafer may cause symptoms and villous atrophy, suggesting a wide range of interindividual sensitivity. The detection of gluten intake and the monitoring of the GFD will be discussed in Chapter 12.

14.3.1.2 Slow histological recovery

Slow histological recovery may be defined, in retrospect, as the lack of significant improvement of small bowel mucosal lesions after at least a 12-month GFD, in patients in whom a recovery has been proven later (Al-Toma et al., 2019). In fact, though clinical response may appear after just a few weeks of gluten withdrawal, histological recovery may require more than one year. In a prospective study, only 65% of adult patients reached a histological recovery, as defined by the regression of more severe lesions (Marsh-Oberhuber 3a-c) to milder lesions (Marsh-Oberhuber 0–2) after two years of a rigorous GFD, while another 20% of patients reached this goal only after five years (Wahab et al., 2002). It is, of course, difficult to assess in such real-life studies whether gluten contamination of the diet or other environmental factors may contribute to slow recovery. For certain, patients who best adhered to a GFD had a greater likelihood to reach complete histological remission (Hutchinson et al., 2010). Globally, these studies showed that the time to histological recovery is rather long and this could mimic a permanent refractoriness.

To conclude, we would suggest that, in patients with symptoms, duodenal mucosa should be reassessed after at least 12 months of a GFD and after a careful dietician review. Documenting a substantial regrowth of villi certainly excludes a diagnosis of RCD, emphasizing once again that deep mucosal healing only happens in a minority of cases and only after much more time of treatment.

14.3.1.3 Misinterpretation of the original biopsy

There are two leading sources of misinterpretation: (1) having mistaken for CD an atrophic enteropathy of different etiology, (2) having mistaken for villous atrophy a histological artifact mimicking CD. In both settings, we deal with non-gluten sensitive conditions that do not respond to a GFD, and that may, therefore mimic RCD.

Non-CD atrophic enteropathies have been the object of recent reviews (Schiepatti et al., 2021). They include a series of uncommon or rare conditions, namely autoimmune enteropathy, common variable immune deficiency, graft-vs-host disease, HIV enteropathy, tropical sprue, chronic *Giardia lamblia* infection, and some drug-related enteropathies (e.g., olmesartan and other sartans, immune checkpoint inhibitors, mycophenolate mofetil, chemotherapy). In all these CD serology is always negative, and all of them are not HLA-restricted. Some of their pathological and clinical features may be of help for addressing the correct diagnosis, as discussed in Chapter 11.

The assessment of small bowel biopsies is not straightforward, as it is hampered by several potential sources of mistake. A proper handling and orientation of size-adequate duodenal biopsies are mandatory for making a correct diagnosis (Chapter 9). It is widespread experience that a poor sample quality and tangential sectioning are common sources of biopsy misinterpretation, to the point that they may lead to the diagnosis of false RCD, even in case of a normal biopsy (Shidrawi et al., 1994).

14.3.2 Refractory celiac disease type I and type II

Both RCD type I and type II are marked by persistent or recurrent villous atrophy, but they are distinguished by immunophenotypic differences of the intraepithelial lymphocytes (IELs). RCD type I is characterized by a polyclonal expansion of IELs having a normal immunophenotype, indistinguishable from that of active CD, while RCD type II is characterized by a clonal proliferation of aberrant IELs in the absence of overt lymphoma (Cellier et al., 1998; Green et al., 2022; Soderquist et al., 2020). Thus, a follow-up duodenal biopsy, including the thorough evaluation of IELs phenotype, is mandatory, providing that all circumstances of false RCD have been excluded. The difference between type I and type II does not only regard immunophenotypic characteristics, but it translates into important differences regarding diagnosis, clinical aspects, treatment, and prognosis. Regarding RCD type II, in most cases a continuum of worsening lesions has been reported, starting from active CD to RCD type II, and overt EATL (Cellier et al., 2000; Meresse et al., 2009); hence, RCD type II should be considered as a severe, preneoplastic, complication of CD. Table 14.1 shows the main features and characteristics of RCD type I and II.

The true prevalence of RCD is largely unknown, due to the cumulation of both type I and II with large heterogeneity of results among centers, the lack of registries, and the lack of data from underrepresented populations. The cumulative incidence of RCD (including both type I and II) has been estimated to range between 0.3% and 1.5% in the celiac population (Malamut & Cellier, 2019; Roshan et al., 2011; Rowinski & Christensen, 2016). This figure could even be

TABLE 14.1 Main features of refractory celiac disease type I and type II.

	Refractory CD type I	Refractory CD type II
Unresponsive villous atrophy	In all cases	In all cases
CD serology	Negative in many cases	Negative in many cases
Potential risk factors	Older age and major symptoms at diagnosis, poor adherence to a GFD; HLA DQ2 homozygosity (~25%–30% of cases)	Older age and major symptoms at diagnosis, poor adherence to a GFD, putative environmental factors; HLA DQ2 homozygosity (~60% of cases)
Signs and symptoms	Major malabsorption symptoms in ~50% of the cases	Major malabsorption symptoms in >80% of the cases
Laboratory alterations	Anemia, hypoalbuminemia, and electrolyte imbalance are present in roughly 50% of the cases	Anemia, hypoalbuminemia, and electrolyte imbalance are present in almost all cases
Endoscopic features	Duodenal/jejunal ulcers or strictures are rare	Duodenal/jejunal ulcers or strictures are common
Immunohistochemistry aberrant IELs (iCD3 +, CD8-, CD4-)	Only a small proportion of cells	>50% of cells
Flow-cytometry aberrant IELs (sCD3-, CD8-, CD4-)	Only a small proportion of cells	>25% of cells
Clonal TCR-γ gene rearrangement	Virtually absent	In most of the cases
Association with autoimmunity	Very common	Less common
Response to steroids or immunosuppressants	Good	Initially good to fair; some immunosuppressants may increase the risk of developing EATL
Risk of developing EATL	Very rare	~50% within 5 years
5-year survival rate	>80%	≤50%

Abbreviations: celiac disease, CD; enteropathy-associated T-cell lymphoma, EATL; gluten-free diet, GFD; human leukocyte antigen, HLA; intraepithelial lymphocyte; intracellular CD, iCD; IEL; superficial CD, sCD; T-cell receptor, TCR.

smaller, as found in an Italian study, reporting an incidence of complicated CD (also including EATL) of only 0.2% after the exclusion of patients referred from other hospitals (Biagi et al., 2014). Further, there is no clear predominance between type I and type II (Al-Toma, Verbeek et al., 2007; Cellier et al., 2000; Malamut et al., 2009; Roshan et al., 2011; Rowinski & Christensen, 2016; Rubio-Tapia, Kelly et al., 2009; Weber et al., 2021). Indeed, the presumptive lower prevalence of RCD type II reported in the United States compared to European series may be attributed to the less or more stringent diagnostic criteria applied (Malamut & Cellier, 2011).

14.3.2.1 Clinical aspects of refractory celiac disease

The persistence or recurrence of malabsorption symptoms in a nominally treated patient should always raise the suspicion of RCD. In this setting, testing for serum CD antibodies is of little help, as in RCD they are negative in most cases (Cellier et al., 2000; Malamut et al., 2009; Roshan et al., 2011; Rubio-Tapia, Kelly et al., 2009). The causes of this low CD seroprevalence are unknown and might be attributed to the strict avoidance of gluten at the time of the study (Malamut et al., 2009) or to antigen exhaustion, as in Hashimoto's thyroiditis and autoimmune gastritis (Lenti et al., 2022). Conversely, also in this setting, seropositivity may reflect a poor adherence to a GFD.

The main clinical features of RCD type I and II reported in six previous series are cumulatively shown in Table 14.2 (Al-Toma, Verbeek et al., 2007; Branchi et al., 2022; Malamut et al., 2009; Roshan et al., 2011; Rubio-Tapia, Kelly et al., 2009; Weber et al., 2021). We should underline that Table 14.2 shows the mean values of heterogeneous studies, and hence should be cautiously interpreted. What is undoubtful, is that these clinical and laboratory alterations are more frequent and severe in type II compared to type I, and the former may often lead to hospitalization and to the need for parenteral nutrition. Further, again in type II, refractoriness usually occurs soon after the diagnosis of CD, suggesting that it is the onset of the complication itself to clinically unleash CD. The mean age of RCD diagnosis is around 45–50 years for type I and 50–60 years for type II. Regarding sex, in almost all series, female patients are usually more commonly affected by RCD type I as compared to males, with a ratio between sexes which is similar to uncomplicated CD (female to male ratio 2.5:1).

TABLE 14.2 Cumulative mean of the main features of patients with refractory celiac disease type I and II as reported in previous series. Percentage calculation was made after the exclusion of patients in whom that specific clinical feature was not reported.

Clinical features	RCD type I ($n = 155$)	RCD type II ($n = 135$)
Sex (female)	74.9%	46.7%
Sex (male)	25.1%	53.3%
Age (years) at CD diagnosis (mean)	45	55
Age (years) at RCD diagnosis (mean)	50	55
Antibody positivity at RCD diagnosis	7.0%	12.0%
Diarrhea	78.5%	81.5%
Weight loss	26.0%	68.5%
Abdominal pain	68.5%	68.3%
Anemia	50.5%	78.1%
Low albuminemia	25.9%	70.2%
Liver dysfunction	54.5%	79.0%
Microscopic colitis	40.0%	25.3%
Ulcerative jejunitis	4.0%	36.3%
Development of EATL	14.3%a	46.2%

Percentage calculation was made after the exclusion of patients in whom that specific clinical feature was not reported.
[a]*This percentage refers to the paper by Malamut et al., the only reporting this finding.*

In RCD type II, this trend is very different, and the female to male ratio tends to reduce (Al-Toma, Verbeek et al., 2007; Rubio-Tapia, Kelly et al., 2009), until reaching 1:1 (Malamut et al., 2009). In the German registry, males affected by RCD type II were even more common than females, confirming their greater predisposition to this complication (Weber et al., 2021). Notably, microscopic colitis has been reported as a rather frequent comorbidity in both RCD subtypes, although much more frequent in type I (Al-Toma, Verbeek et al., 2007; Weber et al., 2021). This is in line with the known, greater predisposition to autoimmunity in RCD type I patients. Regardless of the subtype, patients with RCD usually complain of major malabsorption symptoms, including diarrhea and abdominal pain in all cases, as well as weight loss, peripheral edema, and fatigue (Malamut et al., 2009; Roshan et al., 2011; Rubio-Tapia, Kelly et al., 2009; Weber et al., 2021). Blood tests frequently reveal anemia, hypoalbuminemia, and electrolyte imbalance; in some cases, hypertransaminasemia may coexist (Malamut et al., 2009). Regarding the development of EATL, it is much more common in type II, although these figures are highly influenced by the follow-up length. To note, there are no distinct endoscopic features of CD refractoriness; the duodenal mucosa usually has an atrophic appearance, with flattened or scalloped folds (Malamut et al., 2009), as in untreated CD (see Chapter 9). Similarly, there are no specific radiological features of CD refractoriness. A scoring system based on magnetic resonance enteroclysis to assess RCD has been proposed (Van Weyenberg et al., 2011), by evaluating the number and appearance of jejunal and ileal folds. The median number of jejunal folds per 5 cm was lower in patients with RCD type II, while fold jejunal pattern reversal, diffuse bowel wall thickening were more frequent in type II refractoriness compared to type I.

14.3.2.2 Possibly associated conditions

RCD may be associated to other clinical conditions. In particular, *ulcerative jejunoileitis* (UJI) predominantly associated with RCD type II (Biagi et al., 2000) (Table 14.2), is a condition characterized by multiple ulcerations and strictures of the small bowel, burdened by a raised mortality rate due to bleeding or perforation. In this setting, both double balloon enteroscopy and videocapsule endoscopy are useful for the diagnosis: the ulcerations are usually annular or transverse and scarring, and may be misdiagnosed as other conditions; otherwise, irregular, focal, and nodular ulcers should raise the suspicion of malignancy and EATL. UJI may rarely have a favorable outcome and a good long-term prognosis, unless the strictures have been surgically treated (Biagi et al., 1998).

Adult CD is known to be one of the most frequent conditions associated to *defective spleen function and/or atrophy* (Di Sabatino, Carsetti, et al., 2011), and it may predispose these patients to the risk of major infections (Lenti, Luu, et al., 2022) (see also Chapter 15). Interestingly, a significantly higher risk for hyposplenism was found in RCD than in uncomplicated CD (Di Sabatino, Rosado, et al., 2006). A small spleen volume, which usually implies a more severe spleen hypofunction, has also been more commonly reported with magnetic resonance in RCD type II compared to type I or uncomplicated CD (Van Weyenberg et al., 2011).

Finally, *collagenous sprue* was described for the first time in a patient with an atrophic enteropathy and a primary resistance to a GFD (Weinstein et al., 1970). The full-blown form is represented by thickened and irregular collagen subepithelial band with entrapped lamina propria inflammatory and stromal cells (Maguire et al., 2009; Vakiani et al., 2010). The collagen band could reach more than 200 μm in thickness and, although collagenous sprue has been mostly associated with CD and RCD (Maguire et al., 2009; Vakiani et al., 2010), in more recent years, it has been described in other enteropathies, such as that induced by olmesartan and by other drugs (Kung et al., 2018). Despite being CD serology negativity and HLADQ2/DQ8 positivity rather frequent, the prognosis is favorable in most cases, and the evolution into EATL has never been noticed (Kung et al., 2018). Hence, the relation between collagenous sprue and uncomplicated or RCD remains elusive.

14.3.2.3 Pathological features of refractory celiac disease

As already stated, the diagnosis of RCD relies on histopathology and on the molecular analysis of the duodenal IELs. In particular, immunohistochemistry, flow cytometry, and assessment of gamma chain T cell receptor (TCR) rearrangement should always be included. A combination of these three techniques is always needed, since by only immunohistochemistry, some cases of RCD type II could be overlooked (Malamut & Cellier, 2011).

The definition of an aberrant IEL phenotype is not straightforward, as there is no universal cut-off for their density, nor univocal marker. As shown in Table 14.1, in RCD type II, duodenal mucosa is characterized by unresponsive villous atrophy and increased CD103 + , CD45 + IELs, with >25% of aberrant IELs lacking s(surface)CD3 without CD8 or CD4 on flow cytometry, or >50% aberrant IELs expressing i(intracellular)CD3 without CD8 or CD4 in formol-fixed sections, along with detectable monoclonal TCR gamma chain rearrangement (Malamut et al., 2009; Soderquist & Bhagat, 2021). It has also been suggested to assess the TCR beta chain, as this may improve the diagnostic yield (Perfetti et al., 2012; Ritter et al., 2018).

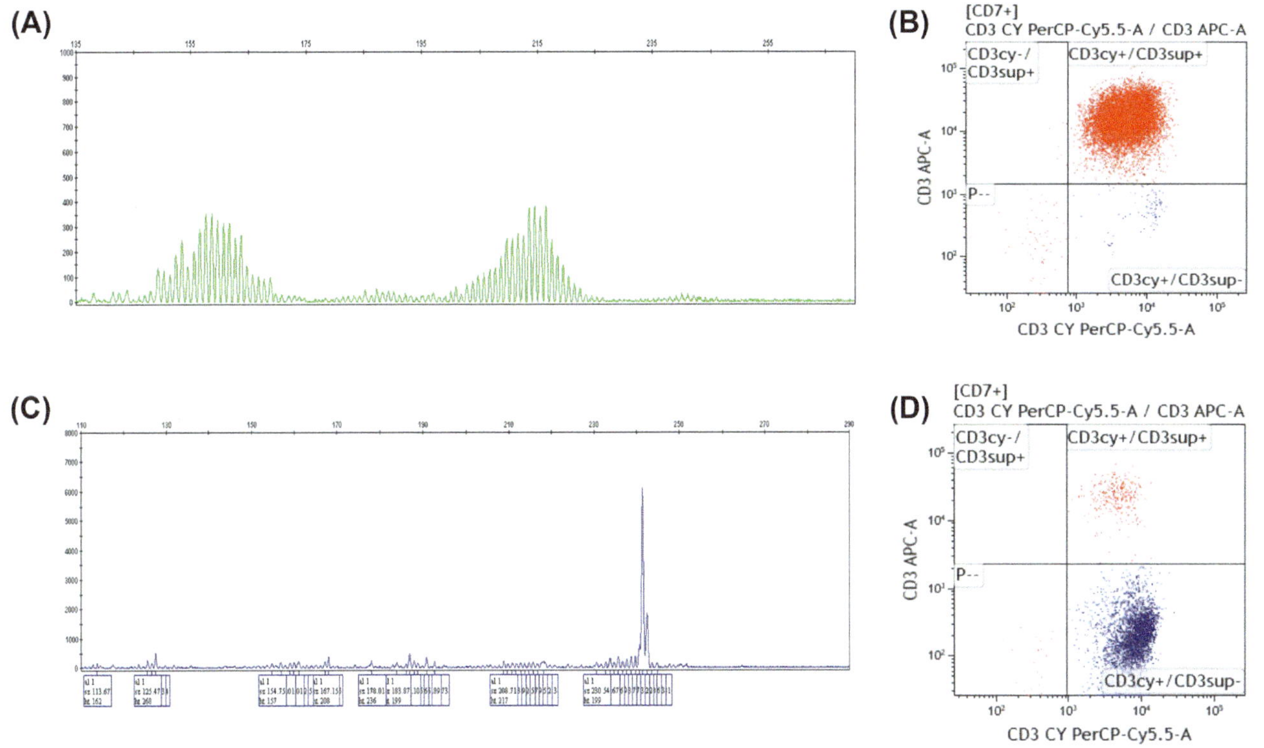

FIGURE 14.2 Panels showing the TCR gamma rearrangement (PCR) and the flow cytometric analysis of IELs of a patient with RCD type I (A, B) compared to a patient with type II (C, D). In RDC type I, polyclonal TCR gamma rearrangements can be noticed (A; multiple, hence polyclonal, peaks expressed as relative fluorescent units). At flow cytometry (B), the immunophenotypic analysis highlighted a normal CD7 + , CD45 + IEL population (red dots), with only a mild aberrant co-expression of surface (s)CD3- and intracellular (i)CD3 + elements (1.3%) (blue dots). In RDC type II, monoclonal TCR gamma rearrangement can be noticed (C; single peak expressed as relative fluorescent units, Vγ1−8 + Jγ1.1/2.1 sequences). At flow cytometry (D), the immunophenotypic analysis highlighted in CD7 + , CD45 + IELs a marked aberrant coexpression of sCD3-/iCD3 + elements (85%) (blue dots), while only a minority of IELs were normal (red dots).

In RCD type I, histology is similar to what is found in untreated CD, with villous atrophy and increased CD103 + , CD45 + IELs, with only <25% of aberrant IELs lacking sCD3 on flow cytometry, or <50% aberrant iCD3 + , CD8-/CD4- cells in formol-fixed sections, in the absence of clonality (i.e., polyclonal TCR) (Malamut et al., 2009) However, in RCD type I, recent updates indicate a lower cut-off (15%−20%) for aberrant IELs (Green et al., 2022). Finally, there may be a discrepancy among the three techniques and a critical interpretation of all the tests is needed (Green et al., 2022). For example, a clonal rearrangement of the TCR, either gamma or beta, in the absence of other immunohistochemical or flow cytometric IEL abnormalities, can be sporadically found in RCD type I and even in uncomplicated CD patients. Conversely, the absence of TCR monoclonal rearrangement does not preclude a diagnosis of RCD type II either, if aberrant IELs are present (Soderquist et al., 2020). In these discrepant cases, clinical features and disease evolution should be taken into account for differentiating between type I and II (Branchi et al., 2022). Fig. 14.2 shows the findings of TCR gamma rearrangement and flow cytometry of a patient with RCD type I (polyclonal rearrangements, normal IEL population) versus type II (monoclonal rearrangement, aberrant IEL population).

There are some additional features that characterize RCD type II. Aberrant IELs also express cytotoxic granule proteins such as granzyme or perforins (Soderquist et al., 2021), and they can be detected in the duodenal *lamina propria* and in extraintestinal sites (e.g., peripheral blood, colon, stomach, skin, and mesenteric lymphnodes) (Malamut et al., 2009; Soderquist et al., 2021; Verbeek et al., 2009; Verkarre, Asnafi, et al., 2003). Additionally, the number of Paneth cells seems to be reduced in duodenal biopsies of complicated CD (including both RCD and EATL), suggesting that their function may be necessary for the integrity of crypts and surface epithelium (Creamer & Pink, 1967; Di Sabatino et al., 2008)

14.3.2.4 Mechanisms of lesions in refractory celiac disease and lymphomagenesis

Some potential risk factors for RCD have been reported. We have already mentioned the increased susceptibility to type II in males (Weber et al., 2021). Regarding age, it was found that in CD patients older than 60 years at diagnosis, the risk of developing complications (RCD type I and II, EATL, UJI) was 18 times higher than in patients diagnosed at 18–40 years, and 9 times higher than in patients diagnosed at 40–60 years (Biagi et al., 2018). Hence, gender, older age and allegedly cumulative longer gluten intake may be important risk factors. In the same study, a major presentation with overt malabsorption and severe malnutrition increased the risk of complications by 7 times. HLA haplotypes may also have a role. DQ2 homozygosity has been shown to increase the severity of mucosal lesions, being present in roughly 25% of the cases of RCD type I, in 44% of type II, and in 50% of EATL in a US series (Al-Toma, Goerres, Meijer, Peña et al., 2006), and the same trend was confirmed in Europe (Biagi et al., 2012; Hrdlickova et al., 2018). Some non-HLA genetic factors, including genetic rearrangements or aberrations affecting crucial immunological pathways may also play a pivotal role in RCD type II and lymphomagenesis, as discussed later (Cording et al., 2022; Ettersperger et al., 2016; Hrdlickova et al., 2018; Levescot et al., 2022; Soderquist & Bhagat, 2021; Verkarre, Romana, et al., 2003).

Regarding RCD type I, the mechanisms that induce resistance to a GFD are largely unknown. However, its frequent association with extraintestinal autoimmunity (Sharma et al., 2018) and its good response to oral budesonide or other immunosuppressants (Green et al., 2022; Malamut et al., 2009; Rubio-Tapia, Kelly et al., 2009) suggest the importance of autoreactive mechanisms. In this perspective, it has been hypothesized that autoreactive, circulating, CD8 + T cells are recruited and activated in chronically inflamed intestinal mucosa, possibly initiating the immunological cascade leading to RCD type I (Levescot et al., 2022). Other authors suggested that RCD type I merely encompasses a subgroup of patients very sensitive to gluten, and once the diet is carefully reviewed and corrected, they improve clinically and histologically (Holmes & Ciacci, 2018). Although the very frequent antibody seronegativity found in these patients argues against this assumption, a preliminary and uncontrolled study showed evidence of raised urinary gluten peptides in patients with RCD type I (Penny, Coleman et al., 2022). Nonetheless, it should be reminded that cases of transitions from RCD type I to type II and EATL, as well as an increased overall mortality in RCD type I in comparison to uncomplicated CD, have been reported (Malamut et al., 2013). Overall, it is likely that RCD type I encompasses a heterogeneous population of patients, and hence the current classification should be partially revised.

Regarding RCD type II, the sequence of events leading to this form and subsequently to EATL has been much more studied. This condition, which is a naturally occurring prelymphoma prototype, has represented an ideal model for studying the mechanisms of EATL development. For sure, interleukin (IL) 15 has been the most studied cytokine in this setting and it plays a major role in determining refractoriness via multiple pathways (Levescot et al., 2022). IL15 promotes the development, differentiation, proliferation, and survival of multiple immune cells, especially T cells and NK cells (Di Sabatino, Calarota, et al., 2011). It can be expressed by various cell types, including immune cells (e.g., monocytes, macrophages, dendritic cells), but also by intestinal epithelial cells and myeloid cells in active CD (Mention et al., 2003). IL15 exerts a proliferative effect on CD8 + T IELs, in both untreated and RCD (type I and II), stimulating the production of gamma interferon (ɣIFN) and tumor necrosis factor (TNF) α, promoting IEL cytotoxicity via perforin/granzyme, Fas-Fas ligand, MIC-NKGD2 ligand against the intestinal epithelium, and inhibiting IEL apoptosis (Abadie et al., 2020; Di Sabatino, Ciccocioppo, et al., 2006; Di Sabatino et al., 2001; Mention et al., 2003; Meresse et al., 2009). In other terms, IL15 sustains and perpetrates the intestinal damage with a waterfall effect, and additional cytokines, inflammatory pathways and genetic aberrations further amplify its detrimental effects. Consistently, an in vitro study showed that a fully humanized anti-IL15 antibody, able to block Jak3 and STAT5 phosphorylation, reinduced a massive IEL apoptosis in untreated CD and in RCD type II (Malamut et al., 2010), as further detailed later. There is evidence that aberrant IELs usually arise from a subset of IL15-dependent innate-lymphoid cells displaying characteristics of both NK and T cells, expressing NKR, NKG2D and NKp46, all boosting IL15 functions (Ettersperger et al., 2016; Levescot et al., 2022). Of note, it was found that NKp46 -but not other NK-related markers- was significantly more expressed by aberrant IELs in RCD type II as compared to type I and uncomplicated CD (Cheminant et al., 2019). Moreover, at immunohistochemistry, the presence of >25 NKp46 + IEL per 100 epithelial cells was able to discriminate RCD type II from type I and uncomplicated CD. Indeed, NKp46 is also expressed in overt EATL.

Besides IL15, other cytokines and inflammatory pathways have been implicated into the pathogenesis of type II refractoriness, even if the evidence is mostly derivative. IL2 acts via the same pathway as IL15 by activating CD8 + T cells (Ring et al., 2012). IL21 is also overexpressed in patients with untreated CD, and its production depends on IL15 (Sarra et al., 2013). The CD40/CD40 ligand pathway, which is a costimulatory pairing crucial for sustaining T cell activation, was upregulated in untreated and RCD; furthermore, blockade of CD40-CD40L interaction abolished gliadin-induced T cells activation and upregulation of γIFN and IL17 (Di Sabatino, Rovedatti, et al., 2011). Finally, neuroendocrine cells, namely serotonin-positive cells, were found to be significantly more expressed in RCD, with no differences between type I and II; in vitro, serotonin was able to upregulate ɣIFN production, at levels comparable to those induced by peptic tryptic digest of gliadin, thus sustaining the hypothesis that neuroendocrine cells may maintain inflammation (Di Sabatino et al., 2014).

While IL15 has an established role in perpetrating intestinal lesions, the mechanisms leading to its upregulation are still under study. Certainly, gluten acts as a trigger, as demonstrated by the increased production of IL15 in the duodenal mucosa of untreated, but not treated, CD (Di Sabatino et al., 2006). It has been proposed that the p31−49 gluten peptide is the fraction specifically responsible for the increased production of IL15 in CD (Hüe et al., 2004). Other environmental factors have been called into question. Infections may activate multiple inflammatory pathways via an augmented production of γIFN, a chronic stimulation and proliferation of CD8 + T cells, and an increased production of IL15 (Levescot et al., 2022; Meresse et al., 2009). In duodenal biopsies of RCD patients (both type I and II), an active Epstein-Barr virus infection was found in 70% of the cases (Perfetti et al., 2016), but other confirmatory studies are needed. Similarly, hepatis B and C infections have also been found in a relevant proportion of refractory patients, though their causal effect is unproven (Malamut et al., 2009).

Some genetic mutations and chromosomal abnormalities have been described in RCD type II and EATL. In RCD type II, pathologic IELs thrive both for gain of function mutations, such as p.G1097 of the JAK1 or the D661 in STAT3, that enhance their response to IL15, and for the impairment of the Smad3 pathway of TGF-β signaling, interfering with the functions of FoxP3 + , CD4 + , CD25 + regulating T cells (Cording et al., 2022; Ettersperger et al., 2016; Meresse et al., 2009; Soderquist et al., 2021). Again, this results in an increased resistance of IEL apoptosis. Apart from the mutations of the JAK-STAT signaling pathway, the mutated cells also harbored recurrent alterations in epigenetic regulators, such as TET2 and KMT2D, nuclear factor-κB, DNA damage repair, and immune evasion pathway genes (Soderquist et al., 2021). The same mutations can also be found in overt EATL, confirming the shared pathogenetic bases. A locus at 7p14.3 was found to be a potential driver of immunological dysregulation in CD that may bring to the development of RCD type II and EATL (Hrdlickova et al., 2018). The presence of a recurrent partial trisomy 1q22-q44 has been observed in the IELs of some cases of RCD type II patients, and this may trigger the early stages of development of overt EATL (Verkarre, Romana, et al., 2003). In EATL, compared to RCD type II, additional mutations can amplify aberrant IELs replication, becoming uncontrolled, and leading to mucosal invasiveness. Among these mutations, KMT2D, TET2, DDX3X, TNFAIP3, and POT1 were the most reported. Moreover, JAK1-STAT3 double mutations, as well as TNFAIP3/A20 mutations, were predictive of EATL transformation (Cording et al., 2022). Finally, it has been postulated that different pathogenetic pathways may underlie EATL arising in patients with a known RCD type II compared to the so called “de novo” EATL, that is, EATL diagnosed in a patient with no history of CD/RCD and whose symptoms have a rapid onset (Chander et al., 2018). In fact, a much more favorable prognosis of this latter has been observed in a large series of EATL (Malamut et al., 2013). However, additional studies are needed to confirm this assumption.

To conclude, in RCD type II, several genetic mutations and chromosome aberrations cumulating in IELs, along with defective IEL apoptosis, constitute the bases of subsequent EATL. Fig. 14.3 schematically summarizes the pathogenic mechanisms of refractoriness and lymphomagenesis.

14.3.2.5 Treatment of refractory celiac disease

Our knowledge regarding the treatment of RCD is mostly based on retrospective data of small series, case reports, and expert opinions. Only few prospective, open-label studies are available. There have been no randomized controlled trials with positive results.

According to the latest guidelines, the first line therapy for both RCD type I and IIis represented by corticosteroids, more specifically open-capsule *budesonide* or, if not available, oral prednisone (Green et al., 2022). Particularly, according to the only available open-label study (Mukewar et al., 2017), open-capsule budesonide (3 mg three times daily; two capsules emptied and stirred into a viscous medium, the third swallowed as a whole) was able to induce a clinical response in 92% of the cases and a histological response in 89% of the cases, even refractory to other immunosuppressants, with no significant differences between type I and type II. In half of the cases of RCD type II, aberrant IELs and TCR clonal rearrangements have been reported to be cleared at follow-up biopsy (Mukewar et al., 2017). Differently, in a previous study using intact budesonide capsules, patients with RCD type II still had aberrant IELs at follow-up biopsy (Brar et al., 2007). According to our clinical experience, patients are likely to need a long-term therapy as they may relapse upon tapering. It is unknown whether intravenous steroids or systemic oral steroids should be given in more clinically compromised patients, but they could represent a bridge option to open-capsule budesonide. Oral *prednisone*, at a dose of 0.5−1 mg/kg, was effective in inducing a clinical response in 90% of RCD type I and in 77% of patients with RCD II, but it is burdened by several side effects (Malamut et al., 2009).

Escalation therapy to immunosuppressants, although not standardized and not evidenced by clinical trials, should be considered in case of steroid refractoriness or dependency. In type I, the addition of *azathioprine* or *6-mercaptopurine* to steroids proved effective in inducing a clinical response in many cases of the available series (Al-Toma, Verbeek et al., 2007; Malamut et al., 2009; Rubio-Tapia et al., 2009. Differently, the use of azathioprine in patients with RCD

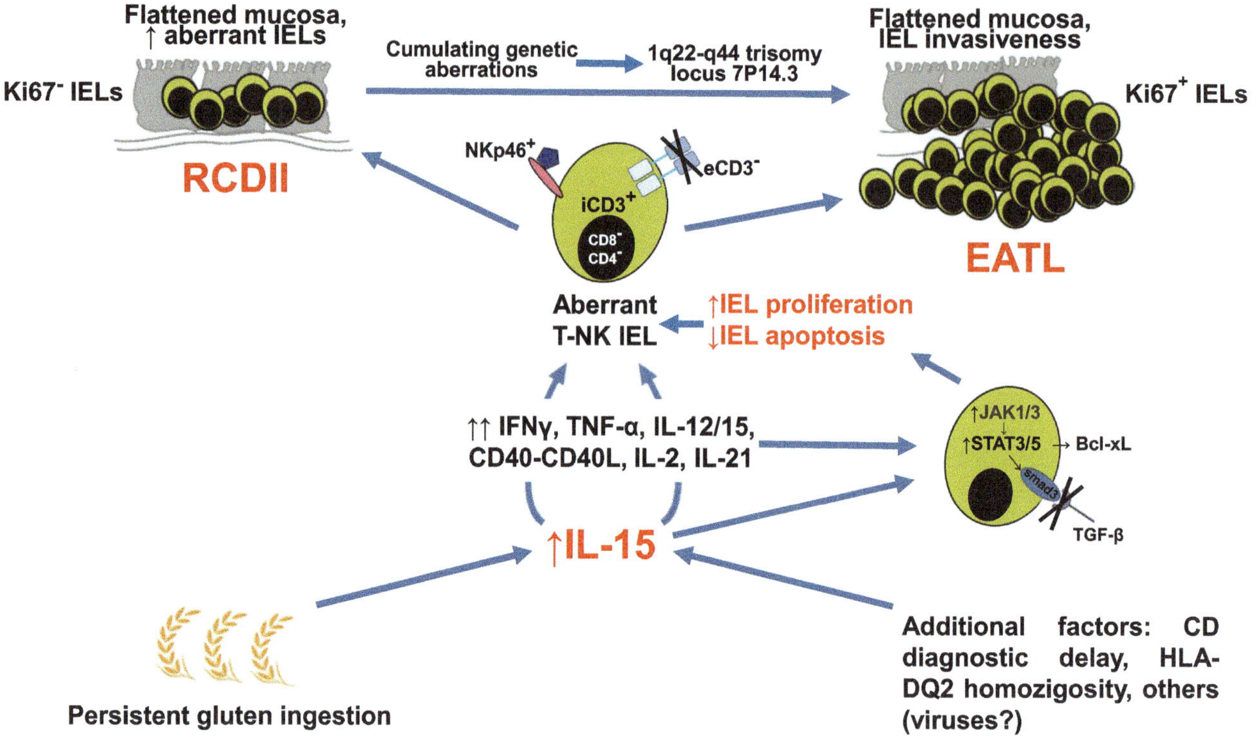

FIGURE 14.3 **Schematic representation of the pathogenesis of RCD type II and of EATL.** It is a multi-step process encompassing multiple, progressively cumulating, immunological and genetic aberrations. In particularly, IL15 increases due to persistent gluten ingestion and/or additional risk factors and exerts a proliferative effect towards IELs, along with an antiapoptotic effect. The effect of IL15 is further amplified by other cytokines, and all together exert an antiapoptotic and proliferative effect on IEL through the amplification of the JAK/STAT signaling which causes an over expression of the antiapoptotic transmembrane molecule B-cell lymphoma-extra large (Bcl-xl). Parallelly, the inhibition of the Smad3-TGF-β interaction occurs. All these changes induce a transformation of the IELs into a NK-T phenotype with aberrant features (iCD3 + , eCD3-; expression of NK markers). When genetic aberrations cumulate, EATL occurs. Indeed, there are many similarities between RCD type II and EATL, such as the expression of the NKp46. However, the two entities can be distinguished through the ki67 expression (absent in RCD type II, markedly increased in EATL).

type II is not currently recommended due to the potential risk of increasing the likelihood of developing EATL (Green et al., 2022). *Vedolizumab*, an antiintegrin monoclonal antibody approved for the treatment of inflammatory bowel disease, has been used in cases of RCD type I who did not respond to budesonide, inducing a clinical and histological remission (from Marsh 3 to Marsh 1) (Lundin et al., 2022).

For RCD type II, other therapies have been attempted in small series or in single case reports; given the small sample size, the safety concerns, and the uncertain risk-benefit profile, the results cannot be considered as conclusive. Among these, *cladribine*, a purine analog, (Al-Toma, Goerres, Meijer, von Blomberg et al., 2006), *alemtuzumab*, an anti-CD52 monoclonal antibody, (Malamut et al., 2009) *infliximab*, (Gillett et al., 2002; Malamut et al., 2009), *autologous hematopoietic stem cell transplantation*, (Al-Toma, Visser et al., 2007; Tack et al., 2011) and *mesenchymal stem cell infusion* (Ciccocioppo et al., 2016) should be mentioned. Despite some positive results, a rapid progression rate to overt EATL has been noticed in some cases.

Finally, the most logical and appropriate candidate for the treatment of RCD type II is the *anti-IL15 monoclonal antibody* AMG714, that has been studied in a phase 2 randomized controlled trial in RCD type II (Cellier et al., 2019). Despite its strong rationale, this trial did not reach the primary endpoint, namely the reduction of aberrant IELs at follow-up duodenal biopsy. Nonetheless, in those patients treated with AMG714, the improvement of diarrhea and the reduction in TCR clonality were noticed; hence, a larger study with a longer follow-up is warranted. An expanded access program is currently available (NCT03439475). Given the JAK/STAT pathways alterations in RCD type II, *tofacitinib*, an oral JAK 1/3 inhibitor, may be of potential interest. A patient suffering from UJI and RCD type II and failing to multiple therapies was treated with tofacitinib and she reached clinical remission and histological improvement (from Marsh 3b-c to Marsh 1−2) at 2-year follow-up (Grewal et al., 2022). Similar results have been obtained in a phase 2a pilot study in three patients out of six (Dieckman et al., 2022). Indeed, more studies are needed for confirming the safety and effectiveness of this drug in this setting.

14.4 Enteropathy-associated T-cell lymphoma

EATL is a rare, aggressive form of primary intestinal non-Hodgkin lymphoma of T/NK origin which arises in patients with a previous -or concomitant- diagnosis of CD (Alaggio et al., 2022; Di Sabatino et al., 2012; Swerdlow et al., 2016). In virtually all cases, it is the end-stage of a continuum represented by active CD → RCD type II → EATL (Malamut et al., 2013). In a few cases, EATL may arise in patients in whom a clear diagnosis of RCD could not be made as death occurred before the completion of the diagnostic work-up (Gale et al., 2000). The classification of EATL has changed over the last years, as it was formerly categorized into type I and type II. Since the 2016 WHO classification, EATL type I was simply renamed into EATL, while EATL type II was renamed into monomorphic epitheliotropic intestinal T-cell lymphoma (MEITL) (Swerdlow et al., 2016).

Data regarding EATL epidemiology are scarce. Similar incidences were found in the general population in the Netherlands and in Scotland with a crude incidence of roughly 0.1/100.000 cases/year (Sieniawski et al., 2010; Verbeek et al., 2008). Notably, the overall incidence of EATL has been found to increase in the United States over the last decades (Sharaiha et al., 2012). Considering the cumulative data from the latest and largest real-life studies in RCD patients (Table 14.2), EATL occurs in about 50% of the cases of RCD type II and in 15% of the cases of RCD type I, meaning a prevalence of <0.5% in the whole CD population. Despite its rarity, EATL constitutes the major mortality burden in CD patients 2001. Overall, the standardized mortality ratio for EATL, was reported to be 69 in CD patients; most deaths occurred within 3 years after CD diagnosis (Abdul Sultan et al., 2015; Corrao et al., 2001; Holmes & Muirhead, 2018; Koskinen et al., 2020). A more recent epidemiological study confirmed that the main determinant of mortality in CD is represented by lymphoproliferative disorders, with a hazard ratio of 2.36; mortality is higher within the first two years after diagnosis (Koskinen et al., 2020).

14.4.1 Clinical aspects of enteropathy-associated T-cell lymphoma

A clinical suspicion of EATL should be raised in two different scenarios. First, in case of a patient with a known history of CD, especially RCD type II and/or UJI, who has further worsening of malabsorption symptoms (e.g., diarrhea, weight loss, severe malnutrition, subocclusive abdominal pain, hypoalbuminemia, anemia) or whose symptoms do not improve after therapy. The second scenario is that of a patient presenting with either an abdominal mass, gastrointestinal bleeding, or acute small-bowel perforation, regardless of a previous history of CD or RCD. Other rarer presentations have been described, including dermatological manifestations and distant metastases (e.g., peripheral lymphnodes, liver, lung) (Malamut et al., 2013). Lactate dehydrogenase, an unspecific marker of cytolysis, and beta2-microglobulin can be increased (Di Sabatino et al., 2012); other laboratory test would typically show anemia, hypoalbuminemia, and electrolyte imbalance. There are no specific onco-markers of EATL. Apart from the classical malabsorption syndrome, "B symptoms" (i.e., fever, weight loss, night sweats) may also occur, as for other lymphomas.

When EATL is suspected, the diagnostic work-up should always include an upper gastrointestinal endoscopy with duodenal biopsies (also including any visible small-bowel lesion), abdomen computed tomography (CT) or magnetic resonance, and 18 FDG-PET (Di Sabatino et al., 2012; Hadithi et al., 2006; van de Water et al., 2010; Van Weyenberg et al., 2011). A videocapsule endoscopy and a double-balloon enteroscopy can detect the multifocal involvement of the small-bowel, distal to the Treitz ligament. Notably, the diagnostic yield of double balloon enteroscopy for malignancy, anterograde in most cases, is greater (roughly 80%) compared to videocapsule only (50%) (Ferretti et al., 2022). At enteroscopy, EATL may manifest as ulcerating masses that could penetrate the mucosa, and this is why perforation at presentation is common (van de Water et al., 2010). Less common presentations are instead as a bulky mass or as a nonexophytic mucosal ulceration. The jejunum is the most common affected site, whereas the stomach and the colon involvement has only been anecdotally reported (Delabie et al., 2011). Abdominal CT scan or magnetic resonance are useful in detecting solid lesions along the small-bowel, as well as metastases in other abdominal sites, such as the mesenteric, paraaortic, and iliac lymphnodes, and the liver (Delabie et al., 2011; Malamut et al., 2013). Finally, fluorodeoxyglucose F18 DG-PET scan is useful for excluding a newly onset EATL, as it is significantly more accurate (100% sensitivity, 90% specificity) than abdominal CT (87% sensitivity, 53% specificity) in detecting EATL (Hadithi et al., 2006); indeed, the definitive diagnosis needs to be confirmed by histopathology. Mesenteric lymph nodes cavitation may also be noticed.

Both endoscopy and imaging techniques are needed for staging EATL. However, there are no specific tools for this purpose, and staging systems validated for other types of lymphoma may be used, such as the Ann Arbor or the Lugano system (Di Sabatino et al., 2012; Malamut et al., 2013).

14.4.2 Pathological features of enteropathy-associated T-cell lymphoma

Histologic examination is the gold-standard for making a definitive diagnosis of EATL. EATL is characterized by a pleomorphic population of lymphocytes (i.e., from small to large immunoblastic/anaplastic cells) with a high mitotic index, angulated or round vesicular nuclei, prominent nucleoli, and abundant, pale cytoplasm (De Leval & Gaulard, 2011). Additionally, these neoplastic lymphocytes are surrounded by a moderate to abundant eosinophilic infiltrate, histiocytes, and other small lymphocytes; mitotic figures and necrosis are common. Indeed, neoplastic lymphocytes, which are $\alpha\beta$ T cells having a cytotoxic profile, display aberrant features (CD3 + , CD4-, CD8-, CD5-, CD7 +) as those noticed in RCD type II, with monoclonal rearrangement of the gamma/beta TCR (De Leval & Gaulard, 2011). The neoplastic cells also express the mucosal homing receptor CD103, which is a feature of the IEL population. EATL are in most cases CD56-, while roughly 50%–60% of them express CD30 (Delabie et al., 2011). CD56 is an NK marker which indicates a non-B-cell lineage, while CD30 is a neoplastic marker of some forms of lymphoma. Additionally, the NK receptor NKp46 is almost invariably expressed in EATL, as well as in RCD type II, and can be considered as a marker of malignancy, allowing discrimination with other indolent T-cell lymphomas in which this marker is negative (Cheminant et al., 2019). The ki67 is highly expressed in neoplastic lymphocytes, allowing discrimination with RCD type II in which IELs are ki67 negative (Mention et al., 2003). The mucosa neighboring EATL typically shows villous atrophy, crypt hyperplasia, and increased IELs, as in untreated or RCD. Fig. 14.4 shows the histopathological features of EATL.

14.4.3 Treatment of enteropathy-associated T-cell lymphoma

To date, there is no validated treatment for EATL. Standard induction chemotherapy, along with the surgical mass debulking, with or without consolidating stem cell transplantation, has been the most commonly reported treatment for EATL (Gale et al., 2000; Jantunen et al., 2013; Malamut et al., 2013; Nijeboer, de Baaij, et al., 2015; Sieniawski et al., 2010; Wöhrer et al., 2004). The overall response rate is 40%–60%, while complete remission appears to be less than

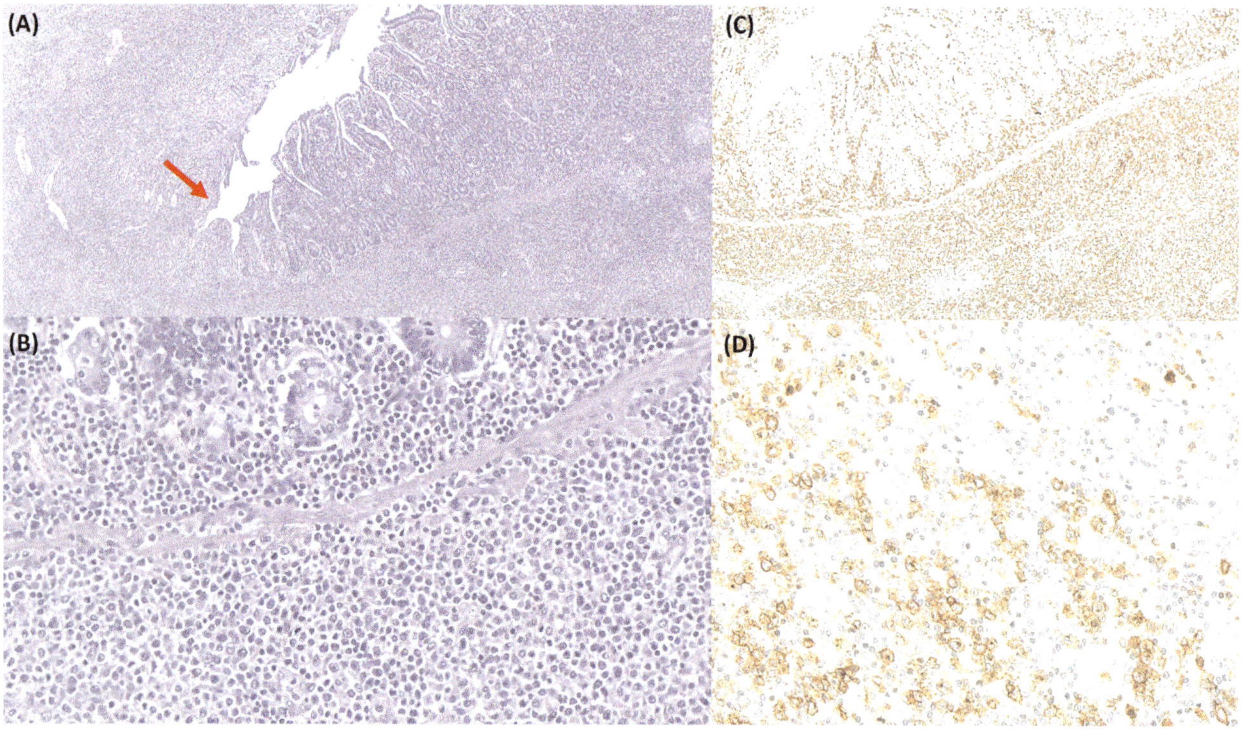

FIGURE 14.4 Histopathological aspects of EATL. (A) Lymphoma infiltrate involving the whole intestinal wall, with mucosal ulceration (red arrow) (hematoxylin/eosin staining; 2.5 ×). (B) Lymphoma infiltrate consists of pleomorphic medium to large sized cells, with vesicular nuclei and prominent nucleoli; some reactive small lymphocytes and eosinophils may be found; the infiltrate invades and crosses the *muscolaris mucosae* (hematoxylin/eosin staining; 20 ×). (C) CD3 immunostaining shows intense positivity on lymphoma cells (5 ×). (D) CD30 expression may be documented in most cases (40 ×).

20% and not strictly related to the EATL stage(Nijeboer, Malamut, et al., 2015). Intensified regimens have shown to lead to a better response rate; however, they are burdened by severe adverse events, including neutropenic sepsis and death. Part of the disappointing results noticed with the available treatment options can be attributed to the poor performance status of the patient, who is often malnourished, anemic, and displays a high rate of perforation and major bleeding (Wierdsma et al., 2016). For these reasons, less than 50% of patients with EATL are eligible to high-dose chemotherapy, while the others have a rapid clinical deterioration until death. In more fit patients, intensified regimens followed by autologous stem cell transplantation were able to prolong survival compared to standard regimens, although the evidence is confined to few cases. The CHOP regimen (cyclophosphamide, doxorubicin, vincristine, prednisolone) has been the most used, sometimes with the addition of etoposide, methotrexate or alemtuzumab (Nijeboer, Malamut, et al., 2015). Other regimens include BACOP (bleomycin, doxorubicin, cyclophosphamide, vincristine, and prednisone), ProMACEMOPP (prednisone, doxorubicin, cyclophosphamide, etoposide, mechlorethamine, vincristine, and procarbazine), VAMP (vincristine, doxorubicin, high-dose methotrexate, and prednisolone), PEACE-BOM (prednisolone, etoposide, doxorubicin, cyclophosphamide-bleomycin, vincristine, and methotrexate) (Di Sabatino et al., 2012). Radiotherapy has been seldomly reported and has little role in the treatment of EATL.

As for other types of neoplasia, monoclonal antibodies could be an ideal treatment option for EATL. In this regard, CD30 is an ideal target, as it is expressed in a substantial proportion of EATL and has a very limited expression in non-neoplastic tissues (Nijeboer, Malamut, et al., 2015). Brentuximab vedotin is a monoclonal antibody which targets CD30 and is conjugated with monomethyl auristatin E, an antimicrotubule agent (van de Donk & E, 2012). There have been only few cases of CD30 + EATL treated with this drug, after chemotherapy, with overall good results (complete remission at 9-month follow-up) (Khalaf et al., 2013). However, larger, prospective studies are needed. The importance of CD30 as a potential target has also been exploited through chimeric antigen receptor (CAR) modified T cells, which are becoming a novel therapeutic option in several neoplastic disorders. There is one case report of a patient with EATL treated with CD30 CAR-T cells, after allogenic stem-cell transplantation, reaching long-term remission (Voorhees et al., 2020).

Other treatments that could ideally be applied to EATL, but have not yet been studied in this setting, are bortezomib and anti-Nkp46 agents. Bortezomib is a proteasome inhibitor which is approved for the treatment of multiple myeloma. Its mechanism of action is able to interfere with the STAT3 signaling, thus enhancing the apoptosis of neoplastic cells (Bao et al., 2017). In an in vitro study, bortezomib was able to inhibit STAT3 phosphorylation in four RCD type II lines, also triggering a pro-apoptotic effect on aberrant cells with no effect on normal lymphocytes (Cording et al., 2022). There are no available data regarding its use in EATL patients. Anti-NK receptor agents, including NKp46, could be ideal agents for stimulating antitumor immunity. However, their clinical application has been hampered by major safety concerns and has not been further studied (Gauthier et al., 2019). Regardless of the treatment, EATL is highly lethal, with an estimated 5-year survival of 8%–20% (Di Sabatino et al., 2012). A clinical prognostic model for EATL has been proposed, able to stratify the risk of death into three categories. The high-risk group is characterized by the presence of B symptoms (overall survival 2 months); the intermediate-risk group encompasses patients without B symptoms and with an International Prognostic Index (IPI) score ≥ 2 (overall survival 7 months); the low-risk group encompasses patients without B symptoms and an IPI score of 0–1 (overall survival 34 months) (De Baaij et al., 2015).

14.4.4 Enteropathy-associated T-cell lymphoma in conditions other than celiac disease

When mentioning EATL, it is implied that CD is the actual underlying enteropathy. However, EATL has also been noticed in other enteropathies which have similar lesions, such as the nongluten sensitive autoimmune enteropathy (Ciccocioppo et al., 2018; Malamut et al., 2012) and the gluten-sensitive dermatitis herpetiformis (Hervonen et al., 2005).

The only two cases of EATL complicating autoimmune enteropathy were characterized by impressive clinical similarities. In both cases, they were 53-year-old women, both with a long history of refractory villous atrophy initially responding to steroids, both with TCR monoclonality, present years before the diagnosis of EATL, and both dying from central nervous system metastases (Ciccocioppo et al., 2018; Malamut et al., 2012).

EATL may also occur in patients with dermatitis herpetiformis, which is a skin manifestation of CD (Chapter 16) (Hervonen et al., 2005). The clinical course and outcome did not differ from that of EATL arising in CD patients, although the most reported affected site was the ileum; a poor adherence to a GFD was the only risk factor found (Hervonen et al., 2005).

14.5 Monomorphic epitheliotropic intestinal T-cell lymphoma and other small bowel lymphomas in celiac disease

According to the latest classification, *MEITL*, initially known as EATL type II (Swerdlow et al., 2016), is an aggressive form of extranodal, small bowel, lymphoma arising from intraepithelial T cells (Alaggio et al., 2022). MEITL is considered as a sporadic and rare form of lymphoma which has not been associated to CD and mostly described in Asian series (Alaggio et al., 2022; Swerdlow et al., 2016). Contrary to EATL, it is characterized by a small-sized, markedly monomorphic, neoplastic lymphocyte infiltrate with marked epitheliotropic features; neoplastic cells are typically CD8 + /CD56 + and CD30-. We have described two cases of MEITL with a confirmed diagnosis of CD (which was certainly refractory in one case) and both characterized by undoubtful histopathological features of a MEITL (Lenti et al., 2019). Notably, MEITL was multifocal, and in both cases the ileum was involved. The prognosis was poor, and chemotherapy was not feasible due to the poor performance status. In one case we were able to study genetic aberrations, and we found 9q gains and 9p losses, which have also been reported in EATL (Roberti et al., 2016). Finally, it should be noted that NKp46 positivity has also been described in MEITL, thus suggesting a shared cell lineage derivation (Cheminant et al., 2019). Fig. 14.5 shows the histopathological features of a MEITL arising in a CD patient.

Several studies have found that people with CD have a higher risk of developing non-EATL, T- or B-cell lymphomas compared to the general population (Halfdanarson et al., 2010; Leslie et al., 2012; Malamut et al., 2014; Smedby et al., 2005), and the risk remains elevated even after long-term adherence to a GFD (Pelizzaro et al., 2021). *Indolent CD4 + T-cell lymphoma* has been the most commonly non-EATL T-cell lymphoma associated with RCD. The largest case series of this rare malignant CD4 + , CD103- condition included 10 patients (five females, mean age at diagnosis 54 years), all having a history of chronic diarrhea and malnutrition, and only one having a known, previous diagnosis of CD (Malamut et al., 2014). In all cases, villous atrophy and TCR gamma monoclonal rearrangement were present; CD serology was positive in 4/10 cases. Proximal small intestinal ulcers were the most frequent endoscopic finding. Histologically, a dense infiltration of the epithelium and lamina propria by a polymorphous population dominated by monomorphic small CD3 + , CD4 + , βF1 + and Vβ lymphoid cells was noticed. A GFD had no impact on symptoms or lesions, but steroid therapy was given in half of the cohort, with a full response in 80% of the cases; the anti-CD52 monoclonal antibody alemtuzumab was given in

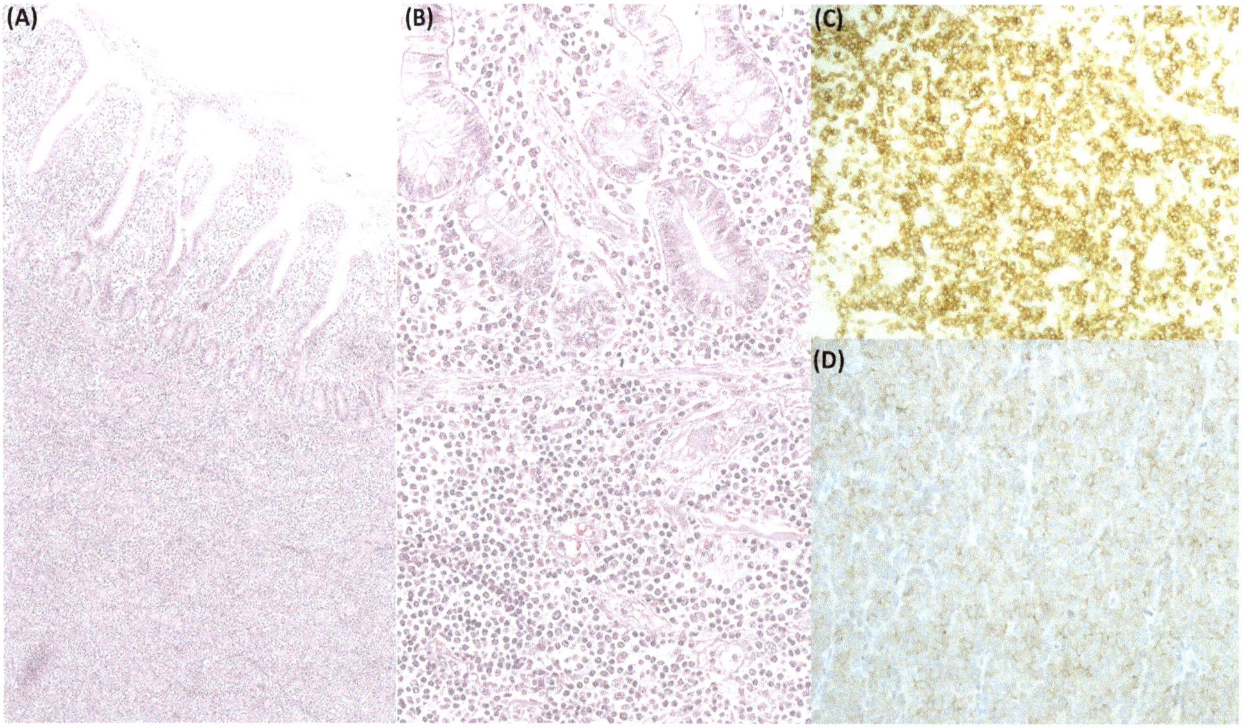

FIGURE 14.5 Histopathological aspects of MEITL. (A) Lymphoma infiltrate involving the whole intestinal wall; villi are shortened and broadly expanded (hematoxylin/eosin staining; 2,5 ×). (B) Lymphoma cells are medium sized and show round nuclei with finely dispersed chromatin; nucleoli are nonprominent. A marked epitheliotropism is evident (hematoxylin/eosin staining; 20 ×). (C) and (D). Tumor cells are usually positive for both CD8 (C, 40 ×) and CD56 (D, 40 ×).

two cases, with a complete resolution of villous atrophy and a normalization of the abnormal infiltration. Globally, the prognosis was good, with no deaths at long-term follow-up. Other less common T-cell lymphomas have been described in CD -not refractory in most cases-, including anaplastic large cell lymphoma and unspecified peripheral T-cell lymphoma (Halfdanarson et al., 2010; Leslie et al., 2012; Smedby et al., 2005).

Patients with CD also carry an increased risk of developing B-cell lymphomas, particularly *diffuse large B-cell lymphoma* (DLBCL) (Halfdanarson et al., 2010; Leslie et al., 2012; Smedby et al., 2005). This condition, having an overall good prognosis, may present in CD with lymphadenopathy, fever, weight loss, and night sweats, as in DLBCL arising in the general population. However, patients with CD-associated DLBCL may also experience abdominal pain, diarrhea, and malabsorption. It is important for clinicians to consider the possibility of DLBCL in CD patients presenting with these symptoms, especially if they are refractory to treatment. A notable aggregation of autoimmune/inflammatory disorders, female sex, and DLBCL in CD has been noticed, pointing at shared pathogenic pathways (Smedby et al., 2005).

References

Abadie, V., Kim, S. M., Lejeune, T., Palanski, B. A., Ernest, J. D., Tastet, O., Voisine, J., Discepolo, V., Marietta, E. V., Hawash, M. B. F., Ciszewski, C., Bouziat, R., Panigrahi, K., Horwath, I., Zurenski, M. A., Lawrence, I., Dumaine, A., Yotova, V., Grenier, J. C., ... Jabri, B. (2020). IL-15, gluten and HLA-DQ8 drive tissue destruction in coeliac disease. *Nature*, *578*(7796), 600–604. Available from https://doi.org/10.1038/s41586-020-2003-8.

Abdulkarim, A. S., Burgart, L. J., See, J., & Murray, J. A. (2002). Etiology of nonresponsive celiac disease: Results of a systematic approach. *American Journal of Gastroenterology*, *97*(8), 2016–2021. Available from https://doi.org/10.1016/S0002-9270(02)04274-0.

Abdul Sultan, A., Crooks, C. J., Card, T., Tata, L. J., Fleming, K. M., & West, J. (2015). Causes of death in people with coeliac disease in England compared with the general population: a competing risk analysis. *Gut*, *64*(8), 1220–1226. Available from https://doi.org/10.1136/gutjnl-2014-308285.

Alaggio, R., Amador, C., Anagnostopoulos, I., Attygalle, A. D., Araujo, I. B. d O., Berti, E., Bhagat, G., Borges, A. M., Boyer, D., Calaminici, M., Chadburn, A., Chan, J. K. C., Cheuk, W., Chng, W. J., Choi, J. K., Chuang, S. S., Coupland, S. E., Czader, M., Dave., ... Xiao, W. (2022). The 5th edition of the World Health Organization Classification of Haematolymphoid Tumours: Lymphoid Neoplasms. *Leukemia*, *36*(7), 1720–1748. Available from https://doi.org/10.1038/s41375-022-01620-2.

Al-Toma, A., Goerres, M. S., Meijer, J. W. R., Peña, A. S., Crusius, J. B. A., & Mulder, C. J. J. (2006). Human leukocyte antigen-DQ2 homozygosity and the development of refractory celiac disease and enteropathy-associated T-cell lymphoma. *Clinical Gastroenterology and Hepatology*, *4*(3), 315–319. Available from https://doi.org/10.1016/j.cgh.2005.12.011.

Al-Toma, A., Goerres, M. S., Meijer, J. W. R., von Blomberg, B. M. E., Wahab, P. J., Kerckhaert, J. A. M., & Mulder, C. J. J. (2006). Cladribine therapy in refractory celiac disease with aberrant T cells. *Clinical Gastroenterology and Hepatology*, *4*(11), 1322–1327. Available from https://doi.org/10.1016/j.cgh.2006.07.007.

Al-Toma, A., Verbeek, W. H. M., Hadithi, M., Von Blomberg, B. M. E., & Mulder, C. J. J. (2007). Survival in refractory celiac disease and enteropathy-associated T-cell lymphoma: Retrospective evaluation of single-centre experience. *Gut*, *56*(10), 1373–1378. Available from https://doi.org/10.1136/gut.2006.114512.

Al-Toma, A., Visser, O. J., Van Roessel, H. M., Von Blomberg, B. M. E., Verbeek, W. H. M., Scholten, P. E. T., Ossenkoppele, G. J., Huijgens, P. C., & Mulder, C. J. J. (2007). Autologous hematopoietic stem cell transplantation in refractory celiac disease with aberrant T cells. *Blood*, *109*(5), 2243–2249. Available from https://doi.org/10.1182/blood-2006-08-042820.

Al-Toma, A., Volta, U., Auricchio, R., Castillejo, G., Sanders, D. S., Cellier, C., Mulder, C. J., & Lundin, K. E. A. (2019). European Society for the Study of Coeliac Disease (ESsCD) guideline for coeliac disease and other gluten-related disorders. *United European Gastroenterology Journal*, *7*(5), 583–613. Available from https://doi.org/10.1177/2050640619844125.

Bao, X., Ren, T., Huang, Y., Ren, C., Yang, K., Zhang, H., & Guo, W. (2017). Bortezomib induces apoptosis and suppresses cell growth and metastasis by inactivation of Stat3 signaling in chondrosarcoma. *International Journal of Oncology*, *50*(2), 477–486. Available from https://doi.org/10.3892/ijo.2016.3806.

Biagi, F., Bianchi, P. I., Vattiato, C., Marchese, A., Trotta, L., Badulli, C., De Silvestri, A., Martinetti, M., & Corazza, G. R. (2012). Influence of HLA-DQ2 and DQ8 on severity in celiac disease. *Journal of Clinical Gastroenterology*, *46*(1), 46–50. Available from https://doi.org/10.1097/MCG.0b013e318221077e.

Biagi, F., Campanella, J., Martucci, S., Pezzimenti, D., Ciclitira, P. J., Ellis, H. J., & Corazza, G. R. (2004). A milligram of gluten a day keeps the mucosal recovery away: A case report. *Nutrition Reviews*, *62*(9), 360–363. Available from https://doi.org/10.1301/nr.2004.sept.360-363.

Biagi, F., & Corazza, G. R. (2001). Defining gluten refractory enteropathy. *European Journal of Gastroenterology and Hepatology*, *13*(5), 561–565. Available from https://doi.org/10.1097/00042737-200105000-00016.

Biagi, F., Gobbi, P., Marchese, A., Borsotti, E., Zingone, F., Ciacci, C., Volta, U., Caio, G., Carroccio, A., Ambrosiano, G., Mansueto, P., & Corazza, G. R. (2014). Low incidence but poor prognosis of complicated coeliac disease: A retrospective multicentre study. *Digestive and Liver Disease*, *46*(3), 227–230. Available from https://doi.org/10.1016/j.dld.2013.10.010.

Biagi, F., Lorenzini, P., & Corazza, G. R. (2000). Literature review on the clinical relationship between ulcerative jejunoileitis, coeliac disease, and enteropathy-associated T-cell lymphoma. *Scandinavian Journal of Gastroenterology*, *35*(8), 785–790. Available from https://doi.org/10.1080/003655200750023129.

Biagi, F., Poggioli, G., Mazzoni, G., & Corazza, G. R. (1998). Intestinal strictures. *The Lancet*, *352*(9131), 876. Available from https://doi.org/10.1016/S0140-6736(98)06170-4.

Biagi, F., Schiepatti, A., Maiorano, G., Fraternale, G., Agazzi, S., Zingone, F., Ciacci, C., Volta, U., Caio, G., Tortora, R., Klersy, C., & Corazza, G. R. (2018). Risk of complications in coeliac patients depends on age at diagnosis and type of clinical presentation. *Digestive and Liver Disease*, *50*(6), 549–552. Available from https://doi.org/10.1016/j.dld.2017.12.001.

Branchi, F., Wiese, J. J., Heldt, C., Manna, S., Dony, V., Loddenkemper, C., Bojarski, C., Siegmund, B., Schneider, T., Daum, S., Hummel, M., Moos, V., & Schumann, M. (2022). The combination of clinical parameters and immunophenotyping of intraepithelial lymphocytes allows to assess disease severity in refractory celiac disease. *Digestive and Liver Disease*, *54*(12), 1649–1656. Available from https://doi.org/10.1016/j.dld.2022.06.024.

Brar, P., Lee, S., Lewis, S., Egbuna, I., Bhagat, G., & Green, P. H. R. (2007). Budesonide in the treatment of refractory celiac disease. *American Journal of Gastroenterology*, *102*(10), 2265–2269. Available from https://doi.org/10.1111/j.1572-0241.2007.01380.x.

Catassi, C., Fabiani, E., Iacono, G., D'Agate, C., Francavilla, R., Biagi, F., Volta, U., Accomando, S., Picarelli, A., De Vitis, I., Pianelli, G., Gesuita, R., Carle, F., Mandolesi, A., Bearzi, I., & Fasano, A. (2007). A prospective, double-blind, placebo-controlled trial to establish a safe gluten threshold for patients with celiac disease. *American Journal of Clinical Nutrition*, *85*(1), 160–166. Available from https://doi.org/10.1093/ajcn/85.1.160.

Cellier, C., Bouma, G., van Gils, T., Khater, S., Malamut, G., Crespo, L., Collin, P., Green, P. H. R., Crowe, S. E., Tsuji, W., Butz, E., Cerf-Bensussan, N., Macintyre, E., Parnes, J. R., Leon, F., Hermine, O., Mulder, C. J., Jabri, B., Murray, J., ... Raymond, R. (2019). Safety and efficacy of AMG 714 in patients with type 2 refractory coeliac disease: A phase 2a, randomised, double-blind, placebo-controlled, parallel-group study. *The Lancet Gastroenterology and Hepatology*, *4*(12), 960–970. Available from https://doi.org/10.1016/S2468-1253(19)30265-1.

Cellier, C., Delabesse, E., Helmer, C., Patey, N., Matuchansky, C., Jabri, B., MacIntyre, E., Cerf-Bensussan, N., & Brousse, N. (2000). Refractory sprue, coeliac disease, and enteropathy-associated T-cell lymphoma. *The Lancet*, *356*(9225), 203–208. Available from https://doi.org/10.1016/S0140-6736(00)02481-8.

Cellier, C., Patey, N., Mauvieux, L., Jabri, B., Delabesse, E., Cervoni, J. P., Burtin, M. L., Guy-Grand, D., Bouhnik, Y., Modigliani, R., Barbier, J. P., Macintyre, E., Brousse, N., & Cerf-Bensussan, N. (1998). Abnormal intestinal intraepithelial lymphocytes in refractory sprue. *Gastroenterology*, *114*(3), 471–481. Available from https://doi.org/10.1016/s0016-5085(98)70530-x.

Chander, U., Leeman-Neill, R. J., & Bhagat, G. (2018). Pathogenesis of enteropathy-associated T cell lymphoma. *Current Hematologic Malignancy Reports*, *13*(4), 308–317. Available from https://doi.org/10.1007/s11899-018-0459-5.

Cheminant, M., Bruneau, J., Malamut, G., Sibon, D., Guegan, N., Van Gils, T., Cording, S., Trinquand, A., Verkarre, V., Lhermitte, L., Brousse, N., Jannot, A. S., Khater, S., Frenzel, L., Delarue, R., Suarez, F., Marçais, A., Mulder, C. J. J., Macintyre, E., ... Hermine, O. (2019). NKp46 is a diagnostic biomarker and may be a therapeutic target in gastrointestinal T-cell lymphoproliferative diseases: A CELAC study. *Gut*, *68*(8), 1396–1405. Available from https://doi.org/10.1136/gutjnl-2018-317371.

Ciccocioppo, R., Croci, G. A., Biagi, F., Vanoli, A., Alvisi, C., Cavenaghi, G., Riboni, R., Arra, M., Gobbi, P. G., Paulli, M., & Corazza, G. R. (2018). Intestinal T-cell lymphoma with enteropathy-associated T-cell lymphoma-like features arising in the setting of adult autoimmune enteropathy. *Hematological Oncology*, *36*(2), 481–488. Available from https://doi.org/10.1002/hon.2494.

Ciccocioppo, R., Gallia, A., Avanzini, M. A., Betti, E., Picone, C., Vanoli, A., Paganini, C., Biagi, F., Maccario, R., & Corazza, G. R. (2016). A refractory celiac patient successfully treated with mesenchymal stem cell infusions. *Mayo Clinic Proceedings*, *91*(6), 812–819. Available from https://doi.org/10.1016/j.mayocp.2016.03.001.

Corazza, G. R., Strocchi, A., & Gasbarrini, G. (1987). Fasting breath hydrogen in celiac disease. *Gastroenterology*, *93*(1), 53–58. Available from https://doi.org/10.1016/0016-5085(87)90313-1.

Cording, S., Lhermitte, L., Malamut, G., Berrabah, S., Trinquand, A., Guegan, N., Villarese, P., Kaltenbach, S., Meresse, B., Khater, S., Dussiot, M., Bras, M., Cheminant, M., Tesson, B., Bole-Feysot, C., Bruneau, J., Molina, T. J., Sibon, D., MacIntyre, E., ... Cerf-Bensussan, N. (2022). Oncogenetic landscape of lymphomagenesis in coeliac disease. *Gut*, *71*(3), 497–508. Available from https://doi.org/10.1136/gutjnl-2020-322935.

Corrao, G., Corazza, G. R., Bagnardi, V., Brusco, G., Ciacci, C., Cottone, M., Sategna Guidetti, C., Usai, P., Cesari, P., Pelli, M. A., Loperfido, S., Volta, U., Calabró, A., & Certo, M. (2001). Mortality in patients with coeliac disease and their relatives: A cohort study. *The Lancet*, *358*(9279), 356–361.

Creamer, B., & Pink, I. J. (1967). Paneth-cell deficiency. *The Lancet*, *289*(7485), 304–306. Available from https://doi.org/10.1016/s0140-6736(67)91239-1.

De Baaij, L. R., Berkhof, J., Van De Water, J. M. W., Sieniawski, M. K., Radersma, M., Verbeek, W. H. M., Visser, O. J., Oudejans, J. J., Meijer, C. J. L. M., Mulder, C. J. J., Lennard, A. L., & Cillessen, S. A. G. M. (2015). A new and validated clinical prognostic model (EPI) for enteropathy-associated T-cell lymphoma. *Clinical Cancer Research*, *21*(13), 3013–3019. Available from https://doi.org/10.1158/1078-0432.CCR-14-2195.

Delabie, J., Holte, H., Vose, J. M., Ullrich, F., Jaffe, E. S., Savage, K. J., Connors, J. M., Rimsza, L., Harris, N. L., Müller-Hermelink, K., Rüdiger, T., Coiffier, B., Gascoyne, R. D., Berger, F., Tobinai, K., Au, W. Y., Liang, R., Montserrat, E., Hochberg, E. P., ... Weisenburger, D. D. (2011). Enteropathy-associated T-cell lymphoma: Clinical and histological findings from the international peripheral T-Cell lymphoma project. *Blood*, *118*(1), 148–155. Available from https://doi.org/10.1182/blood-2011-02-335216.

De Leval, L., & Gaulard, P. (2011). Pathology and biology of peripheral T-cell lymphomas. *Histopathology*, *58*(1), 49–68. Available from https://doi.org/10.1111/j.1365-2559.2010.03704.x.

Dewar, D. H., Donnelly, S. C., McLaughlin, S. D., Johnson, M. W., Ellis, H. J., & Ciclitira, P. J. (2012). Celiac disease: Management of persistent symptoms in patients on a gluten-free diet. *World Journal of Gastroenterology*, *18*(12), 1348–1356. Available from https://doi.org/10.3748/wjg.v18.i12.1348.

Dieckman, Beaumont, H., Bontkes, H., Hayat, Byrnes, V., Lundin, K., Koning, F., Schumann, M., & Bouma. (2022). Tofacitinib: a new treatment option for refractory celiac disease type II. Abstract Book 19th International Celiac Disease Symposium, Sorrento, 175.

Di Sabatino, A., Biagi, F., Gobbi, P. G., & Corazza, G. R. (2012). How I treat enteropathy-associated T-cell lymphoma. *Blood*, *119*(11), 2458–2468. Available from https://doi.org/10.1182/blood-2011-10-385559.

Di Sabatino, A., Calarota, S. A., Vidali, F., MacDonald, T. T., & Corazza, G. R. (2011). Role of IL-15 in immune-mediated and infectious diseases. *Cytokine and Growth Factor Reviews*, *22*(1), 19–33. Available from https://doi.org/10.1016/j.cytogfr.2010.09.003.

Di Sabatino, A., Carsetti, R., & Corazza, G. R. (2011). Post-splenectomy and hyposplenic states. *The Lancet*, *378*(9785), 86–97. Available from https://doi.org/10.1016/S0140-6736(10)61493-6.

Di Sabatino, A., Ciccocioppo, R., Cupelli, F., Cinque, B., Millimaggi, D., Clarkson, M. M., Paulli, M., Cifone, M. G., & Corazza, G. R. (2006). Epithelium derived interleukin 15 regulates intraepithelial lymphocyte Th1 cytokine production, cytotoxicity, and survival in coeliac disease. *Gut*, *55*(4), 469–477. Available from https://doi.org/10.1136/gut.2005.068684.

Di Sabatino, A., Ciccocioppo, R., D'Alò, S., Parroni, R., Millimaggi, D., Cifone, M. G., & Corazza, G. R. (2001). Intraepithelial and lamina propria lymphocytes show distinct patterns of apoptosis whereas both populations are active in Fas based cytotoxicity in coeliac disease. *Gut*, *49*(3), 380–386. Available from https://doi.org/10.1136/gut.49.3.380.

Di Sabatino, A., & Corazza, G. R. (2009). Coeliac disease. *The Lancet*, *373*(9673), 1480–1493. Available from https://doi.org/10.1016/S0140-6736(09)60254-3.

Di Sabatino, A., Giuffrida, P., Vanoli, A., Luinetti, O., Manca, R., Biancheri, P., Bergamaschi, G., Alvisi, C., Pasini, A., Salvatore, C., Biagi, F., Solcia, E., & Corazza, G. R. (2014). Increase in neuroendocrine cells in the duodenal mucosa of patients with refractory celiac disease. *American Journal of Gastroenterology*, *109*(2), 258–269. Available from https://doi.org/10.1038/ajg.2013.426.

Di Sabatino, A., Miceli, E., Dhaliwal, W., Biancheri, P., Salerno, R., Cantoro, L., Vanoli, A., De Vincenzi, M., Blanco, C. D. V., MacDonald, T. T., & Corazza, G. R. (2008). Distribution, proliferation, and function of paneth cells in uncomplicated and complicated adult celiac disease. *American Journal of Clinical Pathology*, *130*(1), 34–42. Available from https://doi.org/10.1309/5ADNAR4VN11TTKQ6.

Di Sabatino, A., Rosado, M. M., Cazzola, P., Riboni, R., Biagi, F., Carsetti, R., & Corazza, G. R. (2006). Splenic hypofunction and the spectrum of autoimmune and malignant complications in celiac disease. *Clinical Gastroenterology and Hepatology*, *4*(2), 179–186. Available from https://doi.org/10.1016/S1542-3565(05)00982-1.

Di Sabatino, A., Rovedatti, L., Vetrano, S., Vidali, F., Biancheri, P., Rescigno, M., Danese, S., MacDonald, T. T., & Corazza, G. R. (2011). Involvement of CD40-CD40 Ligand in uncomplicated and refractory celiac disease. *American Journal of Gastroenterology*, *106*(3), 519–527. Available from https://doi.org/10.1038/ajg.2010.450.

Ettersperger, J., Montcuquet, N., Malamut, G., Guegan, N., Lopez-Lastra, S., Gayraud, S., Reimann, C., Vidal, E., Cagnard, N., Villarese, P., Andre-Schmutz, I., Gomes Domingues, R., Godinho-Silva, C., Veiga-Fernandes, H., Lhermitte, L., Asnafi, V., Macintyre, E., Cellier, C., Beldjord, K., ... Meresse, B. (2016). Interleukin-15-dependent T-cell-like Innate intraepithelial lymphocytes develop in the intestine and transform into lymphomas in celiac disease. *Immunity*, *45*(3), 610–625. Available from https://doi.org/10.1016/j.immuni.2016.07.018.

Ferretti, F., Branchi, F., Orlando, S., Roncoroni, L., Barigelletti, G., Fabiano, S., Vecchi, M., Penagini, R., Doneda, L., & Elli, L. (2022). Effectiveness of capsule endoscopy and double-balloon enteroscopy in suspected complicated celiac disease. *Clinical Gastroenterology and Hepatology*, *20*(4), 941–949. Available from https://doi.org/10.1016/j.cgh.2020.11.010, e3.

Gale, J., Simmonds, P. D., Mead, G. M., Sweetenham, J. W., & Wright, D. H. (2000). Enteropathy-type intestinal T-cell lymphoma: Clinical features and treatment of 31 patients in a single center. *Journal of Clinical Oncology*, *18*(4), 795–803. Available from https://doi.org/10.1200/jco.2000.18.4.795.

Gauthier, L., Morel, A., Anceriz, N., Rossi, B., Blanchard-Alvarez, A., Grondin, G., Trichard, S., Cesari, C., Sapet, M., Bosco, F., Rispaud-Blanc, H., Guillot, F., Cornen, S., Roussel, A., Amigues, B., Habif, G., Caraguel, F., Arrufat, S., Remark, R., ... Vivier, E. (2019). Multifunctional natural killer cell engagers targeting NKp46 trigger protective tumor immunity. *Cell*, *177*(7), 1701–1713. Available from https://doi.org/10.1016/j.cell.2019.04.041, e16.

Gillett, H. R., Arnott, I. D. R., McIntyre, M., Campbell, S., Dahele, A., Priest, M., Jackson, R., & Ghosh, S. (2002). Successful infliximab treatment for steroid-refractory celiac disease: A case report. *Gastroenterology*, *122*(3), 800–805. Available from https://doi.org/10.1053/gast.2002.31874.

Green, P. H. R., & Cellier, C. (2007). Celiac disease. *The New England Journal of Medicine*, *357*(17), 1731–1743.

Green, P. H. R., Paski, S., Ko, C. W., & Rubio-Tapia, A. (2022). AGA clinical practice update on management of refractory celiac disease: Expert review. *Gastroenterology*, *163*(5), 1461–1469. Available from https://doi.org/10.1053/j.gastro.2022.07.086.

Green, P. H. R., Yang, J., Cheng, J., Lee, A. R., Harper, J. W., & Bhagat, G. (2009). An association between microscopic colitis and celiac disease. *Clinical Gastroenterology and Hepatology*, *7*(11), 1210–1216. Available from https://doi.org/10.1016/j.cgh.2009.07.011.

Grewal, J. K., Kassardjian, A., & Weiss, G. A. (2022). Successful novel use of tofacitinib for type II refractory coeliac disease. *BMJ Case Reports*, *15*(4), e244692. Available from https://doi.org/10.1136/bcr-2021-244692.

Hadithi, M., Mallant, M., Oudejans, J., Van Waesberghe, J. H. T. M., Mulder, C. J., & Comans, E. F. I. (2006). 18F-FDG PET versus CT for the detection of enteropathy- associated T-cell lymphoma in refractory celiac disease. *Journal of Nuclear Medicine*, *47*(10), 1622–1627. Available from http://jnm.snmjournals.org/cgi/reprint/47/10/1622.

Halfdanarson, T. R., Rubio-Tapia, A., Ristow, K. M., Habermann, T. M., Murray, J. A., & Inwards, D. J. (2010). Patients with celiac disease and B-Cell lymphoma have a better prognosis than those with T-cell lymphoma. *Clinical Gastroenterology and Hepatology*, *8*(12), 1042–1047. Available from https://doi.org/10.1016/j.cgh.2010.09.007.

Hall, N. J., Rubin, G. P., & Charnock, A. (2013). Intentional and inadvertent non-adherence in adult coeliac disease. A cross-sectional survey. *Appetite*, *68*, 56–62. Available from https://doi.org/10.1016/j.appet.2013.04.016.

Hervonen, K., Vornanen, M., Kautiainen, H., Collin, P., & Reunala, T. (2005). Lymphoma in patients with dermatitis herpetiformis and their first-degree relatives. *British Journal of Dermatology*, *152*(1), 82–86. Available from https://doi.org/10.1111/j.1365-2133.2005.06345.x.

Holmes, G., & Ciacci, C. (2018). The serological diagnosis of coeliac disease - A step forward. *Gastroenterology and Hepatology from Bed to Bench*, *11*(3), 209–215. Available from http://journals.sbmu.ac.ir/ghfbb/index.php/ghfbb/article/download/1311/796.

Holmes, G. K. T., & Muirhead, A. (2018). Mortality in coeliac disease: A population-based cohort study from a single centre in Southern Derbyshire, UK. *BMJ Open Gastroenterology*, *5*(1), e000201. Available from https://doi.org/10.1136/bmjgast-2018-000201.

Hrdlickova, B., Mulder, C. J., Malamut, G., Meresse, B., Platteel, M., Kamatani, Y., Ricaño-Ponce, I., Van Wanrooij, R. L. J., Zorro, M. M., Jan Bonder, M., Gutierrez-Achury, J., Cellier, C., Zhernakova, A., Nijeboer, P., Galan, P., Withoff, S., Lathrop, M., Bouma, G., Xavier, R. J., ... Kumar, V. (2018). A locus at 7p14.3 predisposes to refractory celiac disease progression from celiac disease. *European Journal of Gastroenterology and Hepatology*, *30*(8), 828–837. Available from https://doi.org/10.1097/MEG.0000000000001168.

Hüe, S., Mention, J. J., Monteiro, R. C., Zhang, S. L., Cellier, C., Schmitz, J., Verkarre, V., Fodil, N., Bahram, S., Cerf-Bensussan, N., & Caillat-Zucman, S. (2004). A direct role for NKG2D/MICA interaction in villous atrophy during celiac disease. *Immunity*, *21*(3), 367–377. Available from https://doi.org/10.1016/j.immuni.2004.06.018.

Hutchinson, J. M., West, N. P., Robins, G. G., & Howdle, P. D. (2010). Long-term histological follow-up of people with coeliac disease in a UK teaching hospital. *QJM*, *103*(7), 511–517. Available from https://doi.org/10.1093/qjmed/hcq076.

Jantunen, E., Boumendil, A., Finel, H., Luan, J. J., Johnson, P., Rambaldi, A., Haynes, A., Duchosal, M. A., Bethge, W., Biron, P., Carlson, K., Craddock, C., Rudin, C., Finke, J., Salles, G., Kroschinsky, F., Sureda, A., & Dreger, P. (2013). Autologous stem cell transplantation for enteropathy-associated T-cell lymphoma: A retrospective study by the EBMT. *Blood*, *121*(13), 2529–2532. Available from https://doi.org/10.1182/blood-2012-11-466839.

Kaukinen, K., Peräaho, M., Lindfors, K., Partanen, J., Woolley, N., Pikkarainen, P., Karvonen, A. L., Laasanen, T., Sievänen, H., Mäki, M., & Collin, P. (2007). Persistent small bowel mucosal villous atrophy without symptoms in coeliac disease. *Alimentary Pharmacology and Therapeutics*, *25*(10), 1237–1245. Available from https://doi.org/10.1111/j.1365-2036.2007.03311.x.

Khalaf, W. F., Caldwell, M. E., & Reddy, N. (2013). Brentuximab in the treatment of CD30-positive enteropathy-associated T-cell lymphoma. *JNCCN Journal of the National Comprehensive Cancer Network*, *11*(2), 137–140. Available from https://doi.org/10.6004/jnccn.2013.0021.

Koskinen, I., Virta, L. J., Huhtala, H., Ilus, T., Kaukinen, K., & Collin, P. (2020). Overall and cause-specific mortality in adult celiac disease and dermatitis herpetiformis diagnosed in the 21st century. *The American Journal of Gastroenterology*, *115*(7), 1117–1124. Available from https://doi.org/10.14309/ajg.0000000000000665.

Kung, V. L., Liu, T. C., & Ma, C. (2018). Collagenous enteritis is unlikely a form of aggressive celiac disease despite sharing HLA-DQ2/DQ8 genotypes. *American Journal of Surgical Pathology*, *42*(4), 545–552. Available from https://doi.org/10.1097/PAS.0000000000001022.

Lebwohl, B., Murray, J. A., Rubio-Tapia, A., Green, P. H. R., & Ludvigsson, J. F. (2014). Predictors of persistent villous atrophy in coeliac disease: A population-based study. *Alimentary Pharmacology and Therapeutics*, *39*(5), 488–495. Available from https://doi.org/10.1111/apt.12621.

Leeds, J. S., Höroldt, B. S., Sidhu, R., Hopper, A. D., Robinson, K., Toulson, B., Dixon, L., Lobo, A. J., McAlindon, M. E., Hurlstone, D. P., & Sanders, D. S. (2007). Is there an association between coeliac disease and inflammatory bowel diseases? A study of relative prevalence in comparison with population controls. *Scandinavian Journal of Gastroenterology*, *42*(10), 1214–1220. Available from https://doi.org/10.1080/00365520701365112.

Leffler, D. A., Dennis, M., Hyett, B., Kelly, E., Schuppan, D., & Kelly, C. P. (2007). Etiologies and predictors of diagnosis in nonresponsive celiac disease. *Clinical Gastroenterology and Hepatology*, *5*(4), 445–450. Available from https://doi.org/10.1016/j.cgh.2006.12.006.

Lenti, M., Aronico, n, Bianchi, p, Neri, M., Volta, u, Mumolo, M., Astegiano, M., Calabrò, A., Zingone, F., Latella, G., Di Sario, A., Carroccio, A., Ciacci, C., Luzza, F., Bagnato, C., Fantini, M., Elli, L., Cammarota, g, Gasbarrini, A., ... Di Sabatino, A. (2022). Diagnostic delay in adult coeliac disease: An Italian multicentre study. *Digestive and Liver Disease*, *55*.

Lenti, M., Luu, S., Carsetti, R., Osier, F., Ogwang, R., Nnodu, O., Wiedermann, U., Spencer, J., Locatelli, F., & Corazza. (2022). Di Sabatino A. Asplenia and spleen hypofunction. *Nature Reviews Disease Primers*, *8*.

Lenti, M. V., Biagi, F., Lucioni, M., Di Sabatino, A., Paulli, M., & Corazza, G. R. (2019). Two cases of monomorphic epitheliotropic intestinal T-cell lymphoma associated with coeliac disease. *Scandinavian Journal of Gastroenterology*, *54*(8), 965–968. Available from https://doi.org/10.1080/00365521.2019.1647455.

Lenti, M. V., Rossi, C. M., Melazzini, F., Gastaldi, M., Bugatti, S., Rotondi, M., Bianchi, P. I., Gentile, A., Chiovato, L., Montecucco, C., Corazza, G. R., & Di Sabatino, A. (2022). Seronegative autoimmune diseases: A challenging diagnosis. *Autoimmunity Reviews*, *21*(9). Available from https://doi.org/10.1016/j.autrev.2022.103143.

Leslie, L. A., Lebwohl, B., Neugut, A. I., Gregory Mears, J., Bhagat, G., & Green, P. H. R. (2012). Incidence of lymphoproliferative disorders in patients with celiac disease. *American Journal of Hematology*, *87*(8), 754–759. Available from https://doi.org/10.1002/ajh.23237.

Levescot, A., Malamut, G., & Cerf-Bensussan, N. (2022). Immunopathogenesis and environmental triggers in coeliac disease. *Gut*, *71*(11), 2337–2349. Available from https://doi.org/10.1136/gutjnl-2021-326257.

Losurdo, G., Marra, A., Shahini, E., Girardi, B., Giorgio, F., Amoruso, A., Pisani, A., Piscitelli, D., Barone, M., Principi, M., Di Leo, A., & Ierardi, E. (2017). Small intestinal bacterial overgrowth and celiac disease: A systematic review with pooled-data analysis. *Neurogastroenterology and Motility*, *29*(6). Available from https://doi.org/10.1111/nmo.13028.

Ludvigsson, J. F., Bai, J. C., Biagi, F., Card, T. R., Ciacci, C., Ciclitira, P. J., Green, P. H. R., Hadjivassiliou, M., Holdoway, A., Van Heel, D. A., Kaukinen, K., Leffler, D. A., Leonard, J. N., Lundin, K. E. A., McGough, N., Davidson, M., Murray, J. A., Swift, G. L., Walker, M. M., ... Sanders, D. S. (2014). Diagnosis and management of adult coeliac disease: Guidelines from the British society of gastroenterology. *Gut*, *63*(8), 1210–1228. Available from https://doi.org/10.1136/gutjnl-2013-306578.

Ludvigsson, J. F., Leffler, D. A., Bai, J. C., Biagi, F., Fasano, A., Green, P. H. R., Hadjivassiliou, M., Kaukinen, K., Kelly, C. P., Leonard, J. N., Lundin, K. E. A., Murray, J. A., Sanders, D. S., Walker, M. M., Zingone, F., & Ciacci, C. (2013). The Oslo definitions for coeliac disease and related terms. *Gut*, *62*(1), 43–52. Available from https://doi.org/10.1136/gutjnl-2011-301346.

Lundin, K., Digernes, S., Grzyb, K., & Paulsen, J. (2022). Vedolizumab treatment of refractory celiac disease type I – Presentation of two cases. In: Proceedings of the ninteenth international celiac disease symposium, Sorrento, 189.

Maguire, A. A., Greenson, J. K., Lauwers, G. Y., Ginsburg, R. E., Williams, G. T., Brown, I. S., Riddell, R. H., O'Donoghue, D., & Sheahan, K. D. (2009). Collagenous sprue: A clinicopathologic study of 12 cases. *American Journal of Surgical Pathology*, *33*(10), 1440–1449. Available from https://doi.org/10.1097/PAS.0b013e3181ae2545.

Malamut, G., Afchain, P., Verkarre, V., Lecomte, T., Amiot, A., Damotte, D., Bouhnik, Y., Colombel, J. F., Delchier, J. C., Allez, M., Cosnes, J., Lavergne-Slove, A., Meresse, B., Trinquart, L., Macintyre, E., Radford-Weiss, I., Hermine, O., Brousse, N., Cerf-Bensussan, N., & Cellier, C. (2009). Presentation and long-term follow-up of refractory celiac disease: Comparison of type I with type II. *Gastroenterology*, *136*(1), 81–90. Available from https://doi.org/10.1053/j.gastro.2008.09.069.

Malamut, G., & Cellier, C. (2019). Refractory celiac disease. *Gastroenterology Clinics of North America*, *48*(1), 137–144. Available from https://doi.org/10.1016/j.gtc.2018.09.010.

Malamut, G., Chandesris, O., Verkarre, V., Meresse, B., Callens, C., Macintyre, E., Bouhnik, Y., Gornet, J. M., Allez, M., Jian, R., Berger, A., Châtellier, G., Brousse, N., Hermine, O., Cerf-Bensussan, N., & Cellier, C. (2013). Enteropathy associated T cell lymphoma in celiac disease: A large retrospective study. *Digestive and Liver Disease*, *45*(5), 377–384. Available from https://doi.org/10.1016/j.dld.2012.12.001.

Malamut, G., El Machhour, R., Montcuquet, N., Martin-Lannerée, S., Dusanter-Fourt, I., Verkarre, V., Mention, J. J., Rahmi, G., Kiyono, H., Butz, E. A., Brousse, N., Cellier, C., Cerf-Bensussan, N., & Meresse, B. (2010). IL-15 triggers an antiapoptotic pathway in human intraepithelial lymphocytes that is a potential new target in celiac disease-associated inflammation and lymphomagenesis. *Journal of Clinical Investigation*, *120*(6), 2131–2143. Available from https://doi.org/10.1172/JCI41344.

Malamut, G., Meresse, B., Kaltenbach, S., Derrieux, C., Verkarre, V., Macintyre, E., Ruskone-Fourmestraux, A., Fabiani, B., Radford-Weiss, I., Brousse, N., Hermine, O., Cerf-Bensussan, N., & Cellier, C. (2014). Small intestinal CD4 + T-cell lymphoma is a heterogenous entity with common pathology features. *Clinical Gastroenterology and Hepatology*, *12*(4), 599–e1. Available from https://doi.org/10.1016/j.cgh.2013.11.028.

Malamut, G., & Cellier, C. (2011). Is refractory celiac disease more severe in old Europe? *American Journal of Gastroenterology*, *106*(5), 929–932. Available from https://doi.org/10.1038/ajg.2011.105.

Malamut, G., Verkarre, V., Callens, C., Colussi, O., Rahmi, G., Macintyre, E., Haïoun, C., Meresse, B., Brousse, N., Romana, S., Hermine, O., Cerf–Bensussan, N., & Cellier, C. (2012). Enteropathy-associated T-cell lymphoma complicating an autoimmune enteropathy. *Gastroenterology*, *142*(4), 726–729. Available from https://doi.org/10.1053/j.gastro.2011.12.040, e3.

Mention, J. J., Ahmed, M. B., Bègue, B., Barbe, U., Verkarre, V., Asnafi, V., Colombel, J. F., Cugnenc, P. H., Ruemmele, F. M., McIntyre, E., Brousse, N., Cellier, C., & Cerf-Bensussan, N. (2003). Interleukin 15: A key to disrupted intraepithelial lymphocyte homeostasis and lymphomagenesis in celiac disease. *Gastroenterology*, *125*(3), 730–745. Available from https://doi.org/10.1016/S0016-5085(03)01047-3.

Meresse, B., Ripoche, J., Heyman, M., & Cerf-Bensussan, N. (2009). Celiac disease: From oral tolerance to intestinal inflammation, autoimmunity and lymphomagenesis. *Mucosal Immunology*, *2*(1), 8–23. Available from https://doi.org/10.1038/mi.2008.75.

Mukewar, S. S., Sharma, A., Rubio-Tapia, A., Wu, T. T., Jabri, B., & Murray, J. A. (2017). Open-capsule budesonide for refractory celiac disease. *American Journal of Gastroenterology*, *112*(6), 959–967. Available from https://doi.org/10.1038/ajg.2017.71.

Nijeboer, P., de Baaij, L. R., Visser, O., Witte, B. I., Cillessen, S. A. G. M., Mulder, C. J., & Bouma, G. (2015). Treatment response in enteropathy associated T-cell lymphoma; survival in a large multicenter cohort. *American Journal of Hematology*, *90*(6), 493–498. Available from https://doi.org/10.1002/ajh.23992.

Nijeboer, P., Malamut, G., Mulder, C. J., Cerf-Bensussan, N., Sibon, D., Bouma, G., Cellier, C., Hermine, O., & Visser, O. (2015). Enteropathy-associated T-cell lymphoma: Improving treatment strategies. *Digestive Diseases*, *33*(2), 231–235. Available from https://doi.org/10.1159/000369542.

Pelizzaro, F., Marsilio, i, Fassan, M., Piazza., Barberio, B., Savarino, A., Farinati, E., & Zingone. (2021). The risk of malignancies in celiac disease-A literature review. *Cancers (Basel)*, *13*.

Penny, H. A., Baggus, E. M. R., Rej, A., Snowden, J. A., & Sanders, D. S. (2020). Non-responsive coeliac disease: A comprehensive review from the NHS england national centre for refractory coeliac disease. *Nutrients*, *12*(1). Available from https://doi.org/10.3390/nu12010216.

Penny, H., Coleman, Rej, A., Wild, G., & Sanders, D. (2022). Evidence of raised urinary gluten immunogenic peptides in patients with type 1 refractory celiac disease. In: Proceedings of the 19th international celiac disease symposium, Sorrento, 101.

Penny, H. A., Rej, A., Baggus, E. M. R., Coleman, S. H., Ward, R., Wild, G., Bouma, G., Trott, N., Snowden, J. A., Wright, J., Cross, S. S., Hadjivassiliou, M., & Sanders, D. S. (2022). Non-responsive and refractory coeliac disease: Experience from the NHS England National Centre. *Nutrients*, *14*(13). Available from https://doi.org/10.3390/nu14132776.

Perfetti, V., Baldanti, F., Lenti, M. V., Vanoli, A., Biagi, F., Gatti, M., Riboni, R., Dallera, E., Paulli, M., Pedrazzoli, P., & Corazza, G. R. (2016). Detection of active Epstein–Barr virus infection in duodenal mucosa of patients with refractory celiac disease. *Clinical Gastroenterology and Hepatology*, *14*(8), 1216–1220. Available from https://doi.org/10.1016/j.cgh.2016.03.022.

Perfetti, V., Brunetti, L., Biagi, F., Ciccocioppo, R., Bianchi, P. I., & Corazza, G. R. (2012). TCRβ clonality improves diagnostic yield of TCRγ clonality in refractory celiac disease. *Journal of Clinical Gastroenterology*, *46*(8), 675–679. Available from https://doi.org/10.1097/MCG.0b013e31823eff20.

Rana, S. S., Dambalkar, A., Chhabra, P., Sharma, R., Nada, R., Sharma, V., Rana, S., & Bhasin, D. K. (2016). Is pancreatic exocrine insufficiency in celiac disease related to structural alterations in pancreatic parenchyma. *Annals of Gastroenterology*, *29*(3), 363–366. Available from https://doi.org/10.20524/aog.2016.0042.

Ring, A. M., Lin, J. X., Feng, D., Mitra, S., Rickert, M., Bowman, G. R., Pande, V. S., Li, P., Moraga, I., Spolski, R., Özkan, E., Leonard, W. J., & Christopher Garcia, K. (2012). Mechanistic and structural insight into the functional dichotomy between IL-2 and IL-15. *Nature Immunology*, *13* (12), 1187–1195. Available from https://doi.org/10.1038/ni.2449.

Ritter, J., Zimmermann, K., Jöhrens, K., Mende, S., Seegebarth, A., Siegmund, B., Hennig, S., Todorova, K., Rosenwald, A., Daum, S., Hummel, M., & Schumann, M. (2018). T-cell repertoires in refractory coeliac disease. *Gut*, *67*(4), 644–653. Available from https://doi.org/10.1136/gutjnl-2016-311816.

Roberti, A., Dobay, M. P., Bisig, B., Vallois, D., Boéchat, C., Lanitis, E., Bouchindhomme, B., Parrens, M. C., Bossard, C., Quintanilla-Martinez, L., Missiaglia, E., Gaulard, P., & De Leval, L. (2016). Type II enteropathy-associated T-cell lymphoma features a unique genomic profile with highly recurrent SETD2 alterations. *Nature Communications*, *7*. Available from https://doi.org/10.1038/ncomms12602.

Roshan, B., Leffler, D. A., Jamma, S., Dennis, M., Sheth, S., Falchuk, K., Najarian, R., Goldsmith, J., Tariq, S., Schuppan, D., & Kelly, C. P. (2011). The incidence and clinical spectrum of refractory celiac disease in a north american referral center. *American Journal of Gastroenterology*, *106* (5), 923–928. Available from https://doi.org/10.1038/ajg.2011.104.

Rowinski, S. A., & Christensen, E. (2016). Epidemiologic and therapeutic aspects of refractory coeliac disease - A systematic review. *Danish Medical Journal*, *63*(12). Available from http://www.danmedj.dk/portal/pls/portal/!PORTAL.wwpob_page.show?_docname = 11379500.PDF.

Rubio-Tapia, A., Barton, S. H., Rosenblatt, J. E., & Murray, J. A. (2009). Prevalence of small intestine bacterial overgrowth diagnosed by quantitative culture of intestinal aspirate in celiac disease. *Journal of Clinical Gastroenterology*, *43*(2), 157–161. Available from https://doi.org/10.1097/MCG.0b013e3181557e67.

Rubio-Tapia, A., Hill, I. D., Kelly, C. P., Calderwood, A. H., & Murray, J. A. (2013). ACG clinical guidelines: Diagnosis and management of celiac disease. *American Journal of Gastroenterology*, *108*(5), 656–676. Available from https://doi.org/10.1038/ajg.2013.79.

Rubio-Tapia, A., Hill, I. D., Semrad, C., Kelly, C. P., & Lebwohl, B. (2023). American College of Gastroenterology Guidelines Update: Diagnosis and Management of Celiac Disease. *American Journal of Gastroenterology*, *118*(1), 59–76. Available from https://doi.org/10.14309/ajg.0000000000002075.

Rubio-Tapia, A., Kelly, D. G., Lahr, B. D., Dogan, A., Wu, T.-T., & Murray, J. A. (2009). Clinical staging and survival in refractory celiac disease: A single center experience. *Gastroenterology*, *136*(1), 99–107. Available from https://doi.org/10.1053/j.gastro.2008.10.013.

Sainsbury, A., Sanders, D. S., & Ford, A. C. (2013). Prevalence of irritable bowel syndrome-type symptoms in patients with celiac disease: A meta-analysis. *Clinical Gastroenterology and Hepatology*, *11*(4), 359–365. Available from https://doi.org/10.1016/j.cgh.2012.11.033, e1.

Sarra, M., Cupi, M. L., Monteleone, I., Franzè, E., Ronchetti, G., Di Sabatino, A., Gentileschi, P., Franceschilli, L., Sileri, P., Sica, G., Del Vecchio Blanco, G., Cretella, M., Paoluzi, O. A., Corazza, G. R., Pallone, F., & Monteleone, G. (2013). IL-15 positively regulates IL-21 production in celiac disease mucosa. *Mucosal Immunology*, *6*(2), 244–255. Available from https://doi.org/10.1038/mi.2012.65.

Schiepatti, A., Cincotta, M., Biagi, F., & Sanders, D. S. (2021). Enteropathies with villous atrophy but negative coeliac serology in adults: Current issues. *BMJ Open Gastroenterology*, *8*(1), e000630. Available from https://doi.org/10.1136/bmjgast-2021-000630.

Scotta, M. S., de Giacomo, C., Maggiore, G., Siena, S., & Ugazio, A. G. (1982). Eucharistic problems for celiac patients. *New England Journal of Medicine*, *307*(14), 898. Available from https://doi.org/10.1056/NEJM198209303071423.

Sharaiha, R. Z., Lebwohl, B., Reimers, L., Bhagat, G., Green, P. H., & Neugut, A. I. (2012). Increasing incidence of enteropathy-associated T-cell lymphoma in the United States, 1973-2008. *Cancer*, *118*(15), 3786–3792. Available from https://doi.org/10.1002/cncr.26700.

Sharma, A., Choung, R. S., Wang, X. J., Russo, P. A., Wu, T. T., Nehra, V., & Murray, J. A. (2018). Features of adult autoimmune enteropathy compared with refractory celiac disease. *Clinical Gastroenterology and Hepatology*, *16*(6), 877–883. Available from https://doi.org/10.1016/j.cgh.2017.12.044, e1.

Shidrawi, R. G., Przemioslo, R., Davies, D. R., Tighe, M. R., & Ciclitira, P. J. (1994). Pitfalls in diagnosing coeliac disease. *Journal of Clinical Pathology*, *47*(8), 693–694. Available from https://doi.org/10.1136/jcp.47.8.693.

Sieniawski, M., Angamuthu, N., Boyd, K., Chasty, R., Davies, J., Forsyth, P., Jack, F., Lyons, S., Mounter, P., Revell, P., Proctor, S. J., & Lennard, A. L. (2010). Evaluation of enteropathy-associated T-cell lymphoma comparing standard therapies with a novel regimen including autologous stem cell transplantation. *Blood*, *115*(18), 3664–3670. Available from https://doi.org/10.1182/blood-2009-07-231324.

Smedby, K. E., Åkerman, M., Hildebrand, H., Glimelius, B., Ekbom, A., & Askling, J. (2005). Malignant lymphomas in coeliac disease: Evidance of increased risks for lymphoma types other than enteropathy-type T cell lymphoma. *Gut*, *54*(1), 54–59. Available from https://doi.org/10.1136/gut.2003.032094.

Soderquist, C. R., & Bhagat, G. (2021). Cellular and molecular bases of refractory celiac disease. *International Review of Cell and Molecular Biology*, *358*, 207–240. Available from https://doi.org/10.1016/bs.ircmb.2020.12.001, Elsevier Inc.

Soderquist, C. R., Hsiao, S., Mansukhani, M. M., Alobeid, B., Green, P. H., & Bhagat, G. (2020). Refractory celiac disease type II: An atypical case highlighting limitations of the current classification system. *Hematological Oncology*, *38*(3), 399–405. Available from https://doi.org/10.1002/hon.2720.

Soderquist, C. R., Lewis, S. K., Gru, A. A., Vlad, G., Williams, E. S., Hsiao, S., Mansukhani, M. M., Park, D. C., Bacchi, C. E., Alobeid, B., Green, P. H., & Bhagat, G. (2021). Immunophenotypic spectrum and genomic landscape of refractory celiac disease type II. *American Journal of Surgical Pathology*, *45*(7), 905–916. Available from https://doi.org/10.1097/PAS.0000000000001658.

Swerdlow, S. H., Campo, E., Pileri, S. A., Lee Harris, N., Stein, H., Siebert, R., Advani, R., Ghielmini, M., Salles, G. A., Zelenetz, A. D., & Jaffe, E. S. (2016). The 2016 revision of the World Health Organization classification of lymphoid neoplasms. *Blood*, *127*(20), 2375–2390. Available from https://doi.org/10.1182/blood-2016-01-643569.

Tack, G. J., Wondergem, M. J., Al-Toma, A., Verbeek, W. H. M., Schmittel, A., MacHado, M. V., Perri, F., Ossenkoppele, G. J., Huijgens, P. C., Schreurs, M. W. J., Mulder, C. J. J., & Visser, O. J. (2011). Auto-SCT in refractory celiac disease type II patients unresponsive to cladribine therapy. *Bone Marrow Transplantation*, *46*(6), 840–846. Available from https://doi.org/10.1038/bmt.2010.199.

Testa, A., Imperatore, N., Rispo, A., Rea, M., Tortora, R., Nardone, O. M., Lucci, L., Accarino, G., Caporaso, N., & Castiglione, F. (2018). Beyond irritable bowel syndrome: The efficacy of the low fodmap diet for improving symptoms in inflammatory bowel diseases and celiac disease. *Digestive Diseases*, *36*(4), 271–280. Available from https://doi.org/10.1159/000489487.

Tye-Din, J. A. (2022). Review article: Follow-up of coeliac disease. *Alimentary Pharmacology and Therapeutics*, *56*(1), S49–S63. Available from https://doi.org/10.1111/apt.16847.

Vakiani, E., Arguelles-Grande, C., Mansukhani, M. M., Lewis, S. K., Rotterdam, H., Green, P. H., & Bhagat, G. (2010). Collagenous sprue is not always associated with dismal outcomes: A clinicopathological study of 19 patients. *Modern Pathology*, *23*(1), 12–26. Available from https://doi.org/10.1038/modpathol.2009.151.

van de Donk, N. W., & E, D. (2012). Brentuximab vedotin. *MAbs*, *4*, 458–465. Available from https://doi.org/10.4161/mabs.20230.

van de Water, J. M. W., Cillessen, S. A. G. M., Visser, O. J., Verbeek, W. H. M., Meijer, C. J. L. M., & Mulder, C. J. J. (2010). Enteropathy associated T-cell lymphoma and its precursor lesions. *Best Practice and Research: Clinical Gastroenterology*, *24*(1), 43–56. Available from https://doi.org/10.1016/j.bpg.2009.11.002.

Vanoli, A., Di Sabatino, A., Furlan, D., Klersy, C., Grillo, F., Fiocca, R., Mescoli, C., Rugge, M., Nesi, G., Fociani, P., Sampietro, G., Ardizzone, S., Luinetti, O., Calabrò, A., Tonelli, F., Volta, U., Santini, D., Caio, G., Giuffrida, P., ... Corazza, G. R. (2017). Small bowel carcinomas in coeliac or Crohn's disease: Clinico-pathological, molecular, and prognostic features. A study from the small bowel cancer Italian consortium. *Journal of Crohn's and Colitis*, *11*(8), 942–953. Available from https://doi.org/10.1093/ecco-jcc/jjx031.

Van Weyenberg, S. J. B., Meijerink, M. R., Jacobs, M. A. J. M., Van Kuijk, C., Mulder, C. J., & Van Waesberghe, J. H. T. M. (2011). MR enteroclysis in refractory celiac disease: Proposal and validation of a severity scoring system. *Radiology*, *259*(1), 151–161. Available from https://doi.org/10.1148/radiol.11101808.

Verbeek, W. H. M., Van De Water, J. M. W., Al-Toma, A., Oudejans, J. J., Mulder, C. J. J., & Coupé, V. M. H. (2008). Incidence of enteropathy - Associated T-cell lymphoma: A nation-wide study of a population-based registry in the Netherlands. *Scandinavian Journal of Gastroenterology*, *43*(11), 1322–1328. Available from https://doi.org/10.1080/00365520802240222.

Verbeek, W. H. M., Von Blomberg, B. M. E., Coupe, V. M. H., Daum, S., Mulder, C. J. J., & Schreurs, M. W. J. (2009). Aberrant T-lymphocytes in refractory coeliac disease are not strictly confined to a small intestinal intraepithelial localization. *Cytometry Part B - Clinical Cytometry*, *76*(6), 367–374. Available from https://doi.org/10.1002/cyto.b.20481.

Verkarre, V., Asnafi, V., Lecomte, T., Patey Mariaud-de Serre, N., Leborgne, M., Grosdidier, E., Le Bihan, C., Macintyre, E., Cellier, C., Cerf-Bensussan, N., & Brousse, N. (2003). Refractory coeliac sprue is a diffuse gastrointestinal disease. *Gut*, *52*(2), 205–211. Available from https://doi.org/10.1136/gut.52.2.205.

Verkarre, V., Romana, S. P., Cellier, C., Asnafi, V., Mention, J. J., Barbe, U., Nusbaum, S., Hermine, O., MacIntyre, E., Brousse, N., Cerf-Bensussan, N., & Radford-Weiss, I. (2003). Recurrent partial trisomy 1q22-q44 in clonal intraepithelial lymphocytes in refractory celiac sprue. *Gastroenterology*, *125*(1), 40–46. Available from https://doi.org/10.1016/S0016-5085(03)00692-9.

Voorhees, T. J., Ghosh, N., Grover, N., Block, J., Cheng, C., Morrison, K., Ivanova, A., Dotti, G., Serody, J., Savoldo, B., & Beaven, A. W. (2020). Long-term remission in multiply relapsed enteropathy-associated T-cell lymphoma following CD30 CAR T-cell therapy. *Blood Advances*, *4*(23), 5925–5928. Available from https://doi.org/10.1182/bloodadvances.2020003218.

Wahab, P. J., Meijer, J. W. R., & Mulder, C. J. J. (2002). Histologic follow-up of people with celiac disease on a gluten-free diet: Slow and incomplete recovery. *American Journal of Clinical Pathology*, *118*(3), 459–463. Available from https://doi.org/10.1309/EVXT-851X-WHLC-RLX9.

Weber, M., Wolf, N., Branchi, F., Tangermann, P., Itzlinger, A., Poralla, L., Preiß, J. C., Grunert, P., Daum, S., Siegmund, B., Stallmach, A., & Schumann, M. (2021). Results from the German registry for refractory celiac disease. *Zeitschrift Für Gastroenterologie*, *59*(09), 944–953. Available from https://doi.org/10.1055/a-1540-7476.

Weinstein, W. M., Saunders, D. R., Tytgat, G. N., & Rubin, C. E. (1970). Collagenous sprue — An unrecognized type of malabsorption. *New England Journal of Medicine*, *283*(24), 1297–1301. Available from https://doi.org/10.1056/NEJM197012102832401.

Wierdsma, N. J., Nijeboer, P., de van der Schueren, M. A. E., Berkenpas, M., van Bodegraven, A. A., & Mulder, C. J. J. (2016). Refractory celiac disease and EATL patients show severe malnutrition and malabsorption at diagnosis. *Clinical Nutrition*, *35*(3), 685–691. Available from https://doi.org/10.1016/j.clnu.2015.04.014.

Wöhrer, S., Chott, A., Drach, J., Püspök, A., Hejna, M., Hoffmann, M., & Raderer, M. (2004). Chemotherapy with cyclophosphamide, doxorubicin, etoposide, vincristine and prednisone (CHOEP) is not effective in patients with enteropathy-type intestinal T-cell lymphoma. *Annals of Oncology*, *15*(11), 1680–1683. Available from https://doi.org/10.1093/annonc/mdh427.

Yoosuf, S., Barrett, C. G., Papamichael, K., Madoff, S. E., Kurada, S., Hansen, J., Silvester, J. A., Therrien, A., Singh, P., Dennis, M., Leffler, D. A., & Kelly, C. P. (2023). Pancreatic enzyme supplementation versus placebo for improvement of gastrointestinal symptoms in non-responsive celiac disease: A cross-over randomized controlled trial. *Frontiers in Medicine*, *9*. Available from https://doi.org/10.3389/fmed.2022.1001879.

Chapter 15

Infectious and other oncological complications

Rachele Ciccocioppo[1], Alessandro Vanoli[2], Antonio Di Sabatino[3] and Gino Roberto Corazza[3]
[1]*Gastroenterology Unit, Department of Medicine, AOUI Policlinico GB Rossi & University of Verona, Verona, Italy,* [2]*Pathology Unit, Department of Molecular Medicine, University of Pavia, Fondazione IRCCS Policlinico San Matteo, Pavia, Italy,* [3]*Department of Internal Medicine, University of Pavia, Fondazione IRCCS Policlinico San Matteo, Pavia, Italy*

15.1 Introduction

In addition to the possible development of refractoriness to a gluten-free diet and of enteropathy-associated T-cell lymphoma (EATL) (Chapter 14), a number of studies on the mortality risk in celiac disease (CD) (Corrao et al., 2001; Peters et al., 2003) and a series of case reports (Johnston & Robinson, 1998; O'donoghue, 1986) clearly showed that infections, mostly caused by *Streptococcus pneumoniae*, and other tumors (Askling et al., 2002; Chow et al., 1996), including those with an epithelial origin or outside the gastrointestinal tract, can further complicate adult CD.

15.2 Major infections in celiac disease

An increased frequency of severe infections, mainly by encapsulated bacteria, has been described in adult CD (Ludvigsson et al., 2008; Röckert Tjernberg et al., 2017; Thomas et al., 2008; Zingone et al., 2016). This susceptibility was originally described in an ad hoc study carried out in Sweden on 15,325 CD patients (Ludvigsson et al., 2008), where a higher risk of sepsis (HR = 2.6, $p < 0.001$), mostly from *Pneumococcus* (HR = 3.9, $p < 0.001$), was found with respect to the general population. In contrast, individuals with CD diagnosed in childhood did not show an increased risk of subsequent sepsis. These data were echoed by a study performed in England, where an increased risk of invasive pneumococcal disease (HR = 1.6), including sepsis, pneumonia and meningitis, was found in CD patients with respect to a reference cohort of cases admitted to hospitals (Thomas et al., 2008). Subsequently, similar results were found in a British cohort of unvaccinated CD patients in comparison to unvaccinated control subjects (HR = 1.3) (Zingone et al., 2016), and in a population-based cohort of Swedish CD patients (HR = 1.5) (Röckert Tjernberg et al., 2017). Additionally, in an Italian study, an increased risk of bacterial pneumonia, although not of pneumococcal infection, was shown in children and adolescents with CD (HR = 1.8), with the highest value found especially during the year before the diagnosis (HR = 6.0) (Canova et al., 2019).

Other than infections by encapsulated bacteria, further evidence showed that CD in adulthood is also associated with an increased risk, although modest, of serious infection by nonencapsulated microbes (Röckert Tjernberg et al., 2022). This study was carried out in Sweden by contacting 28 pathology units through the ESPRESSO (Epidemiology Strengthened by histoPathology Reports in Sweden) cohort, and 20,088 CD patients diagnosed in 2002–17 were identified and matched with 80,152 general population comparators, and then followed until the end of 2019. During the follow-up period, CD patients showed a slightly increased risk for serious infections (adjusted HR = 1.3). When examining the causes, the highest HR was found for protozoal infections (HR = 3.3), gastrointestinal infections (HR = 1.6) and sepsis (HR = 1.3). Notably, the risk persisted beyond the first year after CD diagnosis, although it decreased over time.

A few additional studies reported an increased risk of hospital admission for influenza (HR = 2.1) (Mårild et al., 2010), tuberculosis (HR = 2.0) (Ludvigsson et al., 2011), and *Clostridium difficile* infection (HR = 2.0) (Lebwohl et al., 2017) as well. In the latter two, the risk was highest in the first 12 months after CD diagnosis (HR = 3.5 and 5.2,

Pediatric and Adult Celiac Disease. DOI: https://doi.org/10.1016/B978-0-443-13359-6.00023-6

respectively) and remained high even after 5 years (HR = 1.9 and 1.8, respectively) as compared to controls. The association with tuberculosis might be explained by the frequent vitamin D deficiency found in CD, which is recognized to reduce bactericidal activity against mycobacteria through impaired cathelicidin production (Krutzik et al., 2008).

Finally, during the recent severe acute respiratory syndrome coronavirus (SARS-COV)-2 global pandemic, a number of studies have been carried out to investigate whether CD patients are also at increased risk of Coronavirus Disease 2019 (COVID-19) (Gasbarrini et al., 2021; Schiepatti et al., 2021; Uche-Anya et al., 2021). However, since the findings of observational studies are prone to bias due to confounding factors, especially in the case of a new emerging condition, a Mendelian randomized study was carried out to minimize this weakness, and a low susceptibility to both COVID-19 (OR = 0.96, $p = 0.006$) and severe COVID-19 (OR = 0.92, $p = 0.049$) was found in an adult CD cohort of European ancestry (Li et al., 2022). Therefore it may be unnecessary to take extra-precautions in CD.

15.2.1 Predisposing conditions to major infections in celiac disease

Splenic hypofunction/atrophy may be the main culprit of the increased risk of severe infections by encapsulated bacteria, mostly by *Pneumococcus*, in adult CD (Di Sabatino et al., 2011). In addition, a number of further predisposing conditions may also play a role, such as malnutrition, increased intestinal permeability, gut dysbiosis, concomitant immunodeficiency, and possible association with other diseases predisposing to infections, such as type 1 diabetes (Walters et al., 2008).

15.2.1.1 Splenic hypofunction in celiac disease

With the exception of sickle-cell anemia, adult CD is the most frequent medical condition associated with splenic hypofunction/atrophy (Di Sabatino et al., 2011; Lenti et al., 2022), with a prevalence ranging from 33% to 77% (Corazza et al., 1999; O'grady et al., 1984). Interestingly, the prevalence of impaired splenic function, as assessed by counting of pitted red cells or IgM-memory B cells on peripheral blood (see later), ranged from 19% in uncomplicated celiacs, to 59% in those with associated autoimmune conditions, and to 80% in complicated patients with refractory sprue, ulcerative jejuno-ileitis and/or EATL (Di Sabatino et al., 2006). This is why hyposplenism is recognized as an indicator of poor prognosis in adult CD. Interestingly, in a retrospective study, splenic volume was found to be significantly reduced in patients with type II refractory CD in comparison to uncomplicated CD (Van Gils et al., 2017).

At variance with adult patients, splenic hypofunction does not seem to complicate CD in infancy (Corazza et al., 1982), while its occurrence in adult CD correlates with the increasing age at diagnosis (Corazza et al., 1999). Worthy of note, a gluten-free diet is able to improve splenic function except in those patients who had already developed splenic atrophy (Corazza et al., 1983).

In adult CD, the histologic examination of the atrophic spleen showed severe atrophy of the marginal zone of the white pulp and a marked reduction of the expression of IgM-memory B cells (Di Sabatino et al., 2007), a population that displays the unique ability to produce natural antibodies, including those directed against *Streptococcus pneumoniae*, *Neisseria meningitidis*, and *Haemophilus influenzae*, and to initiate T-cell-independent immune responses on infection or vaccination with capsular polysaccharide antigens (Kruetzmann et al., 2003). As a consequence, an increased susceptibility to infections by encapsulated bacteria because of their inability to initiate a protective antibody response, including that towards polysaccharide vaccines, has been reported in patients with hyposplenism (Carsetti et al., 2006; Kruetzmann et al., 2003). Further immunological consequences of impaired splenic function are listed in Table 15.1 (Di Sabatino et al., 2011; Lenti et al., 2022; Walters et al., 2008).

Similarly to CD, splenic hypofunction, although at a lesser degree, has been also found in dermatitis herpetiform, a related gluten-sensitive condition (Corazza et al., 1981).

A series of hypotheses on the possible origin of splenic hypofunction/atrophy in CD have been suggested so far, that is, reticuloendothelial atrophy caused by either lymphocyte egress or failure of splenic recirculation due to loss or their trapping in the gut, reticuloendothelial blockade secondary to circulating immune complexes or pathogens, autoimmune damage leading to atrophy of the marginal zone of the white pulp where IgM-memory B cells reside (Mathur et al., 2022). However, no firm evidence is currently available on the truthfulness of these hypotheses and the possible superiority of one over the others.

15.2.1.2 Assessment of splenic function in celiac disease

The British Society of Gastroenterology recommends adult CD patients to be vaccinated against *Streptococcus pneumoniae*, whereas no definitive indications are given on whether additional vaccinations against other capsulated bacteria,

TABLE 15.1 Immunological consequences of hyposplenism.

Immunologic parameter	Dysfunction	Consequence
IgM-memory B-cell depletion	Defective clearance of encapsulated bacteria Impaired immune response upon vaccination with capsular polysaccharide antigens	Increased susceptibility to infections by encapsulated bacteria Inability to initiate a protective antibody response to polysaccharide vaccines
Reduction of tuftsin/properdin activity	Reduced opsonization of bacteria	Increased susceptibility to infections
Reduction of splenic phagocytes	Impaired pathogen clearance	Increased bacteremia
Reduction of secretory IgA	Altered mucosal barrier	Deficit of blocking bacteria adhesion to mucosal surface
Reduction of regulatory T-cells	Defective immune tolerance	Increased susceptibility to autoimmune disorders

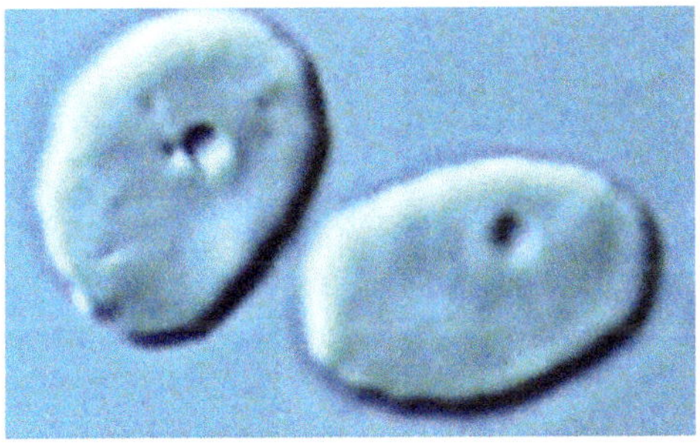

FIGURE 15.1 Blood film from a celiac patient with splenic hypofunction. Both red cells show characteristic membrane excavations clearly visible under the interference phase microscopy known as "pits".

such as *Haemophilus influenzae type B* and *Meningococcus*, should be considered due to the lack of conclusive evidence (Ludvigsson et al., 2014). Disappointingly, the compliance is poor: in one large study of at-risk individuals carried out in 2005, only 4.3% of the 3584 patients with either sickle-cell anemia or CD had been vaccinated in the previous 12 months and 15.9% in the past 5 years (Pebody et al., 2008). A possible alternative strategy might be to offer vaccination only to those adult patients with additional autoimmune disorders and/or CD complications (Di Sabatino et al., 2006), or to those with evidence of splenic hypofunction/atrophy. The last possibility raises the problem of the availability of a method to assess splenic function that would be reliable, accurate, repeatable, minimally invasive, learned through a brief training and affordable. The time-honored Howell-Jolly body, that is, basophilic red blood cell inclusions, counting (Corazza et al., 1990) does not meet all these requirements. By contrast, counting of "pitted cells" that are red cells bearing pits (vacuoles beneath the erythrocyte plasma membrane which are evident as crater-like depressions) visible under interference-phase microscopy on blood film (Fig. 15.1) is regarded as the gold standard for the assessment of splenic function in clinical practice (Corazza et al., 1981).

This method is more sensitive than the Howell-Jolly body counting in recognizing low degrees of splenic hypofunction, and the results have been found to correlate with those obtained by using more sophisticated, costly and invasive methods, such as scintigraphy upon the injection of radiolabeled tracers, or the clearance of heat-damaged, autologous erythrocytes (Corazza et al., 1981; Peters et al., 1981; Spencer et al., 1989). A pitted red cell count of $\geq 4\%$ has been associated with hyposplenism (Corazza et al., 1981), and the inverse correlation between pitted red cells and the rate of IgM memory B-cells indicates that the impairment of the filtering function, as defined by pitted erythrocyte counting, mirrors the parallel impairment of the immunological compartment (Di Sabatino et al., 2006). Although the equipment for pit counting is usually available in research rather than clinical settings, conventional microscopes can be cheaply converted into interference phase contrast mode through Nomarski optics (Corazza et al., 1981).

On the other hand, the incidental finding of a small spleen during an abdominal echotomography should always prompt clinicians to undertake one of the appropriate tests for quantification of splenic function and related disease (Di Sabatino et al., 2013).

15.2.1.3 Splenic atrophy and mesenteric lymph node cavitation

Small bowel mucosa atrophy, splenic atrophy and mesenteric lymph node cavitation (MLNC) is a rare triad related to adulthood CD that largely worsens its prognosis (Freeman, 2010; Holmes, 1986). In fact, the vast majority of reported cases died of cachexia, malnutrition, infections or neoplasms (Howat et al., 1995; Matuchansky et al., 1984), though spontaneous regression of MLNC following a strict gluten-free diet was reported in a case (Arotcarena et al., 2000). MLNC may be seen on abdominal imaging, even though the diagnosis is frequently made on laparotomy or autopsy where enlarged lymph nodes with central cystic degeneration containing acidophilic whitish fluid surrounded by fibrotic tissue are found (Freeman, 2010; Holmes, 1986; Howat et al., 1995; Matuchansky et al., 1984). The mechanisms underlying MLNC are still unclear but are in some way related to the severity of hyposplenism. It has been hypothesized that immune complexes derived from the small bowel mucosa may induce endothelial lesions in mesenteric lymph nodes which in turn trigger complement activation, resulting in localized intravascular coagulation, internal bleeding, and cavitation (Schmitz et al., 2002).

15.2.2 Management of major infections in celiac disease

The management of infections in CD patients with splenic hypofunction/atrophy relies on three complementary interventions, that is, education, therapy, and vaccination.

15.2.2.1 Patient education

Despite the clear association of CD in adulthood with an increased risk of serious infections (Ludvigsson et al., 2008; Röckert Tjernberg et al., 2017; Thomas et al., 2008; Zingone et al., 2016), the vast majority of patients are unaware of this potentially life-threatening condition. In patients with suspected or documented splenic hypofunction/atrophy, an increased awareness may be achieved through information leaflets distributed for instance by patients' associations and the systematic development of registries as well (Arnott et al., 2018). Patients and their relatives should be educated to notify their physicians of any acute febrile illness, especially if associated with systemic symptoms, and of visits to tropical countries because of the high risk of parasitic infections (Davies et al., 2011). Moreover, it is recommended that travelers take preventive measures such as antimalarial prophylaxis, mosquito repellents, and other barrier precautions (Di Sabatino et al., 2011). Disappointingly, these recommendations are often ignored by doctors themselves (Waghorn, 2001), whereas constant monitoring is essential because the risk of severe infections persists over time in patients with defective splenic function.

15.2.2.2 Therapy

At variance to those patients who have developed splenic atrophy, a gluten-free diet is expected to restore splenic function in those with functional hyposplenism (Corazza et al., 1983; O'grady et al., 1984). In these cases, therefore episodic short-term antibiotics at the time of febrile conditions might be a useful strategy (Di Sabatino et al., 2011). In cases of sepsis, treatment with empirical antibiotics is crucial to prevent the development of a life-threatening condition, and includes third-generation cephalosporins (cefotaxime 2 g IV every 8 hours or ceftriaxone 2 g IV every 12–24 hours), combined with gentamicin (5–7 mg/kg IV every 24 hours) or ciprofloxacin (400 mg IV every 12 hours) if a possible urinary or intestinal focus is suspected, or vancomycin (1–1.5 g IV every 12 hours) in case of resistance to penicillin (Brigden & Pattullo, 1999). This empirical approach can be eventually switched to a more specific one if the causative pathogen is identified by means of real-time polymerase chain reaction assay.

15.2.2.3 Vaccine prophylaxis

As already mentioned, the British Society of Gastroenterology (Ludvigsson et al., 2014), as well as other authors (Casella et al., 2019; Di Sabatino et al., 2013), recommend adults with CD to be vaccinated against *Streptococcus pneumoniae*. The choice of the type of vaccine in hypo/asplenic subjects is paramount, and it is based on the use of protein-conjugated vaccines (Di Sabatino et al., 2011). These, by eliciting a T-cell-dependent immune response, can restore the pool of antipneumococcal IgG (and not IgM)-memory B lymphocytes (Rosado et al., 2013). More in depth, the antipneumococcal vaccines available are the 23-valent pneumococcal polysaccharide vaccine (PPV-23; serotypes included: 1–5, 6B, 7 F, 8, 9 N, 9 V, 10 A, 11 A, 12 F, 14, 15B, 17 F, 18 C, 19 A, 19 F, 20, 22 F, 23 F, and 33 F), the 7-valent diphtheria cross-reactive material 197 (CRM197) protein-conjugate pneumococcal vaccine (PCV-7; serotypes included: 4, 6B, 9 V, 14, 18 C, 19 F, and 23 F), the PCV-13 (serotypes included: 1, 3, 4, 5, 6 A, 6B, 7 F, 9 V, 14, 18 C, 19 A, 19 F, and 23 F) (Di Sabatino et al., 2011), the new 15-valent pneumococcal polysaccharide-conjugated vaccine (serotypes included: 1, 3, 4, 5, 6 A, 6B, 7 F, 9 V, 14, 18 C, 19 A, 19 F, 22 F, 23 F, and 33 F) and, finally, the 20-valent pneumococcal polysaccharide-conjugated vaccine (serotypes included: 1, 3, 4, 5, 6 A, 6B, 7 F, 8, 9 V, 10 A, 11 A, 12 F, 14, 15B, 18 C, 19 A, 19 F, 22 F, 23 F, and 33 F) that has recently received approval from the United States Food and Drug Administration. The protective action of PPV-23 is based on the production of opsonizing anticapsular antibodies, however, since no firm evidence on the duration of protective antibody titer persistence are available, yearly monitoring of antibody titers might be a helpful strategy to establish the better timing of revaccination (French, 2003). In this regard, it should be pointed out that measurement of functional antibody titers by opsono-phagocytic assay is highly accurate to establish the protection and the eventual retreatment schedule (Vernacchio et al., 2000), nonetheless its availability is limited to research centers because of technical difficulty. The less expensive conjugate vaccine PCV-7 protects against fewer serotypes than PPV-23 does, but is more immunogenic (Chan et al., 1996). Specifically, conjugation of polysaccharides to the cross-reactive material-197 switches the antipolysaccharide response from a T-independent to a T-dependent one. This property makes this vaccine particularly suitable for those patients with splenic atrophy or functional hyposplenism. In these cases, the presence of an IgM-memory B cell deficit might be responsible for the ineffectiveness of PPV-23 (Carsetti et al., 2006). Finally, the protein-conjugate vaccine PCV-13 (Scott et al., 2007) offers coverage against several additional pneumococcal strains in comparison to PCV-7, and includes all seven serotypes most frequently involved in antibiotic resistance (Dagan & Klugman, 2008).

Current guidelines for splenectomized patients (Waghorn, 2001), that may be extended to those CD patients with spleen atrophy, recommend immunization also against *H. influenzae* type b and *N. meningitidis*. A single vaccination for *H. influenzae* seems protective, whereas revaccination increases the effective antibody response upon the use of conjugate vaccine against meningococcal serogroup C (Borrow et al., 2001). Finally, the availability of a new tetravalent conjugate antimeningococcal vaccine (against strains A, C, Y, and W135 capsular polysaccharides) (Revised Recommendations Of The Advisory Committee On Immunization Practices To Vaccinate All Persons Aged 11–18 Years With Meningococcal Conjugate Vaccine, 2007) might be useful, especially for hyposplenic or asplenic patients who travel to countries where *N. meningitidis* is endemic/epidemic.

In conclusion, patients with adult CD have been shown to be susceptible to major infections, mainly by encapsulated bacteria. Since splenic hypofunction/atrophy may represent a leading predisposing condition, it should be recommended that particularly patients diagnosed later in life with associated autoimmune conditions or refractory to a gluten-free diet be screened for splenic hypofunction or directly undergo vaccine prophylaxis.

15.3 Other oncological diseases

The overall risk of developing malignancies in individuals with CD has been widely investigated in recent decades. In some, but not all studies, this risk appears to be slightly higher compared to the general population, although it was essentially restricted to the first year after CD diagnosis and in those patients diagnosed with CD after the age of 40 (Askling et al., 2002; Card et al., 2004; Elfström et al., 2012; Han et al., 2015; Lebwohl et al., 2022; Tio et al., 2012; West et al., 2004). Further, referral and detection biases may have led to an overestimation of cancer risk in CD.

Beyond the association between CD and the increased risk of developing T cell lymphomas (Chapter 14), another relevant neoplastic complication of CD is small bowel adenocarcinoma (SBA). Moreover, an increased risk of some tumor types arising outside of the small bowel has been recorded.

15.3.1 Small bowel adenocarcinoma

SBA is a malignant epithelial tumor with glandular differentiation arising in the small intestine. Although SBA is a rare malignancy, accounting for less than 5% of all gastrointestinal cancers, it accounts for about 40% of all small bowel cancers and its incidence is rising in recent decades in the United States and Europe (Aparicio et al., 2022; Bouvier et al., 2020; Legué et al., 2016; Lepage et al., 2006). In Europe, an incidence rate of 3600 new SBAs/year has been estimated and most of them turned out to be sporadic (Faivre et al., 2012). Predisposing conditions for SBA development include hereditary cancer syndromes, such as Lynch syndrome, familial adenomatous polyposis and Peutz-Jeghers syndrome, and immune mediated intestinal disorders, mainly Crohn's disease and CD (Giuffrida et al., 2019; Pedersen et al., 2019). In a French prospective cohort study, such predisposing conditions have been identified in around of 20% of SBA cases (Aparicio et al., 2020).

15.3.1.1 The increased occurrence of small bowel adenocarcinoma in celiac disease

The first case report of an SBA associated with CD was published in 1972 (Kenwright, 1972). More recently, several epidemiological studies showed that CD patients have a significantly higher risk for SBA development compared to the general population (Askling et al., 2002; Card et al., 2004; Elfström et al., 2012; Emilsson et al., 2020; Green et al., 2003; Ilus et al., 2014; Silano et al., 2007; Swinson et al., 1983; Van Gils et al., 2018). The few studies that did not find a significantly increased risk suffered from a relatively small sample size (Anderson et al., 2007; Grainge et al., 2012). A metaanalysis of 8 studies reported a pooled odd ratio of 14.4 [95% confidence interval (CI): 5.53−37.60] (Han et al., 2015). In addition, this study found that, although the risk of SBA is higher in the first 2−4 years after CD diagnosis (odd ratio: 17.08, 95% CI: 3.59−81.20), it remains significant also after the fourth year (odd ratio: 4.64, 95% CI: 1.06−20.26). A Swedish nationwide cohort study, specifically designed to reduce the risk of detection bias in the peridiagnostic period, estimated the hazard risk for SBA to be 3.05 (95% CI: 1.86−4.99) in CD patients in comparison with matched controls (Emilsson et al., 2020). An additional study found that cumulative absolute risk estimates for SBAs were lower than those for T-cell lymphomas (Van Gils et al., 2018). Finally, it should be stressed that the absolute risk of SBA in CD patients (0.06%) is, anyway, very low, thus not implying a need for surveillance (Emilsson et al., 2020).

On the other side, the percentage of SBA cases associated with a diagnosis of CD ranges from 1.7% to 13%, depending on the study population (Aparicio et al., 2020; Howdle et al., 2003; Hänninen et al., 2018). Therefore CD serology should be systematically performed after any SBA diagnosis to identify an underlying previously undetected CD.

15.3.1.2 Risk factors for small bowel adenocarcinoma in celiac disease

Risk factors for the development of SBA in CD patients are poorly known. It is noteworthy that the median age at the time of CD diagnosis for patients with SBA was significantly higher than that for CD patients without malignancies, suggesting a potential role of diagnostic delay (Silano et al., 2007). A study of the Small Bowel Cancer Italian Consortium found a median age of 49 years at CD diagnosis in SBA patients, with a median time lapse between CD diagnosis and SBA detection of 23.5 months, whereas in five cases (19%) CD diagnosis was concomitant with that of SBA (Vanoli, Di Sabatino, Furlan, et al., 2017). A recent systematic review confirmed that (1) age at CD diagnosis >50 years is a risk factor for SBA in CD, (2) although the SBA onset time after CD diagnosis is widely variable, most of the cases are diagnosed within 1−3 years, and (3) SBA and CD are diagnosed simultaneously in approximately one-quarter of cases (Boulanger et al., 2022).

In addition to age at CD diagnosis, other putative risk factors for SBA are the severity and the persistence of mucosal lesions. The severity of histologic lesions at the time of CD diagnosis was correlated with the magnitude of SBA risk in a large epidemiological study including more than 30,000 CD patients (Elfström et al., 2012), whereas mucosal healing (defined as Marsh 0−2) was associated with a lower, but not statistically significant, risk of SBA in their cohort (Emilsson et al., 2020). In this regard, it is important to note that diffuse or patchy mucosal lesions have been described in both the cancer-adjacent and cancer-distant small bowel mucosa of most CD-SBA cases (Vanoli et al., 2017). The persistence of mucosal lesions (either in patients on a strict gluten-free diet or in patients with poor compliance to the diet) may contribute to explaining the continuously higher risk of SBA, and even in those cases arising after a strict gluten-free diet (Han et al., 2015; Howdle et al., 2003; Van Gils et al., 2018).

Although two cases of SBA complicating refractory CD have been reported (Chatelanat et al., 2022; Vanoli et al., 2017), the risk of SBA in this condition has not yet been estimated.

The role of epithelial dysplasia and of the classic "adenoma-to-carcinoma sequence" in CD-SBA development is still highly debated, also considering the rarity of residual mass-forming adenomatous dysplastic growths adjacent to the invasive CD-SBAs (Bruno et al., 1997; Vanoli et al., 2017). In addition, while a significantly increased risk of small intestinal adenomas in CD has been reported (Emilsson et al., 2020), another study did not confirm this association (Devi Rampertab et al., 2003). The lack of an overt adenomatous component in most CD-SBA cases suggests that they may arise from hyperproliferative, hyperplastic crypts in inflamed mucosal areas, either directly or through focal, invisible/flat, dysplastic changes, which are rarely seen next to CD-SBAs (Bruno et al., 1997; Vanoli et al., 2017) (Fig. 15.2).

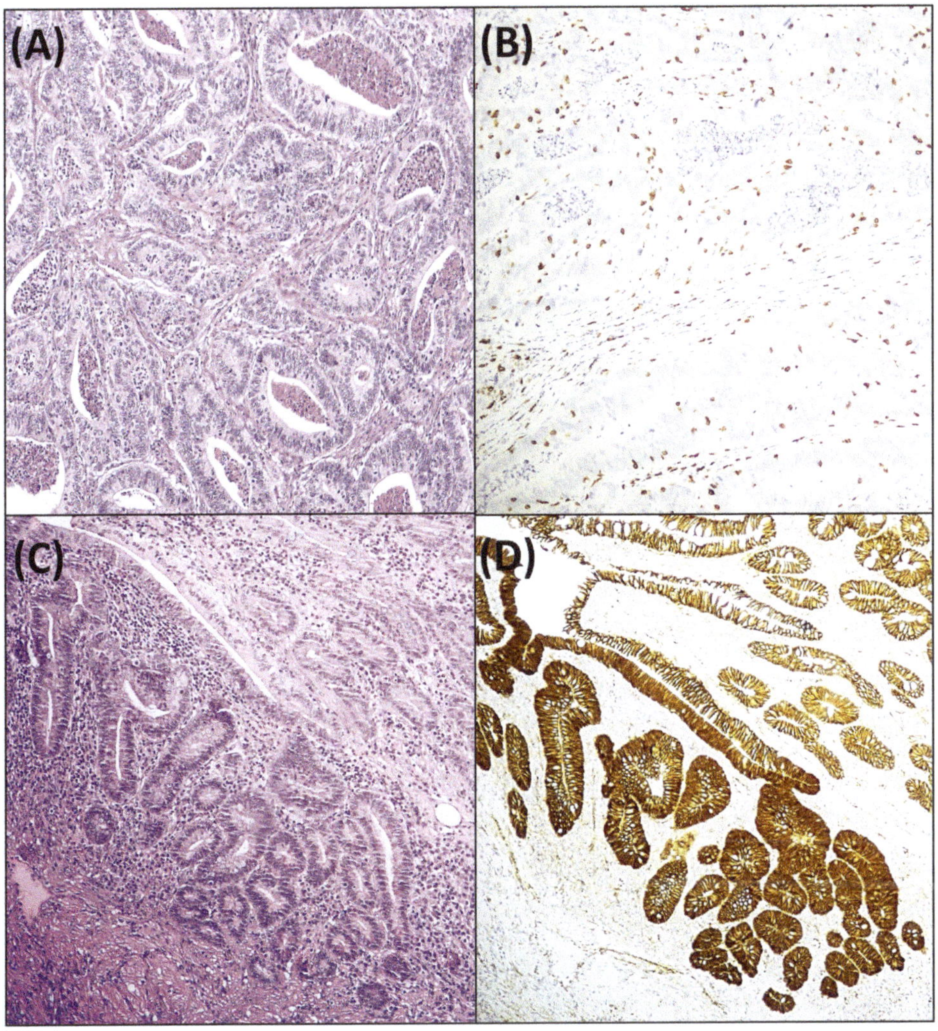

FIGURE 15.2 Representative histologic images of a jejunal adenocarcinoma arising in a patient with celiac disease. (A-B) The neoplasm shows a glandular architecture (A, H&E, ×100) and a high density of $CD3^+$ tumor-infiltrating lymphocytes (B, CD3 immunohistochemistry, ×100). (C) The mucosa adjacent to the adenocarcinoma features a flat low-grade dysplasia (H&E, ×100). The dysplastic lesion is composed of cells arranged in tubular structures with more hyperchromatic and larger nuclei compared to nondysplastic epithelium (on the upper right corner). (D) Several dysplastic cells show aberrant nuclear expression of β-catenin, while normal membranous immunoreactivity for β-catenin is seen in nondysplastic epithelial cells (β-catenin immunohistochemistry, ×100).

15.3.1.3 Clinicopathological, molecular and prognostic features of celiac disease-associated small bowel adenocarcinomas

CD-SBA represents a distinctive entity, showing several differences in clinicopathological and molecular features in comparison with sporadic and Crohn's disease-associated SBAs (Potter et al., 2004; Vanoli et al., 2017; Vanoli et al., 2017).

Studies of the Italian Small Bowel Cancer Consortium found that the median age at SBA diagnosis was significantly lower in CD patients in comparison with sporadic cases (Giuffrida et al., 2020; Vanoli et al., 2017);however, other series did not report significant differences in patient age between the two groups (Emilsson et al., 2020). Although a male preponderance has been reported in some series of CD-SBAs (Howdle et al., 2003; Potter et al., 2004), female sex was slightly more common in others (Giuffrida et al., 2020; Van Gils et al., 2018; Vanoli et al., 2017), likely reflecting the female predominance in uncomplicated CD.

The vast majority of CD-SBAs arise in the proximal small intestine, again reflecting the distribution of small bowel mucosal lesions in CD (Chapter 9), while they are very rare in the ileum, in contrast to SBAs complicating Crohn's disease (Giuffrida et al., 2020; Howdle et al., 2003; Potter et al., 2004). Interestingly, the percentage of jejunal cancers exceeds that of duodenal cases in a US cohort (Potter et al., 2004) and, even more strongly, in a relatively large Italian multicentric cohort of CD-SBA patients (Giuffrida et al., 2020). The reasons for the latter intriguing finding deserve further attention.

Regarding clinical aspects, CD-SBA patients present predominantly with obstructive/subobstructive symptoms or signs, such as abdominal pain, nausea, and vomiting, like sporadic SBA cases. Other possible presentations are overt (e.g., hematemesis or melena), or occult (e.g., otherwise unexplained iron-deficiency anemia) signs of intestinal bleeding, unexplained weight loss, or, more rarely, small intestinal intussusception or perforation (Howdle et al., 2003; Potter et al., 2004). In patients with an established diagnosis of CD, the presence of the above mentioned symptoms or signs should raise suspicion of a CD-related complication (i.e., SBA or lymphoma or ulcerative jejuno-ileitis) and should lead to a prompt diagnostic work-up (Pelizzaro et al., 2021) (see also Chapter 14). Duodenal SBAs may be diagnosed by esophagogastroduodenoscopy, while enteroscopy, computed tomography or magnetic resonance are often needed to detect more distal cancers (Marafini et al., 2020). A potential role of small bowel capsule endoscopy in SBA has been suggested, but it is still debated, due to the risk of capsule retention and the impossibility of taking biopsies (Rondonotti et al., 2018). Capsule endoscopy and double-balloon enteroscopy could be particularly considered for CD patients diagnosed after age of 50 years, with low compliance/poor response to gluten-free diet or alarm symptoms, such as obscure gastrointestinal bleeding, for early detection of small bowel malignancies (Tomba et al., 2014).

Histologically, the majority (56%) of CD-SBAs are conventional (glandular) adenocarcinomas (Fig. 15.2) or medullary-type carcinomas (17%), whereas poorly cohesive carcinomas and other histologic subtypes are very rare in CD patients (Giuffrida et al., 2020; Vanoli et al., 2022; Vanoli et al., 2017). Medullary-type carcinoma, an uncommon SBA histotype characterized by poorly differentiated growth of syncytial neoplastic cells, a prominent tumor infiltration by T lymphocytes, and a pushing margin, appears to be more frequently seen in CD-associated patients compared to sporadic cases (Brcic et al., 2016; Giuffrida et al., 2020; Vanoli et al., 2017). Among CD-SBAs, retained expression of intestinal differentiation markers (e.g., caudal type homeobox 2) is frequent, while the aberrant expression of gastropancreato-biliary markers (e.g., cytokeratin 7) is rarer compared to SBAs associated with Crohn's disease (Vanoli et al., 2017). The tumor microenvironment of CD-SBAs is characterized by a higher density of tumor-infiltrating lymphocytes in comparison to the sporadic form, a finding indicating a greater anticancer host immune response and likely contributing to explaining the higher rate of programmed death ligand 1 (PD-L1) positivity found in CD-SBAs (Giuffrida et al., 2020; Vanoli et al., 2017). These microenvironmental features are, to some extent, linked to distinctive molecular features of CD-SBAs (Diosdado et al., 2010; Potter et al., 2004; Rizzo et al., 2020; Vanoli et al., 2017a, 2017b; Hänninen et al., 2018). In fact, the high rate (50%−73%) of mismatch repair deficiency (MMR-D), which leads to high-level microsatellite instability, is the most striking and well-established molecular feature of CD-SBAs (Giuffrida et al., 2020; Potter et al., 2004; Hänninen et al., 2018). Almost all CD-SBAs with MMR-D show loss of expression of the MMR protein MLH1 due to promoter methylation of the *MLH1* gene (Bergmann et al., 2010; Diosdado et al., 2010; Vanoli et al., 2017). A recent next-generation RNA sequencing investigation on CD-SBAs confirmed that the "microsatellite instability-immune" transcriptomic subtype was the most common, followed by the "mesenchymal subtype" (Rizzo et al., 2020). In addition, while *APC* mutations are rarer in CD-SBAs (<10%) than in non-CD-related cancers (42%), *APC* gene hypermethylation has been reported to be more common in CD-SBAs (73%) (Diosdado et al., 2010). This finding is in keeping with the significantly higher percentage of nuclear immunohistochemical expression of beta-catenin, a surrogate marker of Wnt canonical pathway activation, in CD-SBAs (92%) compared to both sporadic (40%) and Crohn's disease (25%) SBAs (Vanoli et al., 2017). Interestingly, such aberrant nuclear expression of beta-catenin appears to be an early event in CD-related epithelial carcinogenesis, as it has also been seen in the rare dysplastic lesions found adjacent to full-blown invasive carcinomas (Vanoli et al., 2017). On the other hand, the reported frequencies of *KRAS* (31%), *BRAF V600E* (0%), and *PIK3CA* (15%) mutations, as well as of p53 protein overexpression (about 50%) in CD-SBAs are not significantly different from sporadic SBAs (Vanoli et al., 2017).

Importantly, among surgically resected SBA patients, CD-SBAs showed a more favorable prognosis compared to others, and, especially, to SBAs related to Crohn's disease (Emilsson et al., 2020; Potter et al., 2004; Vanoli et al., 2017; Vanoli et al., 2017). The higher rates of MMR-D and the higher density of tumor-infiltrating lymphocytes may contribute to explaining these differences. The favorable prognosis of CD-SBA was also proved among stage II (pT3−4, N0, M0) SBA patients, with potential implications on adjuvant chemotherapeutic choices (Vanoli et al., 2021).

15.3.1.4 Therapy and surveillance for small bowel adenocarcinoma in celiac patients

Treatment recommendations for CD-SBA patients are extrapolated from those of sporadic SBA. Recently, specific National Comprehensive Cancer Network guidelines have been created to establish an evidence-based standard of care for SBA patients (Benson et al., 2019). Surgical resection with regional lymphadenectomy is the only curative treatment for localized resectable SBAs, while stage IV or locally advanced SBAs require systemic chemotherapy. Despite the frequent local or distant recurrences, the clinical benefit of adjuvant chemotherapy for stage II or III disease is still debated. In particular, guidelines do not recommend adjuvant chemotherapy for stage II MMR-D SBAs, which have been found to be associated with better cancer-specific postoperative patient survival compared to stage II MMR-proficient SBAs (Vanoli et al., 2021). In advanced MMR-D SBAs, immunotherapy must be considered as an additional therapeutic option (Le et al., 2017; Pedersen et al., 2021). The *HER2* gene, which has been reported to be a promising targetable alteration in SBA, has been found to be amplified in 8% of CD-SBAs (Vanoli et al., 2017).

To date, no guidelines for SBA surveillance are available and random endoscopic biopsies of small bowel mucosa for SBA surveillance in CD individuals are not specifically recommended.

15.3.2 Other solid malignancies in celiac disease

An increased risk of developing oropharyngeal and esophageal squamous cell carcinomas in CD has been reported (Askling et al., 2002; Elfström et al., 2012; Green et al., 2003; Selby & Gallagher, 1979; Van Gils et al., 2018). A meta-analysis of eight studies reported a pooled odd ratio of 3.72 (95% CI: 1.90−7.28) for esophageal cancer in CD patients and that esophageal cancer risk was higher in the peridiagnostic period (Han et al., 2015). Possible explanations of this increased risk include the reported association of CD with esophagitis, nutritional deficiencies, as well as a higher prevalence of low socioeconomic status reported in CD patients (Choung et al., 2015), which is a risk factor for esophageal cancer (Askling et al., 2002; Han et al., 2015).

Among other gastrointestinal carcinomas, the risk of hepatobiliary and pancreatic adenocarcinomas has been reported to be higher in CD patients, even after the first year from CD diagnosis, in some large population-based Swedish studies (Askling et al., 2002; Elfström et al., 2012; Lebwohl et al., 2022). The association of CD with autoimmune biliary diseases and type 1 diabetes may underlie the observed risk increase for pancreato-biliary cancers (Askling et al., 2002). An increase in colorectal adenocarcinoma was also reported in the same population (Elfström et al., 2012); however, subsequent studies did not confirm this association (Pereyra et al., 2013; Volta et al., 2014).

Among extragastrointestinal malignancies, there are some data suggesting higher risks of melanoma and thyroid papillary carcinomas in CD (Green et al., 2003; Kent et al., 2006; Volta et al., 2011); again, these associations have not been confirmed by other studies (Askling et al., 2002; Card et al., 2004; Lebwohl et al., 2014; Ludvigsson et al., 2013). Therefore there is not sufficient evidence to confirm a formal association between CD and either melanoma or thyroid cancer.

References

Anderson, L. A., Mcmillan, S. A., Watson, R. G. P., Monaghan, P., Gavin, A. T., Fox, C., & Murray, L. J. (2007). Malignancy and mortality in a population-based cohort of patients with coeliac disease or "gluten sensitivity. *World Journal of Gastroenterology*, *13*(1), 146−151. Available from https://doi.org/10.3748/wjg.v13.i1.146.

Aparicio, T., Henriques, J., Manfredi, S., Tougeron, D., Bouché, O., Pezet, D., Piessen, G., Coriat, R., Zaanan, A., Legoux, J. L., Terrebone, E., Pocard, M., Gornet, J. M., Lecomte, T., Lombard-Bohas, C., Perrier, H., Lecaille, C., Lavau-Denes, S., Vernerey, D., & Afchain, P. (2020). Small bowel adenocarcinoma: Results from a nationwide prospective ARCAD-NADEGE cohort study of 347 patients. *International Journal of Cancer*, *147*(4), 967−977. Available from https://doi.org/10.1002/ijc.32860.

Aparicio, T., Pachev, A., Laurent-Puig, P., & Epidemiology, S. M. (2022). Risk factors and diagnosis of small bowel adenocarcinoma. *Cancers (Basel)*, *14*, 2268. Available from https://doi.org/10.3390/cancers14092268, PMID: 35565398; PMCID: PMC9103761.

Arnott, A., Jones, P., Franklin, L. J., Spelman, D., Leder, K., & Cheng, A. C. (2018). A registry for patients with asplenia/hyposplenism reduces the risk of infections with encapsulated organisms. *Clinical Infectious Diseases*, *67*(4), 557−561. Available from https://doi.org/10.1093/cid/ciy141.

Arotcarena, R., Hammel, P., Terris, B., Guth, A., & Bernades, R. (2000). Regression of mesenteric lymph node cavitation syndrome complicating celiac disease after a gluten free diet. *Gastroentérologie Clinique et Biologique*, *24*, 579−581.

Askling, J., Linet, M., Gridley, G., Halstensen, T. S., Ekström, K., & Ekbom, A. (2002). Cancer incidence in a population-based cohort of individuals hospitalized with celiac disease or dermatitis herpetiformis. *Gastroenterology*, *123*(5), 1428−1435. Available from https://doi.org/10.1053/gast.2002.36585.

Benson, A. B., Venook, A. P., Al-Hawary, M. M., Arain, M. A., Chen, Y.-J., Ciombor, K. K., Cohen, S. A., Cooper, H. S., Deming, D. A., Garrido-Laguna, I., Grem, J. L., Hoffe, S. E., Hubbard, J., Hunt, S., Kamel, A., Kirilcuk, N., Krishnamurthi, S., Messersmith, W. A., Meyerhardt, J., & Gurski, L. A. (2019). Small bowel adenocarcinoma, Version 1.2020, NCCN Clinical Practice Guidelines in Oncology. *Journal of the National Comprehensive Cancer Network*, *17*(9), 1109−1133. Available from https://doi.org/10.6004/jnccn.2019.0043.

Bergmann, F., Singh, S., Michel, S., Kahlert, C., Schirmacher, P., Helmke, B., Doeberitz, M. V. K., Kloor, M., & Bläker, H. (2010). Small bowel adenocarcinomas in celiac disease follow the CIM-MSI pathway. *Oncology Reports*, *24*(6), 1535−1539. Available from https://doi.org/10.3892/or-00001015.

Borrow, R., Andrews, N., Goldblatt, D., & Miller, E. (2001). Serological basis for use of meningococcal serogroup C conjugate vaccines in the United Kingdom: Reevaluation of correlates of protection. *Infection and Immunity*, *69*(3), 1568−1573. Available from https://doi.org/10.1128/IAI.69.3.1568-1573.2001.

Boulanger, Meckley, L., Pinto-Sanchez, M., Petrakis, I., Goodwin, B., Pigeon, & Leffler, D.A. (2022). Risk of small bowel cancer in celiac disease: A systematic review of the literature. In: Proceedings of the ninteenth international celiac disease symposium.

Bouvier, A. M., Robaszkiewicz, M., Jooste, V., Cariou, M., Drouillard, A., Bouvier, V., Nousbaum, J. B., Amadeo, B., Arveux, P., Baldi, I., Bara, S., Clavel, J., Colonna, M., Coureau, G., Cowppli-Bony, A., Dabakuyo, S., Dalmeida, T., Daubisse-Marliac, L., Defossez, G., & Woronoff, A. S. (2020). Trends in incidence of small bowel cancer according to histology: A population-based study. *Journal of Gastroenterology*, *55*(2), 181–188. Available from https://doi.org/10.1007/s00535-019-01636-z.

Brcic, I., Cathomas, G., Vanoli, A., Jilek, K., Giuffrida, P., & Langner, C. (2016). Medullary carcinoma of the small bowel. *Histopathology*, *69*(1), 136–140. Available from https://doi.org/10.1111/his.12908.

Brigden, M. L., & Pattullo, A. L. (1999). Prevention and management of overwhelming postsplenectomy infection - An update. *Critical Care Medicine*, *27*(4), 836–842. Available from https://doi.org/10.1097/00003246-199904000-00050.

Bruno, C. J., Batts, K. P., & Ahlquist, D. A. (1997). Evidence against flat dysplasia as a regional field defect in small bowel adenocarcinoma associated with celiac sprue. *Mayo Clinic Proceedings*, *72*(4), 320–322. Available from https://doi.org/10.4065/72.4.320.

Canova, C., Ludvigsson, J., Baldo, V., Barbiellini Amidei, C., Zanier, L., & Zingone, F. (2019). Risk of bacterial pneumonia and pneumococcal infection in youths with celiac disease—A population-based study. *Digestive and Liver Disease*, *51*(8), 1101–1105. Available from https://doi.org/10.1016/j.dld.2019.02.010.

Card, T. R., West, J., & Holmes, G. K. T. (2004). Risk of malignancy in diagnosed coeliac disease: A 24-year prospective, population-based, cohort study. *Alimentary Pharmacology and Therapeutics*, *20*(7), 769–775. Available from https://doi.org/10.1111/j.1365-2036.2004.02177.x.

Carsetti, R., Pantosti, A., & Quinti, I. (2006). Impairment of the antipolysaccharide response in splenectomized patients is due to the lack of immunoglobulin M memory B cells [4]. *Journal of Infectious Diseases*, *193*(8), 1189–1190. Available from https://doi.org/10.1086/501375.

Casella, G., Ingravalle, F., Abbate, G., Monti, C., Bonetti, F., Bassotti, G., Mansueto, P., Villanacci, V., & Carroccio, A. (2019). Pneumococcal vaccination in celiac disease. *Expert Review of Gastroenterology and Hepatology*, *13*(6), 541–546. Available from https://doi.org/10.1080/17474124.2019.1607295.

Chan, C. Y., Molrine, D. C., George, S., Tarbell, N. J., Mauch, P., Diller, L., Shamberger, R. C., Phillips, N. R., Goorin, A., & Ambrosino, D. M. (1996). Pneumococcal conjugate vaccine primes for antibody responses to polysaccharide pneumococcal vaccine after treatment of Hodgkin's disease. *Journal of Infectious Diseases*, *173*(1), 256–258. Available from https://doi.org/10.1093/infdis/173.1.256.

Chatelanat, K., Bruneau, J., Molina, Lhermitte, L., Brouzes, C., Rosenbaum, B., Sibon, D., Hermine, & Cellier. (2022). Clinical presentation and prognosis of small bowel adenocarcinoma in celiac disease patients. In: Proceedings of the ninteenth international celiac disease symposium.

Choung, R. S., Ditah, I. C., Nadeau, A. M., Rubio-Tapia, A., Marietta, E. V., Brantner, T. L., Camilleri, M. J., Rajkumar, S. V., Landgren, O., Everhart, J. E., & Murray, J. A. (2015). Trends and racial/ethnic disparities in gluten-sensitive problems in the United States: Findings from the national health and nutrition examination surveys from 1988 to 2012. *American Journal of Gastroenterology*, *110*(3), 455–461. Available from https://doi.org/10.1038/ajg.2015.8.

Chow, J. S., Chen, C. C., Ahsan, H., & Neugut, A. I. (1996). A population-based study of the incidence of malignant small bowel tumours: SEER, 1973-1990. *International Journal of Epidemiology*, *25*(4), 722–728. Available from https://doi.org/10.1093/ije/25.4.722.

Corazza, G. R., Bullen, A. W., Hall, R., Robinson, P. J., & Losowsky, M. S. (1981). Simple method of assessing splenic function in coeliac disease. *Clinical Science*, *60*(1), 109–113. Available from https://doi.org/10.1042/cs0600109.

Corazza, G. R., Frisoni, M., Vaira, D., & Gasbarrini, G. (1983). Effect of gluten-free diet on splenic hypofunction of adult coeliac disease. *Gut*, *24*(3), 228–230. Available from https://doi.org/10.1136/gut.24.3.228.

Corazza, G. R., Ginaldi, L., Zoli, G., Frisoni, M., Lalli, G., Gasbarrini, G., & Quaglino, D. (1990). Howell-Jolly body counting as a measure of splenic function. A reassessment. *Clinical and Laboratory Haematology*, *12*(3), 269–275.

Corazza, G. R., Lazzari, R., Frisoni, M., Collina, A., & Gasbarrini, G. (1982). Splenic function in childhood coeliac disease. *Gut*, *23*(5), 415–416. Available from https://doi.org/10.1136/gut.23.5.415.

Corazza, G. R., Zoli, G., Di Sabatino, A., Ciccocioppo, K., & Gasbarrini, G. (1999). A reassessment of splenic hypofunction in celiac disease. *American Journal of Gastroenterology*, *94*(2), 391–397. Available from https://doi.org/10.1111/j.1572-0241.1999.00865.x.

Corrao, G., Corazza, G. R., Bagnardi, V., Brusco, G., Ciacci, C., Cottone, M., Sategna Guidetti, C., Usai, P., Cesari, P., Pelli, M. A., Loperfido, S., Volta, U., Calabró, A., & Certo, M. (2001). Mortality in patients with coeliac disease and their relatives: A cohort study. *Lancet*, *358*(9279), 356–361. Available from https://doi.org/10.1016/S0140-6736(01)05554-4.

Dagan, R., & Klugman, K. P. (2008). Impact of conjugate pneumococcal vaccines on antibiotic resistance. *The Lancet Infectious Diseases*, *8*(12), 785–795. Available from https://doi.org/10.1016/S1473-3099(08)70281-0.

Davies, J. M., Lewis, M. P. N., Wimperis, J., Rafi, I., Ladhani, S., & Bolton-Maggs, P. H. B. (2011). Review of guidelines for the prevention and treatment of infection in patients with an absent or dysfunctional spleen: Prepared on behalf of the British Committee for Standards in Haematology by a Working Party of the Haemato-Oncology Task Force. *British Journal of Haematology*, *155*(3), 308–317. Available from https://doi.org/10.1111/j.1365-2141.2011.08843.x.

Devi Rampertab, S., Fleischauer, A., Neugut, A. I., & Green, P. H. R. (2003). Risk of duodenal adenoma in celiac disease. *Scandinavian Journal of Gastroenterology*, *38*(8), 831–833. Available from https://doi.org/10.1080/00365520310004515.

Di Sabatino, A., Brunetti, L., Maffè, G. C., Giuffrida, P., & Corazza, G. R. (2013). Is it worth investigating splenic function in patients with celiac disease. *World Journal of Gastroenterology*, *19*(15), 2313–2318. Available from https://doi.org/10.3748/wjg.v19.i15.2313.

Di Sabatino, A., Carnevale Maffè, G., Brunetti, L., Guerci, M., & Corazza, G. R. (2013). Splenic hypofunction in patients with an incidental finding of small-sized spleen at abdominal ultrasound. *Internal and Emergency Medicine*, *8*(4), 361–362. Available from https://doi.org/10.1007/s11739-012-0869-7.

Di Sabatino, A., Carsetti, R., & Corazza, G. R. (2011). Post-splenectomy and hyposplenic states. *The Lancet*, *378*(9785), 86–97. Available from https://doi.org/10.1016/S0140-6736(10)61493-6.

Di Sabatino, A., Rosado, M. M., Cazzola, P., Riboni, R., Biagi, F., Carsetti, R., & Corazza, G. R. (2006). Splenic hypofunction and the spectrum of autoimmune and malignant complications in celiac disease. *Clinical Gastroenterology and Hepatology*, *4*(2), 179–186. Available from https://doi.org/10.1016/S1542-3565(05)00982-1.

Di Sabatino, A., Rosado, M. M., Miele, L., Capolunghi, F., Cazzola, P., Biancheri, P., Carsetti, R., Gasbarrini, G., & Corazza, G. R. (2007). Impairment of splenic IgM-memory but not switched-memory B cells in a patient with celiac disease and splenic atrophy. *Journal of Allergy and Clinical Immunology*, *120*(6), 1461–1463. Available from https://doi.org/10.1016/j.jaci.2007.07.054.

Diosdado, B., Buffart, T. E., Watkins, R., Carvalho, B., Ylstra, B., Tijssen, M., Bolijn, A. S., Lewis, F., Maude, K., Verbeke, C., Nagtegaal, I. D., Grabsch, H., Mulder, C. J. J., Quirke, P., Howdle, P., & Meijer, G. A. (2010). High-resolution array comparative genomic hybridization in sporadic and celiac disease-related small bowel adenocarcinomas. *Clinical Cancer Research*, *16*(5), 1391–1401. Available from https://doi.org/10.1158/1078-0432.CCR-09-1773.

Elfström, P., Granath, F., Ye, W., & Ludvigsson, J. F. (2012). Low risk of gastrointestinal cancer among patients with celiac disease, inflammation, or latent celiac disease. *Clinical Gastroenterology and Hepatology*, *10*(1), 30–36. Available from https://doi.org/10.1016/j.cgh.2011.06.029.

Emilsson, L., Semrad, C., Lebwohl, B., Green, P. H. R., & Ludvigsson, J. F. (2020). Risk of small bowel adenocarcinoma, adenomas, and carcinoids in a nationwide cohort of individuals with celiac disease. *Gastroenterology*, *159*(5), 1686–1694. Available from https://doi.org/10.1053/j.gastro.2020.07.007, e2.

Faivre, J., Trama, A., De Angelis, R., Elferink, M., Siesling, S., Audisio, R., Bosset, J. F., Cervantes, A., Lepage, C., Zielonk, N., Van Eycken, E., Sundseth, H., Hedelin, G., Bouvier, A. M., Woronoff, A. S., Buemi, A., Tretarre, B., Colonna, M., Bara, S., & White, C. (2012). Incidence, prevalence and survival of patients with rare epithelial digestive cancers diagnosed in Europe in 1995-2002. *European Journal of Cancer*, *48*(10), 1417–1424. Available from https://doi.org/10.1016/j.ejca.2011.10.038.

Freeman, H. J. (2010). Mesenteric lymph node cavitation syndrome. *World Journal of Gastroenterology*, *16*(24), 2991–2993. Available from https://doi.org/10.3748/wjg.v16.i24.2991.

French, N. (2003). Use of pneumococcal polysaccharide vaccines: No simple answers. *Journal of Infection*, *46*(2), 78–86. Available from https://doi.org/10.1053/jinf.2002.1113.

Gasbarrini, G., Dionisi, T., Corazza, G. R., Aronico, N., Cammarota, G., Ianiro, G., De Vitis, I., Candelli, M., Mancarella, F. A., Simeoni, S., Addolorato, G., & Bonvicini, F. (2021). COVID-19 in celiac disease: A multicentric retrospective cohort study. *European Review for Medical and Pharmacological Sciences*, *25*(12), 4400–4404. Available from https://doi.org/10.26355/eurrev_202106_26150.

Giuffrida., Arpa, G., Grillo, F., Klersy, C., Sampietro, G., Ardizzone., Fociani, P., Fiocca, R., Latella, G., Sessa, F., Malvi., Mescoli, d, Rugge, c, Nesi., Ferrero, G., Furlan, S., Poggioli, D., Rizzello, G., Macciomei, F., & Di Sabatino, A. (2020). PD-L1 in small bowel adenocarcinoma is associated with etiology and tumor-infiltrating lymphocytes, in addition to microsatellite instability. *Modern Pathology: an Official Journal of the United States and Canadian Academy of Pathology, Inc*, *33*(7), 1398–1409. Available from https://doi.org/10.1038/s41379-020-0497-0, Erratum in: Mod Pathol. 2020.

Giuffrida, P., Vanoli, A., Arpa, G., Bonometti, A., Luinetti, O., Solcia, E., Corazza, G., Paulli, M., & Di Sabatino, A. (2019). Small bowel carcinomas associated with immune-mediated intestinal disorders: The current knowledge. *Cancers*, *11*(1), 31. Available from https://doi.org/10.3390/cancers11010031.

Grainge, M. J., West, J., Solaymani-Dodaran, M., Card, T. R., & Logan, R. F. A. (2012). The long-term risk of malignancy following a diagnosis of coeliac disease or dermatitis herpetiformis: A cohort study. *Alimentary Pharmacology and Therapeutics*, *35*(6), 730–739. Available from https://doi.org/10.1111/j.1365-2036.2012.04998.x.

Green, P. H. R., Fleischauer, A. T., Bhagat, G., Goyal, R., Jabri, B., & Neugut, A. I. (2003). Risk of malignancy in patients with celiac disease. *American Journal of Medicine*, *115*(3), 191–195. Available from https://doi.org/10.1016/S0002-9343(03)00302-4.

Han, Y., Chen, W., Li, P., & Ye, J. (2015). Association between coeliac disease and risk of any malignancy and gastrointestinal malignancy: A meta-analysis. *Medicine (United States)*, *94*(38), e1612. Available from https://doi.org/10.1097/MD.0000000000001612.

Hänninen, U., Katainen, R., Tanskanen, T., Plaketti, Laine, R., Hamberg, J., Ristimäki, P., Taipale, M., Mecklin, J., Forsström, Pitkänen, E., Palin, K., Välimäki, N., Mäkinen, N., & Aaltonen, L.A. (2018.). Exome-wide somatic mutation characterization of small bowel adenocarcinoma. PLoS Genet, 14(3), e1007200. https://doi.org/10.1371/journal.pgen.1007200. PMID: 29522538; PMCID: PMC5871010.

Holmes, G. K. T. (1986). Mesenteric lymph node cavitation in coeliac disease. *Gut*, *27*(6), 728–733. Available from https://doi.org/10.1136/gut.27.6.728.

Howat, A. J., Mcphie, J. L., Smith, D. A., Aqel, N. M., Taylor, A. K., Cairns, S. A., Thomas, W. E., & Underwood, J. C. (1995). Cavitation of mesenteric lymph nodes: A rare complication of coeliac disease, associated with a poor outcome. *Histopathology*, *27*(4), 349–354.

Howdle, P. D., Jalal, P. K., Holmes, G. K. T., & Houlston, R. S. (2003). Primary small-bowel malignancy in the UK and its association with coeliac disease. *QJM - Monthly Journal of the Association of Physicians*, *96*(5), 345–353. Available from https://doi.org/10.1093/qjmed/hcg058.

Ilus, T., Kaukinen, K., Virta, L. J., Pukkala, E., & Collin, P. (2014). Incidence of malignancies in diagnosed celiac patients: A population-based estimate. *American Journal of Gastroenterology*, *109*(9), 1471–1477. Available from https://doi.org/10.1038/ajg.2014.194.

Johnston, S. D., & Robinson, J. (1998). Fatal pneumococcal septicaemia in a coeliac patient. *European Journal of Gastroenterology and Hepatology*, *10*(4), 353–354. Available from https://doi.org/10.1097/00042737-199804000-00014.

Kent, L., Mcbride, R., Mcconnell, R., Neugut, A. I., Bhagat, G., & Green, P. H. R. (2006). Increased risk of papillary thyroid cancer in celiac disease. *Digestive Diseases and Sciences*, *51*(10), 1875–1877. Available from https://doi.org/10.1007/s10620-006-9240-z.

Kenwright, S. (1972). Coeliac disease and small bowel carcinoma. *Postgraduate Medical Journal*, *48*(565), 673–677. Available from https://doi.org/10.1136/pgmj.48.565.673.

Kruetzmann, S., Rosado, M. M., Weber, H., Germing, U., Tournilhac, O., Peter, H. H., Berner, R., Peters, A., Boehm, T., Plebani, A., Quinti, I., & Carsetti, R. (2003). Human immunoglobulin M memory B cells controlling Streptococcus pneumoniae infections are generated in the spleen. *Journal of Experimental Medicine*, *197*(7), 939–945. Available from https://doi.org/10.1084/jem.20022020.

Krutzik, S. R., Hewison, M., Liu, P. T., Robles, J. A., Stenger, S., Adams, J. S., & Modlin, R. L. (2008). IL-15 links TLR2/1-induced macrophage differentiation to the vitamin D-dependent antimicrobial pathway. *Journal of Immunology*, *181*(10), 7115–7120. Available from https://doi.org/10.4049/jimmunol.181.10.7115.

Le, D. T., Durham, J. N., Smith, K. N., Wang, H., Bartlett, B. R., Aulakh, L. K., Lu, S., Kemberling, H., Wilt, C., Luber, B. S., Wong, F., Azad, N. S., Rucki, A. A., Laheru, D., Donehower, R., Zaheer, A., Fisher, G. A., Crocenzi, T. S., Lee, J. J., & Diaz, L. A. (2017). Mismatch repair deficiency predicts response of solid tumors to PD-1 blockade. *Science (New York, N.Y.)*, *357*(6349), 409–413. Available from https://doi.org/10.1126/science.aan6733.

Lebwohl, B., Eriksson, H., Hansson, J., Green, P. H. R., & Ludvigsson, J. F. (2014). Risk of cutaneous malignant melanoma in patients with celiac disease: A population-based study. *Journal of the American Academy of Dermatology*, *71*(2), 245–248. Available from https://doi.org/10.1016/j.jaad.2014.03.029.

Lebwohl, B., Green, P. H. R., Emilsson, L., Mårild, K., Söderling, J., Roelstraete, B., & Ludvigsson, J. F. (2022). Cancer risk in 47,241 individuals with celiac disease: A nationwide cohort study. *Clinical Gastroenterology and Hepatology*, *20*(2), e111–e131. Available from https://doi.org/10.1016/j.cgh.2021.05.034.

Lebwohl, B., Nobel, Y. R., Green, P. H. R., Blaser, M. J., & Ludvigsson, J. F. (2017). Risk of clostridium difficile infection in patients with celiac disease: A population-based study. *American Journal of Gastroenterology*, *112*(12), 1878–1884. Available from https://doi.org/10.1038/ajg.2017.400.

Legué, L. M., Bernards, N., Gerritse, S. L., Van Oudheusden, T. R., De Hingh, I. H. J. T., Creemers, G. J. M., Ten Tije, A. J., & Lemmens, V. E. P. P. (2016). Trends in incidence, treatment and survival of small bowel adenocarcinomas between 1999 and 2013: A population-based study in The Netherlands. *Acta Oncologica*, *55*(9–10), 1183–1189. Available from https://doi.org/10.1080/0284186X.2016.1182211.

Lenti, M. V., Luu, S., Carsetti, R., Osier, F., Ogwang, R., Nnodu, O. E., Wiedermann, U., Spencer, J., Locatelli, F., Corazza, G. R., & Di Sabatino, A. (2022). Asplenia and spleen hypofunction. *Nature Reviews Disease Primers*, *8*(1), 71. Available from https://doi.org/10.1038/s41572-022-00399-x.

Lepage, C., Bouvier, A. M., Manfredi, S., Dancourt, V., & Faivre, J. (2006). Incidence and management of primary malignant small bowel cancers: A well-defined French population study. *American Journal of Gastroenterology*, *101*(12), 2826–2832. Available from https://doi.org/10.1111/j.1572-0241.2006.00854.x.

Li, J., Tian, A., Yang, D., Zhang, M., Chen, L., Wen, J., & Chen, P. (2022). Celiac disease and the susceptibility of COVID-19 and the risk of severe COVID-19: A Mendelian randomization study. *Clinical and Translational Gastroenterology*, *13*(5), e00480. Available from https://doi.org/10.14309/ctg.0000000000000480.

Ludvigsson, J. F., Bai, J. C., Biagi, F., Card, T. R., Ciacci, C., Ciclitira, P. J., Green, P. H. R., Hadjivassiliou, M., Holdoway, A., Van Heel, D. A., Kaukinen, K., Leffler, D. A., Leonard, J. N., Lundin, K. E. A., Mcgough, N., Davidson, M., Murray, J. A., Swift, G. L., Walker, M. M., & Sanders, D. S. (2014). Diagnosis and management of adult coeliac disease: Guidelines from the British society of gastroenterology. *Gut*, *63*(8), 1210–1228. Available from https://doi.org/10.1136/gutjnl-2013-306578.

Ludvigsson, J. F., Lebwohl, B., Kämpe, O., Murray, J. A., Green, P. H., & Ekbom, A. (2013). Risk of thyroid cancer in a nationwide cohort of patients with biopsy-verified celiac disease. *Thyroid: Official Journal of the American Thyroid Association*, *23*(8), 971–976. Available from https://doi.org/10.1089/thy.2012.0306.

Ludvigsson, J. F., Olén, O., Bell, M., Ekbom, A., & Montgomery, S. M. (2008). Coeliac disease and risk of sepsis. *Gut*, *57*(8), 1074–1080. Available from https://doi.org/10.1136/gut.2007.133868.

Ludvigsson, J. F., Sanders, D. S., Maeurer, M., Jonsson, J., Grunewald, J., & Wahlström, J. (2011). Risk of tuberculosis in a large sample of patients with coeliac disease - A nationwide cohort study. *Alimentary Pharmacology and Therapeutics*, *33*(6), 689–696. Available from https://doi.org/10.1111/j.1365-2036.2010.04572.x.

Marafini, I., Monteleone, G., & Stolfi, C. (2020). Association between celiac disease and cancer. *International Journal of Molecular Sciences*, *21*(11), 4155. Available from https://doi.org/10.3390/ijms21114155.

Mårild, K., Fredlund, H., & Ludvigsson, J. F. (2010). Increased risk of hospital admission for influenza in patients with celiac disease: A nationwide cohort study in Sweden. *American Journal of Gastroenterology*, *105*(11), 2465–2473. Available from https://doi.org/10.1038/ajg.2010.352.

Mathur, A., Mclean, M. H., Cao, H., & Vickers, M. A. (2022). Hyposplenism and gastrointestinal diseases: Significance and mechanisms. *Digestive Diseases*, *40*(3), 290–298. Available from https://doi.org/10.1159/000517338.

Matuchansky, C., Colin, R., Hemet, J., Touchard, G., Babin, P., Eugene, C., Bergue, A., Zeitoun, P., & Barboteau, M. A. (1984). Cavitation of mesenteric lymph nodes, splenic atrophy, and a flat small intestinal mucosa. *Report of six cases. Gastroenterology*, *87*(3), 606–614. Available from https://doi.org/10.1016/0016-5085(84)90533-x.

O'donoghue, D. J. (1986). Fatal pneumococcal septicaemia in coeliac disease. *Postgraduate Medical Journal*, *62*(725), 229–230. Available from https://doi.org/10.1136/pgmj.62.725.229.

O'grady, J. G., Stevens, F. M., Harding, B., O'gorman, T. A., Mcnicholl, B., & Mccarthy, C. F. (1984). Hyposplenism and gluten-sensitive enteropathy. *Natural history, incidence, and relationship to diet and small bowel morphology. Gastroenterology*, *87*(6), 1326–1331. Available from https://doi.org/10.1016/0016-5085(84)90199-9.

Pebody, R. G., Hippisley-Cox, J., Harcourt, S., Pringle, M., Painter, M., & Smith, G. (2008). Uptake of pneumococcal polysaccharide vaccine in at-risk populations in England and Wales 1999-2005. *Epidemiology and Infection*, *136*(3), 360–369. Available from https://doi.org/10.1017/S0950268807008436.

Pedersen, K. S., Foster, N. R., Overman, M. J., Boland, P. M., Kim, S. S., Arrambide, K. A., Jaszewski, B. L., Bekaii-Saab, T., Graham, R. P., Welch, J., Wilson, R. H., & Mcwilliams, R. R. (2021). ZEBRA: A multicenter phase II study of pembrolizumab in patients with advanced small-bowel adenocarcinoma. *Clinical Cancer Research*, *27*(13), 3641–3648. Available from https://doi.org/10.1158/1078-0432.CCR-21-0159.

Pedersen, K. S., Raghav, K., & Overman, M. J. (2019). Small bowel adenocarcinoma: Etiology, presentation, and molecular alterations. *JNCCN Journal of the National Comprehensive Cancer Network*, *17*(9), 1135–1141. Available from https://doi.org/10.6004/jnccn.2019.7344.

Pelizzaro, F., Marsilio, I., Fassan, M., Piazza, F., Barberio, B., D'odorico, A., Savarino, E. V., Farinati, F., & Zingone, F. (2021). The risk of malignancies in celiac disease—A literature review. *Cancers*, *13*(21), 5288. Available from https://doi.org/10.3390/cancers13215288.

Pereyra, L., Gonzalez, R., Mohaidle, A., Fischer, C., Mella, J. M., Panigadi, G. N., Manazzoni, D., Matoso, M. D., Lasa, J. S., Novillo, A., De Paula, J., Soifer, L., Nadales, A., Cimmino, D. G., Pedreira, S., & Boerr, L. (2013). Risk of colorectal neoplasia in patients with celiac disease: A multicenter study. *Journal of Crohn's and Colitis*, *7*(12), e672–e677. Available from https://doi.org/10.1016/j.crohns.2013.06.005.

Peters, A. M., Ryan, P. F. J., Klonizakis, I., Elkon, K. B., Lewis, S. M., & Hughes, G. R. V. (1981). Measurement of splenic function in humans using heat damaged autologous red blood cells. *Scandinavian Journal of Haematology*, *27*(5), 374–380. Available from https://doi.org/10.1111/j.1600-0609.1981.tb00499.x.

Peters, U., Askling, J., Gridley, G., Ekbom, A., & Linet, M. (2003). Causes of death in patients with celiac disease in a population-based Swedish cohort. *Archives of Internal Medicine*, *163*(13), 1566–1572. Available from https://doi.org/10.1001/archinte.163.13.1566.

Potter, D. D., Murray, J. A., Donohue, J. H., Burgart, L. J., Nagorney, D. M., Van Heerden, J. A., Plevak, M. F., Zinsmeister, A. R., & Thibodeau, S. N. (2004). The role of defective mismatch repair in small bowel adenocarcinoma in celiac disease. *Cancer Research*, *64*(19), 7073–7077. Available from https://doi.org/10.1158/0008-5472.CAN-04-1096.

Revised Recommendations Of The Advisory Committee On Immunization Practices To Vaccinate All Persons Aged 11-18 Years With Meningococcal Conjugate Vaccine. (2007). MMWR. Morbidity and Mortality Weekly Report, 56.

Rizzo, F., Vanoli, A., Sahnane, N., Cerutti, R., Trapani, D., Rinaldi, A., Sellitto, A., Ciacci, C., Volta, U., Villanacci, V., Calabrò, A., Arpa, G., Luinetti, O., Paulli, M., Solcia, E., Di Sabatino, A., Sessa, F., Weisz, A., & Furlan, D. (2020). Small-bowel carcinomas associated with celiac disease: Transcriptomic profiling shows predominance of microsatellite instability-immune and mesenchymal subtypes. *Virchows Archiv*, *476*(5), 711–723. Available from https://doi.org/10.1007/s00428-019-02675-w.

Röckert Tjernberg, A., Bonnedahl, J., Inghammar, M., Egesten, A., Kahlmeter, G., Nauclér, P., Henriques-Normark, B., & Ludvigsson, J. F. (2017). Coeliac disease and invasive pneumococcal disease: A population-based cohort study. *Epidemiology and Infection*, *145*(6), 1203–1209. Available from https://doi.org/10.1017/S0950268816003204.

Röckert Tjernberg, A., Mårild, K., Söderling, J., Lebwohl, B., Roelstraete, B., Bonnedahl, J., Green, P. H. R., & Ludvigsson, J. F. (2022). Celiac disease and serious infections: A Nationwide Cohort Study From 2002 to 2017. *The American Journal of Gastroenterology*, *117*(10), 1675–1683. Available from https://doi.org/10.14309/ajg.0000000000001877.

Rondonotti, E., Spada, C., Adler, S., May, A., Despott, E. J., Koulaouzidis, A., Panter, S., Domagk, D., Fernandez-Urien, I., Rahmi, G., Riccioni, M. E., Van Hooft, J. E., Hassan, C., & Pennazio, M. (2018). Small-bowel capsule endoscopy and device-assisted enteroscopy for diagnosis and treatment of small-bowel disorders: European Society of Gastrointestinal Endoscopy (ESGE) Technical Review. *Endoscopy*, *50*(4), 423–446. Available from https://doi.org/10.1055/a-0576-0566.

Rosado, M. M., Gesualdo, F., Marcellini, V., Di Sabatino, A., Corazza, G. R., Smacchia, M. P., Nobili, B., Baronci, C., Russo, L., Rossi, F., Vito, R. D., Nicolosi, L., Inserra, A., Locatelli, F., Tozzi, A. E., & Carsetti, R. (2013). Preserved antibody levels and loss of memory B cells against pneumococcus and tetanus after splenectomy: Tailoring better vaccination strategies. *European Journal of Immunology*, *43*(10), 2659–2670. Available from https://doi.org/10.1002/eji.201343577.

Schiepatti, A., Alimenti, E., Maimaris, S., Nicolardi, M. L., Manzella La Barbera, F., Baiardi, P., & Biagi, F. (2021). Prevalence, incidence and clinical features of SARS-CoV-2 infection in adult coeliac patients. *European Journal of Gastroenterology & Hepatology*, *33*(11), 1361–1366. Available from https://doi.org/10.1097/MEG.0000000000001969.

Schmitz, F., Herzig, K. H., Stüber, E., Tiemann, M., Reinecke-Lüthge, A., Nitsche, R., & Fölsch, U. R. (2002). On the pathogenesis and clinical course of mesenteric lymph node cavitation and hyposplenism in coeliac disease. *International Journal of Colorectal Disease*, *17*(3), 192–198. Available from https://doi.org/10.1007/s00384-001-0361-8.

Scott, D. A., Komjathy, S. F., Hu, B. T., Baker, S., Supan, L. A., Monahan, C. A., Gruber, W., Siber, G. R., & Lockhart, S. P. (2007). Phase 1 trial of a 13-valent pneumococcal conjugate vaccine in healthy adults. *Vaccine*, *25*(33), 6164–6166. Available from https://doi.org/10.1016/j.vaccine.2007.06.004.

Selby, W. S., & Gallagher, N. D. (1979). Malignancy in a 19-year experience of adult celiac disease. *Digestive Diseases and Sciences*, *24*(9), 684–688. Available from https://doi.org/10.1007/BF01314465.

Silano, M., Volta, U., Mecchia., Dessì, M., Benedetto, D., & Vincenzi, D. (2007). Collaborating centers of the Italian registry of the complications of coeliac disease. Delayed diagnosis of coeliac disease increases cancer risk. *BMC Gastroenterology*, *7*. Available from https://doi.org/10.1186/1471-230X-7-8, PMID: 17349035; PMCID: PMC2140155.

Spencer, R. P., Gupta, S. M., & Hansell, J. R. (1989). The spleen: Diagnosis of splenic diseases using radiolabeled tracers. *Critical Reviews in Clinical Laboratory Sciences*, *27*(4), 299–318. Available from https://doi.org/10.3109/10408368909105717.

Swinson, C. M., Coles, E. C., Slavin, G., & Booth, C. C. (1983). COELIAC DISEASE AND MALIGNANCY. *The Lancet*, *321*(8316), 111–115. Available from https://doi.org/10.1016/S0140-6736(83)91754-3.

Thomas, H. J., Wotton, C. J., Yeates, D., Ahmad, T., Jewell, D. P., & Goldacre, M. J. (2008). Pneumococcal infection in patients with coeliac disease. *European Journal of Gastroenterology and Hepatology*, *20*(7), 624–628. Available from https://doi.org/10.1097/MEG.0b013e3282f45764.

Tio, M., Cox, M. R., & Eslick, G. D. (2012). Meta-analysis: Coeliac disease and the risk of all-cause mortality, any malignancy and lymphoid malignancy. *Alimentary Pharmacology and Therapeutics*, *35*(5), 540–551. Available from https://doi.org/10.1111/j.1365-2036.2011.04972.x.

Tomba, C., Elli, L., Bardella, M. T., Soncini, M., Contiero, P., Roncoroni, L., Locatelli, M., & Conte, D. (2014). Enteroscopy for the early detection of small bowel tumours in at-risk celiac patients. *Digestive and Liver Disease*, *46*(5), 400–404. Available from https://doi.org/10.1016/j.dld.2013.12.009.

Uche-Anya, E., Husby, S., Kaplan, G. G., Underwood, F. E., Green, P. H. R., & Lebwohl, B. (2021). An International Reporting Registry of Patients With Celiac Disease and COVID-19: Initial Results From SECURE-CELIAC. *Clinical Gastroenterology and Hepatology*, *19*(11), 2435–2437. Available from https://doi.org/10.1016/j.cgh.2021.06.016, e4.

Van Gils, T., Nijeboer, P., Overbeek, L. I. H., Hauptmann, M., Castelijn, D. A. R., Bouma, G., Mulder, C. J. J., Van Leeuwen, F. E., & De Jong, D. (2018). Risks for lymphoma and gastrointestinal carcinoma in patients with newly diagnosed adult-onset celiac disease: Consequences for follow-up: Celiac disease, lymphoma and GI carcinoma. *United European Gastroenterology Journal*, *6*(10), 1485–1495. Available from https://doi.org/10.1177/2050640618800540.

Van Gils, T., Nijeboer, P., Van Waesberghe, J. H. T., Coupé, V. M. H., Janssen, K., Zegers, J. A., Nurmohamed, S. A., Kraal, G., Jiskoot, S. C. I., Bouma, G., & Mulder, C. J. J. (2017). Splenic volume differentiates complicated and non-complicated celiac disease. *United European Gastroenterology Journal*, *5*(3), 374–379. Available from https://doi.org/10.1177/2050640616663571.

Vanoli, A., Di Sabatino, A., Furlan, D., Klersy, C., Grillo, F., Fiocca, R., Mescoli, C., Rugge, M., Nesi, G., Fociani, P., Sampietro, G., Ardizzone, S., Luinetti, O., Calabrò, A., Tonelli, F., Volta, U., Santini, D., Caio, G., Giuffrida, P., & Corazza, G. R. (2017). Small bowel carcinomas in coeliac or Crohn's disease: Clinico-pathological, molecular, and prognostic features. A study from the small bowel cancer Italian consortium. *Journal of Crohn's and Colitis*, *11*(8), 942–953. Available from https://doi.org/10.1093/ecco-jcc/jjx031.

Vanoli, A., Di Sabatino, A., Martino, M., Klersy, C., Grillo, F., Mescoli, C., Nesi, G., Volta, U., Fornino, D., Luinetti, O., Fociani, P., Villanacci, V., D'armiento, F. P., Cannizzaro, R., Latella, G., Ciacci, C., Biancone, L., Paulli, M., Sessa, F., & Usai, P. (2017). Small bowel carcinomas in celiac or Crohn's disease: Distinctive histophenotypic, molecular and histogenetic patterns. *Modern Pathology*, *30*(10), 1453–1466. Available from https://doi.org/10.1038/modpathol.2017.40.

Vanoli, A., Grillo, F., Guerini, C., Neri, G., Arpa, G., Klersy, C., Nesi, G., Giuffrida, P., Sampietro, G., Ardizzone, S., Fociani, P., Fiocca, R., Latella, G., Sessa, F., D'errico, A., Malvi, D., Mescoli, C., Rugge, M., Ferrero, S., & Di Sabatino, A. (2021). Prognostic role of mismatch repair status, histotype and high-risk pathologic features in Stage II small bowel adenocarcinomas. *Annals of Surgical Oncology*, *28*(2), 1167–1177. Available from https://doi.org/10.1245/s10434-020-08926-4.

Vanoli, A., Guerini, C., Grillo, F., Klersy, C., Fassan, M., Arpa, G., Neri, G., Luinetti, O., Lenti, M. V., Ulivi, P., Tedaldi, G., Furlan, D., Quaquarini, E., Ardizzone, S., Sampietro, G., Biancone, L., Monteleone, G., Solcia, E., Sessa, F., & Di Sabatino, A. (2022). Poorly cohesive carcinoma of the nonampullary small intestine: A distinct histologic subtype with prognostic significance. *American Journal of Surgical Pathology*, *46*(4), 498–508. Available from https://doi.org/10.1097/PAS.0000000000001821.

Vernacchio, L., Romero-Steiner, S., Martinez, J. E., Macdonald, K., Barnard, S., Pilishvili, T., Carlone, G. M., Ambrosino, D. M., & Molrine, D. C. (2000). Comparison of an opsonophagocytic assay and IgG ELISA to assess responses to pneumococcal polysaccharide and pneumococcal conjugate vaccines in children and young adults with sickle cell disease. *Journal of Infectious Diseases*, *181*(3), 1162–1166. Available from https://doi.org/10.1086/315307.

Volta, U., Vincentini, O., Quintarelli, F., Felli, C., Silano, M., & Caio, G. (2014). Low risk of colon cancer in patients with celiac disease. *Scandinavian Journal of Gastroenterology*, *49*(5), 564–568. Available from https://doi.org/10.3109/00365521.2014.893012.

Volta, U., Vincentini, O., & Silano, M. (2011). Collaborating Centers of the Italian Registry of Celiac Disease. Papillary cancer of thyroid in celiac disease. *Journal of Clinical Gastroenterology*, *45*(5), 44–46. Available from https://doi.org/10.1097/MCG.0b013e3181ea11cb.

Waghorn, D. J. (2001). Overwhelming infection in asplenic patients: Current best practice preventive measures are not being followed. *Journal of Clinical Pathology*, *54*(3), 214–218. Available from https://doi.org/10.1136/jcp.54.3.214.

Walters, J. R. F., Bamford, K. B., & Ghosh, S. (2008). Coeliac disease and the risk of infections. *Gut*, *57*(8), 1034–1035. Available from https://doi.org/10.1136/gut.2008.151571.

West, J., Logan, R. F. A., Smith, C. J., Hubbard, R. B., & Card, T. R. (2004). Malignancy and mortality in people with coeliac disease: Population based cohort study. *BMJ (Clinical Research ed.)*, *329*(7468), 716–719. Available from https://doi.org/10.1136/bmj.38169.486701.7c.

Zingone, F., Abdul Sultan, A., Crooks, C. J., Tata, L. J., Ciacci, C., & West, J. (2016). The risk of community-acquired pneumonia among 9803 patients with coeliac disease compared to the general population: A cohort study. *Alimentary Pharmacology and Therapeutics*, *44*(1), 57–67. Available from https://doi.org/10.1111/apt.13652.

Chapter 16

Dermatitis herpetiformis

Marzia Caproni, Alberto Corrà, Elena Biancamaria Mariotti and Cristina Aimo
Rare Diseases Unit, Section of Dermatology, Department of Health Sciences, USL Toscana Centre (European Reference Network-Skin Member), University of Florence, Florence, Italy

16.1 Introduction

Dermatitis herpetiformis (DH) is an inflammatory, autoimmune skin disease typically presenting with highly pruritic and polymorphic lesions, including papules, vesicles, and small blisters in typical anatomic sites such as the extensor surfaces of the limbs, elbows, sacral regions, and the buttocks. It was named after Louis A. Duhring who firstly described it in 1884 and underlined its clinical similarity with herpes virus infection, being its lesions generally grouped. In the 1950s Pierard described the associated recurrent histological pattern, characterized by the presence of clusters of neutrophils and eosinophils in the dermal papillae, while in 1969 Van der Meer detected the presence of granular IgA deposits at the tips of dermal papillae by immunofluorescence microscopy. Despite being initially often confused with bullous pemphigoid or linear IgA dermatosis, DH has been clearly related to gluten intolerance only in the 60s, when the theory of DH as the specific cutaneous manifestation of celiac disease (CD) began spreading among the scientific community. In fact, DH and CD share the same Human Leukocyte Antigen (HLA) haplotypes (DQ2 and DQ8) and they both improve following the start of a gluten-free diet (GFD). Diagnosis of DH relies on two major criteria: the presence of consistent clinical manifestations and granular IgA deposits at the dermal papillae and/or the dermal epidermal junction. Small bowel biopsy is generally recommended in adults, showing that almost all DH patients have typical CD alterations, even in absence of gastrointestinal symptoms. Further studies on DH etiology and pathogenesis have been conducted during the subsequent decades, until the identification of epidermal transglutaminase (eTG) as the major autoantigen of DH by Sardy et al. in 2002. The production of anti-eTG autoantibodies has often been explained with the phenomenon of epitope spreading, while their detection in peripheral blood revealed a helpful tool for the diagnosis. More recent insights into DH pathogenesis suggested a role of T cells in skin damage which is supported by the detection of T cell lymphocytes (TCLs) cross-reacting with both epidermal and tissue transglutaminases (tTG). Since DH is the specific cutaneous manifestation of CD, the affected patients require a GFD. In fact, it is the only causative therapy and it is fundamental in preventing complications, such as, in rare cases, gastrointestinal non-Hodgkin lymphomas. However, while intestinal symptoms resolution usually takes a few weeks after the start of a GFD, cutaneous manifestations may take up to 2 years to be controlled. Therefore, in many cases an additional pharmacological treatment is required to control the cutaneous manifestations. Dapsone is the drug of first choice for symptomatic DH treatment and it usually induces an improvement within 3–4 days. The activity of serum glucose-6-phosphate dehydrogenase (G6PD) is recommended to be determined before administration, because dapsone side effects (e.g., methemoglobinemia and hemolytic anemia) are more frequent in patients with G6PD deficiency.

16.2 Epidemiology

DH is a rare skin disease affecting primarily Caucasian individuals with an incidence varying from 0.4 to 2.6 per 100.000 people per year, while the prevalence ranges from 11.2 to 75.3 per 100.000 in Europe and United States, with the highest value reported in Finland. DH is even rarer among Asians and African Americans, probably because of the lower prevalence of the predisposing haplotypes HLA DQ2 and DQ8 (Antiga et al., 2019). Interestingly, the association between DH and CD seems to be weaker in Asians, including Chinese and Japanese populations, compared to the Western countries. DH typically occurs among adults between the third and fourth decades, while childhood DH seems

Pediatric and Adult Celiac Disease. DOI: https://doi.org/10.1016/B978-0-443-13359-6.00002-9

to be more frequent in Mediterranean countries, but the exact incidence is still unknown. Some Authors suggested a possible underestimation of pediatric DH prevalence due to overlapping clinical features with atopic dermatitis. Although CD is more common in females, males appear to have a higher prevalence of DH, with a male:female ratio of 3:2. However, female cases seem to predominate under 20 years of age (Görög et al., 2021). While the incidence of CD showed an increasing trend during the last decades, DH incidence seems to be in constant decrease. This difference may be explained with the increased awareness of CD leading to a wider prescription of screening tests and thus to an earlier identification of patients with latent or potential CD (Reunala et al., 2021).

16.3 Pathogenesis

While the existence of a complex inflammatory network along the gut-skin axis is well known, the pathogenesis of DH remains only partially understood. Genetic predisposition plays a pivotal role in DH pathogenesis, making it a paradigmatic model of an HLA-mediated autoimmune disease. As a confirmation, a familiar history of DH or CD is reported in about 5%–10% of the cases. The closest associations occur with class I and class II major histocompatibility complex regions, including HLA-A1, HLA-B8, HLA-DPB1, HLA-Dr2 and HLA-Dr3 (Katz et al., 1977; Sun et al., 2016). Interestingly, HLA-DRB1*0301 allele (component of HLA-Dr3) may represent a protective factor for gastrointestinal symptoms, while the presence of villous atrophy may correlate with HLA-B8 allele. However, HLA-DQ2 and DQ8 can be found respectively in 85% and 15% of the patients, resulting in the most frequent associations with both DH and CD. HLA-DQ2 and DQ8 genes map on the chromosome 6 and they seem to be crucially involved in gliadin processing (Antiga et al., 2019). Significant evidence supports the unavoidability of predisposing HLA-DQ antigens together with gluten exposure for the development of DH in Caucasians, while clinical and immunological features of Japanese DH population may present some peculiar characteristics, including the absence of HLA DQ2/DQ8 haplotypes, the infrequent association with CD, a higher clinical involvement of atypical sites such as the extremities and trunk, a fibrillar rather than granular IgA immune deposits in the papillary dermis and a lower incidence of CD-associated autoimmune diseases (Ohata et al., 2012). All these observations raised speculations about the existence of a subtype of DH not elicited by gluten. Besides the simple interaction between HLA-DQ antigens and gluten, the inflammatory response developed in DH may be impacted by endocrine and immunologic factors which may act as modulators of the pathogenetic mechanism. As an example, it seems that potassium iodide, a common compound of expectorants, may induce or worsen DH. The first reported observation dates back to the year 1891, when iodine and its compounds were found to trigger DH; subsequently, oral and topical iodine have been adopted for years for DH diagnosis. In fact, the topical application of potassium iodide can elicit vesicular lesions onset on healthy skin of DH patients, with a concentration-dependent severity (Douglas & Alexander, 1975; From et al., 1974; Michaelsson & Svensson, 1975; Reitamo et al., 1981). Interestingly, it seems that high concentrations of potassium iodide may alter the active site of eTG in the IgA deposits due to steric changes, increasing the enzymatic activity (Taylor & Zone, 2018). Hormonal factors may also play a role in DH, in particular the hypothalamic-pituitary-gonadal axis: a few studies, in fact, reported DH onset following progesterone contraception and analogs of gonadotropin-releasing hormone (GnRH) such as leuprolide, with resolution of the skin lesions after drug discontinuation (Hassan et al., 2007; Yu et al., 2006). Accordingly, robust evidence supports the influence of the hypothalamic-pituitary-gonadal pathway as a modulator of the immune system functions, with androgens playing the role of natural immunosuppressors (Tanriverdi et al., 2003). Androgen-deficient states are therefore associated with increased incidence of autoimmunity, as for systemic lupus erythematosus in patients with Klinefelter's syndrome (Bizzarro et al., 1987; Stern et al., 1977) or in patients treated with GnRH analogs for prostate cancer (Ohata et al., 2012), precocious puberty, endometriosis (Miyagawa et al., 1994) or leiomyomas (Metcalfe & Boulton-Jones, 1997). Some other drugs have been reported as potential triggers of DH, such as infliximab (Sonmezoglu Marakli et al., 2008) or ipilimumab, a monoclonal antibody targeting CTLA-4 administered for metastatic melanoma (Mochel et al., 2016). Regarding the target of autoimmunity in DH, eTG has been identified as the main autoantigen in 2002 (Sárdy et al., 2002). eTG belongs to a nine-member Ca2 + -dependent enzyme family which is believed to have a role in intermolecular cross-linking of cytoplasmic structural components. eTG is encoded by the TGM3 gene on the chromosome 20q11–12 (Wang et al., 1994), being physiologically expressed in the spinous and granular layers of the epidermis and localized in the cytoplasm, contributing to epidermal terminal differentiation, assembly of the cornified envelope and protection of keratinocytes against UVB-induced apoptosis. CD and DH patients develop autoantibodies targeting both the intestinal and epidermal isoforms of transglutaminase; tTG binds to the gliadin in the intestine, catalyzing its deamidation which may result in new epitopes recognized by HLA-DQ2 and -DQ8 on intestinal T cells. The consequent immune response leads to intestinal mucosa damage and the production of IgA antibodies to tTG. It is thought that in patients in which DH will occur, the intestinal IgA-tTG immune-complexes may drive the production of low-affinity anti-eTG IgA antibodies. After a long exposure to gluten, it is likely that the development of a cluster of autoantibodies with low affinity for tTG and high for eTG. In fact, IgA binding to eTG with high avidity can be found only in DH patients (Salmi & Hervonen, 2020). The shift of autoimmune response against eTG can mainly be explained by the phenomenon of epitope spreading, since the sequence homology between tTG and eTG can result in eTG autoimmunity and development of circulating IgA-eTG complexes, leading to DH onset in a subset of CD patients. According to this theory, autoimmunity against eTG would start in the gut (Hietikko et al., 2018). Evidence in support of this hypothesis includes the already mentioned sequence homology between tTG and eTG and the presence of autoimmunity also against the neuronal isoform of transglutaminase, highly similar to tTG and eTG (Kárpáti et al., 2018). Moreover, small bowel secretion of eTG-targeting IgA seems to be significantly higher in active DH compared to CD patients (Hietikko et al., 2018). The role of anti-eTG antibodies in the pathogenesis of DH was supported by the study of Zone et al.: they transferred goat and human anti-eTG IgG and IgA into mice, detecting the development of DH typical granular deposits at the dermal papillae. However, no clinical signs and symptoms of DH were found on mice, suggesting that other mechanisms are involved in the genesis of the typical skin lesions (Zone et al., 2011). According to the aforementioned pathogenetic hypothesis, traumatic epidermal damage may lead to eTG shedding from the spinous and granular layers to the upper dermis, where it binds to circulating anti-eTG IgA, thus explaining the lesions' distribution which suggests the influence of Koebner's phenomenon. Alternatively, eTG/IgA aggregates may exist as circulating immune complexes, with subsequent deposition on dermal vessels wall; this theory is supported by the finding of such asymptomatic immune complexes in the kidney of DH patients and by rare vascular manifestations of DH which may represent the expression of a small vessels vasculitis, including digital purpura and ecchymosis. Moreover, these circulating eTG/IgA immune complexes have already been detected in patients with DH and their concentrations seem to decrease under the GFD (Görög et al., 2016). Besides the central pathogenic activity of the humoral immunity, the involvement of T cells in DH skin lesions has been studied since Garioch et al. found the presence of activated CD4 + T cells and dendritic cells in DH involved skin in 1994 (Garioch et al., 1994). More recent studies found peripheral blood mononuclear cells (PBMCs) collected from DH patients which specifically reacted to gliadin peptides in an IFNg ELISpot, similarly to patients with CD. However, while in CD Th0 and Th1 responses seems to be the main sources of the intestinal damage (Nilsen et al., 1995), the Th2 immunity may play a major or marginal role in the genesis of skin lesions as suggested by immunohistochemical and serological analyses (Caproni et al., 1998; Makino et al., 2017). Moreover, an impaired suppressive activity of lesional Foxp3 + Tregs cells was found in the skin of DH patients compared to healthy controls (Antiga et al., 2015), as well as for intestinal and peripheral Treg cells in CD (Cook et al., 2017; Hmida et al., 2012). Interestingly, a recent study showed that DH patients have increased rates of skin-derived T cells producing TNFα and IL-17A compared to CD patients. Furthermore, the frequency of circulating TNFα- and IL-17A-producing T cells was shown to positively correlate with disease activity and to reduce after GFD in DH patients. Another notable finding is the isolation of T-cell clones responding to both eTG and tTG hysoforms from DH and CD patients, confirming a real cross-reactivity between the two antigens (Caproni et al., 2021). Authors suggested that a certain number of tTG-specific T cells originated in the gut-associated lymphoid tissue (GALT) may also show a high specificity for eTG and thus the potential to drive CD patients to develop DH. Accordingly, all CD subjects may have the risk to develop DH, since tTG/eTG-specific lymphocytes can be obtained also by CD patients in the absence of skin manifestations, but some regulatory mechanisms may prevent the development of the skin involvement in some celiac patients. This points out the possible role of the down-regulation of T-reg cells in DH-skin. The mechanism of blister formation is based mainly on fibrinolysis and neutrophils, which represent the major component of the dermal inflammatory infiltrate in DH. eTG/IgA complexes were demonstrated to be able to activate fibrinogen which is often found at the tips of dermal papillae along with IgA aggregates (Taylor et al., 2015). Fibrin is also thought to represent a chemoattractant and adhesive stimulus for neutrophils (Loike et al., 1991) via CD11c/CD18 receptor, stimulating their migration in the dermal papillae. Neutrophil attraction is enhanced by the eTG/IgA aggregates through an increased activity of FcIgA receptor, while a plethora of other cytokines may reinforce this phenomenon: indeed, the lesional levels of GM-CSF and serum IL-8 and TNFa are augmented in DH patients and correlate with disease activity (Graeber et al., 1993; Hall et al., 2006, 2007). The altered expression of adhesive molecules such as E-selectin and CD11b (Smith et al., 2002) along with the increased ability to bind IgA by circulating neutrophils suggests a sort of "priming" mechanism which may take part at the gut level where the local production of IL-8 is augmented consistently with the gluten intake. The enrollment and activation of these immune cells may also be impacted by the high serum levels of IL-17 and IL-36, as pointed out by recent studies (Zebrowska et al., 2005). Once migrated to the papillary dermis, neutrophils produce elastase and granzyme B resulting in the disruption of basement membrane and thus in vesicles and blister formation. The detachment occurs at the level of lamina lucida (Smith et al., 1992; Suzuki et al., 2014), probably due to the destruction of laminin 332 and to the concomitant production of collagenase, stromelysin-1 and urokinase-type plasminogen activator by keratinocytes (Airola et al., 1995). A local imbalance between matrix metalloproteinases (MMPs) and their inhibitors was also advocated to explain the blistering mechanism in DH (Zebrowska et al., 2005). The pathogenesis of pruritus seems to be the result of a complex interplay between different pathways which to date are only partially explored. The involvement of several neuropeptides was hypothesized, including corticotropin releasing factor and the receptor of endothelin B (Cynkier et al., 2012) which showed high levels in lesional skin samples. Recently, the involvement of IL-31 (Bonciani et al., 2017) was advocated following the finding of high lesional and circulating levels, which would support the previously described role of Th2 immunity in DH, since IL-31 belongs to the family of IL-6 and is produced by Th2 lymphocytes. In humans the enhanced expression of IL-31 is well known to be associated with the induction and persistence of pruritus in other chronic inflammatory skin diseases, such as atopic dermatitis, allergic contact dermatitis, and prurigo nodularis. Finally, in the last years the study of the intestinal microbiota attracted much interest, being suggested to act as a trigger in several autoimmune diseases. In CD microbiota imbalances were hypothesized to trigger or perpetuate the intestinal inflammatory processes, playing a primary or secondary role in the pathogenesis of all the disease phases. Apart from the insights on how a condition of gut dysbiosis would affect mucosal permeability and contribute to gluten intolerance by increasing the production of proinflammatory cytokines, to date very scarce data are available for DH. According to recent works, microbiota may play a role in the clinical manifestation of DH, since its composition seems to be different from CD patients without DH (Cenit et al., 2015; Wacklin et al., 2013). However, further studies are advisable to shed light on this field.

16.4 Clinical manifestations

DH usually manifests with a typical symmetrical rash which involves elbows, knees, shoulders and buttocks. More rarely, the scalp, the neck and the sacral region are affected. The involvement of extensor surfaces exposed to chronic mechanical stress suggests the influence of Koebner's phenomenon on the localization of the disease, supporting the idea that traumatic shedding of eTG from upper epidermal layers has a fundamental role for the onset of the clinical manifestations (Fig. 16.1).

Another hallmark of DH is lesions' polymorphism, as the clinical picture includes erythema, papules and vesicles, which are often grouped resembling herpetic skin infection. Lesion size is variable and rarely small blisters are appreciable (Fig. 16.2).

The rash is invariably, extremely itchy, so that the clinical picture derived by the chronic and intense scratching is often largely represented by crusts, erosions, lichenification and postinflammatory pigmentations. This should be noted since itch is often one of the initial signs of DH and it anticipates the classic lesions by months (Olson et al., 2015). Moreover, Nikolsky's sign is negative. Besides the typical manifestations of the disease, other uncommon presentations have been described (Fig. 16.3), alone or accompanying the characteristic lesions. Indeed, asymptomatic petechiae and purpura have been reported in adult and pediatric cases of DH, suggesting that small vessel inflammation may occur following immune complexes deposition. Accordingly, signs of vascular damage are mainly described in acral and palmoplantar sites with predilection for dominant limbs, supporting the role of repeated microtraumatism (Marks & Jones, 1971; McCleskey et al., 2002; McGovern & Bennion, 1994; Tu et al., 2013). In some cases, microscopic findings of true vasculitis were observed (Hofmann et al., 2009; Naylor et al., 2011). Finally, isolated reports of atypical

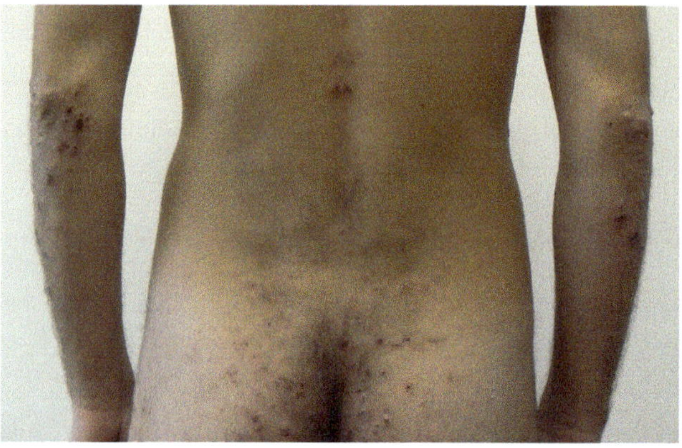

FIGURE 16.1 Classic distribution of DH lesions. Clinical pictures showing the typical distribution of DH lesions involving the extensor surfaces of the limbs and the buttocks with the involvement of the lumbo-sacral region.

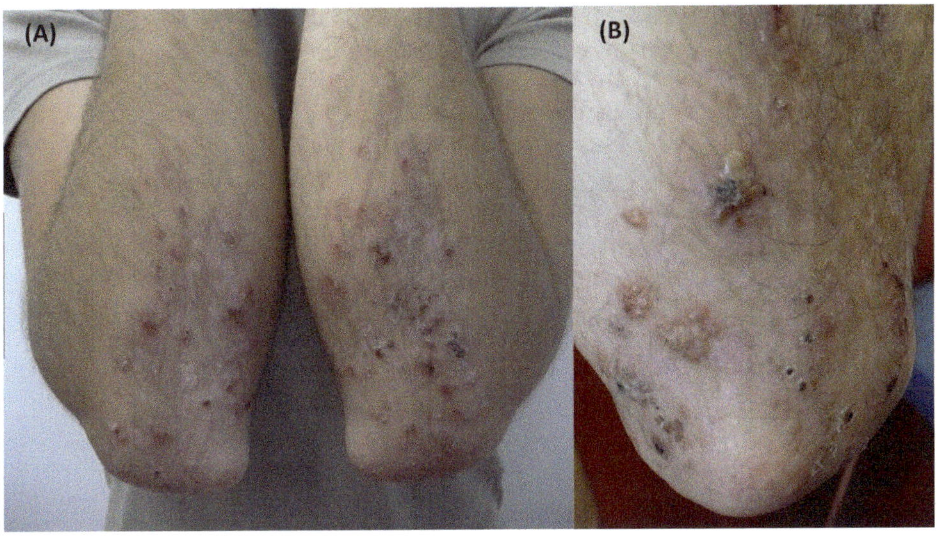

FIGURE 16.2 Polymorphism of grouped DH lesions. Clinical pictures showing erythema and grouped papules and vesicles resembling herpetic skin infection with a symmetrical distribution over the elbows (A). Detail showing the polymorphism of DH with the coexistence of erythema, vesicles, pustules, excoriated and crusted papules (B).

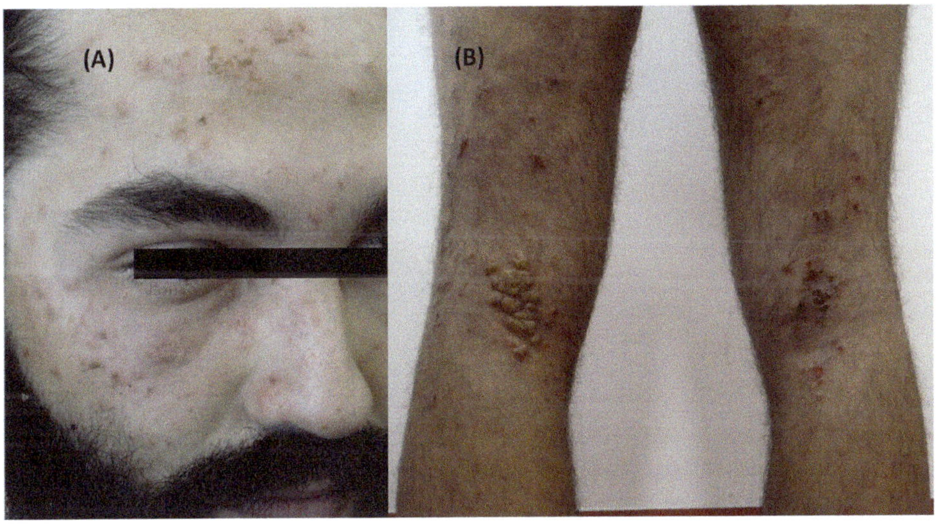

FIGURE 16.3 Rarer clinical presentation of DH. Clinical pictures of DH involving the face (A) and the flexural regions of the limbs (B).

presentations of DH include palmoplantar keratosis, chronic urticaria, intertriginous dermatitis and prurigo pigmentosa-like lesions (Ohshima et al., 2003; Pavlović, 2006; Powell et al., 2004; Saito et al., 2006). The frequency of the mucosal involvement in DH is not clear, with variable reported rates of oral lesions, ranging from 0% to 70% (Economopoulou & Laskaris, 1986; Fraser et al., 1973; Lahteenoja et al., 1998; Lodin & Stigell, 1965). Vesicles, small blisters, erythema and erosions affecting lips, oral mucosa or tongue are described, however in most cases signs and symptoms of underlying CD were not documented. Since papular, erosive or aphthoid lesions have long been associated with CD, it is difficult to ascertain whether these manifestations are due to CD or DH, requiring further and more detailed studies. Dental and enamel abnormalities have been detected in both childhood and adult patients with DH, with similar clinical features to those described in CD cohorts, namely delayed eruption or color and structural defects with horizontal grooves and vertical pits (Aine, 1996; Aine et al., 1992). Gastrointestinal complaints should be considered part of the clinical spectrum since DH represents the specific manifestation of CD. Despite this, the frequency and severity of gastrointestinal symptoms such as diarrhea, constipation or bloating seems to be minor when compared to CD populations. However, mild celiac-type small bowel biopsy alterations are usually found in DH, even in asymptomatic patients, and the grade of inflammation does not affect the prognosis of DH (Mansikka et al., 2017; Mansikka, Hervonen, et al., 2018; Salmi & Hervonen, 2020). DH is well-known to be associated with other autoimmune disorders (Collin et al., 1994; Reunala & Collin, 1997), with autoimmune thyroid disease and type 1 diabetes leading the list for their frequency of association. In fact, type 1 diabetes mellitus seems to affect from 1.0% to more than 5% of DH patients, while the incidence of thyroid autoimmunity ranges from 4.3% to 5.4%. Also, Sjogren syndrome, Addison disease, systemic lupus erythematosus, multiple sclerosis, rheumatoid arthritis, vitiligo and alopecia areata were shown to occur in concomitance with DH, but the association appears weaker (Collin et al., 2017). Recently, a modest but significant association between DH and atopic dermatitis (Kauppi et al., 2021) was described in the pediatric population, while an interesting association was found with bullous pemphigoid (BP), with a 22-fold increased risk among patients with previous DH diagnosis compared to controls (Varpuluoma et al., 2019). It is possible that the dermoepidermal junction damage occurring in the setting of DH may lead to the exposure of BP antigens and consequently to the development of specific autoimmunity. As the specific manifestation of CD, DH should be correlated with an increased incidence of cancer, particularly hematological malignancies, but in fact this association is still debated (Collin et al., 1996; Viljamaa et al., 2006), even if the risk may be considerable during the first 5 years from DH diagnosis. Interestingly, DH seems to carry a reduced mortality rate compared to general population with a significant lower mortality due to cerebrovascular disease and ischemic heart disease, lower incidence of hypercholesterolemia and fewer rate of smokers (Hervonen et al., 2012; Viljamaa et al., 2006).

16.5 Diagnosis

The diagnostic process of DH includes clinical, histological and immunopathological investigations. Being a low prevalent disease and heterogeneous from a clinical and immunological point of view, the diagnosis may be challenging, being further complicated by the wide spectrum of differential diagnoses. This may also result in a considerable diagnostic delay, which has been estimated to vary from 6 months to several years. It has been proposed that the delay may impact more females than males, although not yet clarified. Nevertheless, the time between DH onset and diagnosis seems to be decreasing during the last decades, probably for a higher awareness about CD and increasing prescription of diagnostic serological tests (Antiga et al., 2019; Mansikka et al., 2018; Zhang et al., 2012). In 2021 the new European Guidelines for the diagnosis and management of DH set two major diagnostic criteria to be fulfilled in order to obtain the correct diagnosis (Table 16.1), that is, compatible clinical manifestations and the positivity of direct immunofluorescence on perilesional skin biopsy, which must show granular, micrgranular o fibrillar IgA deposits at the tips of dermal papillae or at the dermoepidermal junction, or a combination of these patterns which involve both IgA1 and IgA2 (Fig. 16.4). Deposits at dermal papillae usually show a typical disposition in vertical lines, the so-called "falling snow" pattern (Dmochowski, 2003; Gornowicz-Porowska et al., 2019; Görög et al., 2021; Ko et al., 2010).

Nonspecific, ancillary findings may be detected, including granular junctional deposition of complement factor 3 (C3) and deposits of fibrinogen or fibrin, but they are nondiagnostic. Notably, positivity of direct immunofluorescence microscopy may occur in CD patients with a healthy skin (Cannistraci et al., 2007) or with non-DH skin diseases (Bonciolini et al., 2018). Direct immunofluorescence represents the gold-standard procedure for DH diagnosis, but in case of repeated negativity and compatible clinical manifestations, the diagnosis can be made with the support of a combination of the following minor criteria: compatible histology; a positive serological test (tTG, eTG or EMA) of high value; CD-compatible duodenal biopsy; compatible HLA testing; positive iodine patch test or oral iodine challenge; response to dapsone (partial recovery in 1 week of 100 mg daily); response to a long term GFD.

TABLE 16.1 Major and minor criteria for the diagnosis of DH.

Major criteria
Compatible clinical manifestations
Perilesional direct immunofluorescence showing granular, microgranular or fibrillar IgA deposits at the tips of dermal papillae or at the DEJ or a combination of these patterns
Minor criteria
Compatible histology
Positive serological test of high value (tTG, eTG or EMA)
CD-compatible duodenal biopsy
Compatible HLA testing
Positive iodine patch test or oral iodine challenge
Response to dapsone (partial recovery in 1 week of 100 mg daily)
Response to a long term GFD

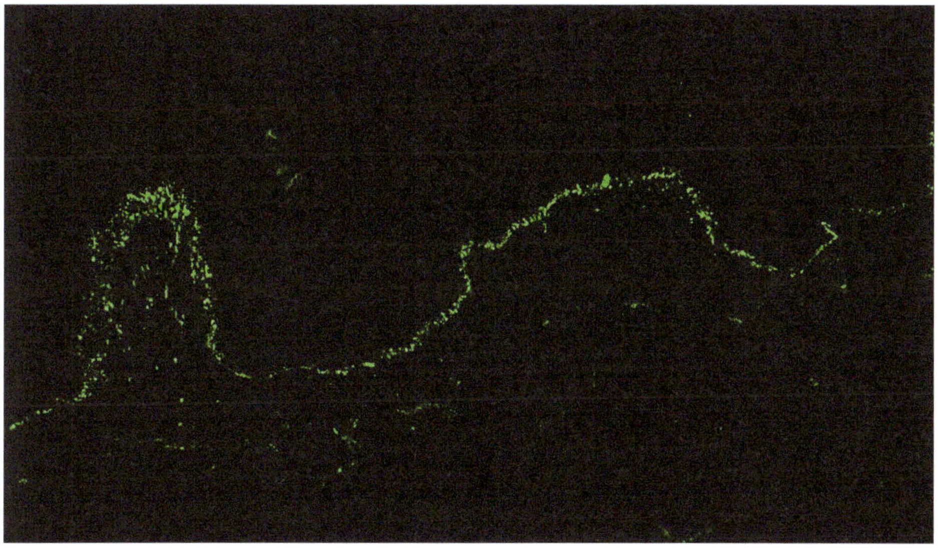

FIGURE 16.4 Typical DH perilesional direct immunofluorescence microscopy. Direct immunofluorescence microscopy showing granular deposits of IgA with reinforcing at the tip of dermal papillae.

16.5.1 Histology

For histological examination, a 4–5 mm lesional punch biopsy is recommended. The pathological hallmarks are accumulation of neutrophils at the papillary tips, resulting in the so-called neutrophilic microabscesses. Edema of the dermal papillae is frequent and subepidermal vesicles or blisters may be found, filled with fibrinogen and polymorphonuclear cells. Variable dermal inflammatory infiltrate may also be present, including neutrophils, eosinophils, and lymphocytes. However, in up to one third of patients, histological findings are nonspecific (Mendes et al., 2013; Warren & Cockerell, 2002).

16.5.2 Serology

Serological analyses are recommended for the first diagnosis of DH when direct immunofluorescence is inconclusive or it cannot be performed, for differential diagnosis or to monitor dietary adherence during follow-up. Circulating IgA antibodies directed against eTG, tTG, and antiendomysium antibodies (EMA) can be detected through indirect immunofluorescence (IIF) microscopy or ELISA. In CD patients with selective IgA deficiency, IgG-based assays may be useful, but since DH is an IgA-mediated disease the use of IgG testing does not provide its diagnosis. Anti-tTG antibodies are considered the most useful marker in CD and high serum concentrations correlate with more severe villous atrophy in both DH and CD (Dahlbom et al., 2010). Anti-tTG alone are not sufficient for the diagnosis of DH, while negative results do not exclude it. EMA can be detected through IIF with a high sensitivity (52%–100%) and specificity (100%) for DH, and they are recommended as a confirmation test for anti-tTG (Alonso-Llamazares et al., 2007; Antiga & Caproni, 2015; Hull et al., 2008; Jaskowski et al., 2009; Lytton et al., 2013). Circulating antibodies against eTG can be found in most DH patients, showing a sensitivity of 52%–100% and a specificity higher than 90%. The sensitivity of anti-eTG sometimes may be higher than the one of anti-tTG (Rose et al., 2009), however, eTG positivity alone is not sufficient for DH diagnosis as it may be positive also in CD cases without skin manifestations (Ziberna et al., 2021). Finally, despite the good sensitivity, antibodies against gliadin or deamidated gliadin are not currently recommended as a diagnostic test for DH since they may be detected also in healthy people or subjects with different conditions.

16.5.3 Small bowel biopsy

According to the European guidelines, a small bowel biopsy is recommended in DH patients to evaluate the degree of enteropathy. At least four specimens from the distal duodenum and one from the duodenal bulb should be taken in adults (see also Chapter 9), while in children with levels of serum tTG antibodies 10-folds above the normal limit and EMA positivity from a separate blood sample, the enteropathy can be diagnosed without biopsy. The histological evaluation of the small intestine should not be conceived as a tool to assess the need for GFD, since DH diagnosis is sufficient to confirm the gluten dependency of the disease, even if enteropathy appears absent in a 10%–30% proportion of DH patients (Collin et al., 2017; Salmi & Hervonen, 2020). The possibility that milder degrees of enteropathy may have been reported as negative should be considered. Small bowel biopsy may also bring clinical information in case of symptoms suggestive of malabsorption, but the grade of enteropathy does not impact the prognosis of DH.

16.5.4 HLA haplotypes genotyping

HLA-DQ2/DQ8 typing is not recommended to be used as a first line in the diagnosis of DH for its low specificity. In fact, all patients with CD carry one of the haplotypes along with a great proportion of the general population. However, HLA testing has a high negative predictive value (99%) so it may serve the purpose of excluding DH diagnosis (Salmi et al., 2016), although it is known that in some populations HLA typing may turn out negative due to different, nondetected polymorphisms. Moreover, in the last years other haplotypes have been hypothesized to predispose to the development of CD and DH, such as HLA-DQ7 or -DQ9 (Araya et al., 2015; Bodd et al., 2012; Ohata et al., 2012; Ramakrishna et al., 2021).

16.6 Differential diagnosis

Many conditions characterized by itching, excoriated, blistering or eroded rash may be included among the differential diagnosis of DH. Therefore a diagnosis of DH may be considered when facing primarily autoimmune bullous diseases. Moreover, due to its polymorphic presentation, it could be misdiagnosed as other chronic non autoimmune pruritic dermatoses (atopic dermatitis, scabies, papular urticaria, nodular or subacute prurigo). The main differential diagnoses of DH are listed below (Tables 16.2 and 16.3).

TABLE 16.2 Differential diagnosis of DH. DEJ = Dermo-epidermal junction; BMZ = basement membrane zone.

Condition	Clinical Manifestations	Histologic report	Direct immunofluorescence microscopy	Screening for celiac disease	Response to GFD diet
Dermatitis Herpetiformis	Papules, excoriated and/or crusted vesicles with a typical distribution	Neutrophilic micro abscesses at the papillary tips	Granular, microgranular or fibrillar IgA deposits at the tips of dermal papillae and/or at the DEJ	Positive	Proved
Linear IgA dermatosis	Annular blisters with a "string of pearls" configuration	Neutrophilic infiltrate	Predominant or exclusive linear deposition of IgA at the DEJ	Negative	Unproven
Bullous pemphigoid	Bullous, vesicular, or urticarial lesions in elderly	Eosinophilic infiltrate	Predominant or exclusive linear deposition of IgG at the DEJ	Negative	Unproven
Epidermolysis bullosa acquisita	Vesicular-bullous lesions at the site of trauma	Mixed infiltrate	Linear deposition of IgG or IgA at the DEJ	Negative	Unproven
Bullous systemic lupus erythematosus	From small clusters of vesicles to large, tense blisters	Interface dermatitis with dermo-epidermal detachment	IgG and IgM deposits at the BMZ	Negative	Unproven
Pemphigus herpetiformis	Vesicular-bullous lesions with arcuate, urticarial and circinate plaques	Focal intraepidermal split with acantholysis and eosinophilic spongiosis	Intraepidermal IgG and C3 deposits with an intercellular pattern	Negative	Unproven
Atopic dermatitis	Itchy grouped papules and vesicles	Eosinophilic spongiosis	Nonconclusive	Negative	Unproven
Scabies	Multiple erythematous and excoriated micro papules	Eosinophilic infiltrate	Nonconclusive	Negative	Unproven

(Continued)

TABLE 16.2 (Continued)

Condition	Clinical Manifestations	Histologic report	Direct immunofluorescence microscopy	Screening for celiac disease	Response to GFD diet
Papular urticarial	Itchy erythematous papules on exposed areas of the skin	Eosinophilic infiltrate	Nonconclusive	Negative	Unproven
Nodular or subacute prurigo	Excoriated non grouped papules or nodules	Eosinophilic infiltrate	Nonconclusive	Negative	Unproven
Nonceliac gluten sensitivity	Pruritus generally associated with excoriated vesicles	Eosinophilic infiltrate	Granular C3 deposits at the DEJ	Negative	Observed

TABLE 16.3 Conditions entering in the differential diagnosis of DH in adults with hallmarks for their clinical recognition.

Condition	Clinical Manifestations	Screening for CD	Response to GFD diet
Dermatitis Herpetiformis	Type of lesions Erythema, papules, excoriated and/or crusted vesicles, pustules Distribution Grouped among the extensor surfaces of the limbs, buttocks, lumbosacral region	Positive	Proved
Linear IgA dermatosis	Type of lesions Annular blisters with a "string of pearls" configuration Distribution Variable, with a predilection for the extensor extremities, the face and the trunk	Negative	Unproven
Bullous pemphigoid	Type of lesions Erythema, urticaria, bullae, vesicles Distribution Initial involvement of the lower abdomen and proximal regions of the limbs, then extension to the entire skin	Negative	Unproven
Epidermolysis bullosa acquisita	Type of lesions Vesicles, bullae Distribution Variable, with a predilection for the site of repeated traumatism	Negative	Unproven
Bullous systemic lupus erythematosus	Type of lesions Vesicles, blisters, erosion Distribution Any site can be involved	Negative	Unproven
Pemphigus herpetiformis	Type of lesions Vesicles-bullae and arcuate, urticarial and circinate plaques Distribution Extensor surfaces of the limbs, the face, the abdomen and the buttocks	Negative	Unproven
Atopic dermatitis	Type of lesions Vesicles, papules, lichen simplex Distribution Variable according to age with a predilection for the flexor regions of the limbs in adults	Negative	Unproven

(Continued)

TABLE 16.3 (Continued)

Condition	Clinical Manifestations	Screening for CD	Response to GFD diet
Scabies	Type of lesions Multiple erythematous and excoriated micropapules Distribution Variable, with a predilection for sides and webs of the fingers, wrists, peri areolar skin and male genitalia	Negative	Unproven
Papular urticarial	Type of lesions Erythematous papules Distribution Area of the skin noncovered by the clothes	Negative	Unproven
Nodular or subacute prurigo	Type of lesions Excoriated papules or nodules Distribution Variable, with a predilection for the trunk and the limbs. Nongrouped	Negative	Unproven
Nonceliac gluten sensitivity	Type of lesions Excoriated vesicles Distribution Extensor surfaces of the limbs, with predilection for elbows and dorsum of the hands, followed by knees, buttocks, chest and neck	Negative	Observed

16.6.1 Linear IgA dermatosis

Linear IgA bullous dermatosis (LABD) is a rare idiopathic or drug induced autoimmune blistering disease occurring with two peaks of incidence according to age: one in adults aged 60 and one during childhood. Its hallmark is represented by the predominant or exclusive linear deposition of IgA at the dermoepidermal junction. In children LABD classically present with widespread annular blisters with a predilection for the lower abdomen, thighs and groin and for perioral and anogenital regions (Fortuna & Marinkovich, 2012). In adults, annular lesions are less common, and the extensor extremities, face and trunk are frequently affected making it difficult to distinguish LABD from DH. However, the "string of pearls" configuration of the vesicles, the possible occurrence of large blisters and the involvement of the oral mucous, characteristics frequently found in LABD (Fortuna & Marinkovich, 2012; Wojnarowska et al., 1988) may help in differentiating the two conditions. Moreover, forms of LABD with a subacute prurigo-like appearance have been reported (Antiga et al., 2012; Caproni et al., 2013; Torchia et al., 2006, 2007). Although histology may be similar in DH and LABD, the distinct immunopathologic findings in LABD characterized by the finding at direct immunofluorescence microscopy of linear IgA deposits along the basement membrane (Shin et al., 2021) and the absence of an associated gluten-sensitive enteropathy allow a correct differential diagnosis.

16.6.2 Bullous pemphigoid

Bullous pemphigoid (BP) may mimic DH especially when presenting as vesicular, urticaria-like or eczema-like variants (Fig. 16.5). However, BP usually occurs in the elderly and is not associated with CD. From an histologic point of view, BP is characterized by a lymphocytic and eosinophilic infiltrate in the absence of the typical neutrophilic microabscesses at the top of the dermal papillae (Mendes et al., 2013) of DH. The diagnosis is confirmed by direct immunofluorescence microscopy revealing linear deposits of IgG and C3 at the dermal-epidermal junction in BP, while ELISA test shows circulating anti-BP180 and/or BP230 antibodies (Miyamoto et al., 2019).

16.6.3 Epidermolysis bullosa acquisita

Epidermolysis bullosa acquisita (EBA) may mimic the configuration of DH lesions in the classic variant where vesicles and blisters are displayed at the sites of trauma but generally lack the symmetrical distribution of DH. The rarer BP-like presentation of EBA may also recall the clinical picture of DH but there is usually associated mucous membrane

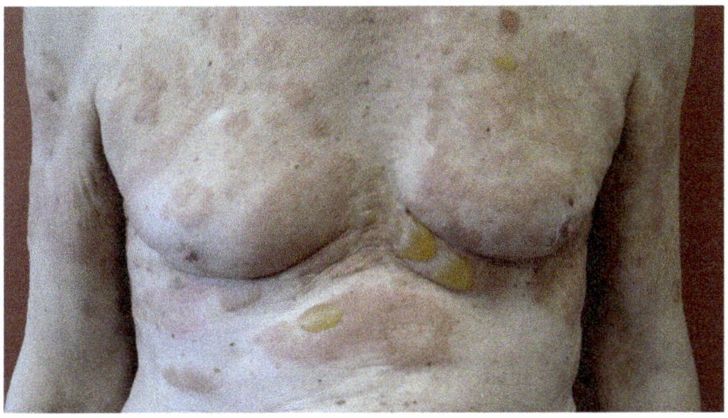

FIGURE 16.5 Bullous pemphigoid. Clinical picture of a man with tense bullous lesions with serous content above urticaroid-erythematous lesions.

involvement, notably of the oral cavity, generally spared in DH. Moreover, unlike DH, vesiculobullous lesions of EBA heal leaving often superficial scars and milia. Eventually, IIF microscopy on salt-split skin displays linear IgG deposits along the floor of the bulla and ELISA reveals the presence of autoantibodies against collagen VII, which is the target autoantigen in EBA (Sticherling & Erfurt-Berge, 2012).

16.6.4 Bullous systemic lupus erythematosus

Bullous systemic lupus erythematosus (BLSE) is a rare presentation of lupus erythematosus occurring in SLE patients that may share clinical and histopathological features with DH. However, DH can be differentiated by the presence of laboratory markers of gluten sensitivity and the absence of serum immunological findings such as antinuclear and other autoantibodies. Furthermore, BSLE is hallmarked by IgG and IgM deposits at the basement membrane zone (BMZ), allowing this condition to be easily differentiated from DH, where only IgA deposits are detected (Barbosa et al., 2011). Similarly to EBA, IIF on salt-split skin displays dermal binding of IgG anti-BMZ antibodies and ELISA is positive for autoantibodies against collagen type VII.

16.6.5 Pemphigus herpetiformis

Pemphigus herpetiformis might be considered another possible differential diagnosis of DH although extremely rare. It is clinically characterized by a vesiculobullous eruption with arcuate, urticarial and circinate plaques with peripheral vesicles mainly involving the trunk. Skin biopsies demonstrate focal intraepidermal split with acantholysis, eosinophilic spongiosis and intraepidermal IgG and C3 deposits with an intercellular pattern using direct immunofluorescence microscopy, differentiating this condition from DH (Kridin, 2018).

16.6.6 Atopic dermatitis

Atopic dermatitis and eczema may present with itchy grouped papulo-vesicular lesions entering in the differential diagnosis of DH. While the localization of atopic dermatitis involves the flexor regions of the limbs in adolescents and adults, the differential diagnosis can be challenging in children where according to age, AD may involve the extensor regions, generally interested by DH. Histology showing a prominent eosinophilic rather than neutrophilic infiltrate may help in differentiating these conditions.

16.6.7 Scabies

The infestation of the skin by the mite *Sarcoptes scabiei*, known as scabies, is predominantly characterized by pruritus, often severe and usually worse at night. Typical cutaneous findings are multiple small, erythematous papules, often excoriated, involving sides and webs of the fingers, flexor aspects of the wrists, extensor aspects of the elbows, peri areolar skin and male genitalia. Burrows may be visible as thin serpiginous lines when not masked by excoriation or secondary infection. Lesions are usually more scattered than DH except in cases of crusted scabies, where crusted lesions may localize on the extensors parts of the limbs leading to misdiagnosis. Therefore because of its high prevalence, scabies must be excluded when considering a diagnosis of DH.

16.6.8 Papular urticaria

Papular urticaria is a hypersensitivity disorder characterized by recurrent and sometimes chronic itchy papules on exposed areas of skin after insects' bites. Children are mainly involved and underwear, genital, perianal and axillary regions are usually spared. These lesions may be urticarial at the start then becoming persistent and papular and/or nodular with time. They are generally discrete, differentiating from DH lesions that are grouped.

16.6.9 Nodular or subacute prurigo

Prurigo may present with excoriations in its subacute phase entering in differential diagnosis with DH, however the lesions are generally distinct and not grouped as in the latter and there is not an association with CD.

16.6.10 Nonceliac gluten sensitivity

Nonceliac gluten sensitivity (NCGS) is one of the most recently described gluten-related disorders (see Chapter 17). Among the wide spectrum of clinical manifestations, several types of cutaneous lesions have been associated with NCGS, including eczematous dermatitis with or without excoriated vesicles, psoriasiform dermatitis and urticaria-angioedema (Faina et al., 2021). Generalized itching in the absence of primitive skin lesions was also described. When present, the lesions usually arise on extensor surfaces, with predilection for elbows and dorsum of the hands, followed by knees, buttocks, chest and neck (Bonciolini et al., 2015). In cases with excoriated papulovesicular lesions, the clinical picture may resemble DH, but perilesional skin direct immunofluorescence is negative, and the pathological picture is often nonspecific. Interestingly, up to 80% of the patients may show junctional granular C3 deposition, a pattern which has been suggested as a potential marker of "cutaneous gluten sensitivity." Moreover, as well as for the intestinal symptoms of NCGS, the associated dermatological manifestations usually go through a faster resolution following GFD compared to DH, sometimes even within a month (Graziano & Rossi, 2018).

16.7 Treatment

The only causative treatment option for DH is a lifelong GFD; therefore it should be adopted after an appropriate confirmatory diagnosis of DH is made. As GFD may take time to demonstrate its effects, other treatments can be associated until clinical remission is reached. Upon strong consensus rather than randomized clinical trials, the drug of first choice is dapsone. Due to its adverse effects, other treatments (sulfasalazine, cyclosporin A, azathioprine, colchicine, tetracyclines, nicotinamide, and mycophenolate) may be preferred in some circumstances. Topical corticosteroids may be associated but are not effective alone in controlling the cutaneous manifestations. Oral antihistamines may also serve the purpose of reducing pruritus in the acute phase. Lately, biologics have also been addressed among the treatments that may be associated to GFD in difficult-to-treat cases or in case of contraindications to conventional therapies. As a few drugs have been implicated in the development/worsening of DH manifestations (Hassan et al., 2007; Mochel et al., 2016; Sonmezoglu Marakli et al., 2008; Yu et al., 2006), a close time-relationship between drug introduction and cutaneous manifestations may lead to drug discontinuation with potential benefits. Patients should be informed of the existence of associations and reference centers where recommendations related to this chronic condition and its management are provided.

16.7.1 Gluten free diet

On GFD the resolution of the skin lesions is usually slow, and it may take up to 2 years while the disappearance of the deposits of IgA and eTG from the papillary dermis may take even longer. On the contrary, gastrointestinal symptoms usually resolve within a few weeks, with the disappearance of circulating tTG autoantibodies in 1–3 years (Reunala et al., 2021). Further benefit of a proper GFD is the prevention of CD associated manifestations, including osteoporosis and gastro-intestinal T lymphoma (see also Chapters 7 and 14).

16.7.2 Dapsone

Dapsone (4,4-diaminodiphenyl sulfone) is an antibiotic and antiinflammatory compound commonly used for the treatment of neutrophil-mediated dermatoses (Urosevic-Maiwald et al., 2013), in fact its efficacy relies on its effects on neutrophil function and recruitment. It is often needed in patients with extensive rash and/or intense itch and its beneficial effects are usually seen within 3–4 days with the clearance of the skin rash. On the contrary, no effect is evidenced on enteropathy or IgA deposits in the skin. Oral dapsone can be administered at a starting dose of 25–50 mg/day, then adapted to the therapeutic response and to be maintained until the disease is controlled (Ingen-Housz-Oro et al., 2011). If a low starting dose is chosen, to minimize potential side effects, then the dose can be increased by 25 mg/day at weekly intervals to a maintenance dose of 100–200 mg/day (Vale et al., 2019). In the maintenance phase 0,5–1 mg/kg/day can generally control the disease. When using dapsone hematological side effects may occur, including hemolysis, methemoglobinemia and agranulocytosis. Hence, blood cell count should be checked weekly for the first month, monthly for the next 5 months, and semiannually thereafter. Patients with a glucose-6-phosphate deficiency (G6PD) are more likely to experience dose-related adverse effects of dapsone, thus G6PD activity should be assessed before treatment (Antiga et al., 2019). In patients with comorbidities or abnormal laboratory values, more frequent follow-ups, dose reduction of dapsone or interruption of therapy should be considered (Görög et al., 2021). Other less frequent side effects include a sensory and motor neuropathy, toxic hepatitis and cholestatic jaundice, and hypoalbuminemia. A chemistry profile should be repeated every 6 months to monitor for possible hepatotoxicity, changes in renal function, and hypoalbuminemia (Egan and Zone, 1999). Idiosyncratic side effects may present as psychiatric symptoms, an infectious mononucleosis-like syndrome, exfoliative dermatitis, erythema multiforme, erythema nodosum, and urticaria. Dapsone hypersensitivity syndrome is an idiosyncratic adverse reaction that manifests in the clinical triad of fever, rash and internal organ involvement, the last of which may be fatal. In a similar fashion, unpredictable agranulocytosis can occur (Antiga et al., 2019).

16.7.3 Corticosteroids

The application of topical super potent corticosteroids may help in alleviating pruritus. However, these agents alone are not sufficient for the management of DH and should be used in conjunction with dapsone and a GFD. Systemic glucocorticoids are usually ineffective for DH.

16.7.4 Alternatives to dapsone

Alternative drugs to dapsone are represented by sulphonamides, such as sulfasalazine. The combination therapy with dapsone and sulfasalazine may be particularly helpful in patients who do not tolerate high doses of dapsone monotherapy. Several other pharmacologic treatments have been proposed for DH, with case reports showing that cyclosporin A, azathioprine, colchicine, tetracyclines, nicotinamide, and mycophenolate may be used to control the acute rash in DH patients. Among them, colchicine has been proposed as a second choice after dapsone, due to its antineutrophilic activity (Görög et al., 2021).

16.7.5 Biologics and small molecules

Biological drugs use in DH has been advocated since 2017 when the correlation between IL-36 and IL-17 and titers of specific antibodies was studied in DH, BV, and PV (Żebrowska et al., 2017). Expression of IL-36 and IL-17 was detected in the serum of all DH, BP and PV samples, and their levels were statistically higher compared to the healthy control group. Moreover, IL-36a levels were statistically higher in DH patients, in comparison to BP and PV patients. IL-36a can up-regulate IL-17a, IL-23 and TNFα expression, building a pro-inflammatory network in inflammatory diseases. This study showed that IL-36 may be helpful in the diagnostic and monitoring of the disease and could be considered as a target for biological therapy (Shin et al., 2021). Recently, the off-label use of JAK inhibitors in inflammatory dermatosis, including DH, has been reported showing positive outcomes (Kahn et al., 2021).

References

Aine, L. (1996). Coeliac-type permanent-tooth enamel defects. *Annals of Medicine*, *28*(1), 9–12. Available from https://doi.org/10.3109/07853899608999067.

Aine, L., Maki, M., & Reunala, T. (1992). Coeliac-type dental enamel defects in patients with dermatitis herpetiformis. *Acta Dermato-Venereologica*, *72*(1), 25–27.

Airola, K., Vaalamo, M., Reunala, T., & Saarialho-Kere, U. K. (1995). Enhanced expression of interstitial collagenase, stromelysin-1, and urokinase plasminogen activator in lesions of dermatitis herpetiformis. *Journal of Investigative Dermatology*, *105*(2), 184–189. Available from https://doi.org/10.1111/1523-1747.ep12317093.

Alonso-Llamazares, J., Gibson, L. E., & Rogers, R. S. (2007). Clinical, pathologic, and immunopathologic features of dermatitis herpetiformis: Review of the Mayo Clinic experience? *International Journal of Dermatology*, *46*(9), 910–919. Available from https://doi.org/10.1111/j.1365-4632.2007.03214.x.

Antiga, E., Bellandi, S., Bianchi, B., Del Bianco, E., Pierini, I., Cozzani, E., Massi, D., Fabbri, P., & Caproni, M. (2012). A further case of subacute prurigo-like linear IgA bullous dermatosis: Growing evidence of a new subset. *International Journal of Dermatology*, *51*(12), 1500–1501. Available from https://doi.org/10.1111/j.1365-4632.2011.05138.x.

Antiga, E., Bonciolini, V., Cazzaniga, S., Alaibac, M., Calabrò, A. S., Cardinali, C., Cozzani, E., Marzano, A. V., Micali, G., Not, T., Quaglino, P., Vassallo, C., Naldi, L., & Caproni, M. (2019). Female patients with Dermatitis Herpetiformis show a reduced diagnostic delay and have higher sensitivity rates at autoantibody testing for celiac disease. *BioMed Research International*, *2019*, 1–7. Available from https://doi.org/10.1155/2019/6307035.

Antiga, E., & Caproni, M. (2015). The diagnosis and treatment of dermatitis herpetiformis. *Clinical, Cosmetic and Investigational Dermatology*, *8*, 257. Available from https://doi.org/10.2147/CCID.S69127.

Antiga, E., Maglie, R., Quintarelli, L., Verdelli, A., Bonciani, D., Bonciolini, V., & Caproni, M. (2019). Dermatitis Herpetiformis: Novel perspectives. *Frontiers in Immunology*, *10*. Available from https://doi.org/10.3389/fimmu.2019.01290.

Antiga, E., Quaglino, P., Pierini, I., Volpi, W., Lami, G., Bianchi, B., Del Bianco, E., Renzi, D., Baroni, G., Novelli, M., Ponti, R., Papini, M., Di Lollo, S., Calabrò, A. S., Fabbri, P., & Caproni, M. (2015). Regulatory T cells as well as IL-10 are reduced in the skin of patients with dermatitis herpetiformis. *Journal of Dermatological Science*, *77*(1), 54–62. Available from https://doi.org/10.1016/j.jdermsci.2014.11.003.

Araya, M., Oyarzun, A., Lucero, Y., Espinosa, N., & Pérez-Bravo, F. (2015). DQ2, DQ7 and DQ8 distribution and clinical manifestations in celiac cases and their first-degree relatives. *Nutrients*, *7*(6), 4955–4965. Available from https://doi.org/10.3390/nu7064955.

Barbosa, W. S., Rodarte, C. M., Guerra, J. G., Maciel, V. G., Fleury Júnior, L. F. F., & Costa, M. B. (2011). Lúpus eritematoso sistêmico bolhoso: Diagnóstico diferencial com dermatite herpetiforme. *Anais Brasileiros de Dermatologia*, *86*(4 suppl 1), 92–95. Available from https://doi.org/10.1590/S0365-05962011000700024.

Bizzarro, A., Valentini, G., Martino, G. D., Daponte, A., De Bellis, A., & Iacono, G. (1987). Influence of testosterone therapy on clinical and immunological features of autoimmune diseases associated with Klinefelter's Syndrome. *The Journal of Clinical Endocrinology & Metabolism*, *64*(1), 32–36. Available from https://doi.org/10.1210/jcem-64-1-32.

Bodd, M., Tollefsen, S., Bergseng, E., Lundin, K. E. A., & Sollid, L. M. (2012). Evidence that HLA-DQ9 confers risk to celiac disease by presence of DQ9-restricted gluten-specific T cells. *Human Immunology*, *73*(4), 376–381. Available from https://doi.org/10.1016/j.humimm.2012.01.016.

Bonciani, D., Quintarelli, L., Del Bianco, E., Bianchi, B., & Caproni, M. (2017). Serum levels and tissue expression of interleukin-31 in dermatitis herpetiformis and bullous pemphigoid. *Journal of Dermatological Science*, *87*(2), 210–212. Available from https://doi.org/10.1016/j.jdermsci.2017.04.008.

Bonciolini, V., Antiga, E., Bianchi, B., Del Bianco, E., Ninci, A., Maio, V., Pimpinelli, N., & Caproni, M. (2018). Granular IgA deposits in the skin of patients with Coeliac disease: Is it always dermatitis herpetiformis? *Acta Dermato-Venereologica*, *99*(1), 78–83. Available from https://doi.org/10.2340/00015555-3001.

Bonciolini, V., Bianchi, B., Del Bianco, E., Verdelli, A., & Caproni, M. (2015). Cutaneous manifestations of non-celiac gluten sensitivity: Clinical histological and immunopathological features. *Nutrients*, *7*(9), 7798–7805. Available from https://doi.org/10.3390/nu7095368.

Cannistraci, C., Lesnoni La Parola, I., Cardinali, G., Bolasco, G., Aspite, N., Stigliano, V., & Picardo, M. (2007). Co-localization of IgA and TG3 on healthy skin of coeliac patients. *Journal of the European Academy of Dermatology and Venereology*, *21*(4), 509–514. Available from https://doi.org/10.1111/j.1468-3083.2006.02017.x.

Caproni, M., Bonciani, D., & Fabbri, P. (2013). Prurigo even as a symptom of a linear IgA bullous disease. *Journal of the European Academy of Dermatology and Venereology*, *27*(6), 794–795. Available from https://doi.org/10.1111/j.1468-3083.2012.04645.x.

Caproni, M., Capone, M., Rossi, M. C., Santarlasci, V., Maggi, L., Mazzoni, A., Rossettini, B., Renzi, D., Quintarelli, L., Bianchi, B., Ninci, A., Lami, G., Calabrò, A., Cosmi, L., Annunziato, F., & Liotta, F. (2021). T cell response toward tissue-and epidermal-transglutaminases in coeliac disease patients developing Dermatitis Herpetiformis. *Frontiers in Immunology*, *12*. Available from https://doi.org/10.3389/fimmu.2021.645143.

Caproni, M., Feliciani, C., Fuligni, A., Salvatore, E., Atani, L., Bianchi, B., Mohammad Pour, S., Proietto, G., Toto, P., Coscione, G., Amerio, P., & Fabbri, P. (1998). Th2-like cytokine activity in dermatitis herpetiformis. *British Journal of Dermatology*, *138*(2), 242–247. Available from https://doi.org/10.1046/j.1365-2133.1998.02068.x.

Cenit, M., Olivares, M., Codoñer-Franch, P., & Sanz, Y. (2015). Intestinal microbiota and celiac disease: Cause, consequence or co-evolution? *Nutrients*, *7*(8), 6900–6923. Available from https://doi.org/10.3390/nu7085314.

Collin, P., Pukkala, E., & Reunala, T. (1996). Malignancy and survival in dermatitis herpetiformis: A comparison with coeliac disease. *Gut*, *38*(4), 528–530. Available from https://doi.org/10.1136/gut.38.4.528.

Collin, P., Reunala, T., Pukkala, E., Laippala, P., Keyriläinen, O., & Pasternack, A. (1994). Coeliac disease - Associated disorders and survival. *Gut*, *35*(9), 1215–1218. Available from https://doi.org/10.1136/gut.35.9.1215.

Collin, P., Salmi, T. T., Hervonen, K., Kaukinen, K., & Reunala, T. (2017). Dermatitis herpetiformis: A cutaneous manifestation of coeliac disease. *Annals of Medicine*, *49*(1), 23–31. Available from https://doi.org/10.1080/07853890.2016.1222450.

Cook, L., Munier, C. M. L., Seddiki, N., van Bockel, D., Ontiveros, N., Hardy, M. Y., Gillies, J. K., Levings, M. K., Reid, H. H., Petersen, J., Rossjohn, J., Anderson, R. P., Zaunders, J. J., Tye-Din, J. A., & Kelleher, A. D. (2017). Circulating gluten-specific FOXP3 + CD39 + regulatory T cells have impaired suppressive function in patients with celiac disease. *Journal of Allergy and Clinical Immunology*, *140*(6), 1592–1603. Available from https://doi.org/10.1016/j.jaci.2017.02.015, e8.

Cynkier, A., Zebrowska, A., Wagrowska-Danilewicz, M., Danilewicz, M., Erkiert-Polguj, A., Stasikowska-Kanicka, O., Sysa-Jedrzejowska, A., & Waszczykowska, E. (2012). Expression of selected neuropeptides in pathogenesis of bullous pemphigoid and dermatitis herpetiformis. *Polish Journal of Pathology*, *63*(1), 31–39. Available from http://www.termedia.pl/Journal/-55/pdf-18480-10?filename = Expression%20of%20selected.pdf.

Dahlbom, I., Korponay-Szabó, I. R., Kovács, J. B., Szalai, Z., Mäki, M., & Hansson, T. (2010). Prediction of clinical and mucosal severity of coeliac disease and Dermatitis Herpetiformis by quantification of IgA/IgG serum antibodies to tissue transglutaminase. *Journal of Pediatric Gastroenterology & Nutrition*, *50*(2), 140–146. Available from https://doi.org/10.1097/mpg.0b013e3181a81384.

Dmochowski, M. (2003). On patterns of IgA depositis in the skin of patients with Dermatitis Herpetiformis. *Advances in Dermatology and Allergology/Postępy Dermatologii i Alergologii*, *20*(1), 46–48.

Douglas, W. S., & Alexander, J. O. D. (1975). Dermatitis herpetiformis, iodine compounds and thyrotoxicosis. *British Journal of Dermatology*, *92*(5), 596–598. Available from https://doi.org/10.1111/j.1365-2133.1975.tb03134.x.

Economopoulou, P., & Laskaris, G. (1986). Dermatitis herpetiformis: Oral lesions as an early manifestation. *Oral Surgery, Oral Medicine, Oral Pathology*, *62*(1), 77–80. Available from https://doi.org/10.1016/0030-4220(86)90074-5.

Egan, C. A., & Zone, J. J. (1999). Linear IgA bullous dermatosis. *International Journal of Dermatology*, *38*(11), 818–827. Available from https://doi.org/10.1046/j.1365-4362.1999.00813.x.

Faina, V., Paolino, G., Bavastrelli, M., Calvieri, S., & Grieco, T. (2021). Classification of cutaneous manifestations in patients with nonceliac gluten sensitivity and wheat allergy. *Journal of the American Academy of Dermatology*, *85*(2), 469–472. Available from https://doi.org/10.1016/j.jaad.2017.12.013.

Fortuna, G., & Marinkovich, M. P. (2012). Linear immunoglobulin A bullous dermatosis. *Clinics in Dermatology*, *30*(1), 38–50. Available from https://doi.org/10.1016/j.clindermatol.2011.03.008.

Fraser, N. G., Kerr, N. W., & Donald, D. (1973). Oral lesions in dermatitis herpetiformis. *British Journal of Dermatology*, *89*(5), 439–450. Available from https://doi.org/10.1111/j.1365-2133.1973.tb03003.x.

From, E., Iodine, k, & Thomsen. (1974). Dermatitis Herpetiformis. A case provoked by iodine. *The British Journal of Dermatology*, *91*(2). Available from https://doi.org/10.1111/J.1365-2133.1974.TB15869.X.

Garioch, J. J., Baker, B. S., Leonard, J. N., & Fry, L. (1994). T lymphocytes in lesional skin of patients with dermatitis herpetiformis. *British Journal of Dermatology*, *131*(6), 822–826. Available from https://doi.org/10.1111/j.1365-2133.1994.tb08584.x.

Gornowicz-Porowska, J., Bowszyc-Dmochowska, M., & Dmochowski, M. (2019). An update on direct immunofluorescence for diagnosing dermatitis herpetiformis. *Postepy Dermatologii i Alergologii*, *36*(6), 655–658. Available from https://doi.org/10.5114/ada.2019.91415.

Görög, A., Antiga, E., Caproni, M., Cianchini, G., De, D., Dmochowski, M., Dolinsek, J., Drenovska, K., Feliciani, C., Hervonen, K., Lakos Jukic, I., Kinyó., Koltai, T., Korponay-Szabó, I., Marzano, A. V., Patsatsi, A., Rose, C., Salmi, T., Schmidt, E., & Sárdy, M. (2021). S2k guidelines (consensus statement) for diagnosis and therapy of dermatitis herpetiformis initiated by the European Academy of Dermatology and Venereology (EADV). *Journal of the European Academy of Dermatology and Venereology*, *35*(6), 1251–1277. Available from https://doi.org/10.1111/jdv.17183.

Görög, A., Németh, K., Kolev, K., Zone, J. J., Mayer, B., Silló, P., Bognár, P., & Kárpáti, S. (2016). Circulating transglutaminase 3-immunoglobulin A immune complexes in Dermatitis Herpetiformis. *Journal of Investigative Dermatology*, *136*(8), 1729–1731. Available from https://doi.org/10.1016/j.jid.2016.03.039.

Graeber, M., Baker, B. S., Garioch, J. J., Valdimarsson, H., Leonard, J. N., & Fry, L. (1993). The role of cytokines in the generation of skin lesions in dermatitis herpetiformis. *British Journal of Dermatology*, *129*(5), 530–532. Available from https://doi.org/10.1111/j.1365-2133.1993.tb00479.x.

Graziano, M., & Rossi, M. (2018). An update on the cutaneous manifestations of coeliac disease and non-coeliac gluten sensitivity. *International Reviews of Immunology*, *37*(6), 291–300. Available from https://doi.org/10.1080/08830185.2018.1533008.

Hall, R. P., Benbenisty, K. M., Mickle, C., Takeuchi, F., & Streilein, R. D. (2007). Serum IL-8 in patients with Dermatitis Herpetiformis is produced in response to dietary gluten. *Journal of Investigative Dermatology*, *127*(9), 2158–2165. Available from https://doi.org/10.1038/sj.jid.5700929.

Hall, R. P., Takeuchi, F., Benbenisty, K. M., & Streilein, R. D. (2006). Cutaneous endothelial cell activation in normal skin of patients with Dermatitis Herpetiformis associated with increased serum levels of IL-8, sE-selectin, and TNF-α. *Journal of Investigative Dermatology*, *126*(6), 1331–1337. Available from https://doi.org/10.1038/sj.jid.5700277.

Hassan, S., Dalle, S., Descloux, E., Balme, B., & Thomas, L. (2007). Dermatitis herpetiformis associated with progesterone contraception. *Annales de Dermatologie et de Venereologie*, *134*, 89198. Available from https://doi.org/10.1016/S0151-9638(07.

Hervonen, K., Alakoski, A., Salmi, T. T., Helakorpi, S., Kautiainen, H., Kaukinen, K., Pukkala, E., Collin, P., & Reunala, T. (2012). Reduced mortality in dermatitis herpetiformis: A population-based study of 476 patients. *British Journal of Dermatology*, *167*(6), 1331–1337. Available from https://doi.org/10.1111/j.1365-2133.2012.11105.x.

Hietikko, M., Hervonen, K., Ilus, T., Salmi, T., Huhtala, H., Laurila, K., Rauhavirta, T., Reunala, T., Kaukinen, K., & Lindfors, K. (2018). Ex vivo culture of duodenal biopsies from patients with dermatitis herpetiformis indicates that transglutaminase 3 antibody production occurs in the gut. *Acta Dermato-Venereologica*, *98*(3), 366–372. Available from https://doi.org/10.2340/00015555-2849.

Hmida, N. B., Ahmed, M. B., Moussa, A., Rejeb, M. B., Said, Y., Kourda, N., Meresse, B., Abdeladhim, M., Louzir, H., & Cerf-Bensussan, N. (2012). Impaired control of effector T cells by regulatory T cells: A clue to loss of oral tolerance and autoimmunity in celiac disease? *American Journal of Gastroenterology*, *107*(4), 604–611. Available from https://doi.org/10.1038/ajg.2011.397.

Hofmann, S. C., Nashan, D., & Bruckner-Tuderman, L. (2009). Petechiae on the fingertips as presenting symptom of dermatitis herpetiformis Duhring. *Journal of the European Academy of Dermatology and Venereology*, *23*(6), 732–733. Available from https://doi.org/10.1111/j.1468-3083.2009.03200.x.

Hull, C. M., Liddle, M., Hansen, N., Meyer, L. J., Schmidt, L., Taylor, T., Jaskowski, T. D., Hill, H. R., & Zone, J. J. (2008). Elevation of IgA anti-epidermal transglutaminase antibodies in dermatitis herpetiformis. *British Journal of Dermatology*, *159*(1), 120–124. Available from https://doi.org/10.1111/j.1365-2133.2008.08629.x.

Ingen-Housz-Oro, S., Bernard, P., Bedane, C., Prost, C., & Joly, P. (2011). Linear IgA dermatosis. Guidelines for the diagnosis and treatment. *Annales de Dermatologie et de Venereologie*, *138*(3), 267–270. Available from https://doi.org/10.1016/j.annder.2011.01.018.

Jaskowski, T. D., Hamblin, T., Wilson, A. R., Hill, H. R., Book, L. S., Meyer, L. J., Zone, J. J., & Hull, C. M. (2009). IgA anti-epidermal transglutaminase antibodies in Dermatitis Herpetiformis and pediatric celiac disease. *Journal of Investigative Dermatology*, *129*(11), 2728–2730. Available from https://doi.org/10.1038/jid.2009.142.

Kahn, J. S., Moody, K., & Rosmarin, D. (2021). Significant improvement of dermatitis herpetiformis with tofacitinib. *Dermatology Online Journal*, *27*(7). Available from https://doi.org/10.5070/d327754365.

Kárpáti, S., Sárdy, M., Németh, K., Mayer, B., Smyth, N., Paulsson, M., & Traupe, H. (2018). Transglutaminases in autoimmune and inherited skin diseases: The phenomena of epitope spreading and functional compensation. *Experimental Dermatology*, *27*(8), 807–814. Available from https://doi.org/10.1111/exd.13449.

Katz, S. I., Hertz, K. C., Rogentine, G. N., & Strober, W. (1977). HLA-B8 and Dermatitis Herpetiformis in patients with IgA deposits in skin. *Archives of Dermatology*, *113*(2), 155–156. Available from https://doi.org/10.1001/archderm.1977.01640020027002.

Kauppi, S., Jokelainen, J., Timonen, M., Tasanen, K., & Huilaja, L. (2021). Atopic dermatitis is associated with Dermatitis Herpetiformis and celiac disease in children. *Journal of Investigative Dermatology*, *141*(1), 191–193. Available from https://doi.org/10.1016/j.jid.2020.05.091, e2.

Ko, C. J., Colegio, O. R., Moss, J. E., & McNiff, J. M. (2010). Fibrillar IgA deposition in dermatitis herpetiformis - An underreported pattern with potential clinical significance. *Journal of Cutaneous Pathology*, *37*(4), 475–477. Available from https://doi.org/10.1111/j.1600-0560.2009.01472.x.

Kridin, K. (2018). 'Pemphigus group: Overview, epidemiology, mortality, and comorbidities. *Immunologic Research*, *66*(2). Available from https://doi.org/10.1007/S12026-018-8986-7.

Lahteenoja, H., Irjala, K., Viander, M., Vainio, E., Toivanen, A., & Syrjanen, S. (1998). Oral mucosa is frequently affected in patients with dermatitis herpetiformis. *Archives of Dermatology*, *134*(6), 756–758. Available from https://doi.org/10.1001/archderm.134.6.756.

Lodin, A., & Stigell, P. O. (1965). Dermatitis herpetiformis. Cases observed at the dermatologic clinic, Karolinska Sjukhuset, 1954–1963. *Acta Dermato-Venereologica*, *45*(5), 355–365.

Loike, J. D., Sodeik, B., Cao, L., Leucona, S., Weitz, J. I., Detmers, P. A., Wright, S. D., & Silverstein, S. C. (1991). 'CD11c/CD18 on neutrophils recognizes a domain at the N terminus of the A alpha chain of fibrinogen. *Proceedings of the National Academy of Sciences of the United States of America*, *88*(3). Available from https://doi.org/10.1073/PNAS.88.3.1044.

Lytton, S. D., Antiga, E., Pfeiffer, S., Matthias, T., Szaflarska-Poplawska, A., Ulaganathan, V. K., Placek, W., Fabbri, P., Hall, R., & Caproni, M. (2013). Neo-epitope tissue transglutaminase autoantibodies as a biomarker of the gluten sensitive skin disease—Dermatitis herpetiformis. *Clinica Chimica Acta*, *415*, 346–349. Available from https://doi.org/10.1016/j.cca.2012.10.051.

Makino, T., Yoshihisa, Y., Mizawa, M., Tsutsui, K., Nishijima, C., Inaoki, M., & Shimizu, T. (2017). Increased serum levels of th2-type cytokines and eotaxin in fibrillar-type dermatitis herpetiformis. *Acta Dermato-Venereologica*, *97*(5), 642–643. Available from https://doi.org/10.2340/00015555-2604.

Mansikka, E., Salmi, T., Kaukinen, K., Collin, P., Huhtala, H., Reunala, T., & Hervonen, K. (2018). Diagnostic delay in Dermatitis Herpetiformis in a high-prevalence Area. *Acta Dermato Venereologica*, *98*(2), 195–199. Available from https://doi.org/10.2340/00015555-2818.

Mansikka, E., Hervonen, K., Kaukinen, K., Collin, P., Huhtala, H., Reunala, T., & Salmi, T. (2018). Prognosis of Dermatitis Herpetiformis patients with and without villous atrophy at diagnosis. *Nutrients*, *10*(5), 641. Available from https://doi.org/10.3390/nu10050641.

Mansikka, E., Hervonen, K., Salmi, T. T., Kautiainen, H., Kaukinen, K., Collin, P., & Reunala, T. (2017). The decreasing prevalence of severe villous atrophy in Dermatitis Herpetiformis: A 45-year experience in 393 patients. *Journal of Clinical Gastroenterology*, *51*(3), 235–239. Available from https://doi.org/10.1097/MCG.0000000000000533.

Marks, R., & Jones, E. W. (1971). Purpura in dermatitis herpetiformis. *British Journal of Dermatology*, *84*(4), 386–388.

McCleskey, P. E., Erickson, Q. L., David-Bajar, K. M., & Elston, D. M. (2002). Palmar petechiae in dermatitis herpetiformis: A case report and clinical review. *Cutis; Cutaneous Medicine for the Practitioner*, *70*(4), 217–223.

McGovern, T. W., & Bennion, S. D. (1994). Palmar Purpura: An atypical presentation of childhood Dermatitis Herpetiformis. *Pediatric Dermatology*, *11*(4), 319–322. Available from https://doi.org/10.1111/j.1525-1470.1994.tb00096.x.

Mendes, F. B. R., Hissa-Elian, A., Abreu, M. A. M. M. D., & Gonçalves, V. S. (2013). Review: Dermatitis herpetiformis. *Anais Brasileiros de Dermatologia*, *88*(4), 594–599. Available from https://doi.org/10.1590/abd1806-4841.20131775.

Metcalfe, W., & Boulton-Jones, J. M. (1997). Exacerbation of lupus nephritis in association with leuprorelin injection for uterine leiomyoma. *Nephrology Dialysis Transplantation*, *12*(8), 1699–1700. Available from https://doi.org/10.1093/ndt/12.8.1699.

Michaelsson, G., & Svensson, L. (1975). Abnormal cutaneous reactions in dermatitis herpetiformis. *Acta Dermato-Venereologica*, *55*(5), 351–358.

Miyagawa, S., Shirai, T., Shimamota, I., Ichijo, M., & Ueki, H. (1994). Worsening of systemic Lupus Erythematosus-associated thrombocytopenia after administration of gonadotropin-releasing hormone analog. *Arthritis & Rheumatism*, *37*(11), 1708–1709. Available from https://doi.org/10.1002/art.1780371123.

Miyamoto, D., Santi, C. G., Aoki, V., & Maruta, C. W. (2019). Bullous pemphigoid. *Anais Brasileiros de Dermatologia*, *94*(2), 133–146. Available from https://doi.org/10.1590/abd1806-4841.20199007.

Mochel, M. C., Ming, M. E., Imadojemu, S., Gangadhar, T. C., Schuchter, L. M., Elenitsas, R., Payne, A. S., & Chu, E. Y. (2016). Cutaneous autoimmune effects in the setting of therapeutic immune checkpoint inhibition for metastatic melanoma. *Journal of Cutaneous Pathology*, *43*(9), 787–791. Available from https://doi.org/10.1111/cup.12735.

Naylor, E., Atwater, A., Selim, M. A., Hall, R., & Puri, P. K. (2011). Leukocytoclastic vasculitis as the presenting feature of dermatitis herpetiformis. *Archives of Dermatology*, *147*(11), 1313–1316. Available from https://doi.org/10.1001/archdermatol.2011.293.

Nilsen, E. M., Lundin, K. E. A., Krajči, P., Scott, H., Sollid, L. M., & Brandtzaeg, P. (1995). Gluten specific, HLA-DQ restricted T cells from coeliac mucosa produce cytokines with Th1 or Th0 profile dominated by interferon γ. *Gut*, *37*(6), 766–776. Available from https://doi.org/10.1136/gut.37.6.766.

Ohata, C., Ishii, N., Hamada, T., Shimomura, Y., Niizeki, H., Dainichi, T., Furumura, M., Tsuruta, D., & Hashimoto, T. (2012). Distinct characteristics in Japanese Dermatitis Herpetiformis: A review of All 91 Japanese patients over the last 35 years. *Clinical and Developmental Immunology*, *2012*, 1–9. Available from https://doi.org/10.1155/2012/562168.

Ohshima, Y., Tamada, Y., Matsumoto, Y., & Hashimoto, T. (2003). Dermatitis herpetiformis Duhring with palmoplantar keratosis. *British Journal of Dermatology*, *149*(6), 1300–1302. Available from https://doi.org/10.1111/j.1365-2133.2003.05660.x.

Olson, M. A., Wieland, C. N., & Newman, C. C. (2015). Diffuse vesicular rash with intractable pruritus. *JAMA: The Journal of the American Medical Association*, *313*(11), 1158. Available from https://doi.org/10.1001/jama.2014.17575.

Pavlović, M. D. (2006). Dermatitis herpetiformis presenting as intertriginous dermatitis. *Acta Dermatovenerologica Alpina, Pannonica et Adriatica*, *15*(1), 52–54.

Powell, G. R., Bruckner, A. L., & Weston, W. L. (2004). Dermatitis Herpetiformis presenting as chronic urticaria. *Pediatric Dermatology*, *21*(5), 564–567. Available from https://doi.org/10.1111/j.0736-8046.2004.21509.x.

Ramakrishna, B. S., Venugopal, G., Singh, A., Pugazhendhi, S., Dutta, S., Ahuja, V., & Makharia, G. K. (2021). Human Leukocyte Antigen DQ (HLA-DQ) genotypes and haplotypes and their association with phenotype in patients with celiac disease in India. *JGH Open*, *5*(10), 1190–1196. Available from https://doi.org/10.1002/jgh3.12651.

Reitamo, S., Reunala, T., Konttinen, Y. T., Saksela, O., & Salo, O. P. (1981). Inflammatory cells, IgA, C3, fibrin and fibronectin in skin lesions in dermatitis herpetiformis. *British Journal of Dermatology*, *105*(2), 167–177. Available from https://doi.org/10.1111/j.1365-2133.1981.tb01202.x.

Reunala, T., & Collin, P. (1997). Diseases associated with dermatitis herpetiformis. *British Journal of Dermatology*, *136*(3), 315–318. Available from https://doi.org/10.1111/j.1365-2133.1997.tb14935.x.

Reunala, T., Hervonen, K., & Salmi, T. (2021). Dermatitis Herpetiformis: An update on diagnosis and management. *American Journal of Clinical Dermatology*, *22*(3), 329–338. Available from https://doi.org/10.1007/s40257-020-00584-2.

Rose, C., Armbruster, F. P., Ruppert, J., Igl, B.-W., Zillikens, D., & Shimanovich, I. (2009). Autoantibodies against epidermal transglutaminase are a sensitive diagnostic marker in patients with dermatitis herpetiformis on a normal or gluten-free diet. *Journal of the American Academy of Dermatology*, *61*(1), 39–43. Available from https://doi.org/10.1016/j.jaad.2008.12.037.

Saito, M., Böer, A., Ishiko, A., & Nishikawa, T. (2006). Atypical dermatitis herpetiformis: A Japanese case that presented with initial lesions mimicking prurigo pigmentosa. *Clinical and Experimental Dermatology*, *31*(2), 290–291. Available from https://doi.org/10.1111/j.1365-2230.2005.02009.x.

Salmi, T. T., Kurppa, K., Hervonen, K., Laurila, K., Collin, P., Huhtala, H., Saavalainen, P., Sievänen, H., Reunala, T., & Kaukinen, K. (2016). Serum transglutaminase 3 antibodies correlate with age at celiac disease diagnosis. *Digestive and Liver Disease*, *48*(6), 632–637. Available from https://doi.org/10.1016/j.dld.2016.03.003.

Salmi, T., & Hervonen, K. (2020). Current concepts of dermatitis herpetiformis. *Acta Dermato-Venereologica*, *100*(100-), 115–121. Available from https://doi.org/10.2340/00015555-3401.

Sárdy, M., Kárpáti, S., Merkl, B., Paulsson, M., & Smyth, N. (2002). Epidermal transglutaminase (TGase 3) Is the autoantigen of Dermatitis Herpetiformis. *Journal of Experimental Medicine*, *195*(6), 747–757. Available from https://doi.org/10.1084/jem.20011299.

Shin, L., Gardner, J. T., & Dao, H. (2021). Updates in the diagnosis and management of linear IgA disease: A systematic review. *Medicina*, *57*(8), 818. Available from https://doi.org/10.3390/medicina57080818.

Smith, A. D., Streilein, R. D., & Hall, R. P. (2002). Neutrophil CD11b, L-selectin and Fc IgA receptors in patients with dermatitis herpetiformis. *British Journal of Dermatology*, *147*(6), 1109–1117. Available from https://doi.org/10.1046/j.1365-2133.2002.05004.x.

Smith, J. B., Taylor, T. B., & Zone, J. J. (1992). The site of blister formation in dermatitis herpetiformis is within the lamina lucida. *Journal of the American Academy of Dermatology*, *27*(2), 209–213. Available from https://doi.org/10.1016/0190-9622(92)70172-c.

Sonmezoglu Marakli, S., Uzun, S., Ozbek, S., & Tuncer, I. (2008). Dermatitis herpetiformis in a patient receiving infliximab for ankylosing spondylitis. *European Journal of Dermatology*, *18*(1), 88–89.

Stern, R., Fishman, J., Brusman, H., & Kunkel, H. G. (1977). Systemic lupus erythematosus associated with klinefelter's syndrome. *Arthritis & Rheumatism*, *20*(1), 18–22. Available from https://doi.org/10.1002/art.1780200103.

Sticherling, M., & Erfurt-Berge, C. (2012). Autoimmune blistering diseases of the skin. *Autoimmunity Reviews*, *11*(3), 226–230. Available from https://doi.org/10.1016/j.autrev.2011.05.017.

Sun, Y., Lin, Y., Yang, B., Wang, C., Fu, X., Bao, F., Yu, G., Zhang, Q., Wang, Z., Zhou, G., Chen, X., Wu, M., Wang, G., Liu, H., Liu, J., & Zhang, F. (2016). The HLA Alleles B*0801 and DRB1*0301 are associated with Dermatitis Herpetiformis in a Chinese population. *Journal of Investigative Dermatology*, *136*(2), 530–532. Available from https://doi.org/10.1016/j.jid.2015.10.057.

Suzuki, M., Hosoda, S., Yamada, T., Komine, M., Murata, S., Yokokura, H., & Ohtsuki, M. (2014). A case of dermatitis herpetiformis with blister formation between laminin-332 and type 7 collagen. *The Journal of Dermatology*, *41*(11). Available from https://doi.org/10.1111/1346-8138.12630, n/a-n/a.

Tanriverdi, F., Silveira, L. F. G., MacColl, G. S., & Bouloux, P. M. G. (2003). The hypothalamic-pituitary-gonadal axis: Immune function and autoimmunity. *Journal of Endocrinology*, *176*(3), 293–304. Available from https://doi.org/10.1677/joe.0.1760293.

Taylor, T. B., & Zone, J. J. (2018). Sensitivity of transglutaminase 3 in the IgA Aggregates in Dermatitis Herpetiformis skin to potassium iodide. *Journal of Investigative Dermatology*, *138*(9), 2066–2068. Available from https://doi.org/10.1016/j.jid.2018.03.1497.

Taylor, T. B., Schmidt, L. A., Meyer, L. J., & Zone, J. J. (2015). Transglutaminase 3 present in the IgA aggregates in Dermatitis Herpetiformis skin is enzymatically active and binds soluble fibrinogen. *Journal of Investigative Dermatology*, *135*(2), 623–625. Available from https://doi.org/10.1038/jid.2014.368.

Torchia, D., Caproni, M., Cozzani, E., Ketabchi, S., & Fabbri, P. (2007). Subacute prurigo-like linear IgA disease. *International Journal of Dermatology*, *46*(10), 1101–1103. Available from https://doi.org/10.1111/j.1365-4632.2007.03314.x.

Torchia, D., Caproni, M., Del Bianco, E., Cozzani, E., Ketabchi, S., & Fabbri, P. (2006). Linear IgA disease presenting as prurigo nodularis. *British Journal of Dermatology*, *155*(2), 479–480. Available from https://doi.org/10.1111/j.1365-2133.2006.07315.x.

Tu, H., Parmentier, L., Stieger, M., Spanou, Z., Horn, M., Beltraminelli, H., & Borradori, L. (2013). Acral purpura as leading clinical manifestation of Dermatitis Herpetiformis: Report of two adult cases with a review of the literature. *Dermatology (Basel, Switzerland)*, *227*(1), 1–4. Available from https://doi.org/10.1159/000347108.

Urosevic-Maiwald, M., Kerl, K., Harr, T., & Allemann, I. B. (2013). Dapsone-induced erythema multiforme with neutropenia in a patient with linear IgA dermatosis: case report and review of the literature. *International Journal of Dermatology*, *52*(11), 1369–1371. Available from https://doi.org/10.1111/j.1365-4632.2011.05413.x.

Vale, E. C. S. D., Dimatos, O. C., Porro, A. M., & Santi, C. G. (2019). Consensus on the treatment of autoimmune bullous dermatoses: Dermatitis herpetiformis and linear IgA bullous dermatosis - Brazilian Society of Dermatology. *Anais Brasileiros de Dermatologia*, *94*(2 suppl 1), 48–55. Available from https://doi.org/10.1590/abd1806-4841.2019940208.

Varpuluoma, O., Jokelainen, J., Försti, A.-K., Timonen, M., Huilaja, L., & Tasanen, K. (2019). Dermatitis Herpetiformis and celiac disease increase the risk of bullous pemphigoid. *Journal of Investigative Dermatology*, *139*(3), 600–604. Available from https://doi.org/10.1016/j.jid.2018.10.010.

Viljamaa, M., Kaukinen, K., Pukkala, E., Hervonen, K., Reunala, T., & Collin, P. (2006). Malignancies and mortality in patients with coeliac disease and dermatitis herpetiformis: 30-year population-based study. *Digestive and Liver Disease*, *38*(6), 374–380. Available from https://doi.org/10.1016/j.dld.2006.03.002.

Wacklin, P., Kaukinen, K., Tuovinen, E., Collin, P., Lindfors, K., Partanen, J., Mäki, M., & Mättö, J. (2013). The duodenal microbiota composition of adult celiac disease patients is associated with the clinical manifestation of the disease. *Inflammatory Bowel Diseases*, *19*(5), 934–941. Available from https://doi.org/10.1097/MIB.0b013e31828029a9.

Wang, M., Kim, I.-G., Steinert, P. M., & McBride, O. W. (1994). Assignment of the human transglutaminase 2 (TGM2) and transglutaminase 3 (TGM3) genes to chromosome 20q11.2. *Genomics*, *23*(3), 721–722. Available from https://doi.org/10.1006/geno.1994.1571.

Warren, S. J. P., & Cockerell, C. J. (2002). Characterization of a subgroup of patients with Dermatitis Herpetiformis with nonclassical histologic features. *The American Journal of Dermatopathology*, *24*(4), 305–308. Available from https://doi.org/10.1097/00000372-200208000-00003.

Wojnarowska, F., Marsden, R. A., Bhogal, B., & Black, M. M. (1988). Chronic bullous disease of childhood, childhood cicatricial pemphigoid, and linear IgA disease of adults: A comparative study demonstrating clinical and immunopathologic overlap. *Journal of the American Academy of Dermatology*, *19*(5), 792–805. Available from https://doi.org/10.1016/S0190-9622(88)70236-4.

Yu, S. S., Connolly, M. K., Berger, T. G., & McCalmont, T. H. (2006). Dermatitis herpetiformis associated with administration of a gonadotropin-releasing hormone analog. *Journal of the American Academy of Dermatology*, *54*(2), S58–S59. Available from https://doi.org/10.1016/j.jaad.2005.08.024.

Zebrowska, A., Narbutt, J., Sysa-Jedrzejowska, A., Kobos, J., & Waszczykowska, E. (2005). The imbalance between metalloproteinases and their tissue inhibitors is involved in the pathogenesis of Dermatitis Herpetiformis. *Mediators of Inflammation*, *2005*(6), 373–379. Available from https://doi.org/10.1155/mi.2005.373.

Żebrowska, A., Woźniacka, A., Juczyńska, K., Ociepa, K., Waszczykowska, E., Szymczak, I., & Pawliczak, R. (2017). Correlation between IL-36α and IL-17 and activity of the disease in selected autoimmune blistering diseases. *Mediators of Inflammation*, *2017*, 1–10. Available from https://doi.org/10.1155/2017/8980534.

Zhang, F., Yang, B., Lin, Y., Chen, S., Zhou, G., Wang, G., Chen, X., Zhang, Y., Tian, H., Yu, M., Shi, Z., & Zhang, D. (2012). Dermatitis herpetiformis in China: A report of 22 cases. *Journal of the European Academy of Dermatology and Venereology*, *26*(7), 903–907. Available from https://doi.org/10.1111/j.1468-3083.2011.04204.x.

Ziberna, F., Sblattero, D., Lega, S., Stefani, C., Dal Ferro, M., Marano, F., Gaita, B., De Leo, L., Vatta, S., Berti, I., Caproni, M., Bonciani, D., Lindfors, K., Salmi, T., Reunala, T., Kaukinen, K., Kalliokoski, S., Kurppa, K., Ura, B., & Not, T. (2021). A novel quantitative ELISA as accurate and reproducible tool to detect epidermal transglutaminase antibodies in patients with Dermatitis Herpetiformis. *Journal of the European Academy of Dermatology and Venereology*, *35*(1), e78–e80. Available from https://doi.org/10.1111/jdv.16822.

Zone, J. J., Schmidt, L. A., Taylor, T. B., Hull, C. M., Sotiriou, M. C., Jaskowski, T. D., Hill, H. R., & Meyer, L. J. (2011). Dermatitis Herpetiformis sera or goat anti–transglutaminase-3 transferred to human skin-grafted mice mimics Dermatitis Herpetiformis immunopathology. *The Journal of Immunology*, *186*(7), 4474–4480. Available from https://doi.org/10.4049/jimmunol.1003273.

Chapter 17

Nonceliac gluten-related disorders

Aurelio Seidita, Pasquale Mansueto, Alessandra Giuliano and Antonio Carroccio
Department of Health Promotion Sciences, Maternal and Infant Care, University of Palermo, Internal Medicine and Medical Specialties (PROMISE), Palermo, Italy

17.1 Introduction

In the last two decades, alongside celiac disease (CD) and wheat allergy (WA) various other wheat/gluten-related clinical entities have been discovered, including nonceliac gluten/wheat sensitivity (NCGS/NCWS), which is probably the most controversial, and whose "dark" sides still seem to predominate over the knowledge acquired.

17.2 Definition

The first clinical reports on the existence of a condition of "gluten intolerance" without the typical immunological and histological features of CD date back to the late 1970s (Ellis & Linaker, 1978). Just a few years later, Cooper et al. reported 8 women who complained of "a gluten-sensitive diarrhea but had no evidence of celiac disease" (Cooper et al., 1980). Following these isolated reports, the number of articles referring to such condition have multiplied exponentially over the years (Kakar et al., 2003; Troncone et al., 2003), revealing its clinical similarity to irritable bowel syndrome (IBS) (Wahnschaffe et al., 2001, 2007).

In 2012, our group proved the existence of gluten sensitivity (GS) by analyzing a large series of subjects with IBS-like symptoms who were diagnosed as having NCWS by means of a double-blind placebo-controlled challenge (DBPCC) with wheat. A comparison of their clinical, serological and histological features was made with control populations of CD and IBS (Carroccio et al., 2012).

In the same year an international consensus provided the first definition of NCGS, namely "...cases of gluten reactions in which neither allergic nor autoimmune mechanisms can be identified. These are generally defined as nonceliac GS or more simply, GS" (Sapone et al., 2012). The authors emphasized that these patients reported "distress when eating gluten-containing products and improvement when following a GFD," presenting "with clinical symptoms that can overlap with CD or WA symptoms," and that it "is not accompanied by the concurrence of antitransglutaminase (tTG) autoantibodies" and that "the small intestine of GS patients is usually normal." They also underlined that a diagnosis can be considered only in those patients "in which both allergic and autoimmune mechanisms have been ruled out" (Sapone et al., 2012). In this fragmentary framework, it was also hypothesized that proteins other than gluten could be responsible for this condition, and therefore that the term NCGS may be replaced by NCWS (Carroccio et al., 2014).

It was only in 2015 that a panel of experts suggested the NCGS diagnostic criteria (the "Salerno Experts" criteria') and elaborated the definition still in use today: "...a syndrome characterized by intestinal and extra-intestinal symptoms related to the ingestion of gluten-containing food, in subjects that are not affected by either CD or WA." Nevertheless, the panel underlined that "The terminology 'NCGS' is still a matter of debate" and that "Although NCGS is triggered by gluten-containing cereals, the offending dietary protein has not been identified," thus "The term 'NCGS' could be changed into 'Nonceliac Wheat Sensitivity' (NCWS) in the near future, although this would exclude other relevant cereals like barley and rye"(Catassi et al., 2015).

It is evident from a quick analysis of this definition that many questions have yet to find an adequate answer, including: what is (or are) the component(s) capable of triggering the onset of symptoms?; if allergic and autoimmune origins are excluded, what are the pathogenetic bases of NCWS?; is it possible to identify a diagnostic biomarker to simplify the diagnosis?; what is the best therapeutic approach to this condition (elimination of gluten vs elimination of wheat vs elimination of other foods)?; is it a chronic or a transient condition that can completely disappear after an adequate period of dieting?

Pediatric and Adult Celiac Disease. DOI: https://doi.org/10.1016/B978-0-443-13359-6.00022-4

17.3 Epidemiology

Considering the many open questions about NCWS, it is a very hard challenge to define the prevalence of this condition.

Data from the United States showed that an increasing number of patients without CD regularly consumed a GFD, from 0.5% in 2009–10 to 1.0% in 2011–12, to 1.7% in 2013–14 ($p = 0.005$) (Choung et al., 2016; DiGiacomo et al., 2013). Moreover, a nonspecialist investigation reported that about 20% of Americans consumed gluten-free foods (Igbinedion et al., 2017), and, recently, a Canadian study estimated that 1.9% of individuals followed a GFD and women had a 2 times higher odds ratio (OR 2.08) of being on a GFD than men (Mudryj et al., 2021). Accordingly, a growth in the US market for gluten-free products has been highlighted, from $1.0 billion in 2003, to $1.31 billion in 2011, $1.68 billion in 2015 (Mansueto et al., 2014), and $6.7 billion in 2022, with an expected further rise to $14.0 billion by 2032 (Gluten-free foods market in the U.S. - statistics & facts, 2022). Even considering in addition to a CD prevalence of 1% in the general population, there are another 0.5%–1.7% of subjects who regularly follow a GFD.

To date, the NCWS prevalence in the general population ranges between 0.5% and 14% (Araya et al., 2020; Aziz et al., 2014; Potter et al., 2020; Volta et al., 2014), the considerable variability depending on the methods of ascertaining the diagnosis and the geographical area. As shown in Table 17.1, most studies carried out on the general population were conducted only by administering and analyzing the results of questionnaires, in the absence of a confirmed diagnosis (Araya et al., 2020; Aziz et al., 2014; Carroccio et al., 2017; Cha et al., 2022; Ontiveros et al., 2021; Potter et al., 2018, 2020; Volta et al., 2014). The variability in the perception of symptoms reported by individual patients may also contribute to the aforementioned discrepancies.

Contrarily, other studies have assessed NCWS prevalence based on the resolution of symptoms during GFD (Sapone et al., 2012) and on their reappearance during a Double-Blind Placebo-Controlled (DBPC) gluten/wheat challenge (Barone et al., 2020; Biesiekierski et al., 2011; Carroccio et al., 2012; Elli et al., 2016; Shahbazkhani et al., 2020). To add more complexity, the challenge, which is certainly more reliable from a methodological point of view, as it minimizes any subjective/psychological component of the disorder, was carried out on subpopulations with IBS or with other functional gastrointestinal diseases in referral centers. This adds an additional selection bias, with prevalence ranging from 6.5% to 38.2% (Biesiekierski et al., 2011; Shahbazkhani et al., 2020). A metanalysis of 10 studies, in which a DBPC challenge was used to estimate the prevalence of NCWS, partially clarified this variability, showing that NCWS was confirmed only in 16% of the patients who self-reported GS. Additionally, this analysis also confirmed a considerable heterogeneity between the methodologies used (Molina-Infante & Carroccio, 2017). Finally, some studies have analyzed the prevalence of NCWS in subsets of patients suffering from fibromyalgia (5.6%) (Almirall et al., 2023), various systemic autoimmune diseases (4.6%) (Vives et al., 2012), inflammatory bowel diseases (IBD) (4.9%–27.6%) (Aziz, Branchi, et al., 2015; Limketkai et al., 2018), and allergies (29.4%) (Massari et al., 2011), or by diagnosing the presence of GS with methodologies that differ from the international criteria (Massari et al., 2011; Vives et al., 2012).

Despite these heterogeneous results, both self-reported questionnaire and DBPC challenge studies show a predominance of NCWS in females (F:M = 6:1) aged 30–50 years (Carroccio et al., 2012; Catassi et al., 2015; Potter et al., 2020; Sapone et al., 2012; Volta et al., 2014), although some cases have also been self-reported in adolescents (Carroccio et al., 2017) and children (Camhi et al., 2019; Francavilla et al., 2018). In these, some important epidemiological and clinical differences compared to adult patients have been reported: a higher prevalence in males and a lower predominance of extra-intestinal manifestations (Francavilla et al., 2014).

Taken all together, these data explain the absolute need to identify specific diagnostic markers of NCWS, in order to simplify its diagnosis and quantify its epidemiology.

17.4 Etiology and pathogenesis

17.4.1 Etiology

Very little is known about the pathogenesis of NCWS. It is still to be defined what component of wheat is the primary etiological agent responsible for symptoms. The various hypotheses have focused on two main macronutrients of wheat, namely gluten or nongluten proteins, such as α-Amylase Trypsin Inhibitors (ATIs), and carbohydrates, such as FODMAPs (Fasano et al., 2015). Some hypotheses argue that these agents, together with the presence of intestinal dysbiosis and increased permeability (Leccioli et al., 2017), have a role in the onset of the intestinal and extra-intestinal symptoms (Mumolo et al., 2020). In view of this uncertainty, some years ago we proposed that the name NCGS should be avoided, considering NCWS a more appropriate and less confusing term (Carroccio et al., 2014).

TABLE 17.1 Studies reporting NCWS prevalence (inverse chronological order).

Authors	Year	Country	Population	Number of patients	Assessment of NCWS/ NCGS	Prevalence
Almirall M. et al. (1)	2022	Spain	Fibromyalgia patients	142	Salerno criteria	5.6%
Cha RR. et al. (2)	2022	South Korea	Patients attending 9 GE outpatient clinics	386	Self-report NCWS using questionnaire	20.7%
Dhoble P. et al. (3)	2021	India	Patients with IBS and healthy subjects	204 IBS patients 400 Healthy subjects	Self-report NCWS using questionnaire	11.3% in IBS 0% in healthy
Shahbazkhani B. et al. (4)	2021	Iran	Patients with refractory functional dyspepsia	77	GFD followed by DBPC gluten challenge with crossover	6.5%
Ontiveros N. et al. (5)	2021	Paraguay	General population	1058	Self-report NCWS using questionnaire	10.3%
Potter M. et al. (6)	2020	Australia	Same general population studied in 2015 and 2018	1322	Self-report NCWS using questionnaire	13.8% in 2015 13.9% in 2018
Barone M. et al. (7)	2020	Italy	IBS patients	30	Low FODMAP-GFD followed by gluten/placebo challenge	46.1%
Arámburo-Gálvez JG. et al. (8)	2020	Brazil	General population	1630	Self-report NCWS using questionnaire	2.3%
Araya M. et al. (9)	2020	Chile	General population	1203	Self-report NCWS using questionnaire	0.5%
Croal ID. Et al. (10)	2019	United Kingdom	General population	1004	Self-report NCWS using questionnaire	32.8%
Camhi SS. et al. (11)	2019	United States	Pediatric patients analyzed retrospectively	500	Salerno criteria	5.2%
Potter MDE. Et al. (12)	2018	Australia	General population	3542	Self-report NCWS using questionnaire	14.9%
Francavilla R. et al. (13)	2018	Italy	Children with chronic gastrointestinal symptoms (negative CD and WA)	1114	GFD followed by DBPC gluten challenge	1.0%
Ontiveros N. et al. (14)	2018	El Salvador	General population	1326	Self-report NCWS using questionnaire	3.1%
Cabrera-Chávez F. et al. (15)	2017	Argentina	General population	1209	Self-report NCWS using questionnaire	6.3%
Carroccio A. et al. (16)	2017	Italy	High-school students	555	Self-report NCWS using questionnaire	12.2%

(Continued)

TABLE 17.1 (Continued)

Authors	Year	Country	Population	Number of patients	Assessment of NCWS/ NCGS	Prevalence
Limketkai BN. et al. (17)	2017	United States	Subjects with IBD	55 Crohn's disease 46 Ulcerative colitis	Self-report NCWS using questionnaire	23.6% in Crohn's Disease, 27.3% in Ulcerative colitis
Elli L. et al. (18)	2016	Italy	Patients with functional gastroenterological symptoms	140	GFD followed by DBPC gluten challenge	20.0%
Cabrera-Chávez F. et al. (19)	2016	Colombia	General population	1207	Self-report NCWS using questionnaire	4.6%
van Gils T. et al. (20)	2016	Netherlands	General population	785	Self-report NCWS using questionnaire	6.2%
Lis DM. et al. (21)	2015	Worldwide	Non-CD athletes	924	Self-report NCWS using questionnaire	22.8%
Aziz I. et al. (22)	2015	United Kingdom	Patients suffering from various organic or functional gastrointestinal diseases	145 IBD 59 IBS 109 Dyspepsia	Self-report NCWS using questionnaire	27.6% in IBD 42.4% in IBS 17.4% in dyspepsia
Sharma H. et al. (23)	2015	India	Patients suffering from IBS or dyspepsia	362 IBS 358 Dyspepsia	Anti-gliadin antibody positivity but CD negative	28.7% in IBS 19.0% in dyspepsia
Capannolo A. et al. (24)	2015	Italy	Subjects complaining of gluten-related symptoms	392	GFD followed by open gluten challenge	6.88%
Shahbazkhani B. et al. (25)	2015	Iran	IBS patients	72	GFD followed by DBPC gluten challenge	43.0%
Ontiveros N. et al. (26)	2015	Mexico	General population	1238	Self-report NCWS using questionnaire	1.0%
Herfarth HH. et al. (27)	2104	United States	IBD patients	1647	Self-report NCWS using questionnaire	4.9%
Lu W. at al. (28)	2014	China	IBS patients	186	Anti-gliadin antibody positivity but CD negative, symptoms improving on a GFD	18.3%
Volta U. et al. (29)	2014	Italy	Consecutive patients without CD and WA from 38 Italian Centers	12255	Self-report NCWS using questionnaire	3.2%
Aziz I. et al. (30)	2014	United Kingdom	Healthy subjects	1002	Self-report NCWS using questionnaire	12.9%
Sapone A et al. (31)	2012	United States	CD outpatient clinic patients	5896	Improvement of symptoms on GFD and recurrence on gluten challenge; exclusion of CD and WA.	5.9%

(Continued)

TABLE 17.1 (Continued)

Authors	Year	Country	Population	Number of patients	Assessment of NCWS/ NCGS	Prevalence
Tanpowpong P. et al. (32)	2012	New Zealand	Healthy children	916	Improvement of symptoms after 6 weeks of GFD	5.0%
Vives MJ. et al. (33)	2012	Spain	Patients with systemic autoimmune diseases	109	Non-CD lymphocytic enteropathy	4.6%
Carroccio A. et al. (34)	2012	Italy	Patients with IBS-like symptoms	920	Elimination diet, with the exclusion of wheat, cow's milk, eggs, tomato, and chocolate followed by DBPC wheat challenge	30.0%
Biesiekierski JR. et al. (35)	2011	Australia	IBS patients	34	GFD followed by DBPC gluten challenge	38.2%
Massari S. et al. (36)	2011	Italy	Allergic patients	236	Non-CD duodenal mucosal lesion	29.4%

CD: Celiac disease; DPBC: double-blind placebo-controlled; FODMAPs: fermentable oligosaccharides, disaccharides, monosaccharides and polyols; GE: gastroenterology; GFD: gluten-free diet; IBD: inflammatory bowel diseases; IBS: irritable Bowel syndrome; NCWS: nonceliac wheat sensitivity; WA: wheat allergy.

Gluten represents 80%–85% of the total proteins in wheat; it is composed of gliadins and glutenins, which are resistant to gastrointestinal proteolytic enzymes (Biesiekierski, 2017; Tatham & Shewry, 2008). Contrary to CD, several studies in NCWS have proved that many patients do not have gluten-related symptoms after a gluten challenge (Biesiekierski et al., 2013; Lionetti et al., 2017; Molina-Infante & Carroccio, 2017; Zanini et al., 2015). A family of proteins reported to be involved in NCWS pathogenesis are the ATIs, which are plant-derived proteins with a pesticide function, mainly found in wheat, rye, rice, potatoes, and peas. They are very resistant to both heat and gastrointestinal proteases and may have a leading role in activating the innate immune response of monocytes, macrophages, and dendritic cells (Schuppan & Zevallos, 2015; Zevallos et al., 2017).

Studies have also focused their attention on FODMAPs, widely found in cereals, fruit, vegetables, milk, and honey (Varney et al., 2017). After fermentation by intestinal bacteria, they could be responsible for an increase in water and gas in the gut, without causing inflammation (Hill et al., 2017), but resulting in abdominal distention and pain (Khan et al., 2020). A low-FODMAPs diet was found to be able to induce the remission of both intestinal and extra-intestinal symptoms in some NCWS patients (Dieterich et al., 2019), with a higher relapse percentage after fructose challenge than with gluten or placebo (Skodje et al., 2018). Consequently, a precise border between FODMAPs intolerance and a NCWS has not yet been established. However, the coexistence of extra-intestinal symptoms and autoimmune or atopic diseases may point at NCWS rather than FODMAPs intolerance.

17.4.2 Pathogenesis

Intestinal permeability seems to be altered in NCWS (Hollon et al., 2015; Leccioli et al., 2017; Uhde et al., 2016). Increased intestinal permeability could be caused by endogenous or exogenous factors. In fact, intestinal dysbiosis occurs in IBD and IBS (Rajilić-Stojanović et al., 2011), and also in NCWS there could be low concentrations of butyrate-producing *Firmicutes* and/or *Bifidobacteria*, which could cause intestinal barrier damage, as proved by the high levels of fatty acid binding protein-2 (FABP2), and a higher paracellular permeability in the mucosa (Leccioli et al., 2017). This mucosal damage may promote the translocation of microbial products, such as lipopolysaccharide (LPS) (Leccioli et al., 2017; Uhde et al., 2016). The translocation of LPS might activate both the innate and adaptive immune systems, as demonstrated by increased serum levels of soluble CD14 (sCD14) and LPS-binding protein (LBP), and by the presence of antibodies against LPS and flagellin. A complete normalization of FABP2 and immune marker levels was observed in some NCWS patients after a wheat-free diet (WFD) (Uhde et al., 2016).

As an exogenous factor, gluten/gliadin has been reported as a possible culprit. In vitro studies have shown that gliadin may cause direct damage to the intestinal barrier, without immune-mediated mechanisms being triggered (Clemente et al., 2003). Other findings have suggested a gliadin-related impaired intestinal permeability: as in active CD subjects, a reduced intestinal transepithelial electrical resistance (TEER) has been proved (Hollon et al., 2015); gliadin binding to C-X-C motif chemokine receptor 3 (CXCR3) might cause the increase in the release of zonulin and therefore consequently greater intestinal permeability (Barbaro et al., 2020; Lammers et al., 2008); gliadin can cause increased apoptosis of the gut epithelial cells (Mazzarella et al., 2008). The increased intestinal permeability could allow gliadin peptides to permeate the lamina propria and activate the innate immune system by the Toll-Like receptors (TLRs) TLR-2 and TLR-4, inducing the release of proinflammatory cytokines (Uhde et al., 2016).

ATIs can also be responsible for the TLR-4-mediated activation of innate immunity. The inflammatory effect appeared to be dose-dependent and could cause an increase in chronic intestinal inflammation (Zevallos et al., 2017). The direct activation of innate immunity by ATIs could cause the indirect activation of the adaptative response, both in the gut and in other organs, resulting in the onset of extra-intestinal symptoms (Schuppan et al., 2015). In this context, cytokines that regulate innate immunity have been found to be increased in the serum of some NCWS patients (Valerii et al., 2015), as has been in mucosal biopsies. On the contrary, an increase has been found in the RNA transcripts of azurocidin 1 (AZU1) and bone morphogenetic protein-7 (BMP7), implicated in the activation of innate immunity. At the rectal site there is a greater production of TNF-α by CD45 + , CD3 + , CD4 + and CD8 + cells and of IL-17 by CD4 + cells compared to healthy controls, suggesting that the adaptive immune system could be involved (Mansueto et al., 2020). In addition, IFN-γ levels have been reported to be markedly increased both in the intestinal and rectal mucosa of NCWS patients after gluten challenge (Brottveit et al., 2013; Di Liberto et al., 2016), but their role in NCWS pathogenesis has been questioned. Finally, intraepithelial CD3 + T lymphocyte, were found to be higher than in a healthy population, but lower than in CD (Carroccio et al., 2019). Similarly, increased levels of mast cells in duodenum biopsies have been reported (Losurdo et al., 2017).

Other mechanisms could be hypothesized, such as a non-IgE-mediated allergic response to gluten/wheat proteins as supported by the finding of high levels of eosinophils in intestinal biopsies. Eosinophils can be involved in both IgE- and non-IgE-mediated reactions; in the case of NCWS there could be a non-IgE food allergy-mediated mechanism (Carroccio et al., 2019; Giancola et al., 2020).

Fig. 17.1 summarizes the putative pathogenic mechanisms of NCWS.

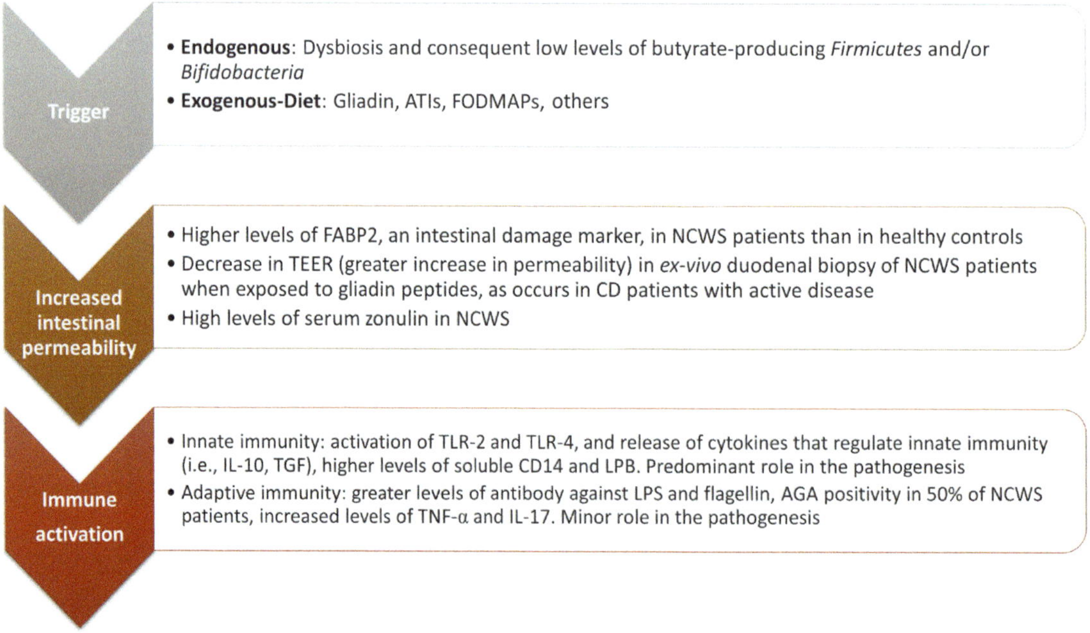

FIGURE 17.1 Putative pathogenic mechanisms of NCWS. ATIs: Amylase/trypsin-inhibitors; AGA: antigliadin antibodies; CD: celiac disease; FABP2: fatty acid binding protein 2; FODMAPs: fermentable oligo-, di-, mono- saccharides and polyols; IL: interleukin; LPS: lipopolysaccharide; NCWS: nonceliac wheat sensitivity; TEER: transepithelial electrical resistance; TGF: transforming growth factor; TLR: toll-like receptor; TNF: tumor necrosis factor.

17.5 Clinical manifestations

Patients suffering from NCWS report a large variety of gastrointestinal and extraintestinal symptoms (Fig. 17.2).

The classic gastrointestinal manifestations of NCWS largely overlap with those reported by IBS individuals and are often associated with functional dyspepsia-like symptoms. A typical feature of all these patients is the rather rapid onset of symptoms, usually ranging from a few minutes to a few hours after gluten/wheat intake (Carroccio et al., 2012; Catassi et al., 2015). Unlike IBS, NCWS patients often complain of several extra-intestinal symptoms, sometimes as the only manifestation of their condition, thus bringing them closer to CD; in most cases, two or more symptoms are present simultaneously (Volta et al., 2012). Of these, the most frequent are brain fog (40%), fatigue (36%), and joint pain (44%) (Carroccio et al., 2012; Catassi et al., 2015; Mansueto et al., 2022; Verdu et al., 2009; Volta et al., 2014).

Four extra-intestinal clusters deserve a specific mention. *Neuropsychiatric symptoms* are heterogeneous in frequency and type. We recently compared the prevalence of neurological symptoms in patients with NCWS, CD and IBS. Half of NCWS patients reported neuropsychiatric symptoms versus only 25% and 28% of patients with CD and IBS. Among the symptoms, the most frequent was headache (33%), followed by fatigue (22%) and brain fog (9%) (Carroccio et al., 2021). NCGS patients were not found to exhibit a tendency for general somatization (Brottveit et al., 2012). In any case, it cannot be forgotten that many subjects with purely neuropsychiatric manifestations tend to self-identify as suffering from food intolerance/allergy, despite double-blind challenge disproves this finding. Thus, being the association among diet, intestinal/extraintestinal symptoms, and self-deception so complex, this relationship must be studied individually, focusing on patients' background and clinical framework (Ferguson, 1990). Fig. 17.3 shows the main neuropsychiatric manifestations in NCWS patients.

Uro-gynecological symptoms seem to worsen the quality of life in NCWS women much more than they do in healthy subjects ($p = 0.04$), and in both CD ($p = 0.02$) and IBS ($p = 0.01$) patients. More than 25% of NCWS patients complained of menstrual cycle disorders, 16% suffered from recurrent vaginitis, 6% from dyspareunia, and 29% from recurrent cystitis. Of note, at 1-year follow-up on a WFD, 46% of the patients with menstrual disorders and 36% with recurrent vaginitis reported the resolution of symptoms (Soresi et al., 2021). Many authors have also reported a high prevalence (10%–22%) of *iron-deficiency anemia* (Carroccio et al., 2012; Losurdo et al., 2018; Volta et al., 2014). A subgroup of NCWS anemic patients were prospectively reevaluated after at least 12 months on a "strict" WFD. In this study, the prevalence of anemia was 34.8%. Again, a long-term WFD was found to significantly improve anemia and

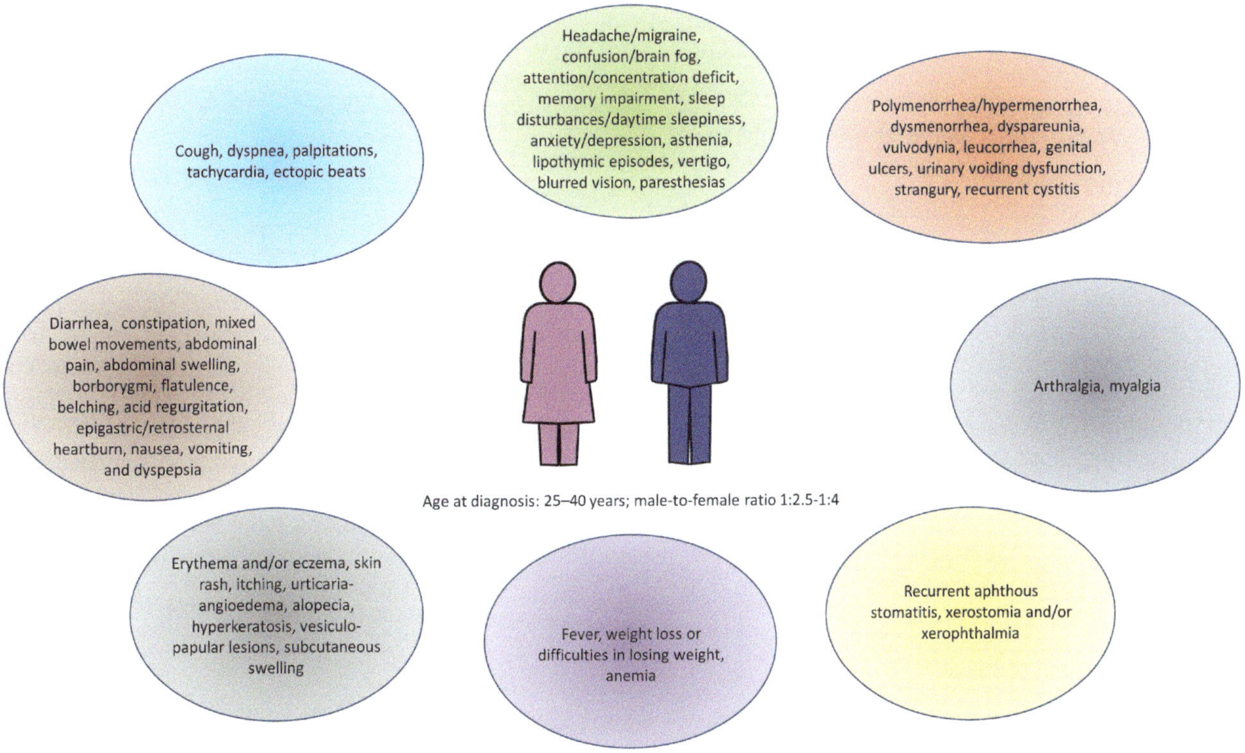

FIGURE 17.2 Summary of gastrointestinal and extraintestinal clinical manifestations of NCWS patients. NCWS: nonceliac wheat sensitivity.

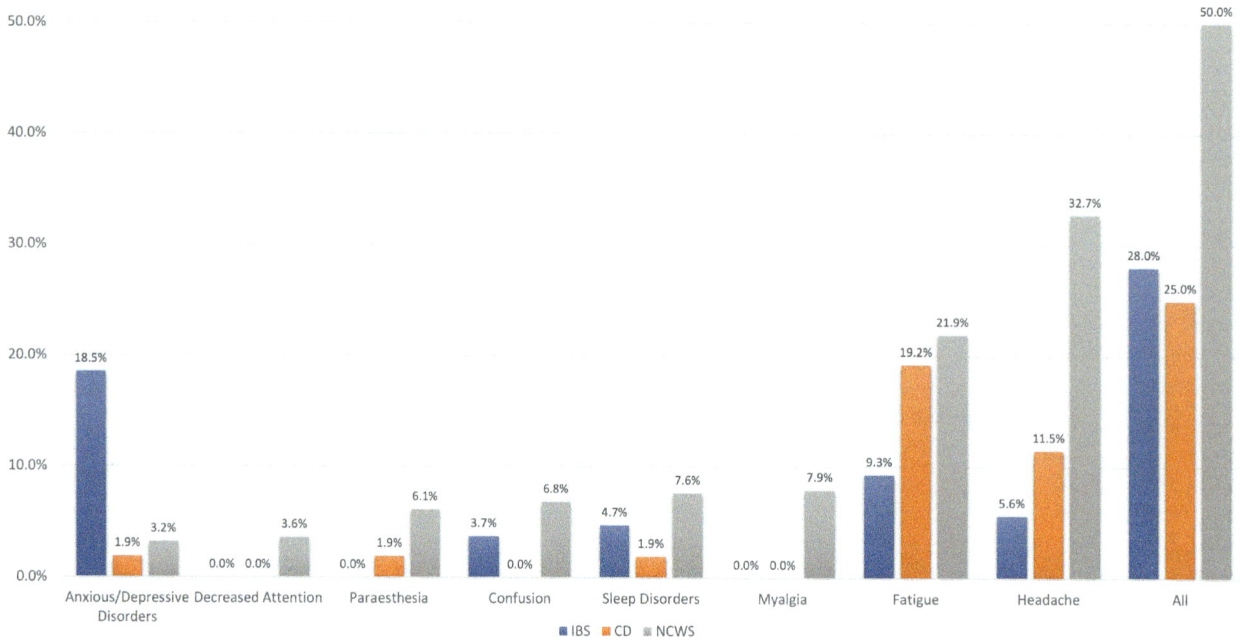

FIGURE 17.3 Summary of neuropsychiatric clinical manifestations of NCWS, CD, and IBS patients. CD: Celiac disease; IBS: irritable Bowel syndrome; NCWS: nonceliac wheat sensitivity. *Modified from Carroccio et al. (2021). Frequency and Clinical Aspects of Neurological and Psychiatric Symptoms in Patients with Non-Celiac Wheat Sensitivity. Nutrients, 13(6). https://doi.org/10.3390/nu13061971.*

iron metabolism. Authors concluded that anemia in NCWS is generally mild, likely due to nutritional imbalances and/or gynecological disorders (Mansueto et al., 2022). *Dermatologic manifestations* are rather common (18%−65%), albeit heterogeneous, feature of NCWS. Among these, the most frequently described are eczema (30%−45%) and nonspecific dermatitis/skin rashes (18%−30%) (Sapone et al., 2012; Volta et al., 2012). Faina et al. reported a 65.6% prevalence of dermatological manifestations, of which 45% eczematous dermatitis, 36% urticaria-angioedema, 10% generalized pruritus, and 9% psoriasis-like lesions (Faina et al., 2021). Other authors have reported the possible presence of psoriasis-like hyperkeratotic lesions, as well as of vesiculo-papular lesions similar to dermatitis herpetiformis (Bonciolini et al., 2015).

17.6 Natural history and complications

17.6.1 Natural history

To date, it is still not possible to establish whether NCWS is a persistent condition or not (Fasano et al., 2015). The only study to have evaluated the long-term effects of NCWS had a median follow-up time of 99 months (Carroccio, D'Alcamo et al., 2017). Applying a modified version of the Pavia score to assess adherence to a GFD in CD (Biagi et al., 2009), 74% of patients had been following a strict WFD for a median of 8 years after diagnosis. All the patients on a strict diet reported a recurrence of both gastrointestinal and extraintestinal symptoms in the case of accidental wheat intake, while the remaining patients had reintroduced wheat without symptoms, or had discovered they were suffering from another disease and reintroduced wheat, or did not adhere to the diet by choice or due to economic problems, reporting, however, the onset of symptoms after eating wheat. Additionally, 88% of patients improved after NCWS diagnosis, especially those who strictly adhered to the WFD. Regarding asymptomatic patients on a strict WFD, some underwent a new DBPC challenge with wheat for two weeks, which confirmed NCWS in 20/22 of them. Although these data should be cautiously interpreted, due to the study limitations, at least in a part of patients with NCWS, the condition is persistent on a long-term.

17.6.2 Complications

While little is known about the natural history of the disease, even less can be said about its potential long-term complications. Various hypotheses have been proposed for two subgroups of patients, that is, those with NCWS alone and those with Multiple Food Hypersensitivity (MFH) including NCWS (Carroccio et al., 2012).

NCWS patients appear to be characterized by a higher frequency of the HLA DQ2/DQ8 haplotype and duodenal lymphocytosis. They therefore appear to be much more like CD than IBS or MFH patients. On this basis, it could be speculated that some NCWS patients could develop villous atrophy. In this respect, the presence of anti-tTG antibody deposits in the duodenal mucosa, could identify those 'preceliac subjects' who may develop villous atrophy (Salmi et al., 2006).

In contrast, patients with MFH including NCWS showed a higher frequency of a family and personal history of allergy, as well as a histological finding of eosinophilic infiltration on duodenal and colonic biopsy, although wheat-specific IgE tests were always negative, suggesting the existence of a non-IgE-mediated hypersensitivity (Carroccio et al., 2012). It could be speculated that in the long term these subjects might develop other clinical manifestations of a non-IgE-mediated hypersensitivity status (e.g., food-protein enteropathy).

17.7 Associated diseases

17.7.1 Multiple food hypersensitivity

In several NCWS subjects there is a quite frequent food allergy history in infancy and/or coexistence of atopic diseases. We showed that NCWS subjects compared to IBS controls are characterized by a personal history of food allergy in the pediatric age ($p = 0.01$), coexistence of atopic diseases ($p = 0.0001$), positive serum anti-betalactoglobulin antibody assay ($p = 0.001$), food-antigen-induced positive in vitro basophil activation test by flow cytometry ($p = 0.0001$), and the presence of eosinophils in the intestinal mucosa biopsies ($p = 0.0001$) (Carroccio et al., 2013). Furthermore, these patients often complained of intolerance to multiple foods, including cow's milk, egg, and others (Carroccio et al., 2012).

17.7.2 Neuropsychiatric disorders

Many authors have reported a variegated pattern of neuropsychiatric manifestations among the NCWS extraintestinal symptoms. Some conditions belonging to the spectrum of autism and schizophrenia, deserve to be considered separately.

The association between CD and autism spectrum disorders has been reported in international studies for some time (Croall et al., 2021). In 2010 "The ScanBrit study," suggested the existence of a certain correlation between autism and gluten exposure, the elimination of which could improve the behavioral spectrum of subjects with autism (Whiteley et al., 2010). Another study identified a high prevalence of AGA in children with autism, which, however, did not correlate with HLA-DQ2/DQ8 (Lau et al., 2013).

In 2014 Jackson et al. demonstrated a higher prevalence of GS and higher mean AGA IgG levels in schizophrenia patients (Jackson et al., 2014). More recently, Dzikowski M. et al. found higher titers of AGA IgA in first-episode schizophrenia (11.5%) and chronic schizophrenia (26.0%) patients compared to healthy individuals (5.0%) (Dzikowski et al., 2020).

17.7.3 Autoimmune diseases

The association between autoimmune conditions and NCWS has been known for years, with a prevalence of about 5% (Losurdo et al., 2018). The correlation between gluten exposure and rheumatic disease has been suggested by other papers (Conti et al., 2015; Elkan et al., 2008; Lidén et al., 2007; Osborne, 2016; Perlmutter, 2018).

The prevalence of autoimmune diseases in a NCWS group was found to be 29%, Hashimoto thyroiditis being the most common. Furthermore, serum samples tested positive for anti-nuclear antibodies (ANA) in 46.0% of subjects with NCWS, significantly higher than in CD and IBS controls (Carroccio et al., 2015). In a prospective study, 25.3% of NCWS patients developed autoimmune diseases, compared with 6.6% of healthy subjects and 11% of IBS patients. In addition, both ANA (71.4%) and extractable nuclear antigen antibody (ENA) (21,9%) positivity were significantly higher in patients with NCWS than in the two control groups (Mansueto et al., 2021). Taken together, these two studies confirmed what had been previously noted (Losurdo et al., 2018; Volta et al., 2016). Hence, according to these works, 1 in 4 NCWS patients has already developed or is at risk of developing an autoimmune disease.

17.7.4 Dermatological disorders

Psoriasis has frequently been reported as connected to food intake, specifically to gluten (Korman, 2020). Some authors have reported psoriasis-like lesions in NCWS patients (Bonciolini et al., 2015; Faina et al., 2021), however, the studies analyzing the role of gluten in psoriasis are far more extensive. The National Psoriasis Foundation, reported the disappearance or improvement of skin lesions after eliminating gluten from their diet in 53.0% of psoriasic patients (Afifi et al., 2017). A 16% of AGA IgA positivity was found in a large psoriasic Swedish population (Michaëlsson et al., 1993). The same authors reported that after a 3-month GFD period, 72.7% of the AGA-positive patients showed a significant decrease in the mean psoriasis area and severity index, 24.2% had their treatment regimen decreased, and 81.8% had a marked reduction in AGA titers. In addition, in 54.5% the index worsened after resuming a gluten-containing diet (Michaëlsson et al., 2000). Finally, the authors proved that GFD reduced CD4 + T lymphocytes and Ki67 + cells in the involved dermis (Michaëlsson et al., 2003). To conclude, AGA positivity in psoriasis is a highly debated topic, with conflicting results reported by the different study groups (Bell et al., 2021; Bhatia et al., 2014).

17.7.5 Other conditions: the case of Familial Mediterranean Fever and fibromyalgia

Familial Mediterranean Fever (FMF) is the most common hereditary autoinflammatory disease, usually occurring in populations from the eastern Mediterranean region (Alghamdi, 2017). Among possible stressors, recently, some physicians have focused on wheat (Ekinci et al., 2020; Yenokyan & Armenian, 2012). A pilot study compared six female volunteers with FMF reporting a self-diagnosed NCWS, 12 age-matched non-FMF NCWS females, and eight sex- and age-matched healthy subjects. After a 2-week DBPC cross-over challenge the patients reported a significant worsening of the FMF symptoms with wheat, but not with rice (Carroccio et al., 2020).

There is a surprising overlap between the clinical manifestation of fibromyalgia and NCWS (Heidari et al., 2017), and several authors have reported the possible co-existence of these conditions (Aziz et al., 2015; Losurdo et al., 2018). An improvement of symptoms after a GFD has been reported, but not confirmed (Isasi et al., 2014; Slim et al., 2017). In a Spanish study, an improvement of intestinal symptoms after a GFD was noticed in fibromyalgia patients (Almirall et al., 2023). Recently, we showed that 65.1% of primary fibromyalgia women self-reported NCWS. Among the patients who underwent WFD, a reduction of fibromyalgia-related symptoms was observed, as well as a reduction of gastrointestinal symptoms (Schinocca et al., 2021).

17.8 Diagnosis

Diagnosing NCWS is challenging. No diagnostic methodologies, no serum markers, and no histological alterations, individually or in combination, proved reliable and reproducible. Therefore NCWS diagnosis can only be made after excluding other organic diseases.

17.8.1 The Salerno Experts' criteria

"Given the lack of a NCGS biomarker, there is the need for standardizing the procedure leading to the diagnosis confirmation"; with these words the Salerno experts clarify the purpose of their work (Catassi et al., 2015).

NCWS suspected patients are those who report "persistent intestinal and/or extraintestinal complaints" with a "worsening of symptoms after eating gluten-rich food." The idea of a diagnosis by exclusion is immediately clarified; indeed, the authors explain that only patients "showing a normal result of the CD and WA serological markers on a gluten-containing diet" should be considered. Therefore it is necessary to preliminarily test all patients for CD antibodies (Al-Toma et al., 2019), and to ensure they have negative IgE-mediated immune-allergy tests to wheat (skin prick tests and serum specific IgE detection) (Carroccio et al., 2012; Sapone et al., 2012).

However, these exclusion criteria do not take into consideration patients with suspected "seronegative" CD who require a dedicated diagnostic process. In this context, it is mandatory to perform duodenal biopsies whilst on a gluten-containing diet in all seronegative patients with gluten-related symptoms and a positive HLA DQ2/DQ8 haplotype (Schiepatti et al., 2022).

Once the patients have been allegedly identified, the proposed diagnostic protocol includes two subsequent steps:

1. Evaluation of the clinical response after the introduction of a GFD;
2. Evaluation of the clinical response after the reintroduction of gluten into the diet.

To evaluate the symptoms and their fluctuations with the dietary regimes, a modified version of the Gastrointestinal Symptom Rating Scale (GSRS) has been proposed (Kulich et al., 2008), which investigates both the intestinal symptoms and the main NCWS extraintestinal manifestations.

17.8.1.1 Evaluation of the clinical response after the introduction of a GFD

Before eliminating gluten from the diet, all patients should follow a gluten-containing diet for at least six weeks. In the two weeks preceding and just before starting a GFD, patients should be asked to complete the revised version of the GSRS. The patient can then start the GFD, which, despite often leading to a consistent improvement in symptoms within a few days, must be continued for at least six weeks to confirm the causal relationship between symptoms and gluten ingestion. Throughout this phase the patient will have to fill in the modified version of the GSRS weekly. Patients will be considered as "responders" to the GFD if they report a "$>30\%$ reduction of one to three main symptoms or at least 1 symptom with no worsening of others for at least 50% of the observation time." A diagnosis of NCWS can be ruled out in patients considered "nonresponders."

Indeed, most patients who self-report gluten intolerance are already following a GFD. For these patients, the authors propose starting the diagnostic protocol directly from step 2, if patients have experienced a significant benefit from a GFD. This choice is, in our opinion, entirely acceptable, the only exception being the need to exclude CD with certainty, especially seronegative CD, if the serological and biopsy investigations have not been performed during a gluten-containing diet.

Finally, although not specified by the Salerno experts, it should be verified whether patients really follow a "strict" GFD. For this purpose, a GFD adherence questionnaire can be used, even if it was specifically designed for CD and not NCWS (Biagi et al., 2009).

17.8.1.2 Evaluation of the clinical response after the reintroduction of gluten/wheat into the diet

The second step is required to confirm the actual gluten sensitivity of these subjects by a challenge in which the patient is unaware of whether gluten or placebo has been introduced. To this end, two different methods are suggested: (1) the DBPC gluten challenge with crossover, which is more rigorous, but more complex to implement (Biesiekierski et al., 2011; Carroccio et al., 2012; Elli et al., 2016; Francavilla et al., 2018; Shahbazkhani et al., 2020); (2) the single-blind placebo-controlled gluten challenge with crossover, which is easier to perform, but potentially biased. The Salerno criteria suggests that the DBPC gluten challenge should be the only one to be used for experimental purposes, while in clinical practice the single-blind placebo-controlled gluten challenge might be sufficient.

The methodology to perform the challenge has been carefully described, suggesting the addition of 8 grams/day of gluten, containing at least 0.3 grams of ATIs/8 grams of gluten, to the GFD. Furthermore, the use of muesli bars, bread, or muffins as vehicles for gluten intake is encouraged, thus avoiding the use of gelatin capsules. Obviously, whatever the vehicle used, the organoleptic and packaging features of the gluten and placebo doses must be completely indistinguishable. In this regard, it seems likely to affirm that the best results are obtained by directly supplying the flours (with and without gluten, indistinguishable from each other) to the patients, leaving them free to cook (Carroccio et al., 2012, 2019, 2021; Mansueto et al., 2020, 2021, 2022; Seidita et al., 2022).

The overall duration of the challenge is a minimum of 3 weeks: a first week of challenge, with the intake of gluten or placebo, a second week of wash-out, and, finally, a third week of challenge with cross-over. The authors suggest extending the duration of the challenge if the symptoms are blurred. Throughout the challenge period, including wash-out, patients should complete the modified version of the GSRS daily. In the challenge-negative patients ($<30\%$ worsening between placebo and gluten) or in patients experiencing a "nocebo effect" ($>30\%$ worsening between gluten and placebo) a NCWS diagnosis can be ruled out.

17.8.1.3 General considerations

This diagnostic approach, which has tried to bring an element of clarity and reproducibility to this challenging issue, raises at least three considerations.

17.8.1.3.1 Difficulty of application in real-life clinical practice

As the Salerno experts point out, proposing a similar diagnostic protocol to patients with suspected NCWS is challenging. First, a large percentage of cases come to the first visit already following a GFD, and categorically refuse to reintroduce gluten. Furthermore, even when they decide to undergo the challenge, they are introduced directly in the second step, being considered "responders" to the GFD. Although this is acceptable in clinical practice, it is, however, not correct from a methodological point of view, as it is not possible to be certain either of the composition of the pre-GFD diet, or of the methods of application of the GFD itself. Additionally, this may represent just a placebo effect.

Finally, many of these patients eliminate multiple foods from their diet at the same time, as they are probably suffering from MFH, thus making it very difficult to identify what the real trigger of their symptoms is (Carroccio et al., 2012, 2013). In addition to the above, the long timescales involved (6 weeks of run-in, followed by at least 6 weeks of strict GFD and 3 weeks of challenge) and the complexity of performing a double- or single-blind challenge discourage patients and physicians from taking this path.

17.8.1.3.2 Choice of products to be tested in the challenge

The proposed diagnostic protocol involves the use of gluten only, even if it is normally balanced with at least 0.3 grams of ATIs per 8 grams of gluten, and the vehicle should be FODMAPs free. Since the pathogenic aspects of NCWS are uncertain and gluten may not be the real culprit, the question arises whether performing a challenge with gluten is sufficient or not. To overcome this problem, we suggest to carry out the challenge with wheat flour rather than gluten (Carroccio et al., 2019, 2021; Mansueto et al., 2020, 2021, 2022).

17.8.1.3.3 Poor reproducibility of the diagnostic protocol

In the light of the many difficulties, it is extremely unlikely that the patient will agree to repeat it, although this depends on the individual patient, and there is no doubt that the poor repeatability of the protocol reduces its diagnostic value.

These difficulties increasingly lead to a self-diagnosis of NCWS being accepted as sufficient to classify a patient as truly suffering from NCWS. While this approach has been accepted in clinical practice, it certainly cannot be considered suitable for research purposes. Therefore the Salerno diagnostic protocol remains to date the only one capable of guaranteeing a relatively certain diagnosis of NCGS, although not of NCWS.

17.8.2 The role of AGA in NCWS diagnosis

Exposure to gluten may activate an immunological process, as this seems to be confirmed by the finding of AGA positivity in a large subgroup of NCWS patients. In our study population of 276 NCWS subjects diagnosed by DBPC wheat challenge, 39.9% were positive for AGA IgA and 55% for AGA IgG (Carroccio et al., 2012). Another study demonstrated a positivity for AGA IgG in 56.4% of cases, but only 7.7% of AGA IgA (Volta et al., 2012). Again, the same latter group evaluated AGA IgA and IgG in 44 NCGS patients compared with 40 CD, before and after 6 months of GFD. After 6 months of GFD, in 93.2% of patients with NCGS there was a negativization of AGA IgG, while this occurred only in 60.0% of CD patients. Of note, AGA IgG persistence in NCGS was correlated with a low GFD adherence and with a mild clinical response (Caio et al., 2014).

Differently, Biesiekierski et al. analyzed AGA in 34 NCGS patients, finding no significant difference before and after a DBPC gluten challenge with crossover, thus concluding that AGA were not induced in this condition (Biesiekierski et al., 2011).

Although these data are inconclusive, and AGA certainly cannot be used as a diagnostic marker of NCWS, it seems that these patients develop this antibody response more often than patients with IBS (Wahnschaffe et al., 2007), or with various autoimmune conditions (Pellegrini et al., 1991), or the general population (Volta et al., 2001).

17.8.3 Intestinal permeability in NCWS diagnosis

It has been suggested that alterations in intestinal permeability could represent one of the main pathologic mechanisms of NCWS, and consequently they might be studied as a potential diagnostic (although not specific) marker of NCWS.

The lactulose/mannitol (La/Ma) test is a noninvasive method to study intestinal permeability. A recent US study evaluated the reliability of the La/Ma test in differentiating between CD patients, NCGS, and healthy volunteers during a DBPC gluten challenge. La/Ma test was not only unable to differentiate subjects with NCGS from CD and healthy controls, but also it did not change between baseline and challenge in any of the groups (Cartee et al., 2022). Another study, analyzing NCGS, CD and functional dyspepsia patients, showed that the La/Ma test values were significantly higher in CD than in NCGS patients, the latter presenting even lower values than functional dyspepsia (Sapone et al., 2011). Thus, although some authors have reported impaired intestinal permeability in NCWS (Clemente et al., 2003; Hollon et al., 2015; Uhde et al., 2016), the La/Ma ratio test does not seem able to identify it with sufficient accuracy, as also confirmed in the Salerno Experts' criteria (Catassi et al., 2015).

Among other potential serological markers of intestinal permeability, Barbaro et al. evaluated the diagnostic performance of serum zonulin in 86 NCWS patients, comparing them with 59 IBS, 15 CD and 25 healthy controls. In this study, NCWS and CD subjects had significantly higher serum zonulin levels than IBS and healthy controls, but no significant differences were found between them. In 44 of the 86 NCWS patients zonulin levels were reassessed after 6 months of WFD, demonstrating a significant reduction only in the subgroup with HLA DQ2/DQ8 positivity. The diagnostic accuracy of zonulin in distinguishing NCWS from IBS and healthy controls was 81% and 87% respectively, whereas it was not adequate in discriminating NCWS from CD patients (Barbaro et al., 2020).

17.8.4 Histological findings in NCWS diagnosis

Since the first studies on NCWS, it has been evident that some NCWS patients have alterations of the intestinal mucosa. Out of 276 NCWS patients, 89.6% presented intraepithelial lymphocyte (IEL) infiltration >25/100 epithelial cells on duodenal biopsy, which was significantly higher than in patients with IBS; moreover, the duodenal eosinophil count was significantly higher in NCWS patients than in both CD and IBS. On rectal biopsy, IEL infiltration was found in 24.6% of NCWS patients (not significantly different from IBS), but eosinophilic infiltration was found much more frequently both intraepithelial (63.0%) and in the lamina propria (59.8%), with a significantly higher frequency than in IBS patients (Carroccio et al., 2012).

Other studies have reported an increased number of IELs in the duodenal mucosa of NCWS patients, although in a lower percentage of cases. In a large multicenter study, the authors highlighted that 53% of NCGS patients were Marsh 0, and 47% were Marsh I-II, whereas all control participants were Marsh 0. Median IELs in NCGS patients with Marsh 0 was higher than in controls. Therefore, the authors assessed the utility of IEL density in distinguishing Marsh 0 NCGS from controls, reporting that an IEL cut-off of 14 had a 79% sensitivity and 55% specificity (Rostami et al., 2022). Of course, due to the lack of specificity, these findings cannot help in NCWS diagnosis (Volta et al., 2020).

17.8.5 Gut microbiota assessment in NCWS diagnosis

Recently, the role of the intestinal microbiota in NCWS has attracted increasingly wider attention (Barbara et al., 2021; Dieterich et al., 2019; Nobel et al., 2021). The search for a possible characteristic microbiota profile in NCWS is, however, still far from providing clear and reproducible indications; nevertheless, it should be a target of future investigations.

17.9 Treatment

The therapeutic approach consists of eliminating gluten/wheat from the diet (Catassi et al., 2015). It might therefore appear superfluous to emphasize how a GFD/WFD is able to lead, if not to the complete remission, at least to a consistent reduction in both the intestinal and extra-intestinal symptoms. Nonetheless, more than one perplexity remains concerning the most appropriate dietary approach. According to some researchers, it is not sufficient to exclude only gluten from the diet since the etiological agent could be one of the nongluten proteins or the glucidic fractions of wheat (Catassi et al., 2017). In addition, several patients with self-reported NCWS have been shown to fail the DBPC gluten challenge, which suggests that gluten might be tolerated, and instead ATIs or FODMAPs might be the harmful, at least in a subgroup of patients (Biesiekierski et al., 2013; Skodje et al., 2018).

Finally, some study groups are focusing attention on the possible benefits of "ancient" grains, which were widely cultivated in the past, before intensive plant breeding, and are today only grown on a small scale because of their low yields (Geisslitz et al., 2022). These wheat strains seem to show a lower ATI activity compared to the modern hexaploid wheats (Geisslitz et al., 2020; Zevallos et al., 2017). However, there are conflicting data about this issue and recommending an "ancient wheat-based diet" in NCWS is far from being accepted (Dieterich et al., 2022). In this context, we recently evaluated the consumption of ancient Sicilian wheat cultivars in NCWS subjects. More than half of patients reported consuming ancient grains before NCWS diagnosis; of note, the diagnostic delay in this group was significantly higher than in nonconsumers. A subgroup of patients, with optimal adherence to a WFD reported ancient wheat intake after NCWS diagnosis. Among them, some reported milder symptoms than those caused by modern wheat intake, and a few reported an excellent tolerability. Consequently, this study showed that in a low percentage of NCWS patients ancient wheat strains might be tolerated (Seidita et al., 2022). Nevertheless, other authors showed that no differences could be proved in symptoms, signs of inflammation or markers for intestinal barrier integrity after wheat or spelt products intake (Zimmermann et al., 2022).

Indeed, other questions remain open. Is there a general threshold value of gluten/wheat, ATIs, FODMAPs, etc. that can be tolerated, or is a patient's clinical response dose-independent? Is it therefore conceivable to recommend that these patients follow a diet containing low doses of the trigger element, rather than completely depriving affected patients of some otherwise nutritionally beneficial foods? Furthermore, the current absence of long-term follow-up studies on these patients does not allow us to establish whether they are able to regain tolerance over the years, although isolated reports seem to indicate this possibility (Carroccio, D'Alcamo, et al., 2017; Roncoroni et al., 2019).

Finally, we do not know what the clinical consequences are of adopting such a restrictive diet for a long time and the implications for a patient's quality of life. Indeed, the likelihood of developing nutritional deficiencies/malnutrition following a nonsupplemented and unsupervised GFD, is well-known (Al-Toma et al., 2019). It is plausible that similar conditions might also be valid for NCWS subjects. Specific studies are required to confirm this hypothesis.

17.10 Different points of view

Although many studies support the idea that some patients might suffer of a wheat sensitivity not related to CD or WA, some authors have questioned the real existence of NCGS/NCWS.

Both the Australian and the Norwegian study groups focused on a prevalent role of FODMAPs (Biesiekierski et al., 2013; Skodje et al., 2018), proving that patients who self-reported NCGS had symptom's improvement/worsening regardless of the gluten intake, so that they concluded: "That NCGS, as currently defined, might not be a discrete entity or that this entity might be confounded by FODMAP restriction" (Biesiekierski et al., 2013). In addition, an Italian randomized DBPC cross-over trial with small amount of gluten, showed that most of self-reported NCGS patients showed approximately equal degrees of symptoms with either gluten or placebo (Di Sabatino et al., 2015).

Even more conflicting data come from studies aimed at identifying the pathogenetic basis of NCWS. Di Sabatino et al. analyzed the innate and adaptive immunity pattern in self-reported NCGS subjects on an unrestricted diet, proving no differences in expression of IL-1β, IL-4, IL-5, IL-6, IL-8, IL-10, IL-12p70, IL13, IL-15, IL-17A IL-23, IL-27, IL-32α, thymic stromal lymphopoietin, TNF-α, IFN-α, IFN-γ, CCL1, CCL2, CCL3, CCL4, CCL5, CXCL1, CXCL10, granulocyte colony stimulating factor and granulocyte-macrophage colony stimulating factor, compared to CD and healthy controls (Di Sabatino et al., 2016).

However, we should stress that these studies considered only gluten as the putative trigger of NCWS symptoms, so the hypothesis that wheat challenge might prove different results remains entirely intact.

17.11 Future perspectives

As of March 2023, more than 15 trials related to NCWS were in progress. These range from epidemiological studies, aimed at identifying the prevalence of NCWS in pediatric and adolescent populations, to studies analyzing the association between NCWS and some autoimmune or neuropsychiatric diseases, and to studies designed to identify potential diagnostic biomarkers or different therapeutic approaches. It is patently clear that a lot of questions about NCWS still need to be answered. Once the way to discriminate patients with NCWS will be identified, as happened for CD about 40 years ago, the details of the epidemiological, pathogenic and immunohistochemical characteristics will be unraveled with more certain results. The final goal of these efforts will be to identify an effective and safe therapeutic approach, either dietary or not.

17.12 Gluten/wheat-dependent neurological manifestations

Outside of the neuropsychiatric manifestation of NCWS, a group of patients with neurological manifestations and positive AGA antibodies have been deeply described, mainly by the Sheffield research group (Hadjivassiliou et al., 1996, 2003, 2014, 2016, 2017; Rouvroye et al., 2020).

17.12.1 Definition

Gluten/wheat-dependent neurological manifestations are symptoms not attributable to other neurological diseases and which improve on a GFD/WFD and reappear after DPBC challenge with wheat/gluten. Such manifestations have been found not only in CD patients, but also in patients with GS without gluten enteropathy (Hadjivassiliou et al., 2016).

Patients with gluten-related neurological manifestations are considered GS only in the presence of positive AGA IgG and/or IgA, with or without positivity of anti-tTG and EMA, but without evidence of enteropathy (Hadjivassiliou et al., 1996). In addition, it must be remembered that neurological symptoms can be an extraintestinal manifestation in patients with NCWS.

17.12.2 Epidemiology and pathogenesis

There are no accurate data on the prevalence of gluten/wheat-related neurological disorders in the general population (Rouvroye et al., 2020). Very little is known about the pathogenesis of gluten/wheat-related neurological manifestations, but an immune-mediated basis has been hypothesized (Rouvroye et al., 2020). The intestinal microbiota could favor the accumulation of wheat neurotoxic peptides so that these could first reach the bloodstream due to the increased intestinal permeability, and subsequently cross the blood-brain barrier, influencing the endogenous opiate system and cerebral neurotransmission (Cryan & Dinan, 2012; Dehhaghi et al., 2019). Neurotoxic peptides could also have an immunogenic potential, therefore they could activate the innate and adaptive immunity, accounting for extraintestinal and neurological symptoms (Sapone et al., 2010).

17.12.3 Clinical manifestations

Several retrospective studies have shown that there are no differences in neurological manifestations between patients with GS and those with CD. The most frequent gluten-related neurological syndromes are encephalopathy, ataxia, peripheral neuropathy, and myopathy. Epilepsy, chorea, myelopathy, and stiff person syndrome are rarer (Hadjivassiliou et al., 2014). Encephalopathy was shown to be more frequent in patients with CD, while gangliopathy and neuropathy were more frequent in patients with GS (Hadjivassiliou et al., 2014, 2016). As far as ataxia is concerned, it is not clear whether it is a more frequent manifestation in CD (Hadjivassiliou et al., 2016) or GS (Rodrigo et al., 2016).

17.12.4 Diagnosis

The diagnosis of gluten/wheat-dependent neurological manifestations is made only by excluding all potential organic conditions, both overt neurological and psychiatric diseases. Furthermore, it must be ruled out that symptoms are a complication of a patient's comorbidity, such as diabetes, exposure to neurotoxic agents, or vitamin deficiency. Diagnosis is currently based on the results of serological tests. AGA IgG/IgA must be always detectable, while anti-tTG IgA/IgG and EMA are not, because they are specific to CD. In some cases, studies have demonstrated the presence of magnetic resonance imaging abnormalities in patients with neurological symptoms and gluten-related disorders. In ataxia, for example, cerebellar atrophy is present, while the presence of white matter abnormalities in a patient with GS and headache might suggest a condition of gluten encephalopathy (Hadjivassiliou et al., 2014). In the case of patients with GS, where antibody positivity is not present, it is necessary to define the cause-effect correlation between the gluten/wheat intake and symptoms, based on their improvement during the GFD/WFD and subsequent worsening during the DBPC challenge with gluten/wheat.

17.12.5 Treatment

Treatment should include the use of drugs for neurological symptom control and an etiological treatment based on a GFD/WFD (Zis & Hadjivassiliou, 2019). Several studies have demonstrated not only the clinical improvement of symptoms during a strict GFD, but also a radiological (Hadjivassiliou et al., 2017), and an electrophysiological amelioration (Hadjivassiliou et al., 2003). A different study also highlighted that the efficacy of a GFD was inversely related to the diagnostic delay, therefore an early initiation of the diet was more frequently associated with the remission of neurological symptoms. If the neurological damage was already irreversible, a GFD prevented its worsening (Zis & Hadjivassiliou, 2019). In addition, rigorously following a GFD led to an improvement in the patients' quality of life due to the consequent better pain control (Zis et al., 2018).

References

Afifi, L., Danesh, M. J., Lee, K. M., Beroukhim, K., Farahnik, B., Ahn, R. S., Yan, D., Singh, R. K., Nakamura, M., Koo, J., & Liao, W. (2017). Dietary behaviors in psoriasis: Patient-reported outcomes from a U.S. National Survey. *Dermatology and Therapy*, *7*(2), 227–242. Available from https://doi.org/10.1007/s13555-017-0183-4.

Alghamdi, M. (2017). Familial Mediterranean Fever, review of the literature. *Clinical Rheumatology*, *36*(8), 1707–1713. Available from https://doi.org/10.1007/s10067-017-3715-5.

Almirall, M., Casellas, F., Dot, J., de Torres, I., Segurola, H., & Marsal, S. (2023). Prevalence of non-coeliac gluten sensitivity and assessment of the response to gluten-free diet in a cohort of patients with fibromyalgia. *Rheumatology (Oxford, England)*, *62*(3), 1069–1077. Available from https://doi.org/10.1093/rheumatology/keac434.

Al-Toma, A., Volta, U., Auricchio, R., Castillejo, G., Sanders, D. S., Cellier, C., Mulder, C. J., & Lundin, K. E. A. (2019). European Society for the Study of Coeliac Disease (ESsCD) guideline for coeliac disease and other gluten-related disorders. *United European Gastroenterology Journal*, *7*(5), 583–613. Available from https://doi.org/10.1177/2050640619844125.

Araya, M., Bascuñán, K. A., Alarcón-Sajarópulos, D., Cabrera-Chávez, F., Oyarzún, A., Fernández, A., & Ontiveros, N. (2020). Living with gluten and other food intolerances: Self-reported diagnoses and management. *Nutrients*, *12*(6). Available from https://doi.org/10.3390/nu12061892.

Aziz, I., Branchi, F., Pearson, K., Priest, J., & Sanders, D. S. (2015). A study evaluating the bidirectional relationship between inflammatory bowel disease and self-reported non-celiac gluten sensitivity. *Inflammatory Bowel Diseases*, *21*(4), 847–853. Available from https://doi.org/10.1097/MIB.0000000000000335.

Aziz, I., Hadjivassiliou, M., & Sanders, D. S. (2015). The spectrum of noncoeliac gluten sensitivity. *Nature Reviews. Gastroenterology & Hepatology*, *12*(9), 516–526. Available from https://doi.org/10.1038/nrgastro.2015.107.

Aziz, I., Lewis, N. R., Hadjivassiliou, M., Winfield, S. N., Rugg, N., Kelsall, A., Newrick, L., & Sanders, D. S. (2014). A UK study assessing the population prevalence of self-reported gluten sensitivity and referral characteristics to secondary care. *European Journal of Gastroenterology & Hepatology*, *26*(1), 33–39. Available from https://doi.org/10.1097/01.meg.0000435546.87251.f7.

Barbara, G., Barbaro, M. R., Fuschi, D., Palombo, M., Falangone, F., Cremon, C., Marasco, G., & Stanghellini, V. (2021). Inflammatory and Microbiota-related regulation of the intestinal epithelial barrier. *Frontiers in Nutrition*, *8*, 718356. Available from https://doi.org/10.3389/fnut.2021.718356.

Barbaro, M. R., Cremon, C., Morselli-Labate, A. M., Di Sabatino, A., Giuffrida, P., Corazza, G. R., Di Stefano, M., Caio, G., Latella, G., Ciacci, C., Fuschi, D., Mastroroberto, M., Bellacosa, L., Stanghellini, V., Volta, U., & Barbara, G. (2020). Serum zonulin and its diagnostic performance in non-coeliac gluten sensitivity. *Gut*, *69*(11), 1966–1974. Available from https://doi.org/10.1136/gutjnl-2019-319281.

Barone, M., Gemello, E., Viggiani, M. T., Cristofori, F., Renna, C., Iannone, A., Di Leo, A., & Francavilla, R. (2020). Evaluation of non-celiac gluten sensitivity in patients with previous diagnosis of irritable bowel syndrome: A randomized double-blind placebo-controlled crossover trial. *Nutrients*, *12*(3). Available from https://doi.org/10.3390/nu12030705.

Bell, K. A., Pourang, A., Mesinkovska, N. A., & Cardis, M. A. (2021). The effect of gluten on skin and hair: A systematic review. *Dermatology Online Journal*, *27*(4).

Bhatia, B. K., Millsop, J. W., Debbaneh, M., Koo, J., Linos, E., & Liao, W. (2014). Diet and psoriasis, part II: Celiac disease and role of a gluten-free diet. *Journal of the American Academy of Dermatology*, *71*(2), 350–358. Available from https://doi.org/10.1016/j.jaad.2014.03.017.

Biagi, F., Andrealli, A., Bianchi, P. I., Marchese, A., Klersy, C., & Corazza, G. R. (2009). A gluten-free diet score to evaluate dietary compliance in patients with coeliac disease. *The British Journal of Nutrition*, *102*(6), 882–887. Available from https://doi.org/10.1017/S0007114509301579.

Biesiekierski, J. R. (2017). What is gluten? *Journal of Gastroenterology and Hepatology*, *32*(Suppl 1), 78–81. Available from https://doi.org/10.1111/jgh.13703.

Biesiekierski, J. R., Newnham, E. D., Irving, P. M., Barrett, J. S., Haines, M., Doecke, J. D., Shepherd, S. J., Muir, J. G., & Gibson, P. R. (2011). Gluten causes gastrointestinal symptoms in subjects without celiac disease: A double-blind randomized placebo-controlled trial. *The American Journal of Gastroenterology*, *106*(3), 508–514. Available from https://doi.org/10.1038/ajg.2010.487, quiz 515.

Biesiekierski, J. R., Peters, S. L., Newnham, E. D., Rosella, O., Muir, J. G., & Gibson, P. R. (2013). No effects of gluten in patients with self-reported non-celiac gluten sensitivity after dietary reduction of fermentable, poorly absorbed, short-chain carbohydrates. *Gastroenterology*, *145*(2), 320–328. Available from https://doi.org/10.1053/j.gastro.2013.04.051, e1–3.

Bonciolini, V., Bianchi, B., Del Bianco, E., Verdelli, A., & Caproni, M. (2015). Cutaneous manifestations of non-celiac gluten sensitivity: Clinical histological and immunopathological features. *Nutrients*, *7*(9), 7798–7805. Available from https://doi.org/10.3390/nu7095368.

Brottveit, M., Beitnes, A.-C. R., Tollefsen, S., Bratlie, J. E., Jahnsen, F. L., Johansen, F.-E., Sollid, L. M., & Lundin, K. E. A. (2013). Mucosal cytokine response after short-term gluten challenge in celiac disease and non-celiac gluten sensitivity. *The American Journal of Gastroenterology*, *108*(5), 842–850. Available from https://doi.org/10.1038/ajg.2013.91.

Brottveit, M., Vandvik, P. O., Wojniusz, S., Løvik, A., Lundin, K. E., & Boye, B. (2012). Absence of somatization in non-coeliac gluten sensitivity. *Scandinavian Journal of Gastroenterology*, *47*(7), 770–777. Available from https://doi.org/10.3109/00365521.2012.679685.

Caio, G., Volta, U., Tovoli, F., & De Giorgio, R. (2014). Effect of gluten free diet on immune response to gliadin in patients with non-celiac gluten sensitivity. *BMC Gastroenterology*, *14*, 26. Available from https://doi.org/10.1186/1471-230X-14-26.

Camhi, S. S., Sangal, K., Kenyon, V., Lima, R., Fasano, A., & Leonard, M. M. (2019). Pediatric nonceliac gluten sensitivity: A gluten-related disorder treatment center experience. *Journal of Pediatric Gastroenterology and Nutrition*, *69*(2), 200–205. Available from https://doi.org/10.1097/MPG.0000000000002335.

Carroccio, A., D'Alcamo, A., Cavataio, F., Soresi, M., Seidita, A., Sciumè, C., Geraci, G., Iacono, G., & Mansueto, P. (2015). High proportions of people with nonceliac wheat sensitivity have autoimmune disease or antinuclear antibodies. *Gastroenterology*, *149*(3), 596–603. Available from https://doi.org/10.1053/j.gastro.2015.05.040, e1.

Carroccio, A., D'Alcamo, A., Iacono, G., Soresi, M., Iacobucci, R., Arini, A., Geraci, G., Fayer, F., Cavataio, F., La Blasca, F., Florena, A. M., & Mansueto, P. (2017). Persistence of Nonceliac wheat sensitivity, based on long-term follow-up. *Gastroenterology*, *153*(1), 56–58. Available from https://doi.org/10.1053/j.gastro.2017.03.034, e3.

Carroccio, A., Giambalvo, O., Blasca, F. L., Iacobucci, R., D'Alcamo, A., & Mansueto, P. (2017). Self-Reported non-celiac wheat sensitivity in high school students: Demographic and clinical characteristics. *Nutrients*, *9*(7). Available from https://doi.org/10.3390/nu9070771.

Carroccio, A., Giannone, G., Mansueto, P., Soresi, M., La Blasca, F., Fayer, F., Iacobucci, R., Porcasi, R., Catalano, T., Geraci, G., Arini, A., D'Alcamo, A., Villanacci, V., & Florena, A. M. (2019). Duodenal and rectal mucosa inflammation in patients with non-celiac wheat sensitivity. *Clinical Gastroenterology and Hepatology: The Official Clinical Practice Journal of the American Gastroenterological Association*, *17*(4), 682–690. Available from https://doi.org/10.1016/j.cgh.2018.08.043, e3.

Carroccio, A., Mansueto, P., D'Alcamo, A., & Iacono, G. (2013). Non-celiac wheat sensitivity as an allergic condition: Personal experience and narrative review. *The American Journal of Gastroenterology*, *108*(12), 1845–1852. Available from https://doi.org/10.1038/ajg.2013.353, quiz 1853.

Carroccio, A., Mansueto, P., Iacono, G., Soresi, M., D'Alcamo, A., Cavataio, F., Brusca, I., Florena, A. M., Ambrosiano, G., Seidita, A., Pirrone, G., & Rini, G. B. (2012). Non-celiac wheat sensitivity diagnosed by double-blind placebo-controlled challenge: Exploring a new clinical entity. *The American Journal of Gastroenterology*, *107*(12), 1898–1906. Available from https://doi.org/10.1038/ajg.2012.236, quiz 1907.

Carroccio, A., Mansueto, P., Soresi, M., Fayer, F., Di Liberto, D., Monguzzi, E., Lo Pizzo, M., La Blasca, F., Geraci, G., Pecoraro, A., Dieli, F., & Schuppan, D. (2020). Wheat consumption leads to immune activation and symptom worsening in patients with Familial Mediterranean Fever: A pilot randomized trial. *Nutrients*, *12*(4). Available from https://doi.org/10.3390/nu12041127.

Carroccio, A., Rini, G., & Mansueto, P. (2014). Non-celiac wheat sensitivity is a more appropriate label than non-celiac gluten sensitivity. *Gastroenterology*, *146*(1), 320–321. Available from https://doi.org/10.1053/j.gastro.2013.08.061.

Carroccio, A., Soresi, M., Chiavetta, M., La Blasca, F., Compagnoni, S., Giuliano, A., Fayer, F., Mandreucci, F., Castellucci, D., Seidita, A., Affronti, A., Florena, A. M., & Mansueto, P. (2021). Frequency and clinical aspects of neurological and psychiatric symptoms in patients with non-celiac wheat sensitivity. *Nutrients*, *13*(6). Available from https://doi.org/10.3390/nu13061971.

Cartee, A. K., Choung, R. S., King, K. S., Wang, S., Dzuris, J. L., Anderson, R. P., Van Dyke, C. T., Hinson, C. A., Marietta, E., Katzka, D. A., Nehra, V., Grover, M., & Murray, J. A. (2022). Plasma IL-2 and symptoms response after acute gluten exposure in subjects with celiac disease or nonceliac gluten sensitivity. *The American Journal of Gastroenterology*, *117*(2), 319–326. Available from https://doi.org/10.14309/ajg.0000000000001565.

Catassi, C., Alaedini, A., Bojarski, C., Bonaz, B., Bouma, G., Carroccio, A., Castillejo, G., De Magistris, L., Dieterich, W., Di Liberto, D., Elli, L., Fasano, A., Hadjivassiliou, M., Kurien, M., Lionetti, E., Mulder, C. J., Rostami, K., Sapone, A., Scherf, K., & Sanders, D. S. (2017). The overlapping area of non-celiac gluten sensitivity (NCGS) and wheat-sensitive irritable bowel syndrome (IBS): An update. *Nutrients*, *9*(11). Available from https://doi.org/10.3390/nu9111268.

Catassi, C., Elli, L., Bonaz, B., Bouma, G., Carroccio, A., Castillejo, G., Cellier, C., Cristofori, F., de Magistris, L., Dolinsek, J., Dieterich, W., Francavilla, R., Hadjivassiliou, M., Holtmeier, W., Körner, U., Leffler, D. A., Lundin, K. E. A., Mazzarella, G., Mulder, C. J., & Fasano, A. (2015). Diagnosis of non-celiac gluten sensitivity (NCGS): The Salerno Experts' Criteria. *Nutrients*, *7*(6), 4966–4977. Available from https://doi.org/10.3390/nu7064966.

Cha, R. R., Kim, J. H., Koo, H. S., Jung, K. W., Min, Y. W., Choi, C. H., Ryu, H. S., Kwon, Y. H., Cho, D. H., Kwon, J. G., Park, K. S., & Kim, H. J. (2022). Self-reported non-celiac gluten sensitivity in the Korean population: Demographic and clinical characteristics. *Journal of Neurogastroenterology and Motility*, *28*(2), 283–290. Available from https://doi.org/10.5056/jnm21108.

Choung, R.S., Unalp-Arida, A., Ruhl, C.E., Brantner, T.L., Everhart, J.E., & Murray, J.A. (2016). Less hidden celiac disease but increased gluten avoidance without a diagnosis in the United States: Findings from the National Health and Nutrition Examination Surveys From 2009 to 2014. Mayo Clinic Proceedings. https://doi.org/10.1016/j.mayocp.2016.10.012.

Clemente, M. G., De Virgiliis, S., Kang, J. S., Macatagney, R., Musu, M. P., Di Pierro, M. R., Drago, S., Congia, M., & Fasano, A. (2003). Early effects of gliadin on enterocyte intracellular signalling involved in intestinal barrier function. *Gut*, *52*(2), 218–223.

Conti, V., Leone, M. C., Casato, M., Nicoli, M., Granata, G., & Carlesimo, M. (2015). High prevalence of gluten sensitivity in a cohort of patients with undifferentiated connective tissue disease. *European Annals of Allergy and Clinical Immunology*, *47*(2), 54–57.

Cooper, B. T., Holmes, G. K., Ferguson, R., Thompson, R. A., Allan, R. N., & Cooke, W. T. (1980). Gluten-sensitive diarrhea without evidence of celiac disease. *Gastroenterology*, *79*(5 Pt 1), 801–806.

Croall, I. D., Hoggard, N., & Hadjivassiliou, M. (2021). Gluten and autism spectrum disorder. *Nutrients*, *13*(2). Available from https://doi.org/10.3390/nu13020572.

Cryan, J. F., & Dinan, T. G. (2012). Mind-altering microorganisms: The impact of the gut microbiota on brain and behaviour. *Nature Reviews. Neuroscience*, *13*(10), 701–712. Available from https://doi.org/10.1038/nrn3346.

Dehhaghi, M., Kazemi Shariat Panahi, H., & Guillemin, G. J. (2019). Microorganisms, tryptophan metabolism, and kynurenine pathway: A complex interconnected loop influencing human health status. *International Journal of Tryptophan Research: IJTR*, *12*. Available from https://doi.org/10.1177/1178646919852996, 1178646919852996.

Di Liberto, D., Mansueto, P., D'Alcamo, A., Lo Pizzo, M., Lo Presti, E., Geraci, G., Fayer, F., Guggino, G., Iacono, G., Dieli, F., & Carroccio, A. (2016). Predominance of Type 1 innate lymphoid cells in the rectal mucosa of patients with non-celiac wheat sensitivity: Reversal after a wheat-free diet. *Clinical and Translational Gastroenterology*, *7*(7), e178. Available from https://doi.org/10.1038/ctg.2016.35.

Di Sabatino, A., Giuffrida, P., Fornasa, G., Salvatore, C., Vanoli, A., Naviglio, S., De Leo, L., Pasini, A., De Amici, M., Alvisi, C., Not, T., Rescigno, M., & Corazza, G. R. (2016). Innate and adaptive immunity in self-reported nonceliac gluten sensitivity versus celiac disease. *Digestive and Liver Disease: Official Journal of the Italian Society of Gastroenterology and the Italian Association for the Study of the Liver*, *48*(7), 745–752. Available from https://doi.org/10.1016/j.dld.2016.03.024.

Di Sabatino, A., Volta, U., Salvatore, C., Biancheri, P., Caio, G., De Giorgio, R., Di Stefano, M., & Corazza, G. R. (2015). Small amounts of gluten in subjects with suspected nonceliac gluten sensitivity: A randomized, double-blind, placebo-controlled, cross-over trial. *Clinical Gastroenterology and Hepatology: The Official Clinical Practice Journal of the American Gastroenterological Association*, *13*(9), 1604–1612. Available from https://doi.org/10.1016/j.cgh.2015.01.029, e3.

Dieterich, W., Schuppan, D., Schink, M., Schwappacher, R., Wirtz, S., Agaimy, A., Neurath, M. F., & Zopf, Y. (2019). Influence of low FODMAP and gluten-free diets on disease activity and intestinal microbiota in patients with non-celiac gluten sensitivity. *Clinical Nutrition (Edinburgh, Scotland)*, *38*(2), 697–707. Available from https://doi.org/10.1016/j.clnu.2018.03.017.

Dieterich, W., Schuster, C., Gundel, P., Scherf, K. A., Pronin, D., Geisslitz, S., Börner, A., Neurath, M. F., & Zopf, Y. (2022). Proteins from modern and ancient wheat cultivars: impact on immune cells of healthy individuals and patients with NCGS. *Nutrients*, *14*(20). Available from https://doi.org/10.3390/nu14204257.

DiGiacomo, D. V., Tennyson, C. A., Green, P. H., & Demmer, R. T. (2013). Prevalence of gluten-free diet adherence among individuals without celiac disease in the USA: Results from the Continuous National Health and Nutrition Examination Survey 2009–2010. *Scandinavian Journal of Gastroenterology*, *48*(8), 921–925. Available from https://doi.org/10.3109/00365521.2013.809598.

Dzikowski, M., Juchnowicz, D., Dzikowska, I., Rog, J., Próchnicki, M., Kozioł, M., & Karakula-Juchnowicz, H. (2020). The differences between gluten sensitivity, intestinal biomarkers and immune biomarkers in patients with first-episode and chronic schizophrenia. *Journal of Clinical Medicine*, *9*(11). Available from https://doi.org/10.3390/jcm9113707.

Ekinci, R. M. K., Balci, S., Bisgin, A., Cetin, F. T., & Tumgor, G. (2020). The contribution of diet preference to the disease course in children with familial Mediterranean fever: A cross-sectional study. *Reumatologia*, *58*(2), 81–86. Available from https://doi.org/10.5114/reum.2020.95361.

Elkan, A.-C., Sjöberg, B., Kolsrud, B., Ringertz, B., Hafström, I., & Frostegård, J. (2008). Gluten-free vegan diet induces decreased LDL and oxidized LDL levels and raised atheroprotective natural antibodies against phosphorylcholine in patients with rheumatoid arthritis: a randomized study. *Arthritis Research & Therapy*, *10*(2), R34. Available from https://doi.org/10.1186/ar2388.

Elli, L., Tomba, C., Branchi, F., Roncoroni, L., Lombardo, V., Bardella, M. T., Ferretti, F., Conte, D., Valiante, F., Fini, L., Forti, E., Cannizzaro, R., Maiero, S., Londoni, C., Lauri, A., Fornaciari, G., Lenoci, N., Spagnuolo, R., Basilisco, G., & Buscarini, E. (2016). Evidence for the presence of non-celiac gluten sensitivity in patients with functional gastrointestinal symptoms: Results from a multicenter randomized double-blind placebo-controlled gluten challenge. *Nutrients*, *8*(2), 84. Available from https://doi.org/10.3390/nu8020084.

Ellis, A., & Linaker, B. D. (1978). Non-coeliac gluten sensitivity? *Lancet (London, England)*, *1*(8078), 1358–1359.

Faina, V., Paolino, G., Bavastrelli, M., Calvieri, S., & Grieco, T. (2021). Classification of cutaneous manifestations in patients with nonceliac gluten sensitivity and wheat allergy. *Journal of the American Academy of Dermatology*, *85*(2), 469–472. Available from https://doi.org/10.1016/j.jaad.2017.12.013.

Fasano, A., Sapone, A., Zevallos, V., & Schuppan, D. (2015). Nonceliac gluten sensitivity. *Gastroenterology*, *148*(6), 1195–1204. Available from https://doi.org/10.1053/j.gastro.2014.12.049.

Ferguson, A. (1990). Food sensitivity or self-deception? *The New England Journal of Medicine*, *323*(7), 476–478.

Francavilla, R., Cristofori, F., Verzillo, L., Gentile, A., Castellaneta, S., Polloni, C., Giorgio, V., Verduci, E., D'Angelo, E., Dellatte, S., & Indrio, F. (2018). Randomized double-blind placebo-controlled crossover trial for the diagnosis of non-celiac gluten sensitivity in children. *The American Journal of Gastroenterology*, *113*(3), 421–430. Available from https://doi.org/10.1038/ajg.2017.483.

Francavilla, R., Cristofori, F., Castellaneta, S., Polloni, C., Albano, V., Dellatte, S., Indrio, F., Cavallo, L., & Catassi, C. (2014). Clinical, serologic, and histologic features of gluten sensitivity in children. *The Journal of Pediatrics*, *164*(3), 463–467. Available from https://doi.org/10.1016/j.jpeds.2013.10.007, e1.

Geisslitz, S., Longin, C. F. H., Koehler, P., & Scherf, K. A. (2020). Comparative quantitative LC-MS/MS analysis of 13 amylase/trypsin inhibitors in ancient and modern Triticum species. *Scientific Reports*, *10*(1), 14570. Available from https://doi.org/10.1038/s41598-020-71413-z.

Geisslitz, S., Weegels, P., Shewry, P., Zevallos, V., Masci, S., Sorrells, M., Gregorini, A., Colomba, M., Jonkers, D., Huang, X., De Giorgio, R., Caio, G. P., D'Amico, S., Larré, C., & Brouns, F. (2022). Wheat amylase/trypsin inhibitors (ATIs): Occurrence, function and health aspects. *European Journal of Nutrition*, *61*(6), 2873–2880. Available from https://doi.org/10.1007/s00394-022-02841-y.

Giancola, F., Volta, U., Repossi, R., Latorre, R., Beeckmans, D., Carbone, F., Van den Houte, K., Bianco, F., Bonora, E., Gori, A., Costanzini, A., Boschetti, E., Caio, G., Vanuytsel, T., Stanghellini, V., Tack, J., & De Giorgio, R. (2020). Mast cell-nerve interactions correlate with bloating and abdominal pain severity in patients with non-celiac gluten/wheat sensitivity. *Neurogastroenterology and Motility: The Official Journal of the European Gastrointestinal Motility Society*, *32*(6), e13814. Available from https://doi.org/10.1111/nmo.13814.

Gluten-free foods market in the U.S. - statistics & facts. (2022). https://www.statista.com/topics/2067/gluten-free-foods-market/#topicHeader__wrapper.

Hadjivassiliou, M., Davies-Jones, G. A. B., Sanders, D. S., & Grünewald, R. A. (2003). Dietary treatment of gluten ataxia. *Journal of Neurology, Neurosurgery, and Psychiatry*, *74*(9), 1221–1224.

Hadjivassiliou, M., Gibson, A., Davies-Jones, G. A., Lobo, A. J., Stephenson, T. J., & Milford-Ward, A. (1996). Does cryptic gluten sensitivity play a part in neurological illness? *Lancet (London, England)*, *347*(8998), 369–371.

Hadjivassiliou, M., Grünewald, R. A., Sanders, D. S., Shanmugarajah, P., & Hoggard, N. (2017). Effect of gluten-free diet on cerebellar MR spectroscopy in gluten ataxia. *Neurology*, *89*(7), 705–709. Available from https://doi.org/10.1212/WNL.0000000000004237.

Hadjivassiliou, M., Duker, A. P., & Sanders, D. S. (2014). Gluten-related neurologic dysfunction. *Handbook of Clinical Neurology*, *120*, 607–619. Available from https://doi.org/10.1016/B978-0-7020-4087-0.00041-3.

Hadjivassiliou, M., Rao, D. G., Grìnewald, R. A., Aeschlimann, D. P., Sarrigiannis, P. G., Hoggard, N., Aeschlimann, P., Mooney, P. D., & Sanders, D. S. (2016). Neurological dysfunction in coeliac disease and non-coeliac gluten sensitivity. *The American Journal of Gastroenterology, 111*(4), 561–567. Available from https://doi.org/10.1038/ajg.2015.434.

Heidari, F., Afshari, M., & Moosazadeh, M. (2017). Prevalence of fibromyalgia in general population and patients, a systematic review and meta-analysis. *Rheumatology International, 37*(9), 1527–1539. Available from https://doi.org/10.1007/s00296-017-3725-2.

Hill, P., Muir, J. G., & Gibson, P. R. (2017). Controversies and recent developments of the low-FODMAP diet. *Gastroenterology & Hepatology, 13* (1), 36–45.

Hollon, J., Puppa, E. L., Greenwald, B., Goldberg, E., Guerrerio, A., & Fasano, A. (2015). Effect of gliadin on permeability of intestinal biopsy explants from celiac disease patients and patients with non-celiac gluten sensitivity. *Nutrients, 7*(3), 1565–1576. Available from https://doi.org/10.3390/nu7031565.

Igbinedion, S. O., Ansari, J., Vasikaran, A., Gavins, F. N., Jordan, P., Boktor, M., & Alexander, J. S. (2017). Non-celiac gluten sensitivity: All wheat attack is not celiac. *World Journal of Gastroenterology, 23*(40), 7201–7210. Available from https://doi.org/10.3748/wjg.v23.i40.7201.

Isasi, C., Colmenero, I., Casco, F., Tejerina, E., Fernandez, N., Serrano-Vela, J. I., Castro, M. J., & Villa, L. F. (2014). Fibromyalgia and non-celiac gluten sensitivity: A description with remission of fibromyalgia. *Rheumatology International, 34*(11), 1607–1612. Available from https://doi.org/10.1007/s00296-014-2990-6.

Jackson, J., Eaton, W., Cascella, N., Fasano, A., Santora, D., Sullivan, K., Feldman, S., Raley, H., McMahon, R. P., Carpenter, W. T., Demyanovich, H., & Kelly, D. L. (2014). Gluten sensitivity and relationship to psychiatric symptoms in people with schizophrenia. *Schizophrenia Research, 159* (2–3), 539–542. Available from https://doi.org/10.1016/j.schres.2014.09.023.

Kakar, S., Nehra, V., Murray, J. A., Dayharsh, G. A., & Burgart, L. J. (2003). Significance of intraepithelial lymphocytosis in small bowel biopsy samples with normal mucosal architecture. *The American Journal of Gastroenterology, 98*(9), 2027–2033.

Khan, A., Suarez, M. G., & Murray, J. A. (2020). Nonceliac gluten and wheat sensitivity. *Clinical Gastroenterology and Hepatology: The Official Clinical Practice Journal of the American Gastroenterological Association, 18*(9), 1913–1922. Available from https://doi.org/10.1016/j.cgh.2019.04.009, e1.

Korman, N. J. (2020). Management of psoriasis as a systemic disease: What is the evidence. *The British Journal of Dermatology, 182*(4), 840–848. Available from https://doi.org/10.1111/bjd.18245.

Kulich, K. R., Madisch, A., Pacini, F., Piqué, J. M., Regula, J., Van Rensburg, C. J., Ujszászy, L., Carlsson, J., Halling, K., & Wiklund, I. K. (2008). Reliability and validity of the Gastrointestinal Symptom Rating Scale (GSRS) and Quality of Life in Reflux and Dyspepsia (QOLRAD) questionnaire in dyspepsia: A six-country study. *Health and Quality of Life Outcomes, 6*, 12. Available from https://doi.org/10.1186/1477-7525-6-12.

Lammers, K. M., Lu, R., Brownley, J., Lu, B., Gerard, C., Thomas, K., Rallabhandi, P., Shea-Donohue, T., Tamiz, A., Alkan, S., Netzel-Arnett, S., Antalis, T., Vogel, S. N., & Fasano, A. (2008). Gliadin induces an increase in intestinal permeability and zonulin release by binding to the chemokine receptor CXCR3. *Gastroenterology, 135*(1), 194–204. Available from https://doi.org/10.1053/j.gastro.2008.03.023, e3.

Lau, N. M., Green, P. H. R., Taylor, A. K., Hellberg, D., Ajamian, M., Tan, C. Z., Kosofsky, B. E., Higgins, J. J., Rajadhyaksha, A. M., & Alaedini, A. (2013). Markers of celiac disease and gluten sensitivity in children with autism. *PLoS One, 8*(6), e66155. Available from https://doi.org/10.1371/journal.pone.0066155.

Leccioli, V., Oliveri, M., Romeo, M., Berretta, M., & Rossi, P. (2017). A new proposal for the pathogenic mechanism of non-coeliac/non-allergic gluten/wheat sensitivity: Piecing together the puzzle of recent scientific evidence. *Nutrients, 9*(11). Available from https://doi.org/10.3390/nu9111203.

Lidén, M., Kristjánsson, G., Valtýsdóttir, S., & Hällgren, R. (2007). Gluten sensitivity in patients with primary Sjögren's syndrome. *Scandinavian Journal of Gastroenterology, 42*(8), 962–967.

Limketkai, B. N., Sepulveda, R., Hing, T., Shah, N. D., Choe, M., Limsui, D., & Shah, S. (2018). Prevalence and factors associated with gluten sensitivity in inflammatory bowel disease. *Scandinavian Journal of Gastroenterology, 53*(2), 147–151. Available from https://doi.org/10.1080/00365521.2017.1409364.

Lionetti, E., Pulvirenti, A., Vallorani, M., Catassi, G., Verma, A. K., Gatti, S., & Catassi, C. (2017). Re-challenge studies in non-celiac gluten sensitivity: A systematic review and meta-analysis. *Frontiers in Physiology, 8*, 621. Available from https://doi.org/10.3389/fphys.2017.00621.

Losurdo, G., Piscitelli, D., Pezzuto, F., Fortarezza, F., Covelli, C., Marra, A., Iannone, A., Amoruso, A., Principi, M., Ierardi, E., & Di Leo, A. (2017). T helper lymphocyte and mast cell immunohistochemical pattern in nonceliac gluten sensitivity. *Gastroenterology Research and Practice, 2017*, 5023680. Available from https://doi.org/10.1155/2017/5023680.

Losurdo, G., Principi, M., Iannone, A., Amoruso, A., Ierardi, E., Di Leo, A., & Barone, M. (2018). Extra-intestinal manifestations of non-celiac gluten sensitivity: An expanding paradigm. *World Journal of Gastroenterology, 24*(14), 1521–1530. Available from https://doi.org/10.3748/wjg.v24.i14.1521.

Losurdo, G., Principi, M., Iannone, A., Giangaspero, A., Piscitelli, D., Ierardi, E., Di Leo, A., & Barone, M. (2018). Predictivity of autoimmune stigmata for gluten sensitivity in subjects with microscopic enteritis: A retrospective study. *Nutrients, 10*(12). Available from https://doi.org/10.3390/nu10122001.

Mansueto, P., Di Liberto, D., Fayer, F., Soresi, M., Geraci, G., Giannone, A. G., Seidita, A., D'Alcamo, A., La Blasca, F., Lo Pizzo, M., Florena, A. M., Dieli, F., & Carroccio, A. (2020). TNF-α, IL-17, and IL-22 production in the rectal mucosa of nonceliac wheat sensitivity patients: role of adaptive immunity. *American Journal of Physiology. Gastrointestinal and Liver Physiology, 319*(3), G281–G288. Available from https://doi.org/10.1152/ajpgi.00104.2020.

Mansueto, P., Seidita, A., D'Alcamo, A., & Carroccio, A. (2014). Non-celiac gluten sensitivity: Literature review. *Journal of the American College of Nutrition*, *33*(1), 39–54. Available from https://doi.org/10.1080/07315724.2014.869996.

Mansueto, P., Seidita, A., Soresi, M., Giuliano, A., Riccio, G., Volta, U., Caio, G., La Blasca, F., Disclafani, R., De Giorgio, R., & Carroccio, A. (2022). Anemia in non-celiac wheat sensitivity: Prevalence and associated clinical and laboratory features. *Digestive and Liver Disease: Official Journal of the Italian Society of Gastroenterology and the Italian Association for the Study of the Liver*. Available from https://doi.org/10.1016/j.dld.2022.11.022.

Mansueto, P., Soresi, M., Candore, G., Garlisi, C., Fayer, F., Gambino, C. M., La Blasca, F., Seidita, A., D'Alcamo, A., Lo Sasso, B., Florena, A. M., Geraci, G., Caio, G., Volta, U., De Giorgio, R., Ciaccio, M., & Carroccio, A. (2021). Autoimmunity features in patients with non-celiac wheat sensitivity. *The American Journal of Gastroenterology*, *116*(5), 1015–1023. Available from https://doi.org/10.14309/ajg.0000000000000919.

Massari, S., Liso, M., De Santis, L., Mazzei, F., Carlone, A., Mauro, S., Musca, F., Bozzetti, M. P., & Minelli, M. (2011). Occurrence of nonceliac gluten sensitivity in patients with allergic disease. *International Archives of Allergy and Immunology*, *155*(4), 389–394. Available from https://doi.org/10.1159/000321196.

Mazzarella, G., Stefanile, R., Camarca, A., Giliberti, P., Cosentini, E., Marano, C., Iaquinto, G., Giardullo, N., Auricchio, S., Sette, A., Troncone, R., & Gianfrani, C. (2008). Gliadin activates HLA class I-restricted CD8 + T cells in celiac disease intestinal mucosa and induces the enterocyte apoptosis. *Gastroenterology*, *134*(4), 1017–1027. Available from https://doi.org/10.1053/j.gastro.2008.01.008.

Michaëlsson, G., Ahs, S., Hammarström, I., Lundin, I. P., & Hagforsen, E. (2003). Gluten-free diet in psoriasis patients with antibodies to gliadin results in decreased expression of tissue transglutaminase and fewer Ki67 + cells in the dermis. *Acta Dermato-Venereologica*, *83*(6), 425–429.

Michaëlsson, G., Gerdén, B., Hagforsen, E., Nilsson, B., Pihl-Lundin, I., Kraaz, W., Hjelmquist, G., & Lööf, L. (2000). Psoriasis patients with antibodies to gliadin can be improved by a gluten-free diet. *The British Journal of Dermatology*, *142*(1), 44–51.

Michaëlsson, G., Gerdén, B., Ottosson, M., Parra, A., Sjöberg, O., Hjelmquist, G., & Lööf, L. (1993). Patients with psoriasis often have increased serum levels of IgA antibodies to gliadin. *The British Journal of Dermatology*, *129*(6), 667–673.

Molina-Infante, J., & Carroccio, A. (2017). Suspected nonceliac gluten sensitivity confirmed in few patients after gluten challenge in double-blind, placebo-controlled trials. *Clinical Gastroenterology and Hepatology: The Official Clinical Practice Journal of the American Gastroenterological Association*, *15*(3), 339–348. Available from https://doi.org/10.1016/j.cgh.2016.08.007.

Mudryj, A., Waugh, A., Slater, J., Duerksen, D. R., Bernstein, C. N., & Riediger, N. D. (2021). Dietary gluten avoidance in Canada: A cross-sectional study using survey data. *CMAJ Open*, *9*(2), E317–E323. Available from https://doi.org/10.9778/cmajo.20200082.

Mumolo, M. G., Rettura, F., Melissari, S., Costa, F., Ricchiuti, A., Ceccarelli, L., de Bortoli, N., Marchi, S., & Bellini, M. (2020). Is gluten the only culprit for non-celiac gluten/wheat sensitivity? *Nutrients*, *12*(12). Available from https://doi.org/10.3390/nu12123785.

Nobel, Y. R., Rozenberg, F., Park, H., Freedberg, D. E., Blaser, M. J., Green, P. H. R., Uhlemann, A.-C., & Lebwohl, B. (2021). Lack of effect of gluten challenge on fecal microbiome in patients with celiac disease and non-celiac gluten sensitivity. *Clinical and Translational Gastroenterology*, *12*(12), e00441. Available from https://doi.org/10.14309/ctg.0000000000000441.

Ontiveros, N., Real-Delor, R. E., Mora-Melgem, J. A., Beltrán-Cárdenas, C. E., Figueroa-Salcido, O. G., Vergara-Jiménez, M., de, J., Cárdenas-Torres, F. I., Flores-Mendoza, L. K., Arámburo-Gálvez, J. G., & Cabrera-Chávez, F. (2021). Prevalence of wheat/gluten-related disorders and gluten-free diet in paraguay: An online survey-based study. *Nutrients*, *13*(2). Available from https://doi.org/10.3390/nu13020396.

Osborne, P. (2016). No grain, No pain. A 30-day diet for eliminating the root cause of chronic pain. Atria Books.

Pellegrini, G., Scotta, M. S., Soardo, S., Avanzini, M. A., Ravelli, A., Burgio, G. R., & Martini, A. (1991). Elevated IgA anti-gliadin antibodies in juvenile chronic arthritis. *Clinical and Experimental Rheumatology*, *9*(6), 653–656.

Perlmutter, D. (2018). Grain brain: The surprising truth about wheat, carbs, and sugar—Your brain's silent killers. Little, Brown and Company.

Potter, M. D. E., Walker, M. M., Jones, M. P., Koloski, N. A., Keely, S., & Talley, N. J. (2018). Wheat intolerance and chronic gastrointestinal symptoms in an Australian population-based study: Association between wheat sensitivity, celiac disease and functional gastrointestinal disorders. *The American Journal of Gastroenterology*, *113*(7), 1036–1044. Available from https://doi.org/10.1038/s41395-018-0095-7.

Potter, M. D., Jones, M. P., Walker, M. M., Koloski, N. A., Keely, S., Holtmann, G., & Talley Ac, N. J. (2020). Incidence and prevalence of self-reported non-coeliac wheat sensitivity and gluten avoidance in Australia. *The Medical Journal of Australia*, *212*(3), 126–131. Available from https://doi.org/10.5694/mja2.50458.

Rajilić-Stojanović, M., Biagi, E., Heilig, H. G. H. J., Kajander, K., Kekkonen, R. A., Tims, S., & de Vos, W. M. (2011). Global and deep molecular analysis of microbiota signatures in fecal samples from patients with irritable bowel syndrome. *Gastroenterology*, *141*(5), 1792–1801. Available from https://doi.org/10.1053/j.gastro.2011.07.043.

Rodrigo, L., Hernández-Lahoz, C., Lauret, E., Rodriguez-Peláez, M., Soucek, M., Ciccocioppo, R., & Kruzliak, P. (2016). Gluten ataxia is better classified as non-celiac gluten sensitivity than as celiac disease: a comparative clinical study. *Immunologic Research*, *64*(2), 558–564. Available from https://doi.org/10.1007/s12026-015-8750-1.

Roncoroni, L., Bascuñán, K. A., Vecchi, M., Doneda, L., Bardella, M. T., Lombardo, V., Scricciolo, A., Branchi, F., & Elli, L. (2019). Exposure to different amounts of dietary gluten in patients with non-celiac gluten sensitivity (NCGS): An exploratory study. *Nutrients*, *11*(1). Available from https://doi.org/10.3390/nu11010136.

Rostami, K., Ensari, A., Marsh, M. N., Srivastava, A., Villanacci, V., Carroccio, A., Asadzadeh Aghdaei, H., Bai, J. C., Bassotti, G., Becheanu, G., Bell, P., Di Bella, C., Bozzola, A. M., Cadei, M., Casella, G., Catassi, C., Ciacci, C., Apostol Ciobanu, D. G., Cross, S. S., & Derakhshan, M. (2022). Gluten induces subtle histological changes in duodenal mucosa of patients with non-coeliac gluten sensitivity: A multicentre study. *Nutrients*, *14*(12). Available from https://doi.org/10.3390/nu14122487.

Rouvroye, M. D., Zis, P., Van Dam, A.-M., Rozemuller, A. J. M., Bouma, G., & Hadjivassiliou, M. (2020). The neuropathology of gluten-related neurological disorders: A Systematic review. *Nutrients*, *12*(3). Available from https://doi.org/10.3390/nu12030822.

Salmi, T. T., Collin, P., Korponay-Szabó, I. R., Laurila, K., Partanen, J., Huhtala, H., Király, R., Lorand, L., Reunala, T., Mäki, M., & Kaukinen, K. (2006). Endomysial antibody-negative coeliac disease: Clinical characteristics and intestinal autoantibody deposits. *Gut*, *55*(12), 1746–1753.

Sapone, A., Bai, J. C., Ciacci, C., Dolinsek, J., Green, P. H. R., Hadjivassiliou, M., Kaukinen, K., Rostami, K., Sanders, D. S., Schumann, M., Ullrich, R., Villalta, D., Volta, U., Catassi, C., & Fasano, A. (2012). Spectrum of gluten-related disorders: Consensus on new nomenclature and classification. *BMC Medicine*, *10*, 13. Available from https://doi.org/10.1186/1741-7015-10-13.

Sapone, A., Lammers, K. M., Casolaro, V., Cammarota, M., Giuliano, M. T., De Rosa, M., Stefanile, R., Mazzarella, G., Tolone, C., Russo, M. I., Esposito, P., Ferraraccio, F., Cartenì, M., Riegler, G., de Magistris, L., & Fasano, A. (2011). Divergence of gut permeability and mucosal immune gene expression in two gluten-associated conditions: Celiac disease and gluten sensitivity. *BMC Medicine*, *9*, 23. Available from https://doi.org/10.1186/1741-7015-9-23.

Sapone, A., Lammers, K. M., Mazzarella, G., Mikhailenko, I., Cartenì, M., Casolaro, V., & Fasano, A. (2010). Differential mucosal IL-17 expression in two gliadin-induced disorders: Gluten sensitivity and the autoimmune enteropathy celiac disease. *International Archives of Allergy and Immunology*, *152*(1), 75–80. Available from https://doi.org/10.1159/000260087.

Schiepatti, A., Sanders, D. S., Baiardi, P., Caio, G., Ciacci, C., Kaukinen, K., Lebwohl, B., Leffler, D., Malamut, G., Murray, J. A., Rostami, K., Rubio-Tapia, A., Volta, U., & Biagi, F. (2022). Nomenclature and diagnosis of seronegative coeliac disease and chronic non-coeliac enteropathies in adults: The Paris consensus. *Gut*, *71*(11), 2218–2225. Available from https://doi.org/10.1136/gutjnl-2021-326645.

Schinocca, C., Di Liberto, D., Mansueto, P., Lo Pizzo, M., Grasso, G., Messana, E., Dieli, F., Ciccia, F., Carroccio, A., & Guggino, G. (2021). Effect of the gluten-free diet on quality of life, gastrointestinal symptoms and immune system in patients with fibromyalgia and non-celiac wheat sensitivity. *Journal of Biotechnology and Biomedicine*, *4*(4), 221–227. Available from https://doi.org/10.26502/jbb.2642-912800x46.

Schuppan, D., & Zevallos, V. (2015). Wheat amylase trypsin inhibitors as nutritional activators of innate immunity. *Digestive Diseases (Basel, Switzerland)*, *33*(2), 260–263. Available from https://doi.org/10.1159/000371476.

Schuppan, D., Pickert, G., Ashfaq-Khan, M., & Zevallos, V. (2015). Non-celiac wheat sensitivity: Differential diagnosis, triggers and implications. *Best Practice & Research. Clinical Gastroenterology*, *29*(3), 469–476. Available from https://doi.org/10.1016/j.bpg.2015.04.002.

Seidita, A., Mansueto, P., Giuliano, A., Chiavetta, M., Mandreucci, F., Soresi, M., Pistone, M., Compagnoni, S., Castellucci, D., Bisso, G., Faraci, F., Maestri, S., Disclafani, R., Sapone, A., Fasano, A., & Carroccio, A. (2022). Potential tolerability of ancient grains in non-celiac wheat sensitivity patients: A preliminary evaluation. *Frontiers in Medicine*, *9*, 995019. Available from https://doi.org/10.3389/fmed.2022.995019.

Shahbazkhani, B., Fanaeian, M. M., Farahvash, M. J., Aletaha, N., Alborzi, F., Elli, L., Shahbazkhani, A., Zebardast, J., & Rostami-Nejad, M. (2020). Prevalence of non-celiac gluten sensitivity in patients with refractory functional dyspepsia: A randomized double-blind placebo controlled trial. *Scientific Reports*, *10*(1), 2401. Available from https://doi.org/10.1038/s41598-020-59532-z.

Skodje, G. I., Sarna, V. K., Minelle, I. H., Rolfsen, K. L., Muir, J. G., Gibson, P. R., Veierød, M. B., Henriksen, C., & Lundin, K. E. A. (2018). Fructan, rather than gluten, induces symptoms in patients with self-reported non-celiac gluten sensitivity. *Gastroenterology*, *154*(3), 529–539. Available from https://doi.org/10.1053/j.gastro.2017.10.040, e2.

Slim, M., Calandre, E. P., Garcia-Leiva, J. M., Rico-Villademoros, F., Molina-Barea, R., Rodriguez-Lopez, C. M., & Morillas-Arques, P. (2017). The effects of a gluten-free diet versus a hypocaloric diet among patients with fibromyalgia experiencing gluten sensitivity-like symptoms: A pilot, open-label randomized clinical trial. *Journal of Clinical Gastroenterology*, *51*(6), 500–507. Available from https://doi.org/10.1097/MCG.0000000000000651.

Soresi, M., Incandela, S., Mansueto, P., Incandela, G., La Blasca, F., Fayer, F., D'Alcamo, A., Florena, A. M., & Carroccio, A. (2021). Gynecological disorders in patients with non-celiac wheat sensitivity. *Digestive Diseases and Sciences*, *66*(1), 167–174. Available from https://doi.org/10.1007/s10620-020-06184-8.

Tatham, A. S., & Shewry, P. R. (2008). Allergens to wheat and related cereals. *Clinical and Experimental Allergy: Journal of the British Society for Allergy and Clinical Immunology*, *38*(11), 1712–1726. Available from https://doi.org/10.1111/j.1365-2222.2008.03101.x.

Troncone, R., Franzese, A., Mazzarella, G., Paparo, F., Auricchio, R., Coto, I., Mayer, M., & Greco, L. (2003). Gluten sensitivity in a subset of children with insulin dependent diabetes mellitus. *The American Journal of Gastroenterology*, *98*(3), 590–595.

Uhde, M., Ajamian, M., Caio, G., De Giorgio, R., Indart, A., Green, P. H., Verna, E. C., Volta, U., & Alaedini, A. (2016). Intestinal cell damage and systemic immune activation in individuals reporting sensitivity to wheat in the absence of coeliac disease. *Gut*, *65*(12), 1930–1937. Available from https://doi.org/10.1136/gutjnl-2016-311964.

Valerii, M. C., Ricci, C., Spisni, E., Di Silvestro, R., De Fazio, L., Cavazza, E., Lanzini, A., Campieri, M., Dalpiaz, A., Pavan, B., Volta, U., & Dinelli, G. (2015). Responses of peripheral blood mononucleated cells from non-celiac gluten sensitive patients to various cereal sources. *Food Chemistry*, *176*, 167–174. Available from https://doi.org/10.1016/j.foodchem.2014.12.061.

Varney, J., Barrett, J., Scarlata, K., Catsos, P., Gibson, P. R., & Muir, J. G. (2017). FODMAPs: Food composition, defining cutoff values and international application. *Journal of Gastroenterology and Hepatology*, *32*(Suppl 1), 53–61. Available from https://doi.org/10.1111/jgh.13698.

Verdu, E. F., Armstrong, D., & Murray, J. A. (2009). Between celiac disease and irritable bowel syndrome: The "no man's land" of gluten sensitivity. *The American Journal of Gastroenterology*, *104*(6), 1587–1594. Available from https://doi.org/10.1038/ajg.2009.188.

Vives, M.-J., Esteve, M., Mariné, M., Fernández-Bañares, F., Alsina, M., Salas, A., Loras, C., Carrasco, A., Almagro, P., Viver, J. M., & Rodriguez-Carballeira, M. (2012). Prevalence and clinical relevance of enteropathy associated with systemic autoimmune diseases. *Digestive and Liver Disease: Official Journal of the Italian Society of Gastroenterology and the Italian Association for the Study of the Liver*, *44*(8), 636–642. Available from https://doi.org/10.1016/j.dld.2012.02.013.

Volta, U., Bardella, M. T., Calabrò, A., Troncone, R., & Corazza, G. R. (2014). An Italian prospective multicenter survey on patients suspected of having non-celiac gluten sensitivity. *BMC Medicine*, *12*, 85. Available from https://doi.org/10.1186/1741-7015-12-85.

Volta, U., Bellentani, S., Bianchi, F. B., Brandi, G., De Franceschi, L., Miglioli, L., Granito, A., Balli, F., & Tiribelli, C. (2001). High prevalence of celiac disease in Italian general population. *Digestive Diseases and Sciences*, *46*(7), 1500–1505.

Volta, U., Caio, G., & De Giorgio, R. (2016). Is autoimmunity more predominant in nonceliac wheat sensitivity than celiac disease? *Gastroenterology*, *150*(1), 282. Available from https://doi.org/10.1053/j.gastro.2015.08.058.

Volta, U., Caio, G., Ghirardi, C., Lungaro, L., Mansueto, P., Carroccio, A., & De Giorgio, R. (2020). Minimal lesions of the small intestinal mucosa: More than morphology. *Digestive Diseases and Sciences*, *65*(10), 2761–2768. Available from https://doi.org/10.1007/s10620-020-06571-1.

Volta, U., Tovoli, F., Cicola, R., Parisi, C., Fabbri, A., Piscaglia, M., Fiorini, E., & Caio, G. (2012). Serological tests in gluten sensitivity (nonceliac gluten intolerance). *Journal of Clinical Gastroenterology*, *46*(8), 680–685. Available from https://doi.org/10.1097/MCG.0b013e3182372541.

Wahnschaffe, U., Schulzke, J.-D., Zeitz, M., & Ullrich, R. (2007). Predictors of clinical response to gluten-free diet in patients diagnosed with diarrhea-predominant irritable bowel syndrome. *Clinical Gastroenterology and Hepatology: The Official Clinical Practice Journal of the American Gastroenterological Association*, *5*(7), 844–850.

Wahnschaffe, U., Ullrich, R., Riecken, E. O., & Schulzke, J. D. (2001). Celiac disease-like abnormalities in a subgroup of patients with irritable bowel syndrome. *Gastroenterology*, *121*(6), 1329–1338.

Whiteley, P., Haracopos, D., Knivsberg, A.-M., Reichelt, K. L., Parlar, S., Jacobsen, J., Seim, A., Pedersen, L., Schondel, M., & Shattock, P. (2010). The ScanBrit randomised, controlled, single-blind study of a gluten- and casein-free dietary intervention for children with autism spectrum disorders. *Nutritional Neuroscience*, *13*(2), 87–100. Available from https://doi.org/10.1179/147683010X12611460763922.

Yenokyan, G., & Armenian, H. K. (2012). Triggers for attacks in familial Mediterranean fever: Application of the case-crossover design. *American Journal of Epidemiology*, *175*(10), 1054–1061. Available from https://doi.org/10.1093/aje/kwr460.

Zanini, B., Basché, R., Ferraresi, A., Ricci, C., Lanzarotto, F., Marullo, M., Villanacci, V., Hidalgo, A., & Lanzini, A. (2015). Randomised clinical study: Gluten challenge induces symptom recurrence in only a minority of patients who meet clinical criteria for non-coeliac gluten sensitivity. *Alimentary Pharmacology & Therapeutics*, *42*(8), 968–976. Available from https://doi.org/10.1111/apt.13372.

Zevallos, V. F., Raker, V., Tenzer, S., Jimenez-Calvente, C., Ashfaq-Khan, M., Rüssel, N., Pickert, G., Schild, H., Steinbrink, K., & Schuppan, D. (2017). Nutritional wheat amylase-trypsin inhibitors promote intestinal inflammation via activation of myeloid cells. *Gastroenterology*, *152*(5), 1100–1113. Available from https://doi.org/10.1053/j.gastro.2016.12.006, e12.

Zimmermann, J., Longin, F. H., Schweinlin, A., Basrai, M., & Bischoff, S. C. (2022). No difference in tolerance between wheat and spelt bread in patients with suspected non-celiac wheat sensitivity. *Nutrients*, *14*(14). Available from https://doi.org/10.3390/nu14142800.

Zis, P., & Hadjivassiliou, M. (2019). Treatment of neurological manifestations of gluten sensitivity and coeliac disease. *Current Treatment Options in Neurology*, *21*(3), 10. Available from https://doi.org/10.1007/s11940-019-0552-7.

Zis, P., Sarrigiannis, P. G., Rao, D. G., & Hadjivassiliou, M. (2018). Quality of life in patients with gluten neuropathy: A case-controlled study. *Nutrients*, *10*(6). Available from https://doi.org/10.3390/nu10060662.

Chapter 18

Normative framework and public health interventions for the protection of celiac disease patients

Marco Silano[1], Roberto Copparoni[2] and Giuseppe Plutino[2]

[1]*Department of Cardiovascular and Endocrine-Metabolic Diseases and Ageing, Istituto Superiore di Sanità, Rome, Italy,* [2]*Office V—Nutrition and Information to Consumers, Directorate General for Food Safety and Nutrition, Italian Ministry of Health, Rome, Italy*

18.1 Introduction

At the moment, the only known treatment for celiac disease (CD) is a strict and lifelong withdrawal of gluten from a diet. Although a gluten-free diet (GFD) may heavily affect the social life of a celiac patient, this dietary regimen is able to normalize the serum level of antitransglutaminase and intestinal lesions, regress CD-related signs and symptoms, so preventing the life-threatening complications of CD.

A GFD is based on the consumption of three groups of food:

1. naturally gluten-free foods, such as meat, fish, eggs, pulses, milk and unprocessed cheeses, vegetables, fruit, potato, maize, rice, nuts, and pseudo-cereals, such as amaranth and buckwheat quinoa;
2. processed foods formulated to avoid cross-contamination with gluten and/or gluten-containing cereals (wheat, rye and barley);
3. foods consisting of or made only from one or more cereals that do not contain gluten (maize, rice and pseudo-cereals) or that have been specially processed to remove gluten, by physical or chemical means. In this group falls staple foods, such as gluten-free (GF) substitutes of pasta, bread, crackers, pizzas and salty and sugary biscuits.

The quantitative determination of gluten in foods and ingredients shall be based on an immunologic method or other method providing at least equal sensitivity and specificity. A robust and effective normative framework has been gradually developed, first at the national level and then at the European level, with a twofold aim, namely to ensure that GF foods are safe for CD patients and to make GF foods widely available to the public and to private catering. In this chapter, this normative framework is described and discussed (Fig. 18.1).

18.2 Why the limit of 20 ppm of gluten in gluten-free food?

Contrary to the IgE-mediated food allergy whose life-threatening symptoms might be elicited in susceptible individuals, even with traces of the trigger food protein, CD patients can safely tolerate traces of dietary gluten. Therefore the foods labeled GF and safe in a GFD are not required to contain "zero" gluten residues, but, currently, it is widely established from a scientific point of view and then set by the European regulation that the gluten content in the labeled GF food products has to be below 20 ppm (equivalent to 20 mg/kg). This threshold was officially established in 1979 by the Codex Alimentarius Commission (FAO/WHO Codex Alimentarius standard for foods for special dietary use for persons intolerant to gluten codex standard 118, 1979). The Codex Alimentarius Commission is a FAO–WHO joint commission created in 1963 to develop food standards and guidelines to protect consumers' health. Since then, the decisions and opinions of the Codex Alimentarius Commission have been a reference for consumers, food producers and processors, and food agencies.

Pediatric and Adult Celiac Disease. DOI: https://doi.org/10.1016/B978-0-443-13359-6.00016-9

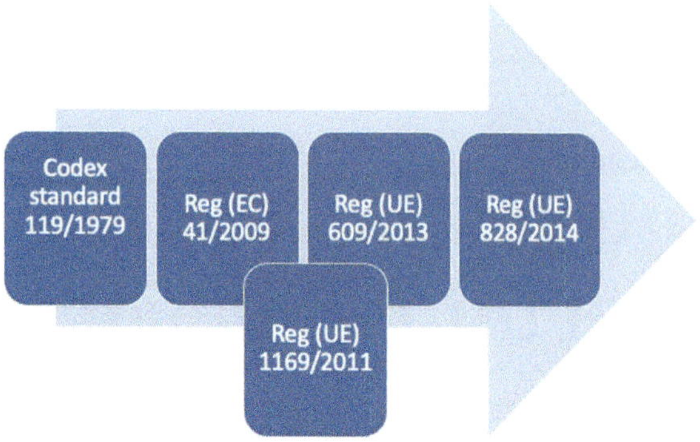

FIGURE 18.1 Schematic representation of the timeline of the EU normative on CD. Timeline of the EU normative on CD.

This limit of 20 ppm was decided at that time because it was the lowest gluten amount that could be measured by the available methods in a reliable, reproducible and sustainable way. Codex Standard 118–1979, in a later amendment issued in 2008, provided a general outline of the method of gluten determination in a food matrix:

1. the antibody used should react with the cereal protein fractions that are toxic for persons intolerant to gluten and should not cross-react with other cereal proteins or constituents of the foods or ingredients;
2. methods used for determination should be validated and calibrated against a certified reference material, if available;
3. the detection limit has to be appropriate according to the state of the art and technical standards. It should be 10 mg gluten/kg or below;
4. the qualitative analysis that indicates the presence of gluten shall be based on relevant methods (e.g., ELISA-based methods, DNA methods).

It also specified that the method for determining gluten was the Enzyme-linked Immunoassay (ELISA) based on the R5 antibody developed by Mendez (Hernando et al., 2008; Valdés et al., 2003).

Codex Standard 118–1979, also provided a number of definitions in Section 2.1.1. According to this normative, gluten-free (GF) foods are those "consisting of or made only from one or more ingredients which do not contain wheat (rye, barley, oats or their crossbred varieties), and the gluten level does not exceed 20 mg/kg in total, based on the food as sold or distributed to the consumer; and/or consisting of one or more ingredients from wheat (i.e., all Triticum species, such as durum wheat, spelt and khorasan wheat, which is also marketed under different trademarks such as kamut), which have been specially processed to remove gluten, and the gluten level does not exceed 20 mg/kg in total, based on the food as sold or distributed to the consumer."

The section 2.1.2 of the above-mentioned Standard also specified that low-gluten (LG) foods are "foods consisting of one or more ingredients from wheat (i.e., all Triticum species, such as durum wheat, spelt and khorasan wheat, which is also marketed under different trademarks such as kamut), rye, barley, oats, and their crossbred varieties, which have been specially processed to reduce the gluten content to a level above 20 up to 100 mg/kg in total, based on the food as sold or distributed to the consumer. Decisions on the marketing of products described in this section may be determined at the national level."

It should be said that the market of LG foods has not been developed as extensively as that of GF products. LG products might contain up to five times the gluten compared to the GF equivalents, which significantly reduces the daily amount of GF foods that a person with CD can consume without exceeding the safe overall gluten dietary intake of 50 mg per day (Gibert et al., 2013).

18.3 The national Italian regulations

Following the first publication of Codex Standard 118–1979, several European member states (MSs) adopted a normative framework to regulate the national marketing of GF products. Italy was one of the first MSs to have a dedicated legislation about these products. Since 1982, this MS has had a normative body aimed at protecting CD patients. The decree of the Italian Ministry of Health on 4 August 1982 was the first act to regulate the labeling of GF food products, which at that time fell in the category of *dietotherapeutics*. A few years later, the decree of the Italian Ministry of Health on 8 June 2001 confirmed the gluten threshold of 20 ppm for foods labeled GF; in Italy, the standard for LG food was never implemented by law. Then, the *National Register of the Foods for Special Needs* was established, with the aim of keeping a record of all the prescription GF foods in the Italian regions and autonomous provinces of Trento and Bolzano (Italian Minister of Health, 2001) (Fig. 18.2).

FIGURE 18.2 Italian label identyfing a gluten-free product. Sign to identify the products notified in the National Register of Foods for special needs.

18.4 The harmonization at the European level of the normative body for gluten-free foods

The European Commission harmonized across the UE MSs legislation on the labeling of GF in 2009, with Regulation (EC) 41/2009, whose Article 3 stated that: "foodstuffs for people intolerant to gluten, consisting of or containing one or more ingredients which substitute wheat, rye, barley, oats or their crossbred varieties shall not contain a level of gluten exceeding 20 mg/kg in the food as sold to the final consumer. The labeling, presentation and advertising of those products shall bear the term "gluten-free," whereas foodstuffs for people intolerant to gluten, consisting of or containing one or more ingredients made from wheat, rye, barley, oats or their crossbred varieties which have been especially processed to reduce gluten, shall not contain a level of gluten exceeding 100 mg/kg in the food as sold to the final consumer" (European Commission, 2009).

The above-mentioned regulation was implemented in all of the EU MSs as of 1 January 2012, thus creating for the first time a European common market for GF foods and ensuring that individuals with CD could find safe and clearly labeled GF food products suitable for a GFD across the union.

This normative act also clarified the safe use of oats in a GFD, as Article 3.3 reported the following: "Oats contained in foodstuffs for people intolerant to gluten must have been specially produced, prepared and/or processed in a way to avoid contamination by wheat, rye, barley, or their crossbred varieties and the gluten content of such oats must not exceed 20 mg/kg."

The horizontal European normative framework on GF foodstuff was completely revised in the provision of Regulation (EU) 609/2013 on food intended for infants and young children, food for special medical purposes and total diet replacement for weight control (European Commission, 2013). This regulation repealed, among others, the Commission Regulations (EC) 41/2009 and demanded the transfer of the rules on GF labels to a future regulation. This to-be-implemented legislation should consider methods for ensuring that people with CD are adequately informed of the differences between foods that are specially produced, prepared and/or processed in order to reduce the gluten content of one or more gluten-containing ingredients and other foods made exclusively from ingredients naturally free of gluten.

The most important novelty introduced by the abrogation of Regulation (CE) 41/2009 by Regulation (UE) 609/2013 was that the category of *dietotherapeutics* was terminated, and, since then, the labeling of GF must be considered as information for consumers. This implied, among others, that the authorization of the production site and notifications on the label are no longer required in marketing GF foods.

What is more, Regulation (UE) 609/2013 posed a significant obstacle to the current Italian legislative framework, which the above-mentioned protection of CD patients relied on at the time. With the abrogation of the concept of "dietotherapeutics" and the inclusion of GF-labeled products in the category of "normal" foods, there was no longer a normative basis for the inclusion of GF foods in the *Italian National Register of the Foods for Special Needs*. At that point, Italy was one of the MSs that drove the discussion at the Standing Committee on Plants, Animals, Food and Feed of the European Commission, aimed at implementing regulation that might have allowed the continuation of the register and the consequent monthly wage for the purchase of GF foods (see below). Thus, the European Commission set up Regulation (EU) 828/2014, which currently regulates the labeling of GF food in EU MSs. According to this, the information to consumers of GF may be accompanied by the statements "suitable for people intolerant to gluten" or "suitable for celiacs."

The food information referred to in paragraph 1 may be accompanied by the statements "specifically formulated for people intolerant to gluten" or "specifically formulated for celiacs" if the food is specially produced, prepared and/or processed to:

1. reduce the gluten content of one or more gluten-containing ingredients; or
2. substitute the gluten-containing ingredients with other ingredients naturally free of gluten.

It is noteworthy that Regulation (EU) 828/2014 set out, for the first time, that an ingredient with reduced gluten content, obtained by means of physical and/or chemical methods, could be used in GF-labeled products (European Commission, 2014).

18.5 The declaration of gluten presence in the foods as allergen

On this premise, Regulation (UE) 1169/2011 provides a framework for delivering consumers information on the presence of cereals containing gluten (European Commission, 2011). This information is intended for people with allergies and not those with CD, although it could also be of interest to people with CD. When a cereal containing gluten is used as a food ingredient, its name must be reported in the ingredient list in bold and in a larger font. No indications about the amount in grams or the percentage of the ingredient used and/or present in the food product on the shelf are required on the label; if this does not occur, the conditions are specified in Article 22 of Regulation (UE) 1169/2011 itself.

In case the food business operator is unable to maintain contamination control of the raw materials and/or the production chain of gluten-containing cereals, the following voluntary precautionary labeling might be used: contains cereal containing gluten. As stated before, this information is of limited usefulness for persons with CD because it does not imply information about the amount of gluten in the final food product; thus it is not an indication of whether the food could be included safely in a GFD.

18.6 The regulation in countries outside Europe

Outside Europe, the United States Food and Drug Administration (FDA) has stipulated since 2013 that GF-labeled foods must have a gluten content lower than 20 ppm (Labeling of GF foods in USA, 2013). Different from what the Codex Alimentarius Commission and European legislation have stated, the US FDA allows for a naturally GF food to be labeled GF. In 2020, the FDA issued a final rule on GF labels for fermented or hydrolyzed foods. This rule does not change the definition of GF food but establishes compliance requirements for these hydrolyzed and fermented foods, in the absence of a scientifically valid analytical method effective in precisely detecting and quantifying the gluten protein content in fermented or hydrolyzed foods in terms of equivalent amounts of intact gluten proteins (https://www.federalregister.gov/documents/2020/08/13/2020–17088/food-labeling-gluten-free-labeling-of-fermented-or-hydrolyzed-foods, 2020). Canada also adheres to the Codex Alimentarius Commission's regulations on GF foods (Labeling of GF foods in Canada, 2020).

The Australiancompetent authority for nutrition and food safety, the Food Standards Australia New Zealand, defines a GF claim as a nutrition content claim, and it specifies that any business making such a claim needs to be confident that their product or menu item contains no detectable gluten. This is in contrast with other countries, whose regulations allow GF products to contain up to 20 ppm of gluten (Labeling of GF foods in Australia, 2018).

In fact, contrary to Europe and the United States, in Australia, a GF food must not contain the following:

any detectable gluten;
oats or oat products;
cereals containing gluten that have been malted, or products of such cereals.
The labeling of food products in these countries in relation to gluten includes the following:
gluten free: contains no detectable gluten;
low gluten: contains no more than 200 ppm of gluten;
contains/is high in gluten.

Chile set 3 ppm and Japan 10 ppm for GF-labeled food products (Labeling of GF foods in Chile, 2015) (Table 18.1).

It is evident that the approach to risk management for a GFD varies from country to country. If, at first glance, the approach most cautionary for individuals with CD seems to be the lowest threshold implemented in Australia and Chile, some aspects should be considered.

TABLE 18.1 Gluten limit for GF-labeled foods in different countries.

Country	Limit for GF-labeled packed foods (ppm—mg/kg)
Australia	0
Canada	20
Chile	3
The European Union	20
Japan	10
The United Kingdom	20
The United States	20

First, there is a cost to ensure the gluten content in LG food products is much higher than that of GF foods with a gluten limit of 20 ppm. This higher cost necessarily affects the final consumers, jeopardizing the adherence to a GFD.

Second, at the moment, the limit of the analytical methods, specifically the ELISA kit based on the R5 antibody (see below), allows the official control to have a reasonable "zero gluten" equal to the limit of quantification (LoQ) of 3–5 ppm.

Third, a seminal paper has demonstrated that CD patients can tolerate up to 10 mg/die and that the dietary consumption of gluten below this amount does not trigger any mucosal changes in treated individuals with CD (Catassi et al., 2007). This means that, considering the most conservative approach, in a GFD, 500 mg of GF foods with a limit of 20 ppm might be safely present. This quantity is more than sufficient for a balanced diet, where the remaining recommended daily energy and nutrient requirements should be provided by naturally or minimally processed GF foods, such as rice, maize, pseudo-cereals, meat, eggs, vegetables and fruits (Bascuñán et al., 2017).

Finally, in Italy, the 20-ppm limit has been implemented for roughly 45 years, and no significant burden of symptoms and signs or CD-related complications have been reported in individuals with CD consuming a GFD containing a large proportion of GF-labeled foods (Aljada et al., 2021; Catassi et al., 2022).

18.7 Official control of gluten-free products

In Italy, the National Plan of Control is the tool by which the competent authority—the Ministry of Health—programs and conducts the annual official controls of food products to check eventual nonconformities, according to the enforced legislation (microbiological contamination, ingredients and allergens not declared in the labeling and residue levels higher than those set by the regulations). A systematic control on GF-labeled products was last carried out in 2016. Out of the 28 food items controlled, as well as in the audits on the production and distribution of GF foods, zero nonconformity was found (Italian Ministry of Health, 2016).

One scientific study also measured the gluten content in 200 food products available on the Italian market, including both naturally GF and GF-labeled foods. Of the 18 (9%) products determined to have a gluten content higher than 20 ppm, only two were labeled GF, while the remaining 16 were made with GF ingredients, but were not labeled GF. In four of these the precautionary allergen label *may contain cereals containing gluten* (Verma et al., 2017) was added.

Out of the 1471 recalls for allergen contamination recorded in the US in the period 2013–19, only 34 were related to an undeclared gluten presence (Sharma et al., 2023).

A similar study performed in India showed that, out of 794 GF products tested, 80 (10.1%), including 38 of the 360 labeled and 42 of the 354 nonlabeled/naturally GF food products, had gluten content >20 ppm. It is noteworthy that none of the imported GF products had a gluten content more than the recommended limit (Mehtab et al., 2021)

A meta-analysis in 2020 concluded there was an overall contamination prevalence of 13.2% in industrialized GF products (Falcomer et al., 2020).

There are some discrepancies in the results about gluten contamination of the official control and those of the scientific papers. It is to be noted that the official control by competent authorities has a more rigorous approach than that of scientific studies. In fact, the official control is carried out by laboratories that are under an External Control Quality System, with validated and accredited tests and instruments and trained personnel. Despite these limitations,

TABLE 18.2 Amount of monthly wage of Italians with CD (Euros).

Age range	Male	Female
6 months–5 years	56	
6–9 years	70	
10–13 years	100	90
14–17 years	124	99
18–59 years	110	90
>60 years	89	75

the above-mentioned results show that the GF foods available on the market are safe for CD patients. Of course, a healthy and balanced GFD should not include only GF food products specifically formulated for individuals with CD, mostly including naturally GF foods, such as fruits, vegetables, meats and fish (see Chapter 13).

18.8 The subsidies from the national health systems to patients with celiac disease

Some European and extra-European countries support persons with CD in purchasing GF foods specifically formulated for them, with the aim of improving their compliance with a GFD, because these products have shown to be more expensive than those with gluten. In Italy, CD patients certified at one of the tertiary regional medical centers of the regional health systems, receive a monthly voucher to be used for purchasing GF processed foods, regardless of their individual or household income. The amount of this wage varies according to sex and age (Table 18.2). This wage can be used to purchase only GF foods specifically formulated for people with CD and listed in the *National Register of the Foods for Special Needs*, maintained and regularly updated by the Italian Ministry of Health. The GF foods listed in the register can be easily recognized by consumers with CD from an appropriate labeling (Fig. 18.2) (Report on celiac disease to the Italian Parliament, 2021).

The food categories that can be listed in the *National Register of the Foods for Special Needs* are:

1. breads and similar, salty baked products and similar;
2. pastas and similar, pizzas and similar, ready to eat pasta;
3. sweet baked products and other sweet products;
4. mixed flours, formulation for cakes, breads, pizzas and others;
5. breakfast cereals.

Monthly amounts have been calculated, assuming that 30% of the recommended daily allowance for carbohydrates comes from GF pasta, bread and other GF foods for people with CD and considering the average price of GF pasta, bread, crackers, and sweet biscuits. In 2021, in Italy, GF foods, for a total of €233.349.439, have been purchased by people with CD using their own wages, for an average expense of €965 per CD patients (Italian Minister of Health, 2018).

In the United States and Canada, a GF tax deduction is offered for those diagnosed with CD and on a GFD for medical reasons. In the United States, the deduction for GF food is added to the total deduction for medical expenses, which is eligible to be written off for the amount that exceeds 7.5% of the personal annual income. In Canada, differently from the United States, the deduction for GF foods is independent of that for other medical expenses, and there is not a threshold to access the tax deduction. In Canada, food eligible for tax deduction includes GF foods specifically formulated for people with CD and some naturally GF products, which are usually ingredients for the home preparation of GF recipes, such as rice/maize flour and spices (Tax deduction for GF foods in USA, 2022; Tax return for CD patients in Canada, 2022).

In the United Kingdom, individuals with CD receive free staple GF food under a prescription plan of the general practitioner, whose quantity in units is determined according to age, sex and other factors, such as pregnancy and breastfeeding. GF foods that can be prescribed are listed in the Food and Drink Directory (GF food prescription in UK, 2018).

In New Zealand, GF products are partially subsidized by the health care system. The patient can purchase GF foods at 50% of the retail price, the other 50% being funded by the country (Subsidy for GF foods in New Zealand, 2022). In Spain and Serbia, the corresponding national health systems provide CD patients with a monthly supply of specific staple GF food, such as GF flour (Pinto-Sanchez et al., 2015).

18.9 Interventions to ensure an early diagnosis and the prevention of complications

Italy is the sole country worldwide that has set a normative framework for the protection of CD patients. The main structure of this normative framework, as described in the above paragraphs, is law 123/2005 (Law 123/2015, 2015).

This law, in Article 2, legally binds the Italian health system to:

1. promote the early diagnosis of CD;
2. promote the prevention of CD complications;
3. facilitate for CD patients fair and safe access to collective food services;
4. develop an information campaign on CD for the general population and for individuals with CD;
5. deliver periodic CME programs on CD diagnosis and treatment for health personnel;
6. support scientific research on CD;
7. provide blood and instrumental tests to all citizens with clinical signs and symptoms of CD for the early diagnosis of CD and the prevention of its complications.

Article 3 of this law sets out that, to ensure early CD diagnosis and the prevention of CD complications, national guidelines for the diagnosis and follow-up of CD should be developed and that a national network of secondary and tertiary referral centers with recognized experience and excellence in the diagnosis and treatment of CD should be defined.

With the aim of facilitating CD patients' full adherence to a GFD, Article 4 of law 123/2015 requires the canteens of public schools, hospitals and offices to provide GF meals. The same article allocates a yearly fund to support canteens in the procurement of GF foods and for fitting kitchens to prepare safe GF meals (Report on celiac disease to the Italian Parliament, 2021).

What is more, this article reaffirms the monthly wage for CD patients to purchase GF food products. To ensure that CD patients have access to safe GF meals when away from home, Article 5 allocates funding to provide information to workers in charge of the private collective food service.

Regarding the obligations set by Article 3 of Law 123/2005, flowcharts for the diagnosis and follow-up of CD in adults and children, respectively, have been published in the Italian Official Gazette n 191–19 August 2015. These guidelines, one for children and one for adults, have twofold importance. First, they represent up-to-date, evidence-based diagnostic tools that physicians of the network's referral centers and the general practitioners/pediatricians should follow for a correct and early diagnosis (Husby et al., 2019, 2020). Second, the presence of blood and instrumental tests in these flowcharts shows that the secondary and tertiary medical centers are obliged to deliver these tests to all citizens suspected of having CD or diagnosed with CD.

A normative act on 17 May 2017 established a network of secondary and tertiary referral clinical centers that are appointed to certify the diagnosis of CD and to ensure CD patients have access to a monthly allowance for GF products and other rights (Agreement between the State and the Region and Provincies Autonomous, 2017). The minimum criteria for a hospital/clinical center to be appointed as a secondary referral center have been set as follows:

Capability to perform the following diagnostic tests

1. measurement of serological antibodies: EMA, anti-TG and antideamidated gliadin peptides;
2. determination at high definition of the haplotype DQ2/8;
3. duodenoscopy (with narcosis for pediatric patients);
4. histological examination of duodenal mucosal specimens (with hematoxylin/eoxin staining and immunohistochemistry).

Expertise and education of the personnel

1. medical staff with at least one of the following specialties: internal medicine, gastroenterology, immunology (adults), pediatrics (children), dermatology (for dermatitis herpetiform only) and extensive, proven experience in the diagnosis and treatment of CD and dermatitis herpetiform;
2. pathologists with extensive, proven experience in the diagnosis of CD;
3. dietitians.

For the tertiary referral centers, all of the above-mentioned requirements are necessary, along with the following:

1. proven experience in the diagnosis and treatment of refractory CD;
2. organization of CME events;
3. scientific excellence, as proven by papers and participant speakers at international meetings and conferences on CD.

18.10 Law on secreening on celiac disease and diabetes mellitus type 1

In May 2023, the Italian Parliament approved a legislative act that binds the Minister of Health and the Regions and Provincies Autonomous to set a mandatory serogical screening on a national basis among the pediatric population, starting from the year 2024, aiming at the early diagnosis of CD and the prevention of diabetic chetoacidosis.

For the implementation of this multiannual program, a funding of 3.85 million euros has been established for each of the years 2024 and 2025 and of 2.85 million euros per year starting from the year 2026 (Law n 130/2023, 2023).

Italy is at the moment the only country worldwide that will have such a screening program on CD, that will allow to monitor yearly the incidence and the prevalence of CD and clarify the epidemiological and clinical profiles of newly diagnosed CD cases.

References

Agreement between the State and the Region and Provincies Autonomous. (2017). https://www.gazzettaufficiale.it/eli/id/2017/11/23/17A07862/sg.

Aljada, B., Zohni, A., & El-Matary, W. (2021). The gluten-free diet for celiac disease and beyond. *Nutrients*, *13*(11), 3993. Available from https://doi.org/10.3390/nu13113993.

Bascuñán, K. A., Vespa, M. C., & Araya, M. (2017). Celiac disease: Understanding the gluten-free diet. *European Journal of Nutrition*, *56*(2), 449–459. Available from https://doi.org/10.1007/s00394-016-1238-5.

Catassi, C., Fabiani, E., Iacono, G., D'Agate, C., Francavilla, R., Biagi, F., Volta, U., Accomando, S., Picarelli, A., De Vitis, I., Pianelli, G., Gesuita, R., Carle, F., Mandolesi, A., Bearzi, I., & Fasano, A. (2007). A prospective, double-blind, placebo-controlled trial to establish a safe gluten threshold for patients with celiac disease. *American Journal of Clinical Nutrition*, *85*(1), 160–166. Available from https://doi.org/10.1093/ajcn/85.1.160.

Catassi, C., Verdu, E., Baj, J., & E, L. (2022). Coeliac disease. *Lancet*, *399*(10344), 2413. Available from https://doi.org/10.1016/S0140-6736(22)00794-2.

European Commission. (2009). Regulation (EC) 41/2009.

European Commission. (2011). Regulation (EU) 1169/2011.

European Commission. (2013). Regulation (EU) 609/2013.

European Commission. (2014). Regulation (EU) 828/2014.

Falcomer, A. L., Santos Araújo, L., Farage, P., Santos Monteiro, J., Yoshio Nakano, E., & Puppin Zandonadi, R. (2020). Gluten contamination in food services and industry: A systematic review. *Critical Reviews in Food Science and Nutrition*, *60*(3), 479–493. Available from https://doi.org/10.1080/10408398.2018.1541864.

FAO/WHO Codex Alimentarius standard for foods for special dietary use for persons intolerant to gluten codex standad 118. (1979).

GF food prescription in UK. (2018). https://www.england.nhs.uk/publication/prescribing-gluten-free-foods-in-primary-care-guidance-for-ccgs/.

Gibert, A., Kruizinga, A. G., Neuhold, S., Houben, G. F., Canela, M. A., Fasano, A., & Catassi, C. (2013). Might gluten traces in wheat substitutes pose a risk in patients with celiac disease? A population-based probabilistic approach to risk estimation. *American Journal of Clinical Nutrition*, *97*(1), 109–116. Available from https://doi.org/10.3945/ajcn.112.047985.

Hernando, A., Mujico, J. R., Mena, M. C., Lombardía, M., & Méndez, E. (2008). Measurement of wheat gluten and barley hordeins in contaminated oats from Europe, the United States and Canada by Sandwich R5 ELISA. *European Journal of Gastroenterology and Hepatology*, *20*(6), 545–554. Available from https://doi.org/10.1097/MEG.0b013e3282f46597.

https://www.federalregister.gov/documents/2020/08/13/2020-17088/food-labeling-gluten-free-labeling-of-fermented-or-hydrolyzed-foods. (2020). Labeling of GF hydrolyzed of fermented GF foods in USA.

Husby, S., Koletzko, S., Korponay-Szabó, I., Kurppa, K., Mearin, M. L., Ribes-Koninckx, C., Shamir, R., Troncone, R., Auricchio, R., Castillejo, G., Christensen, R., Dolinsek, J., Gillett, P., Hróbjartsson, A., Koltai, T., Maki, M., Nielsen, S. M., Popp, A., Størdal, K., & Wessels, M. (2020). European Society Paediatric Gastroenterology, Hepatology and Nutrition Guidelines for Diagnosing Coeliac Disease 2020. *Journal of Pediatric Gastroenterology and Nutrition*, *70*(1), 141–156. Available from https://doi.org/10.1097/MPG.0000000000002497.

Husby, S., Murray, J. A., & Katzka, D. A. (2019). AGA clinical practice update on diagnosis and monitoring of celiac disease—Changing utility of serology and histologic measures: Expert review. *Gastroenterology*, *156*(4), 885–889. Available from https://doi.org/10.1053/j.gastro.2018.12.010.

Italian Minister of Health. (2001). Decree.

Italian Ministry of Health. (2016). National Plan of Official Control. https://www.salute.gov.it/relazioneAnnuale2016/dettaglioRA2016.jsp?cap = capitolo1&id = 1447&sez = ra16-1-alimenti.

Italian Minister of Health. (2018). Decree.

Labeling of GF foods in Australia. (2018). https://www.foodstandards.gov.au/code/Pages/default.aspx.

Labeling of GF foods in Canada. (2020). https://www.canada.ca/content/dam/hc-sc/documents/services/food-nutrition/reports-publications/food-safety/gluten-pamphlet/gluten-pamphlet-eng.pdf.

Labeling of GF foods in Chile. (2015). https://www.isl.gob.cl/wp-content/uploads/2015/04/D.S-N-977actualizado-2013.pdf.

Labeling of GF foods in USA. (2013). http://10. https://www.federalregister.gov/documents/2013/08/05/2013-18813/food-labeling-gluten-free-labeling-of-foods.

Law 123/2015. (2015). https://www.normattiva.it/uri-res/N2Ls?urn:nir:stato:legge:2005;123.

Law n 130/2023. (2023). https://www.normattiva.it/uri-res/N2Ls?urn:nir:stato:legge:.

Mehtab, W., Sachdev, V., Singh, A., Agarwal, S., Singh, N., Malik, R., Malhotra, A., Ahuja, V., & Makharia, G. (2021). Gluten content in labeled and unlabeled gluten-free food products used by patients with celiac disease. *European Journal of Clinical Nutrition*, *75*(8), 1245–1253. Available from https://doi.org/10.1038/s41430-020-00854-6.

Pinto-Sanchez, M. I., Verdu, E. F., Gordillo, M. C., Bai, J. C., Birch, S., Moayyedi, P., & Bercik, P. (2015). Tax-deductible provisions for gluten-free diet in Canada compared with systems for gluten-free diet coverage available in various countries. *Canadian Journal of Gastroenterology and Hepatology*, *29*(2), 104–110. Available from https://doi.org/10.1155/2015/508156.

Report on celiac disease to the Italian Parliament. (2021). https://www.salute.gov.it/imgs/C_17_pubblicazioni_3308_allegato.pdf.

Sharma, G. M., Ma, Y., & Luccioli, S. (2023). Recalls associated with food allergens and gluten in FDA-regulated foods from fiscal years 2013 to 2019. *Journal of Food Protection*, *86*(4), 100069. Available from https://doi.org/10.1016/j.jfp.2023.100069.

Subsidy for GF foods in New Zealand. (2022). https://www.tewhatuora.govt.nz/our-health-system/claims-provider-payments-and-entitlements/special-authority.

Tax deduction for GF foods in USA. (2022). https://www.irs.gov/taxtopics/tc502.

Tax return for CD patients in Canada. (2022). https://www.canada.ca/en/revenue-agency/services/tax/individuals/topics/about-your-tax-return/tax-return/completing-a-tax-return/deductions-credits-expenses/lines-33099-33199-eligible-medical-expenses-you-claim-on-your-tax-return/details-medical-expenses.html.

Valdés, I., García, E., Llorente, M., & Méndez, E. (2003). Innovative approach to low-level gluten determination in foods using a novel sandwich enzyme-linked immunosorbent assay protocol. *European Journal of Gastroenterology & Hepatology*, *15*(5), 465–747. Available from https://doi.org/10.1097/01.meg.0000059119.41030.df.

Verma, A., Gatti, S., Galeazzi, T., Monachesi, C., Padella, L., Baldo, G., Annibali, R., Lionetti, E., & Catassi, C. (2017). Gluten contamination in naturally or labeled gluten-free products marketed in Italy. *Nutrients*, *9*(2), 115. Available from https://doi.org/10.3390/nu9020115.

Chapter 19

The Role of Patients' Associations: the Italian Celiac Association as a case of best practice

Caterina Pilo, Susanna Neuhold and Luisa Novellino
Associazione Italiana Celiachia AIC—APS, Genoa, Italy

19.1 The Italian Celiac Association: building a link between patients and institutions

Patients' associations drive their advocacy action to protect a category of individuals in a disadvantaged situation with respect to their health state. They are nonprofit and work as a connection point between patients, families, and caregivers, and also between the scientific community and the institutions. By creating networks of associations to strengthen their role, they advocate the needs of patients in an organized and unified way in the frame of the institutions that are responsible for health policies. Moreover, they assist patients and families by improving information and knowledge about their disease, provide them with assistance, sharing good practices and experience, thus promoting awareness, ability to manage the disease, and improve the quality of life. In the precious relationship with healthcare professionals, they offer support to the community of patients and families by conveying correct information, complementary to that provided during the diagnosis and follow-up. Furthermore, starting from the requests and needs of patients, the associations stimulate research and often contribute to its financing.

The role of associations is increasingly recognized as subjects capable of contributing to the definition of the best health policies. In the growing, formal, recognition of the role of patients and associations, there is a need to strengthen the tools and methods of involvement in dialog with institutions and in decision-making processes.

In Italy, this activity is regulated by the Third Sector (TS) Code which rules the broadest and most general category of all forms of association and business which pursue civic, solidarity, and social utility purposes, on a nonprofit basis. The Reform of the TS is recognized as the best implementation tool for the constitutional principle of *subsidiarity*. The principle of subsidiarity, recently included into the constitutional provisions, is the result of an extensive project to rethink the organizational structure of the powers of the State and of a modern view of the general interest, namely the instances coming from the society, through a participative and collaborative approach to the decision-making process (Terzjus & Pastorino Olmi, 2021).

The Third Sector Reform impacted on the Italian Celiac Association (AIC) giving rise to a new governance of the association, which in years 2022 and 2023 returned to a one and united organization by mean of two mergers by incorporation, thus becoming legally a Third Sector Entity (ETS) and an Association for Social Promotion (APS), capable of implementing its complex statutory purposes.

As described in the first part of the article, in fact, the Third Sector Code has integrated the nonprofit regulations into a single Code, identifying, among others, the list of 26 activities which can be carried out by all the ETSs.

Before the Reform, many activities were precluded to Nonprofit Organizations (NPOs), both by their intrinsic nature and by the commercial nature of the activities, which themselves excluded the involvement of NPOs.

Conforming to its statutory purposes, AIC has carried out programs, activities and services such as the AIC Food Handbook, the granting of the license for the Crossed Grain Trademark to companies, the funding of the scientific research, the creation of the Celiac News house organ, the development of the Gluten-Free Eating Out program, etc. All activities which, in order to comply with both civil law and fiscal legislation, had to be managed by suitable organizations.

Pediatric and Adult Celiac Disease. DOI: https://doi.org/10.1016/B978-0-443-13359-6.00014-5

Between 2004 and 2005, in compliance with the rules in force at the time, AIC had first created the Celiac Foundation ONLUS, which held the aims of financing scientific research in the area of Celiac Disease (CD), Dermatitis Herpetiformis (DH) and the gluten-free diet. AIC then also founded the company SBS Spiga Barrata Service, which became a social enterprise in 2011 and which managed all the commercial activities.

Three entities, notwithstanding their legal independence, operated synergistically in compliance with the AIC mission, thanks to a complex governance. In fact, they adhered to the AIC Federation, as well as the local AIC Associates. A strategic political body (the Group Executive Council) allowed for synergistic and complementary strategic choices, guaranteeing full adherence to the AIC mission and vision.

Today, AIC ETS APS is legally capable and suitable for carrying out all the activities previously managed by the three national entities. The merger by incorporation has allowed the internalization of each independent entity's purposes, including all the legal agreements (research, brand licensing, collaborations), allowing for continuity of relationships and activities, without any interruption. All these will allow AIC to improve its performance in assisting patients and families, achieving the defined strategic objectives and responding more efficiently to changes in the context in which we operate.

19.2 AIC with respect to lifestyle, health and therapy needs of celiacs: goals and present activities

19.2.1 The history and the goals achieved by AIC

AIC (http://www.celiachia.it) was founded in 1979 by the families of some celiac children in a period when CD was scarcely known and it was mainly considered a pediatric disease. Moreover, the overall scenario provided no social care (e.g., school meals for celiac children) and no clinical management of the disease. This setting was the driving force yielding these celiac families to associate: AIC was established as the first Italian CD patient advocacy group dedicated to driving diagnosis and enabling access to gluten-free food. Later, AIC pivots to research for proximity to high QoL, treatments and cure, especially diffusing and supporting the concept of a strict long-life gluten-free (GF) diet as the only evidence-based therapy of CD. Primary goals of the nascent AIC were especially: (1) understanding the clinical history and the etiopathogenesis of CD; (2) safely driving and shortening the diagnosis course of patients; (3) patient and institution empowerment about the concept of "gluten-free diet"; (4) supporting the identification and the validation of safe gluten-free products suitable for celiacs; (5) facilitating the distribution of such products to celiac citizens and their families, including pharmacists and, recently, the retail sector (grocery stores and supermarkets).

Later, AIC has further grown up, empowering the rights of patients in Italy, contributing to Italian specific Laws dedicated to: (1) the free delivery of GF products to diagnosed patients, as a therapy; (2) public services for celiacs such as school meals; (3) dissemination of awareness about CD and its implications in the frame of Institutions, leading to the Report on Celiac Disease to the Italian Parliament, cured by the Ministry of Health and the Italian NIH (ISS, Istituto Superiore di Sanità); (4) dissemination of awareness about CD and its implications in the public opinion, physicians, hospitals, diagnosis centers, dietitians, and all the operators in the food sector (including manufacturers).

In the 1980s, the first "Food Handbook" (*Prontuario*) was created as a collection of food products whose safety for celiacs was verified, also including the absence of gluten contaminations. In the meantime, the first companies were also born that specialized in the production of dietetic foods for celiacs, alternatives to bread, pasta, and biscuits. At first the products were not particularly tasty, the costs were prohibitive, but they represented an important response from the commercial field which recognized celiacs as consumers with specific needs. In Italy, today, various scientific studies have demonstrated that not only the organoleptic quality, but also the nutritional profile of substitute foods specifically formulated for celiacs is similar to those of traditional products (Fiori et al., 2022; Cornicelli et al., 2018). This dispels the false myth that these products are more caloric or rich in fats and sugars with respect to the conventional food, a fake news that refers to outdated data, at least as far as concerns Italy.

Today, AIC is a Federation of 21 associated Regional Associations (local groups) which represents celiac patients and their needs all over Italy and abroad, in the frame of all the most important institutional contexts and working groups. AIC is both the promoter and the institutional partner of projects aimed at improving the patient's daily life, initiatives and communication plans aimed at informing and sensitizing the public, the media and the physicians as well, especially GPs dedicated to the primary care, who have the pivot role of identifying new CD patients and start the diagnose process. Moreover AIC supports and funds the best scientific research through national calls for proposals aiming at the funding of projects about the high quality proximity to patients' QoL, therapy and management. Finally, AIC controls and manages, in Italy, the registered Crossed Grain Trademark (*Spiga Barrata*), the top-grade guarantee of safety and suitability of GF-products for celiacs, in the frame of the European Scheme, the European licensing system, of the AOECS (the Association of European Celiac Societies).

19.2.2 The advocacy and lobbying action of AIC in the first decades of activity

In the first decades, AIC gave top priority to the lobbing actions toward the institutions, aiming at recognizing the needs and the rights of a fragile group of the population, affected by a chronic disease that requires rigorous lifelong therapy: the GF diet which was not only very expensive, but it also involved a strong impact on people's social life.

Originating from this continuous lobbing activity with the Italian institutions, a series of laws, decrees and regulations were set to protect patients, and this set of regulations still remains today a unique case worldwide.

Among the most important regulations, we would like to mention the Italian Celiac Disease System, an evidence-based assistance model, which provides for the provision of substitute products to celiacs according to maximum monthly expenditure limits, which are calculated on average market prices and nutritional needs, therefore yielding different cost/nutritional settings according to the age and the sex. In Italy, the products supplied to celiacs are authorized by the Ministry of Health, which reports them in the National Register of GF foods. The procedures for accessing this Register, far from being purely formal, represent an important form of control of the highest national authority responsible for the protection of Italian health. The law not only verifies the safety of the ingredients used and the production processes, but also strictly regulates the use of misleading claims or information to celiac patients. The intervention of the Ministry of Health is therefore also aimed at protecting the general consumers from misleading claims, as well as from the increasing phenomenon of the GF fashion, mainly intended as a general healthy life style or a "slimming diet" practice.

Furthermore, the Italian Celiac System recognizes CD, thanks to the Law 123/2005, as a "social disease" with the aim of promoting the integration of celiacs into social life (school, sports, and work activities through fair and safe access to catering services), also focusing on the fundamental importance of the early diagnosis and the consequent prevention of the complications originating from wrong/delayed diagnoses.

19.2.3 AIC today mission

AIC main purposes are as follow:

1. Improving the health and well-being of persons affected by CD or DH, through our strategic investments in research, advocacy, and education (dissemination and empowerment);
2. Supporting people with CD and DH to get a full and correct awareness of their condition. We work each day to promote understanding of CD, empower our patient and provider community, and collaborate with partners and stakeholders to carry our mission forward;
3. Disseminating a correct and broad knowledge of CD and DH in society. We invest in programs developed by leading academic research centers in CD to engage healthcare professionals, researchers, and the patient community.

19.2.4 AIC: scientific activities, educational programs, diagnosis and research

AIC promoted the definition of a Health Ministry official protocol (ACCORDO 30 luglio 2015, Accordo, ai sensi dell'articolo 4, del decreto legislativo 28 agosto 1997, n. 281, tra il Governo, le regioni e le province autonome di Trento e di Bolzano sul «Protocollo per la diagnosi e il follow up della celiachia». Rep. Atti n. 125/CSR. 15A06387, Pag. 147 della Gazzetta Ufficiale della Repubblica Italiana, Anno 156° - Numero 191, 19 agosto 2015 (2015) for the diagnosis and the follow up of CD, by mean of its National Scientific Board. Moreover, AIC develops and supports editorial, digital and conference tools to healthcare professionals, patients, and family members through initiatives such as (1) the annual AIC Scientific Congress, the only Symposium yearly held in Italy completely dedicated to CD and DH; (2) editorial products aimed at the general population, such as the brochures dedicated to the CD in women (Donna & Celiachia. Raccomandazioni Cliniche, 2016), and to sports and physical activity in relation to a GF diet (Guida Sport&Celiachia. La Dieta Senza Glutine e La Performance Sportiva, 2017); (3) editorial products aimed at educating/updating physicians, being aware of the critical role of primary care and GPs in avoiding the missing of diagnosis due to the wide range of possible symptoms connected to this "camaleontic" disease. (Celiachia e Dermatite Erpetiforme, 2022; Celiachia e Test Genetico HLA, 2022; CELIAC DISEASE IN CHILDREN AND ADULTS: FOR PHYSICIANS OF GENERAL MEDICINE AND PRIMARY CARE PAEDIATRICIANS, 2018)

Most of these initiatives are under the supervision of the National Scientific Board, comprising Italian internationally renowned physicians and researchers in the field of CD.

The local AICs also rely on the support of local Scientific Committees and Referents which both organize local events and joint the educational/informative national AIC programs (e.g., the annual AIC Scientific Congress).

As for mission, AIC also supports and funds the best scientific research through national calls.

19.2.5 AIC and the gluten-free diet

As CD is a chronic disease that has the GF diet as the only available therapy, information and services related to nutrition are of critical importance. For this reason, different activities have been developed over the years by AIC and replicated in many other countries, also thanks to the great willingness and generosity shown by AIC sharing its models and projects internationally.

The educational program on nutrition developed by AIC for the celiac patient-consumer takes into consideration several objectives:

- education about healthy eating and lifestyle;
- GF diet education, with particular reference to the issue of accidental contamination (cross-contact); (Le Contaminazioni Nella Dieta Senza Glutine, 2022).
- education on the economic sustainability of the diet, promoting a naturally GF diet, and in any case discouraging the sole use of commercial GF products.

In order to achieve these aims, AIC implements several actions on food education as well as on nutrition and food safety topics: paper and digital publications, use of social networks and dissemination of information and position papers on the AIC communication channels, publications of recipes also reporting the nutritional profiles and some 'healthy' tips, and dedicated webinars. Nutritional education is also carried out by local groups, such as the psychological and nutritional counseling services, which are now even more accessible thanks to the possibility of remote consultancy, empowerment meetings and self-help groups, cooking classes for adults and children to introduce people to a varied, tasty and healthy GF diet. The AIC monitors the adherence of patients to the GF diet through specific studies carried out through the years in collaboration with university research centers, in order to identify the most effective strategies to guarantee adherence to this life-saving therapy (Corposanto et al., 2015).

Since 2007, AIC has also developed a permanent program, entitled "Escape from Gluten," reserved to the schools and pre-schools, which has spread throughout Italy and has being replicated in other European countries (Hungary, Switzerland, and others are joining).

The "Escape from Gluten" program includes specific didactic material, and aims to inform, in a playful way, all children, young people and the school community, what it means to be celiac and eat GF, without this being seen as a different. Another item on which the program is focused is responsibility: particular needs are indeed an opportunity to act collectively in the school environment on the *three levels of responsibility*, as also reported by the Council of Europe:

- responsibility for one's own physical, social, and mental health;
- interpersonal relationships;
- responsibility for the environment in its broadest sense.

From a psychological point of view, the program sets CD as one of the various existing conditions of diversity, which can drive sharing and co-participation dynamics, and it also proposes games and activities in a playful nature according to the school level.

19.2.6 Eating out and traveling

With the exception of school and hospital canteens and public facilities, which guarantee by law GF meals for celiacs since 2005 complying to the Italian law no. 123, eating out remains the most troublesome issue of celiacs' life, with strong impact on both job and job traveling, and their social life.

For this reason, since 2000 the AIC has developed a specific program (called "*Gluten-free eating outside home*") dedicated to the dissemination of the correct GF training to the catering services. The program includes materials and procedures to train all kind of catering services (restaurants, pizzerias, hotels, bars, ice cream parlors, pastry shops) and make them aware and trained on CD and the technical requirements to make a service suitable for the dietary needs of celiac patients.

Today the catering services joining the AIC Network are around 4000 and distributed throughout Italy. This program starts with a specific course, held by specialized and qualified trainers, on CD and GF diet. The course covers both theoretical and practical sessions. The theoretical session clarifies the specific technical requirements (rules and procedures) to be applied in the preparation of GF meals with the aim of guaranteeing the safety of the meal for celiacs. The practical session of the course shows the technological characteristics of basic GF products, such as flour and pasta.

The local groups also offer consultancy to the catering services joining the Network: a "training completion" which takes place through meetings with the manufacturing/catering facility. AIC also helps the facility to defining the menu, the storage methods and the preparation areas for GF meals. In terms of follow up of the training program, AIC offers constant availability for the entire duration of the facility's membership to the Network. Every year, each facility joining the Network is given an AIC window sticker indicating the current year, which must be displayed in the window of the venue. AIC carries out periodic audits of all the accredited venues through qualified personnel.

The list of the venues joining the Network is published in the AIC Eating Out Guide, which is distributed to all AIC members, and published on the AIC website. The digital website version of the Guide, as well as the Mobile smartphone app version (increasingly used by consumers, tourists and members), are updated in real time. Taking advantage of the GPS geolocation system, the accredited venues closest to the user is located and indicated by the app. The app also contains a series of useful information on diet, on shops where you can find GF food, news on AIC activities and on everything that may be of interest to celiacs.

19.2.7 The crossed grain trademark and the "Food Handbook"

With the help of scientific experts, AIC creates and maintains an updated "classification" of food categories with reference to the risk of gluten contamination. This classification is specific for the Italian market, so it is possible to find slight differences from similar lists of other countries.

Moreover, the market is constantly evolving to follow the needs of consumers and the production processes change and improve. For these reasons, it is possible that the ranking might undergo changes.

In general, in relation to CD, foods can be classified as:

ALWAYS SAFE. These foods are always safe for celiacs as naturally GF or manufactured through processes that do not involve any risk of contamination;

UNSAFE-CHECK THE LABEL. These foods may contain gluten as ingredient or traces due to contamination. It is necessary to check the label.

FORBIDDEN. These food products always contain gluten. They are NOT suitable for celiacs. These food products are -of course- NOT listed in AIC's Food & Drink Italian Directory.

These classifications and the corresponding lists (foods and beverages) are published on the AIC website either in Italian and English. Also the Celiac Associations in other countries publish these lists on their own website. These lists are the result of constant analyses, research and surveys carried out by the Celiac Associations in collaboration with Manufacturers associations and national Food Agencies.

With some practice, celiacs learn to identify the foods suitable for their GF diet, also because the "gluten-free" claim guarantees the absence of ingredients and traces of gluten, according to the European law.

In addition to reading the label, there are tools that help celiacs, families and restaurant operators to choose safe foods. The registered Crossed Grain Trademark (*Spiga Barrata*) (Fig. 19.1), consisting of a crossed wheat ear, is now shown on the label of an increasing number of food products. *The Crossed Grain logo* is a trademark owned by AIC for the Italian market and by the Federation Of European Celiac Associations (www.aoecs.org) and other foreign celiac associations for their reference markets. *The Crossed Grain logo* is therefore a globally registered trademark and can only be used under license on prepackaged foods and beverages that meet a rigorous technical standard, created by the AOECS. For celiac patients *The Crossed Grain logo* is considered as the most important means of recognizing the suitability of products, and it represents the gold standard for CD food safety.

On the Italian market there are more than 2500 products licensed by *Crossed Grain* Trademark; AOECS member societies have licensed over 20,000 GF products that are safe to date. Use of this trademark is strictly monitored by AOECS and its member societies to ensure its integrity.

All the production plants of the licensed *Crossed Grain* products are yearly checked by qualified technical personnel who conduct the necessary inspections to assess the compliance of the producing/factory systems to the AOECS standard.

FIGURE 19.1 The registered Crossed Grain Trademark (*Spiga Barrata*) The registered Crossed Grain Trademark (*Spiga Barrata*) consists of an imagination design recalling a crossed wheat.

The European standard is also periodically reviewed to improve the quality and safety of its procedures and technical requirements. Since the beginning, AIC has been a member of the Technical Committee and Working Groups which deal with the revision and updating activity.

The "Food Handbook" (Prontuario) is an editorial and digital tool, annually published and updated, intending to help celiacs in selecting suitable GF foods. As in other countries, the Italian "Food Handbook," especially the digital version, represents a quick and effective consultation tool for families and operators. Approximately 30,000 paper copies of the "Food Handbook" are distributed to all AIC members, catering services, canteens, restaurants and dieticians. In Italy there are about 500 member companies and almost 20,000 products controlled and verified by AIC.

19.2.8 AIC and the advocacy action

As one of the leading aim, AIC's mission also comprises advocacy programs to promote understanding of CD, empower patients and provider community, and collaborate with institutions and stakeholders to carry AIC mission forward.

AIC is present on the main social media (Facebook, Instagram, Twitter, Youtube, LinkedIn) with institutional accounts with over 100,000 followers, and it is often present on national and local media with articles, interviews, insights. Among the many initiatives, the most important is certainly the National CD Awareness Week, an initiative that takes place every year on May, since 2015, in connection with the International CD Awareness Day (May 16th). A week to provide correct information, making the public and the media aware of a disease that affects about 600,000 persons in Italy, almost 400,000 of whom have not been diagnosed yet. Moreover, the National CD Awareness Week is an important chance to turn the spotlight on the false myths and fake news that often contribute to generate bad information, poor knowledge of CD and consequently a false perception of the needs and the rights of patients.

Apart the National CD Awareness Week, AIC carries out many other advocacy/information/dissemination events throughout the year, such as conferences and meetings dedicated to the general public or to specific professional health operator/physicians managing CD and the GF diet.

19.3 Networking European Associations: the activities of AOECS in Europe and worldwide

AOECS (aoecs.org) brings together the Celiac Associations of the European continent, therefore not only the European Union, but also those of Norway, Russia, Switzerland, Ukraine, and the United Kingdom, as well as some Associations outside the European continent, which can join as "affiliate members." To date it has 38 "full members" and represents celiacs and people affected by DH at an international level: the approximately 300,000 members of the various national associations and over 7 million European celiacs.

Born in 1988, AIC was one of its founding members and has always contributed very actively to its growth.

AOECS is an independent NPO and is involved in many international initiatives to raise awareness of CD and encourage research, essential for its diagnosis and treatment. From a European umbrella organization, created to encourage exchange between the first Celiac Associations born on our continent, and to represent celiac patients at a supranational level, it is increasingly becoming the global reference point for those suffering from CD, far beyond the European borders.

The main aims of AOECS are:

1. the best possible safety, availability and labeling of products suitable for gluten intolerant people;
2. encouraging more multinational research projects on CD and DH;
3. a more open exchange of information on CD, nonceliac wheat sensitivity (NCWS), DH and their suitable diets;
4. developing and sharing skills and knowledge, particularly among newer national celiac societies, to support the social welfare of those affected by CD, NCWS and DH.

AOECS regularly participates in sessions of the Codex Alimentarius Commission, a FAO/WHO organization responsible for the development and adoption of standards and guidelines for food law around the world. This participation has given, over the years, important results on various Codex Standards and the consequent national regulations to protect the health of celiac people.

The General Assembly is the most important body of AOECS and has responsibility for decisions and actions of AOECS. It is made up of the so-called "full members" of the AOECS, who must meet specific requirements to join our Federation. In fact, European Celiac Associations can join the AOECS if they are registered as NPOs in their respective countries. In addition to the General Assembly, AOECS is made up of a Board of seven members, including a President and a Vice-President, elected among the national delegates, which operationally implements the decisions of the General Assembly, a small but very competent staff, and develops its business through "Working Groups." Much of the work of the AOECS is carried out through the Working Groups, composed of voluntary members of the National Associations, who lend their work for free to develop the different objectives of AOECS according to specific themes, such as lobbying initiatives on the issue of food labeling, relations with the world of medical-scientific research, promotion of a European scheme aimed at catering, drafting of position statements on subjects of interest and cooperation with the Scientific Societies of doctors and dieticians, other patient associations and the trade associations of producers of foods.

19.4 Focus on scientific research: AIC funding activities of the research projects

AIC started promoting scientific research since 1995 by mean of the "Free of Gluten" Task Force and one of the greatest AIC fundraising activities which enrolled 8000 members all over Italy and yielded about 500,000 €. Telethon Foundation also contributed to fund raising.

According to the Italian nonprofit laws, on 2004 AIC established the Italian Foundation for Celiac Disease (FC) with the aim of supporting advanced biomedical research toward the treatment of CD.

Since 2013 FC has promoted high quality Italian research by granting Projects by mean of the national call for proposals system, thus becoming the first Italian charity in the field of CD. (FC Grants for the Scientific Research. An Overlook of a 6-Year Period, 2018)

The development of the national call for proposals system led to two main issues: (1) the improvement of the peer-review procedure, and (2) the creation of a web platform for researchers and reviewers to manage the funding process.

Accomplishing the current Italian nonprofit laws, FC has been merged by incorporation into AIC that is now the sole resulting entity. Upon the merger, the incorporating company, AIC, took the rights and obligations of FC, continuing in all its activities, including the funding of Italian research on CD, DH and other gluten-related disorders by the national call for proposals system.

19.4.1 Proposal types

Two kinds of national calls for proposals were developed with different aims and funding procedures (https://www.celiachia.it/ricerca/fc-call/)

19.4.2 The investigator call for proposals

These Calls are aimed at awarding excellent Research Projects presented by highly qualified senior researchers with proved and recognized international expertise in the field of CD and/or DH. The Project Proposals selected for funding are carried out in a public or a private nonprofit Italian research institute/organization where the Principal Investigators are affiliated.

In the first three editions, the Grant covered all the costs of the projects, not exceeding € 70,000 per year. In the last 2020 edition, the Grant was redesigned as a cofunding program to allow AIC for funding a higher number of projects. In all cases, annual, biennial, triennial and also pilot studies were invited to join the call. Partnership as well as multicenter studies are encouraged to participate.

19.4.3 The fellowships call for proposals

These Calls are aimed at awarding Triennial Fellowships to highly qualified young graduates and/or postdoctoral fellows who wish to broaden or acquire experience in scientific research within the field of CD and/or DH in public or private nonprofit Italian research institutes/organizations.

Each Triennial Fellowship Program is granted for a total of € 85,000, of which € 75,000 for the triennial gross salary of the Fellow and € 10,000 to be used during the Triennial research program exclusively for the purchase of consumables and supplies.

The Fellowships Call for Proposals is a co-funding program, as long as all the other costs involved in the accomplishment of the research project are at the expense of the Head of the Hosting Lab and/or of the Institute.

Both types of calls are open-theme, and all biomedical, food-tech, nutritional, clinical and basic research topics are included.

With eight editions of the national call for proposals, AIC has funded 29 Research Projects and almost the same number of Institutes, for total financing € 3,200,000, yielding about 70 publications.

19.5 The peer review methodology

The AIC national call for proposals system shares with other great worldwide charities the Peer Review process which represents an essential part in the promotion of high quality research.

AIC can count on a group of International Peer Reviewers all over the world, especially United States and Europe, who guarantee that the funded research is the most relevant in the field of CD and related topics.

Before the launch of the call, the availability of the reviewers is asked for confirmation, and the reviewers receive the Peer Review Regulation, the time-schedule of the selection/evaluation steps, the guidelines for the evaluation and the use of the web platform to make the online evaluation, and the Call for Proposals containing the rules and the eligibility criteria.

In this phase, the composition of the peer reviewers for each research area is evaluated to assure the proper covering of the projects focusing on each area. In case, and however, periodically for turnover reasons, reviewers are also asked to provide references of other international researchers to be contacted and to increase the group of the International Peer Reviewers.

Submitted Project Proposal Applications undergo the steps reported in Table 19.1.

TABLE 19.1 The processing steps of submitted Project Proposal Applications.

	Steps
1	Administrative conformity check;
2	Peer review evaluation of project proposal applications complying the administrative conformity check, and formulation of the ranking list;
3	Approval of the ranking list by the AIC Board of Directors.

TABLE 19.2 Priority criteria.

Priority criteria of the triaging evaluation of the project proposal application
Adherence to the AIC's aims priorities
Consistency with the duration declared by the PI applicant
Consistency with the research topic declared by the PI applicant
Consistency with the research area declared by the PI applicant
Consistency with the typology of study declared by the PI applicant

19.5.1 Administrative conformity check

Submitted Project Proposal Applications first undergo an administrative check for assessing the compliance to the *Call for Proposals* eligibility criteria and regulations. Project Proposal Applications that meet all the formal requirements undertake the Peer Review Evaluation. All the PI Applicants will be notified of the outcome of the Administrative Conformity Check.

19.5.2 Peer review evaluation of project proposal applications, and formulation of the ranking list

Each admitted Project Proposal is evaluated individually by three independent international Peer Reviewers. The Peer Review evaluation of each Project Proposal Application is performed by a two steps procedure, an IN/OUT triage and a scoring with a ranking list. The Peer Reviewers triage the Project Proposal as anIN Application or as an OUT Application on the basis of the priority criteria reported in Table 19.2.

Project Proposals which do not meet only one of the above Priority Criteria are triaged OUT by the Peer Reviewers. In the second step of the Evaluation, for each Project Proposal, the Peer Reviewers assess the scientific value and the feasibility. The Ranking List is prepared by positioning the Project Proposals on the basis of their *Final Overall Score*.

19.6 Research projects funded by AIC

Project Proposals, focused on either biomedical and food/food tech topics, must mandatorily have aims and purposes that fully adhere to one or more of the following AIC Aims Priorities:

1. Proximity to find new pathogenic mechanisms
2. Proximity to cure
3. Natural history of the disease by linking different phases of the disease to specific biological/genetic profiles
4. Interactions between environmental risk factors, genetic profiles and biomarkers
5. Innovation of clinical methodologies
6. Evaluation of the efficacy of diagnostic and therapeutic approaches in clinical practice, in terms of outcome and quality of life
7. New therapeutic targets, procedures or strategies (pilot clinical studies)

In general, submissions are mainly focused on the natural history of CD and cover basic research areas, while clinical studies are less represented. In particular, a substantial number of projects are focused on the development of new biomarkers for either an alternative diagnosis of CD or to measure the adherence to a GF diet. Some of these studies involved miRNA differential signatures to be evaluated in the peripheral blood, while other studies developed new urine and fecal biomarkers to assess the adherence. Another emerging topic copes with the correlation between HLA alleles and the risk to develop CD, with the aim of supporting an early diagnosis, especially in subjects at risk. Finally, an increasing importance is given to studies investigating the role of intestinal microbiota on the development of CD.

References

ACCORDO 30 luglio 2015, Accordo, ai sensi dell'articolo 4, del decreto legislativo 28 agosto 1997, n. 281, tra il Governo, le regioni e le province autonome di Trento e di Bolzano sul «Protocollo per la diagnosi e il follow up della celiachia». (Rep. Atti n. 125/CSR). (15A06387), Pag. 147 della Gazzetta Ufficiale della Repubblica Italiana, Anno 156° - Numero 191, 19 agosto 2015. (2015). Rep. Atti n. 125/CSR). (15A06387), Pag. 147 Della Gazzetta Ufficiale Della Repubblica Italiana.

CELIAC DISEASE IN CHILDREN AND ADULTS: FOR PHYSICIANS OF GENERAL MEDICINE AND PRIMARY CARE PAEDIATRICIANS. (2018). Associazione Italiana Celiachia APS (AIC). https://celiachia.b-cdn.net/assets/uploads/2019/12/Booklet-Celiachia-ENG.pdf.

Celiachia e Dermatite Erpetiforme. (2022). Associazione Italiana Celiachia APS (AIC). https://celiachia.b-cdn.net/assets/uploads/2022/05/Raccomandazioni_Celiachia-e-dermatite-erpetiforme.pdf.

Celiachia e Test Genetico HLA. (2022). Associazione Italiana Celiachia APS (AIC). https://celiachia.b-cdn.net/assets/uploads/2022/05/Raccomandazioni_Celiachia-e-test-genetico-HLA.pdf.

Cornicelli, M., et al. (2018). Nutritional composition of gluten-free food versus regular food sold in the Italian market. Digestive and Liver Disease: Official Journal of the Italian Society of Gastroenterology and the Italian Association for the Study of the Liver.

Corposanto, C., Molinari, B., & Neuhold, S. (2015). Celiac Disease Seen with the Eyes of the Principle Component Analysis and Analyse Des Donn'es. *Open Journal of Statistics*, *05*(03), 211–222. Available from https://doi.org/10.4236/ojs.2015.53023.

Donna & Celiachia. Raccomandazioni Cliniche. (2016). Associazione Italiana Celiachia APS (AIC). https://celiachia.b-cdn.net/assets/uploads/2020/10/DonnaeCeliachia_Raccomandazioni.pdf.

FC Grants for the Scientific Research. An overlook of a 6-year period. (2018). Associazione Italiana Celiachia APS (AIC). https://celiachia.b-cdn.net/assets/uploads/2019/12/TriennioBandi_ENG.pdf.

Fiori, F., Parpinel, M., Morreale, F., & Pellegrini, N. (2022). The Update of the Italian Food Composition Database of Gluten-Free Products and Its Application In Food-Based Dietary Guidelines Menus. *Nutrients*, *14*(19). Available from https://doi.org/10.3390/nu14194171.

Guida Sport&Celiachia. La Dieta Senza Glutine e la Performance Sportiva. (2017). Associazione Italiana Celiachia APS (AIC). https://celiachia.b-cdn.net/assets/uploads/2020/10/Sport-e-Celiachia-Guida-Web.pdf.

Le Contaminazioni nella Dieta Senza Glutine. (2022). Associazione Italiana Celiachia APS (AIC). https://celiachia.b-cdn.net/assets/uploads/2022/02/A.05.Sito-Istituzionale-AIC-Supporto-editoriale-Contaminazioni.pdf.

Terzjus, & Pastorino Olmi, P. (2021). Gli enti del Terzo Settore come soggetti protagonisti del disegno Costituzionale: Un nuovo orizzonte per la promozione della ricerca scientifica e per la tutela del diritto alla salute. https://terzjus.it/articoli/note-e-commenti/gli-enti-del-terzo-settore-come-soggetti-protagonisti-del-disegno-costituzionale-un-nuovo-orizzonte-per-la-promozione-della-ricerca-scientifica-e-per-la-tutela-del-diritto-alla-salute/.

Chapter 20

Putative therapeutic targets in celiac disease

Antonio Di Sabatino, Giovanni Santacroce, Paola Ilaria Bianchi and Nicola Aronico
Department of Internal Medicine, University of Pavia, Fondazione IRCCS Policlinico San Matteo, Pavia, Italy

20.1 Introduction

A strict and long-term gluten-free diet (GFD) is the cornerstone of celiac disease (CD) management and, at the moment, is the only available effective treatment for this condition (Lebwohl et al., 2018). However, patients may experience persistent symptoms and some degree of intestinal damage despite GFD, largely due to unintentional or deliberate dietary noncompliance that it has been reported to occur in up to 50% of cases (Rubio-Tapia et al., 2010). Moreover, gluten restriction requires extensive patient effort, impacting personal, social, and economic aspects, and affecting their quality of life (Martínez-Martinez et al., 2019; Shah et al., 2014). Not least, an inadequately balanced GFD may be associated with several nutritional disadvantages (Cardo et al., 2021; Marciniak et al., 2021). More specifically, the lower proportion of whole grains and fibers in gluten-free products, together with the higher consumption of refined carbohydrates, have been implicated in increasing in treated CD patients the rates of metabolic-related disorders, including liver steatosis, coronary syndrome, and cardiovascular events (Larussa et al., 2021; Potter et al., 2018). Thus the need to develop new strategies in order to improve clinical and social outcomes is actually an issue of the highest concern (Pinto-Sanchez et al., 2021). Furthermore, the willingness for therapies beyond GFD is increasing in adult patients and in CD children's parents (Andrew et al., 2022; Norsa et al., 2015). If these alternative therapies could fully replace GFD or should be considered as additional to GFD to counteract occasional and inadvertent exposure to gluten is still a matter of debate.

The exponential development of biological therapies designed for targeting specific inflammatory molecules implicated in the pathogenesis of many chronic inflammatory disorders, including inflammatory bowel diseases (Danese et al., 2015), and the significant advances in understanding the molecular mechanisms implicated in the mucosal damage of CD (Di Sabatino & Corazza, 2009; Sollid & Jabri, 2013) (see also Chapter 5), has favored the proliferation of numerous biologic or immune-modulatory agents aimed at dampening mucosal inflammation and inducing mucosal healing or restoring intestinal barrier. Several putative endogenous therapeutic targets have been investigated through a translational approach, including intercellular adhesion complexes, such as tight junctions (TJs), enzymes, such as tissue transglutaminase (tTG)-2 or cathepsin S, immune cells, such as lamina propria B and T lymphocytes, modulators of the interaction between antigen presenting cells (APCs) and T cells, such as HLA-DQ2 molecules, proinflammatory cytokines, such as interleukin (IL)-15, intracellular signal transduction pathways, such as Janus kinase (JAK) and Signal Transducer and Activator of Transcription (STAT), adhesion molecules, such as $\alpha4\beta7$ integrin, and chemokines, such as C-chemokine receptor (CCR)9.

Apart from the therapeutic neutralization of endogenous inflammatory targets, another therapeutic strategy is that based on the detoxification of gliadin peptides. A number of molecules capable of binding, neutralizing or degrading gluten peptides are currently under investigation, some of them synthetic, others derived from natural sources, such as plants, bacteria, and fungi.

Aim of the present chapter is to explore and critically revise all the therapeutic strategies alternative to GFD, which are currently under investigation in clinical trials in uncomplicated CD patients (Fig. 20.1). Although some of these molecules are promising and have reached advanced stages in clinical trials, none of them is yet approved for the treatment of uncomplicated CD patients in clinical practice.

Pediatric and Adult Celiac Disease. DOI: https://doi.org/10.1016/B978-0-443-13359-6.00013-3

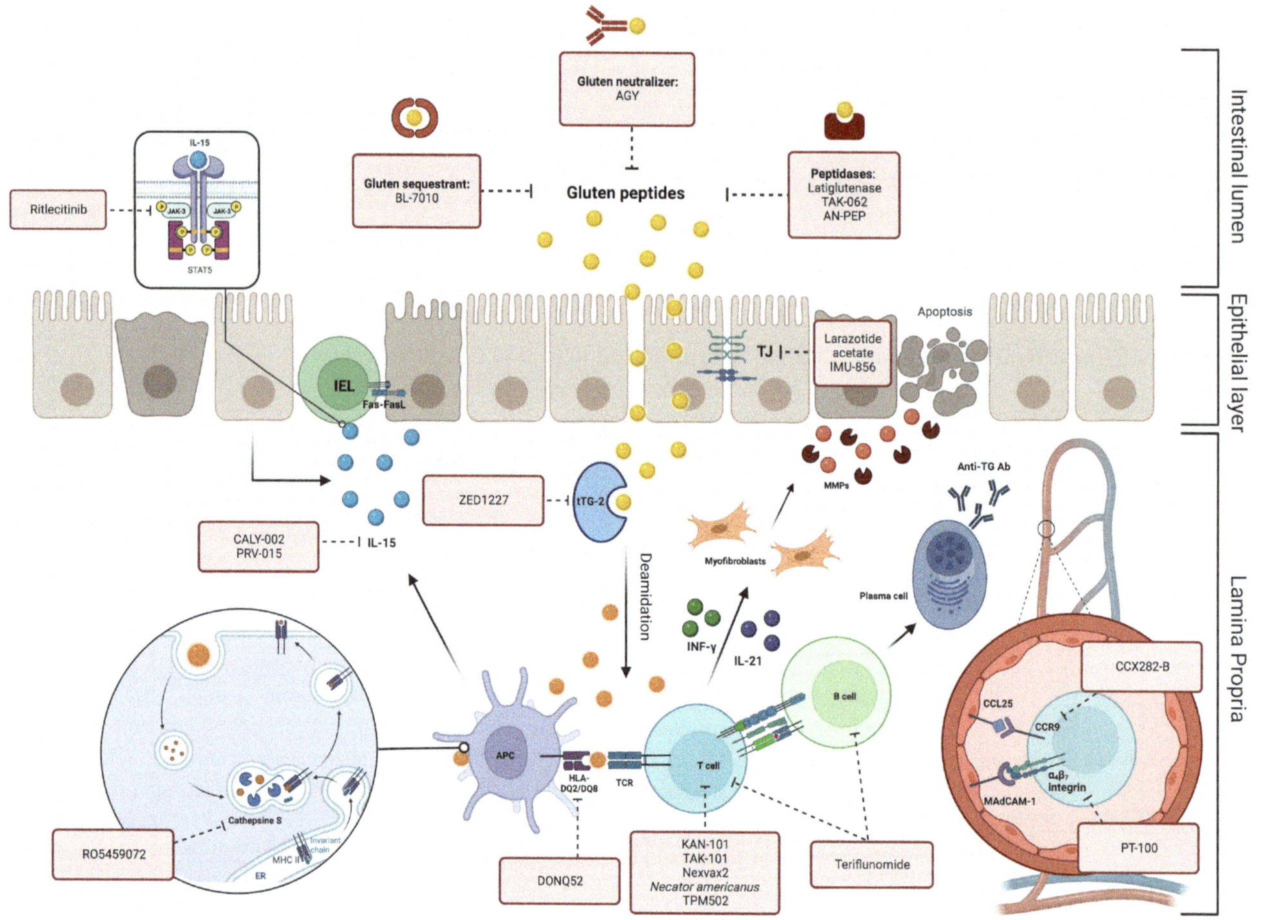

FIGURE 20.1 Therapeutic targets for compounds investigated in clinical trials for celiac disease. This figure shows the main pathogenetic mechanisms underlying celiac disease (CD), focusing on putative targets for therapeutics. The compounds are shown in red squares in the figure and are correlated with their target by a dashed line. Gluten peptides, etiological factors of CD, reach the intestinal lumen. At this level, some molecules explored as therapeutic options act by sequestering [BL-7010], neutralizing [AGY] or degrading gluten [Latiglutenase; TAK-062; AN-PEP]. By paracellular and transcellular pathways, gluten peptides cross the intestinal barrier, whose integrity is maintained by tight junctions (TJs), possible targets of barrier integrity modulators [Larazotide acetate and IMU-856]. Once in the lamina propria, gluten peptides are deamidated by tissue transglutaminase-2, target of transglutaminase inhibitors [ZED1227]. The deamidated peptides are phagocytized into antigen-presenting cells (APCs). In the cytoplasm of APCs (see circle in the lower left of the figure), gluten peptides form a complex with the human leukocyte antigen (HLA) DQ2-DQ8, after its activation by removal of the invariant chain via cathepsin S. A synthetic cathepsin S inhibitor [RO5459072] has been evaluated as therapeutic agent for CD. The complex HLA-DQ2—gluten epitope is another therapeutic target [DONQ52]. The interaction between APCs and lymphocytes determines the activation of T cells and the release of several cytokines, such as interferon gamma and interleukin (IL)-21. These molecules activate myofibroblasts, leading to the production of metalloproteases that induce epithelial apoptosis. In addition, the interaction between activated T cells and B cells stimulates the production of plasma cells secreting antitransglutaminase antibodies. T cells and B cells are target of tolerogenic molecules [KAN-101; TAK-101; Nexvax2; *Necator americanus*; TPM502; teriflunomide]. Activated APCs produce IL -15 (which is also secreted by the epithelium), which stimulates intraepithelial lymphocytes (IEL) through the activation of Janus kinase (JAK)-3—signal transducer and activator of transcription (STAT)5 pathway. IL-15 and JAK-3 are possible therapeutic targets [CALY-002; PRV-015; ritlecitinib]. The activated IEL interacts with enterocytes via the Fas-Fas Ligand pathway, determining cellular damage. In the lower right of the figure is represented a section of an intestinal vessel in which activated T lymphocytes circulate and interact with endothelial cells via the MAdCAM-1—$\alpha4\beta7$ integrin system and the C-chemokine receptor 9—C-chemokine ligand 25 system. The lymphocyte homing can be regulated through selective antagonists of these trafficking systems [PT-100; CCX282-B]. *Ab, antibody; AGY, egg yolk antigliadin antibody; AN-PEP, Aspergillus niger- prolyl endopeptidase; APC, antigen presenting cell; CCL, C-chemokine ligand; CCR, C-chemokine receptor; ER, endoplasmic reticulum; HLA, human leukocyte antigen; IEL, intraepithelial lymphocyte; IL, interleukin; INF, interferon; JAK, Janus kinase; MAdCAM, mucosal vascular addressing cell adhesion molecule; MHC, major histocompatibility complex; MMP, metalloprotease; STAT, signal transducer and activator of transcription; TCR, T cell receptor; tTG, tissue transglutaminase; TJ, tight junction; –| modulation. Created with 'Biorender.com'.*

20.2 Endogenous therapeutic targets

20.2.1 Tight junctions

TJs are interenterocyte adhesion complexes able to control the paracellular transport of molecules and the intestinal permeability. In physiological conditions, TJs prevent the passage of harmful molecules and pathogenic bacteria into the lamina propria. Epithelial junctional proteins are altered in the mucosa of untreated CD patients (Ciccocioppo et al., 2006), and the functional impairment of these junctions due to gliadin-mediated zonulin up-regulation is considered one of the main and early event in CD pathogenesis (Schuppan et al., 2009). More in depth, in genetically predisposed subjects, the gliadin binding with the chemokine receptor CXCR3 induces the MyD88-dependent release of zonulin, leading to TJ impairment, gluten peptide paracellular translocation into the lamina propria and activation of the inflammatory cascade leading to mucosal damage (Fasano, 2011). Hence, the modulation of TJs has been evaluated as a therapeutic option for CD.

20.2.1.1 Larazotide acetate

A first TJ modulator is the oral synthetic zonulin receptor antagonist called larazotide acetate (formerly AT-1001), whose sequence was derived from a zonula-occludens toxin secreted by *Vibrio cholerae*. In vitro studies in Madin-Darby canine kidney type II and Caco-2 cells, and in vivo preclinical studies in HLA-HCD4/DQ8 and zonulin transgenic mice have shown that larazotide acetate is able to restore TJ integrity and barrier function, by reducing permeability to gliadin peptides (Gopalakrishnan et al., 2012; Gopalakrishnan et al., 2012; Sturgeon et al., 2017).

In the clinical setting, the phase I double-blind randomized controlled trial (DBRCT) evaluating the safety and efficacy of larazotide acetate 12 mg vs placebo in 21 CD patients on at least 6-month GFD and undergoing gluten challenge for 3 days, showed a good safety profile of this compound. Also, the results of the trial highlighted a good efficacy of larazotide acetate, as suggested by the reduction of intestinal barrier dysfunction -evaluated through the urinary lactulose/mannitol fractional excretion ratio-, down-regulation of interferon-γ production, and amelioration of gastrointestinal symptoms (Paterson et al., 2007). Later phase IIa and phase IIb trials, exploring respectively the effect of different doses of larazotide acetate on 86 and 184 CD patients, both under at least 6-month GFD, did not confirm the effect of the compound on intestinal permeability, but showed a good safety profile and efficacy in reducing gluten-induced symptoms, especially for low doses (Kelly et al., 2013; Leffler et al., 2012). The effect of larazotide acetate on gastrointestinal symptoms was further investigated in another phase II DBRCT, enrolling 342 CD patients nonresponsive to GFD, who were randomized to receive either different doses of the synthetic peptide or placebo for 12 weeks. The primary endpoint, which consisted in the reduction of signs and symptoms, was met just for the lower dose (0.5 mg), but not for the doses of 1 and 2 mg, with a safety comparable to placebo (Leffler et al., 2015). The peptide aggregation at higher doses, with a consequent reduced activity in vivo, was claimed by the authors as a possible explanation for their results.

Given the promising results of phase II trials, a phase III trial was started to confirm the efficacy of 0.25 or 0.50 mg of larazotide acetate on symptomatic treated CD patients. However, the *interim* analysis showed that the additional number of patients needed to determine a significant clinical outcome between larazotide acetate and placebo was too large to support trial continuation (accesswire.com, 2022). Therefore the trial was discontinued for sponsor decision in 2022. Further studies are needed to determine if specific subgroups of symptomatic treated CD patients might benefit from treatment with larazotide acetate. However, the conflicting data on the structure and function of this molecule (Sollid & Koning, 2021), as well as the lack of proven efficacy on intestinal barrier dysfunction raised some doubts on its real benefit. Also, the lack of data on duodenal histology represents a major limitation of these trials.

20.2.1.2 IMU-856

A new synthetic oral molecule called IMU-856, which acts as a modulator of TJs, has recently been evaluated in 71 healthy subjects in the first part of a phase I trial, showing that a single ascending dose and multiple ascending doses of this compound are safe, well-tolerated and with a good pharmacokinetic profile (Polasek et al., 2022). The second part of this trial is ongoing and recruiting CD patients under GFD to evaluate safety and efficacy of 80 and 160 mg of IMU-856.

20.2.2 Tissue transglutaminase-2

tTG-2 is an enzyme playing a crucial role in CD pathogenesis through the deamidation of gluten peptides reaching the lamina propria. tTG deamidation generates specific negatively charged gliadin peptides that bind with a higher affinity to CD-associated HLA-DQ2 or DQ8 molecules expressed on the surface of lamina propria APCs, thus enhancing T-cell activation and proinflammatory cytokine production (Schuppan et al., 2009). Hence, tTG2 inhibition has been considered an optimal strategy in preventing the inflammatory cascade leading to mucosal damage in CD. A first proof-of-concept study investigated whether two tTG2 inhibitors, cell-impermeable R281 and cell-permeable R283, could prevent the gliadin damage in vitro, using Caco-2 cells, and ex vivo, using organ culture of duodenal biopsies collected from CD patients (Rauhavirta et al., 2013). These inhibitors showed protective effect against gliadin-induced detrimental effect in vitro, with a more potent action of R281. The latter also showed a preventive action on gluten-induced modifications ex vivo, including the increase in lamina propria CD25- and IL-15-positive lymphocytes. However, the action of these molecules has not yet been evaluated in vivo due to their potential side effects. The possible cross-inhibition of other members of the transglutaminase family, ubiquitously expressed in human tissues, could lead to an impairment of the coagulation cascade and the wound healing process (Dickneite et al., 2015; Telci & Griffin, 2006). Moreover, tTG2 knock-out mice showed a defective clearance of apoptotic cells, predisposing to autoimmunity, splenomegaly, and immune complex glomerulonephritis (Szondy et al., 2003).

20.2.2.1 ZED1227

To overcome the previously discussed safety issues, a new generation of soluble highly selective intestinal tTG2 inhibitors have been developed. Among them, ZED1227 showed a high affinity for intestinal tTG2 in a mouse model of small bowel inflammation (Ventura et al., 2018). Further studies on transgenic mouse model of CD (NOD-DQ8 mice) highlighted the ability of ZED1227 to prevent gluten-induced enteropathy and immune activation (Ventura et al., 2021). This oral peptidomimetic has been also evaluated in a phase II DBRCT on 163 CD patients on at least 12-month GFD, who were randomized to receive either ZED1227 (10, 50 or 100 mg) or placebo after gluten challenge. Treatment with all the three dosage levels of the compound was able to attenuate the gluten-induced duodenal mucosal injury, measured through the villus height to crypt depth ratio, and the 100 mg dose improved symptom and quality of life scores (Schuppan et al., 2021), with a good safety profile. A new phase II trial started on 2021, and it is still ongoing to evaluate ZED1227 efficacy on treated symptomatic CD patients (EudraCT 2020−004612-97, 2023).

20.2.2.2 GSK3915393

The first in-human phase I DBRCT evaluating safety, tolerability and pharmacokinetics of single ascending dose and repeated oral doses of GSK3915393, a new tTG2 inhibitor, has been recently completed in healthy subjects. Results are awaited to eventually plan trials in CD patients (NCT04604795, 2021).

20.2.3 T cells

The immunological activation by deamidated gluten peptides represents the cornerstone of CD pathogenesis, and T cells are the major players of this process (Schuppan et al., 2009). Therefore one of the most attractive options for the therapeutic management of CD has been that aimed at restoring tolerance to ingested gluten by remodulating T-cell function (Di Sabatino et al., 2018). Although the mechanisms underlying vaccine-mediated reinduction of immune tolerance are only partially understood, deletion of naïve T cells specific for the gliadin peptides composing the vaccine, and subsequent induction of regulatory T cell differentiation has been hypothesized as a possible therapeutic option in CD (Anderson & Jabri, 2013). Also, a vaccine direct action on memory T cells promoting anergy and causing deletion has been postulated.

20.2.3.1 Nexvax2

Nexvax2 is a peptide-based vaccine composed of three immunodominant gliadin peptides named NPL001, NPL002 and NPL003, identified as the major activators of gluten-specific T cells in CD patients (Tye-Din et al., 2010). Nexvax2 targets the HLA-DQ2.5 epitope-T cell receptor complex, which allows the stimulation of T cells by APCs. After successful studies in HLA-DQ2.5 transgenic mice (Anderson & Jabri, 2013), Nexvax2 was tested in a phase I DBRCT on 34 HLA-DQ2.5-positive CD patients under GFD, randomized to receive either weekly intradermal injection of 9, 30, 60 or 90 μg vaccine or placebo for 3 weeks (Brown et al., 2011). This study showed a good safety profile for the tolerizing agent, with gastrointestinal adverse events occurring more frequently at the highest doses. Furthermore, the time course of symptoms and gluten-specific T-cell mobilization after vaccine administration was similar to those triggered by acute oral gluten exposure. Two additional phase I DBRCTs showed a maximum tolerated dose of 150 μg for twice-weekly intradermal administration. The transient acute gastrointestinal adverse events, similar to those caused by gluten ingestion, occurred after the initial vaccine doses (Goel et al., 2017). The symptoms appeared in concomitance with the elevation in plasma levels of a number of cytokines, including interleukin (IL)-2, IL-6, IL-8, IL-10, monocyte chemoattractant protein-1 and interferon γ-induced protein 10, starting from 2-hour postdose (Goel et al., 2016). However, at the maximum tolerated dose the recall immune response to gluten was modified, being T cells specific for Nexvax2 epitopes unresponsive to further antigenic stimulation, and no evidence of worsening in duodenal histology was detected. A further phase I trial found that gradual escalation of vaccine administration allowed safer administration of higher doses (up to 900 μg), without immune activation (Daveson et al., 2017).

Given the encouraging data from the phase I studies, a phase II study (RESET CeD) was conducted to evaluate the Nexvax2 subcutaneous administration safety and efficacy in improving patient reported outcome scores in adult CD patients after bolus exposure to gluten (Tye-Din et al., 2023). The study was discontinued by the sponsor since the *interim* analysis showed that Nexvax2 did not reduce acute gluten-induced symptoms and immune activation. However, in an exploratory posthoc endpoint, duodenal histology showed significant improvement after an 18-week treatment period. Despite the disappointing results of RESET CeD study, the strategy adopted by Nexvax2 seems to be interesting and thus requires further investigation. The primary endpoint previously pursued, namely the amelioration of acute symptoms induced by heavy gluten challenge, is one of the hardest to achieve in therapeutic trials. Hence, new studies with revised design and endpoints are needed to address different purposes of the vaccine, such as its ability to control low gluten exposure and serve as supplement to dietary restriction more than a replacement to GFD. Also, the best route of administration and the most appropriate dosage of Nexvax2 has yet to be defined.

20.2.3.2 KAN-101

KAN-101 is a novel compound aimed at re-educating T cells driving the mucosal damage in CD, so they do not respond to gluten peptides, while leaving the other components of the immune system intact. The phase I DBRCT enrolled a total of 41 HLA-DQ2.5-positive CD patients under GFD, treated with intravenous KAN-101 single-ascending dose or multiascending doses versus placebo, followed by oral gluten challenge. The results showed that KAN-101 was both safe and tolerated in CD patients at all dose levels tested, while a maximally tolerated dose was not identified (anokion.com, 2023). Also, KAN-101 administration led to a reduction of gliadin peptide-specific T cell response following gluten challenge, as measured by both whole blood restimulation with gliadin and by ELISpot analysis, compared to placebo. Furthermore, KAN-101 controlled the immune response to gluten challenge, as demonstrated by the prevented gluten-induced plasma IL-2 response at higher doses, and blunted gut-homing CD8-positive T-cell response. A new phase I/II clinical trial recruiting CD patients has recently started to confirm the safety, tolerability, pharmacodynamics and pharmacokinetics of KAN-101 (NCT05574010, 2023).

20.2.3.3 TAK-101

Another compound targeting T cells and currently under investigation is TAK-101, formerly known as TIMP-GLIA. The safety and efficacy of this poly(lactic-co-glycolic acid)-gliadin nanoparticle has been previously investigated in three mouse models of CD, namely C57BL/6, RAG1$^{-/-}$ and HLA-DQ8, huCD4 transgenic Ab0 NOD mice. The intravenous injection of TAK-101 in mice significantly decreased gliadin-specific T-cell proliferation, proinflammatory cytokine secretion, circulating gliadin-specific IgG/IgG2c and gluten-dependent enteropathy (Freitag et al., 2020). TAK-101 was later evaluated in a phase I dose escalation safety study and a phase IIa DBRCT. In the phase I study 23 CD patients under GFD were treated with a single ascending dose followed by repeated doses, while for the phase II trial, 34 CD patients under GFD were randomized to either TAK-101 8 mg/kg or placebo on days 1 and 8, followed by 14 days of gluten challenge (Kelly et al., 2021). TAK-101 was well tolerated and no serious adverse events occurred following its administration. Moreover, in CD patients who completed the 14-day gluten challenge, the infusion of TAK-101 induced a significant reduction in circulating gliadin-specific interferon-γ producing cells and in circulating $\alpha_4\beta_7^+$ CD4$^+$, $\alpha E\beta 7^+$ CD8$^+$ and effector memory $\gamma\delta$ T cells, compared to placebo. No effect of the compound was found on the number of intraepithelial lymphocytes (IELs). In addition, a trend toward a reduction in gluten-induced deterioration of duodenal histology was observed with TAK-101 compared to placebo, but this did not reach a statistical significance. These results suggested an effect of this compound in preventing gluten-induced immune activation, and have paved the way for a new phase II dose-ranging DBRCT that has started in 2022 but is not yet recruiting patients (NCT04530123, 2023).

20.2.3.4 TPM502

A new peptide-encapsulated nanoparticle called TPM502, able to induce immune tolerance modulating T-cell function, is currently under evaluation in a phase II DBRCT. This trial aims at assessing the safety and the pharmacodynamics of

TPM502 intravenous infusion in HLA-DQ2.5 CD patients under GFD (EudraCT 2022−001656-41, 2023). Results are expected in 2024.

20.2.3.5 Necator americanus

Based on the "hygiene hypothesis," according to which the lack of exposure to intestinal microorganisms during early childhood could predispose to autoimmunity (Bach, 2002), a possible immunomodulatory role has been hypothesized for the gastrointestinal nematode *N. americanus*, based on its potential ability in shifting the Th1 immune response triggered by gluten into Th2 response. Phase I and II trials showed a good tolerance to the intestinal hookworm infection, without negative impact on gut microbiota (Cantacessi et al., 2014; Croese et al., 2015, 2020; Daveson et al., 2011). However, the efficacy results of the intradermal inoculation of *N. americanus* are controversial. The first DBRCT, evaluating the effect of *N. americanus* in 20 HLA-DQ2-positive CD patients under GFD, did not show a beneficial effect of the inoculation on the degree of villous atrophy following gluten challenge (Cantacessi et al., 2014; Daveson et al., 2011). Conversely, a subsequent single group assignment trial, evaluating the effect of small incremental doses of gluten over 16 weeks in 12 CD patients previously infected with *N. americanus*, showed reduced mucosal damage and no rise in CD antibodies after gluten microchallenge compared to historical controls (Croese et al., 2015). These results were associated with an increase of regulatory T cells and a decrease of interferon-γ-producing T cells. Also, there was an improvement in the quality of life of patients, who did not experience gastrointestinal symptoms after gluten challenge. In a more recent trial assessing the effect of *N. americanus* on 54 HLA-DQ2-positive CD patients under GFD, the parasite inoculation did not restore tolerance to sustained moderate consumption of gluten (2 g/day), but it was associated with improved symptom scores after intermittent consumption of lower gluten doses (Croese et al., 2020). When considering the conflicting results of this therapeutic approach, which involves the inoculation of larvae, it is difficult to imagine its routine clinical use in CD management.

20.2.3.6 Teriflunomide

Teriflunomide is an immunomodulator used for multiple sclerosis and with a cytostatic effect on T cells due to its inhibition of dihydro-orotate dehydrogenase (Bar-Or et al., 2014). A phase II DBRCT evaluating the effect of an oral administration of 14 mg/day teriflunomide in HLA-DQ2.5 CD patients is currently underway (NCT04806737, 2023).

20.2.4 Cathepsin S

Cathepsin S is a lysosomal protease predominantly expressed in immune cells, including B cells, dendritic cells and macrophages (Smyth et al., 2022). Aside from degrading proteins within the lysosome, this enzyme plays a key role in HLA maturation, by degrading the invariant chain/CD74, and in antigen processing and presentation. Hence, the modulation of cathepsin S has recently aroused interest as a novel therapeutic approach of several immune-mediated disorders (Smyth et al., 2022), including CD. RO5459072 is an orally administered competitive inhibitor of cathepsin S which has been evaluated in a phase I DBRCT (NCT02679014, 2016). This trial investigated the pharmacokinetics, pharmacodynamics, safety and tolerability of RO5459072 in 19 CD patients on at least 12-month GFD. Although the study was completed, the results are not yet available.

20.2.5 HLA-DQ2.5

CD patients carry one of two HLA class II allele variants, namely HLA-DQ2 or DQ8, which represent the genetic background of the disease (Schuppan et al., 2009). Indeed, HLA-DQ2/DQ8 molecules, expressed on APCs, bind with high affinity gliadin-derived peptides, leading to the activation of T cells, which are main effector player of the intestinal inflammation leading to villous atrophy. Hence, agents able to block HLA-DQ2/DQ8 molecules have been developed. Preclinical studies were conducted on modified gluten peptides able to bind with high affinity HLA-DQ2 molecules, thus blocking the subsequent gluten-induced immune response (Kapoerchan et al., 2008, 2010). However, these compounds are still at the preclinical phase. DONQ52, an antibody targeting the HLA-DQ2.5-gluten peptide complex, is actually under evaluation in a phase I trial, the so-called LILY study, aimed at assessing the safety, tolerability, pharmacokinetics, and immunogenicity of DONQ52 in HLA-DQ2.5 CD patients under GFD (NCT05425446, 2023). The trial is still recruiting patients. However, a potential limit of this approach is that the compounds studied so far target only DQ2 molecules, thus being unable to work on HLA-DQ8-positive CD patients. Finally, the safety of these compounds needs to be more thoroughly evaluated, as the ubiquitous requirement of HLA in several immune responses could result in severe immunosuppression.

20.2.6 Interleukin-15

IL-15 is a cytokine mostly produced by intestinal epithelial cells and dendritic cells/macrophages, which modulates both the innate and adaptative immune response in the gut, by inducing the differentiation and proliferation of T cells, B cells and natural killer cells. Several studies found a gluten-dependent IL-15 overexpression, coupled with increased expression of IL-15 receptor and lower IL-15 response threshold in active CD patients (Di Sabatino et al., 2006). More in depth, IL-15 up-regulates the NKG2D receptor on IELs and enhances IEL cytotoxic activity thus favoring an exaggerated enterocyte apoptosis which, in turn, causes villous atrophy. Furthermore, IL-15 can lead to the impairment of Smad-3 dependent transforming growth factor-γ signaling, promoting intestinal inflammation. Most of the attempts in targeting IL-15 have been conducted in complicated CD patients, due to the role of this cytokine in decreasing IEL apoptosis thus favoring the emergence of aberrant IELs in refractory CD (see Chapter 14). In the context of uncomplicated CD patients, the therapeutic experience is scarce and the few data available are described below.

20.2.6.1 CALY-002

CALY-002 is a humanized monoclonal anti-IL-15 antibody, under actual investigation in a recruiting phase I DBRCT (NCT04593251, 2023). Adult CD patients on at least 12-month GFD are randomized to either intravenous multiascending dose of CALY-002 or placebo. The primary outcome of the trial is the incidence of treatment-emergent adverse events. Results of this trial are not yet available.

20.2.6.2 PRV-015

Another potential agent is the anti-IL15 human monoclonal antibody PRV-015 (previously known as AMG714). It was investigated as an alternative to GFD in a phase II DBRCT, which was conducted by randomizing 64 CD patients to receive six subcutaneous 150 or 300 mg doses of PRV-015 or placebo for 10 weeks and a subsequent 2.5 g/day gluten challenge (Lähdeaho et al., 2019). The agent did not show any effect in reducing the degree of villous atrophy, but a significant trend in gluten-related symptom improvement and a reduction in IEL density was found in CD patients treated with 300 mg. A new phase II DBRCT has started to evaluate the efficacy and safety of three dose regimens of this agent in adult CD patients nonresponsive to GFD (NCT04424927, 2023). The primary outcome is represented by symptom attenuation. The effect of PRV-015 on IEL infiltration, the emergence of antidrug antibodies and the safety are secondary outcome of this trial.

20.2.7 Janus kinases

The IL-15 activation of cytotoxic IELs is mediated by the activation of the JAK-STAT signal transduction pathway (Schuppan et al., 2009). For this reason, the antiinflammatory action of tofacitinib, a pan-JAK inhibitor, has been explored in a transgenic mouse model of CD overexpressing human IL-15, that is, T3(b)-hIL-15 Tg mice (Yokoyama et al., 2013). This study demonstrated a lasting reversal of pathologic manifestations in treated mice. However, clinical studies on tofacitinib are at the moment focused only on refractory CD patients. Ritlecitinib, a synthetic JAK-3 inhibitor, is actually under a phase II clinical trial recruiting CD patients under GFD to establish its safety and efficacy in preventing gluten-induced enteropathy and symptoms (NCT05636293, 2023).

20.2.8 Integrins

Gut-selective $\alpha4\beta7$ integrins expressed on T cells, through the binding with the mucosal addressing cellular adhesion molecule 1 (MAdCAM-1) on endothelial cells, are responsible for lymphocyte gut-homing, with subsequent spread of inflammatory-driven tissue injury. The $\alpha4\beta7$ integrin—MAdCAM-1 pathway has been investigate as a possible determinant of mucosal damage in CD (Di Sabatino et al., 2009). More in depth, MAdCAM-1 overexpression in the duodenum, associated with depletion of circulating $\alpha4\beta7$ integrin-positive blood T cells, was found in active CD patients. Nonetheless, agents targeting the $\alpha4\beta7$ integrin, such as the humanized monoclonal antibody vedolizumab, have been found to be effective and safe in the treatment of other Th1-mediated diseases, namely Crohn's disease and ulcerative colitis, and are currently employed in clinical practice (Danese et al., 2015). PTG-100, a selective $\alpha4\beta7$ integrin antagonist, has been recently evaluated in a phase I DBRCT in CD (NCT04524221, 2022). Twelve CD patients were randomized to receive either 600 mg of PTG-100 or placebo for 42 days, together with gluten challenge, to determine the safety and efficacy of this agent in preventing gluten-induced inflammatory injury. The trial has been concluded in 2022 but results are still awaited.

20.2.9 Chemokines

Intestinal lymphocyte trafficking is also regulated by the chemokine CCR9, expressed by T cells, and its endothelial ligand C-chemokine ligand (CCL)25. Several studies have investigated the role of CCR9-CCL25 axis in CD pathogenesis. In particular, $CCR9^+$ dendritic cells from CD patients, when elicited by gliadin-derived peptides, showed a proinflammatory cytokine profile and a higher capacity to promote gut-homing CCR9 expression in autologous T cells (Escudero-Hernández et al., 2020). Also, an impairment in CCR9-expression on epithelial and lamina propria T cells in untreated CD patients was found (Olaussen et al., 2007). Hence, CCR9 antagonists have been considered as a potential therapy for CD. CCX282-B is a selective CCR9 antagonist which was evaluated in a phase II DBRCT (EudraCT 2007−003450-28, 2008; NCT00540657, 2008). The efficacy of 250 mg of oral CCX282-B compared to placebo in mitigating the effects of gluten ingestion in 90 CD patients under GFD was tested. The outcomes measured included the degree of villous atrophy, markers of intestinal inflammation, serology, and gluten-induced gastrointestinal symptoms. However, results showed no benefit in CD patients who were challenged with gluten. In addition, an increase in liver enzymes was found in both the treatment and placebo arms, probably due to gluten challenge, but was more pronounced in patients treated with CCX282-B.

Compounds directed against endogenous targets and investigated in clinical trials in CD are summarized in Table 20.1.

20.3 Gluten as a therapeutic target

Oral endoluminal therapy targeting gluten peptides has been explored as a therapeutic approach in CD. Molecules capable of binding, neutralizing or degrading gluten peptides are currently under investigation. Such compounds need to be active and stable under gastroduodenal conditions, should promptly neutralize T- cell activating gluten peptides and must be safe for human consumption. However, the capability to neutralize the total amount of gluten present in an average wheat-based diet represents the main limitation for these molecules.

20.3.1 Neutralizing strategies

Gluten neutralizing strategies foresee the use of exogenous antigluten antibodies. Among them, one of the most promising is represented by the oral antigliadin antibody AGY, produced by egg yolk of chickens during the immunization period, which has been shown to exert an in vitro and in vivo binding activity against gliadin. A phase I study published in 2017 assessed the safety and efficacy of oral administration of two 500 mg capsules in ten adult CD patients, showing reduction of symptoms, increased quality of life and reduction of serum anti-tTG antibodies (Sample et al., 2017). Thereafter, a DBRCT was started to evaluate AGY safety and efficacy on gastrointestinal symptoms in CD symptomatic patients aged from 10 to 65 year (NCT03707730, 2023). Patients were randomized to either AGY or placebo. The trial was interrupted in March 2020 due to the COVID-19 pandemic, but resumed in October 2020, allowing remote participation. As of 31 March 2022, 114 participants have been enrolled and 66 have completed the study, without significant adverse events (Sample et al., 2022). Efficacy results are not yet available.

20.3.2 Sequestering strategies

Gluten sequestering strategies are based on polymeric molecules able to bind gliadin and prevent its digestion from gastrointestinal enzymes, avoiding the generation of immunogenic peptides. BL-7010 is a copolymer of poly-hydroxyethyl-methacrylate (HEMA) and styrene sulfonate (SS), formerly known as Poly(HEMA-co-SS), which was developed and studied in in vitro. BL-7010 showed to bind α-gliadin and form supramolecular complex particles, inducing α-gliadin structural changes with a decrease of α-helical content and increase of β-turn and β-sheet contents (Liang et al., 2008). The polymer hinders the digestion of wheat gliadin and barley hordein by gastrointestinal enzymes, thus minimizing the formation of immunogenic gliadin peptides. In gluten-sensitized mice, Poly(HEMA-co-SS) reduced paracellular permeability and modulated the systemic immune response to gluten (Pinier et al., 2012). In vitro

TABLE 20.1 Endogenous therapeutic targets investigated in clinical trials in celiac disease.

Target	Compound	RoA	Phase	Type (duration)	Population	Identifier	Status	Main findings	References
Tight junction	*Larazotide Acetate* Synthetic peptide-Zonulin receptor antagonist	Oral	I	DBRCT: larazotide 12 mg QD versus placebo (3-day)	21 adult CD on at least 6-month GFD	NCT00386165	Complete 2006	– ↓ Intestinal barrier dysfunction and proinflammatory cytokine production – ↓gluten-induced symptoms – Good safety profile	Paterson et al. (2007)
			II	DBRCT: larazotide 0.25–1–4–8 mg TID versus placebo (21-day)	86 adult CD on at least 6-month GFD	NCT00362856	Complete 2009	– No effect on intestinal permeability – ↓ gluten-induced symptoms at 0.25 and 4 mg – Good safety profile	Leffler et al. (2012)
			II	DBRCT: larazotide 1–4–8 mg TID versus placebo (6-week)	184 adult CD on at least 6-month GFD	NCT00492960	Complete 2009	– No effect on intestinal permeability – ↓ gluten-induced symptoms and immune reactivity – Good safety profile	Kelly et al. (2013)
			II	DBRCT: larazotide 0.5–1–2 mg TID versus placebo (12-week)	342 symptomatic adult CD on at least 12-month GFD	NCT01396213	Complete 2013	– 0.5 mg effective on signs and symptoms; 1 and 2 mg no effective – Good safety profile	Leffler et al. (2015)
			III	DBRCT: larazotide 0.25–0.5 mg TID versus placebo (12-week)	307 symptomatic adult CD on at least 6-month GFD	NCT03569007	Terminated (sponsor decision) 2022	– Interim analysis: number of patients needed for a significant clinical outcome too large to support trial continuation	accesswire.com (2022)

(Continued)

TABLE 20.1 (Continued)

Target	Compound	RoA	Phase	Type (duration)	Population	Identifier	Status	Main findings	References
	IMU-856 Synthetic molecule- Tight junction modulator	Oral	I	DBRCT: Part A - IMU-856 single ascending dose up to 160 mg OD versus placebo (Over 14-day) Part B - IMU-856 multiple ascending dose up to 160 mg OD versus placebo (Over 14-day) Part C - IMU-856 80–160 mg OD versus placebo (Over 28-day)	71 adult healthy subjects (45 Part A and 26 B) Adult CD on GFD (Part C)	n.a.	• Part A and B complete 2022 • Part C recruiting start 2022	– Part A: single ascending dose safe and well-tolerated; no MTD reached; no ADE; good PK profile – Part B: multiple ascending dose safe and well-tolerated; no MTD reached; no serious ADE; good PK profile	Polasek et al. (2022)
Tissue transglutaminase-2	*ZED1227* Peptidomimetic- Transglutaminase 2 inhibitor	Oral	II	DBRCT: ZED1227 10–50–100 mg OD versus placebo (6-week)	163 adult CD on at least 12-month GFD	EudraCT2017–002241–30	Complete 2020	– Attenuation of gluten-induced duodenal mucosal damage – Possible improvement of gluten-induced symptoms and QoL – Good safety profile	Schuppan et al. (2021)
			II	DBRCT: ZED1227 10–25–50 mg versus placebo (12-week)	Symptomatic adult CD on at least 12-month GFD	EudraCT2020–004612–97	Recruiting start 2021	n.a.	EudraCT 2020–004612-97 (2023)
T cell	*Nexvax2* Mixture of three soluble peptides- Immune tolerance inductor	ID or SC	I	DBRCT: Nexvax2 9–30–60–90 μg ID weekly versus placebo (3-week)	34 adult HLA-DQ2 CD on at least 12-month GFD	NCT00879749	Complete 2010	– Mobilization of gluten-specific T cells similar to acute oral gluten exposure – Well tolerated – GI ADE more common for 60 and 90 μg	Brown et al. (2011)

			I	DBRCTs • 3-dose study: multiple ascending dose of Nexvax2 (60–90–150 μg weekly) ID versus placebo (Over 15-day) • 16-dose study: multiple ascending dose of Nexvax2 (150–300 μg twice weekly) ID versus placebo (Over 53-day)	82 adult HLA-DQ2.5 CD on at least 12-month GFD	ACTRN12613001331729; ACTRN12612000355875	Complete 2017	– MTD is 150 μg for twice weekly ID administration – Modified immune responsiveness to Nexvax2 peptides without deterioration in duodenal histology – Immune activation from 2 h after 1st dose (↑ IL-2, IL-6, IL-8, IL-10, CCL2, CXCL10) – GI symptoms after 1st administration (above 30 μg) resembled oral gluten challenge	Goel et al., 2017 (2016)
			I	DBRCT: Nexvax2 ID gradual escalation versus placebo—group assignment based on HLA-DQ2.5 status (Over 46 days)	36 adult HLA-DQ2.5 CD on at least 12-month GFD	NCT02528799	Complete 2017	– Safe administration at high dose, without immune activation, if preceded by gradual dose escalation – Trends towards improved duodenal histology	Daveson et al. (2017)
			II	DBRCT: Nexvax2 SC twice weekly versus placebo (32-week)	Adult HLA-DQ2.5 CD on at least 12-month GFD	NCT03644069	Discontinued (by sponsor) 2019	– In interim analysis Nexvax2 not provided meaningful protection from gluten exposure – In exploratory posthoc endpoint, intestinal histology improved after 18-week treatment	Tye-Din et al. (2023)

(Continued)

TABLE 20.1 (Continued)

Target	Compound	RoA	Phase	Type (duration)	Population	Identifier	Status	Main findings	References
	KAN-101 Recombinant protein-Immune tolerance inductor	IV	I	DBRCT: Part A – KAN-101 single- ascending dose Part B - KAN-101 multiple ascending dose versus placebo (up to 28-day)	41 adult HLA-DQ2.5 CD on at least 12-month GFD	NCT04248855	Complete 2021	– ↓ gliadin specific T cell response – Dose-dependent ↓ gluten-induced IL-2 – No ↑ gut-homing CD8 T cell response after gluten challenge – Good safety profile	anokion.com (2023)
			I/II	Part A – OL: KAN-101 multiple ascending dose Part B - DBRCT: KAN-101 3 doses cohorts versus placebo (56-day)	Adult HLA-DQ2.5 CD on at least 12-month GFD	NCT05574010	Recruiting start 2022	n.a.	NCT05574010 (2023)
	TAK-101 (formerly TIMP-GLIA) PLGA-gliadin nanoparticle - Immune tolerance inductor	IV	I/II	Phase 1: OL single ascending dose (0.1,0.5,1,2,4,8 mg/kg) followed by repeat dose Phase 2: DBRCT: TAK-101 8 mg/Kg day 1 and 8 versus placebo (Up to 67-day)	Phase 1: 23 adult CD on at least 10-day GFD Phase 2: 34 HLA-DQ2 or DQ8 adult CD on at least 6-month GFD	NCT03486990 NCT03738475	Complete 2019	– Prevention of gluten-induced immune activation – Good safety profile	Kelly et al. (2021)
			II	DBRCT: TAK-101 1–2–4 mg/Kg versus placebo (24-week)	HLA-DQ2 or DQ8 adult CD on at least 6-month GFD	NCT04530123	Not yet recruiting start 2022	n.a.	NCT04530123 (2023)
	TPM502 Peptide-encapsulated nanoparticle - Immune tolerance inductor	IV	II	DBRCT: TPM502 2 infusions (escalating dose) versus placebo (43-day)	Adult HLA-DQ2.5 CD on at least 6-month GFD	NCT05660109	Recruiting start 2022	n.a.	EudraCT 2022–001656-41 (2023)

	Necator americanus (NA) Hookworm-Immune tolerance inductor	ID	I/II	DBRCT: NA inoculation 0- and 12-weeks versus placebo (21-week)	20 adult HLA-DQ2 CD on at least 6-month GFD	NCT00671138	Complete 2009	– No benefit on duodenal mucosa during gluten-challenge – NA infection well-tolerated – No impact on gut microbiota	Cantacessi et al. (2014), Daveson et al. (2011)
			I/II	Single group assignment: NA inoculation (Up to 36-week)	12 adult HLA-DQ2 CD on GFD previously infected with NA (see previous trial)	NCT01661933	Complete 2014	– ↓ mucosal deterioration and no rise in CD serology after gluten microchallenge compared with historical controls – Improvement QoL, no symptoms after gluten-challenge – ↑ CD4 + Foxp3 + T cells, ↓ IFN-γ T cells – NA-induced infection well-tolerated	Croese et al. (2015)
			I	DBRCT with OL extension phase: low vs medium dose NA versus placebo (Up to 94-week)	54 adult HLA-DQ2 CD on GFD	NCT02754609	Complete 2020	– NA not restored tolerance to sustained moderate gluten-challenge (2 g/day) – Improved symptoms after lower intermittent gluten doses	Croese et al. (2020)
	Teriflunomide Enamide-Dihydroorotate dehydrogenase inhibitor	Oral	II	DBRCT: Teriflunomide 14 mg/day versus placebo (14-day)	Adult HLA-DQ2.5 CD on at least 6-month GFD	NCT04806737	Recruiting start 2021	n.a.	NCT04806737 (2023)

(*Continued*)

TABLE 20.1 (Continued)

Target	Compound	RoA	Phase	Type (duration)	Population	Identifier	Status	Main findings	References
Cathepsin S	*RO5459072* Synthetic molecule-Cathepsin S inhibitor	Oral	I	DBRCT: RO5459072 100 mg BID versus placebo (28-day)	19 adult CD on at least 12-month GFD	NCT02679014	Complete 2016	n.a.	NCT02679014 (2016)
HLA-DQ2.5	*DONQ52* Antibody-AntiHLA-DQ2.5/gluten peptide complex	SC	I	DBRCT: Part A - DONQ52 single ascending dose versus placebo (Up to 246-day) Part B - DONQ52 multiple ascending dose versus placebo (Up to 3-month)	Adult HLA-DQ2.5 CD on at least 12-month GFD	NCT05425446	Recruiting start 2022	n.a.	NCT05425446 (2023)
IL-15	*CALY-002* Humanized monoclonal antibody-IL-15 inhibitor	IV	I	DBRCT: CALY-002 1 h infusion multiple ascending dose versus placebo (Up to 3-month)	Adult CD on at least 12-month GFD	NCT04593251	Recruiting start 2020	n.a.	NCT04593251 (2023)
	PRV-015 (formerly AMG714) Human monoclonal antibody-IL-15 inhibitor	SC	II	DBRCT: PRV-015 150,300 mg every 2 weeks versus placebo (Over 10-week)	64 adult CD on at least 12-month GFD	NCT02637141	Complete 2019	– No effect on mucosal injury – 300 mg effective on ↓ IEL density – Effective on gluten-dependant symptoms – Injection site reactions most common ADE; no serious ADE	Lähdeaho et al. (2019)
		SC	II	DBRCT: PRV-015 low, medium, high dose versus placebo (28-week)	Symptomatic adult CD on at least 12-month GFD	NCT04424927	Recruiting start 2020	n.a.	NCT04424927 (2023)
JAK-3	*Ritlecitinib* Synthetic molecule-JAK-3 inhibitor	Oral	II	DBRCT: ritlecitinib 200 mg OD versus placebo (21- day)	Adult CD on at least 6-month GFD	NCT05636293	Recruiting start 2022	n.a.	NCT05636293 (2023)

$\alpha_4\beta_7$ integrin	*PTG-100* Peptide-$\alpha_4\beta_7$ integrin selective antagonist	Oral	I	DBRCT: PTG-100 600 mg BID versus placebo (42-day)	12 adult CD	NCT04524221	Complete 2022	n.a.	NCT04524221 (2022)
C-Chemokine Receptor 9	*CCX282-B* Synthetic molecule-C-Chemokine receptor 9 antagonist	Oral	II	DBRCT: CCX82-B 250 mg BID versus placebo (13-week)	90 adult CD on at least 24-month GFD	NCT00540657; EudraCT2007–003450–28	Complete 2008	– No benefit in patients challenged with gluten – ↑ liver transaminases in treatment group more than placebo	EudraCT 2007–003450-28 (2008), NCT00540657 (2008)

Abbreviations: ADE, adverse event; BID, bis in die; CCL2, monocyte chemoattractant protein-1; CD, celiac disease; CK, cytokine; CXCL10, interferon gamma-induced protein 10; DBRCT, double-blind randomized controlled trial; GFD, gluten free diet; GI, gastrointestinal; HLA, human leukocyte antigen; ID, intradermal; IEL, intraepithelial lymphocyte; IFN, interferon; IL, interleukin; IV, intravenous; JAK, Janus kinase; MTD, maximum tolerated dose; n.a., not available; NA, Necator americanus; OD, once daily; OL, open label; PK, pharmacokinetic; PLGA, Poly (lactic-co-glycolic acid); QD, quarter in die; QoL, quality of life; RoA, route of administration; Ref, reference; SC, subcutaneous; TID, ter in die; ↓, reduction; ↑, increase.

experiments showed high affinity of BL-7010 to gliadin and low interaction to digestive enzymes (McCarville et al., 2014). Moreover, in rats exposed to a dose of 3000 mg/kg/day for 14 days, BL-7010 showed absence of systemic absorption. A phase I/II DBRCT on BL-7010 started in 2013 to evaluate the safety and systemic exposure of single escalating administrations and repeated administration (3 times a day for 14 days) of BL-7010, compared to placebo, in 40 adult CD patients on at least 12-month GFD (NCT001990885, 2017). However, the last update belongs to 2017 and no result are available at the moment.

20.3.3 Degrading strategies

Several enzymes, including prolyl endopeptidases (PEPs), cysteine proteases and subtilisins, have been shown to degrade human digestion-resistant gluten peptides both in vitro and in vivo (Wei et al., 2020). A big amount of these enzymes has been identified in bacteria, fungi and plants. Examples are PEPs derived from bacteria, such as *Flavobacterium meningosepticum*, *Sphingomonas capsulata* and *Myxococcus xanthus*, and from fungi, such as *Aspergillus niger* and *Trichophyton rubrum*. Also, degrading enzymes produced by human oral colonizers, such as *Rothia* bacteria, showed optimized gluten-degrading activities. This chapter will focus on natural or synthetic enzymes under evaluation in clinical trials.

20.3.3.1 AN-PEP

AN-PEP is a glutenase produced by the fungus *A. niger*, which is active against intact gluten for pH values between 4.0 and 5.0. This glutenase, which demonstrated enhanced gluten digestion in the stomach of healthy volunteers (Salden et al., 2015), has been evaluated in a phase I/II DBRCT in 16 CD patients under GFD who underwent gluten challenge (Tack et al., 2013). In the safety phase of the trial, 16 patients consumed gluten (7 g/day) with a concomitant administration of AN-PEP (168 proline protease unit) for 2 weeks. After a 2-week washout period, during which patients were adherent to GFD, 14 patients were randomized to gluten intake with either AN-PEP or placebo for 2-week (efficacy phase). The enzyme showed a satisfactory safety profile but no difference with placebo in preventing gluten-dependent mucosal damage. In 2021, another exploratory phase IV DBRCT started on 40 adult CD patients under GFD for at least two years (NCT04788797, 2022). The study aimed to evaluate the effect of daily administration of AN-PEP at a dose of two capsules/meal (breakfast, lunch and dinner) for 8 weeks compared to placebo. The recruitment phase has been completed, but results are not yet available.

20.3.3.2 Latiglutenase

The combination in a 1:1 ratio of the glutamine-specific endopeptidase (EP-B2) -known as ALV001- and the PEP from *Sphingomonas capsulate* (SC-PEP) -known as ALV002- led to a new peptidase oral product, known as latiglutenase (formerly called ALV003 or IMGX003). Two phase I single-blind randomized controlled trials evaluated the safety, tolerability and activity of latiglutenase administered at escalating doses (100, 300, 900, 1800 mg) via nasogastric tube in healthy controls and CD patients under GFD (Siegel et al., 2012). These studies showed an acceptable tolerance for all doses and no serious adverse events. Also, latiglutenase appeared to be stable in the fed stomach and degrade dietary gluten: 100 and 300 mg doses degraded respectively 75% and 88% of 1 g wheat bread gluten. Latiglutenase showed a good safety profile and a significant protection from gluten-dependent villous atrophy and intraepithelial lymphocytosis in CD patients under GFD. In a further phase II DBRCT,CD patients under GFD were randomized to either latiglutenase or placebo in order to assess both safety and efficacy of the compound (Lähdeaho et al., 2014). However, no significant change was observed in the degree of villous atrophy and IEL infiltration in treated CD patients. The subsequent phase II DBRCT, evaluating the efficacy of different doses of latiglutenase compared to placebo in symptomatic CD patients, did not show histological improvement but a reduction in symptom severity and frequency in a posthoc analysis (Murray et al., 2017; Syage et al., 2017). In 2022, a new phase II DBRCT was completed (Murray et al., 2022). Forty-three adult patients, with a well-controlled CD after at least 12-month GFD, were randomized to either latiglutenase 1200 mg or placebo with a following 2 g gluten challenge for 6 weeks. In the latiglutenase arm a reduction of mucosal and symptom deterioration was observed following gluten ingestion, compared to placebo. However, it should be noted that the mean change in mucosal damage and symptom severity scores reached only a significant trend. Moreover, gluten immunogenic peptide detection in the urine indicated that latiglutenase was able to degrade 95% of gluten in the stomach. Finally, no evidence of safety issues were reported. Further two DBRCTs have been started and are actually recruiting symptomatic CD patients under at least 12-month GFD (NCT04243551, 2023; NCT04839575, 2023). These two studies, in which patients are randomized to either latiglutenase or placebo, have been designed to evaluate as primary outcome the reduction in symptom severity after 6 months. The estimated time for the study completion is 2024.

20.3.3.3 TAK-062

TAK-062 is a novel computationally designed third-generation endopeptidase targeting the proline-glutamine dipeptide motifs. TAK-062 has demonstrated in vitro the capability to degrade more than 99% of gluten (3 and 9 g) within 10 min (Pultz et al., 2021). In a phase I randomized controlled trial healthy participants and CD patients were randomized to either oral TAK-062 (100, 300, 600 and 900 mg) or placebo, followed by gluten challenge (3 or 6 g) (Pultz et al., 2021). In the first part of the study, the safety and tolerability of single ascending doses in healthy controls and CD patients was confirmed. In the second part of the trial, the ability of TAK-062 to degrade gluten in the stomach of healthy participants was studied, resulting in a median gluten degradation ranging from 97% to more than 99% in complex gluten-containing meals at 20–65 min postdose. A phase II DBRCT, aimed at evaluating TAK-062 safety and efficacy in symptomatic adolescent and adult CD patients under GFD, is ongoing (NCT05555446, 2023; Silvester et al., 2022).

Molecules capable of binding, neutralizing or degrading gluten peptides and currently under investigation for CD are summarized in Table 20.2.

20.4 Perspectives

Despite numerous attempts to identify potential therapeutic targets for CD, it is still difficult to find an ideal therapy alternative to GFD. Indeed, GFD is the most effective and safe approach for CD patients, and thus it currently remains the cornerstone of CD management. The compounds evaluated so far in clinical trials, either targeting endogenous proinflammatory cytokines/pathways or neutralizing gluten, still raise several safety concerns and failed to achieve the main outcomes, namely the full protection from gluten-induced mucosal damage and the complete remission of gastrointestinal symptoms under gluten challenge.

Among the compounds directed against endogenous damaging effector mechanisms, those targeting the epithelial barrier did not prove effective in modulating the barrier function in CD patients thus fully counteracting the gluten peptide translocation from the lumen into the lamina propria. As regards the tTG2 inhibitors, the potential immunogenic activity of certain gluten peptides, even in the absence of tTG-induced deamidation, could lead to their inefficacy. The immune tolerizing approach underlying vaccines remains certainly the most attractive strategy from a pathophysiological point of view (Di Sabatino & Corazza, 2017), however, many issues should be clarified. One of the major limitations is that the available vaccines have been developed only with known and previously investigated immunogenic epitopes, and this limits their activity spectrum. Moreover, they are restricted to HLA-DQ2.5-positive CD patients, thus excluding at least one-third of CD population. Finally, molecules targeting down-stream immunological pathways, such as proinflammatory cytokines, might not be sufficient to counteract the damaging mechanisms as the redundant immune response in the gut may bypass the therapeutic blockade through the switch of the mucosal cytokine scenario. The possibility of immunosuppressive systemic adverse events should also be considered when using these approaches.

As regards compounds aimed at neutralizing, sequestering or degrading gluten, they seem to be safer, but their use is hampered by the difficulty in adjusting the dose of the drug according to the amount of gluten ingested. Also, the possible inactivation and/or digestion of these agents in the gastrointestinal lumen could affect their activity. Finally, the complex matrix in which gluten peptides are embedded in foods could interfere with their complete neutralization. Many strategies are currently being developed to increase the efficacy of antigluten agents, such as the combination of different molecules or the optimization of their chemical-physical properties. For example, either protein PEGylation, that consists in covalent or noncovalent attachment of polyethylene-glycol to a molecule, or microencapsulation have been hypothesized as possible strategies to improve the stability and activity of these agents (Darwish et al., 2019).

Even if more evidences from clinical trials are needed, in our opinion restoring the immune tolerance to gluten through vaccines might represent the most promising and safe therapeutic strategy alternative to GFD. Indeed, deceiving the immune system by masking specific tissue antigens or polarizing immune cells, including lymphocytes, dendritic cells, and macrophages, towards a tolerogenic phenotype is an approach currently explored in many immune-mediated disorders. For example, the modulation through vaccines of mucosal dendritic cells, which have been demonstrated to play a crucial pathogenic role in active CD (Di Sabatino et al., 2007), making them tolerogenic for immunogenic gliadin peptides, could be the key in enabling the safe resumption of dietary gluten in CD patients. Indeed, vaccines can polarize specific APC subsets and modulate intracellular antigen trafficking, thus affecting antigen cross-presentation (Cohn & Delamarre, 2014). On the other hand, they could induce anergy or deletion of T cells, thus dampening the consequent

TABLE 20.2 Gluten as a therapeutic target in clinical trials in celiac disease.

Strategy	Compound	RoA	Phase	Type (duration)	Population	Identifier	Status	Main findings	References
Neutralization	*AGY* Egg yolk-derived gliadin antibody	Oral	I	OL: AGY 1 g (6-week)	10 symptomatic adult CD on at least 6-month GFD	NCT01765647	Complete 2015	– ↓ symptoms (tiredness, headache, bloating) – ↑ QoL – ↓ tTG antibodies	Sample et al. (2017)
			II	DBRCT: AGY versus placebo (16-week)	Symptomatic CD aged 10–65 years on at least 12-month GFD	NCT03707730	Recruiting start 2019	– Interim analysis: no significant adverse event	NCT03707730 (2023), Sample et al. (2022)
Sequestration	*BL-7010* Copolymer poly-hydroxyethyl-methacrylate-co-styrene sulfonate		I/II	DBRCT: Part 1- BL-7010 single escalating dose versus placebo (7-week) Part 2—BL-7010 repeated administration versus placebo (4-week)	40 adult CD on at least 6-month GFD	NCT01990885	Complete 2017	n.a.	NCT001990885 (2017)
Degradation	*AN-PEP* *Aspergillus niger* prolyl-endopeptidase enzyme	Oral	I/II	DBRCT: Safety phase-gluten with AN-PEP (2-week) Efficacy phase- gluten with AN-PEP versus placebo (2-week)	16 adult CD on at least 12-month GFD	NCT00810654	Complete 2009	– Well tolerated – No difference with placebo in preventing gluten-induced mucosal damage	Tack et al. (2013)
			IV	DBRCT: AN-PEP 2 capsules/TID versus placebo (8-week)	40 adult CD on at least 2-year GFD	NCT04788797	Complete 2022	n.a.	NCT04788797 (2022)
	Latiglutenase (also known as ALV003 or IMGX003) Synthetic combination: glutamine-specific endopeptidase + prolyl-endopeptidase	Oral	I	Single-blind RCTs: latiglutenase increasing dose (100–300–900–1800 mg) via nasogastric tube versus placebo	Study 1: 24 healthy subjects and 4 adult well-controlled CD Study 2:	NCT00626184 NCT00669825	Complete 2008	– All doses (especially 300 mg) well tolerated – 100 and 300 mg degraded 75 ± 10% and	Siegel et al. (2012)

	from *Sphingomonas capsulate*			Study 1—drug administered in the fasted state Study 2—drug administered with gluten (duration n.a.)	52 healthy subjects and 1 adult CD on at least 8-week GFD			88 ± 5%, respectively, of 1 gram of wheat bread gluten	
			II	DBRCT: latiglutenase versus placebo (6-week)	41 adult CD on GFD	NCT00959114 NCT01255696	Complete2011	↓ gluten-induced mucosal injury	Lähdeaho et al. (2014)
			II	DBRCT: latiglutenase 100–300–450–600–900 mg TID versus placebo (12- or 24-week)	494 symptomatic CD on at least 12-month GFD	NCT01917630	Complete 2015	– No histological improvement – Posthoc analysis: ↓ in symptoms severity and frequency	Murray et al. (2017), Syage et al. (2017)
			II	DBRCT: latiglutenase 1200 mg versus placebo (6 weeks)	43 adult CD on at least 12-month GFD	NCT03585478	Complete 2021	– ↓ gluten-induced mucosal and symptom deterioration – Latiglutenase degraded 95% gluten in the stomach – No safety issues reported	Murray et al. (2022)
			II	DBRCT: latiglutenase versus placebo (duration n.a.)	Symptomatic adult CD on at least 12-month GFD	NCT04243551	Recruiting start 2020	n.a.	NCT04243551 (2023)
			II	DBRCT: latiglutenase versus placebo (duration n.a.)	Symptomatic adult CD on at least 12-month GFD	NCT04839575	Recruiting start 2021	n.a.	NC T04839575 (2023)

(*Continued*)

TABLE 20.2 (Continued)

Strategy	Compound	RoA	Phase	Type (duration)	Population	Identifier	Status	Main findings	References
	TAK-062 Computationally designed endopeptidase	Oral	I	Single-blind RCT: TAK-062 100,300,600,900 mg versus placebo (6-week) Part 1: single ascending dose Part 2: single-dose	Part 1: 9 adult CD on at least 6-month GFD + 24 healthy participants Part 2: 43 healthy partecipants	NCT03701555	Complete 2021	– Well tolerated – Rapid and effective degradation of large amount of gluten	Pultz et al. (2021)
			II	DBRCT: TAK-062 increasing dose versus placebo (24-week)	Symptomatic adolescent/ adult CD on at least 12-month GFD	NCT05353985; EudraCT2020–005438–14	Recruiting start 2022	n.a.	NCT05555446 (2023), Silvester et al. (2022)

Abbreviations: AGY, egg yolk antigliadin antibody; AN-PEP, Aspergillus niger- prolyl endopeptidase; CD, celiac disease; DB, double blind; GFD, gluten free diet; n.a., not available; O.L., open label; PPU, proline protease unit; QoL, quality of life; RCT, randomized controlled trial; Ref, reference; RoA, route of administration; TID, ter in die; tTG: tissue transglutaminase; ↓, reduction; ↑, increase.

T cell-induced mucosal damage (Anderson & Jabri, 2013). The development of new highly engineered vaccines raises hopes for their future use in clinical practice.

From a pathophysiological point of view, the idea of resetting the immune system in CD patients seems to be quite intriguing. This is the case of allogeneic stem cell transplantation, which in a few cases reported in the literature showed long-term gluten tolerance reinduction (Ben-Horin et al., 2013; Zandi et al., 2019). However, allogeneic stem cell transplantation is a high-risk procedure and traditionally reserved for the treatment of life-threatening hematologic malignancies. More feasible could be the mesenchymal stem cell-based therapy. Mesenchymal stem cells are multipotent stromal cells that have the ability to undergo multilineage differentiation and migrate to sites of inflammation and injury where they exert immunoregulatory and repairing effects. Moreover, their transplantation can be achieved without myeloablative conditioning (Moheb-Alian et al., 2016). An interesting in vitro study on bone marrow-derived mesenchymal stem cells showed a strong immunomodulatory effects of these cells on gliadin-specific T-cell lines obtained from mucosal specimens of CD patients. A reduction of $CD4^+$ T cells and an expansion of regulatory $FoxP3^+$ subsets was observed in T cells co-cultured in vitro with mesenchymal stem cells (Ciccocioppo et al., 2014). However, the source of these cells, their dosage and route of administration still represent a point of discussion and research in view of their potential application in the therapy of uncomplicated CD.

In conclusion, although several interesting endogenous targets have been identified in the recent years thanks to the advances in understanding the complex molecular mechanisms underlying CD pathogenesis (Fig. 20.2), none of the

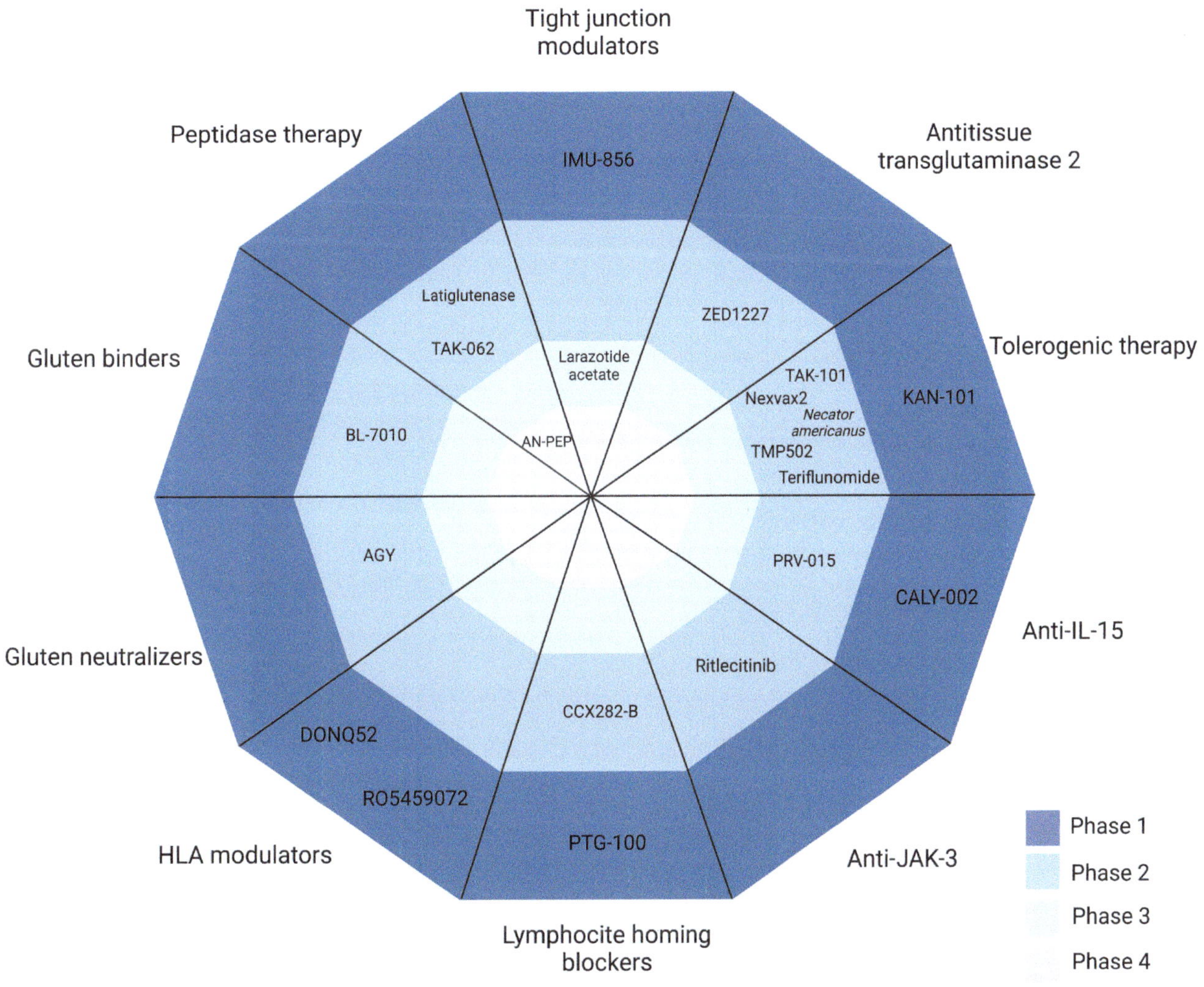

FIGURE 20.2 Compounds investigated for celiac disease, divided according to mechanism of action and clinical trial phase. AGY, egg yolk antigliadin antibody; AN-PEP, Aspergillus niger- prolyl endopeptidase; HLA, human leukocyte antigen; IL, interleukin; JAK, Janus kinase. *Created with "Biorender.com."*

compounds tested so far in clinical trials in uncomplicated CD patients can compare to GFD both in terms of safety and efficacy. Thus the strict and life-long exclusion of gluten from the diet remains the only possible therapeutic option in CD.

References

accesswire.com. (2022). https://www.accesswire.com/705856/9-Meters-Biopharma-Announces-Interim-Analysis-of-Phase-3-Study-of-Larazotide-for-Celiac-Disease-Does-Not-Support-Trial-Continuation.

Anderson, R. P., & Jabri, B. (2013). Vaccine against autoimmune disease: Antigen-specific immunotherapy. *Current Opinion in Immunology*, *25*(3), 410–417. Available from https://doi.org/10.1016/j.coi.2013.02.004.

Andrew, A., Passomore, R., & Urwin, H. (2022). Adults with coeliac disease are interested in therapeutics beyond the gluten free diet; a UK survey. In Proceedings of the ninteenth international celiac disease symposium, Sorrento, 169–170.

anokion.com. (2023). https://anokion.com/press_releases/anokion-announces-positive-data-from-phase-1-aced-clinical-trial-evaluating-kan-101-as-a-treatment-for-celiac-disease-presented-in-a-plenary-session-at-ddw-2022/.

Bach, J. F. (2002). The effect of infections on susceptibility to autoimmune and allergic diseases. *New England Journal of Medicine*, *347*(12), 911–920. Available from https://doi.org/10.1056/NEJMra020100.

Bar-Or, A., Pachner, A., Menguy-Vacheron, F., Kaplan, J., & Wiendl, H. (2014). Teriflunomide and its mechanism of action in multiple sclerosis. *Drugs*, *74*(6), 659–674. Available from https://doi.org/10.1007/s40265-014-0212-x.

Ben-Horin, S., Polak-Charcon, S., Barshack, I., Picard, O., Fudim, E., Yavzori, M., Avivi, C., Mardoukh, C., Shimoni, A., Chowers, Y., & Maor, Y. (2013). Celiac disease resolution after allogeneic bone marrow transplantation is associated with absence of gliadin-specific memory response by donor-derived intestinal T-cells. *Journal of Clinical Immunology*, *33*(8), 1395–1402. Available from https://doi.org/10.1007/s10875-013-9943-9.

Brown, G. J., Daveson, J., Marjason, J. K., Ffrench, R. A., Smith, D., Sullivan, M., Tye-Din, J. A., & Anderson, R. P. (2011). A phase I study to determine safety, tolerability and bioactivity of Nexvax2® in HLA DQ2 + volunteers with celiac disease following a long-term, strict gluten-free diet. *Gastroenterology*, *140*(5), S-437–S-438. Available from https://doi.org/10.1016/s0016-5085(11)61794-0.

Cantacessi, C., Giacomin, P., Croese, J., Zakrzewski, M., Sotillo, J., McCann, L., Nolan, M. J., Mitreva, M., Krause, L., & Loukas, A. (2014). Impact of experimental hookworm infection on the human gut microbiota. *The Journal of Infectious Diseases*, *210*(9), 1431–1434. Available from https://doi.org/10.1093/infdis/jiu256.

Cardo, A., Churruca, I., Lasa, A., Navarro, V., Vázquez-Polo, M., Perez-Junkera, G., & Larretxi, I. (2021). Nutritional imbalances in adult celiac patients following a gluten-free diet. *Nutrients*, *13*(8). Available from https://doi.org/10.3390/nu13082877.

Ciccocioppo, R., Camarca, A., Cangemi, G. C., Radano, G., Vitale, S., Betti, E., Ferrari, D., Visai, L., Strada, E., Badulli, C., Locatelli, F., Klersy, C., Gianfrani, C., & Corazza, G. R. (2014). Tolerogenic effect of mesenchymal stromal cells on gliadin-specific Tlymphocytes in celiac disease. *Cytotherapy*, *16*(8), 1080–1091. Available from https://doi.org/10.1016/j.jcyt.2014.03.002.

Ciccocioppo, R., Finamore, A., Ara, C., Di Sabatino, A., Mengheri, E., & Corazza, G. R. (2006). Altered expression, localization, and phosphorylation of epithelial junctional proteins in celiac disease. *American Journal of Clinical Pathology*, *125*(4), 502–511. Available from https://doi.org/10.1309/DTYRA91G8R0KTM8M.

Cohn, L., & Delamarre, L. (2014). Dendritic cell-targeted vaccines. *Frontiers in Immunology*, *5*(MAY). Available from https://doi.org/10.3389/fimmu.2014.00255.

Croese, J., Giacomin, P., Navarro, S., Clouston, A., McCann, L., Dougall, A., Ferreira, I., Susianto, A., O'Rourke, P., Howlett, M., McCarthy, J., Engwerda, C., Jones, D., & Loukas, A. (2015). Experimental hookworm infection and gluten microchallenge promote tolerance in celiac disease. *Journal of Allergy and Clinical Immunology*, *135*(2), 508–516. Available from https://doi.org/10.1016/j.jaci.2014.07.022, e5.

Croese, J., Miller, G., & Marquart, l (2020). Placebo controlled trial of experimental hookworm infection for improving gluten tolerance in celiac disease. *Clinical and Translational Gastroenterology*, *11*.

Danese, S., Vuitton, L., & Peyrin-Biroulet, L. (2015). Biologic agents for IBD: Practical insights. *Nature Reviews Gastroenterology & Hepatology*, *12*(9), 537–545. Available from https://doi.org/10.1038/nrgastro.2015.135.

Darwish, G., Helmerhorst, E. J., Schuppan, D., Oppenheim, F. G., & Wei, G. (2019). Pharmaceutically modified subtilisins withstand acidic conditions and effectively degrade gluten in vivo. *Scientific Reports*, *9*(1). Available from https://doi.org/10.1038/s41598-019-43837-9.

Daveson, A. J., Ee, H. C., Andrews, J. M., King, T., Goldstein, K. E., Dzuris, J. L., MacDougall, J. A., Williams, L. J., Treohan, A., Cooreman, M. P., & Anderson, R. P. (2017). Epitope-specific immunotherapy targeting CD4-positive T cells in celiac disease: Safety, pharmacokinetics, and effects on intestinal histology and plasma cytokines with escalating dose regimens of Nexvax2 in a randomized, double-blind, placebo-controlled phase 1 study. *EBioMedicine*, *26*, 78–90. Available from https://doi.org/10.1016/j.ebiom.2017.11.018.

Daveson, A. J., Jones, D. M., Gaze, S., McSorley, H., Clouston, A., Pascoe, A., Cooke, S., Speare, R., Macdonald, G. A., Anderson, R., McCarthy, J. S., Loukas, A., Croese, J., & Gluud, L. L. (2011). Effect of hookworm infection on wheat challenge in celiac disease – A randomised double-blinded placebo controlled trial. *PLoS One*, *6*(3). Available from https://doi.org/10.1371/journal.pone.0017366.

Di Sabatino, A., Ciccocioppo, R., Cupelli, F., Cinque, B., Millimaggi, D., Clarkson, M. M., Paulli, M., Cifone, M. G., & Corazza, G. R. (2006). Epithelium derived interleukin 15 regulates intraepithelial lymphocyte Th1 cytokine production, cytotoxicity, and survival in coeliac disease. *Gut*, *55*(4), 469–477. Available from https://doi.org/10.1136/gut.2005.068684.

Di Sabatino, A., & Corazza, G. R. (2009). Coeliac disease. *The Lancet*, *373*(9673), 1480–1493. Available from https://doi.org/10.1016/S0140-6736(09)60254-3.

Di Sabatino, A., & Corazza, G. R. (2017). Epitope-specific immunotherapy: A vaccine for coeliac disease? *The Lancet Gastroenterology and Hepatology*, *2*(7), 464–465. Available from https://doi.org/10.1016/S2468-1253(17)30141-3.

Di Sabatino, A., Lenti, M. V., Corazza, G. R., & Gianfrani, C. (2018). Vaccine immunotherapy for celiac disease. *Frontiers in Medicine*, *5*. Available from https://doi.org/10.3389/fmed.2018.00187.

Di Sabatino, A., Pickard, K. M., Gordon, J. N., Salvati, V., Mazzarella, G., Beattie, R. M., Vossenkaemper, A., Rovedatti, L., Leakey, N. A. B., Croft, N. M., Troncone, R., Corazza, G. R., Stagg, A. J., Monteleone, G., & MacDonald, T. T. (2007). Evidence for the role of interferon-alfa production by dendritic cells in the Th1 response in celiac disease. *Gastroenterology*, *133*(4), 1175–1187. Available from https://doi.org/10.1053/j.gastro.2007.08.018.

Di Sabatino, A., Rovedatti, L., Rosado, M. M., Carsetti, R., Corazza, G. R., & MacDonald, T. T. (2009). Increased expression of mucosal addressin cell adhesion molecule 1 in the duodenum of patients with active celiac disease is associated with depletion of integrin $\alpha 4\beta 7$-positive T cells in blood. *Human Pathology*, *40*(5), 699–704. Available from https://doi.org/10.1016/j.humpath.2008.10.014.

Dickneite, G., Herwald, H., Korte, W., Allanore, Y., Denton, C. P., & Cerinic, M. M. (2015). Coagulation factor XIII: A multifunctional transglutaminase with clinical potential in a range of conditions. *Thrombosis and Haemostasis*, *113*(4), 686–697. Available from https://doi.org/10.1160/TH14-07-0625.

Escudero-Hernández, C., Martín, A., Pedro Andrés, R., Fernández-Salazar, L., Garrote, J. A., Bernardo, D., & Arranz, E. (2020). Circulating dendritic cells from celiac disease patients display a gut-homing profile and are differentially modulated by different gliadin-derived peptides. *Molecular Nutrition & Food Research*, *64*(6). Available from https://doi.org/10.1002/mnfr.201900989.

EudraCT 2007–003450-28. (2008). Available from https://www.clinicaltrialsregister.eu/ctr-search/trial/2007-003450-28/results.

EudraCT 2020–004612-97. (2023). https://www.clinicaltrialsregister.eu/ctr-search/trial/2020-004612-97/IE.

EudraCT 2022-001656-41. (2023). https://www.clinicaltrialsregister.eu/ctr-search/trial/2022-001656-41/NO.

Fasano, A. (2011). Zonulin and its regulation of intestinal barrier function: The biological door to inflammation, autoimmunity, and cancer. *Physiological Reviews*, *91*(1), 151–175. Available from https://doi.org/10.1152/physrev.00003.2008.

Freitag, T. L., Podojil, J. R., Pearson, R. M., Fokta, F. J., Sahl, C., Messing, M., Andersson, L. C., Leskinen, K., Saavalainen, P., Hoover, L. I., Huang, K., Phippard, D., Maleki, S., King, N. J. C., Shea, L. D., Miller, S. D., Meri, S. K., & Getts, D. R. (2020). Gliadin nanoparticles induce immune tolerance to gliadin in mouse models of celiac disease. *Gastroenterology*, *158*(6), 1667–1681. Available from https://doi.org/10.1053/j.gastro.2020.01.045, e12.

Goel, G., King, T., Daveson, A. J., Andrews, J. M., Krishnarajah, J., Krause, R., Brown, G. J. E., Fogel, R., Barish, C. F., Epstein, R., Kinney, T. P., Miner, P. B., Tye-Din, J. A., Girardin, A., Taavela, J., Popp, A., Sidney, J., Mäki, M., Goldstein, K. E., ... Anderson, R. P. (2017). Epitope-specific immunotherapy targeting CD4-positive T cells in coeliac disease: Two randomised, double-blind, placebo-controlled phase 1 studies. *The Lancet Gastroenterology and Hepatology*, *2*(7), 479–493. Available from https://doi.org/10.1016/S2468-1253(17)30110-3.

Goel, G., Mayassi, T., Qiao, S.-W., Ciszewski, C., King, T., Daveson, A. J., Andrews, J. M., Krishnarajah, J., Krause, R., Brown, G. J., Fogel, R., Barish, C. F., Epstein, R., Kinney, T., Miner, P. B., Tye-Din, J. A., Girardin, A., Goldstein, K., Dzuris, J. L., ... Anderson, R. P. (2016). Sa1396 a single intradermal (ID) injection of Nexvax2®, a peptide composition with dominant epitopes for gluten-reactive CD4 + T cells, activates T cells and triggers acute gastrointestinal symptoms in HLA-DQ2.5 + people with celiac disease (CeD). *Gastroenterology*, *150*(4). Available from https://doi.org/10.1016/s0016-5085(16)31065-4.

Gopalakrishnan, S., Durai, M., Kitchens, K., Tamiz, A. P., Somerville, R., Ginski, M., Paterson, B. M., Murray, J. A., Verdu, E. F., Alkan, S. S., & Pandey, N. B. (2012). Larazotide acetate regulates epithelial tight junctions in vitro and in vivo. *Peptides*, *35*(1), 86–94. Available from https://doi.org/10.1016/j.peptides.2012.02.015.

Gopalakrishnan, S., Tripathi, A., Tamiz, A. P., Alkan, S. S., & Pandey, N. B. (2012). Larazotide acetate promotes tight junction assembly in epithelial cells. *Peptides*, *35*(1), 95–101. Available from https://doi.org/10.1016/j.peptides.2012.02.016.

Kapoerchan, V. V., Wiesner, M., Hillaert, U., Drijfhout, J. W., Overhand, M., Alard, P., van der Marel, G. A., Overkleeft, H. S., & Koning, F. (2010). Design, synthesis and evaluation of high-affinity binders for the celiac disease associated HLA-DQ2 molecule. *Molecular Immunology*, *47*(5), 1091–1097. Available from https://doi.org/10.1016/j.molimm.2009.10.036.

Kapoerchan, V. V., Wiesner, M., Overhand, M., van der Marel, G. A., Koning, F., & Overkleeft, H. S. (2008). Design of azidoproline containing gluten peptides to suppress CD4 + T-cell responses associated with celiac disease. *Bioorganic and Medicinal Chemistry*, *16*(4), 2053–2062. Available from https://doi.org/10.1016/j.bmc.2007.10.091.

Kelly, C. P., Green, P. H. R., Murray, J. A., Dimarino, A., Colatrella, A., Leffler, D. A., Alexander, T., Arsenescu, R., Leon, F., Jiang, J. G., Arterburn, L. A., Paterson, B. M., & Fedorak, R. N. (2013). Larazotide acetate in patients with coeliac disease undergoing a gluten challenge: A randomised placebo-controlled study. *Alimentary Pharmacology and Therapeutics*, *37*(2), 252–262. Available from https://doi.org/10.1111/apt.12147.

Kelly, C. P., Murray, J. A., Leffler, D. A., Getts, D. R., Bledsoe, A. C., Smithson, G., First, M. R., Morris, A., Boyne, M., Elhofy, A., Wu, T. T., Podojil, J. R., Miller, S. D., Fogel, R., Freitag, T. L., Gerber, M., Haynes, P. K., Koren, M., Matson, M., ... Turner, M. (2021). TAK-101 nanoparticles induce gluten-specific tolerance in celiac disease: A randomized, double-blind, placebo-controlled study. *Gastroenterology*, *161*(1), 66–80. Available from https://doi.org/10.1053/j.gastro.2021.03.014, e8.

Lähdeaho, M. L., Kaukinen, K., Laurila, K., Vuotikka, P., Koivurova, O. P., Kärjä-Lahdensuu, T., Marcantonio, A., Adelman, D. C., & Mäki, M. (2014). Glutenase ALV003 attenuates gluten-induced mucosal injury in patients with celiac disease. *Gastroenterology*, *146*(7), 1649–1658. Available from https://doi.org/10.1053/j.gastro.2014.02.031.

Lähdeaho, M. L., Scheinin, M., Vuotikka, P., Taavela, J., Popp, A., Laukkarinen, J., Koffert, J., Koivurova, O. P., Pesu, M., Kivelä, L., Lovró, Z., Keisala, J., Isola, J., Parnes, J. R., Leon, F., & Mäki, M. (2019). Safety and efficacy of AMG 714 in adults with coeliac disease exposed to gluten challenge: A phase 2a, randomised, double-blind, placebo-controlled study. *The Lancet Gastroenterology and Hepatology*, *4*(12), 948–959. Available from https://doi.org/10.1016/S2468-1253(19)30264-X.

Larussa, T., Abenavoli, L., Procopio, A. C., Iannelli, C., Polimeni, N., Spagnuolo, R., Doldo, P., & Luzza, F. (2021). The role of gluten-free diet in nonalcoholic fatty liver disease development. *European Review for Medical and Pharmacological Sciences*, *25*(21), 6613–6618. Available from https://doi.org/10.26355/eurrev_202111_27104.

Lebwohl, B., Sanders, D. S., & Green, P. H. R. (2018). Coeliac disease. *The Lancet*, *391*(10115), 70–81. Available from https://doi.org/10.1016/S0140-6736(17)31796-8.

Leffler, D. A., Kelly, C. P., Abdallah, H. Z., Colatrella, A. M., Harris, L. A., Leon, F., Arterburn, L. A., Paterson, B. M., Lan, Z. H., & Murray, J. A. (2012). A randomized, double-blind study of larazotide acetate to prevent the activation of celiac disease during gluten challenge. *American Journal of Gastroenterology*, *107*(10), 1554–1562. Available from https://doi.org/10.1038/ajg.2012.211.

Leffler, D. A., Kelly, C. P., Green, P. H. R., Fedorak, R. N., Dimarino, A., Perrow, W., Rasmussen, H., Wang, C., Bercik, P., Bachir, N. M., & Murray, J. A. (2015). Larazotide acetate for persistent symptoms of celiac disease despite a gluten-free diet: A randomized controlled trial. *Gastroenterology*, *148*(7), 1311–1319. Available from https://doi.org/10.1053/j.gastro.2015.02.008, e6.

Liang, L., Pinier, M., Leroux, J., & Subirade. (2008). Interaction of a-gliadin with Poly(HEMA-co-SS): Structural characterization and biological implication. *Biopolymers*, *2*, 169–177.

Marciniak, M., Szymczak-Tomczak, A., Mahadea, D., Eder, P., Dobrowolska, A., & Krela-Kaźmierczak, I. (2021). Multidimensional disadvantages of a gluten-free diet in celiac disease: A narrative review. *Nutrients*, *13*(2). Available from https://doi.org/10.3390/nu13020643.

Martínez-Martinez, M. I., Alegre-Martínez, A., García-Ibánez, J., & Cauli, O. (2019). Quality of life in people with coeliac disease: Psychological and socio- economic aspects. *Endocrine, Metabolic & Immune Disorders - Drug Targets*, *19*(2), 116–120. Available from https://doi.org/10.2174/1871530318666180723100003.

McCarville, J. L., Nisemblat, Y., Galipeau, H. J., Jury, J., Tabakman, R., Cohen, A., Naftali, E., Neiman, B., Halbfinger, E., Murray, J. A., Anbazhagan, A. N., Dudeja, P. K., Varvak, A., Leroux, J. C., & Verdu, E. F. (2014). BL-7010 demonstrates specific binding to gliadin and reduces gluten-associated pathology in a chronic mouse model of gliadin sensitivity. *PLoS One*, *9*(11). Available from https://doi.org/10.1371/journal.pone.0109972.

Moheb-Alian, A., Forouzesh, F., Rostami-Nejad, M., & Rostami, K. (2016). Mesenchymal stem cells as potential therapeutic approaches in celiac disease. *Gastroenterology and Hepatology from Bed to Bench*, *9*, S1–S7. Available from http://journals.sbmu.ac.ir/ghfbb/index.php/ghfbb/article/download/1054/617.

Murray, J. A., Kelly, C., Green, P., Marcantonio, A., Wu, T., Mäki, M., & Adelman, D. (2017). CeliAction Study Group of Investigators. No difference between latiglutenase and placebo in reducing villous atrophy or improving symptoms in patients with symptomatic celiac disease. *Gastroenterology*, *152*, 787–798.

Murray, J. A., Syage, J. A., Wu, T. T., Dickason, M. A., Ramos, A. G., Van Dyke, C., Horwath, I., Lavin, P. T., Mäki, M., Hujoel, I., Papadakis, K. A., Bledsoe, A. C., Khosla, C., Sealey-Voyksner, J. A., Hinson, C., Loskutov, V., Norum, A., Linberg, S., Goldkind, L., . . . Nezzer, J. (2022). Latiglutenase protects the mucosa and attenuates symptom severity in patients with celiac disease exposed to a gluten challenge. *Gastroenterology*, *163*(6), 1510–1521. Available from https://doi.org/10.1053/j.gastro.2022.07.071, e6.

NCT001990885. (2017). https://clinicaltrials.gov/ct2/show/NCT001990885.

NCT00540657. (2008). https://clinicaltrials.gov/ct2/show/NCT00540657.

NCT02679014. (2016). https://clinicaltrials.gov/ct2/show/results/NCT02679014.

NCT03707730. (2023). https://clinicaltrials.gov/ct2/show/NCT03707730.

NCT04243551. (2023). https://clinicaltrials.gov/ct2/show/NCT04243551.

NCT04424927. (2023). https://clinicaltrials.gov/ct2/show/NCT04424927.

NCT04524221. (2022). https://clinicaltrials.gov/ct2/show/study/NCT04524221.

NCT04530123. (2023). https://clinicaltrials.gov/ct2/show/NCT04530123.

NCT04593251. (2023). https://clinicaltrials.gov/ct2/show/NCT04593251.

NCT04604795. (2021). https://clinicaltrials.gov/ct2/show/results/NCT04604795.

NCT04788797. (2022). https://clinicaltrials.gov/ct2/show/NCT04788797.

NCT04806737. (2023). https://clinicaltrials.gov/ct2/show/NCT04806737.

NCT04839575. (2023). https://clinicaltrials.gov/ct2/show/NCT04839575.

NCT05425446. (2023). https://clinicaltrials.gov/ct2/show/NCT05425446.

NCT05555446. (2023). https://clinicaltrials.gov/ct2/show/NCT05555446.

NCT05574010. (2023). https://clinicaltrials.gov/ct2/show/NCT05574010.

NCT05636293. (2023). https://clinicaltrials.gov/ct2/show/NCT05636293.

Norsa, L., Tomba, C., Agostoni, C., Branchi, F., Bardella, M. T., Roncoroni, L., Conte, D., & Elli, L. (2015). Gluten-free diet or alternative therapy: A survey on what parents of celiac children want. *International Journal of Food Sciences and Nutrition*, *66*(5), 590–594. Available from https://doi.org/10.3109/09637486.2015.1064872.

Olaussen, R. W., Karlsson, M. R., Lundin, K. E. A., Jahnsen, J., Brandtzaeg, P., & Farstad, I. N. (2007). Reduced chemokine receptor 9 on intraepithelial lymphocytes in celiac disease suggests persistent epithelial activation. *Gastroenterology*, *132*(7), 2371–2382. Available from https://doi.org/10.1053/j.gastro.2007.04.023.

Paterson, B. M., Lammers, K. M., Arrieta, M. C., Fasano, A., & Meddings, J. B. (2007). The safety, tolerance, pharmacokinetic and pharmacodynamic effects of single doses of AT-1001 in coeliac disease subjects: A proof of concept study. *Alimentary Pharmacology & Therapeutics*, *26*(5), 757–766. Available from https://doi.org/10.1111/j.1365-2036.2007.03413.x.

Pinier, M., Fuhrmann, G., Galipeau, H., Rivard, N., Murray, J., David, C., Drasarova, H., Tuckova, L., Leroux, J., & Verdu, E. F. (2012). The copolymer P (HEMA-co-SS) binds gluten and reduces immune response in glute-sensitized mice and human tissues. *Gastroenterology*, *142*, 316–326.

Pinto-Sanchez, M. I., Silvester, J. A., Lebwohl, B., Leffler, D. A., Anderson, R. P., Therrien, A., Kelly, C. P., & Verdu, E. F. (2021). Society for the study of celiac disease position statement on gaps and opportunities in coeliac disease. *Nature Reviews Gastroenterology and Hepatology*, *18*(12), 875–884. Available from https://doi.org/10.1038/s41575-021-00511-8.

Polasek, T., Burianel, F., Mihajlovic, K., Peelen, E., Fonseca, J., Schreieck, A., Kehler, I., Vitt, D., Kohlhof, H., & Mähler, A. (2022). First in human trial of IMU-856, an orally available tight-junction regulator for the treatment of celiac disease. Abstract Book 19th International Celiac Disease Symposium, Sorrento, 161.

Potter, M. D. E., Brienesse, S. C., Walker, M. M., Boyle, A., & Talley, N. J. (2018). Effect of the gluten-free diet on cardiovascular risk factors in patients with coeliac disease: A systematic review. *Journal of Gastroenterology and Hepatology (Australia)*, *33*(4), 781–791. Available from https://doi.org/10.1111/jgh.14039.

Pultz, I. S., Hill, M., Vitanza, J. M., Wolf, C., Saaby, L., Liu, T., Winkle, P., & Leffler, D. A. (2021). Gluten degradation, pharmacokinetics, safety, and tolerability of TAK-062, an engineered enzyme to treat celiac disease. *Gastroenterology*, *161*(1), 81–93. Available from https://doi.org/10.1053/j.gastro.2021.03.019, e3.

Rauhavirta, T., Oittinen, M., Kivistö, R., Männistö, P. T., Garcia-Horsman, J. A., Wang, Z., Griffin, M., Mäki, M., Kaukinen, K., & Lindfors, K. (2013). Are transglutaminase 2 inhibitors able to reduce gliadin-induced toxicity related to celiac disease? A proof-of-concept study. *Journal of Clinical Immunology*, *33*(1), 134–142. Available from https://doi.org/10.1007/s10875-012-9745-5.

Rubio-Tapia, A., Rahim, M. W., See, J. A., Lahr, B. D., Wu, T. T., & Murray, J. A. (2010). Mucosal recovery and mortality in adults with celiac disease after treatment with a gluten-free diet. *American Journal of Gastroenterology*, *105*(6), 1412–1420. Available from https://doi.org/10.1038/ajg.2010.10.

Salden, B. N., Monserrat, V., Troost, F. J., Bruins, M. J., Edens, L., Bartholomé, R., Haenen, G. R., Winkens, B., Koning, F., & Masclee, A. A. (2015). Randomised clinical study: Aspergillus niger-derived enzyme digests gluten in the stomach of healthy volunteers. *Alimentary Pharmacology and Therapeutics*, *42*(3), 273–285. Available from https://doi.org/10.1111/apt.13266.

Sample, D.A., Dieleman, L.A., & Turner, J. (2022). Update on the AGY clinical trial for patients with symptomatic celiac disease. In Proceedings of the nineteenth international celiac disease symposium, Sorrento, 160.

Sample, D. A., Sunwoo, H. H., Huynh, H. Q., Rylance, H. L., Robert, C. L., Xu, B. W., Kang, S. H., Gujral, N., & Dieleman, L. A. (2017). AGY, a novel egg yolk-derived anti-gliadin antibody, is safe for patients with celiac disease. *Digestive Diseases and Sciences*, *62*(5), 1277–1285. Available from https://doi.org/10.1007/s10620-016-4426-5.

Schuppan, D., Junker, Y., & Barisani, D. (2009). Celiac disease: From pathogenesis to novel therapies. *Gastroenterology*, *137*(6), 1912–1933. Available from https://doi.org/10.1053/j.gastro.2009.09.008.

Schuppan, D., Mäki, M., Lundin, K. E. A., Isola, J., Friesing-Sosnik, T., Taavela, J., Popp, A., Koskenpato, J., Langhorst, J., Hovde, Ø., Lähdeaho, M. L., Fusco, S., Schumann, M., Török, H. P., Kupcinskas, J., Zopf, Y., Lohse, A. W., Scheinin, M., Kull, K., & Greinwald, R. (2021). A randomized trial of a transglutaminase 2 inhibitor for celiac disease. *New England Journal of Medicine*, *385*(1), 35–45. Available from https://doi.org/10.1056/NEJMoa2032441.

Shah, S., Akbari, M., Vanga, R., Kelly, C. P., Hansen, J., Theethira, T., Tariq, S., Dennis, M., & Leffler, D. A. (2014). Patient perception of treatment burden is high in celiac disease compared with other common conditions. *American Journal of Gastroenterology*, *109*(9), 1304–1311. Available from https://doi.org/10.1038/ajg.2014.29.

Siegel, M., Garber, M. E., Spencer, A. G., Botwick, W., Kumar, P., Williams, R. N., Kozuka, K., Shreeniwas, R., Pratha, V., & Adelman, D. C. (2012). Safety, tolerability, and activity of ALV003: Results from two phase 1 single, escalating-dose clinical trials. *Digestive Diseases and Sciences*, *57*(2), 440–450. Available from https://doi.org/10.1007/s10620-011-1906-5.

Silvester, J., Smithson, G., Siegelman, J., Maxwell, J.R., Meckely, L.M., Elli, L., Lei, Q., Kabilan, S., & Leffler, D.A. (2022). A phase 2, randomized, double-blind, placebo-controlled, dose-ranging study to evaluate the efficacy and safety of TAK-062: study design. In Proceedings of the nineteenth international celiac disease symposium, Sorrento, 165–166.

Smyth, P., Sasiwachirangkul, J., Williams, R., & Scott, C. J. (2022). Cathepsin S (CTSS) activity in health and disease - A treasure trove of untapped clinical potential. *Molecular Aspects of Medicine*, *88*. Available from https://doi.org/10.1016/j.mam.2022.101106.

Sollid, L. M., & Jabri, B. (2013). Triggers and drivers of autoimmunity: Lessons from coeliac disease. *Nature Reviews. Immunology*, *13*(4), 294–302. Available from https://doi.org/10.1038/nri3407.

Sollid, L. M., & Koning, F. (2021). Lack of relationship of AT1001 to zonulin and prehaptoglobin-2: Clinical implications. *Gut*, *70*(11), 2211–2212. Available from https://doi.org/10.1136/gutjnl-2020-323829.

Sturgeon, C., Lan, J., & Fasano, A. (2017). Zonulin transgenic mice show altered gut permeability and increased morbidity/mortality in the DSS colitis model. *Annals of the New York Academy of Sciences*, *1397*(1), 130–142. Available from https://doi.org/10.1111/nyas.13343.

Syage, J. A., Murray, J. A., Green, P. H. R., & Khosla, C. (2017). Latiglutenase improves symptoms in seropositive celiac disease patients while on a gluten-free diet. *Digestive Diseases and Sciences*, *62*(9), 2428–2432. Available from https://doi.org/10.1007/s10620-017-4687-7.

Szondy, Z., Sarang, Z., Molnár, P., Németh, T., Piacentini, M., Mastroberardino, P. G., Falasca, L., Aeschlimann, D., Kovács, J., Kiss, I., Szegezdi, E., Lakos, G., Rajnavölgyi, E., Birckbichler, P. J., Melino, G., & Fésüs, L. (2003). Transglutaminase 2-/- mice reveal a phagocytosis-associated crosstalk between macrophages and apoptotic cells. *Proceedings of the National Academy of Sciences of the United States of America*, *100*(13), 7812–7817. Available from https://doi.org/10.1073/pnas.0832466100.

Tack, G. J., van de Water, J. M. W., Bruins, M. J., Kooy-Winkelaar, E. M. C., van Bergen, J., Bonnet, P., Vreugdenhil, A. C. E., Korponay-Szabo, I., Edens, L., von Blomberg, B. M. E., Schreurs, M. W. J., Mulder, C. J., & Koning, F. (2013). Consumption of gluten with gluten-degrading enzyme by celiac patients: A pilot-study. *World Journal of Gastroenterology*, *19*(35), 5837–5847. Available from https://doi.org/10.3748/wjg.v19.i35.5837.

Telci, D., & Griffin, M. (2006). Tissue transglutaminase (TG2) - A wound response enzyme. *Frontiers in Bioscience*, *11*(1), 867–882. Available from https://doi.org/10.2741/1843.

Tye-Din, J. A., Daveson, A. J. M., Goel, G., Goldstein, K. E., Hand, H. L., Neff, K. M., Popp, A., Taavela, J., Maki, M., Isola, J., Williams, L. J., Truitt, K. E., Anderson, R. P., Adams, A., Andrews, J. M., Behrend, C. E., Brown, G. J. E., Chen Yi Mei, S. L., Coates, A. G., . . . Wilson, S. A. (2023). Efficacy and safety of gluten peptide-based antigen-specific immunotherapy (Nexvax2) in adults with coeliac disease after bolus exposure to gluten (RESET CeD): An interim analysis of a terminated randomised, double-blind, placebo-controlled phase 2 study. *The Lancet Gastroenterology and Hepatology*, *8*(5), 446–457. Available from https://doi.org/10.1016/S2468-1253(22)00428-9.

Tye-Din, J. A., Stewart, J. A., Dromey, J. A., Beissbarth, T., Van Heel, D. A., Tatham, A., Henderson, K., Mannering, S. I., Gianfrani, C., Jewell, D. P., Hill, A. V. S., McCluskey, J., Rossjohn, J., & Anderson, R. P. (2010). Comprehensive, quantitative mapping of T cell epitopes in gluten in celiac disease. *Science Translational Medicine*, *2*(41). Available from https://doi.org/10.1126/scitranslmed.3001012.

Ventura, M. A. E., Hils, M., Tewes, B., Pasternack, R., Greinwald, R., Verdu, E. F., & Schuppan, D. (2021). 245 the oral tissue transglutaminase inhibitor Zed1227 prevents gluten-induced enteropathy in a transgenic mouse model of celiac disease. *Gastroenterology*, *160*(6), S-53–S-54. Available from https://doi.org/10.1016/s0016-5085(21)00873-8.

Ventura, M. A. E., Sajko, K., Hils, M., Pasternack, R., Greinwald, R., Tewes, B., & Schuppan, D. (2018). Su1161 - the oral transglutaminase 2 (TG2) inhibitor Zed1227 blocks TG2 activity in a mouse model of intestinal inflammation. *Gastroenterology*, *154*(6). Available from https://doi.org/10.1016/s0016-5085(18)31861-4.

Wei, G., Helmerhorst, E. J., Darwish, G., Blumenkranz, G., & Schuppan, D. (2020). Gluten degrading enzymes for treatment of celiac disease. *Nutrients*, *12*(7). Available from https://doi.org/10.3390/nu12072095.

Yokoyama, S., Perera, P. Y., Waldmann, T. A., Hiroi, T., & Perera, L. P. (2013). Tofacitinib, a Janus Kinase inhibitor demonstrates efficacy in an IL-15 transgenic mouse model that recapitulates pathologic manifestations of celiac disease. *Journal of Clinical Immunology*, *33*(3), 586–594. Available from https://doi.org/10.1007/s10875-012-9849-y.

Zandi, S., Kanfar, S., Cserti-Gazdewich, C., Lipton, J. H., & Pendergrast, J. (2019). Resolution of celiac disease, IgA deficiency and platelet refractoriness after allogeneic bone marrow transplantation for acute leukemia. *Haematologica*, *104*(3), e121–e123. Available from https://doi.org/10.3324/haematol.2018.203943.

Chapter 21

Future directions

Alessio Fasano[1,2]

[1]*T.H. Chan School of Public Health, Department of Nutrition, Boston, MA, United States,* [2]*Massachusetts General Hospital, Harvard Medical School, Department of Pediatrics, Boston, MA, United States*

21.1 Introduction

Celiac disease (CD) is now considered a chronic inflammatory disease with autoimmune connotations (see Chapter 5 for more details). CD is unique among autoimmune diseases given the high penetrance of the HLA DQ2 and/or DQ8 genes among patients, the fact that the environmental trigger (gluten) is known, and that the disease-specific autoantibodies have been identified as robust biomarkers of the break of tolerance to gluten with the subsequent onset of the autoimmune enteropathy that characterizes this condition. Therefore exposure to the environmental trigger can be carefully traced based on the time of introduction of gluten in the diet, and frequent prospective screening of the autoantibody against tissue transglutaminase can determine precisely when the loss of tolerance to gluten occurs. These unique features related to CD, together with new knowledge on its epidemiology and pathogenesis, amply covered in chapters 1–5 challenge two major paradigms concerning chronic inflammatory diseases in general and, more specifically, autoimmune diseases: 1. genetic predisposition and exposure to environmental triggers are necessary and sufficient to develop these diseases; 2. once the tolerance to antigens triggering the autoimmune disease is broken, the inflammatory process is not reversible.

Based on the first paradigm, in the past we assumed that the exposure to gluten coincident with the introduction of baby food in the infant's diet was the initial step leading to the onset of CD in genetically predisposed subjects. This assumption also implied that CD is a pediatric pathology and that adults with CD would have developed the disease in childhood and remained asymptomatic until the moment of diagnosis. Prospective epidemiological studies following subjects at risk over time (family members of celiac patients) showed that this is not the case. Rather, there are subjects capable to tolerate gluten for years/decades and then, for reason now only partially understood, they break tolerance and develop the disease in adulthood or even in their elderly years (Catassi et al., 2010). Another evidence that challenged this paradigm is the epidemiological observation that the prevalence of CD, like any other chronic inflammatory disease, increased in the past two-three decades (Gatti et al., n.d.; Lee et al., 2023), a period too short to blame human genetic mutations or changes in the organoleptic characteristics of gluten-containing grains as driving forces of this increase in prevalence. Therefore, there must be other factors at play to explain why the onset of CD may occur at any age and its prevalence raise over time, particularly in populations that embrace a Western lifestyle.

Concerning the second paradigm, being the only autoimmune disease in which the environmental trigger is known, CD has proven that if the interplay between genes and environmental triggers is avoided, the inflammatory process can be turned off. Indeed, in most celiac patients, the correct implementation of a strict gluten free diet (GFD) that eliminates the environmental factor from the equation, leads to resolution of symptoms, disappearance of the autoantibodies, and healing of the celiac enteropathy.

For all the aforementioned reasons, CD represents an ideal model to tackle the multifactorial nature of chronic inflammatory diseases with the goal to identify new strategies for patients' stratification (precision medicine) and possibly develop primary prevention strategies based on the identification and validation of potential biomarkers of early steps involved in switching from genetic predisposition to clinical outcome as the consequence of breaking tolerance to gluten.

Pediatric and Adult Celiac Disease. DOI: https://doi.org/10.1016/B978-0-443-13359-6.00008-X

21.2 What triggers chronic inflammation?

As mentioned above, until about thirty years ago, when the Human Genome Project was still in the planning stages, the general hypothesis was that to develop a chronic inflammatory disease, genetic predisposition and exposure to an environmental trigger were both considered necessary and sufficient. While the genes and the environmental triggers may vary from one group of disease to another, what seems to remain constant is the mismanagement by the immune system of the exposure to environmental triggers. Parallel to the early hypothesis of genes plus environment equals chronic inflammatory disease was the epidemiological observation that major epidemics of these conditions coincided with the declining rate of infectious disease in the Western hemisphere. This generated the hygiene hypothesis (Okada et al., 2010), meaning that, because in the Western hemisphere people had become too clean for their own good, they would die slowly of chronic inflammatory diseases instead of succumbing rapidly to infectious disease as still happens in developing countries.

If these two elements, genetics and environment, were the only ones at play (and that was the premise), the interpretation of these "epidemics" of chronic inflammatory diseases, including the rise of CD over time (see Chapter 1 for more details) can only lead to two possible conclusions. If we embrace a pessimistic reading of these epidemiological trends, they should be interpreted as the consequence of the drastic changes in our environment caused by human intervention that occurred too rapidly for us to genetically adapt.

Conversely, if we embrace a more optimistic view, these epidemics of chronic inflammatory diseases teach us a very different lesson. The old concept that genetic mutations carrying the risk for a chronic inflammatory disease will translate by default to clinical outcome over time is challenged by these epidemiological observations. The old concept that if you have the genes for the disease, you have the destiny to develop that disease, no matter how you behave or conduct your lifestyle, would indeed not explain the epidemiological acceleration of these conditions.

What we learned when the Human Genome Project was completed is that we are genetically much more rudimental than we had previously thought (Collins et al., 1995). The premise of one gene, one protein, and one disease cannot explain the complexity of the balance between health and disease. Twenty-seven thousand genes are insufficient to explain all the permutations of human pathophysiology, including if, and when, and why we develop diseases.

Rather, it is the interplay between us as individuals and the environment in which we live that causes epigenetic modulations in a way that we can stay healthy despite unfavorable genetic predisposition. Or else, our poor lifestyle may lead to the shift from genetic predisposition to clinical outcome, including autoimmunity and other chronic inflammatory diseases.

Over time, there have been accumulating growing evidence that, beside genetic predisposition and exposure to environmental triggers, there are at least three other elements at play leading to chronic inflammation (Fig. 21.1).

Indeed, the loss of mucosal barrier (especially gut barrier) function (Furman et al., 2019), a hyperbelligerent immune system, and an unbalanced microbiome all need to be at play to develop chronic inflammation (Valitutti & Fasano, 2019). In the following sections, we will review these additional three domains with the understanding that they are highly interconnected and that they could be exploited for novel diagnostic and/or therapeutic targets for the management of CD.

21.3 Intestinal barrier function

At approximately eight to nine meters in length, the human intestine (both small and large) provides the largest interface between our body and the outside world and the port of entry for the vast majority of environmental triggers of chronic inflammatory diseases. The intestinal epithelium maintains a barrier that segregate the host from the external environment while allowing the absorption of nutrients and defends the host against toxins, antigens, and other invaders. Under physiologic circumstances, a tightly controlled antigen trafficking through the paracellular pathway controlled by several signaling molecules, including zonulin (Fasano, 2011; Sturgeon & Fasano, 2016) (see also Chapter 20 for more details), together with coordination with local mucosal immune cells and micromilieu shaped by the presence of specific antiinflammatory cytokines, provide a tolerogenic state of mucosal anergy (Fig. 21.2).

Changes of this mucosal homeostasis by loss of barrier function (the third element involved in the pathogenesis of chronic inflammatory diseases), with subsequent uncontrolled antigen trafficking, increased mucosal inflammation mediated by the production of proinflammatory cytokines that per se increases permeability, will establish a vicious loop that leads to break of mucosal tolerance and the start of the march from genetic predisposition to clinical outcome whose nature depends on the genetic makeup of the host and the epigenetic environmental pressure (Valitutti & Fasano, 2019) (Fig. 21.2).

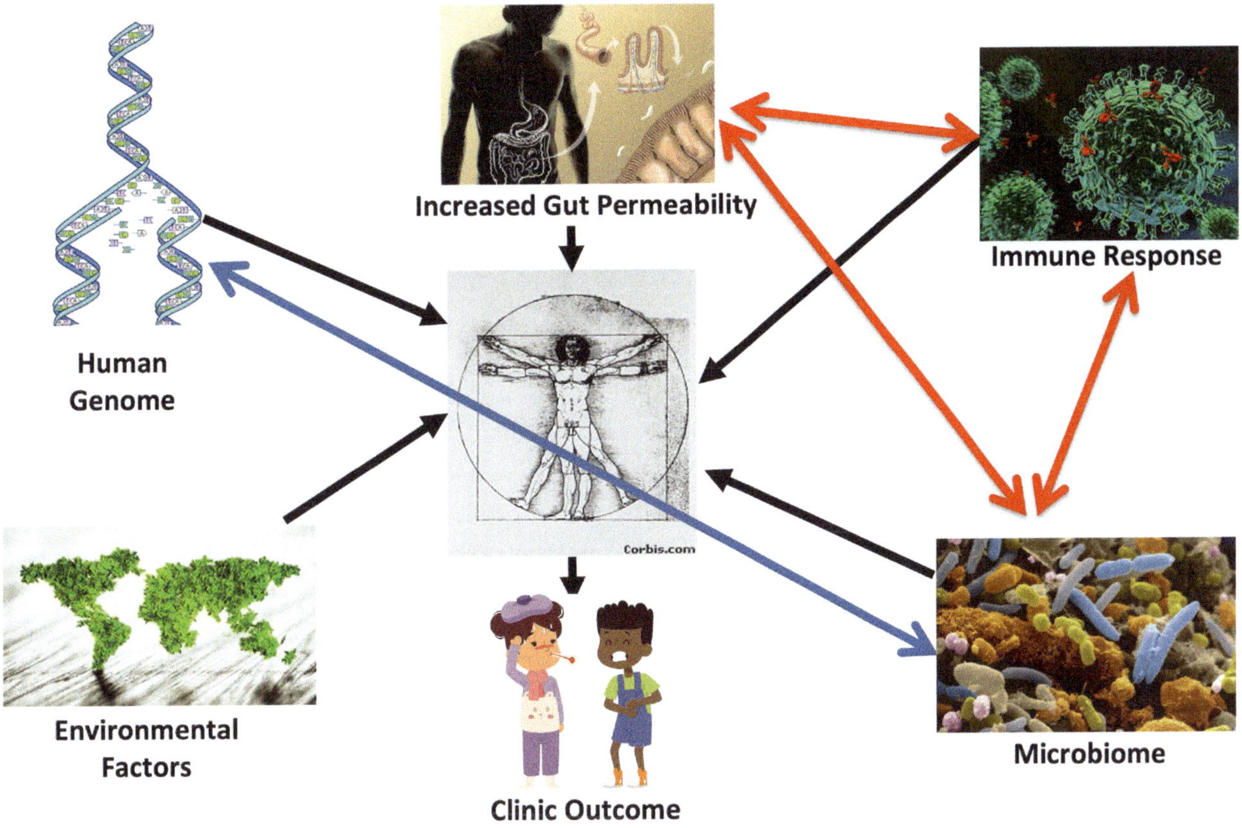

FIGURE 21.1 The five elements involved in the pathogenesis of chronic inflammatory diseases, including celiac disease Beside genetic predisposition and exposure to environmental riggers, loss of gut barrier function, a proinflammatory tone of the immune system and an imbalanced microbiome (dysbiosis) are integral part of the inflammatory process. The last three elements influence each other in a bidirectional manner (*red* lines), while there is growing evidence of the epigenetic role of the human microbiome that dictates the switch from genetic predisposition, the step that start the march toward the clinical outcome (*blue* line).

21.4 The first 1000 days of life to program the immune system

A hyperbelligerent immune system is the fourth element necessary to develop chronic inflammation. In this case, the immune system produces inflammation constantly, not just when it is appropriately unleashed, for example, in response to exposure to a pathogen. Inflammation is not always detrimental. Indeed, it is an essential part of the protection and healing processes when the body is exposed to infection or injury. In an appropriate inflammatory response, factors produced by immune cells also protect our tissues against bacteria and viruses until the microorganisms are defeated, when it is time to turn the inflammation off. How does the immune system decide to unleash inflammation and when is it time to turn it off? This seems the result of a sophisticated programming that occurs during the first 1000 days of our lives, from conception until 2 years of age. There is now growing evidence that it is the microbiome in charge of this programming (Donald & Finlay, 2023). The microbiome, with its diverse composition and balance, trains the immune system about if and when to unleash inflammation to fight danger. Evolutionally speaking, the immune system had as the single goal to defeat infections, the most frequent cause of morbidity and mortality for the human species until the last century. When exposed to microorganisms, the immune system, using both the innate and adaptive immune branches, creates a state of local inflammation to generate a micromilieu inhospitable for microorganisms.

Conversely, if the gut microbiome's composition is negatively influenced by factors that are not aligned with human evolutionary planning, including poor maternal lifestyle, cesarean-section delivery, inappropriate nutrition, or overuse of antibiotics, the engraftment of the microbiome will be off balance and the immune system will be inappropriately programmed in a proinflammatory tone. Based on these considerations, we can hypothesize that an unbalanced microbiota engraftment in infancy secondary to detrimental environmental factors typical of the Western lifestyle may have far-reaching consequences during the entire lifespan of the host (Donald & Finlay, 2023).

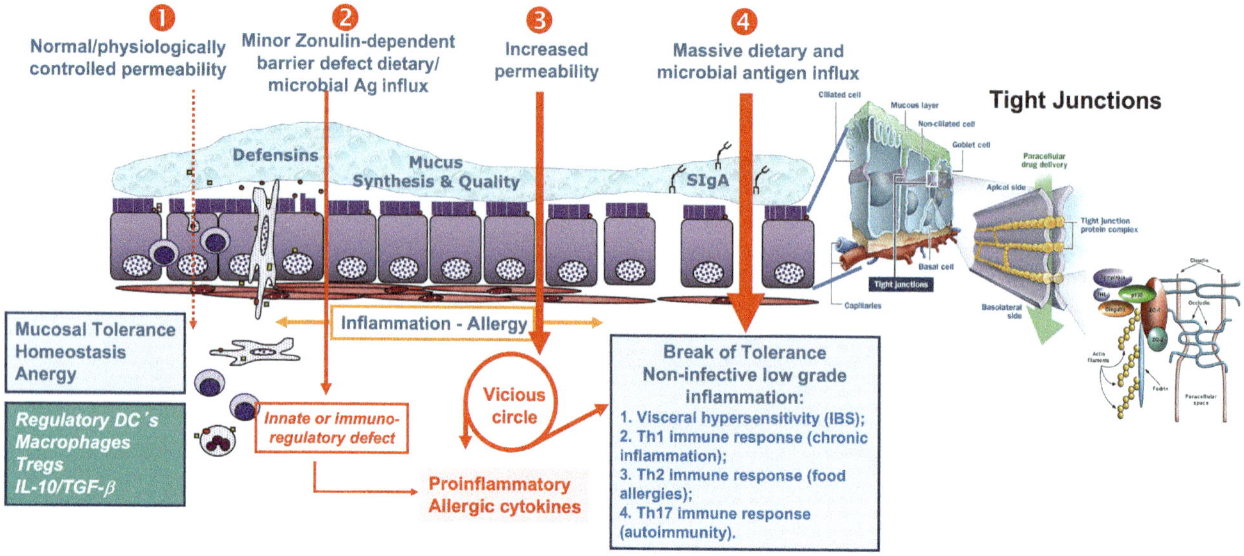

FIGURE 21.2 Proposed mechanisms involved in breaking mucosal tolerance and subsequent onset of noninfective chronic inflammatory diseases (1) Under physiological circumstances, there is a tightly control of antigen trafficking from the gut lumen into the intestinal lamina propria (antigen sampling) that leads to mucosal tolerance, homeostasis and anergy mediated by the involvement of a variety immune cells and antiinflammatory cytokines. (2) Following a variety of environmental stimuli, including gut microbiome imbalance (dysbiosis), paracellular permeability is increased secondary to the upregulation of specific pathways, including the zonulin pathway. (3) As a consequence of prolonged and uncontrolled antigen trafficking, innate immune response or immunoregulatory defects leads to the onset of inflammation with secondary production of proinflammatory cytokines, including TNF-alpha and IFN-gamma that, by themselves, are causing further increase in gut permeability by activating the myosin light chain kinase pathway. (4) This mechanism starts a vicious cycle of further increase in gut permeability and subsequent antigen trafficking that ultimately leads to break of tolerance and onset of chronic inflammatory diseases. *Adapted from Fasano, Alessio (2020). All disease begins in the (leaky) gut: Role of zonulin-mediated gut permeability in the pathogenesis of some chronic inflammatory diseases.* F1000Research, *9, 69. https://doi.org/10.12688/f1000research.20510.1.*

21.5 The microbiome and its possible epigenetic role in chronic inflammatory disease pathogenesis

The fifth pillar of inflammation is the human microbiome, which is perhaps the most important element. Since *Homo sapiens* appeared on the evolution stage approximately three hundred thousand years ago, we have coevolved with the collective microorganisms that inhabit our body. Therefore it should be no surprise that this coevolution requires constant adjustments to establish a pacific, symbiotic relationship, which can be achieved only through communications mediated by epigenetic influence from the microbiome. Indeed, growing evidence suggests that the microbiome dictates the balance between health and disease by deciding epigenetically if and when our genes should be expressed or repressed, and so determines if and when we shift from genetic predisposition to clinical outcome (Furman et al., 2019). It should also be emphasized that the three additional pillars of chronic inflammatory diseases—gut permeability through the loss of barrier function, a hyperimmune system, and the microbiome's composition and function—are constantly interacting and influencing each other. A change in the *status quo* in one domain (gut dysbiosis), may affect another domain (increased gut permeability).

Prenatal, perinatal, and postnatal environmental factors are instrumental influencers of the composition and function of the human microbiome. Most likely, these changes help us to quickly adapt to our surroundings through the microbiome's epigenetic modulation of our genes. However, with the drastic modifications to lifestyle and environment that have occurred in the past two to three decades in the Western hemisphere, this plasticity has shifted from a superb asset to a major liability.

This postulate can explain the epidemiological data of CD occurring at any age, even after decades of the exposure to gluten as the instigator of the autoimmune process. The initial proinflammatory tone may translate in disease onset following changes in lifestyle causing changes in microbiome composition and function. This dysbiosis triggers epigenetic changes that can lead to a series of biological consequences, including alteration of gut permeability, increased antigen trafficking, and abnormal immune response (see Fig. 21.1) that cause break of mucosal tolerance and the start of the march from the genetic predisposition to clinical outcome (see Fig. 21.2).

Taking into consideration all the aforementioned factors and based on our current knowledge of the biology and immunology of the human species as an integral part of a much more complex ecosystem, we can explore novel approaches for the management of chronic inflammatory diseases, including CD.

21.6 Developing therapeutic targets

The five pillars contributing to chronic inflammation that are listed in this chapter provide five different possible targets to mitigate that inflammation. The first approach is the editing of the human genome. As we have seen from events in China, after gene editing was practiced on human embryos before two live births, there are enormous ethical, cultural, political, and social concerns with this approach (Cyranoski, 2020; Li et al., 2019). Even if these issues were satisfactorily resolved, the multifactorial nature of chronic inflammatory diseases (like CD) makes genetic editing unfeasible. Hundreds, if not thousands, of genes are involved, and they can be different in each individual despite exerting the same function even when affected by the same disease.

The second approach, eliminating the environmental trigger, is the one we routinely pursue in CD by avoiding the ingestion of gluten-containing grains, an approach that is not always successful and that has a large burden on the quality of life of celiac patients (for more details, see Chapter 12). Therefore there is a great level of interest in pursuing alternative and/or complementary strategies to the GFD.

The current approaches focused on mitigating inflammation and antigen trafficking (the third and fourth elements involved in the pathogenesis of chronic inflammatory diseases) are covered in Chapter 20. This chapter will focus on future research pipelines aimed at implementing personalized interventions (precision medicine) and primary prevention exploiting novel approaches, including the use of artificial intelligence (AI), synthetic biology, and/or the manipulation of the microbiome composition and function.

21.7 Artificial intelligence to implement personalized medicine and primary prevention

Thirty years after the launch of the Human Genome Project, we continue to struggle with the complex nature of human biology as we seek therapeutic targets to address a variety of diseases. Disappointingly, the map of the human genome led to the identification of clinically relevant mutations involving only 2% of possible therapeutic targets. This finding applies to CD as well, in which, beside the HLA DQ2/8 genes accounting for approximately half of the genetic load, there are more than 100 additional genetic mutations found so far, each contributing minimally to the genetic load of CD (for more details, see Chapter 3). However, the lesson learned from these results was that the remaining 98% of human diseases is driven by changes in gene expression rather than their mutations, suggesting a key function of what was initially thought to be noncoding "junk DNA" as targets for epigenetic regulation of coding DNA.

Understanding how environmental factors modulate this "regulatory genome" represents a transformational advance toward identifying novel therapeutic and preventive targets for many human diseases. Technology and costs are no longer limiting factors to achieve these goals. Rather, the limitation with the explosion of data we can generate by genomic, transcriptomic, metabolomic, and proteomic studies of the human cell lines or of the human microbiome is our inability to deal with a staggering amount of high-dimensional data. Analysis for these data often requires the development of new computational tools and algorithms.

Researchers are turning to AI and machine learning to help turn these data into knowledge and to learn from this knowledge in order to make smarter decisions to identify targets for therapeutic interventions and to improve patient outcomes through precision, personalized medicine. By training machine learning algorithms on sample data that accumulate over time as more data are collected, machine learning algorithms can grow smarter and smarter in performing these tasks in a manner that surpasses human ability. In a model based on human learning, the computer learns from observational data, figuring out its own solution to the problem at hand.

But why do we need these very sophisticated analysis methodologies and not the typical statistical approaches that we have been using for so long in science? Some people may believe that machine learning and statistics are aimed at the same goals, with machine learning being "the next generation" of a more powerful statistical method. However, they are not synonymous. Statistics is the mathematical study of data and, therefore statistics cannot be applied unless there are data available. A statistical model is a model for the data that is used either to infer something about the relationships within the data or to create a model that can predict future values. To make this even more explicit, there are many statistical models that can make predictions, but predictive accuracy is not their strength. Likewise, machine learning models provide various degrees of interpretability, but they generally sacrifice interpretability for predictive power. Typical statistical approaches follow rigid rules dictating which method of analysis is to be applied based on data distribution. Some approaches are used if the data are normally distributed, and others if the data are not normally distributed. Conversely, in machine learning, there is more flexibility because the machines can "learn" from data without having to impose any rigid rules or assumptions about them. Another advantage is that machine learning approaches can simultaneously analyze and learn from multiple, disparate data types and an enormous number of potential predictors, which is not feasible with statistical approaches.

In the case of the use of machine learning or multiomic analysis, there are some interesting questions: are the host genome and RNA sequencing informative enough to infer specific pathways involved in the pathogenesis of the disease studied? Which feature of metagenomic data (microbes or encoded functional pathways) can be linked to a specific disease? Is metatranscriptomic information useful in understanding the possible epigenetic role of the microbiome? Can metabolites produced by the microbiome and/or the host potentially be mechanistically linked to the onset of the disease? Most likely, in the near future machine learning approaches may achieve not only these results, but even more ambitious goals. For example, once we have established links between the composition and function of the microbiome and specific diseases, we might be able to identify the disease of a specific patient simply by analyzing his or her microbiome. Even more exciting, if the machine learning approach is trained on longitudinal data, we may be able to predict the risk of developing a specific disease or the severity of the disease in an individual at a specific point in time. This is the "Holy Grail" of first, disease interception and second, primary prevention by modifying the microbiome composition and function through targeted personalized interventions. These interventions could include changes in diet and the use of specific prebiotics, probiotics, or a combination of the two (i.e., synbiotics). In the next two sections we will cover some of these potential applications in CD.

21.8 Novel therapeutic approaches for celiac disease personalized interventions

With the support of AI we will be able to know a priori which intervention will be effective to a specific subpopulation of patients and which would not. There are many examples that could apply to CD, but for practical reasons, two approaches will be discussed in depth because of their potential to revolutionize our therapeutic approach to CD.

21.8.1 mRNA to negatively select CD4 + gliadin specific T-cells

Vaccines based on messenger RNA (mRNA) have gained primetime attention during the COVID-19 pandemic, particularly for their superb efficacy and the speed of their development and large-scale manufacturing. mRNA vaccines consist of antigen-encoding RNA encapsulated in lipid nanoparticles. Initially, mRNA vaccines were developed to stimulate an immune response against tumors. In those studies, the net negative charge of the mRNA–lipid nanoparticle resulted in their macropinocytosis by splenic dendritic cells, activation of Toll-like receptor 7 (TLR7), production of interferon-α, and triggering of antigen-specific immunity (Kranz et al., 2016). In parallel, other investigators developed a modification of mRNA-based vaccines that incorporated 1-methylpseudouridine (m1Ψ), a naturally occurring component of eukaryotic 18 S ribosomal RNA, in place of uridine, finding that m1Ψ mRNA had a higher translational capacity and lower innate immune activation (Karikó et al., 2005).

The mRNA platform technology is plastic enough to be applied to several conditions, mainly to prime antigen-specific immune response as successfully used to develop a SARS-CoV2 mRNA vaccine. Sahin et al. have harnessed this vaccine technology to suppress, rather than stimulate, antigen-specific immune responses (Krienke et al., 2021; Sahin et al., 2014), reasoning that an antiinflammatory, m1Ψ-based mRNA vaccine could be advantageous in those diseases in which suppression of immunity was desired as it is the case of autoimmunity. Working with mouse models of multiple sclerosis, they show that immunization with a modified m1Ψ mRNA encoding a self-antigen and delivered through noninflammatory nanoparticle carriers led to a dampening of autoimmunity through the activation of antigen-specific regulatory T cells (T_{reg}) (Krienke et al., 2021). A similar approach could be applied to CD by using nanoparticles containing m1Ψ mRNA encoding for immune-dominant deamidated gluten peptides to be taken up by antigen presenting cells, with subsequent activation of gluten-specific FoxP3$^+$ T_{reg} cells and subsequent suppression of CD4 + gliadin specific T-cells.

Sahin et al. have clearly demonstrated the potential of RNA lipoplex vaccines to deliver a noninflammatory form of an mRNA vaccine encoding a self-antigen to prevent and limit autoimmune disease in mice. It is noteworthy that m1Ψ-containing mRNA is also used for the COVID-19 mRNA vaccine, indicating that the proversus antiinflammatory nature of m1Ψ mRNA vaccines can be modulated depending on the specific antigen and specific encapsulating lipid formulation. In the case of the BNT162b2 vaccine for COVID-19, the antigen is a foreign protein formulated in an immunostimulatory lipid nanoparticle. In case of CD, the antigen would be a nonself protein (gluten) that should be delivered in a nonimmunogenic lipoplex formulation, and an extra mRNA purification step to remove any residual immunostimulatory molecules. This method allows antigen presentation in the absence of inflammation and co-stimulation, preferential expansion of preexisting T_{reg} cells, and possibly also their de novo development.

Before m1Ψ mRNA vaccination to treat autoimmune diseases can be translated to humans, key questions related to defining optimal gluten immune-dominant targets and to the longevity and specificity of the effect must be answered. More work will be needed to fully understand the capacity for, and breadth of, bystander suppression. The report by Sahin et al. did not include formal demonstration of tolerance by a later rechallenge with the autoantigen in a nontolerogenic format, or by the adoptive transfer of T_{reg} cells to naive mice (for demonstrating "dominant" tolerance). Active CD implies continuous exposure to gluten; therefore, it will be important to establish the indication of a "celiac vaccine" (protection against cross-contamination, treatment of nonresponsive celiac cases, allowing the return to an unrestricted diet) to establish the length of the immunoregulatory effect and whether repeat vaccination is safe or advantageous. Finally, the fact that similar approaches can be used to stimulate immune responses underlines the need to fully understand the mechanisms that control the proversus antiinflammatory nature of mRNA vaccines to eliminate the risk of deleterious immune stimulation.

A complementary/alternative approach is to use mRNA therapeutics to selectively program cell death of undesirable CD4 + gliadin specific T cell clones. This approach has been originally developed to specifically target tumoral cells in which mRNA encoding for a toxic protein is delivered through nanoparticles taken up by the tumoral cells (Jain et al., 2018). Similarly, nanoparticles decorated externally with immunodominant gliadin peptides and containing mRNA encoding caspase can be used to selectively program gliadin-specific T cells self-disruption. Once the nanoparticles are included within antigen presenting cells or CD4 + T cells, caspase will be expressed and cause cells apoptosis, so eliminating gliadin-committed clones.

21.8.2 Synthetic biology for personalized celiac disease treatment

The definition of synthetic biology is applying engineering techniques to develop genetic circuitry. This circuitry can be implanted into a microorganism to perform a specific metabolic function and record events, which we can use as information to achieve the goal of targeted therapeutic interventions. Synthetic biology is the sophisticated evolution of genetic engineering. A classic example of old-fashioned, traditional genetic engineering is the cloning and production of insulin, which is focused on engineering bacteria to produce abundant amounts of a single protein. This technology has been developed and has been robustly used to manufacture drugs. Synthetic biology engineers want to move this technology to the next level by answering the following question: how do you put together multiple genes, not just a single one, and how do those genes interact with each other in a way that is basically turning the cell into a miniature computer? This goal is obtained by implanting genetic circuitries in microorganisms capable to make specific decisions based on preset programming.

The initial application of genetic engineering to microorganisms was aimed at transforming them into a type of factory with a unidimensional function. The bioengineer would insert genes, perhaps an insulin gene, in the microorganism's genome. The only thing the microbe was required to do, the only "decision" it had to make was to replicate itself, and by doing that it would make a large amount of insulin: that is classical genetic engineering.

Now, we have the opportunity to go to the next level. Instead of asking microorganisms to make a single product, we ask them to make its own decisions based on contingent environmental situations, which dictate if and when to enact a series of actions and gene expressions based on specific programming to obtain a specific outcome. But if the environmental conditions change, then the microorganism's genetic circuitry will activate a different decision-making process leading to a different outcome.

This almost sounds like science fiction because this concept unlocks an unbelievable number of opportunities. But we can now capitalize on these novel technologies to achieve the necessary next step in the world of microbiome research, namely, to move from a scientific proof of concept or a purely theoretical possibility to its applicability in human therapies. Indeed, in collaborations with colleagues at the Massachusetts Institute of Technology in Boston we are developing a patient-tailored platform of therapeutic bacteria for in situ delivery of gluten degrading proteases. We have isolated specific Lactobacillus strains from the duodenal mucosa of celiac patients in remission. We then engineered these strains to secrete microbe-derived gluten-degrading enzymes to dismantle undigestible gluten peptides. This platform can be personalized with different colonizing bacteria engineered to express multiple endopeptidases with complementary cleavage specificities and different pH ranges of activity, all working sequentially and synergistically to eliminate gluten immunogenic uncleavable epitopes during digestion. Specifically, we aim at engineering Lactobacillus strains to secrete propyl endopeptidases (PEPs) (see Chapter 20 for more details) upon sensing bile salts and/or gluten, indicative of food ingestion. We will also incorporate a pH sensor to express the type of therapeutic enzyme needed at each pH during the digestion process, optimizing the protein yield and the fitness of the therapeutic bacteria. Finally, novel PEPs will be developed using a machine learning approach to engineer more efficient gluten-degrading proteases. We will leverage machine learning ability to solve multiobjective design problems to increase PEP specificity for immunogenic peptides and increase activity in the gastric and duodenal environment. Some of these candidates have been already validated in the human organoid model that we have developed and used for other celiac-related projects (Freire et al., 2019; Nickerson et al., 2021).

21.9 Celiac disease primary prevention

Now that we have set the stage for the importance of the human microbiome in epigenetically controlling our clinical destiny and that we have developed the knowhow and technical tools to properly define the microbiome composition and function, we could exploit its manipulation for diagnostic and therapeutic applications. However, the entire field is at a crucial crossroad, since most of the microbiome studies published so far are merely descriptive, while there is a strong need to move from association to causation to clinically capitalize on this wealth of information accumulated on the human microbiome in the past decade (Martin et al., 2016). The major roadblock to achieve this goal is that most of the studies on human microbiome published so far are case-control, cross-sectional studies in which we compare two cohorts, one healthy and the other affected by the disease object of the study, with the goal to show a different microbiome composition. From these results, in general we infer that the cause of the disease is indeed the changes in microbiome composition and function. However, this is a mere assumption, since it can be just the opposite, for example an inflamed intestinal mucosa typical of the CD enteropathy can lead to changes of the microbiota and not vice versa. To overcome this limitation, the most powerful data come from longitudinal prospective cohort studies with at-risk individuals followed over time, as we have done with the Celiac Disease Genomic, Environment, Microbiome, and Metabolomic (CD-GEMM) study (Leonard et al., 2023; Leonard et al., 2015). So far, we have recruited an extremely comprehensive longitudinal prospective birth cohort of more than 500 children at-risk of CD, gathering a unique collection of almost 130,000 metadata and 28,500 biospecimens specifically designed to mechanistically link microbiome composition and function to diseases pathogenesis. Thanks to the CD-GEMM cohort, we are using data before, during, and after the onset of CD for training machine learning algorithms that can result in more accurate predictive models, linking specific microbiome components, ideally at the strain level, at specific time points to CD susceptibility or protection, More specifically, we are analyzing changes in microbiome and metatranscriptomic profiles over time in relation to factors such as mode of delivery, exposure to antibiotics and feeding regimens, including breast feeding and timing of gluten introduction (Leonard et al., 2020). This approach allows for an in-depth characterization of the infants' metabotypes (microbe-derived metabolomes), which will be linked to the microbiome composition, genomic information, and environmental factors to build top-down AI-driven system models of integrated multiomic profiles. These profiles can then be interrogated with respect to outcome (tolerance vs. immune response) to obtain predictive models and mechanistic insight into predisposing factors leading to CD autoimmunity. Ultimately, creating theoretical models of interaction between host genetic makeup, the microbiome, the metabolome, and environmental "pressure" in infants at risk of CD will help to identify and validate biomarkers predictor of late CD onset and specific microbiome and metabolomic profiles that can predict loss of tolerance in children genetically at risk of autoimmunity. The identification of these profiles will also provide possible targets for therapeutic and early preventive interventions through the manipulation of the microbiome composition in order to re-establish gluten tolerance and ultimately prevent CD. Below we will briefly review the identification of biomarkers predictive of subsequent CD onset and the exploitation of microbiome analysis for preventing the break of mucosal tolerance to gluten in the preceliac phase.

21.9.1 Identification of preclinical biomarkers predictor of celiac disease onset

We exploited the CD-GEMM cohort to identify possible serological biomarkers predictor of subsequent onset of CD in at-risk infants. In a first study we included 325 children for which multiple serum samples collected every 6 months were available. Of those, 275 were HLA-DQ2 + or HLA-DQ8 + and 38 had anti-tTG IgA seroconversion and were classified as part of the celiac disease spectrum, with 21 celiac cases subsequently confirmed. Of the 38 children, 28 (73.6%) had a rise in antideamidated antigliadin IgG antibodies (aDGPs) at least 6–12 months before their anti-tTG IgA seroconversion, with median age at IgG aDGPs rise of 18 months (age range 12–48 months) (Valitutti et al., 2023). Rise of aDGPs IgG held a relative risk (RR) of 5.03 (95%CI 3.59 to 7.05; $p < 0.0001$) for CD spectrum after 6 or 12 months. The Area Under the ROC Curve for "predictive" aDGPs IgG among the celiac disease spectrum group was 0.86 (95%CI 0.81 to 0.89), and the best cut-off value for this purpose was >10.0 U (sensitivity 73.7%; specificity 85.4%). The AUC decreased to 0.82 (95%CI 0.77 to 0.86) in children diagnosed with CD and the best cut-off of aDGPs IgG value was >10.3 U (sensitivity 71.4%; specificity 82.2%) (Valitutti et al., 2023). The median rise of these antibodies in children with CD prior to diagnosis was 17.2 U (range 10.4–89.9 U). Therefore these antibodies may represent a reliable biomarker of subsequent onset of CD in at-risk subjects.

In a follow up study using the same cohort we also evaluated the level of serum zonulin used as a biomarker of gut barrier integrity. As we mentioned above and in Chapter 20, increase in zonulin-dependent gut permeability seems to be an integral element of the pathogenesis of several chronic inflammatory diseases, including CD (Fasano, 2020; Sturgeon & Fasano, 2016; Valitutti & Fasano, 2019). Based on the hypothesis outlined in Fig. 21.2, loss of the zonulin-dependent gut barrier function with subsequent uncontrolled antigen (including gluten) trafficking could be an early event in breaking mucosal tolerance. We had evidence of this hypothesis in animal models (Watts et al., 2005) and in patients affected by type 1 diabetes (Sapone et al., 2006), but not in CD. For this reason, we evaluated 52 CD autoimmunity cases (defined as confirmed celiac patients by current protocols or with sustained presence of tTG antibodies but not confirmed yet by biopsy) and matched controls, who were followed for at least 12 months and consumed gluten. We measured serum zonulin from age 12 months to time of CD autoimmunity onset, and the corresponding time point in controls, and examined clinical factors of interest. We ran a mixed-effects longitudinal model with zonulin as a dependent variable. Children who developed CD autoimmunity had a significant increase in zonulin in the 18.3 months (range 6–78) preceding diagnosis compared to those that remained healthy (DaFonte et al., 2023). Among metadata considered, zonulin trajectory was only influenced by increasing number of antibiotic courses, which increased the slope of zonulin trajectory over time. These data show that zonulin levels significantly rise in the months (average 18 months) that precede CD autoimmunity diagnosis, even before the appearance of the aDGP IgG biomarkers (9 months before onset). Therefore, these results suggest (as outlined in Fig. 21.2) that zonulin-dependent loss of gut barrier function is an early event (average 18 months before CD autoimmunity onset), followed by increase gluten peptides trafficking leading to the appearance of antibodies against deamidated gliadin peptide on average 9 months before the onset of CD autoimmunity marked by the first detection of tTG antibodies. Also interesting is the observation that a greater number of previous antibiotic courses were associated with an increase in zonulin levels in CD autoimmunity subjects, is in line with our previous report that dysbiosis is one of the strongest *stimuli* for the upregulation of the zonulin pathway.

21.9.2 Exploiting the gut microbiome to find targets to implement primary prevention

The CD-GEMM cohort was also utilized to examine the infants' microbiome and metabolome profiling during the first six months after birth in relation to environmental risk factors and genetic markers known to increase the susceptibility to CD development (Leonard et al., 2020). In a subsequent study, we investigated changes in microbiome and metabolomic profiles preceding the onset of CD by up to 18 months (Leonard et al., 2021). Among the differences identified, we detected the decreased abundance of several microbial species/strains known to have a potential antiinflammatory, protective effect in cases (defined as at-risk children who developed CD) before disease onset compared to controls (defined as age-matched, at-risk infants who did not develop the disease) (Leonard et al., 2021). Based on these results we further investigated the potential protective role of some of these bacterial strains isolated from children at risk of CD that were protected from the onset of the disease. Among the identified bacterial strains, we used a prototype novel *Bacteroides vulgatus* strain 20220303-A2 to perform epigenetic profiling and functional studies using human patient-derived gut organoids to mechanistically establish a protective effect of bacterial bioproducts against gluten-dependent changes on human gut epithelial functions. Following gliadin exposure, we observed increases in epithelial cell death, epithelial monolayer permeability, and secretion of proinflammatory cytokines. These effects were mitigated upon exposure to *B. vulgatus* 20220303-A2 conditional media which had, relative to reference strains, genetic mutations in gene products whose function matched the observed phenotype. These protective effects were mediated by epigenetic reprogramming of the organoids treated with *B. vulgatus* conditional media as revealed by microRNA expression changes controlling pathways related to cell cycle, immune response, and barrier function. These results confirm that the there are some microbiome species, including *B. vulgatus* that may exert a beneficial role by protecting CD epithelium against gluten-induced break of mucosal tolerance through miRNA reprogramming. Loss of these protective strains due to lifestyle changes affecting microbiome composition and function may start the march from genetic predisposition to CD onset (see Fig. 21.2).

21.10 Conclusions

The future of medicine lies in recognizing the unique paths that individuals take to arise at a common disease state, the so-called Precision Medicine. Equally potentially attainable is the possibility to prevent the onset of chronic inflammatory diseases in subjects genetically at risk (primary prevention). Transformational, clinically applicable discoveries will come from appreciating a complex model of disease in which AI-dependent multiomic analysis is utilized along with longitudinal prospective studies to combine multiomic data with clinical outcomes. CD represents a unique model of chronic inflammatory diseases that may lead to these transformational changes in deploying novel strategies to deliver the health care of the 21st century.

References

Catassi, C., Kryszak, D., Bhatti, B., Sturgeon, C., Helzlsouer, K., Clipp, S. L., Gelfond, D., Puppa, E., Sferruzza, A., & Fasano, A. (2010). Natural history of celiac disease autoimmunity in a USA cohort followed since 1974. *Annals of Medicine*, *42*(7), 530–538. Available from https://doi.org/10.3109/07853890.2010.514285.

Collins, F. S., Ph, M. Dd, & Fink, L. (1995). The human genome project. *Alcohol Health and Research World*, *19*, 190–195.

Cyranoski, D. (2020). What CRISPR-baby prison sentences mean for research. *Nature*, *577*(7789), 154–155. Available from https://doi.org/10.1038/d41586-020-00001-y.

DaFonte, T., Valitutti, F., Kenyon, Locascio, J., Fasano, A., & Leonard, M. (2023). Increased intestinal permeability precedes celiac disease in a birth cohort of at-risk children. Abstract DDW Chicago.

Donald, K., & Finlay, B.B. (2023). Early-life interactions between the microbiota and immune system: impact on immune system development and atopic disease. Nature Reviews. Immunology. Available from https://doi.org/10.1038/s41577-023-00874-w.

Fasano, A. (2011). Zonulin and its regulation of intestinal barrier function: The biological door to inflammation, autoimmunity, and cancer. *Physiological Reviews*, *91*(1), 151–175. Available from https://doi.org/10.1152/physrev.000030.2008.

Fasano, Alessio (2020). All disease begins in the (leaky) gut: Role of zonulin-mediated gut permeability in the pathogenesis of some chronic inflammatory diseases. *F1000Research*, *9*, 69. Available from https://doi.org/10.12688/f1000research.20510.1.

Freire, R., Ingano, L., Serena, G., Cetinbas, M., Anselmo, A., Sapone, A., Sadreyev, R. I., Fasano, A., & Senger, S. (2019). Human gut derived-organoids provide model to study gluten response and effects of microbiota-derived molecules in celiac disease. *Scientific Reports*, *9*(1). Available from https://doi.org/10.1038/s41598-019-43426-w.

Furman, D., Campisi, J., Verdin, E., Carrera-Bastos, P., Targ, S., Franceschi, C., Ferrucci, L., Gilroy, D. W., Fasano, A., Miller, G. W., Miller, A. H., Mantovani, A., Weyand, C. M., Barzilai, N., Goronzy, J. J., Rando, T. A., Effros, R. B., Lucia, A., Kleinstreuer, N., & Slavich, G. M. (2019). Chronic inflammation in the etiology of disease across the life span. *Nature Medicine*, *25*(12), 1822–1832. Available from https://doi.org/10.1038/s41591-019-0675-0.

Gatti, S., Lionetti, E., Balanzoni, L., Verma, Galeazzi, T., Gesuita, R., Scattolo, N., Cinquetti, M., Fasano, A., & Catassi, C. (n.d.). Celiac Screening Team (2020). Increased Prevalence of Celiac Disease in School-age Children in Italy. Clinical Gastroenterology and Hepatology: the Official Clinical Practice Journal of the American Gastroenterological Association, 18, 596–603.

Jain, R., Frederick, J. P., Huang, E. Y., Burke, K. E., Mauger, D. M., Andrianova, E. A., Farlow, S. J., Siddiqui, S., Pimentel, J., Cheung-Ong, K., McKinney, K. M., Köhrer, C., Moore, M. J., & Chakraborty, T. (2018). MicroRNAs enable mRNA therapeutics to selectively program cancer cells to self-destruct. *Nucleic Acid Therapeutics*, *28*(5), 285–296. Available from https://doi.org/10.1089/nat.20180.0734.

Karikó, K., Buckstein, M., Ni, H., & Weissman, D. (2005). Suppression of RNA recognition by Toll-like receptors: The impact of nucleoside modification and the evolutionary origin of RNA. *Immunity*, *23*(2), 165–175. Available from https://doi.org/10.1016/j.immuni.2005.06.008.

Kranz, L. M., Diken, M., Haas, H., Kreiter, S., Loquai, C., Reuter, K. C., Meng, M., Fritz, D., Vascotto, F., Hefesha, H., Grunwitz, C., Vormehr, M., Hüsemann, Y., Selmi, A., Kuhn, A. N., Buck, J., Derhovanessian, E., Rae, R., Attig, S., & Sahin, U. (2016). Systemic RNA delivery to dendritic cells exploits antiviral defence for cancer immunotherapy. *Nature*, *534*(7607), 396–401. Available from https://doi.org/10.1038/nature18300.

Krienke, C., Kolb, L., Diken, E., Streuber, M., Kirchhoff, S., Bukur, T., Akilli-Öztürk, Ö., Kranz, L. M., Berger, H., Petschenka, J., Diken, M., Kreiter, S., Yogev, N., Waisman, A., Karikó, K., Türeci, Ö., & Sahin, U. (2021). A noninflammatory mRNA vaccine for treatment of experimental autoimmune encephalomyelitis. *Science (New York, N.Y.)*, *371*(6525), 145–153. Available from https://doi.org/10.1126/science.aay3638.

Lee, R. U., Stahlman, S. L., & Magee, J. S. (2023). Celiac disease on the rise in the US military population: A 22 year retrospective epidemiologic study. *Digestive Diseases and Sciences*, *68*. Available from https://doi.org/10.1007/s10620-023-07964-8.

Leonard, M. M., Camhi, S., Huedo-Medina, T. B., & Fasano, A. (2015). Celiac disease genomic, environmental, microbiome, and metabolomic (CDGEMM) study design: Approach to the future of personalized prevention of celiac disease. *Nutrients*, *7*(11), 9325–9336. Available from https://doi.org/10.3390/nu7115470.

Leonard, M. M., Karathia, H., Pujolassos, M., Troisi, J., Valitutti, F., Subramanian, P., Camhi, S., Kenyon, V., Colucci, A., Serena, G., Cucchiara, S., Montuori, M., Malamisura, B., Francavilla, R., Elli, L., Fanelli, B., Colwell, R., Hasan, N., Zomorrodi, A. R., & De Villsante, G. C. (2020). Multi-omics analysis reveals the influence of genetic and environmental risk factors on developing gut microbiota in infants at risk of celiac disease. *Microbiome*, *8*(1). Available from https://doi.org/10.1186/s40168-020-00906-w.

Leonard, M. M., Kenyon, V., Valitutti, F., Pennacchio-Harrington, R., Piemontese, P., Francavilla, R., Norsa, L., Passaro, T., Crocco, M., Baldassarre, M., Trovato, C. M., & Fasano, A. (2023). Cohort profile: Celiac disease genomic, environmental, microbiome and metabolome study; a prospective longitudinal birth cohort study of children at-risk for celiac disease. *PLoS One*, *18*(3), e0282739. Available from https://doi.org/10.1371/journal.pone.0282739.

Leonard, M. M., Valitutti, F., Karathia, H., Pujolassos, M., Kenyon, V., Fanelli, B., Troisi, J., Subramanian, P., Camhi, S., Colucci, A., Serena, G., Cucchiara, S., Trovato, C. M., Malamisura, B., Francavilla, R., Elli, L., Hasan, N. A., Zomorrodi, A. R., Colwell, R., & Fasano, A. (2021). Microbiome signatures of progression toward celiac disease onset in at-risk children in a longitudinal prospective cohort study. *Proceedings of the National Academy of Sciences of the United States of America*, *118*(29). Available from https://doi.org/10.1073/pnas.2020322118.

Li, J., Walker, S., Nie, J., & Zhang, X. (2019). Experiments that led to the first gene-edited babies: The ethical failings and the urgent need for better governance. *Journal of Zhejiang University-SCIENCE B*, *20*(1), 32–38. Available from https://doi.org/10.1631/jzus.b1800624.

Martin, V. J., Leonard, M. M., Fiechtner, L., & Fasano, A. (2016). Transitioning from descriptive to mechanistic understanding of the microbiome: The need for a prospective longitudinal approach to predicting disease. *Journal of Pediatrics*, *179*, 240–248. Available from https://doi.org/10.1016/j.jpeds.2016.08.049.

Nickerson, K. P., Llanos-Chea, A., Ingano, L., Serena, G., Miranda-Ribera, A., Perlman, M., Lima, R., Sztein, M. B., Fasano, A., Senger, S., & Faherty, C. S. (2021). A versatile human intestinal organoid-derived epithelial monolayer model for the study of enteric pathogens. *Microbiology Spectrum*, *9*(1), 1–17. Available from https://doi.org/10.1128/Spectrum.00003-21.

Okada, H., Kuhn, C., Feillet, H., & Bach, J. F. (2010). The "hygiene hypothesis" for autoimmune and allergic diseases: An update. *Clinical and Experimental Immunology*, *160*(1), 1–9. Available from https://doi.org/10.1111/j.1365-2249.2010.04139.x.

Sahin, U., Karikó, K., & Türeci, Ö. (2014). MRNA-based therapeutics-developing a new class of drugs. *Nature Reviews. Drug Discovery*, *13*(10), 759–780. Available from https://doi.org/10.1038/nrd4278.

Sapone, A., De Magistris, L., Pietzak, M., Clemente, M. G., Tripathi, A., Cucca, F., Lampis, R., Kryszak, D., Cartenì, M., Generoso, M., Iafusco, D., Prisco, F., Laghi, F., Riegler, G., Carratu, R., Counts, D., & Fasano, A. (2006). Zonulin upregulation is associated with increased gut permeability in subjects with type 1 diabetes and their relatives. *Diabetes*, *55*(5), 1443–1449. Available from https://doi.org/10.2337/db05-1593.

Sturgeon, C., & Fasano, A. (2016). Zonulin, a regulator of epithelial and endothelial barrier functions, and its involvement in chronic inflammatory diseases. *Tissue Barriers*, *4*(4). Available from https://doi.org/10.1080/21688370.2016.1251384.

Valitutti, F., & Fasano, A. (2019). Breaking Down barriers: How understanding celiac disease pathogenesis informed the development of novel treatments. *Digestive Diseases and Sciences*, *64*(7), 1748–1758. Available from https://doi.org/10.1007/s10620-019-05646-y.

Valitutti, F., Leonard, M. M., Kenyon, V., Montuori, M., Piemontese, P., Francavilla, R., Malamisura, B., Norsa, L., Calvi, A., Lionetti, M. E., Baldassarre, M., Trovato, C. M., Perrone, M., Passaro, T., Sansotta, N., Crocco, M., Morelli, A., Raguseo, L. C., Malerba, F., & Fasano, A. (2023). Early antibody dynamics in a prospective cohort of children at risk of celiac disease. *American Journal of Gastroenterology*, *118*(3), 574–577. Available from https://doi.org/10.14309/ajg.0000000000002192.

Watts, T., Berti, I., Sapone, A., Gerarduzzi, T., Not, T., Zielke, R., & Fasano, A. (2005). Role of the intestinal tight junction modulator zonulin in the pathogenesis of type I diabetes in BB diabetic-prone rats. *Proceedings of the National Academy of Sciences of the United States of America*, *102*(8), 2916–2921. Available from https://doi.org/10.1073/pnas.0500178102.

Index

Note: Page numbers followed by "*f*" and "*t*" refer to figures and tables, respectively.

D

G

H

I

J

K

L

M

N

O

P

Q

R

S

T

U

V

W

Z